(a) A warning sign

(b) A mandatory sign

(c) A prohibition sign

(d) A fire equipment sign

(e) Safe condition signs

xamples of fire safety signs illustrating the standard shape and colour pattern for ach class (Courtesy: British Standards Institution)

(a) Symbol alone

(b) and (c) Wording and symbol

xample of a fire safety sign where the standard symbol is used alone or in onjunction with the standard wording (Courtesy: British Standards Institution)

Safety at Work

Safety at Work

Fourth edition

Edited by
John Ridley, BSc(Eng), CEng, MIMechE, FIOSH

mann Ltd
ordan Hill, Oxford OX2 8DP

mber of the Reed Elsevier group

LONDON BOSTON
H NEW DELHI SINGAPORE SYDNEY
TORONTO WELLINGTON

rst published 1983
Second edition 1985
Reprinted 1987
Third edition 1990
Reprinted 1991, 1992, 1993
Fourth edition 1994

British Library Cataloguing in Publication Data
Safety at Work. – 4Rev.ed
I. Ridley, John R.
363.110941

ISBN 0 7506 0746 7

Library of Congress Cataloguing in Publication Data
Safety at work/edited by John Ridley. – 4th ed.
p. cm.
Includes bibliographical references and index.
ISBN 0 7506 0746 7
1. Industrial safety. I. Ridley, John R.
T55.S214 93–46862
363.11–dc20 CIP

Composition by Genesis Typesetting, Laser Quay, Rochester, Kent
Printed and bound in Great Britain by Hartnoll Ltd, Bodmin

Contents

Foreword

Frank J. Davies CBE, O St J, *Chairman, Health and Safety Commission*

I am very pleased to provide the foreword to the fourth edition of this book.

My forty years experience of working in industry have taught me the importance of health and safety. Even so, since becoming Chairman of the Health and Safety Commission in October 1993, I have learned more about the extent to which health and safety issues impact upon so much of our economic activity. An HSE study has brought this sharply into focus. The study estimates that the overall direct cost to British industry of all work accidents and work-related ill health is between £4 billion and £9 billion. Industry cannot afford to overlook these findings when deciding how best to manage health and safety.

In his foreword to the third edition my predecessor, Sir John Cullen, commented on the increasing impact of Europe in the field of health and safety, most notably through European Community Directives. This is borne out by the revisions to this edition, such as updating the section on EC directives and covering new health and safety legislation, much of which has emanated from Europe.

It is, of course, vital and inescapable that an issue as critical as health and safety should be grounded in sound and effective legislation. Despite the progress since the 1974 Health and Safety at Work Act in rationalising and modernising the law, the HSC is well aware that the sheer volume of health and safety legislation and its accompanying guidance can appear daunting, even to experts. That is why we have been pleased to undertake a review of health and safety regulations with the aim of simplifying the regulatory regime for industry. However, the Commission and Ministers are agreed that it is essential to maintain the levels of health and safety protection.

This review, involving employers and trade unionists, typifies the way that the Health and Safety Commission and Executive regularly examine issues that are of direct concern to industry. Similarly, an internal HSE review has examined its efforts on each major class of occupational health risk. The outcome will lead to a programme of action designed to help employers and workers.

This book covers many other important health and safety developments, including environmental and industrial relations law which touch on this area to varying degrees. I welcome the contribution it makes towards the goal of reaching and maintaining effective health and safety policies and practices throughout the workplace.

Preface to fourth edition

In the Preface to the third edition I made a statement that 'a spate of directives have been adopted by the European Community aimed at the approximation of occupational health and safety laws of Member States by the time the free market is established in 1993'. 1993 has come and gone, the free market has been established and the contents of those directives have been incorporated by the UK into its national laws.

Specifically, the Employment Framework Directive with its first five daughters is now incorporated in a series of Regulations (known familiarly as the 'six-pack'). These particular Regulations are concerned with health and safety in the workplace and the protection of the workpeople. The Machinery Directive which was aimed at promoting the free movement of plant and equipment between Member States and which used safety as its criterion has been incorporated in the Supply of Machinery (Safety) Regulations 1992. A number of directives concerned with conditions of employment have been incorporated in the Trade Union Reform and Employment Rights Act 1993, while in other spheres such as those concerning the use of chemicals and the environment, there have been on-going developments as a consequence of EC initiatives.

While much of the content of these new laws encompasses extant requirements, there are also a number of significant changes that reflect the differing attitudes and approaches of other Member States of the EC. These new legislative requirements are covered in the revised text.

In the rapidly changing times in which we live, interests and responsibilities change, drawing those involved into new spheres of activity. This book has had its fair share of changes of contributors with new authors stepping in to revise, update or rewrite chapters. Where this has happened, I am grateful to the original authors for so freely allowing their earlier material to be used by their successors.

We live in a fluid world – and none more so than in the safety and health arena – where the acceptance of change and the ability to adapt are the essentials of survival. The beginning of 1993 brought major changes in health, safety and employment legislation – what is now needed is a period of entrenchment to enable those new laws to be understood and assimilated into current practice. I hope that this edition will help in that process.

John Ridley
1994

Preface to first edition

Since the first welfare Act was put on the Statute Book in 1802 there has been a steady development in safety and health legislation aimed at improving the lot of those who work in mills, factories, and even in offices. In the past two decades official concern has increased, culminating in 1974 in the Health and Safety at Work etc. Act. 'Safety at Work' has now taken on a new and more pertinent meaning for both employer and employee.

Developments in the field of safety extend throughout much of the world, indicating an increasing concern for the quality of working life. In Europe the number of directives promulgated by the European Economic Community are evidence of this growing official awareness of the dangers that the individual faces in his work.

Health and safety laws in the UK are the most complex and comprehensive of all employment laws. Consequently employers are looking to a new breed of specialists, the occupational safety advisers, for expert advice and guidance on the best means for complying with, and achieving the spirit of, the law. These specialists must have the necessary knowledge of a wide range of disciplines extending from safety and related laws to occupational health and hygiene, human behaviour, management and safety techniques, and of course, the hazards inherent in particular industries or pursuits. With this demand for expert advice has grown a need for a nationally recognised qualification in this new industrial discipline. With this in mind the Institution of Occupational Safety and Health (IOSH) published in 1978 a syllabus of subjects for study by those seeking to become professionally qualified in this field. This syllabus now forms the foundation upon which the National Examination Board in Occupational Safety and Health (NEBOSH) sets its examinations.

Prepared in association with IOSH, this book covers the complete syllabus. It is divided into five parts to reflect the spectrum of the five major areas of recommended study. Each part has a number of chapters, which deal with specific aspects of health or safety. To enable readers to extend their study of a particular subject, suitable references are given together with recommendations for additional reading. Further information and details of many of the techniques mentioned can also be obtained through discussions with tutors. A table is given in Appendix II to guide students in their selection of the particular chapters to study for the appropriate levels and parts of the examination.

A major objective of this book has been to provide an authoritative, up-to-date guide in all areas of health and safety. The contributing authors are recognised specialists in their fields and each has drawn on his or her personal knowledge and experience in compiling the text, emphasising those facets most relevant to the safety advisers' needs. In this they have drawn material from many sources and the views they have expressed are their own and must not be construed as representing the opinions or policies of their employers nor of any of the organisations which have so willingly provided material.

It has been common practice to refer to the safety specialist as the 'safety officer', but this implies a degree of executive authority which does not truly indicate the rôle he plays. Essentially that rôle is one of monitoring the conditions and methods of work in an organisation to ensure the maintenance of a safe working environment and compliance with safety legislation and standards. Where performance is found wanting his function is to advise the manager responsible on the corrective action necessary. Reflecting this rôle, the safety specialist is throughout the book referred to as the 'safety adviser', a title that more closely reflects his true function.

The text has been written primarily for the student. However, a great deal of the content is directly relevant to the day-to-day work of practising managers. It will enable them to understand their safety obligations, both legal and moral, and to appreciate some of the techniques by which a high standard of safe working can be achieved. It will also provide an extensive source of reference for established safety advisers.

The text of any book is enhanced by the inclusion of tables, diagrams and figures and I am grateful to the many companies who have kindly provided illustrations. I would also like to acknowledge the help I have received from a number of organisations who have provided information. Particularly I would like to thank the journal *Engineering*, the Fire Prevention Association, the Health and Safety Executive, the British Standards Institution and the International Labour Office.

I also owe grateful thanks to many people for the help and encouragement they have given me during the preparation of this book, in particular Mr J. Barrell, Secretary of IOSH, Mr D.G. Baynes of Napier College of Science and Technology, Mr N. Sanders, at the time a senior safety training adviser to the Road Transport Industry Training Board, Dr Ian Glendon of the Department of Occupational Health and Safety at the University of Aston in Birmingham, Mr R.F. Roberts, Chief Fire and Security Officer at Reed International's Aylesford site and David Miskin, a solicitor, for the time each gave to check through manuscripts and for the helpful comments they offered.

I am also indebted to Reed International P.L.C. for the help they have given me during the editing of this book, a task which would have been that much more onerous without their support.

John Ridley

List of Contributors

E.W. Adrian, FIOSH
L. Bamber, BSc, DIS, MBIM, FIRM, FIOSH, RSP
AMI Occupational Health Ltd
Dr A.J. Boyle, PhD, BSc, ABPsS.
Health and Safety Technology and Management Ltd
Steve Bradley, BSc(Hons), DipSH, MIOSH, RSP, MIWM
Head of Safety and Environment, British Sugar plc.
Dr A.R.L. Clark, MSc, MB, BS, MFOM, DIH, DHMSA
Dr T. Coates, MB, BS, FFOM, DIH, DMHSA
Frank S. Gill, BSc, MSc, CEng, MIMinE, FIOSH, FFOM(Hon), Dip.Occ.Hyg.
Consultant ventilation engineer and occupational hygienist.
Professor Andrew Hale, PhD, CPsychol., MErgS, FIOSH
Professor of General Safety Science, Delft University of Technology.
Chris Hartley, MSc, MIBiol
Senior Lecturer, Health and Safety Unit, Aston University.
R.W. Hodgin, LL.M.
Senior Lecturer in Law, The University of Birmingham.
Consultant, Pinsent & Co
Edwin G. Hooper, MPhil, CEng, FIEE, FIOSH
R. Hudson, FIOSH
Construction safety consultant
Dr R.G. Lawson, LL.M., PhD
Consultant in marketing and advertising law.
E.S. Long, MSc, BA(Hons), PE(USA), HSDip, MIQA
Safety Engineer, British Engine Insurance Ltd
Colin Mackay, BSc, PhD
Principal Psychologist, Health and Safety Executive
R.D. Miskin
A solicitor
John Ridley, BSc(Eng), CEng, MIMechE, FIOSH
Stan Simpson, CEng, MIMechE, ACII, MIOSH
Eric J. Skellett
A solicitor

Ron W. Smith, BSc(Eng), MSc(Noise & Vibration)
Ecomax Acoustics Ltd
Dr Peter Waterhouse, PhD, BSc, FIOSH, FIRM
Brenda Watts, MA, BA
Barrister, Senior Lecturer, Southampton College of Higher Education
Ashton West, BA(Hons), ACII
Claims Deputy Manager, Iron Trades Insurance Group
Dr A.D. Wrixon, DPhil, BSc(Hons)
Principal Scientific Officer, National Radiological Protection Board

Part I

Law

Laws are necessary for the government and regulation of the affairs and behaviour of individuals and communities for the benefit of all. As societies and communities grow and become more complex, so do the laws and the organisation necessary for the enforcement and administration of them.

The industrial society in which we live has brought particular problems relating to the work situation and concerning the protection of the worker's health and safety, his employment and his right to take 'industrial action'.

This section looks at how laws are administered in the UK and the procedures to be followed in pursuing criminal actions and common law remedies through the courts. It also considers the various Acts and Statutes that determine or influence the implementation of safe working in the workplace. Further, the processes are reviewed by which liabilities for damages due to either injury or faulty product are established and settled.

Chapter 1

Explaining the law

Brenda Watts

1.1 Introduction

To explain the law an imaginary incident at work is used which exemplifies aspects of the operation of our legal system. These issues will be identified and explained with differences of Scottish and Irish law being indicated where they occur.

1.2 The incident

Bertha Duncan, an employee of Hazards Ltd, while at work trips over a piece of wire in a badly-lit part of the factory, and breaks her leg. The employer notifies the accident in accordance with his statutory obligations. The investigating factory inspector, Instepp, is dissatisfied with some of the conditions at Hazards, so he issues an improvement notice in accordance with the Health and Safety at Work etc. Act 1974 (HSW), requiring adequate lighting in specified work areas.

1.3 Some possible actions arising from the incident

The *inspector*, in his official capacity, may consider a prosecution in the criminal courts where he would have to show a breach of a relevant provision of the safety legislation. The likely result of a successful safety prosecution is a fine, which is intended to be penal. It is not redress for Bertha.

The *employee*, Bertha, has been injured. She will seek money compensation to try to make up for her loss. No doubt she will receive State industrial injury benefit, but this is intended as support against misfortune rather than as full compensation for lost wages, reduced future prospects or pain and suffering. Bertha will therefore look to her employer for compensation. She may have to consider bringing a civil action, and will then seek legal advice (from a solicitor if she has no union to turn to) about claiming compensation (called damages). To succeed, Bertha must prove that her injury resulted from breach of a legal duty owed to her by Hazards.

For the *employer*, Hazards Ltd, if they wish to dispute the improvement notice, the most immediate legal process will be before an industrial tribunal. The company should, however, be investigating the accident to ensure that they comply with statutory requirements; and also in their own interests, to try to prevent future mishaps and to clarify the facts for their insurance company and for any defence to the factory inspector and/or to Bertha. As a company, Hazards Ltd has legal personality; but it is run by people and if the inadequate lighting and slack housekeeping were attributable to the personal neglect of a senior officer (s. 37 HSW), as well as the company being prosecuted, so too might the senior officer.

1.4 Legal issues of the incident

The preceding paragraphs show that it is necessary to consider:

criminal and civil law,
the organisation of the courts and court procedure,
procedure in industrial tribunals, and
the legal authorities for safety law: legislation and court decisions.

1.5 Criminal and civil law

A *crime* is an offence against the State. Accordingly, in England prosecutions are the responsibility of the Crown Prosecution Service; or, where statute allows, an official such as a factory inspector (ss. 38, 39 HSW). Very rarely may a private person prosecute. In Scotland the police do not prosecute since that responsibility lies with the procurators-fiscal, and ultimately with the Lord Advocate. In Northern Ireland the Director of Public Prosecutions (DPP) initiates prosecutions for more serious offences, and the police for minor cases. The DPP may also conduct prosecutions on behalf of Government Departments in magistrates' courts when requested to do so. The procurators-fiscal, and in England and Northern Ireland the Attorney General acting on behalf of the Crown, may discontinue proceedings; an individual cannot.

Criminal cases in England are heard in the magistrates' courts and in the Crown Court; in Scotland mostly in the Sheriff Court, and in the High Court of Justiciary. In Northern Ireland criminal cases are tried in magistrates' courts and in the Crown Court. In all three countries the more serious criminal cases are heard before a jury, except in Northern Ireland for scheduled offences under the Northern Ireland (Emergency Provisions) Acts of 1978 and 1987.

The burden of proving a criminal charge is on the prosecution; and it must be proved beyond reasonable doubt. However, if, after the incident at Hazards, Instepp prosecutes, alleging breach of, say s. 2 of HSW, then Hazards must show that it was not reasonably practicable for the company to do more than it did to comply (s. 40 HSW). This section puts the burden on the accused to prove, on the balance of probabilities, that he had complied with a practicable or reasonably practicable statutory duty under HSW.

The rules of evidence are stricter in criminal cases, to protect the accused. Only exceptionally is hearsay evidence admissible. In Scotland the requirement of corroboration is stricter than in English law.

The main sanctions of a criminal court are imprisonment and fines. The sanctions are intended as a punishment, to deter and to reform, but not to *compensate* an injured party. A magistrates' court may order compensation to an individual to cover personal injury and damage to property. Such a compensation order is not possible for dependants of the deceased in consequence of his death. At present the upper limit for compensation in the magistrates' court is £5000[1].

A *civil action* is between individuals. One individual initiates proceedings against another and can later decide to settle out of court. Over 90% of accident claims are so settled.

English courts hearing civil actions are the county courts and the High Court; in Scotland the Sheriff Court and the Court of Session. In Northern Ireland the County Court and the High Court deal with civil accident claims. Civil cases rarely have a jury; in personal injury cases only in the most exceptional circumstances.

A civil case must be proved on the balance of probabilities, a lower standard than the criminal one of beyond reasonable doubt.

In civil actions the plaintiff (the pursuer in Scotland) sues the defendant (the defender) for remedies beneficial to him. Often the remedy sought will be damages – that is, financial compensation. Another remedy is an injunction, for example, to prevent a factory committing a noise or pollutant nuisance.

1.6 Branches of law

As English law developed it followed a number of different routes or branches. The diagram in *Figure 1.1* illustrates the main legal sources of English law and some of the branches of English law.

Criminal law is one part of public law. Other branches of public law are constitutional and administrative law, which include the organisation and jurisdiction of the courts and tribunals, and the process of legislation.

Civil law has a number of branches. Most relevant to this book are contract, tort (delict in Scotland) and labour law. A contract is an agreement between parties which is enforceable at law. Most commercial law (for example, insurance) has a basis in contract. A tort is a breach of duty imposed by law and is often called a civil wrong. The two most frequently heard of torts are nuisance and trespass. However, the two most relevant to safety law are the torts of negligence and of breach of statutory duty.

The various branches of law may overlap and interact. At Hazards, Bertha has a *contract* of employment with her employer, as has every employee and employer. An important implied term of such contracts is that an employer should take reasonable care for the safety of employees. If Bertha proves that Hazards were in breach of that duty, and that in consequence she suffered injury, Hazards will be liable in the *tort* of negligence. There

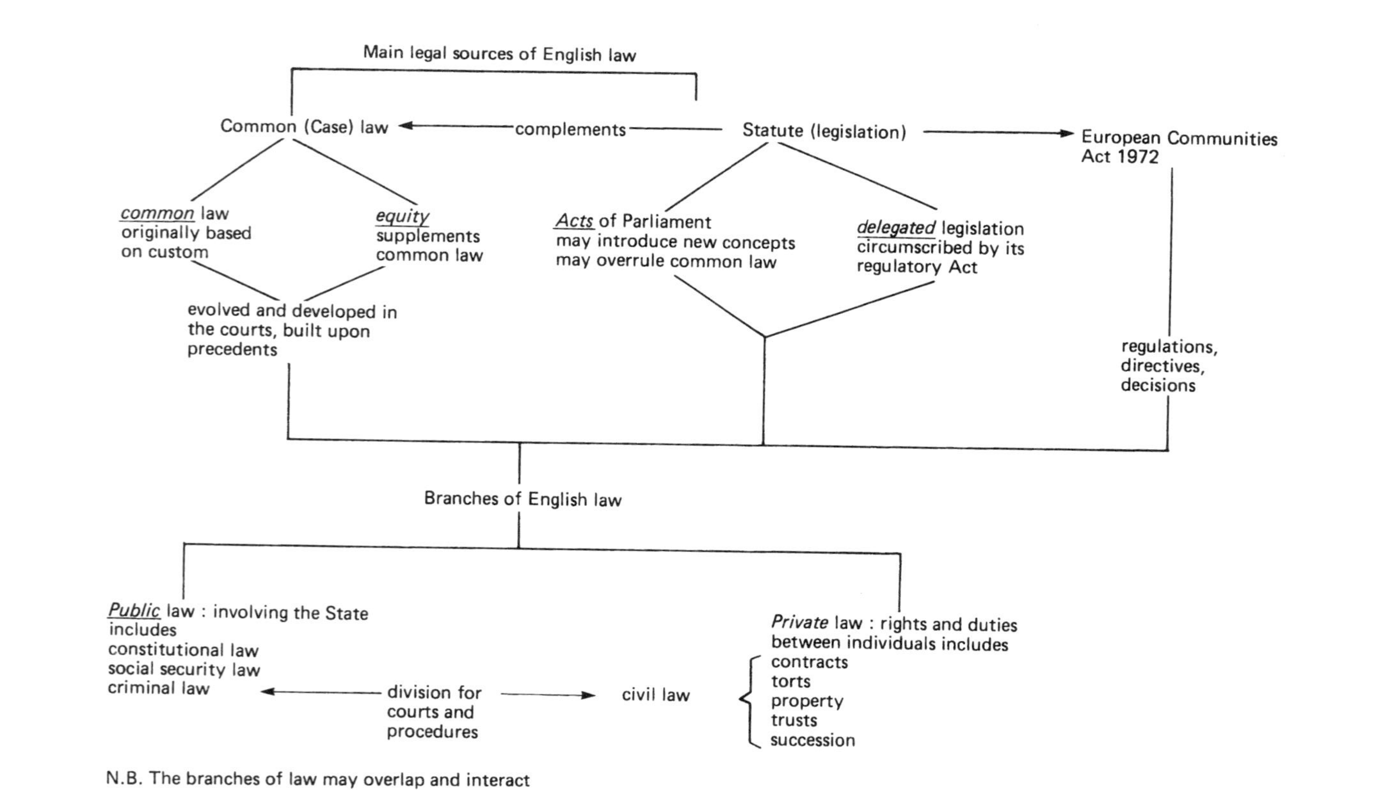

Figure 1.1 Sources and branches of English law

could be potential *criminal* liability under HSW. Again, Hazards might discipline a foreman, or Bertha's workmates might refuse to work in the conditions, taking the situation into the field of *industrial relations* law.

1.7 Law and fact

It is sometimes necessary to distinguish between questions of law and questions of fact.

A jury will decide only questions of fact. *Questions of fact* are about events or the state of affairs and may be proved by evidence. *Questions of law* seek to discover what the law is, and are determined by legal argument. However, the distinction is not always clear-cut. There are more opportunities to appeal on a question of law than on a question of fact.

Regulation 12 of the Workplace (Health, Safety and Welfare) Regulations 1992 (WHSW) requires an employer (and others, to the extent of their control) to keep, so far as reasonably practicable, every floor in the workplace free from obstructions and from any article which may cause a person to slip, trip or fall. In the Hazards incident Bertha's tripping, her injury, the wire being there, the routine of Hazards, are questions of fact. However, the meaning of 'obstruction', of 'floor', of 'reasonably practicable' are questions of law.

1.8 The courts

1.8.1 First instance: Appellate

A court may have *first instance* jurisdiction, which means that it hears cases for the first time; it may have *appellate* jurisdiction which means that a case is heard on appeal; or a court may have both.

1.8.2 Inferior: Superior

Inferior courts are limited in their powers: to local jurisdiction, in the seriousness of the cases tried, in the sanctions they may order, and, in England, in the ability to punish for contempt.

In England the superior courts are the House of Lords, the judicial Committee of the Privy Council, and the Supreme Court of Judicature. Magistrates' and county courts are inferior courts.

For Scotland the Sheriff Court is an inferior court while the superior courts are the House of Lords, the Court of Session and the High Court of Justiciary.

In Northern Ireland the superior courts are the House of Lords and the Supreme Court of Judicature of Northern Ireland. The inferior courts are the magistrates' courts and the county courts.

1.8.3 Criminal proceedings – Trial on indictment; Summary trial

The indictment is the formal document containing the charge(s), and the trial is before a judge and a jury (of 12 in England and N. Ireland, of 15 in Scotland). A summary trial is one without a jury.

The most serious crimes, such as murder, or robbery, must be tried on indictment (or solemn procedure in Scotland). Some offences are triable only summarily (for example, most road traffic offences), others (for example, theft) are triable either way according to their seriousness. Most offences under HSW are triable either way, but in practice are heard summarily.

1.8.4 Representation

A practising lawyer will be a solicitor or a barrister (advocate in Scotland). Traditionally, barristers concentrate on advocacy and provide specialist advice. A qualification for senior judicial appointment is sufficient experience as an advocate. A barrister who has considerable experience and thinks he has attained some distinction may apply to the Lord Chancellor to 'take silk'. A solicitor is likely to be a general legal adviser. Until the Courts and Legal Services Act 1990, a solicitor's right to represent in court was limited to the lower courts. That Act provides for the ending of the barrister's monopoly appearances in the higher courts. Solicitors will be able to appear in the High Court and before juries; and be appointed judges in the High Court. A party may always defend himself, but there are restrictions on an individual personally conducting a private prosecution in the Crown Court or above (*R.* v. *George Maxwell Ltd*[2]). There is no general right of private prosecution in Scotland.

1.8.5 An outline of court hierarchy in England

There is a system of courts for hearing civil actions and a system for criminal actions. These are shown diagrammatically in *Figures 1.2* and *1.3*. However, some courts have both civil and criminal jurisdiction.

The lowest English courts are the magistrates' courts, which deal mainly with criminal matters; and the county courts, which deal only with civil matters.

Magistrates determine and sentence for many of the less serious offences. They also hold preliminary examinations into other offences to see if the prosecution can show a prima facie case on which the accused may be committed for trial. Serious criminal charges (and some not so serious where the accused has the right to jury trial) are heard on indictment in the Crown Court. The Crown Court also hears appeals from magistrates. For civil cases, the Courts and Legal Services Act increases the jurisdiction of the county courts. All personal injury claims for less than £50 000 will start in a county court. District judges attached to the small claims courts may deal with personal injury cases for less than £5000. More important civil matters,

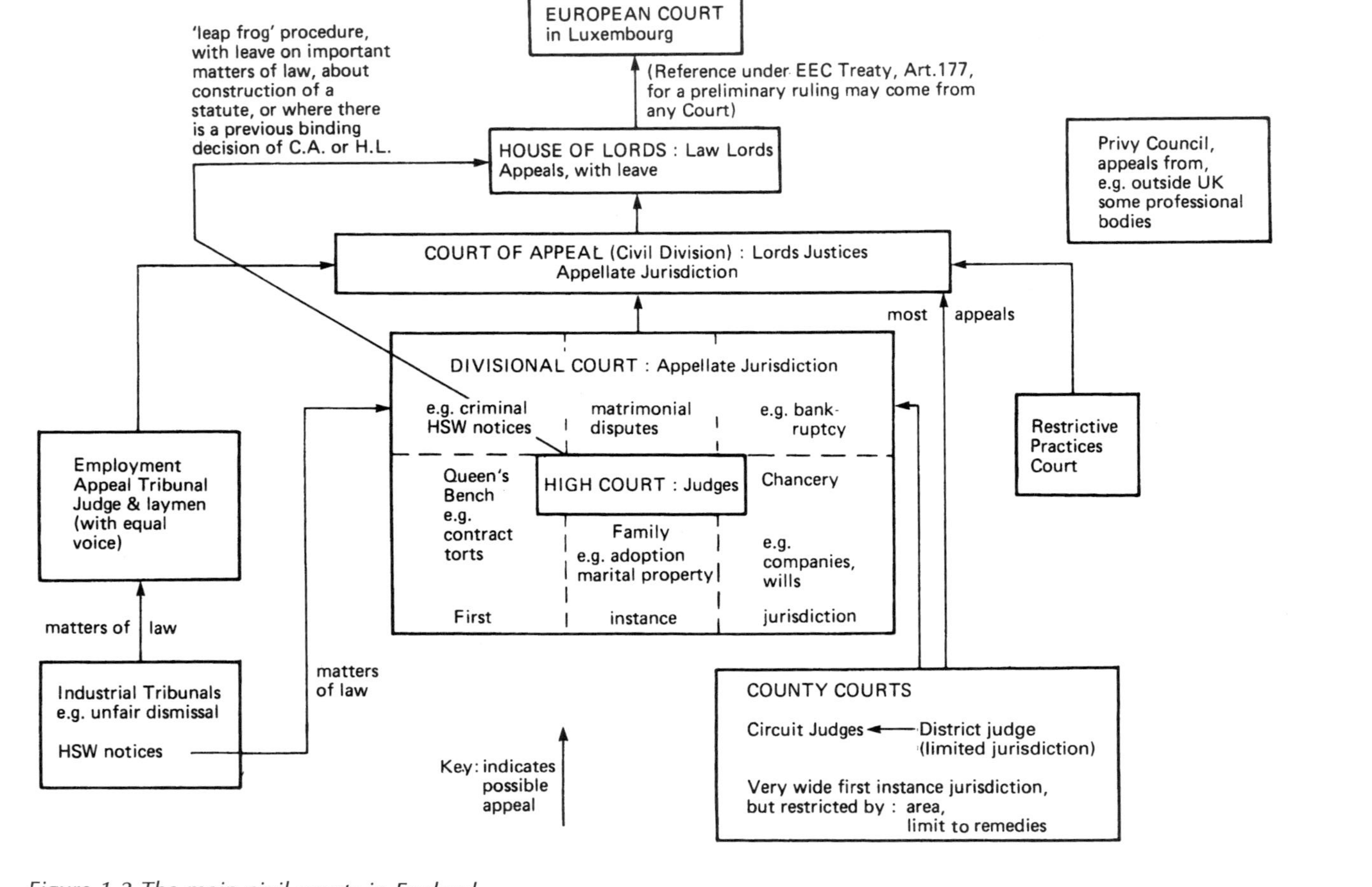

Figure 1.2 The main civil courts in England

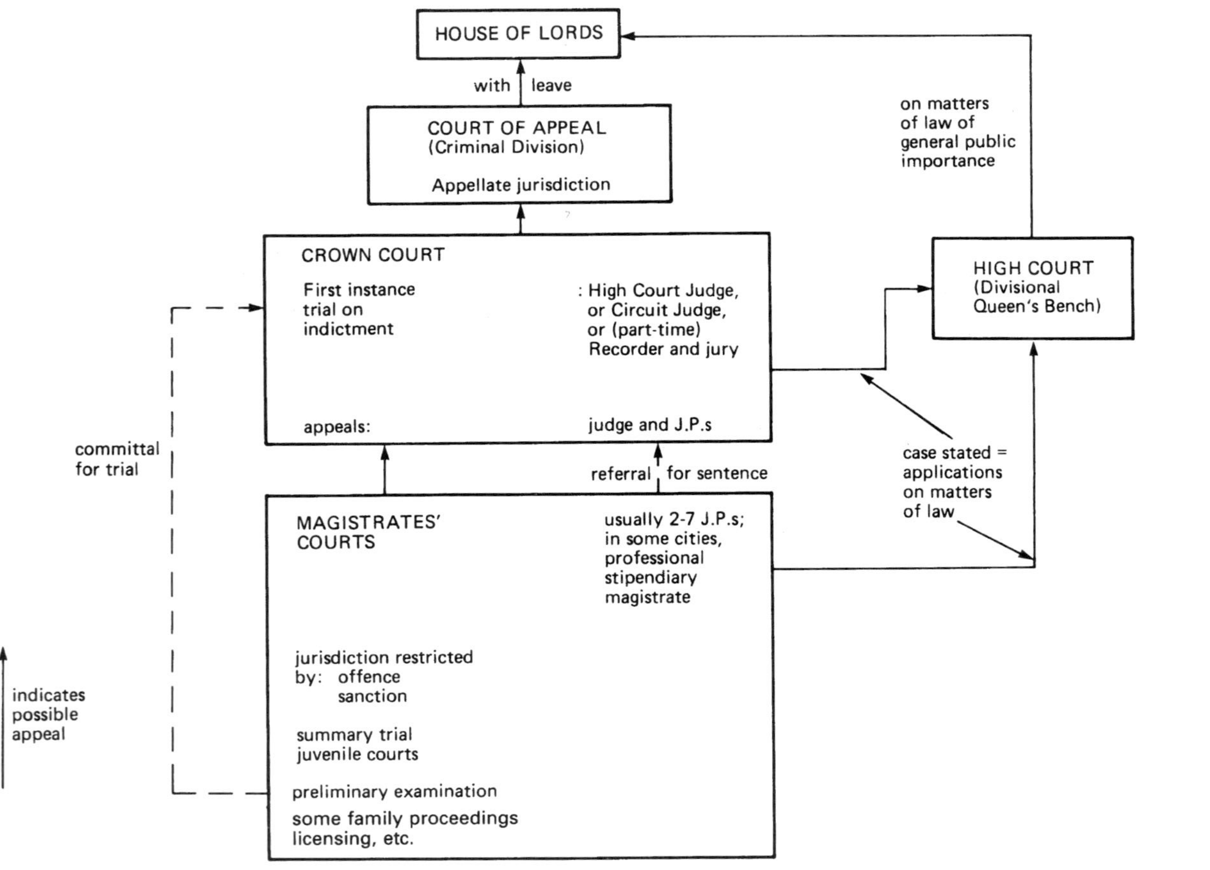

Figure 1.3 The main criminal courts in England

because of the sums involved or legal complexity, will start in the High Court of Justice. The High Court has three divisions:

Queen's Bench (for contract and torts),
Chancery (for matters relating to, for instance, land, wills, partnerships and companies),
Family.

In addition the Queen's Bench Division hears appeals on matters of law:

(1) from the magistrates' courts and from the Crown Court on a procedure called 'case stated', and
(2) from some tribunals, for example the finding of an industrial tribunal on an enforcement notice under HSW.

It also has some supervisory functions over lower courts and tribunals if they exceed their powers or fail to carry out their functions properly, or at all.

The High Court, the Crown Court and the Court of Appeal are known as the Supreme Court of Judicature.

The Court of Appeal has two divisions: the Civil Division which hears appeals from the county courts and the High Court; and the Criminal Division which hears appeals from the Crown Court. Further appeal, in practice on important matters of law only, lies to the House of Lords from the Court of Appeal and in restricted circumstances from the High Court. The Judicial Committee of the Privy Council is not part of the mainstream judicial system, but hears appeals, from, for instance, the Channel Islands, some Commonwealth countries and some disciplinary bodies.

Since our entry into the European Community, our courts must follow the rulings of the European Court of Justice. On an application from a member country, the European Court will determine the effect of European directives on domestic law. Potentially, the involvement is far-reaching in industrial obligations, including safety.

1.8.6 Court hierarchy in Scotland

Scotland also has separate but parallel frameworks for the organisation of its civil and criminal courts. These are shown diagrammatically in *Figures 1.4* and *1.5* and are discussed below.

The court most used is the local Sheriff Court which has wide civil and criminal jurisdiction. Civilly it may sit as a court of first instance or as a court of appeal (to the Sheriff Principal from a sheriff's decision). For criminal cases the sheriff sits with a jury for trials on indictment, and alone to deal with less serious offences prosecuted on complaints, when its jurisdiction encompasses that of the restricted district court.

The Court of Session is the superior civil court. The Outer House, sometimes sitting with a jury, has original jurisdiction; the Inner House hears appeals from the Sheriff Court and from the Outer House. Matters of law may be referred to the Inner House for interpretation, and it also hears appeals on matters of law from some committees and tribunals, such as decisions on HSW enforcement notices. Appeal from the Inner House is to the House of Lords. For criminal cases the final court of appeal is the High

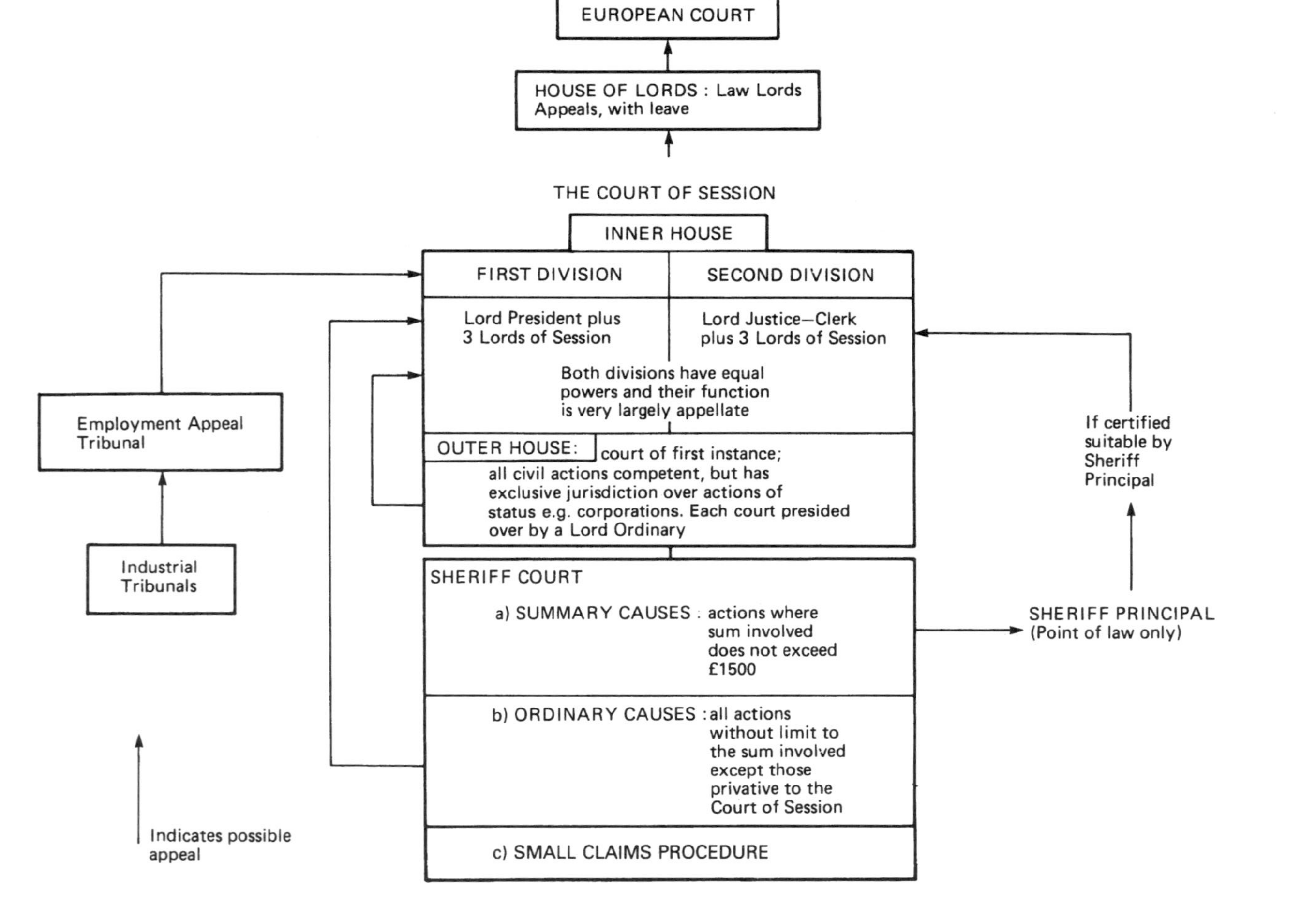

Figure 1.4 The main civil courts in Scotland

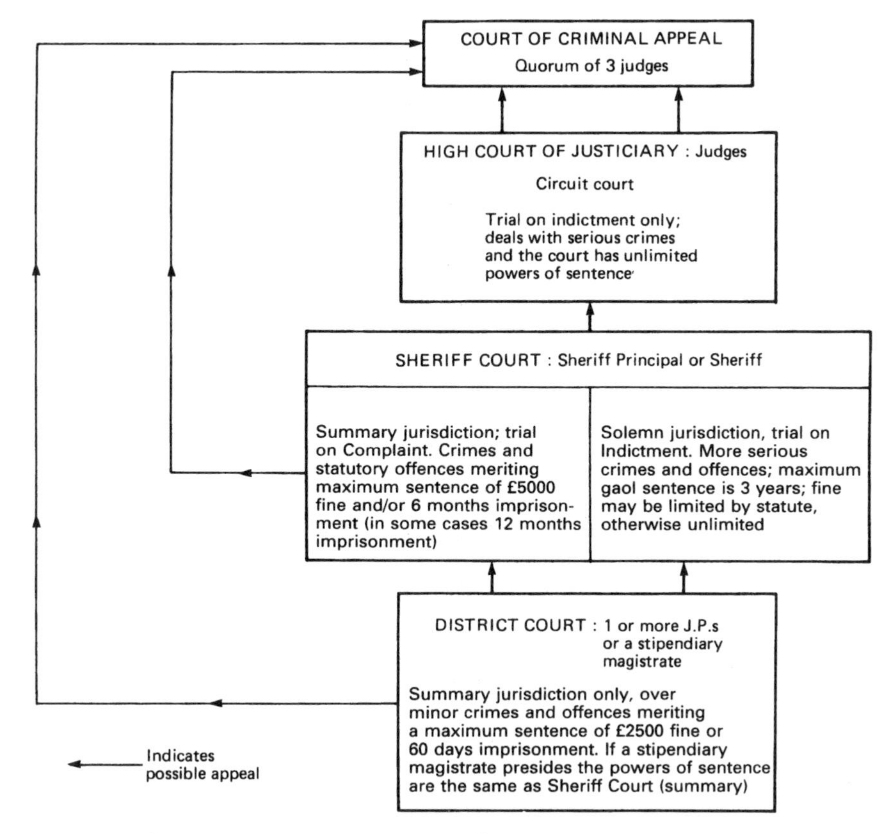

Figure 1.5 The main criminal courts in Scotland

Court of Justiciary, with three or more judges. When sitting with one judge and a jury it is a court of first instance, having exclusive jurisdiction in the most serious criminal matters and unrestricted powers of sentencing. The High Court of Justiciary hears appeals from the first instance courts but only on matters of law in cases tried summarily in the Sheriff Court and the district courts. The judges of the High Court are the same persons as the judges of the Court of Session. They have different titles and wear different robes.

1.8.7 Court hierarchy in Northern Ireland

The hierarchy of courts in Northern Ireland is different from that for the English courts and is shown in *Figures 1.6* and *1.7*.

Most criminal charges are heard in the magistrates' courts. Magistrates try summary accusations or indictable offences being dealt with summarily. They also undertake a preliminary examination of a case to be heard in the Crown Court on indictment (committal proceedings).

Following trial in a magistrates' court, the defendant may appeal to the County Court; or, on matters of law only, by way of 'case stated' to the Court

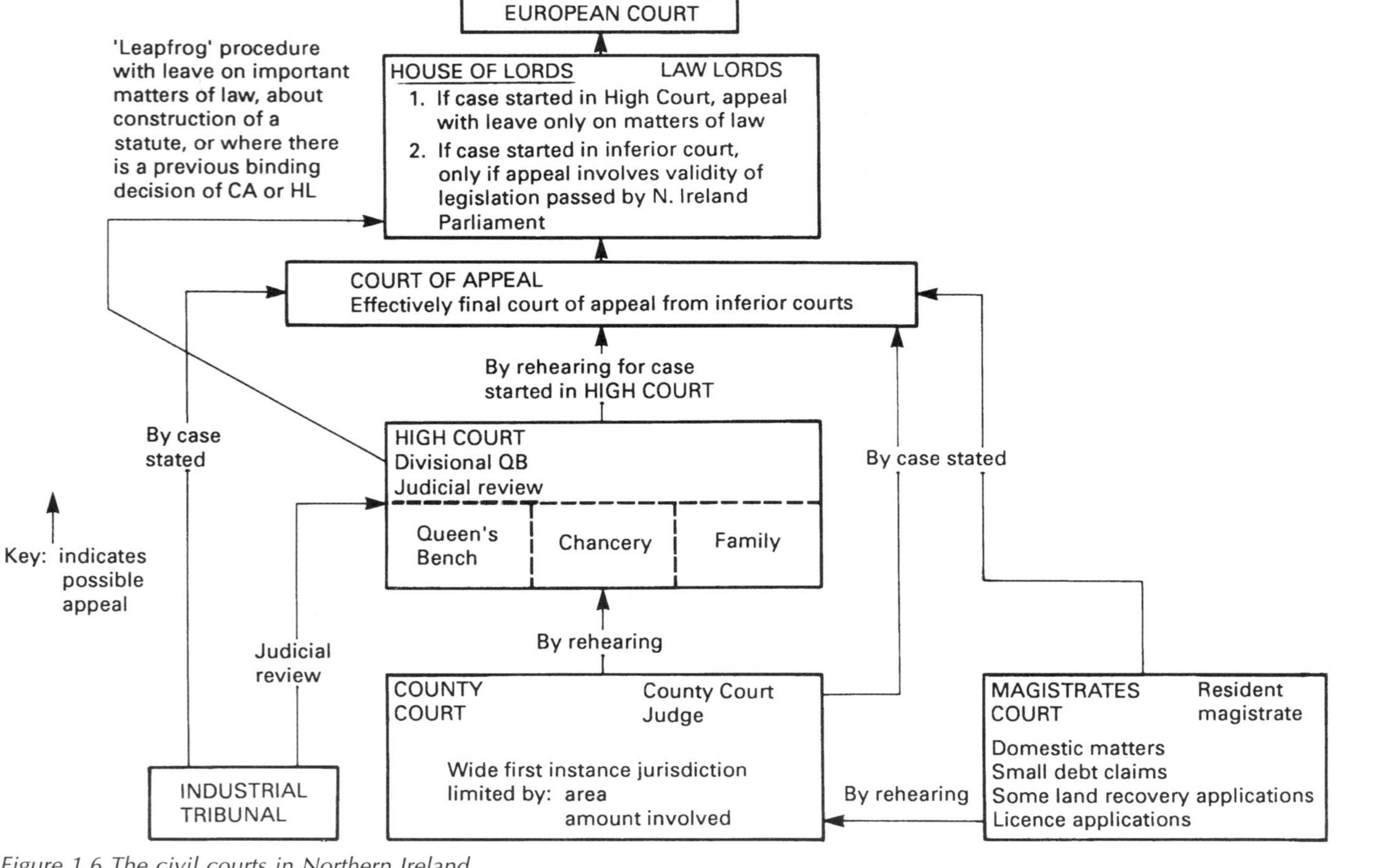

Figure 1.6 The civil courts in Northern Ireland

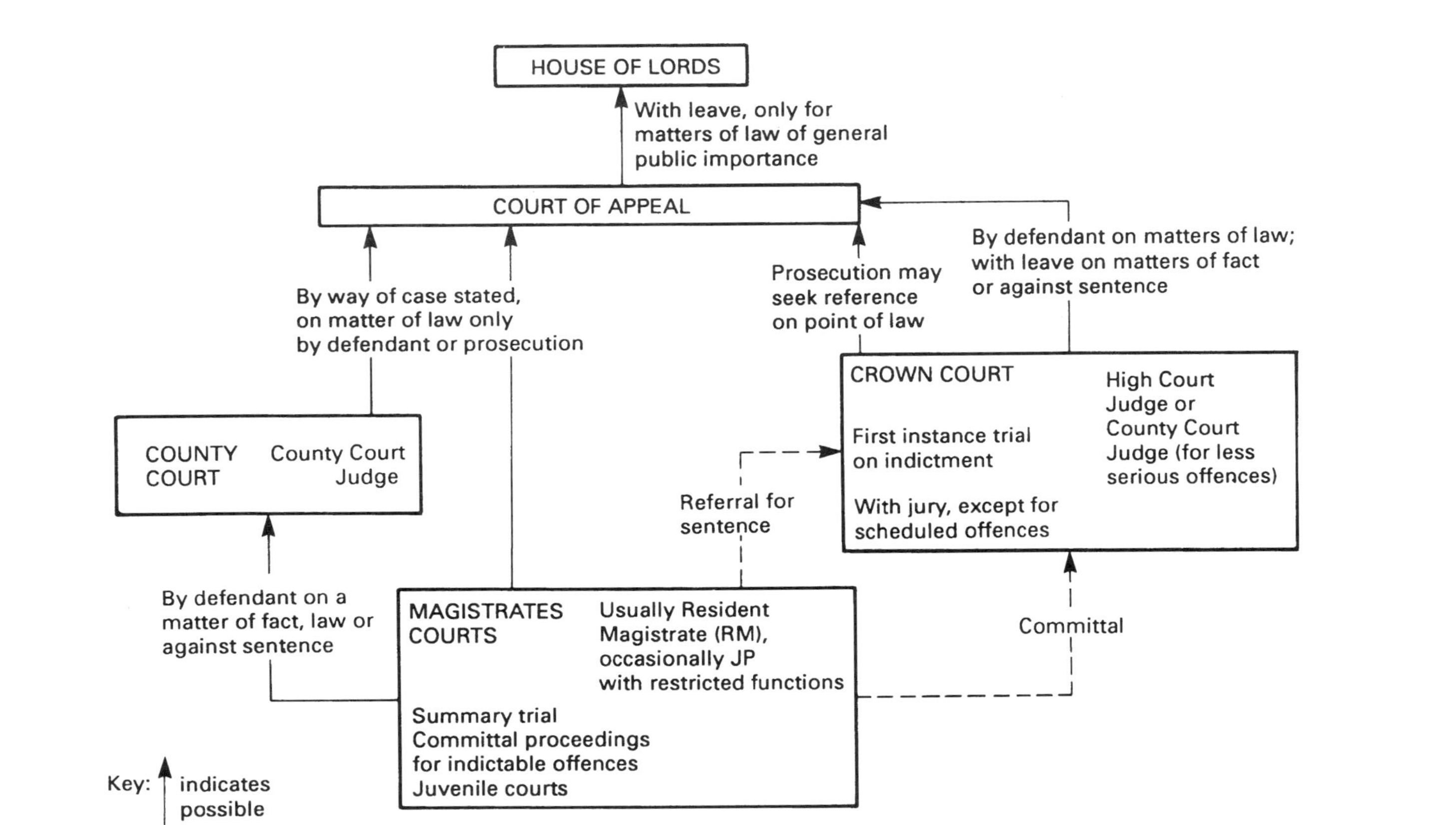

Figure 1.7 The criminal courts in Northern Ireland

of Appeal. The prosecution may appeal only to the Court of Appeal and only on a matter of law by way of 'case stated'. Trial on indictment, for more serious offences, is in the Crown Court, before a judge and jury (except for scheduled offences under the emergency legislation when cases are heard before a judge alone).

Appeal from the Crown Court is to the Court of Appeal. The defendant needs leave unless he is appealing only on a matter of law. The prosecution may refer a matter of law to the Court of Appeal, but this will not affect an acquittal. Final appeal by either side is to the House of Lords, but only with leave and only on matters of law of general public importance. Some civil proceedings take place in a magistrates' court before a resident magistrate (RM). County courts have a wider and almost exclusive civil first instance jurisdiction. The procedure is less formal than in English county courts. Appeal from a County Court is to the High Court for a rehearing, or to the Court of Appeal on a matter of law only.

The High Court has unlimited civil jurisdiction. Appeal by way of rehearing is to the Court of Appeal; or in exceptional circumstances on important matters of law, direct to the House of Lords. Appeal from the Court of Appeal to the House of Lords is possible on matters of law only and with leave.

The Divisional Court hears application for judicial review and habeas corpus in contrast to the wider jurisdiction on 'case stated' of the English court and the English Divisional Courts for Chancery and Family.

1.9 Judicial precedent

Previous court decisions are looked to for guidance. English law has developed a strong doctrine of judicial precedent (sometimes referred to as *stare decesis* – let the decision stand). Some decisions (precedents) **must** be followed in a subsequent case. Other precedents are only persuasive. To operate the doctrine of judicial precedent it is necessary to know:

(1) the legal principle of a judgement, and
(2) when a decision is binding and when persuasive.

Higher courts bind lower courts, and superior courts usually follow their own previous decisions unless there is good reason to depart from them. Only since 1966 has departure been possible for the House of Lords, and the Civil Division of the Court of Appeal is not expected to depart from its own properly made previous decisions. The Criminal Division has more latitude because the liberty of the accused may be affected.

Decisions of the superior courts which are not binding are *persuasive*, judicial decisions of other common law countries or from the Judicial Committee of the Privy Council (see below: 1.14, para. 2) are also persuasive. The judgements of inferior courts are mostly on questions of fact and are not strict precedents. Decisions of the Court of Justice of the European Communities bind English courts on the interpretation of EC legislation.

The legal principle of a judgement, the actual findings on the particular facts, is called the *ratio decidendi*. Any other comments, such as what the likely outcome would have been had the facts been different, or reference to law not directly relevant, are persuasive but not binding. They are called *obiter dicta* – 'comments by the way'. The *obiter dicta* can be so persuasive that they are incorporated into later judgements and become part of the *ratio decidendi*. This happened to the *dicta* in the famous negligence case of *Hedley Byrne* v. *Heller & Partners*[3] (see below: 1.18, para. 4). Also, *obiter* is any dissenting judgement.

A precedent can bind only on similar facts. A court may *distinguish* the facts in a present case from those in an earlier case so that a precedent may not apply. A previous decision which has been distinguished may still be *persuasive*. An appeal court may *approve* or *disapprove* a precedent. A higher court may *overrule* a precedent, i.e. overturn a principle (though not the actual decision) of a lower court in a different earlier case. If a decision of a lower court is taken to appeal, the higher court will confirm or *reverse* the specific original decision.

The English doctrine of judicial precedent has evolved to give certainty and impartiality to a legal system relying upon case law decisions. Other advantages of the doctrine are the range of cases available and the practical information therein is said to provide flexibility for application to new circumstances and at the same time detailed guidance. Criticisms of the doctrine are that it is not always easy to discover the *ratio decidendi* of a judgement. One way in which a court may avoid a previous decision is to hold that it is *dicta* and not *ratio*. Other criticisms are that the doctrine leads to rigid compliance in a later case unless the previous decision can be distinguished; and that trying to avoid or distinguish a precedent can lead to legal deviousness. The doctrine of binding judicial precedent applies similarly in N. Ireland. In Scotland precedent is important, but there is also emphasis on principle. The European Court of Justice regards precedents but is not bound by them.

For the doctrine of precedent to operate there must be reliable law reporting. Important judgements are published in the Weekly Law Reports (WLR), some of which are selected for the Law Reports. Another important series is the All England Reports (All ER). Important Scottish cases are reported in Sessions Cases (SC) and Scots Law Times (SLT). In N. Ireland the two main series of law reports are the Northern Ireland Law Reports (NI) and the Northern Ireland Judgements Bulletin (N.I.J.B.), sometimes called the Bluebook. There are various specialist law reports, to which reference may be made when considering safety cases. A list of their abbreviations is published in *Current Law*[4] which also summarises current developments and current accident awards.

Legal terminology in Law Reports includes abbreviations such as L.J. (Lord Justice), M.R. (Master of the Rolls), per Mr Justice Smith (meaning 'statement by'); *per curiam* means statement by all the court; *per incuriam* means failure to apply a relevant point of law.

A decision of a higher court is a precedent, even though it is not reported in a law report. As well as written law reports, there are computerised data bases. An important example is Lexis[5], which includes unreported judgements of the Civil Division of the Court of Appeal. This very useful development may also accentuate a practical problem of the doctrine of judicial precedent. The volume of cases which may be cited may unnecessarily complicate a submission and lengthen legal hearings. This danger has been recognised in the House of Lords[6].

1.10 Court procedure

English, Irish and Scottish law follow an 'adversary' system, in which each side develops its cases and answers the contentions of the other. The judge's functions are to ensure that the correct procedures are followed, to clarify ambiguities, and to decide the issue. He may question, but he should not 'come down into the arena' and enter into argument.

An indication of the possible proceedings that could arise following an accident to an employee at work are shown in *Figure 1.8* and considered below.

Referring to the incident, should criminal proceedings be instituted against Hazards, in England and Wales any information stating the salient facts is laid before a magistrate.

Section 38 of HSW requires this to be by an inspector or by or with the consent of the Director of Public Prosecutions. The Magistrate will issue a summons to bring the defendant before the court, and this would be served on Hazards at their registered office. Since a company has no physical existence, and therefore cannot represent itself, it would act through a solicitor or barrister.

In Scotland offences are reported to the local procurator-fiscal who decides whether to prosecute (and in what form when offences are triable either way). With serious cases he would consult with the Crown Office. If there is to be a summary trial a complaint is served on the accused stating the details of the charge.

Most HSW prosecutions are heard summarily, and then trial may commence when the accused is before the magistrates (in England and N. Ireland) or the sheriff (in Scotland). In England and N. Ireland, if the trial is to be on indictment, the magistrates will sit as examining justices to see if there is a case to answer before committing the accused for trial at the Crown Court. A magistrate may issue a witness summons and a procurator-fiscal a citation if it appears that a witness will not attend voluntarily.

In a civil claim in the High Court or Court of Session Bertha Duncan, the plaintiff (pursuer), starts her action by obtaining a writ of summons and then serving this on Hazards Ltd. Hazards would consult their solicitors who would acknowledge service and indicate whether they intend to contest proceedings (if they do not, there may be judgement in default).

Then come the pleadings when the plaintiff details the grounds of her claim and the damages she is claiming; and the defendant replies to the specific allegations.

Before trial each side must disclose to the other the existence of documents relevant to its case. The other side is allowed to inspect documents which are not privileged. An important ground of privilege is the protection of communication between a party and his legal advisers. In 1979 the House of Lords in *Waugh* v. *British Railways Board*[7] held that legal advice must be the dominant purpose of a document for it to be privileged. In this case disclosure was ordered of the report of a works accident,

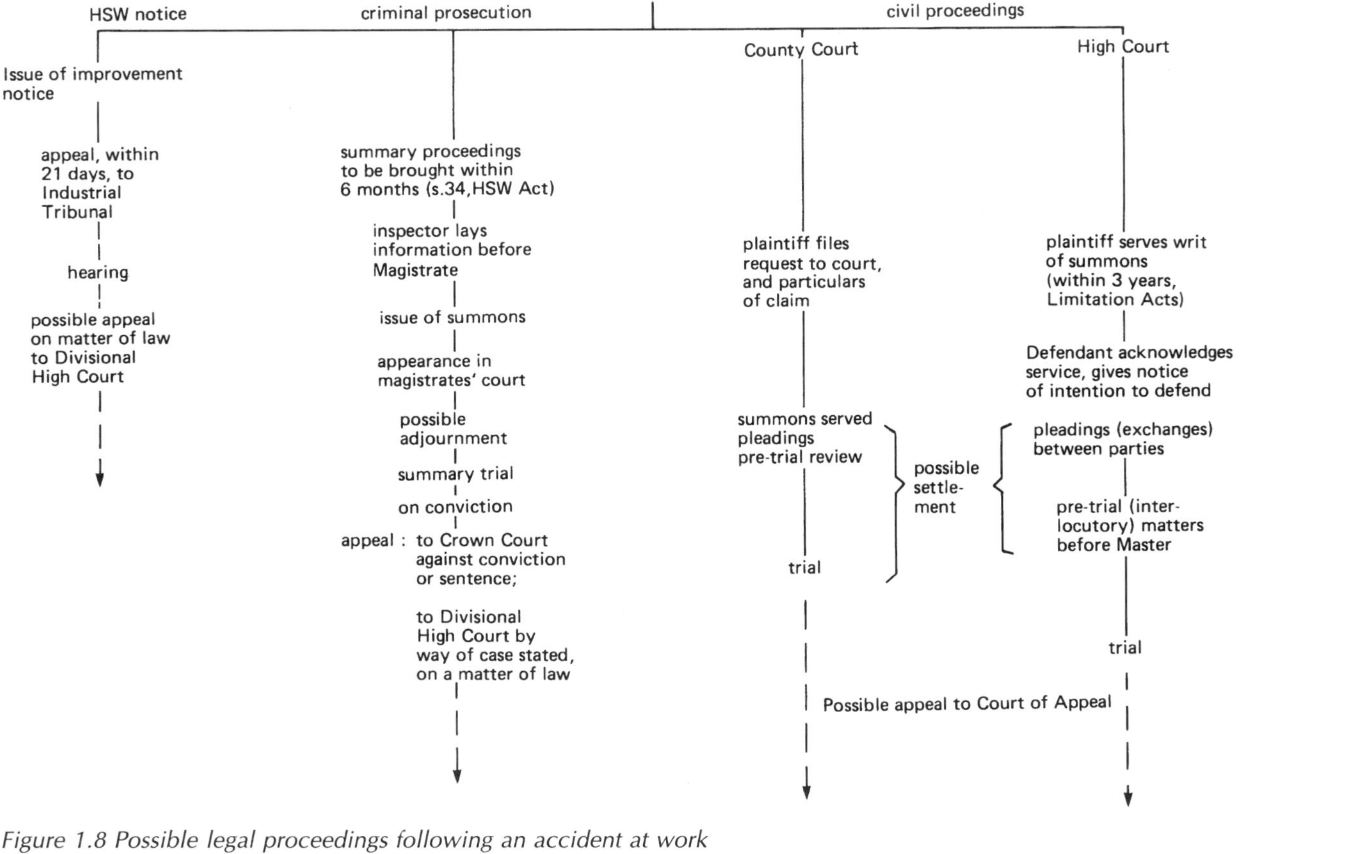

Figure 1.8 Possible legal proceedings following an accident at work

incorporating witnesses' statements, which while intended to establish the cause of the accident was intended also for the Board's solicitors.

An order (subpoena) requiring the attendance of a witness may be obtained. In N. Ireland witnesses may remain in court during the hearing of evidence, unlike England.

Proceedings in the inferior courts are similar to those in the High Court and Court of Session, but quicker, cheaper and more under the direction of the court administrators.

Usually a criminal case is decided before a related civil hearing comes on. The Civil Evidence Act 1968 (1971 for N. Ireland) allows a conviction to be used in subsequent civil proceedings. The conviction and the intention to rely on it must be set out in the formal civil pleadings. If this happened with Hazards then it would be for the company to plead and to prove (on the balance of probabilities) that the conviction is irrelevant or was erroneous.

If Bertha Duncan (*née* Smith) is suing in Scotland her case is referred to as *Smith (or Duncan)* v. *Hazards Ltd*, though for brevity it may be quoted as *Duncan* v. *Hazards Ltd.* The latter is also the English and N. Ireland practice (in speech the case is referred to as Duncan *and* Hazards Ltd).

On appeal, the party appealing, who may have been the defendant in the earlier trial, may be called the appellant and the other party the respondent.

1.11 Identity of court personnel

1.11.1 The English system

Court personnel include the bench, that is judges or magistrates; counsel for either side (see paragraph 1.8.4); and the court usher appointed to keep silence and order in court, and to attend upon the judge. All judges are appointed by the Crown, and the appointment is salaried and pensionable.

In the Magistrates' Court there are 2–7 Justices of the Peace; or, in London and some large cities, possibly a stipendiary magistrate. Justices of the Peace are lay persons appointed by the Lord Chancellor on behalf of the Queen. The office dates back to the thirteenth century, but is now mainly regulated by the Justices of the Peace Act, 1979. Justices sit part-time. They are not paid, but are reimbursed for financial expenses incurred from the office. A stipendiary magistrate is appointed by the Lord Chancellor, and is a qualified solicitor or barrister of at least seven years standing. The office is salaried and full-time.

A Clerk to the Justices advises justices on questions of law, procedure and evidence; but should not be involved in the magistrates' function of trying the case. Legislation specifies the qualifications for justices' clerks.

Officiating in the County Court is a Circuit judge; or a District judge for small claims and interlocutory (pre-trial) matters. A Circuit judge may also sit in the Crown Court. As a result of the Courts and Legal Services Act 1990, eligibility for appointment to the bench is based on having sufficient years of right of audience (qualification) in the courts. A Circuit judge must have 10 years' county court or Crown Court qualification, or be a Recorder, or have held other specified appointments. A District judge requires a 7 year general qualification (i.e. right of audience in any court).

First instance cases in the Crown Court are tried before a judge (to decide on matters of law); and a lay jury (for matters of fact). The Crown Court has three kinds of judge according to the gravity of the offence: a High Court judge, a Circuit judge or a Recorder. A High Court judge (necessary for a serious case) will be a Circuit judge with at least two years experience, or have a 10 year High Court qualification. A Recorder is part-time, with a 10 year county court or High Court qualification. For appeals to the Crown Court, there will be no jury, but possibly the judge will sit with 2–4 justices.

For the Court of Appeal, normally three judges sit. They are called Lord Justices of Appeal. Appointments are normally made from High Court judges. An alternative prerequisite is 10 years High Court qualification. High Court judges may also be asked to assist in the Court of Appeal. The Master of the Rolls is president of the Civil Division of the Court of Appeal. The Lord Chief Justice presides in the Criminal Division.

The Appellate Committee of the House of Lords as the final court of appeal sits with at least three 'Law Lords'. The Law Lords include the Lord Chancellor, the Lords of Appeal in Ordinary (who must have held high judicial office for two years or have 15 years Supreme Court (see para. 1.8.5) qualification), and Peers of Parliament who hold or have held high judicial office.

The head of the judiciary and president of the House of Lords is the Lord Chancellor. He is also a government minister, and the Speaker of the House of Lords. He is exceptional in combining judicial, executive and legislative functions.

The Attorney General is the principal law officer of the Crown. He is usually an M.P. and answers questions on legal matters in the House of Commons. He may appear in court in cases of exceptional public interest. His consent is required to bring certain criminal actions, for example in respect of offences against public order. The Solicitor General is immediately subordinate to the Attorney General.

The Director of Public Prosecutions must have a 10 year general qualification. He undertakes duties in accordance with the directions of the Attorney General. He will prosecute cases of murder and crimes amounting to an interference with justice.

1.11.2 Legal personnel in Scotland

In Scotland the Lord Advocate is the chief law officer of the Crown and has ultimate responsibility for prosecutions. He and the Secretary of State for Scotland undertake the duties which in England and Wales are the responsibility of the Home Secretary, the Lord Chancellor and the Attorney General. The Lord Advocate is assisted by the Solicitor General.

Judicial appointment, to the Supreme Court and the Sheriff Court, is by Royal authority on the recommendation of the Secretary of State. Judges in the District Courts are lay justices of the peace, apart from some stipendiary magistrates in Glasgow.

The two branches of the legal profession are solicitors and advocates. As in England, advocates no longer have exclusive rights of audience in the higher courts. Traditionally a Scottish solicitor is more a manager of his client's affairs than in England.

1.11.3 Legal personnel in Northern Ireland

The Lord Chancellor, and the English Attorney General and Solicitor General act also for Northern Ireland. The Director of Public Prosecutions is appointed by the Attorney General, and has ten years legal practice in Northern Ireland. His chief function is responsibility for prosecutions in serious cases (compare the Crown Prosecution Service in England, and the Lord Advocate and procurators fiscal in Scotland).

As in England, appointment to the bench and advocacy in the superior courts is at present restricted to barristers. A major difference between the legal system of Northern Ireland and England is the appointment of resident magistrates (RM). They are full-time and legally qualified, with responsibility for minor criminal offences, committal proceedings, and some civil matters. The powers of lay Justices of the Peace in Northern Ireland are limited in comparison with JPs in England and Wales.

1.12 Industrial tribunals

Industrial Tribunals were set up in 1964 to deal with matters arising under the Industrial Training Act of that year. Now they have statutory jurisdiction in a range of employment matters, such as unfair dismissal, redundancy payments, equal pay and sex and race discrimination. The Secretary of State may by order confer jurisdiction on industrial tribunals in respect of claims for breach of contract of employment. Such jurisdiction does not include a claim in respect of personal injuries[8]. In the context of HSW they hear appeals against prohibition and improvement notices, and applications by statutory safety representatives about payment for time off for training.

Tribunals sit locally and consist of a legally qualified chairman plus a representative from each side of industry. Proceedings begin with an originating notice of application in which the applicant sets out the name and address of both parties and the facts of the claim. The application must be made within the prescribed time limit. This varies. It is 21 days with enforcement notice; three months for unfair dismissal and paid time off for union duties; six months for redundancy applications.

Proceedings are on oath, but they are more informal than in the courts and the strict rules of evidence are not followed. Legal aid is not available for representation. A friend or union official may represent (which is not possible in the courts). Costs are rarely awarded. Like the courts, tribunal proceedings are open to the public, and visits are the best way to understand their working.

An appeal is possible from an industrial tribunal decision, but only on a matter of law. In respect of enforcement notices it is to the High Court in England; and to the Court of Session in Scotland. In respect of other matters it is to the Employment Appeal Tribunal except in N. Ireland.

The Employment Appeal Tribunal is a superior court associated with the High Court. It sits with a judge and 2–4 lay members, and all have equal voice. Parties may be represented by any person they wish, and legal aid is available. Further appeal is to the Court of Appeal (in Scotland to the Inner House of the Court of Session). In N. Ireland there is no Employment Appeal Tribunal but an Industrial Tribunal's decision may be challenged by review by the Tribunal itself, by judicial review by the High Court, or by way of case stated to the Court of Appeal.

1.13 European Community Courts

1.13.1 The Court of Justice of the European Communities

The European Court is the supreme authority on Community law. Its function is to 'ensure that in the interpretation and application of the EEC Treaty the law is observed' (art. 164). The EEC Treaty and the Single European Act, 1986, are concerned with matters such as freedom of competition between Member States; and aspects of social law, including health and safety at work.

The Court has two types of jurisdiction, direct actions, and reference for preliminary rulings.

Direct actions may be:

- against a Member State for failing to fulfill its obligations under Community Law and be brought by the Commission or by another Member State;
- against a Community institution, for annulment of some action, or for failure to act (judicial review);
- against the Community for damages for injury by its institutions or servants;
- against a Community institution brought by one of its staff.

References for preliminary rulings are requests by national courts for interpretation of a Community provision. Art. 177 provides that any court or tribunal may ask the European Court for a ruling; but only the final court of appeal (the House of Lords in the UK) must ask for a ruling if a party requests it. In the English case of *Bulmer* v. *Bollinger*[9] the Court of Appeal held that the High Court and the Court of Appeal may interpret Community law.

The European Court is based at Luxembourg. There are 13 judges (to include one from each Member State), assisted by six Advocates General. The function of an Advocate General is to assist the Court by presenting submissions, in which he analyses the relevant issues and makes relevant recommendations for the use of the court. The judgement itself is a single decision, thus an odd number of judges is required. With the increase in workload, there is a facility for the Court to sit in sub-divisions called Chambers. Cases brought by a Member State or by a Community institution must still be heard by the full court. Although the Court seeks to have consistency in its findings, precedents are persuasive rather than binding on itself. Decisions are binding on the particular member state.

Referrals to the Court of Justice are requests to it to rule on the interpretation or applicability of particular parts of Community law. Where the Court of Justice makes a decision, it not only settles the particular matter at issue but also spells out the construction to be placed on disputed passages of Community legislation, thereby giving clarification and guidance as to its implementation.

It keeps under review the legality of acts adopted by the Council and the Commission and also can be invited to give its opinion on an agreement which the Community proposes to undertake with a third country, such opinions become binding on the Community.

Through its judgements and interpretations, the Court of Justice is helping to create a body of Community law that applies to all Community Institutions, Member States, national governments and private citizens. Judgements of the European Court of Justice take primacy over those of national courts on the matters referred to it.

Although appointed by the Member States, the Court of Justice is not answerable to any Member State or to any other EC institution. The independence of the judges is guaranteed.

Under the Single European Act, 1986, the Council of Ministers has the power to create a new Court of First Instance. This Court was established by Council decision in 1988 and became effective in September 1989. It has 12 Members, appointed by common accord of the Member States. Members may also be asked to perform the task of an Advocate General. It may sit with three or five judges.

The jurisdiction is:

- disputes between the Community and its staff;
- applications for judicial review against the Council or Commission;
- applications for judicial review in some matters against the European Coal and Steel Community.

There is also a Court of Auditors, which supervises the implementation of the budget.

1.13.2 The European Court of Human Rights

This Court should not be confused with the Court of Justice of the European Communities. The Court of Human Rights sits at Strasbourg. Its function is to interpret the European Convention for the Protection of Human Rights and Fundamental Freedoms, drawn up by the Council of Europe in 1950. The Council of Europe comprises 23 Western European states. It is active on social and cultural fronts rather than economic. The United Kingdom ratified the Convention in 1951, so that it is binding on the UK internationally. However, UK legislation has not yet incorporated the Convention. The articles of the Convention provide for matters such as the right not to be subjected to inhuman or degrading treatment, the right to freedom of peaceful assembly, the right to respect for family life, home and correspondence.

An example of a decision directed to the UK was the 'Sunday Times thalidomide case' in 1981. A drug prescribed for pregnant women caused severe abnormalities in the children. The manufacturers sought an injunction to prevent the Sunday Times publishing an article about the drug. The Court of Human Rights ruled that the House of Lords' confirmation of an injunction was a violation of the right of freedom of expression[10].

1.14 Sources of English law

The two main sources of UK law are legislation, and legal principles developed by court decisions (common or case law).

English common law, based on custom and evolving since the eleventh century, developed indigenous concepts, and unlike most European countries was little influenced by Roman law. In Scotland Roman law was an important influence from the sixteenth to the eighteenth century, particularly on the law of obligations, which includes contract and delict. In Ireland, before the seventeenth century, Brehan law (of early Irish jurists) or English common law predominated according to political control at the time. Since the seventeenth century the law in Ireland and England developed along similar lines in general, with some exceptions such as marriage and divorce. English common law concepts were applied in former British territories. Today most of the United States, Canada (other than Quebec), Australia, New Zealand, India and some African countries remain and are called common law countries.

England, Scotland and N. Ireland do not have codified legal systems. Nearly all of our law of contract and much of the law of tort or delict is case law. This will gradually change with the production and implementation of Law Commission reports.

As with most subjects, law has specific terminology. The historic development of our law is illustrated by the Latin, old French and old English phrases which are sometimes used. This chapter contains some Latin words, for example, *obiter dicta* and *ratio decidendi* (section 1.9); and some coming from the French, such as tort and plaintiff (sections 1.5, 1.6). The most straightforward rule for legal Latin or French is to pronounce words as though they were English. Other words and phrases met with have a particular legal meaning, such as damages, contract of employment, relevant statutory provision; and abbreviations such as J.P. or *v.* (as in Donoghue *v.* Stevenson). There are a number of Law Dictionaries to explain or to translate words and these are listed at the end of this chapter.

1.15 Legislation

1.15.1 Acts of Parliament and delegated legislation

Since the eighteenth century increasing use has been made of legislation. Legislation comprises Acts of Parliament and delegated legislation made by subordinate bodies given authority by Act of Parliament. Examples of delegated legislation are ministerial orders and regulations (Statutory

Instruments), local authority bye-laws and court rules of procedure. All legislation is printed and published by HMSO. Often, but not always, delegated legislation requires the approval of Parliament, for example by negative resolution (that is by not receiving a negative vote of either House); or, more rarely, by affirmative resolution (that is by requiring a positive vote of 'yes').

HSW and its associated regulations is an example of how extensive subordinate legislation may be. HSW is an enabling Act. Section 15, schedule 3 and s. 80 give very wide powers to the Secretary of State to make regulations. The regulations are subject to negative resolution (s. 82). They may be made to give effect to proposals of the Health and Safety Commission (in N. Ireland the Health and Safety Agency); or independently of such proposals, but following consultation with the Commission and such other bodies as appear appropriate (s. 50). The Commission may also issue Approved Codes of Practice (s. 16 HSW) for practical guidance. Such codes are not legislation and s. 17 confirms that failure to observe such codes cannot of itself ground legal proceedings. However, failure to comply is admissible evidence and will be proof of failure to comply with a legislative provision to which the code relates unless the court is satisfied that there is compliance in some other way.

Delegated legislation is suitable for detailed technical matters. By avoiding the formality required for an Act of Parliament the legislation can be adapted, and speedily (for example, the maximum unfair dismissal payment may be increased quickly by an Order).

Long drawn out consultation may slow down any legislation. In 1955 the decision in a famous case of *John Summers & Sons Ltd* v. *Frost*[11] virtually meant that an abrasive wheel was used illegally unless every part of that dangerous machinery was fenced. Regulations were required to allow its legal use. There were drafts and consultations, but it was 1971 before the Abrasive Wheels Regulations came into operation[12].

During its passage through Parliament and before it receives the Royal Assent an intended Act is called a *bill*. Most government bills start in the House of Commons, but non-controversial ones may start in the House of Lords. Ordinary public bills such as that for HSW go through the following process. The bill is introduced and has a formal first reading. At the second reading there is discussion on the general principles and purpose of the bill. It then goes to committee. After detailed consideration the committee reports the bill to the House, which considers any amendments. The House may make further amendments and return the bill to Committee for further consideration. After the report stage the bill is read for the third time. At third reading in the Commons only verbal alterations may be made.

The bill now goes through similar stages in the other House. If the second House amends the bill it is returned to the first House for consideration. If the Lords reject a bill for two sessions it may receive the Royal Assent without the Lords' agreement. Practically, the Lords can delay a bill for a maximum of one year.

After being passed by both Houses the bill receives the Royal Assent, which conventionally is always granted, and thus becomes an Act. A statute will normally provide at the end whether it is to apply in Scotland and N. Ireland as well as in England and Wales. Subsequent legislation may apply provisions to Scotland or N. Ireland, for example the Health and Safety at Work (NI) Order 1978.

Parliament has supreme authority. It may enact any measure, other than binding future Parliaments. It is not answerable to the judiciary.

The United Kingdom is now part of the European Community and subject to the Community's regulations and directives (see paragraph 1.15.4). These require member states to implement agreed standards on, among other concerns, safety and health at work and the environment.

The ultimate sovereignty of the UK Parliament is theoretically retained in that Parliament could repudiate agreement to EC membership[13]. Also, since the Single European Act there has been increased emphasis on *subsidiarity*. This is the principle that decisions should be taken at the most suitable level down the hierarchy of power, that is at national rather than EC level where appropriate.

1.15.2 Statutory interpretation

Inevitably some legislation has to be interpreted by the courts, to clarify uncertainties, for example, and substantial case law may attach to a statute. Judicial consideration of the effect of legislation for the fencing of dangerous machinery is an example of this (see sections 1.16.3 and 1.18.1).

Statutes normally contain an interpretation section. There is also the Interpretation Act 1978 which provides, for example, that unless the contrary is stated, then male includes female, the singular includes the plural, writing includes printing, photography and other modes of representing or reproducing words in visible form. In modern legislation, the detail is often relegated to Schedules at the end of the Act.

Parliamentary discussions are reported verbatim in Hansard. In 1992 the House of Lords decided that if there is an ambiguity, a minister's clear explanation to Parliament, as published in Hansard, may be used to interpret a statute[14].

As a result of Article 5 of the EEC Treaty 1957, which requires member states to 'take all appropriate measures to ensure fulfillment of the obligations arising out of the treaty', UK courts give a *purposive* interpretation where the purpose of UK legislation is to give meaning to a directive. An example is *Pickstone* v. *Freeman plc*[15]. The House of Lords interpreted regulations amending the Equal Pay Act against their literal meaning to allow a female warehouse operative to use as a comparison a man doing a different job of equal value.

1.15.3 White Papers and Green Papers

Proposed legislation may be preceded by documents presented by the government to Parliament for consideration. A green paper is a discussion document. A white paper contains policy statements and explanations for proposed legislation. Such papers are published as Command Papers.

On a narrower basis the Government also consults with outside interests when drafting legislation, bodies such as the CBI and TUC on industrial and economic matters. Legislation may require such consultation, for example s. 50 HSW.

1.15.4 European Community (EC) legislation

Originally known as the European Economic Community (EEC) but now referred to as The European Community (EC), its primary EC legislation is the Treaty of Rome 1957, which established the Community and was incorporated into UK law by the European Communities Act 1972; and the Single European Act 1986, incorporated into UK law by the European Communities (Amendment) Act 1986.

Secondary community legislation takes three forms: Regulations which are binding on Member States, Directives which require national implementation (see section 1.15.1) and Decisions of the Council or Commission. Such a decision is specific rather than general. Its main use is if a State asks permission to depart from the EEC Treaty, for example in respect of competition policy.

Legislation is usually initiated by the European Commission and, after statutory consultation is adopted by the Council of the European Community (CEC). The administration of the EC is in the hands of the Commission which has 17 members, one from each Member State but with the larger Member States having two. The supreme body of the EC is the Council of Ministers with one member from each State but with weighted voting rights according to size. The Council receives proposals from the Commission and consults with the Economic and Social Committee (EcoSoC) and the European Parliament with the aim of reaching a common position on the proposal. When the Council adopts a proposal it places obligations on Member States to incorporate its requirements into national laws within a stated time scale. Adopted legislation is published in the Official Journal of the European Communities.

In outline, the procedure for secondary legislation is that the Commission proposes and consults; the European Parliament considers and may propose amendments; the Council adopts; and Member States implement. *Figure 1.9* illustrates this in more detail. *Figure 1.10* shows internal UK procedure for incorporating a directive into UK law.

The function of the European Parliament is advisory and supervisory rather than legislative and much of its work is done in committees. Parliamentary influence has increased following the introduction of direct elections in 1979 and the Single European Act in 1986.

The Single European Act introduced, for certain matters, a 'co-operative' procedure between Council, Parliament and the Commission which allows Parliament a second chance to comment and suggest changes to proposals once a common position has been adopted by Council. If the Commission accepts these changes then Council can adopt the proposal by qualified (weighted) majority vote. If the Commission and Parliament cannot reconcile their opinions then Council can only adopt the proposal by unanimous vote. This new procedure applies, among other matters, to proposals concerning the health and safety of workers. The introduction of

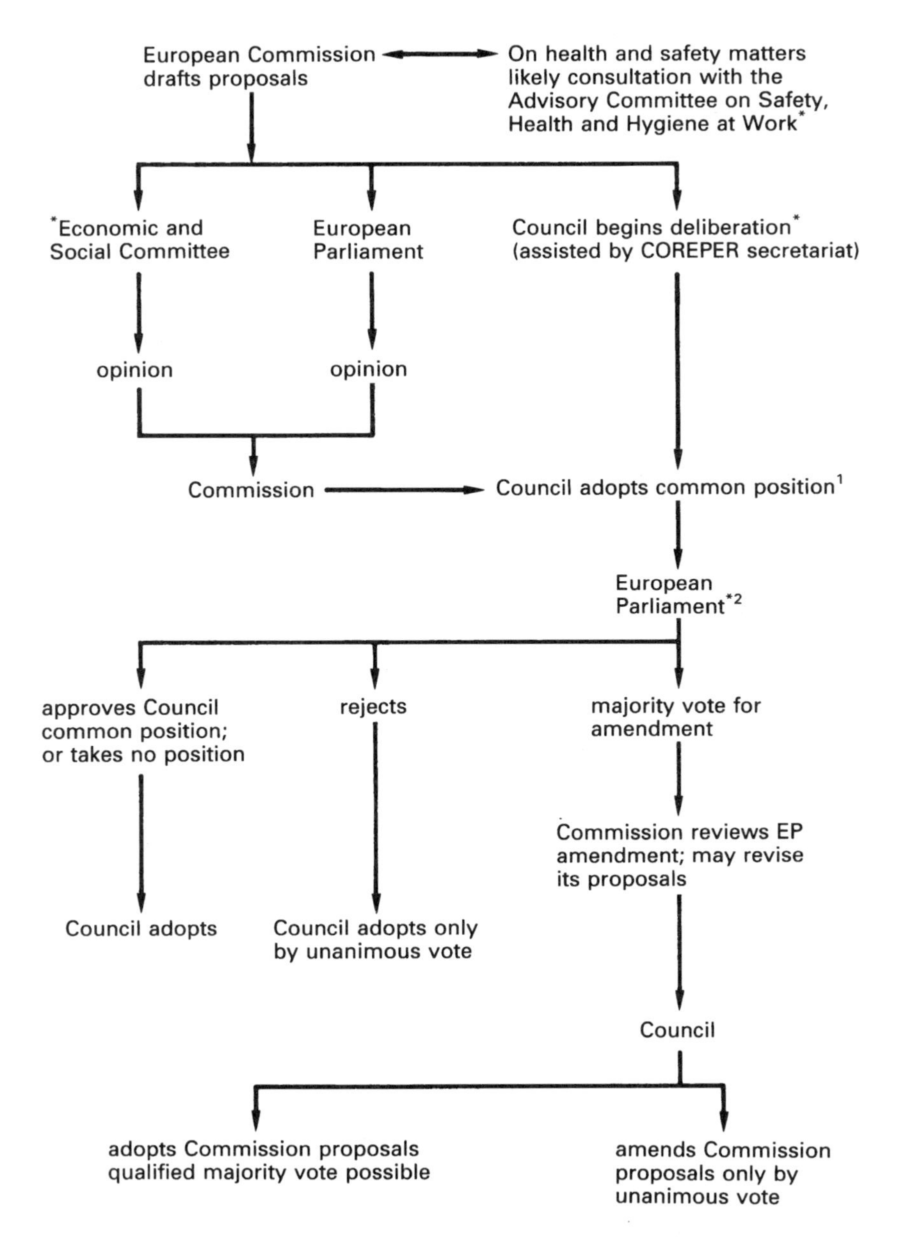

* denotes points at which consultation/lobbying may occur

1. Council may adopt common position by qualified majority vote

2. European Parliament has a second imput.
   Note: The Treaty on European Unity (Maastricht) gives European Parliament veto power for certain proposals, including health

Figure 1.9 EC co-operation procedure for adoption of directives

Community adopts Directive
(member States must implement within time limit)

1. HSE prepare proposals
2. Limited consultations with e.g. CBI, TUC, professional bodies
3. Draft regulations to HSC
4. HSC issues consultative document for public comment (consultation period of some 4 months)
5. Comments co-ordinated by HSE
6. Finalised proposals, taking account of public comment
7. Final proposals submitted to HSC
8. HSC submit proposals to Secretary of State
9. Proposals placed before parliament; negative resolution
10. Put on Statute Book and becomes UK law
11. Effective on date announced

Figure 1.10 Stages of internal UK procedure for implementing a directive

qualified majority voting was the stimulus for a great increase in EC health and safety directives from 1989 onwards.

The restricted power of the elected body, the European Parliament, causes concern that the EC lacks democratic accountability. The Maastricht Treaty (Treaty on European Unity 1991) includes greater power for the European Parliament, allowing it to reject certain proposals, including those on health and consumer affairs (Article 189b). Also, Article 138b provides that the European Parliament may, acting by majority vote of its members, request the Commission to prepare appropriate proposals on any matters on which it considers that a Community act is required for the purposes of implementing the Treaty.

1.15.5 Application of EC legislation to an individual

The Treaty and Community legislation must be recognised in the Member States, but an individual can only enforce it, if at all, in the national courts; and only if it has 'direct effect' for that individual. Community legislation takes two main forms, regulations and directives (see also paragraph 1.15.4). A regulation is a law in the Member States to which it is directed; it is said to be 'directly applicable' to that State. According to its content a Community regulation may impose obligations and confer rights on individuals enforceable in the national courts; it is then said to have 'direct effect'. A directive must be enacted by the Member State, and then, according to how it is enacted, may give enforcement rights to individuals in the national courts. Sometimes a Directive, even before implementation by the Member State, may have 'direct effect' for an individual to rely on it against the State. This could be so if the date of implementation had passed and the existing law of the Member State contravenes the directive[16]. The directive must be sufficiently clear, precise and unconditional.

Any such direct effect of a directive does not give rise to obligations between individuals. However, in *Marshall* v. *Southampton and South West Hampshire Area Health Authority (Teaching)*[17], Mrs Marshall successfully challenged the health authority's compulsory retiring age of 65 for men and

60 for women as being discriminatory. An individual may not enforce such a decision against a private employer but can against a government body[18]. However, The European Court of Justice has required national courts to *interpret* national legislation to be consistent with directives.

1.16 Safety legislation before the Health and Safety at Work etc. Act

1.16.1 Factories

Early factory legislation, in the nineteenth century, concerned the textile and allied industry. It was directed towards the protection of young persons and women and was motivated by concern for moral welfare and sanitation as much as for safety. Between 1875 and 1937 there were attempts to unify the increasing but fragmented legislation, but subsequent inadequacies resulted in patchwork amendments. The Factories Act 1937 was intended as a co-ordinating measure. It brought together health, safety and welfare in all factories; and introduced some new requirements such as those for floors, passages and stairs, and for safe access.

But regulations made under previous legislation continued in force as though made under the 1937 Act. This practice was repeated by the Factories Act 1961 so that some of the provisions and standards were outdated. The HSW and consequent regulations, including those implementing EC directives, have replaced much of the Factories Act and associated legislation.

Similarly, HSW regulations have superseded or augmented other workplace-specific provisions, such as for offices, agriculture, mines and quarries.

1.16.2 Offices

In 1949 the Gower Committee report made recommendations about the health, welfare and safety of employed persons outside the protections of existing legislation. In 1960 an Offices Act was passed. Before it became operative however it was repealed and replaced by the Offices, Shops and Railway Premises Act 1963. This adopted much of the structural content of the Factories Act 1961 but not the regulations, which apply only to factories.

1.16.3 Mines, quarries, etc.

The law relating to safety and management in mines and quarries was examined in the 1950s and the principal Act is now the Mines and Quarries Act 1954. HSW regulations are more likely to augment and update rather than absorb rules for this very particular work environment. There is wide power to make regulations. Other Acts refer to work practices in agriculture, aviation and shipping.

1.17 Safety legislation today

1.17.1 Health and Safety at Work etc. Act 1974

In 1970 the Robens Committee was set up to review the provision made for the safety and health of persons in the course of their employment. At that time safety requirements were contained in a variety of enactments (as the list of relevant statutory provisions in schedule 1 of HSW indicates). An estimated five million employees had no statutory protection. Protection was uneven. Administration was diverse and enforcement powers were considered inadequate. The wording and intent of the legislation were not directed towards personal involvement of the worker; and in parts it was obsolete.

HSW corrects many of these defects. General principles are enacted, to be supplemented by regulations. The provisions apply to employments generally to protect persons at work and those at risk from work activities.

The Act was intended to be wide to facilitate changing circumstances. Examples of development are the sanctions for non-compliance; and the use of the extensive powers to make regulations under section 15 and Schedule 3.

Magistrates may now impose a fine up to £20 000 for breach of sections 2 to 6 HSW or for a breach of an improvement or prohibition notice or a court remedy order. In addition, magistrates may imprison individuals for up to six months for breach of an improvement or prohibition notice or court remedy order[19].

Sections 2 to 6 were selected because they contain the main health and safety duties of those responsible for workplace safety. It was considered that a company charged with breach of one of these sections is probably responsible for a systematic failure to meet these general duties and is putting its employees and possibly others at risk. Failure to comply with a notice indicates a deliberate flouting of health and safety law.

The maximum magistrates' fine for other offences is £5000[20].

The Crown Court has used powers under section 2(1) of the Company Directors Disqualification Act 1986 to disqualify a director for two years. The Act allows the court to make a disqualification order against a person convicted of an indictable offence connected with the *management* of a company. The accused's company was fined under s. 33 HSW for breach of a prohibition notice; and the accused under s. 37 HSW because the company's offence was committed with his consent, connivance or attributable to his neglect[21].

1.17.2 EC influence

The Single European Act 1986, with the objective of a single market by 1st January 1993, has had a dynamic effect on the introduction of health and safety legislation. The implementation of effective common health and safety standards is considered conducive to attaining a 'level playing field' for employers across the Community; and to the participation of the workforce in the intended resulting economic benefits.

Article 118A (introduced by the 1986 Act) provides that Member States shall 'pay particular attention to *encouraging improvements, especially in the working environment, as regards the health and safety of workers,* and shall set as their objective the harmonisation of conditions in this area, while maintaining the improvements made'.

A change in EC approach has been the use of Framework and related 'Daughter' Directives. The Framework Directive on the introduction of measures to encourage improvements in the safety and health of workers at work, with five daughter directives is an example[22]. The Directive has been implemented in the UK as the Management of Health and Safety at Work Regulations 1992 (MHSW). The core of these regulations is the duty to assess the risks to health and safety to employees and anyone who may be affected by the work activity, and to follow through with appropriate measures of planning, care and information.

Implementation has been possible under HSW. Section 1(2) provides for the progressive replacement of existing legislation by a system of regulations and approved codes of practice 'designed to **maintain or improve** the standards of health, safety or welfare established by or under those enactments'.

There are a number of further directives and draft directives relevant to health and safety. The HSC will not negotiate all the implementation. For example, a number of the 'Technical Harmonisation and Standards' directives are co-ordinated by the DTI. The HSC and HSE are involved in consultation.

1.17.3 Standards of duty

A criminal law offence usually requires the prosecution to prove guilty intent. Statute occasionally creates a crime of absolute liability, when the prosecution needs to prove only the facts of the offence. The accused cannot plead a defence. HSW s. 9, the duty not to charge an employee for things provided because of a specific statutory requirement, has been suggested as a rare example.

More often there is statutory strict liability, but with a possible defence, such as that the risk was not foreseeable or that suitable steps had been taken. For example, every employer is required to make a suitable and sufficient risk assessment of their undertaking for employees and others[23]. To undertake a risk assessment is a strict requirement. The approved code of practice[31] suggests risk 'reflects both the likelihood that harm will occur and its severity'. That will affect whether the assessment is suitable and sufficient. There is an apparently strict duty on the employer in reg. 11 of the Provision and Use of Work Equipment Regulations 1992 (PUWER) to prevent access to any *dangerous* part of machinery. 'Dangerous' is not defined in these regulations. It was interpreted in the now repealed s. 14 of FA as being 'a reasonably foreseeable cause of injury to anybody acting in a way in which a human being may reasonably be expected to act in circumstances which may reasonably be expected to occur'[11]. Thus the duty to fence 'every dangerous part of any machinery' (former s. 14 FA), or 'prevent access to any dangerous part of machinery' (reg. 11 PUWER) is strict rather than absolute.

Furthermore, the reg. 11 requirement is *qualified* by steps to be taken when *practicable* (reg. 11(2)). Practicable has been interpreted to mean not as arduous as physically possible. A measure is practicable if it is possible in the light of current knowledge and invention[24].

Practicable is a stricter duty than *reasonably practicable* which is used extensively in HSW and in some of the 1992 'daughter' regulations[25]. Reasonably practicable implies that the risk should be balanced against the 'cost' of the measures necessary to avert the risk (whether in money, time or trouble) to see if there is gross disproportion[26].

Some apparently strict duties of EC health and safety directives have been transposed into UK legislation as being reasonably practicable. The HSE has explained that this is to avoid conflict of two absolute duties. For example Article 3 of the EC manual handling of loads directive[27] requires the employer to use appropriate means to avoid manual handling and to take steps to control manual handling that does take place. European law is accustomed to deal with such conflicts with the doctrine of *proportionality*, that is balancing consequences to see whether an absolute ban is disproportionate to a goal which could be achieved by less restrictive means.

1.18 Principles developed by the courts

1.18.1 Case law interpretation

Case law interpretation has had an adverse effect on some safety legislation. A notorious example is the fencing requirements for dangerous machinery (then s. 14 FA), as illustrated by, for example, *Close* v. *Steel Company of Wales*[28]. With reluctance judges interpreted the statute so that s. 14 could not be used where parts of the machine or of the material being worked on have been ejected at a workman. This interpretation has now been remedied by reg. 12(3) of PUWER.

Such interpretations affect the scope of legislation, and of civil action for breach of statutory duty. Breach of statutory duty and the tort of negligence are the two most frequent grounds for civil claims following accidents at work.

1.18.2 Tort of negligence

Negligence is a relatively modern tort, but today it is probably the most important in number of cases and for the amount of damages which may be awarded for serious injury.

The tort consists of a breach by the defendant of a legal duty to take care not to damage the plaintiff or his property and consequent damage from that breach. From early times the common law has placed on the employer duties towards his employees. In 1932, Lord Atkin, in the leading case of *McAlister (or Donoghue)* v. *Stevenson*[29] suggested a general test for when a duty is owed. It is owed to persons whom one ought reasonably to have in mind as being affected by the particular behaviour. In 1963 the persuasive precedent of *Hedley Byrne* v. *Heller & Partners*[3] extended the duty to include financial loss resulting from some careless statements.

1.18.3 Tort of breach of statutory duty

When a statutory duty is broken there is liability for any penalty stipulated in the statute. In addition a person suffering damage from the breach may sometimes bring a civil action in tort to obtain compensation. Sometimes the Act specifies this (for example, the Consumer Protection Act 1987). Sometimes the Act is silent but the courts allow the action, as happened with FA and related regulations; or the Act is silent but the courts deny a civil action. This happened with the Food and Drugs Act 1955 (which has now been consolidated with other enactments relating to food into the Food Safety Act 1990) when it was decided that the statute was not intended to add to a buyer's civil remedies for breach of contract or of negligence.

Section 47 of HSW provides that breach of the Act will not give rise to a civil action, but breach of any regulation made under the Act is actionable, unless the regulations say otherwise. So far the only regulations to provide otherwise are the MHSW[30].

Negligence and breach of statutory duty are two different torts, but both may be relevant following an incident. Bertha, injured at work because of an obstruction of the factory floor, might allege negligence plus breach of regulation 12 of the Workplace (Health, Safety and Welfare) Regulations 1992 (WHSW), and possibly succeed in both torts. She would not recover double damages because the remedy is compensation for the actual loss suffered.

The fact that an accident has occurred and resulted in legal action being taken is unsatisfactory. An award cannot repair an injury; the outcome of an action is uncertain; and the considerable cost and ingenuity expended in the investigation, developing the pleadings and the trial itself, could have been used more positively in trying to avoid such accidents. Such avoidance is an objective of HSW; and of the EC Directives, which are having increasing importance.

Because of the constraints of space, this chapter can be an outline only. Students are recommended to complement the chapter with further reading (see below) and visits to courts and tribunals.

References and endnotes

1. Powers of Criminal Courts Act 1973, Criminal Justice Act 1991, HMSO, London
2. R. *v.* George Maxwell Ltd (1980) 2 All ER 99
3. Hedley Byrne & Co. Ltd *v.* Heller & Partners Ltd (1964) AC 463
4. *Current Law:* a monthly publication from Sweet and Maxwell
5. Operated by Butterworth (Telepublishing) Ltd
6. For example: Roberts Petroleum Ltd *v.* Bernard Kenny Ltd (1983) 1 All ER 564 HL
7. Waugh *v.* British Railways Board (1979) 2 All ER 1169

8. Employment Protection Consolidation Act 1978; Trade Union Reform and Employment Rights Act 1993
9. Bulmer *v.* Bollinger (1974) 4 All ER 1226
10. AG *v.* Times Newspapers Ltd (1979) 2 EHRR 245, European Court of Human Rights
11. John Summers & Sons Ltd *v.* Frost (1955) AC 740
12. Much of the Abrasive Wheels Regulations 1970 is to be replaced by the Provisions and Use of Work Equipment Regulations 1992 and the Workplace (Health, Safety and Welfare) Regulations 1992
13. McCarthys Ltd *v.* Smith (1979) 3 All ER 325
14. Pepper *v.* Hart (1992) NLJ Vol 143 p. 17
15. Pickstone *v.* Freeman plc (1989) 1 AC 66
16. Van Duyn *v.* Home Office (Case 41/74) (1975) 3 All ER 190
17. Marshall *v.* Southampton and South West Hampshire Area Health Authority (Teaching) (1986) case 152/84 1 CMLR 688; (1986) QB 401
18. Rolls Royce plc *v.* Doughty (1992) ICR 538
19. Effective since 6.3.1992, by section 4 of the Offshore Safety Act 1992
20. Effective since October 1992 by the Criminal Justices Act 1991
21. Rodney Chapman at Lewes Crown Court on 26.6.92. In 1991 an Employment Minister said in Parliament that the potential scope of section 2(1) of HSW was understood to be very broad and that 'management' includes the management of health and safety
22. EC Directive No. 89/391/EEC, adopted 12.6.89 with five daughter directives
23. Regulation 3 of the Management of Health and Safety at Work Regulations 1992
24. For example: Adsett *v.* K & L Steelfounders and Engineers Ltd (1953) 1 All ER 97 and 2 All ER 320
25. For example: The Manual Handling Operations Regulations 1992
26. Edwards *v.* National Coal Board (1949) 1 All ER 743
27. Regulation 4 of the Manual Handling Operations Regulations 1992
28. Close *v.* Steel Company of Wales (1962) AC 367
29. Donoghue *v.* Stevenson (1932) AC 562
30. Regulation 15 of the Management of Health and Safety at Work Regulations 1992
31. Health and Safety Executive, *Approved Code of Practice: Management of Health and Safety at Work Regulations 1992*, HSE Books, Sudbury (1992)

Further reading

Atiyah, P. S., *Accidents, Compensation and the Law*, 4th edn, Weidenfeld & Nicolson, London (1987)
Barnard, D., *The Civil Court in Action*, 2nd edn, Butterworths, London (1985)
Barnard, D., *The Criminal Court in Action*, 3rd edn, Butterworths, London (1988)
Barrett, B. and Howells, R., *Health and Safety Law*, Pitman Publishing, London (1993)
Dane, J. and Thomas, P., *How to Use a Law Library*, 2nd edn, Sweet & Maxwell, London (1987)
Dewis and Stranks, *Tolley's Health and Safety at Work Handbook*, 4th edn, Tolley, London (1991)
Dickson, B., *The Legal System of Northern Ireland*, 2nd edn, SLS Publications (NI), Belfast (1989)
Encyclopaedia of Health and Safety at Work, Sweet and Maxwell, London (loose-leaf)
Hutchins, E. L. and Harrison, A., *History of Factory Legislation*, F. Cass, London (1966)
James, P., *Introduction to English Law*, 12th edn, Butterworths, London (1989)
Munkman, J., *Employer's Liability at Common Law*, 11th edn, Butterworths, London (1990)
Phillips, O. Hood, *A First Book of English Law*, 8th edn, Sweet and Maxwell, London (1989)
Selwyn, N., *The Law of Employment*, 7th edn. Butterworths, London (1991)
Smith, K. C. and Keenan, D. J., *English Law*, 10th edn, Pitman Publishing Ltd, London (1992)
Walker, D. M., *The Scottish Legal System*, 6th edn, W. Green, Edinburgh (1992)
Walker, R. J., *The English Legal System*, 7th edn, Butterworths, London (1992)
Williams, Glanville, *Learning the Law*, 11th edn, Sweet and Maxwell, London (1982)

Law Dictionaries

Curzon, *A Dictionary of Law* 3rd edn, Pitman, London (1988)
Jowitt and Burke, *Dictionary of English Law,* 2nd edn plus supplement, Sweet and Maxwell, London (1985)
Mozley and Whiteley's Law Dictionary, 10th edn, Butterworths, London (1988)
Osborn's Concise Law Dictionary, 7th edn, Sweet and Maxwell, London (1983)
Concise Dictionary of Law, 2nd edn, Oxford University Press, Oxford (1990)
Walker, *The Oxford Companion to Law,* Oxford University Press, Oxford (1980)

Chapter 2

Principal Health and Safety Acts

S. Simpson

UK health and safety legislation consists of a number of main or principal Statutes or Acts which are supported by a great deal of subordinate legislation in the form of Regulations and Orders. This chapter deals with the more commonly applied main Acts that are concerned with protecting the health and safety of the working population and those who may be put at risk from the manner in which the work is carried out.

2.1 The Health and Safety at Work etc. Act 1974

2.1.1 Pre-1974 legislation

For more than a century health and safety legislation for persons at work in the UK had developed a piece at a time, each piece covering a particular class of person and not in a consistent manner each time. Separate legislation with variations in details and in the methods of enforcement would apply to a process or requirement when undertaken in a factory, as opposed to an office, a mine or a quarry. For example, an air receiver situated in a factory would be required to be examined for safety reasons by a competent person at least once every 26 months, but the same receiver moved to a shop would not require examination nor would the same receiver need to be inspected in the factory if, instead of air, another gas at the same or even higher working pressure was substituted.

In the main, the principal Act affecting the particular groups of persons, usually on the basis of the kind of premises in which they worked, was supplemented by regulations. The Act and its regulations would be enforced by a particular inspectorate (e.g. by factory inspectors for factories and notional factories such as construction sites, mines inspectors for mines and quarries and local authority inspectors for offices and shops). Any breach of the appropriate legislation could lead to a prosecution by an inspector which in turn could lead to a fine usually imposed on the company or other organisation rather than an individual.

The major responsibility for observing the requirements of the legislation was that of the employer with some responsibilities falling on the occupier, if he was not the employer, and on the employees. Only in mining legislation was there also a criminal liability placed on managers and other officials. On the whole, legislation tended to look to the protection of plant and equipment as a way of preventing injuries to workers. Visitors, contractors, neighbours and other third parties were mainly ignored in the drafting of these earlier Acts and regulations, as were many employees who did not work on premises (e.g. roadsweepers) or worked in premises not covered (e.g. schools, research establishments, hospitals, etc.).

By 1970 many organisations, especially the trade unions, were questioning whether the existing legislation was either sufficient or effective in providing proper protection for work people.

The effect that workers' organisations could have on workshop safety was limited and large sections of the working population were not covered.

A Private Member's Bill aimed at providing for the compulsory involvement of workers in accident prevention was withdrawn when in 1970 a committee was set up under the chairmanship of Lord Robens to look at safety and health at work. After studying the whole problem in depth the committee reported in 1972[1] making many recommendations of a wide ranging nature.

The essence of the 'Robens Report' recommendations was to:

(1) Replace the mass of existing safety legislation with one Act applying generally to all persons at work.
(2) Replace the mass of detail with a few simple and easily assimilated precepts of general application.
(3) Change methods of enforcement so that prosecution is not always the first resort.
(4) Ensure that occupational safety should also protect visitors and the public.
(5) Place more emphasis on safe systems of work rather than technical standards.
(6) Actively involve the workers in the procedures for accident prevention at their place of work.

In spite of changes of Governments, the main recommendations of the Robens Committee were accepted by Parliament and were incorporated in the Health and Safety at Work etc. Act 1974.

2.1.2 The Health and Safety at Work etc. Act 1974

Drafted as an enabling Act, it permitted the Secretary of State or other Ministers to make regulations with a view to replacing the existing piecemeal legislation, typified by those Acts listed in schedule 1 of HSW, by regulations and codes of practice requiring improved standards of safety, health and welfare. It established a co-ordinating enforcement authority, the Health and Safety Commission, giving its inspectors greater powers than hitherto. It also extended legislative protection for health and safety to everyone who was employed, whether paid or not (except domestic servants) and imposed more general but wider duties on both employer and employee.

The Act makes provision for protecting others against risks to health and safety from the way in which work activities are carried out. It also seeks to control certain emissions into the atmosphere, as does the Control of Pollution Act 1974, and to control the storage and use of dangerous substances. In addition, the Act ensures the continuation of the Employment Medical Advisory Service and in Scotland deals with those parts of Building Regulations that affect the health and safety of those using the buildings.

Although eventually due to be superseded there is still a need to comply with the requirements of parts of the pre-1974 legislation which remain in effect but this applies only to those work activities covered previously.

2.1.3 General duties on employers and others

These duties are outlined in ss. 2–5 where the obligations are qualified by the phrases 'so far as is reasonably practicable' and 'best practicable means'. Interpretations of these phrases have been made[2] which indicate that 'reasonably practicable' implies a balance of the degree of risk against the inconvenience and cost of overcoming it, whereas 'best practicable means' ignores the cost element but recognises possible limitations of current technical knowledge.

In common law, employers have had, and still have, duties of care with regard to the health and safety of their employees, duties which are now incorporated into statute law as part of section 2 of this Act.

The first part of s. 2 contains a general statement of the duties of employers to their employees while at work and is qualified in subsection (2) which instances particular obligations to:

(1) Provide and maintain plant and systems of work that are safe and without risks to health. Plant covers any machinery, equipment or appliances including portable power tools and hand tools.
(2) Ensure that the use, handling, storage and transport of articles and substances is safe and without risk.
(3) Provide such information, instruction, training and supervision to ensure that employees can carry out their jobs safely.
(4) Ensure that any workshop under his control is safe and healthy and that proper means of access and egress are maintained, particularly in respect of high standards of housekeeping, cleanliness, disposal of rubbish and the stacking of goods in the proper place.
(5) Keep the workplace environment safe and healthy so that the atmosphere is such as not to give rise to poisoning, gassing or the encouragement of the development of diseases. Adequate welfare facilities should be provided.

In this section 'work' means any activities undertaken as part of employment and includes extra voluntary jobs for which payment is received or which are accepted as part of the particular job, i.e. part-time firemen, collecting wages etc.

Further duties are placed on the employer by:

s. 2(3) to prepare and keep up-to-date a written safety policy supported by information on the organisation and arrangements for carrying out the policy. The safety policy has to be brought to the notice of employees. Where there are five or less employees this section does not apply.

s. 2(6) to consult with any safety representatives appointed by recognised trade unions to enlist their co-operation in establishing and maintaining high standards of safety.

s. 2(7) to establish a safety committee if requested by two or more safety representatives.

The general duties of employers and self-employed persons include in s. 3 a requirement to conduct their undertakings in such a way that persons other than their employees are not exposed to risks to their health and safety. In certain cases information may have to be given as to what these risks are.

Landlords or owners are required by s. 4 to ensure that means of access or egress are safe for those using their premises and these are defined in s. 53 as any place and, in particular, any vehicle, vessel, aircraft or hovercraft, any installation on land, any offshore installation and any tent or movable structure.

Those in charge of premises are required by s. 5 to use the best practicable means for preventing noxious or offensive fumes or dusts from being exhausted into the atmosphere, or that such exhausts are harmless. Offensive is not defined and may depend upon one individual's opinion.

Duties are placed by s. 6 on everyone in the supply chain, from the designer to the final installer, of articles of plant or equipment for use at work or any article of fairground equipment to:

(1) ensure that the article will be safe and without risks to health at all times when it is being set, used, cleaned or maintained,
(2) carry out any necessary testing and examination to ensure that it will be safe, and
(3) provide adequate information about its safe setting, use, cleaning, maintenance, dismantling and disposal.

There is obligation on designers or manufacturers to do any research necessary to prove safety in use. Erectors or installers have special responsibilities to make sure when handed over that the plant or equipment is safe to use.

Similar duties are placed on manufacturers and suppliers of substances for use at work to ensure that the substance is safe when properly used, handled, processed, stored or transported, to provide adequate information and do any necessary research, testing or examining.

Where articles or substances are imported, the suppliers' obligations outlined above attach to the importer, whether a separate importing business or the user himself.

Often items are obtained through hire purchase, leasing[3] or other financing arrangements with the ownership of the item being vested with the financing organisation. Where the financing organisation's only function is to provide the money to pay for the goods, the suppliers' obligations do not attach to them.

The employees' duties are laid down in ss. 7 and 8 which state that, whilst at work, every employee must take care for the health and safety of himself and of other persons who may be affected by his acts or omissions.

Employees should co-operate with the employer to meet legal obligations and they must not, either intentionally or recklessly, interfere with or misuse anything, whether plant equipment or methods of work, provided by the employer to meet obligations under this or any other related Act.

The employer is not allowed by s. 9 to charge any employee for anything done or provided to meet statutory requirements.

2.1.4 Administration of the Act

The Act through s. 10 caused the establishment of two bodies to direct and enforce legislative matters concerned with health and safety. The Health and Safety Commission (HSC), appointed by the Secretary of State, consists of a chairman and six to nine members. Three of the members are appointed after consultation with the employers' organisations, three after consultation with employees' organisations and two after consulting local authorities.

It is the duty of the Commission (s. 11) to:

(1) assist and encourage persons in furthering safety,
(2) arrange for the carrying out of research and to encourage research and the provision of training and information by others,
(3) provide an information and advisory service,
(4) submit proposals for regulations, and
(5) report to and act on directions given to it by the Secretary of State.

It also liaises with local authority and fire authority organisations to whom it has delegated[4,5] (s. 18) some of its duties.

Whereas the Commission has the function of formulating policies, the Health and Safety Executive (HSE) is responsible for their implementation. The Executive which is appointed by the Commission and consists of three persons, one of whom is the director, has a duty to exercise on behalf of the Commission such functions as the Commission directs. If so requested by a Minister, the Executive shall provide him with information of the activities of the Executive on any matter in which he is concerned and to provide him with advice.

The Commission may direct the Executive or authorise any other person to investigate or make a special report on any accident, occurrence, situation or other matter for a general purpose or with a view to making regulations.

The duties of the Commission and the Executive are contained in ss. 11–14 of HSW Act.

2.1.5 Regulations and Codes of Practice

The enabling powers of this Act are exercised through s. 15 whereby the appropriate Secretary of State or Minister may make regulations without referring the matter to Parliament. The regulations may be drawn up by the Executive and submitted through the Commission to the Secretary or

Minister. Although there is a general requirement for the Commission and Executive to keep interested parties 'informed of and adequately advised on, such matters' (s. 11(2)c) there is no obligation to consult. However, in drafting regulations that affect workplace safety, extensive consultation does occur.

The regulations may repeal or modify any of the existing regulations and matters related to ss. 2–9 of the Act. They can also approve or refer to specified documents, such as British Standard Specifications. A list of 22 subject matters that can be covered by regulations is given in schedule 3 of the Act.

The need to provide guidance on the regulations is recognised in s. 16 which gives the Commission power to prepare and approve Codes of Practice on matters contained not only in the regulations but also in ss. 2–7 of the Act.

Before approving a code, the Executive acting for the Commission must consult with any interested body. The Commission have powers to approve codes prepared by bodies other than themselves, and some British Standards have been approved.

An Approved Code is a quasi-legal document and although non-compliance with it does not constitute a breach, if the contravention of the Act or a regulation is alleged, the fact that the code was not followed will be accepted in court as evidence of failure to do all that was reasonably practicable. A defence would be to prove that something equally as good or better had been done (s. 17(2)). To supplement the Approved Codes of Practice, the Executive issue guidance notes which are purely advisory and have no standing in law.

2.1.6 Enforcement

2.1.6.1 General

The enforcement of the Act (s. 18), with some exceptions in respect of noxious and offensive emissions[6] (s. 5), is the responsibility of the HSE through its constituent inspectorates with certain premises delegated to local authorities[4] and for certain fire matters to the Fire Authority[5].

Actual enforcement is carried out by inspectors (s. 19) who should have suitable qualifications and be authorised by a written warrant outlining the powers he may exercise. An inspector must produce his warrant on request; without it he has no powers of entry or enforcement.

2.1.6.2 Powers of inspectors

By virtue of his warrant an inspector has the powers outlined in s. 20. These powers relate only to the field of the inspectorate authorising him and include:

(1) The right to enter premises and if resisted to enlist the support of a police officer.
(2) To inspect the premises.
(3) To require, following an incident, that plant is not disturbed.

(4) Taking measurements and photographs although in the latter case it is usual to obtain permission first.
(5) Taking samples of suspect substances.
(6) Require tests to be carried out on suspect plant or substances.
(7) Requiring the dismantling of plant.
(8) Require those with possible knowledge relevant to his investigation to give it either verbally or in a written statement. The inspector has discretion to allow another to be present during questioning and the taking of a written statement.
(9) The right to inspect and take copies of books or documents required to be kept by safety legislation or others it is necessary for him to see as part of his investigation but he has no right to examine documents for which legal privilege is claimed.
(10) Requiring assistance within a person's limits of responsibilities.

Where an inspector takes samples of substances he must leave a similar identified sample with a responsible person or leave a conspicuous notice stating that he has taken a sample.

Information contained in an answer to an inspector cannot be used in criminal proceedings against the giver.

A customs officer may seize any imported article or imported substance and detain it for not more than two working days on behalf of an inspector (s. 25A).

Where an employer suffers damage of property or business, as a result of actions of an inspector, the inspector can be sued for recompense against which he may be indemnified by the enforcing authority.

After an inspector has completed his investigation he has a duty to inform representatives of the workpeople of actual matters he has found (s. 28(8)) and must give the employer similar information.

2.1.6.3 Notices

If an inspector is of the opinion that a breach has, or is likely to, occur he may serve an Improvement Notice (s. 21) on the employer or workman. The notice must state which statutory provision the inspector believes has been contravened and the reason for his belief. It should also state a time limit in which the matter should be put right.

However if the activity involves immediate risk of serious personal injury, the inspector may serve a Prohibition Notice (s. 22) requiring immediate cessation of the activity. This notice must state what, in the inspector's opinion, is the cause of the risk and any possible contravention. If the risk is great but not immediate a deferred Prohibition Notice may be served stating a date after which the activity must cease unless the matter has been put right. Where corrective work cannot be completed in time, the inspector may extend the period of the notice. There is no procedure for certifying that a notice has been complied with.

Appeals against a notice may be made to an Industrial Tribunal[7]. On entering an appeal an Improvement Notice is suspended until the appeal is disposed of or withdrawn, whereas a Prohibition Notice continues in effect unless the Tribunal directs otherwise.

2.1.7 Offences

It is an offence to:

(1) fail to discharge a duty imposed by sections 2–7,
(2) contravene sections 8 and 9, any regulation or notice,
(3) make false entries in a register,
(4) obstruct or pretend to be an inspector, and
(5) make false statements etc.

If an inspector decides to institute legal proceedings, he must do so within six months of learning of the alleged contravention (s. 34(3)). Cases can be heard either summarily which attracts a fine not exceeding level 5 on the standard scale on conviction, or on indictment where the penalty can be imprisonment and/or an unlimited fine. The types of offences are outlined in s. 33 which allows those concerned with interfering with the powers or work of an inspector (s. 33(1)(d,f,h and n)) to be dealt with summarily. For all the other offences listed in s. 33(1) plus in certain circumstances contravention of a requirement imposed by an inspector in the exercise of his powers (s. 33(1)e) the case can be tried either summarily or if the offence is serious enough and the parties agree on indictment, when the penalty on conviction can be an unlimited fine.

Responsibility for an offence usually attaches to the employer but may attach to an employee (ss. 7–8). However where the contravention was caused with the consent or knowledge or be due to the neglect of a director, manager, company secretary or other officer (s. 37) then he too can be prosecuted.

In proceedings alleging a failure to use reasonably practicable or best practicable means the prosecution only has to state the suspicion and it is up to the accused to prove that what was done was as good as, if not better than, the duty required (s. 40).

Penalties have recently been increased by the Offshore Safety Act 1992 so that failing to discharge a duty under sections 2 to 6 attracts a liability on summary conviction to a fine not exceeding £20 000 and on conviction on indictment to a fine. For specified offences, a person (such as a director, manager, etc.) found guilty of the offence shall be liable on summary conviction to imprisonment for a term not exceeding six months or a fine not exceeding £20 000 but for conviction on indictment to imprisonment for a term not exceeding two years or a fine or both. Fines for other offences are set at Level 5 (at present, through the Criminal Justices Act 1991, this is a sum not exceeding £5000).

2.1.8 Extensions

Part 1 of the Act has been extended to include:

(1) the protection of the public from danger associated with the transmission and distribution of gas through pipelines,
(2) securing the health, safety and welfare of persons on offshore installations engaged in pipeline works,

(3) securing the safety of such installations and preventing accidents on or near them,
(4) securing the proper construction and safe operation of pipelines and preventing damage to them, and
(5) securing the safe dismantling, removal and disposal of offshore installations or pipelines.

2.1.9 Parts II to IV and Schedules

Part II of the Act allows for the continuation of the Employment Medical Advisory Service, defines the purpose and responsibilities of the service, allows for fees to be charged, for payments to be made and for the keeping of accounts.

Part III modifies the Public Health Acts 1936–61 and the Building (Scotland) Act 1959 with respect of powers to make building regulations and deals with provisions as to the approval of plans, unsuitable materials, type relaxations, approved types of building, carrying out of tests, etc.

Part IV is a miscellaneous and general part amending the Radiological Protection Act 1970, Fire Precautions Act 1971, Companies Act 1967 and stating such matters as the extent and application of the HSW Act.

The following schedules of the Act cover:

(1) Relevant existing enactments.
(2) The constitution etc. of the Commission and Executive.
(3) Subject matter of health and safety regulations.
(4)–(7) Repealed.
(8) Transitional provisions with respect to Fire Certificates.
(9) Minor amendments to the Act.
(10) Repeals.

2.1.10 Definitions

Sections 52 and 53 contain a number of definitions aimed at clarifying part I of the Act:

'Work' means an activity a person is engaged in whether as an employee or as a self-employed person. An employee is considered to be at work all the time he is following his employment whether paid or not and a self-employed person is at work throughout such time as he devotes to work as a self-employed person. Regulations can extend the meaning of 'work' and 'at work' to other situations such as YTS training[8].

Other definitions include:

'Article for use at work' includes any plant designed for use at work and any article designed for use as a component in such plant.
'Code of practice' includes a standard, a specification or any other documentary form of practical guidance.
'Domestic premises' means premises occupied as a private dwelling (including gardens, yards, garages etc.).

'Employee' means an individual who works under a contract of employment.
'Personal injury' includes any disease or any impairment of a person's physical or mental condition.
'Plant' includes any machinery, equipment or appliance used at work.
'Premises' include any place, vehicle, vessel, aircraft, hovercraft, installation on land, offshore installation, installation resting on the sea bed or other land covered by water and any tent or movable structure within territorial waters. This definition has been extended by the Health and Safety at Work etc. Act 1974 (Application outside Great Britain) Order 1989 to include North Sea oil rigs etc.
'Self-employed person' is an individual who works for gain or reward otherwise than under a contract of employment, whether or not he employs others.
'Substance' means any natural or artificial substance whether solid, liquid, gas or a vapour and includes micro-organisms.

2.2 The Factories Act 1961

The Factories Act 1961 was in the main a consolidating Act, bringing together earlier Factories Acts. Many of the major provisions with regard to health, safety and welfare no longer continue in force.

However, those sections that do remain in effect refer to particular safety requirements but apply only to factories as defined in the Act and cover the following matters.

No young person may work at any of the machines prescribed in the Dangerous Machines (Training of Young Persons) Order 1954 unless he has been fully instructed in how to operate them safely (s. 21) or is under the supervision of a person who has a thorough knowledge and experience of the machine.

Hoists and lifts must be of good design and construction, properly maintained and thoroughly examined by a competent person at least once every six months (s. 22). The lift hoistway must be enclosed and fitted with interlocked gates that cannot be opened except when the cage or platform is at the landing and arranged so that the lift cannot move until the gate is closed.

Construction should not enable anyone to become trapped between lift and fixed structure or counterweight. The maximum safe working load must be conspicuously marked and not exceeded.

On lifts used for carrying passengers, automatic overrunning devices must be fitted (s. 23) and the cage fitted with gates that can be opened only at floors and when open the cage cannot be moved. There must be at least two ropes or chains supporting the cage, each capable of carrying the whole load. An efficient brake capable of arresting and supporting a fully loaded cage must be fitted. A hoist or a lift is defined as having the movement of the cage or platform restricted by a guide or guides (s. 25). A number of variations to the above requirements are given in the Hoists Exemption Orders 1962–83.

The equipment used to connect the load or cage to the lifting machine, i.e. support chains, rope and lifting tackle which includes (s. 26) chain slings, rope slings, rings, hooks, shackles and swivels, must be of good construction, sound material, adequate strength and free from patent defect. Each should be thoroughly examined by a competent person every six months and marked with its safe working load. For multiple slings a Table showing safe working loads should be displayed in the store and where the slings are being used. All this equipment should be tested before its first use, except fibre ropes and slings, and, unless exempt (i.e. high tensile chains) or a rope sling, be annealed at least once every 14 months (6 months for chains used in connection with molten metal or molten slag). The plate chains on fork lift trucks should be examined every six months but the truck itself is not considered a lifting machine. A record should be kept of tests and examinations.

Cranes and other lifting machines (crab, winch, teagle, pulley, block, gin wheel, transporter or runway) including their anchoring and fixing devices must be of good construction, sound material, adequate strength, free from patent defect and properly maintained (s. 27). They shall be thoroughly examined by a competent person at least once every 14 months and display a safe working load. Jib cranes should have either an automatic load indicator or a Table to indicate the safe working load at different inclinations of the jib. Before being taken into use for the first time in a factory, the crane must be tested and examined by a competent person and a test certificate issued. Where anyone is working on or near the crane wheel track and is liable to be struck by the crane, the driver must be warned and arrangements made to ensure that the crane does not approach within 6 m (20 ft 0 in). Similarly anyone working where they are liable to be struck by an overhead travelling crane or its load should be warned.

Ladders must be securely constructed and footed at the bottom or lashed at the top when in use.

Where work has to be done in a confined space (s. 30) such as a chamber, vat, pit, pipe or flue, the following precautions must be taken.

(1) If there is no adequate means of egress, a manhole shall be provided of a minimum specified size.
(2) The space must be certified by a responsible person as being safe, for a specified period, for entry without breathing apparatus. A space shall not be certified until effective steps have been taken to prevent ingress of dangerous fumes, all sludge and other deposits have been removed and there is adequate ventilation.
(3) If the space is not certified as safe, no person shall enter or remain in it unless he is wearing breathing apparatus, has been authorised to enter, and, where practicable, is wearing a belt with rope attached and a person outside keeping watch.
(4) A supply of approved breathing apparatus, belts, ropes, reviving apparatus and oxygen shall be kept available and in good working order. They shall be thoroughly examined at least once a month.
(5) Sufficient trained persons to use the apparatus to be employed.

No person shall enter or remain in a confined space in which the proportion of oxygen in the air may be substantially reduced unless he is wearing breathing apparatus or the space is adequately ventilated, tested and certified as safe.

No work is permitted in a boiler flue or furnace until it has been sufficiently cooled.

Where any process gives rise to dust which is liable to explode on ignition (s. 31), the plant should be enclosed, dust should not be allowed to accumulate and possible sources of ignition should be excluded. Unless the plant can withstand an explosion, suitable explosion vents, chokes or baffles should be provided. Explosion suppression systems can be used.

Any part of a plant containing explosive or flammable gas under pressure shall not be opened except when the pressure is reduced to atmospheric and flange joints shall not be broken until the flow of gas to that part is effectively stopped.

No plant, tank or vessel which contains, or has contained, any explosive or flammable substance shall be welded, brazed, soldered or cut, or opened by the application of heat until the substance and any fumes from it have been removed or rendered non-flammable. Before the plant, tank or vessel is refilled with an explosive or flammable substance, it must be cooled sufficiently to prevent ignition.

Entry into boilers (s. 34) for repair or examination may only be made if steam and hot water pipes are disconnected or all valves are shut and the blow-down valve, if it feeds to a common manifold, is shut.

Every water sealed gasholder having a storage capacity of not less than 140 m^3 (5000 ft^3) shall be of sound construction, properly maintained and examined externally every two years (s. 39). Gasholders with a lifting part that has been in use for more than twenty years shall have the internal state of the sheeting examined by a competent person every ten years. No gasholder shall be repaired or demolished, except under the direct supervision of an experienced person who knows the risks of explosion and gassing.

The occupier of a humid factory is required to inform the factory inspector in writing when artificial humidity is first produced (s. 68). Every room in which artificial humidity is produced should have two hygrometers and a humidity table. The readings of the hygrometers shall be taken at certain times of the day and entered on the prescribed form, F48. Artificial humidification is not permitted in a room in which the wet bulb thermometer exceeds 22.5°C (72.5°F) or for certain processes 26.5°C (80°F) or when the difference between the readings of the dry and wet bulb thermometers is less than that indicated in the humidity table. No water which is liable to cause injury to health shall be used for artificial humidification.

Workrooms below ground level (s. 69) present special hazards and an inspector may restrict the work that is carried out there.

Prohibitions are placed on the employment of young girls in processes involving salt (s. 73), and women and young persons in processes involving lead (ss. 74, 75).

2.3 The Fire Precautions Act 1971

The Act furthers the provisions for the protection of persons from fire risks. If any premises are put to use and are designated a certificate is required from the fire authority. Although classes of use cover the provisions of sleeping accommodation; use as an institution; use for the purposes of entertainment, recreation, instruction, teaching, training or research; use involving access to the premises by members of the public and use as a place of work, so far only the provision of sleeping accommodation and use as a place of work have been designated.

Houses occupied as single private dwellings are exempt, but the fire authority have powers to make it compulsory for some dwellings to be covered by a fire certificate.

Applications for fire certificates must be made on the prescribed form and the fire authority must be satisfied that the means of escape in case of fire, means of fire fighting and means of giving persons in the premises warning in case of fire are all adequate. Every fire certificate issued shall specify particular use or uses of the premises, its means of escape, details of the means of fire fighting, and of fire warning and, in the case of factories, particulars of any explosive or highly flammable materials which may be stored or used on the premises. The certificate may impose such restrictions as the fire authority considers appropriate and may cover the instruction or training of persons in what to do in case of fire or it may limit the number of persons who may be in the premises at any one time. In certain circumstances the fire authority may grant exemption from the requirements to have a fire certificate, otherwise a copy of the fire certificate is sent to the occupier and it must be kept on the premises. The owner of the building is also sent a copy of the certificate.

It is an offence not to have or to have applied for a fire certificate for any designated premises. Contravention of any requirement imposed in a fire certificate is also an offence. A person guilty of an offence (with some exceptions) shall be liable on summary conviction to a fine not exceeding level 5 on the standard scale and on conviction on indictment a fine or imprisonment or both.

So long as a certificate is in force, the fire authority may inspect the premises to ascertain whether there has been a change in conditions. Any proposed structural alterations or extensions to the premises, major changes in the layout of furniture or equipment or, in factories, to begin to use or store or increase the extent of explosive or flammable material shall, before the proposals are begun, be notified to the fire authority.

It is also necessary whilst the certificate is in force, or an exemption has been granted under s. 5A, for the occupier to give notice of any proposed material extension or alterations to the premises or its internal arrangement and, in the case of factories, to store or use or to materially increase the amount of explosive or highly flammable materials. Within two months of receiving notice, the fire authority must, if they regard the requirements of the relevant fire certificate as becoming inadequate, inform the occupier, or owner, and give such directions as they consider appropriate. If the directions are duly taken the fire authority will amend the certificate or issue a new one. Not giving suitable notice or contravening a direction are offences that on conviction could lead to a fine or imprisonment, or both. The rights of appeal are detailed in s. 9.

The coming into effect of The Fire Safety and Safety of Places of Sport Act 1987 amended but did not replace the FPA and gave the Fire Authority much wider powers. These include the power to charge a reasonable fee for the initial issue, or the amendment or the issue of a new fire certificate (s. 8B). Even though premises may be exempt from the requirements for a fire certificate there are duties to provide both means of escape and means of fighting fire (s. 9A). In order to assist occupiers to meet these duties the Secretary of State may issue Approved Codes of Practice and the fire authority may serve Improvement Notices if they think a code is not being met (ss. 9A–9F).

Should the fire authority be of the opinion that, in the event of fire, the use of premises involves or will involve so serious a risk to persons on the premises that continuing use ought to be prohibited or restricted, the authority may serve a Prohibition Notice on the occupier. There are rights of appeal against these notices (ss. 10–10B).

The Secretary of State has powers under the Act to make regulations about fire precautions in designated premises other than those in which manufacturing processes are carried on (s. 12).

The Act deals with matters pertaining to building regulations (s. 13), the duties of consultation between local authorities (s. 15 et seq.), fire authorities and other authorities such as the HSE, the enforcement of the Act (s. 18), the powers of inspectors to enter premises (s. 19), offences, penalties and legal proceedings (ss. 22–27) and the amendment of other Acts (s. 29 et seq.).

Schedule 1 has the effect of making of special provisions for factory, office, railway or shop premises, that do not form part of a mine, in relation to leasing, part ownership, the issue of licences under the Explosives Act 1875 and the Petroleum (Consolidation) Act 1928. It also has an effect on the proposed or actual storage or use of explosives or highly flammable material in factory premises.

2.4 The Mines and Quarries Acts 1954–71

The main Acts laying down the general safety duties of mines and quarries personnel (i.e. owners, managers, undermanagers, surveyors and officials) are the Mines and Quarries Act 1954, the Mines and Quarries (Tips) Act 1969 and the Mines Management Act 1971 (the latter Act in particular and some sections of the 1954 Act will be revoked if the draft Management and Administration of Health and Safety in Mines Regulations are accepted).

In the 1954 Act the safety, health and welfare provisions for the mines are given separately from those of quarries whilst common sections deal with workmen's inspections, employment of women and young persons, records and returns, and fencing of abandoned and disused mines and quarries.

Either through the Act or by Regulations provisions have to be made at each mine for:

(a) the keeping of plans of the workings of the mine;
(b) securing safe ingress and egress (including communication between shafts, limitation on numbers of workers, winding and haulage apparatus);
(c) securing shafts and entrances to disused workings;

(d) the construction and maintenance of roadways (including transport rules, refuge holes);
(e) safe operation of winding and rope haulage apparatus and conveyors;
(f) systematic support for roofs and sides;
(g) adequate ventilation;
(h) lighting and lamps;
(i) contraband, electricity and electrical apparatus, blasting materials and devices, fire precautions and rescue, dust precautions and external dangers to workings;
(j) duties of officials and workmen in case of danger;
(k) machinery and apparatus (including construction, maintenance, restrictions, loading of cranes);
(l) surface buildings and structures;
(m) training and discipline;
(n) prohibition of heavy work by women and young persons;
(o) general welfare provisions.

In some cases requirements for coal, shale and fireclay mines are different from other mines.

In all quarries the following requirements apply.

All parts and working gear, including anchoring of all machinery and apparatus, must be of good construction, suitable material, adequate strength and free from patent defect and properly maintained. Flywheels and every other dangerous exposed part of machinery must be securely fenced and fencing properly maintained and kept in position.

All vessels containing or producing air, gas or steam at a pressure greater than atmospheric pressure must be so constructed, installed, maintained and used as to obviate any risk from fire, bursting, explosion or collapse or the production of noxious gases.

The safe working load shall be plainly marked on every crane and winch and it must not be exceeded.

All buildings and structures must be kept in a safe condition.

Safe means of access must be provided and maintained. Fencing is required where a person may fall more than 3 m unless there is a secure foothold and, where necessary, a secure handhold.

No person shall be employed in a quarry unless adequately instructed or trained or is under the instruction and supervision of a person who is competent to instruct and supervise that work.

A person who contravenes tipping rules or directions or rules made by the manager for compliance with the Act shall be guilty of an offence.

A person who negligently or wilfully does anything likely to endanger the safety or health or wilfully omits to do anything necessary for securing safety or without permission removes, alters or tampers with anything provided for safety or health shall be guilty of an offence.

No woman or young person shall lift, carry or move a load so heavy as to be likely to cause injury. Adequate supplies of wholesome drinking water shall be provided and maintained at suitable points.

No quarry shall be worked unless there is a sole manager, or managers with jurisdiction for particular parts, who shall give close and effective supervision of all operations in progress.

Managers must read, or arrange for some other competent person to read, each report, record or other item of information required to be entered in a book by virtue of the Act. The manager's attention must be drawn to anything abnormal or unusual.

It is the duty of every manager to take precautions to avoid danger from falls.

Safety devices have to be provided to prevent vehicles that run on rails from running away. No ropeway shall be used unless it meets the requirements of the Quarries (Ropeways and Vehicles) Regulations 1958.

Where natural light is insufficient suitable and sufficient artificial lighting shall be provided and maintained. Steps shall be taken to protect persons against the inhalation of dust.

Where any person in charge of a part of a quarry is of the opinion that a danger exists he must clear everyone out of that part, inform his immediate superiors and then ascertain the measures necessary to render it safe[9].

2.5 The Environmental Protection Act 1990

To prevent the pollution from emissions to air, land or water from scheduled processes the concept of integrated Pollution Control has been introduced. Authorisation to operate the relevant processes must be obtained from the enforcing authority which, for the more heavily polluting industries, is H M Inspectorate of Pollution. Control for pollution to air from the less heavily polluting processes is through the local authority.

Regulations also place a 'duty of care' on all those involved in the management of waste, be it collecting, disposing of or treating *Controlled Waste* which is subject to licensing.

In addition to extending the Clean Air Acts by including new measures to control nuisances, the Regulations introduce litter control; amend the Radioactive Substances Act 1960; regulate genetically modified organisms; regulate the import and export of waste; regulate the supply, storage and use of polluting substances and allow the setting up of contaminated land registers by the local authority. In 1991 the Water Act 1989 which controlled the pollution and supply of water was replaced by five separate Acts (see section 31.6.5).

2.6 The road traffic acts 1972–91

The road traffic acts, including the Road Traffic Regulation Act 1984, together with the Motor Vehicle (Construction and Use) Regulations 1986, Road Vehicle Lighting Regulations 1981, Goods Vehicles (Plating and Testing) Regulations 1988, the Motor Vehicle (Tests) Regulations 1982 and numerous other regulations, form comprehensive safety legislation not only of the occupants of the vehicles but also for members of the general public who may be affected by the driving and parking of vehicles.

In the construction of vehicles, safety features include the provision of suitable braking systems; burst-proof door latches and hinges; material for fuel tanks; types of lamps and reflectors; the fitting of audible warnings, mirrors, safety glass windscreens, seat belts; acceptable tyres, the driver's view of the road and the lighting of vehicles. Noise and smoke emissions are also topics related to safety and covered by the legislation.

When loading a vehicle care must be taken to ensure that the load is evenly distributed to conform to the vehicle's individual axle weight and where necessary the driver must make suitable corrections on multi-delivery work to ensure that no axle becomes overloaded due to transfer of weight. Since it is an offence to have an insecure load, all loads must be securely fixed and roped and, if necessary, sheeted. Restrictions are placed on projecting loads, extra long or extra wide loads and abnormal indivisible loads. The carriage of dangerous goods, be they toxic, flammable, radioactive or corrosive, is covered by regulations made under other Acts (e.g. Petroleum (Consolidation) Act).

The road traffic acts also deal with offences connected with the driving of motor vehicles and of traffic generally, accidents, road safety, licensing of drivers, driving instruction, restrictions on the use of motor vehicles, periodic testing of vehicles to ensure that they are roadworthy etc.

2.7 The Public Health Act 1936

This is another consolidating Act and in Part III statutory nuisances and offensive trades are dealt with.

Statutory nuisances are any premises in such a state as to be prejudicial to health or a nuisance, likewise the keeping of any animal, allowing any accumulation or deposit and causing any trade, business or process dust or effluvia to affect inhabitants of the neighbourhood. Not ventilating, not keeping clean and not keeping free from noxious effluvia or overcrowding any workplace are also statutory nuisances.

Where a statutory nuisance exists, the local authority can serve an abatement notice on the appropriate person, owner or occupier. If the abatement notice is disregarded, the court has powers to make a nuisance order.

Consent of the local authority is required before specified trades or business can be carried on. These offensive trades include blood boiling and drying, bone boiling, fat extracting and melting, fell mongering, glue making, soap boiling, tripe boiling and dealing in rags and bones.

The Act gives local authorities power to make bye-laws with regard to offensive trades and of fish-frying.

An allied piece of legislation is the Food Hygiene (General) Regulations 1970 made under the Food and Drugs Act 1955 but now covered by the Food Safety Act 1990. These regulations apply in England and Wales only but similar food hygiene regulations also exist for Scotland and for Northern Ireland.

The principal requirements of the Regulations relate to:

(a) the cleanliness of premises and the equipment used for the purpose of a food business;
(b) the hygienic handling of food;
(c) the cleanliness of persons engaged in the handling of food;

(d) the construction of premises used for the purposes of a food business and their repair and maintenance;
(e) the provision of water supply and washing facilities;
(f) the proper disposal of waste;
(g) the temperature at which certain foods are to be kept.

2.8 Petroleum (Consolidation) Act 1928

Little remains of this Act which is now restricted to licensing for the keeping of petroleum spirit, the making of byelaws for filling stations and canals and to testing petroleum. Extant regulations cover compressed gas cylinders, the keeping of petroleum spirit and extending the provisions of the Act to other substances such as carbide of calcium and liquid methane.

2.9 Crown premises

Although section 48 of the HSW makes provision for binding the Crown to the provisions of part of the Act it excepts sections 21 to 25 which deal with prohibition and improvement notices. The Crown immunity exists because in the exercise of justice in the name of the monarch it is not constitutionally possible for one part of the Crown service to pursue another part of the service into the Courts. Nevertheless, the HSE does apply a version of prohibition and improvement notices, called Crown Notices, when deficiencies are found on Crown property. Through the National Health Service (Amendment) Act 1986, Crown immunity for both food legislation and health and safety legislation have been removed from the Health Authority.

References

1. Report of the Committee on Safety and Health at Work. 1970–71 (Robens Report), Cmnd 5034, HMSO, London (1972)
2. Fife, I. and Machin, E.A. *Redgrave Fife and Machin Health and Safety,* Butterworth, London (1990)
3. Health and Safety Executive, *The Health and Safety (Leasing Arrangements) Regulations 1992,* SI 1992 No. 1524, HMSO, London
4. Health and Safety Executive, *The Health and Safety (Enforcing Authority) Regulations 1989,* SI 1989 No 1903, HMSO, London
5. Health and Safety Executive, *The Fire Precautions (Factories, Offices and Railway Premises) Order 1989,* SI 1989 No. 76, HMSO, London
6. Health and Safety Executive, *The Health and Safety (Emissions into the Atmosphere) Regulations 1983,* SI 1983 No. 943, HMSO, London
7. Health and Safety Executive, *The Industrial Tribunals (Improvement and Prohibition Notices Appeals) Regulations 1974,* SI 1974 No. 1925, HMSO, London
8. Health and Safety Executive, *The Health and Safety (Training for Employment) Regulations 1990,* SI 1990 No. 1380, HMSO, London
9. Health and Safety Executive, *The Law relating to Safety and Health in Mines and Quarries, Parts 1–4* HSE Books, Sudbury (1972–1979)

Further reading

Selwyn, N., *Law of Health and Safety at Work*, Croners, London (1993)
Mahaffy and Dodson on Road Traffic, Butterworth, London, (loose-leaf)
Garner, *Environmental Law,* Butterworth, London, (loose-leaf)
Encyclopedia of Environmental Health, Sweet and Maxwell, London (loose-leaf)

Chapter 3

Subordinate safety legislation

S. Simpson

Chapter 2 dealt with the principal Acts aimed at improving the level of safety and health in the workplace. Under those Acts, Regulations and Orders have been made. Codes of Practice approved and Guidance Notes issued, all with the object of giving detail to general requirements, and providing information and guidance needed to ensure that the intentions of the principal legislation are met. Details of some of the Regulations that have common application follow, although The Pressure Systems and Transportable Gas Containers Regulations 1986 is fully dealt with in chapter 29.

Various Regulations made in late 1992 replaced many of the welfare and safety requirements, such as those for the guarding of machinery, detailed in the Factories Act 1961, the Offices, Shops and Railway Premises Act 1963, the Abrasive Wheels Regulations 1970, the Protection of Eyes Regulations 1974, the Woodworking Machines Regulations 1974 and the like.

3.1 The Asbestos (Licensing) Regulations 1983

These regulations cover all work involving asbestos insulation and asbestos coating and relate to any material containing asbestos that is used for thermal, acoustic or other insulation purposes including fire protection. Asbestos cement and asbestos board are outside these regulations. Every operator undertaking work involving the listed types of asbestos materials must possess a licence issued by the HSE. However no licence is required where:

(a) the person carrying out the work spends no more than one hour a week on it; and
(b) the total time so spent in any factory does not exceed two hours per week; or
(c) the work is carried out by employees of the occupier; or
(d) the work consists only of taking air samples or material samples for identification.

Applications for a licence, if not already held, must be made 28 days before any work is started. When issued, the licence may contain restrictions or conditions which if not obeyed, or the licensee is convicted of certain offences, may cause the licence to be revoked.

If the work is to be undertaken by the occupier's own employees, 28 days notice – or shorter period if agreed – must be given to the enforcing authority. The employer is required to tell those carrying out the work about the material they will be handling and give them any necessary training. He must also ensure that any exposure to the dust is kept as low as is reasonably practicable.

Everyone who works with asbestos must be medically examined before commencing the work and re-examined at least once in every two years. Where relevant the HSE may grant exemptions to any of these requirements provided it is satisfied that the health of anyone affected will not be prejudiced as a result.

3.2 The Construction (General Provisions) Regulations 1961

The principal requirements of the General Provisions can be summarised as follows:

(1) Safety supervisors, either full or part-time, must be appointed in writing by contractors employing more than 20 workpeople.
(2) For excavations, shafts, tunnels, coffer-dams and caissons
    (a) suitable timbers or other materials to be provided and used to prevent falls of earth etc.,
    (b) they should be examined by a competent person every day,
    (c) they shall only be erected, altered or dismantled under the direction of a competent person,
    (d) positions of safety to be provided where flooding may occur,
    (e) edges to be safeguarded against collapses.
(3) Explosives shall not be handled or used unless controlled by a competent person.
(4) Adequate ventilation of working places in excavations etc. and other confined spaces must be provided (atmospheres to be tested where appropriate).
(5) Suitable precautions shall be taken when work is on or near water.
(6) Precautions are detailed for the use of locomotives and vehicles.
(7) Demolition work must be supervised by a competent and experienced person and before and during demolition precautions must be taken to prevent fire, gas explosion, flooding and accidental collapse.
(8) Measures shall also be taken to
    (a) prevent steam, smoke or other vapour from obscuring the site,
    (b) protect persons from falling material,
    (c) have working places lit,
    (d) remove timbers with projecting nails,
    (e) keep walkways free of loose material, and
    (f) avoid danger of collapse of a structure.

3.3 The Construction (Health and Welfare) Regulations 1966

Subject to exceptions, a contractor shall provide for persons employed:

(1) accommodation for taking shelter during bad weather,
(2) accommodation for clothing not worn during working hours,

(3) accommodation for protective clothing if it becomes wet,
(4) accommodation for taking meals, including facilities for boiling water and heating food,
(5) a supply of drinking water,
(6) facilities for washing, means of cleaning and drying and hot and cold, or warm, water,
(7) sanitary conveniences, and
(8) safe access to and egress from the above facilities.

3.4 The Construction (Lifting Operations) Regulations 1961

In general the principal provisions of these regulations require:

Lifting appliances, and their components, to:

(1) be in good mechanical condition, of sound material, adequate strength and free from visible defect,
(2) be properly maintained,
(3) be inspected weekly by the driver if competent,
(4) be securely suspended,
(5) have safe platforms for drivers,
(6) have cabins for drivers of power driven appliances that give weather protection and a clear view,
(7) be provided with suitable drums, pulleys, brakes, controls and safety devices,
(8) have safe means of access for the purpose of examination, lubrication or repair,
(9) be stable when used on soft, uneven or sloping surfaces,
(10) be operated by trained and competent persons,
(11) be thoroughly examined by a competent person every 14 months (6 months for hoists) and following any substantial repair or alteration.

Cranes, etc. must be:

(1) securely anchored or properly ballasted to ensure stability,
(2) tested by a competent person before being taken into use and at least once every four years thereafter, or after any substantial repair or alteration,
(3) erected under the supervision of a competent person,
(4) marked with a safe working load which should not be exceeded,
(5) in some cases, fitted with an automatic safe load indicator that should be examined weekly.

Chains, ropes and slings must be:

(1) of good construction, sound material, adequate strength and quality and free from visible defect,
(2) tested and examined before use by a competent person, marked with the appropriate safe working load (with exceptions) and means of identification, and re-examinations should be undertaken every six months,
(3) able to meet the requirements specified for
    (a) wire ropes with regard to broken wires,
    (b) testing after repairs,

(c) prevention of displacement from hooks,
(d) use of slings etc.

Other provisions cover the carrying of persons by means of lifting appliances, hoists carrying persons, secureness of loads, the keeping of records and having suitable and efficient signalling systems where necessary.

3.5 The Construction (Working Places) Regulations 1966

Contractors and employers, so far as reasonably practical, must provide a safe working place with suitable access and egress. This may require the provision of scaffolds, ladders, safety nets or safety belts.

A scaffold is a temporary structure which may include guardrails, toeboards, working platforms, gangways, ladders etc.

Every scaffold should:

(1) be erected, altered or dismantled under the direction of a competent person,
(2) be constructed of suitable and sound material,
(3) be properly maintained with parts secured to prevent displacement,
(4) have all standards or uprights, ledgers, putlogs, ladders erected in accordance with the detailed regulations,
(5) be stable,
(6) be inspected by a competent person every seven days or after exposure to adverse weather conditions and a report issued.

Further provisions apply to slung, cantilever, jib, figure, bracket and other suspended scaffolds, boatswain's chairs, cages and skips, and trestle scaffolds.

Working platforms, gangways and runs shall be:

(1) close boarded, planked or plated with boards etc. secured to prevent movement and of adequate thickness,
(2) fitted with footlaths if slope exceeds 1 in 4 (no run should exceed 1 in 1½),
(3) of adequate width for the purpose provided,
(4) fitted with guardrails and toeboards,
(5) safe underfoot.

Ladders (and folding stepladders) shall:

(1) be of good construction, suitable and sound material,
(2) be securely fixed and properly supported,
(3) be on a level and firm footing,
(4) extend above the highest platform to be reached unless there are adequate handholds,
(5) not be used for risers over 9 m (30 ft 0 in) without an intermediate platform.

Precautions must also be taken to prevent falls due to openings, corners, breaks, edges, open joisting, holes in platforms for ladders, sloping roofs and work on or near fragile roof materials.

3.6 The Control of Asbestos at Work Regulations 1987

In general, the duties of employers with respect to employees apply also to other persons who may be affected by the work activity. Prior to doing any work in which persons would or are liable to be affected by asbestos, the asbestos must be identified or assumed not to be chrysotile alone, treated accordingly and an assessment of the exposure must be made. The assessment shall determine the nature and degree of exposure and the steps to be taken to prevent exposure or to reduce it to the lowest reasonably practical level.

Unless the employer is licensed under the Asbestos (Licensing) Regulations or the exposure will not exceed the action level (that is where the cumulative exposure in a 12 week period to chrysotile is 96 fibre-hours per millilitre of air, or for other types of asbestos 48 fibre-hours per millilitre of air) he shall not commence any work until 28 days after he has notified the enforcing authority in writing of his intention to carry out the work. However, the enforcing authority may agree to a shorter time.

Employees who are liable to be exposed to asbestos or to do work in connection with the employer's duties under the Regulations are to be given adequate information, instruction and training.

Exposure of employees to asbestos should be prevented, but where this is not reasonably practicable exposure should be reduced to the lowest reasonable level and, if necessary, suitable respiratory protective equipment should be provided. Where practicable a less hazardous substance should be substituted for asbestos. Control measures, including personal protective equipment, shall be used and maintained and, in the case of exhaust ventilation equipment, regularly examined and tested at suitable intervals by a competent person.

Adequate and suitable protective clothing shall be provided and shall be cleaned at suitable intervals or disposed of as asbestos waste. If the cleaning is not carried out on the premises, special packing arrangements are required. Employers shall provide washing and changing facilities and storage for both protective clothing and for personal clothing not worn during working hours.

The employer has a duty to prevent the spread of asbestos from the place of work. Premises, or parts of premises, where asbestos work is carried out are to be kept in a clean state. Where a manufacturing process gives rise to asbestos dust the premises shall be designed and constructed to facilitate cleaning and be equipped with a vacuum cleaning system, preferably fixed. Designated areas shall be demarcated and identified by notices, as an 'asbestos area' when exposure exceeds, or would be liable to exceed, the action level, and as a 'respirator zone' when the concentration of asbestos would exceed, or be liable to exceed, the control limit (which for chrysotile is 0.5 fibres per millilitre of air averaged over 4 hours, or 1.5 fibres per millilitre of air over 10 minutes; for other types of asbestos the levels are 0.2 fibres and 0.6 fibres respectively). Employees must not eat, drink or smoke in a designated area. When monitoring is appropriate for protection of the health of those employed, it shall be done and the records of the monitoring kept for at least 5 years.

Employers should prepare a plan of work when asbestos is being removed from any building, plant, skip or the like. The written plan should include the nature and probable duration of the work, the location of the place of work, the methods to be applied when handling the asbestos or materials containing asbestos and the equipment for protection and decontamination.

Employees exposed to asbestos concentrations in excess of the action level shall be medically examined and the records kept for at least 40 years after the date of the last entry. The affected employees must be kept under medical surveillance by an Employment Medical Adviser or appointed doctor. The medical surveillance shall include a medical examination incorporating a specific examination of the chest which should have occurred not more than two years before the beginning of the exposure and thereafter at intervals not exceeding two years. Certificates of examination shall be kept for four years and a copy shall be given to the employee.

When raw asbestos or waste is stored, received into or dispatched from any place of work or is distributed, suitable sealed containers clearly marked to show they contain asbestos must be used. No person is to supply any product or component containing asbestos unless it is labelled in accordance with the Regulations. Approved Codes of Practice have been issued in connection with these Regulations[1,2].

3.7 The Control of Industrial Major Accident Hazards Regulations 1984 (CIMAH)

Subject to listed exceptions these Regulations apply to any operation in an industrial installation listed in the schedule (i.e. producing, processing or treating organic or inorganic chemicals, energy gases, distilling petroleum or its products, etc.) which involves one or more dangerous substances (e.g. that are very toxic, toxic, explosive or flammable), unless that operation is incapable of causing a major accident hazard. It also applies to the storage of specified quantities of listed substances.

Manufacturers must demonstrate that they have identified the major accident hazards, taken steps to prevent or limit the consequences of any major accident and provided suitable information, training and equipment for persons working on the site. Any major accident must be reported to the HSE. Where quantities of dangerous substances exceed the appropriate scheduled amounts, a report containing prescribed particulars (viz. information relating to the substance, quantity, installation, management control system and accident potential) must be sent to the HSE. All reports are to be kept up-to-date. Three months notice must be given before starting a new activity.

Manufacturers are required to prepare and keep up-to-date an on-site emergency plan detailing how major accidents will be dealt with on site, whilst the local authority must have an off-site emergency plan based on information supplied by the manufacturer and others. The local authority may recover from the manufacturer reasonable costs incurred in the preparation of the off-site emergency plans.

3.8 The Control of Lead at Work Regulations 1980

In these Regulations 'lead' means lead (including lead alloys and compounds and lead as a constituent of any substance) which is liable to be inhaled, ingested or absorbed by persons whether employees or not.

Where any work may expose anyone to lead the employer must assess the risk before that work is commenced to determine the nature and degree of exposure.

Adequate information, instruction and training must be given to those likely to be affected.

The first action must be to introduce measures to control materials, plant and processes so that adequate protection is provided against exposure to lead without the need for respiratory protective equipment or protective clothing. If these control measures do not give adequate protection against airborne lead, approved respiratory protection must be provided. Where the exposure to lead is significant each employee must be provided with adequate protective clothing. Adequate washing facilities must be provided and arrangements made for the separate storage of protective clothing and any personal clothing not worn during working hours.

Eating, drinking or smoking is not permitted in places liable to be contaminated by lead, and workplaces, premises, plant and protective equipment must be kept clean.

Employers and employees have a duty to prevent the spread of lead contamination,. Employers must see that any equipment for controlling exposure to lead is properly used and maintained and employees must use those control methods properly, reporting any defects.

Where appropriate suitable air monitoring procedures shall be implemented, medical surveillance and biological testing of exposed employees undertaken and records kept of such monitoring, surveillance and testing.

The Approved Code of Practice[3] suggests that exposure to lead is significant if the level of airborne lead is in excess of half the lead-in-air standard or there is a substantial risk of ingesting lead or there is a risk of skin contact with concentrated lead alkyls. For lead-in-air the standard for an 8 hour time weighting is an average concentration of 0.15 mg/m^3 of air except for tetraethyl lead where the level is 0.10 mg/m^3 of air.

3.9 The Control of Pollution (Special Waste) Regulations 1980

Special waste is controlled waste consisting of or containing listed substances that could be dangerous to life (i.e. where a single dose of not more than 5 cm^3 would cause death or serious damage to tissue if ingested by a child of 20 kg body weight or where exposure to it for 15 minutes or less would be likely to cause serious damage to human tissue by inhalation, skin contact or eye contact) or has a flash point of 21°C or less or is a medical product available only on prescription. A substance that is only dangerous because it is radioactive is not covered by these Regulations.

A person who produces special waste has to prepare six copies of a specified five part consignment note and complete some of the parts before the substance can be disposed of. Copies of the consignment note must go to the people who transfer the special waste from where it is produced to another place, to the disposal authority of the area in which it is to be disposed and to the carrier. The disposer has to complete the consignment note. Exceptions can be made for regular consignments. Special waste for disposal in another area must be referred to that particular authority.

Registers containing copies of the consignment notes must be kept by the producer, by the carrier and by the disposer. Site records of actual deposits must also be kept.

Listed substances include acids and alkalis, various metallic compounds and peroxides.

3.10 The Control of Pollution (Supply and Use of Injurious Substances) Regulations 1980

In these Regulations an 'injurious substance' means:

(a) a polychlorinated biphenyl (PCB);
(b) a polychlorinated terphenyl (PCT);
(c) any preparation with a PCB or PCT content higher than 0.1 per cent by weight.

Effectively, injurious substances are banned in the UK except:

(1) when they are being transported;
(2) where currently they are used in transformers and capacitors;
(3) as a hydraulic fluid in underground mining equipment;
(4) in a process in which the substance is destroyed or converted to a non-injurious substance; or
(5) in the course of analysis, research and development.

3.11 The Control of Substances Hazardous to Health Regulations 1988 (COSHH)

The substances that are hazardous to health include:

(a) those that are listed in the Classification, Packaging and Labelling of Dangerous Substances Regulations as being very toxic, toxic, harmful, corrosive or irritant;
(b) a substance that has a maximum exposure limit;
(c) certain micro-organisms; and
(d) dust in substantial concentration in air.

The Regulations do not apply to lead, asbestos or where the substance is only hazardous to health due to its radioactive, explosive or flammable properties or because it is at high or low temperature, but they do now cover carcinogens. Certain substances, such as benzidine, cannot be imported into the UK.

In essence, an employer shall not carry on any work liable to expose any employee to a substance hazardous to health unless he has made a suitable and sufficient assessment of the risks and the steps to be taken to meet the requirements of the Regulations. The assessment should be repeated if it is no longer regarded as being valid or if sufficient change in the work has taken place.

Every employer shall ensure that the exposure of employees to such substances shall be prevented or, where this is not reasonably practicable, controlled either with or without the use of personal protective equipment. Where control measures or personal protective equipment is provided the employer shall see that it is properly used and the employee is required to make full and proper use of them and report any defects. Control measures are to be maintained in an efficient state, in efficient working order and in good repair. Where engineering controls are provided they shall be thoroughly examined and tested in the case of local exhaust ventilation plant at least once every 14 months (except for some scheduled plants where the intervals are less) and in any other case at suitable intervals. Non-disposable respiratory protective equipment shall be examined, and where appropriate, tested at suitable intervals. The records of examinations and tests are to be kept for at least five years.

The exposure of an employee to substances hazardous to health may have to be monitored in the cases of scheduled substances or processes (viz. vinyl chloride monomer, electrolytic chromium process) or as a requisite for ensuring the maintenance of adequate controls. Suitable records shall be kept, in the case of identifiable employees for 30 years and in other cases five years. Health surveillance of employees is required in scheduled processes involving certain substances (e.g. in the manufacture of potassium or sodium chromate or auramine magenta). A health record in respect of each affected employee, containing approved particulars, must be kept for at least 40 years from the date of the last entry. Health surveillance will be by an appointed doctor or an employment medical adviser and will normally take place every 12 months. Information, instruction and training are to be given to persons who may be exposed to hazardous substances, whether employee or not. Other Regulations deal with fumigation and exemption certificates whilst other schedules detail:

(a) substances assigned maximum exposure limits;
(b) medical surveillance;
(c) excepted fumigations; and
(d) repeals and revocations.

Approved Codes of Practice expand the Regulations[4–8].

3.12 The Electricity at Work Regulations 1989

The general regulations impose requirements with regard to:

(a) construction and maintenance of electrical systems;
(b) carrying out work activities on or near electrical systems;
(c) the provision of protective equipment;
(d) the putting into use of electrical equipment;
(e) the construction of electrical equipment to resist adverse or hazardous environments;

(f) the insulation and protection of conductors;
(g) earthing;
(h) sound connections;
(i) means of protection against excess current and the cutting off of the supply;
(j) work on or near live conductors;
(k) working space, access and lighting; and
(l) persons engaged in electrical work to be competent to prevent danger.

The Regulations place duties, so far as they relate to matters within their control, on employers (with special reference to managers of a mine or a quarry), self-employed persons and employees. Certain of the Regulations place additional duties on managers of mines. Comprehensive guidance is given in a Memorandum[9] and Approved Codes of Practice[10,11].

3.13 The Fire Certificates (Special Premises) Regulations 1976

Most premises at which persons are 'at work' require a fire certificate which is issued by the Fire Authority under the Fire Precautions Act. However, for some premises, usually because of the quantities of hazardous substances being stored or used, the fire certificate will be issued by the Health and Safety Executive.

These latter premises are listed in schedule 1 of the Regulations and include those which:

(1) use over 50 tonnes of highly flammable liquid in the manufacturing process,
(2) manufacture 50 tonnes or more of expanded cellular plastics per week,
(3) can store 100 tonnes or more of liquefied petroleum gas unless required as a fuel or for the heat treatment of metals,
(4) can store 100 tonnes or more of liquefied natural gas unless kept solely as a fuel,
(5) can store 100 tonnes of liquefied flammable gas consisting predominantly of methyl acetylene unless kept solely as a fuel,
(6) manufacture oxygen and have storage facilities for 135 tonnes or more of liquid oxygen,
(7) can store 50 tonnes of chlorine unless kept solely for water purification,
(8) manufacture artificial fertilisers and where there are storage facilities for 250 tonnes or more of ammonia,
(9) process, manufacture, use or store at any one time 5 tonnes or more of phosgene, 20 tonnes or more of ethylene oxide, 50 tonnes of carbon disulphide, acrylonitrile or hydrogen cyanide, 100 tonnes or more of ethylene or propylene or 400 tonnes or more of any highly flammable liquid not otherwise specified,
(10) are licensed explosives factories or magazines,
(11) are surface buildings of mines,

(12) are licensed, or UKAEA and Crown properties that otherwise would have been licensed, nuclear installations,
(13) house particle accelerators of not less than 50 megavolt (hospitals are exempt), and
(14) use in process, manufacture or storage at any one time 10 or more curies of Class 1, 100 curies of Class II and III or 1000 curies of Class IV radionuclides or unsealed radioactive substances.

Temporary buildings on construction sites are also included unless the following conditions are satisfied:

(1) not more than 20 persons are employed at any one time in the building,
(2) not more than 10 persons are employed at any one time elsewhere than on the ground floor,
(3) no explosive or highly flammable material is stored or used in or under the building, and
(4) means of escape and apparatus for fire fighting are provided.

An application for a fire certificate should be made by the 'responsible person' (that is, the person having control of the premises who may or may not be the occupier). The particulars to be included in an application are listed in schedule 2 of the Regulations. Such matters as address and description of premises, nature of processes carried on, nature and quantities of explosives or highly flammable substances and maximum number of persons likely to be in any building are to be included. If required the applicant must furnish plans, not only of the premises to be covered by the certificate but also of adjoining premises.

Where an application for a fire certificate has been made and all necessary plans furnished, it is the duty of the Executive to arrange for an inspection to be carried out. If the Executive is then satisfied that suitable means of escape and means of fire fighting are provided then a fire certificate shall be issued.

The certificate specifying means of escape and fire fighting etc. and imposing any special conditions the Executive consider appropriate shall be sent to the responsible person with a copy to the occupier. The occupier must keep it available for inspection at all reasonable times and keep a notice displayed at a suitable place on the premises which states:

(1) that a certificate has been issued,
(2) where it may be inspected, and
(3) the date of posting of the notice.

When it is proposed to make any material change to the building or process which may affect matters specified in a fire certificate the HSE shall be informed in writing of the proposal before the change is made.

No work should be carried out in these special premises until a certificate has been issued and its conditions complied with or an application has been made for a certificate which has not been refused or withdrawn.

Certificates issued under the Fire Precautions and other Acts will remain valid pending applications for the special certificates required under these Regulations.

A fire certificate is not required for any berth to which The Dangerous Substances in Harbour Areas Regulations 1987 apply.

3.14 The Fire Precautions (Places of Work) Regulations (proposed)

Proposals for these Regulations, which have been prepared by the Home Office to accommodate the requirements of the Workplace Directive, will, if they come into effect, require assessments of the risks to persons in case of fire in the workplace. From the assessment, emergency plans will have to be made covering means of fighting fires, evacuation procedures, etc. Reasonable means of escape in case of fire must be provided and provision must also be made for means of fire fighting; giving warning of and detecting fires; maintaining the means of escape and the fire equipment; storing combustible refuse safely; supervising maintenance and construction work where there are fire risks; and instruction and training of persons who have fire precaution duties in a workplace.

3.15 The Health and Safety at Work etc. Act 1974 (Application outside Great Britain) Order 1989

By this Order in Council, sections 1 to 59 and 80 to 82 of the HSW are extended to cover work in connection with offshore installations and pipelines, together with construction works, diving operations and certain other activities within territorial waters (which are normally three miles beyond low water mark) and to mines extending under the sea. The extension includes any areas designated under the Continental Shelf Act 1964.

Responsibility for enforcement belongs to the HSC and HSE inspectors cover activities on and in the vicinity of offshore platforms as well as activities elsewhere. The structural safety, manning and design of rigs and the navigation of vessels used in connection with installations and pipelines is also the responsibility of the HSE.

3.16 The Health and Safety (Display Screen Equipment) Regulations 1992

Where an employee is a user, as defined (in essence an operator who habitually uses display screen equipment as a significant part of the normal work), of display screen equipment (any alpha-numeric or graphic display on a screen) the work station has to be assessed with regard to the health and safety of the user. The findings of the assessment should be used to reduce the risks identified.

The minimum requirements of the work station (i.e. display screen equipment, keyboard, software, accessories, disc drive, telephone, modem, printer, document holder, work chair, work surface or desk, etc. when provided) are laid down in the Schedule. The scheduling of the work of a user shall be such that the work on the display screen equipment is periodically interrupted by breaks or changes of activity.

Before a person is employed as a user that person has a right to have an appropriate eye and eyesight test carried out by a competent person. Such tests shall be available to users at regular intervals. Where normal corrective appliances cannot be used when the operator is experiencing visual difficulties which reasonably may be considered to be caused by work on display screen equipment, the employer shall ensure special corrective appliances are provided.

Before any person becomes a user he should be provided with appropriate health and safety training in the use of any work station on which he is to work. Users are to be given information on the measures being taken to comply with the Regulations insofar as they relate to the user.

The Schedule covers such matters as:

(1) Equipment – display screen, keyboard, work desk, work chair.
(2) Environmental – space, lighting, reflection and glare, noise, heat, radiation, humidity.
(3) User/computer interface – software, systems.

3.17 The Health and Safety (First-Aid) Regulations 1981

First aid means, so far as the Regulations are concerned,

(a) in cases where a person will need help from a medical practitioner or nurse, such treatment necessary to preserve life and minimise the consequences of injury and illness until they can be treated by a doctor or nurse;
(b) treatment of minor injuries which would otherwise receive no treatment or which do not need treatment by a medical practitioner or nurse.

An employer must provide or ensure that there are available adequate and appropriate equipment and facilities to render first aid to his employees if they are injured or become ill at work and to provide sufficient suitable persons for rendering first aid.

A suitable person will have undergone such training and have such qualifications as the HSE may approve. The temporary or exceptional absence of this person can be covered by appointing someone else to take charge if any injury or illness emergency occurs. In deciding what facilities and type of first aid services to provide the employer must consider the number of employees, the nature of the undertaking, the size and location of the establishment and the distribution and location of employees in the course of their work.

Employees must be informed of the first aid arrangements including the location of those who can give treatment, equipment and facilities. On multi-occupancy sites common facilities and arrangements are acceptable.

A self-employed person must provide suitable equipment to enable him to render first-aid to himself.

The regulations do not apply to diving operations, certain vessels and some types of mine.

Criteria for deciding who should give first aid treatment, what provisions will be adequate and appropriate and the extents of first aid boxes and kits are given in an Approved Code of Practice[12].

3.18 Highly Flammable Liquids and Liquefied Petroleum Gases Regulations 1972

3.18.1 Highly flammable liquids

A flammable liquid is one that at atmospheric pressure gives off a flammable vapour at a temperature of 32°C and a liquefied flammable gas is one which, even though stored under pressure as a liquid, is a flammable vapour at 20°C at atmospheric pressure.

All highly flammable liquids (hfl) shall be stored:

(1) in suitable fixed storage tanks, or
(2) in suitable closed vessels in the open air, or
(3) up to 50 litres (11 gallons) in work rooms if kept in fire resisting cupboards or bins.

When not in use storage cupboards, tanks, vessels, bins shall be kept closed and all reasonably practicable steps shall be taken to contain leaks or to make them safe.

Every store room, cupboard, bin, tank and vessel used to store hfl is to be clearly and boldly marked 'Highly Flammable'; 'Flashpoint below 32°C'; 'Flashpoint in the range of 22°C to 32°C'; or some other appropriate indication. If impracticable to mark the store room etc, the words 'Highly Flammable Liquid' shall be clearly and boldly marked as near as possible.

Where hfl are conveyed in a factory a totally enclosed pipework system should be used where reasonably practicable, otherwise they should be conveyed in vessels with sealed non-spill caps. The quantity of hfl present in a workplace shall be as small as possible and not exceed that required for the shift or day's work.

Empty tanks or vessels should not be kept in the work area but removed without delay to the open air or to a fire resistant store room or cupboard and kept there until next required. They should be kept closed.

Likely sources of ignition of vapour from hfl must be removed from areas where a dangerous concentration of vapour may build up.

Where a dangerous concentration of vapours may reasonably be expected, any cotton waste or other materials contaminated with hfl, or liable to spontaneous combustion, should be put in a metal container and the lid closed or removed to a safe place.

Where cellulose nitrate is present in hfl steps should be taken to prevent any solid residue depositing on any surface which might reach 120°C (248°F).

Where hfl is used in a workplace its vapours must be extracted and exhausted to atmosphere so that vapour cannot escape into the workplace.

Ideally, the process should be carried on within a fume cabinet or other fire resistant enclosure that has mechanical ventilation.

The electric motors in exhaust ventilation systems should not be in the path of the hfl vapours unless flameproofed.

All venting devices of fixed tanks or vessels containing hfl shall discharge to a safe place and where necessary be provided with a wire gauze or other suitable flame arrester.

Provision may be made in fire resisting structures (store room, work room etc.) for pressure relief in the case of explosion, such relief to be arranged so that any pressure will vent to a safe place.

Fire escape routes from rooms in which hfl is manufactured, used or manipulated must be well identified and kept clear.

Where solid waste residues from any process involving hfl is a fire risk suitable precautions should be taken until the residue can be removed to a safe place. In the removal of these residues iron or steel implements must not be used but non-sparking bronze tools may be.

There should be no smoking wherever hfl is used or stored.

Hfl should be burnt in suitable plant in a safe manner, unless the purpose is to dispose of waste or to provide fire fighting training when the burning should be done by a competent person in a safe manner and in a safe place. Fire fighting training using burning hfl must be carried out under the direct and continuous supervision of a competent person.

Where hfl is manufactured or used, appropriate well maintained fire fighting equipment shall be readily available.

3.18.2 Liquefied petroleum gases

All liquefied petroleum gas (LPG) not in use shall be stored:

(1) in underground reservoirs, fixed storage tanks or vessels,
(2) in movable storage tanks or vessels kept in the open air,
(3) in pipelines, or
(4) in suitable containers kept in the open air or where this is not reasonably practicable the LPG may be kept in a well ventilated store room, either in a safe position or of fire resisting structure. Apart from LPG, only acetylene cylinders may be kept in the same store room.

LPG kept in a workroom may only be in cylinders or pipelines whose number must be kept to a minimum having regard to the processes carried on. LPG cylinders are to be stored until required and empty cylinders removed from workplaces as soon as reasonably practicable.

Every tank, vessel, reservoir, cylinder and store room used for storing LPG shall be clearly and boldly marked 'Highly Flammable – LPG' or similar.

It is the duty of every person employed in a factory to which these Regulations apply to comply with the Regulations. If a person discovers any defect in the plant, equipment or appliances it is his duty to report the defect without delay to the occupier, manager or other responsible persons.

3.19 The Ionising Radiations Regulations 1985

These regulations relate to any work involving radioactive substances having an activity greater than 100 Bq g^{-1} and also to the use of machines that generate ionising radiations. They supersede the Ionising Radiations (Unsealed Sources) Regulations 1968 and the Ionising Radiations (Sealed Sources) Regulations 1969.

Twenty eight days before any work involving ionising radiations is undertaken for the first time, the HSE must be informed. Records must be kept of the quantity and location of all radioactive substances and those records retained for at least 2 years.

Any area where the instantaneous dose rate exceeds 7.5 μSv h^{-1} should be made a 'controlled area' and any employee who receives a dose of ionising radiations at or above this level must be designated a 'classified person'. The radiation dose to which a classified person is subjected must be monitored using an approved dosimetry service and that person shall be subject to medical surveillance. A 'radiation protection adviser' should be appointed.

Where the radiation dose rate lies between 2.5 and 7.5 μSv h^{-1} the area shall be designated a 'supervised area' and a radiation protection supervisor appointed. The employer shall monitor the level of ionising radiations using equipment that is well maintained and tested every 14 months. Employers using radioactive substances should prepare and put in writing safe working rules aimed at ensuring compliance with the Regulations and he should ensure the work is properly supervised.

Where any employee is subjected to a radiation dose level thought to be in excess of the stated dose limit, the circumstances must be investigated to ensure all reasonably practicable steps have been taken to keep exposure to a minimum. If overexposure has occurred the HSE must be informed and arrangements made for the individual to be medically examined. Records of such investigations should be kept for 50 years.

Approved Codes of Practice have been published[13,14]

3.20 The Management of Health and Safety at Work Regulations 1992

The broad general duties apply to most kinds of work and require employers to:

(1) assess the risks to the health and safety of employees and others who may be affected by their activities so that the necessary measures for prevention and protection can be identified;
(2) make arrangements for putting into practice measures shown by the risk assessment to be necessary for the management of health and safety. The arrangements to cover planning, organisation, control, monitoring and review;
(3) where necessary, provide health surveillance;
(4) appoint competent persons to assist him in taking the measures needed to comply with the relevant statutory provisions. These persons need not be employees;

(5) set up emergency procedures;
(6) provide employees with information about health and safety;
(7) co-operate with other employers sharing the work site;
(8) provide non-employees working on the site with information;
(9) ensure employees have adequate health and safety training and are capable enough to avoid risks; and
(10) give temporary workers appropriate health and safety information.

Employees must follow health and safety instructions and report dangerous situations[15].

3.21 The Manual Handling Operations Regulations 1992

Where manual handling, which means the transporting or supporting of a load, including lifting, putting down, pushing, pulling, carrying or moving it by hand or bodily force, cannot be avoided, when it involves a risk of injury the operation must be assessed. A suitable and sufficient assessment will consider such factors as:

(1) The task – Does it involve holding the load at a distance from the trunk, unsatisfactory body movement, excessive movement of the load, risk of sudden movement of the load, frequent or prolonged physical effort or insufficient rest or recovery periods?
(2) The load – Is it heavy, bulky, difficult to grasp, unstable, sharp or hot?
(3) The working environment – Are there space constraints, poor floors, extremes of temperature or humidity or poor lighting?
(4) Individual capability – Does the job require unusual strength, create hazards to those who are pregnant or have a health problem or require special knowledge or training?

Where it is not reasonably practicable to avoid the need for employees to do manual handling which could involve a risk of their being injured, they must be trained or provided with suitable equipment and given information about the load.

Assessments should be reviewed when conditions change. Employees must make full and proper use of the equipment provided and inform their employer of any physical conditions suffered by them which could affect their ability to undertake manual handling operations.

3.22 The Noise at Work Regulations 1989

These Regulations require employers to make and review noise assessments and to keep records thereof. The risk of damage to the hearing of employees from exposure to noise must be reduced and at the first action level (a daily personal exposure of 85 dB(A)) suitable personal hearing protection must be provided. At the second action level (a daily personal exposure of 90 dB(A) or above) or at the peak action level (200 pascals or above) hearing protection zones, clearly demarcated and identified, have to be established. Hearing protectors should be worn by employees entering a hearing protection zone. Employees likely to be exposed to any level of noise above the first action level are to be provided with adequate information, instruction and training. There is a specified duty on designers, manufacturers, etc., to provide such information under section 6 of HSW Act.

3.23 The Personal Protective Equipment at Work Regulations 1992

The Regulations, which do not apply to personal protective equipment that is required to be provided under earlier legislation dealing with lead, asbestos, ionising radiations, hazardous substances, etc., cover the provision of equipment designed to be worn or held by a person to protect him against one or more hazards likely to endanger his safety and health at work, but do not extend to ordinary work clothes or uniforms. PPE includes aprons, waterproof clothing, gloves, safety footwear, high visibility waistcoats, eye protectors, respirators, safety harnesses, etc.

Unless a risk is controlled by other means an employer is required to provide suitable PPE to each of his employees who may be exposed to the risk at work. Before choosing a PPE an employer must make an assessment to determine whether the PPE is suitable for the particular use. The assessment should comprise such factors as whether the risk could be avoided by other means and the characteristics which the PPE should have to be effective against the risk.

All the PPE provided should be maintained in a good and clean order and be stored in suitable accommodation when not in use. Employees should be given information, instruction and training on the risks to be avoided, the purpose for which and the manner in which the PPE is to be used and the actions to be taken to keep the PPE in an efficient and clean state.

Employees must use the PPE provided and report any loss or defect. Employees must not be charged for the PPE with which they are issued.

The standards to be met in respect of the manufacture and testing of PPE are detailed in the Personal Protective Equipment (European Community Directive) Regulations 1992.

3.24 The Provision and Use of Work Equipment Regulations 1992

These Regulations pull together and tidy up the laws governing equipment used at work. Instead of piecemeal legislation covering particular kinds of equipment in different industries, they place general duties on employers and list minimum requirements for work equipment to deal with selected hazards whatever the industry.

'Work equipment' is broadly defined to include everything from a hand tool, through machines of all kinds, to complete plant such as a refinery. 'Use' includes starting, stopping, programming, setting, transporting, repairing, modifying, maintaining, servicing and cleaning.

The general duties require employers to:

(1) take into account the working conditions and risks in the workplace when selecting equipment;
(2) make sure that equipment is suitable for the use that will be made of it and it is properly maintained; and
(3) give adequate information, instruction and training.

Specific requirements cover:

(4) protection from dangerous parts of machinery;
(5) maintenance operations;
(6) danger caused by other specified hazards;
(7) parts and materials at high or very low temperatures;
(8) control systems and controls;
(9) isolation of equipment from power sources;
(10) stability of equipment;
(11) lighting; and
(12) warnings and markings.

The measures relating to the design and construction, and to the placing on the market and putting into service, of machinery are contained in the Supply of Machinery (Safety) Regulations 1992. By conforming with these regulations machinery qualifies for the EC mark and can be placed anywhere on the European market.

3.25 The Power Presses Regulations 1965

These regulations require that power presses embodying a flywheel and clutch mechanism must have the tools set and safety devices checked by a properly trained person.

Setters who do setting, adjusting, trying out and fitting safety devices must be over 18 years of age, trained, competent and appointed in writing. They must be given a certificate of appointment. Setters under training must be supervised by a fully trained competent person.

Every power press must be thoroughly examined and tested by a competent person before being put into use, thereafter every 12 months for presses with fixed fencing or 6 months for other presses. Safety devices other than fixed fencing to be examined on installation and then every six months. A report of every examination and test on an approved form to be completed within 14 days.

The competent person must notify both the occupier and the factory inspector of any defect which is, or may become, a cause of danger.

If the defect is dangerous, the press may not be used until the defect is remedied, otherwise the competent person must specify the time within which the defect has to be remedied. Measures taken to remedy defects in presses or safety devices must be recorded.

Safety devices must be inspected and tested after setting or re-setting tools or the adjustment of tools involving alteration or disturbance of safety devices and not later than four hours after the start of each working day or shift. A certificate is to be kept at or near the press, showing particulars identifying the press and its safety devices, the date and time of the inspection and test, the setter's (tester's) signature and that every safety device is in efficient working order. All certificates to be preserved for six months.

Missing, defective or unsuitable safety devices to be reported to the employer immediately.

Every press and safety device to be clearly identified and the maximum permissible speed of the flywheel and its direction of rotation to be conspicuously marked. That speed must not be exceeded and the direction many only be reversed in an emergency.

Registers of appointment of setters and press examinations to be preserved for two years after the final entry.

3.26 The Reporting of Injuries, Diseases and Dangerous Occurrences Regulations 1985

These regulations relate to incidents connected with work and cover not only employees and the self employed but members of the public, pupils and students, hotel residents, prisoners etc. who may suffer injury or death from an accident arising from or connected with work. The responsibility for reporting the accident lies with the employer or the person in charge of the premises where the work was being carried out.

Incidents that result in death, major injury, specified illness, detention in hospital for more than 24 hours or a dangerous occurrence must be reported to the enforcing authority forthwith (usually by telephone) and followed within 7 days by a written report on form F2508. If as a result of an accident an employee dies within a year the enforcing authority must be notified in writing as soon as the employer has knowledge of the death. In addition, if an employee is diagnosed as suffering from one of the diseases listed in the schedule 2 the employer must report the fact on form F2508A. Gas suppliers must report any death or injury resulting from the use of their product.

Records of the details of reportable accidents must be kept by the employer and this requirement can be met by keeping a copy of the reporting form F2508/2508A although the employer's own records that contain the necessary particulars are acceptable. These records must be retained for at least 3 years.

Injuries and illness conditions that must be reported are:

(1) fracture of the skull, spine or pelvis,
(2) fracture of any bone:
  (a) in the arm or wrist but not the hand,
  (b) in the leg or ankle but not the foot,
(3) amputation of a hand or foot, and of a finger, thumb or toe if a bone or joint is severed,
(4) loss of sight in an eye,
(5) burns or loss of consciousness from electric shock,
(6) loss of consciousness through lack of oxygen,
(7) decompression sickness,
(8) acute illness from the inhalation, ingestion or absorption of any substance,

(9) acute illness resulting from exposure to pathogens or infected material,
(10) any other injury that results in detention in hospital for more than 24 hours,
(11) any of the diseases listed in schedule 2 to the regulations.

The dangerous occurrences that must be reported are listed in schedule 2 and include:

(1) collapse or overturning of any hoist, lift, crane, derrick, mobile platform, excavator or pile driver but excluding winches, gin wheels, transporters, runways and lift trucks,
(2) the failure of any passenger carrying equipment,
(3) explosion or collapse of a pressure vessel,
(4) explosion, fire, electrical shortcircuit or overload that stops plant for more than 24 hours,
(5) the escape of more than one tonne of a flammable substance,
(6) collapse of scaffolding or part of a building,
(7) failure of a pipeline.

Chapter 11 deals with the procedures for investigating and reporting accidents.

Additional dangerous occurrences are notifiable in relation to mines or quarries, or railways, and there are also additional provisions relating to mines and quarries.

3.27 The Safety Representatives and Safety Committee Regulations 1977

For the purpose of section 2(4) of the Health and Safety at Work Act recognised trade unions may appoint safety representatives from amongst the employees. An employer should be notified in writing of the names of the persons appointed and the group or groups of employees they represent. The nominee should either have been employed by his employer throughout the preceding two years or have at least two years' experience in similar employment.

In addition to representing employees in consultation with the employer, a safety representative has the following functions:

(1) To investigate potential hazards and dangerous occurrences at the workplace and to examine the cause of accidents at the workplace.
(2) To investigate complaints by any employee he represents relating to that employee's health, safety or welfare.
(3) To make representations to employers on matters arising out of the above.
(4) To make representations to employers on general matters affecting the health, safety and welfare of employees at the workplace.
(5) To carry out inspections of the workplace, following notifiable accidents and, if relevant to the inspection, see certain documents.
(6) To represent the employees in consultation with inspectors.

(7) To receive information from inspectors.
(8) To attend safety committee meetings.

But none of these functions should be construed as imposing any duty on him.

A safety representative may take such time off, with pay, as shall be reasonably necessary for performing his functions or undertaking appropriate training.

Safety representatives shall be entitled to inspect the workplace if they have not inspected it within the previous three months and they have given reasonable notice to the employer. More frequent inspections can be made by agreement with the employer. Where there has been a substantial change in the conditions of the workplace or where new information has been published by the HSE the safety representative, after consultation with the employer, shall be entitled to carry out an inspection notwithstanding that three months have not elapsed since the last inspection. The employer shall provide such facilities and assistance as may be reasonably required by the safety representative to make such inspections. Similar inspections can be carried out by the safety representative following a notifiable accident or dangerous occurrence in a workplace or if a notifiable disease has been contracted there.

If he has given reasonable notice to the employer a safety representative may inspect and take copies of any document relevant to the workplace that is required to be kept by any relevant statutory provision listed in schedule 1 of HSW except documents relating to the health record of an identifiable individual. Subject to listed exceptions, the employer shall make available the information necessary to enable a safety representative to fulfil his functions.

Where there is not one in existence and two or more safety representatives request an employer in writing to establish a safety committee, the employer shall consult with the safety representatives and other representatives for that workplace and establish a committee within three months. The employer shall post a notice stating the composition of the committee and the workplace(s) to be covered.

An employer should keep appropriate safety representatives informed of the steps being taken to meet the duties of the Management of Health and Safety at Work Regulations 1992 (MHSW).

The HSE is given power to grant exemptions to these requirements. A safety representative may appeal to an industrial tribunal if his rights of inspection or time off are refused.

Codes of practice have been issued covering both the functions and time off for the training of safety representatives[16]. Guide notes relating to these regulations have also been published.

3.28 The Safety Signs Regulations 1980

A safety sign giving health or safety information or instruction must comply with Part 1 of BS5378: 1980 as must any strip of alternate colours used to identify a hazard. These requirements do not apply to any sign or colour:

(a) for regulating rail, road, marine or air traffic or relating to any load carried by such traffic or on an aircraft, hovercraft or vessel;
(b) at a coal mine or tip;
(c) relating to fire fighting or rescue equipment or emergency exits;
(d) any label or marking on a package or container or explosive article.

Except for mines or tips, signs used for regulating road traffic are prescribed under the Road Traffic Signs Regulations and General Directive 1981.

Part 1 of BS 5378 describes the safety colours as red, yellow, blue and green. Red is for stop or prohibition with the contrasting colour in white and any symbol black. Yellow is caution or risk of danger with black contrasting colour and symbol. Blue is for mandatory signs with contrasting colour and symbol in white. Green is the safe conditions colour, again with white for contrasting colour and symbols. (See inside covers.)

Prohibition signs have a white background with a red circular band and crossbar and five symbols are detailed:

No smoking;
Smoking or naked flames prohibited;
Do not extinguish with water;
Not drinking water;
Pedestrians prohibited.

Warning signs have black triangular bands, yellow background with symbol or text in black. There are ten signs detailed:

Caution, risk of fire;
Caution, risk of explosion;
Caution, toxic hazard;
Caution, corrosive substances;
Caution, risk of ionising radiations;
Caution, overhead load;
Caution, industrial trucks;
Caution, risks of electric shock;
Caution, laser beam;

and a general warning:

Caution, risk of danger.

Mandatory signs have a blue circle with symbol or text in white. The listed mandatory signs are:

Eye protection must be worn;
Head protection must be worn;
Hearing protection must be worn;
Respiratory protection must be worn;
Hand protection must be worn.

Safe conditions are indicated by a green square or rectangle with white symbol or text, but only first aid and direction signs are specified in part 1 of the specification.

Any supplementary signs must be rectangular or square. Usually they are white with black text but can be the same colour as the background of the associated sign with text in the symbol colour.

Alternative colours of black with fluorescent orange, red or yellow can be used to indicate a place of danger.

3.29 The Workplace (Health, Safety and Welfare) Regulations 1992

In replacing a total of 38 pieces of older law including parts of the Factories Act 1961 and the Offices, Shops and Railway Premises Act 1963, the Regulations cover many aspects of health, safety and welfare in the workplace and apply to all places of work except: means of transport; construction sites; mines and quarries and other extractive sites; and fishing boats. Workplaces on agricultural or forestry land away from main buildings are also exempted from most requirements.

The Regulations set general requirements in:

(1) The working environment including: temperature; ventilation; lighting including emergency lighting; room dimensions; suitability of work stations and seating; and outdoor workstations (e.g. weather protection).
(2) Safety including: safe passage of pedestrians and vehicles; windows and skylights (safe opening, closing and cleaning); glazed doors and partitions (use of safe material and marking); doors, gates and escalators (safety devices); floors (their construction, obstructions, slipping and tripping hazards); falls from heights and into dangerous substances; and falling objects.
(3) Facilities including: toilets; washing, eating and changing facilities; clothing storage; seating; rest areas (and arrangements in them for non-smokers); and rest facilities for pregnant and nursing mothers.
(4) Housekeeping including maintenance of the workplace, its equipment and facilities; cleanliness; and removal of waste materials.

Guidance on the implementation of the Regulations is given in an Approved Code of Practice[17].

Note: To augment the above the Home Office is preparing regulations with regard to fire precautions in the workplace. Assessments will have to be made of the risks to persons in case of fire leading to the drawing up of emergency plans and establishing reasonable means of escape. Fire fighting and detection equipment will have to be provided and maintained and persons in the workplace instructed and trained.

3.30 Status of European Directives

Directives dealing with health and safety matters that stem from the Treaty of Rome as amended emanate mainly from two sources within the Commission of the European Communities (CEC); those concerned with the establishment of a free market (referred to as Article 100A directives) under the aegis of Directorate General III (DG III) which place requirements on new plant and equipment; and those concerned with employment (referred to as Article 118A directives) which are from DGV and lay down standards of health and safety to be achieved in the workplace. In the UK, negotiations for, and development of, regulations to implement Art. 100A directives are the responsibility of the Department of Trade and Industry (DTI) under powers deriving from the European Communities Act 1972, while Art. 118A directive implementation is the responsibility of the Department of Employment through the HSC/HSE and are made under powers authorised by the HSW '74. As the free market in the European Community develops so the number of directives dealing with health and safety matters will grow. Information on the latest position can be obtained from technical journals and publications and from HSC publications.

References

1. Health and Safety Executive, *Approved Code of Practice for the Control of Asbestos at Work*, HSE Books, Sudbury (1988)
2. Health and Safety Executive, *Approved Code of Practice for Work with Asbestos Insulation, Asbestos Coating and Asbestos Insulation Board*, HSE Books, Sudbury (1988)
3. Health and Safety Executive, *Approved Code of Practice for the Control of Lead at Work*, HSE Books, Sudbury (1985)
4. Health and Safety Executive, *Approved Code of Practice for the Control of Substances Hazardous to Health*, HSE Books, Sudbury (1992)
5. Health and Safety Executive, *Approved Code of Practice for the Control of Substances Hazardous to Health in Fumigation Operations*, HSE Books, Sudbury (1988)
6. Health and Safety Executive, *Approved Code of Practice for the Control of Vinyl Chloride at Work*, HSE Books, Sudbury (1988)
7. Health and Safety Executive, *Approved Code of Practice for the Control of Substances Hazardous to Health in the Production of Pottery*, No. COP 41, HSE Books, Sudbury (1990)
8. Health and Safety Executive, *The control of legionellosis including legionnaire's disease*, Guidance booklet HS(G) 70, HSE Books, Sudbury (1991)
9. Health and Safety Executive, *Memorandum of Guidance on the Electricity at Work Regulations 1989*, HSE Books, Sudbury (1989)
10. Health and Safety Executive, *Approved Code of Practice for the Use of Electricity in Mines*, HSE Books, Sudbury (1989)
11. Health and Safety Executive, *Approved Code of Practice for the Use of Electricity in Quarries*, HSE Books, Sudbury (1989)
12. Health and Safety Executive, *First aid at works, Health and Safety (First Aid) Regulations 1981, Approved Code of Practice and Guidance*, No. COP 42, HSE Books, Sudbury (1981)
13. Health and Safety Executive, *Approved Code of Practice for the Protection of Persons against Ionising Radiations arising from any Work Activity*, HSE Books, Sudbury (1985)
14. Health and Safety Executive, *Approved Code of Practice for Exposure to Radon*, HSE Books, Sudbury (1988)
15. Health and Safety Executive, *Approved Code of Practice: Management of Health and Safety at Work Regulations 1992*, No. L21, HSE Books, Sudbury (1992)
16. Health and Safety Executive, *Approved Code of Practice for Safety Representatives and Safety Committees*, HSE Books, Sudbury (1978)
17. Health and Safety Executive, *Approved Code of Practice and Guidance on the Workplace (Health, Safety and Welfare) Regulations 1992*, No. L24, HSE Books, Sudbury (1992)

Some Other Regulations

The Asbestos (Prohibitions) Regulations 1985
The Classification, Packaging and Labelling of Dangerous Substances Regulations 1984

The Construction (Head Protection) Regulations 1989
The Dangerous Substances in Harbour Areas Regulations 1987
The Docks Regulations 1988
The Gas Safety (Installations and Use) Regulations 1984
The Health and Safety (Enforcing Authority) Regulations 1989
The Health and Safety Information for Employees Regulations 1989
The Health and Safety (Training for Employment) Regulations 1990
The Lifting Plant and Equipment (Record of Test and Examination, etc.) Regulations 1992
The Loading and Unloading of Fishing Vessels Regulations 1988
The Notification of Installations Handling Dangerous Substances Regulations 1982
The Notification of New Substances Regulations 1982
The Road Traffic (Carriage of Dangerous Substances in Packages etc.) Regulations 1992
The Road Traffic (Carriage of Dangerous Substances in Road Tankers and Tank Containers) Regulations 1992
The Road Traffic (Training of Drivers of Vehicles carrying Dangerous Goods) Regulations 1992
The Shipbuilding and Ship-repairing Regulations 1960

all available from HMSO, London.

Chapter 4

Law of contract

R. W. Hodgin

4.1 Contracts

4.1.1 Formation of contract

A contract is an agreement between two or more parties and to be legally enforceable it requires certain basic ingredients. It must be certain in its wording and consist of an offer made by one party which must be accepted unconditionally by the other (*Scammell* v. *Ouston*[1]; *Carlill* v. *Carbolic Smoke Ball Co.*[2]). This does not prevent negotiations taking place and alterations being made by both parties during the early stages of the discussions, but in its final stage there must be complete and clear agreement as to the terms of the contract (*Bigg* v. *Boyd Gibbins Ltd*[3]). The great majority of contracts need not be in writing and those made daily by the general public, buying a newspaper or food, clearly show this.

There must however be 'consideration' that flows from one party to the other. Consideration is the legal ingredient that changes an informal agreement into a legally binding contract. It is the exchange of goods for payment, of work for wages, of a journey for the price of a ticket that amounts to consideration (*Dunlop Pneumatic Tyre Co. Ltd* v. *Selfridge & Co. Ltd*[4]). It is possible for future promises, mutually exchanged, to amount to consideration. Thus X agrees to buy a car from Y on January 1st and payment will not be made until that date. The contract dates from the day that these promises are exchanged.

Two other essentials for a valid contract are that the parties must intend to enter into a legally binding agreement and both parties must have the legal capacity to make such a contract. Lack of intention is rarely a problem in reality, but if one party alleges that he never intended to enter into a contract, only the courts can say, having analysed the behaviour of the parties, whether or not a contract was created (*Balfour* v. *Balfour*[5]). Capacity to contract means that the parties must be sane, sober and over the age of 18, although there are obvious exceptions to the age requirement when it comes to contracts of employment; such employment contracts are now largely regulated by statute, e.g. Employment of Children Act 1973.

Although the limits of a contract should be reflected in what the parties to it have expressly agreed upon, it is possible for the courts, or for Parliament, to imply terms into the agreement. The most notable example of statutory implied terms is the Sale of Goods Act 1979 which among other things implies into contracts of sale a condition that goods shall be of merchantable quality. The court's role in implying terms is a more difficult area of the law because the general approach adopted by the courts is one of non-intervention.

It is clear, however, that in certain circumstances they will adopt a more positive role particularly where the implied term is necessary to give business sense to the agreement. (*Matthews v. Kuwait Bechtal Corporation*[6], The Moorcock[7]).

4.1.2 Faults in a contract

Despite the outward appearance of agreement there may be fundamental faults that will affect the validity of the contract. The parties are not always clear and precise in the language they use and it may be that the contract will be void, that is unenforceable, on the grounds of Mistake. This can arise in a number of ways. It may be that the subject matter of the contract is no longer in existence at the time the contract is made, e.g. where X appears to sell a machine to Y but earlier the machine had been destroyed in a fire at the factory.

It is possible that the parties have been negotiating at cross-purposes, e.g. where X intends to sell one particular machine but Y has in mind another machine at X's factory. In this situation the basic requirement of agreement is missing and no contract comes into existence.

A third possibility is where one party is mistaken as to the identity of the other contracting party and can prove that identity was crucial to his entering into the contract. It should be stressed however that the courts will not easily allow the mistaken party to avoid the contract for this would be an easy way for people to escape from contracts that look as though they are about to take a disastrous financial turn.

It may be that one party has been led into the contract by a Misrepresentation made by the other party, e.g. untrue statements about the capabilities of the machine. The remedies available in this situation vary depending on whether the misrepresentation was made innocently, negligently or fraudulently. Under the Misrepresentation Act 1967 where the statement was made innocently the party misled may ask for the contract to be set aside, but the court has the power to refuse this and instead grant damages. If the statement was made negligently then damages can be awarded and the contract may be set aside, but the court may decide that damages alone are sufficient and rule that the contract should continue. Lastly, the most serious misrepresentation is where it is fraudulent and here both damages and setting aside the contract will be ordered.

Even though the formation of the contract meets all the requirements it may still be declared to be an illegal contract and unenforceable. This is a complex topic but one example can be taken from the contract of employment. There are often restraint clauses in such contracts whereby an employer seeks to prevent an employee who leaves the firm from working for a rival company or setting himself up in competition. Such restraints are basically unenforceable because they are not in the public interest and are contrary to the employee's freedom to work. However it is possible to enforce such a restraint if the employer can show that the wording of the clause was reasonable in scope and that he has some interests, such as trade secrets or customer lists, that need protecting (*Fitch* v. *Dewes*[8]). What the employer cannot do however is to seek to prevent competition that the ex-employee may threaten.

Similar restraints are often found in the sale of businesses. Part of the price of a business sale reflects the goodwill that the owner has built up over the years. The buyer obviously would not want the seller to start up in competition with him by opening a similar venture in a location which would pose a financial threat to the newly acquired business. The court approaches the problem in the same way as in employer–employee restraints. Basically the restraint will be struck out as being against the public interest unless the buyer can so word the restraint that the court regards it as reasonable in the circumstances. In both types of restraint the type of work or business, the length of time of the restraint and the geographical area of the restraint will all be taken into account by the court in deciding whether it is reasonable or not (*Nordenfelt* v. *Maxim Nordenfelt*[9]).

4.1.3. Remedies

In the vast majority of cases contracts are satisfactorily concluded with both sides completing their respective obligations. But when one party fails to do so and is in breach of contract then the question of remedies arises. The normal remedy is damages, or monetary compensation.

The aim of damages is to put the innocent party in the position he would have been in if the breach had not occurred (*Parsons* v. *B.N.M. Laboratories Ltd*[10]). It is not for him to profit from the wrongdoer's behaviour and there is in fact a duty on the innocent party to mitigate the loss wherever possible (*Darbishire* v. *Warren*[11]). The claim will also be limited to what the wrongdoer can reasonably have been expected to foresee would be the outcome of his breach. For instance where one party sends a piece of machinery to the other party for repair and the repairer is in breach of contract by not returning it by an agreed date, claims covering loss of production will be allowed if the repairer should have foreseen the likely losses caused by his delay (*Hadley* v. *Baxendale*[12]). The safest thing is to inform the repairer at the time the contract is made of the exact function the machinery plays in the manufacturing process so that he is aware of its importance (*Victoria Laundry (Windsor) Ltd* v. *Newman Industries Ltd*[13]).

Another possible remedy is *quantum meruit.* This arises where the innocent party has completed part of his contract but is prevented from continuing by the other party. His claim is then based on the amount of work he has completed up to that date (*Planche* v. *Colburn*[14]). However this claim cannot be maintained by a party whose failure to complete is of his own doing. A builder cannot build half a garage and refuse to complete and yet claim for the work done. If the work is completed but badly, then the contract price must be paid less a deduction to compensate for the faulty work.

It is possible for the court to grant Specific Performance as a remedy whereby one party is ordered to complete his part of the contract. The remedy is discretionary and little used outside of land sales. It will not be granted where the contract is one of personal services, e.g. in a contract of employment. Although an industrial tribunal may order reinstatement of an employee following an unfair dismissal, such a remedy cannot be enforced against an unwilling employer and his refusal will merely be reflected in the compensation awarded to the former employee.

The above general discussion is obviously of the briefest nature. What follows is a closer look at specific contracts; contracts which depend for their content and form on legislation.

4.2 Contracts of employment

It is important to distinguish between contracts of service and contracts for services. The former describes the relationship between employer and employee while the latter is concerned with employing independent contractors to carry out certain specific tasks. Unfortunately it is not always easy to distinguish between the two and yet it is essential in order to determine the legal liabilities and responsibilities of the parties. This is particularly important in situations involving main contractors and sub-contractors. The wording of the contract can place responsibility on any party but care should be taken to set this out clearly in the various contracts. If this is done then the parties involved can cover their responsibilities by obtaining insurance.

In a contract of service it is said that a man is employed as part of the business; whereas under a contract for services his work, although done for the business, is not integrated into it but is only accessory to it (*Stevenson, Jordan and Harrison* v. *Macdonald and Evans*[15]; *The Ready-mixed Concrete (South East) Ltd.* v. *The Minister of Pensions and National Insurance*[16]). The distinction has serious repercussions on tortious liability for the general rule is that the employer is liable for the torts committed by his employees acting in the course of their employment, but he is not liable for the tortious behaviour of independent contractors. It must be stressed however that there are a number of exceptions to this basic rule. Even where an exception applies and the employer is liable to third parties, it may be that the contract will give the employer rights of reimbursement from the contractor.

4.3 Employment legislation

A contract of employment can be in any form, but the more informal it is the more difficult it may be to define its true scope.

Parliament enacted the first Contract of Employment Act in 1963, requiring that certain basic ingredients be expressed in writing within thirteen weeks of the commencement of employment. Governments since have been active in the area of employment law and much of the present law is to be found in the Employment Protection (Consolidation) Act 1978 and the Employment Act 1982 and the Employment Act 1989. The information to be communicated to the employee must be in writing and include the names of employer and employee, the date of commencement of employment, hours of work, pay, holiday entitlements, incapacity for work, sick pay provisions, length of notice which the employee must give and is entitled to receive, pension provisions and the employee's job title. Any changes in the terms of employment must be notified to the employee within one month of the change although they need not be retained by the employee.

Information regarding disciplinary rules and grievance procedure must also accompany the written particulars. These requirements are not however conclusive evidence of the terms of the contract of employment, but an employee can ask for the contract to be altered to correspond with the terms if he feels there are discrepancies. It is common also for the particulars to refer to other documentation, for instance, collective agreements, and by so doing to incorporate them into the Contract of Employment (*Systems Floors (UK) Ltd.* v. *Daniel*[17]). In all these the written agreement is persuasive but not necessarily conclusive evidence of the relationship between the parties.

In addition to the above terms, the Employment Protection (Consolidation) Act 1978, as amended, creates a large number of rights in favour of employees; the major rights include: ss. 12–18 guarantee that an employee will be paid a minimum wage where he is unable to earn his contractual wage because the employer cannot provide him with work. There are exceptions to this right, in particular where the workless period has been caused by a trade dispute involving fellow employees.

SS. 19–22 compel the employer to pay the employee certain sums for up to 26 weeks where the employee is unable to earn his contractual wage because he has been suspended from work on medical grounds under one of the statutory health and safety enactments.

SS. 23–32 guarantee to the employee the right of membership of trade unions and also the right to be paid time off work to carry out trade union duties and time off without pay to participate in certain other union or public activities.

SS. 33–48, as amended by the Social Security Act 1986, provide a right to maternity pay from the employer and a right to return to work within 29 weeks of confinement. There are no statutory rights for paternity leave.

SS. 49–53 give rights to both employer and employee to a minimum period of notice of termination of employment and, if he has been in that employment for a minimum period of 26 weeks, the employee is entitled to be given written reasons for dismissal within 14 days of his requesting it.

Where an employee alleges unfair dismissal, Part V of the 1978 Act (as amended) sets out to define the scope of unfair dismissal and the remedies that are available together with the methods of calculating compensation. Part VI, as amended, deals with redundancy payments.

Several other recent pieces of legislation also play an important role in determining the relationship between employer and employee. The Equal Pay Act 1970 (see also the Equal Pay (Amendment) Regulations 1983 introducing the concept of equal value) introduces into all contracts an equality clause stating that where people are employed on broadly similar work with people of the opposite sex then any discrepancy in terms or conditions between them must be equalised upwards unless such differences can be explained on grounds other than difference of sex. The Sex Discrimination Act 1975 as amended by the Employment Act 1989 and the Race Relations Act 1976 render it unlawful for a person to treat another less favourably on sexual or racial grounds. Any employment variations must be shown to be justified for non-sexual or non-racial reasons.

Apart from the above legislation certain other terms are implied into the employment contract, having been built up by court decisions over the years. These can be supplanted only by express provisions of an Act so stating.

The most important of the implied conditions are:

that both parties exercise reasonable care in carrying out their duties under the contract. This means that the employer should provide a safe system of work including machinery, a safe place of work and skilled fellow employees;

the employee must take care to act reasonably and not injure others (*Lister* v. *Romford Ice and Cold Storage Co. Ltd.*[18]);

the employee owes his employer a duty of fidelity which prevents him from working for a rival firm or divulging secret information (*Hivac Ltd.* v. *Park Royal Scientific Instruments Ltd.*[19]);

the employee should cooperate fully with his employer to achieve the goal of the employment contract.

4.4 Law of sale

The most common example of a contract is one for the sale and purchase of goods. While the basic common law rules and rules of equity still apply to such sales, many of the rules are to be found in the Sale of Goods Act 1979, which consolidates legislation that began in 1893. The Act covers the whole range of contract topics such as formation, terms, performance, transfer of ownership, rights of unpaid sellers and remedies.

Worthy of particular mention here are the implied conditions and warranties found in ss. 12–15. Implied terms, as we have seen in the previous section, are terms which the courts will read into the contract where parties have failed to mention them. Because of the hesitancy of the courts in implying terms into contracts, the 1979 Act specifically sets out the important terms that must be read into the contract of sale. These terms provide the buyer with a certain basic protection against buying faulty or unsuitable goods.

Section 12(i) states that there is an implied condition that the seller has good title to the goods and is therefore capable of passing true ownership to the buyer (*Rowland* v. *Divall*[20]).

Section 13 covers a situation where goods are sold by description. There is an implied condition that the goods will correspond to the description given (*Beale* v. *Taylor*[21]).

Section 14 contains two important implied conditions where the seller is selling in the course of a business, in contrast to a private sale. The first is that the goods must be of a merchantable quality unless the seller has drawn the buyer's attention to the defect or the buyer has examined the goods and should have detected the faults before the contract was made (*Wilson* v. *Rickett, Cockerell & Co. Ltd*[22]). The second implied condition is that where the buyer has expressly or by implication made known to the seller the particular purpose for which the goods are bought then such goods must be reasonably fit for such purpose (*Henry Kendall & Sons* v. *William Lillico & Sons Ltd*[23]).

Section 15 is concerned with sales by sample and implies a condition that the bulk will correspond to the sample in quality; that the buyer will have a reasonable opportunity of comparing bulk with sample and that the goods will be free from any defect rendering them unmerchantable which would not have been apparent on examination of the sample (*Ashington Piggeries Ltd* v. *Christopher Hill Ltd*[24]).

Certain transactions, i.e. exchange and barter, whereby goods change ownership may not fall within the definition of 'sale of goods'. Also, where a repair is carried out, the transfer of any new part itself is not regarded as a sale of goods.

To give protection to the new owner the Supply of Goods and Services Act 1982 was introduced whereby similar implied conditions to those listed above under the Sale of Goods Act 1979 are incorporated into the contract. As the title of the Act suggests it also covers the service element of the contract. Thus, where the supplier is acting in the course of business there is an implied term that he will exercise reasonable care and skill. Where no time is stipulated for completion of the service then there is an implied term that it must be carried out within a reasonable time and that is determined by the facts of the case. Likewise, where no price is fixed by the contract then there is an implied term that a reasonable price will be paid.

This is an opportune point to mention one of the most important pieces of consumer legislation in recent years, the Unfair Contract Terms Act 1977. The original Sale of Goods Act 1893, following the general principles of the common law contracts, permitted the parties to exclude themselves from legal liability for wrongful performance of their contractual duties by suitably worded clauses or notices. The 1977 Act, together with the 1982 Act now forbids such clauses any legal recognition if the aim is to avoid liability for injury or death caused by negligence, for instance where an electrical appliance is faulty.

Where the clause is aimed at avoiding economic loss caused by one party to the other, for instance by selling unmerchantable goods, the Act recognises two situations. If the seller sells in the course of business and the buyer is not buying in the course of business and the goods are of a type normally supplied for private use or consumption, then the Act prohibits the exclusion of the implied conditions of the 1979 Act. Where the contract is not a consumer sale, then an exclusion clause will be valid but only if it can meet the test of reasonableness as laid down in the 1977 Act, the burden of proof resting on the party who wishes to utilise the exclusion clause (*George Mitchell (Chesterhall) Ltd.* v. *Finney Lock Seeds Ltd*[25]). Section 12 of the 1979 Act cannot be excluded in either category of sale. The 1982 Act adopts similar procedures. Somewhat surprisingly the 1979 Act does not apply to insurance contracts, an area where consumer protection would seem to be a necessary requirement.

The omission from the legislation was allowed in exchange for the insurance industry, through the Association of British Insurers, drawing up their own Statements of Insurance Practice. The purpose of such Statements is to give the private insured greater rights against his insurance company. The Statements do not however carry the force of law and represent only voluntary agreements on the part of those insurers who are members of A.B.I.

4.5 Specialised legislation affecting occupational safety advisers

The responsibility of occupiers of land to those who enter the premises is to be found in the Occupier's Liability Acts of 1957 and 1984. The 1957 Act covers both tortious and contractual liability. The basic obligation is that the occupier owes a common duty of care to see that the premises are reasonably safe for the purpose for which the visitor has been permitted to enter (s. 2). It is possible of course for any contract that may exist between the occupier and the visitor to state a higher duty of care than that of s. 2. But under the Act it was possible for the occupier to exclude this basic duty by a suitably worded exclusion clause in a contract or by a notice (*Ashdown* v. *Samuel Williams & Sons*[26]). But the Unfair Contract Terms Act 1977 now prohibits any attempt to avoid liability for personal injury or death caused by negligence where the premises are used for business purposes and the injured person was a lawful entrant. Even under the 1957 Act where a party entered premises by virtue of a contract to which he was not a party, for example, contractor's workmen, any exclusion of liability in that contract did not affect his rights against the occupier. He was owed the common duty of care unless the contract stated that higher obligations were owed, in which case he could then benefit from the higher duty (s. 3).

The Occupier's Liability Act 1984 makes two important changes to the law. The original Act made no reference to the duty of an occupier to a trespasser. The courts were left to evolve their own rules to cater for this category of person.

Section 1 of the 1984 Act states that an occupier owes a duty to take such care as is reasonable in the circumstances for the safety of a trespasser if he is aware or should have been aware of the existence of the danger and if he knows or should know that a trespasser may come within the vicinity of the danger.

Section 2 of the Act makes a change to the Unfair Contract Terms Act so that where an occupier allows someone to enter his business premises for purposes that are recreational or educational and not connected with the business itself then the occupier may rely on the use of an appropriately worded exclusion clause or notice.

Special reference should be made to two sections of the Health and Safety at Work etc. Act 1974 in a chapter on contract.

Section 4 is concerned with the various duties owed by those who have control of premises to those who are not their employees. Subsection (3) states that where such a person enters non-domestic premises by virtue of a contract or tenancy which creates an obligation for maintenance or repair or responsibility for the safety of or absence from risks to health arising from plant or substances in any such premises, then the person deemed to be in control owes a duty to see that reasonable measures are taken to ensure that such premises, plant or substances are safe and without risks to the health of the person entering. This section therefore would provide that safety standards be extended to someone who enters a cinema (contract) or enters a factory to inspect machinery (licensee) in addition to the other aim of the Act which is concerned with the safety of employees. It should be noted that liability is on the person who has control over premises or who can be described as an occupier and that case law shows that more than one person can be in that position (*Wheat* v. *Lacon & Co. Ltd*[27]).

The Consumer Protection Act 1987 schedule 3 amends and widens the scope of s. 6 of the Health and Safety at Work etc. Act 1974 which is concerned with the general standard of care and safety owed by manufacturers, designers, importers and suppliers of articles for use at work. Such persons must ensure, so far as is reasonably practicable, that the article is so designed and constructed that it will be safe and without risk to health at all times when being set, used, cleaned or maintained by a person at work. To meet these requirements there is a duty to carry out or arrange for such testing or examination as may be necessary in the circumstances. Also adequate information must be given to the person supplied about the use for which the article was designed. Revisions of earlier information must also be given. Similar obligations exist where it is a 'substance' rather than an 'article' that is being supplied. The major alteration here however is that the duties are not restricted to 'for use at work' but cover also when it is being 'used, handled, processed, stored or transported by a person at work'. It must be stressed that liability under the HSW Act 1974 is penal in character and civil remedies are only permissible where the Secretary of State introduces specific regulations. However there may be a contractual relationship between the supplier and the recipient of articles or substances referred to above. A remedy for breach of that contract may therefore be available. Subsection 8 of HSW Act (as amended by the CP Act 1987) allows the originator of a defective article to escape liability if he has obtained a written undertaking from the person supplied that the person will take specified steps sufficient to ensure, so far as is reasonably practicable, that the article is safe and without risk to health when used, set, cleaned or maintained by a person at work. Subsection (9) also has the effect of pinpointing responsibility for contravening the standards imposed by s. 6 on the effective supplier of such goods when another person has in fact been the contracting party with the customer by virtue of a hire-purchase agreement, conditional sale or credit sale agreement. Usually the financing of these arrangements is carried out by means of finance houses. Contractually the goods are sold to the finance house who then in turn enters into his own contract with the customer. This subsection rightly seeks to ensure that the basic obligations of s. 6 remain with the originator of the faulty design, product etc., rather than allowing it to pass to the finance house. Similarly, the Health and Safety (Leasing Arrangements) Regulations 1980 extend to those who, purely as financiers, lease articles for use at work, the immunity from the duties of care imposed by s. 6 and leaves the obligation of this section on the shoulders of the effective supplier rather than the ostensible supplier.

Further reading

Full citation is given below to the cases cited in the text. Extracts of those cases marked with an asterisk can be found in:
H. G. Beale, W. D. Bishop and M. P. Furmston, *Contract Cases and Materials,* Butterworth, London (1985)
For an introduction to the law of contract see:
G. H. Treitel, *An Outline of the Law of Contract,* 4th edn, Butterworth, London (1989).
For more detailed coverage of employment law see:
I. T. Smith and John C. Wood, *Industrial Law,* 4th edn, Butterworth, London (1989).

Reference cases

*1. Scammell *v.* Ouston (1941) 1 All ER 14
*2. Carlill *v.* Carbolic Smoke Ball Co. (1893) 1 QB 256
3. Bigg *v.* Boyd Gibbins Ltd (1971) 2 All ER 183
*4. Dunlop Pneumatic Tyre Co. Ltd *v.* Selfridge & Co. Ltd (1915) AC 847
*5. Balfour *v.* Balfour (1919) 2 KB 571
6. Matthews *v.* Kuwait Bechtal Corporation (1959) 2 QB 57
*7. The Moorcock (1886–1890) All ER Rep 530
8. Fitch *v.* Dewes (1921) 2 AC 158
9. Nordenfelt *v.* Maxim Nordenfelt Guns and Ammunition Co. (1894) AC 535
*10. Parsons *v.* B.N.M. Laboratories Ltd (1963) 2 All ER 658
11. Darbishire *v.* Warren (1963) 3 All ER 310
*12. Hadley *v.* Baxendale (1854) 9 Exch. 341
*13. Victoria Laundry (Windsor) Ltd *v.* Newman Industries Ltd (1949) 2 KB 528
14. Planche *v.* Colburn (1831) 8 Bing 14
15. Stevenson, Jordan and Harrison *v.* Macdonald & Evans (1951) 68 R.P.C. 190
16. The Ready-mixed Concrete (South East) Ltd *v.* The Minister of Pensions and National Insurance (1968) 1 All ER 433
17. Systems Floors (UK) Ltd. *v.* Daniel (1982) ICR 54; (1981) IRLR 475
18. Lister *v.* Romford Ice and Cold Storage Co. Ltd (1957) AC 535
19. Hivac Ltd *v.* Park Royal Scientific Instruments Ltd (1946) 1 All ER 350
20. Rowland *v.* Divall (1923) 2 KB 500
21. Beale *v.* Taylor (1967) 3 All ER 253
22. Wilson *v.* Rickett, Cockerell & Co. Ltd (1954) 1 All ER 868
23. Henry Kendall & Sons *v.* William Lillico & Sons Ltd (1968) 2 All ER 444
24. Ashington Piggeries Ltd *v.* Christopher Hill Ltd (1971) 1 All ER 847
*25. George Mitchell (Chesterhall) Ltd. *v.* Finney Lock Seeds Ltd. (1983) 2 All ER 737
26. Ashdown *v.* Samuel Williams & Sons (1957) 1 All ER 35
27. Wheat *v.* Lacon & Co. Ltd (1966) 1 All ER 582

List of the Acts referred to

Employment of Children Act 1973
The Misrepresentation Act 1967
Contract of Employment Act 1963
Equal Pay Act 1970
Sex Discrimination Act 1975
Race Relations Act 1976
Employment Protection (Consolidation) Act 1978
Sale of Goods Act 1893
Sale of Goods Act 1979
Unfair Contract Terms Act 1977
Occupier's Liability Acts 1957 and 1984
Health and Safety at Work etc. Act 1974
Health and Safety (Leasing Arrangements) Regulations 1980. SI 1980 No. 907

Supply of Goods and Services Act 1982
Consumer Protection Act 1987
Equal Pay (Amendment) Regulations 1983
Employment Acts 1982 and 1989
Social Security Act 1986

Chapter 5

Industrial relations law

R. D. Miskin

5.1 Introduction

Up until the end of the 1960s an employee in the UK had little or no protection at law so far as the continuity of his employment was concerned. The employer was under no obligation to give the employee any specific form of contract or enlighten him as to the terms of his employment. He could thus dismiss him as and when he so decided and was under no obligation to give the employee reasons for such dismissal. The employee was only entitled in the majority of cases to one week's wages and if the employer failed to pay these, his only recourse was to take proceedings at common law, which could often be lengthy and expensive.

It was only in a small minority of cases that more senior employees had some form of agreement which perhaps gave them slightly more protection, but not to any great degree. The scales were very much weighted in favour of the employer, who had no responsibility to look after the welfare of his workers and, in many cases, did not do so. This often resulted in injustice and a person losing his job without cause.

Towards the end of the 1960s the then government decided that an employee deserved proper protection in his employment and passed a series of Acts of Parliament which, although not giving the employee total security of tenure in his job, certainly created the right for him or her not to be unfairly dismissed, which put a much heavier obligation upon the employer to treat his employees fairly and reasonably. An employer could still dismiss where the situation warranted it, but he had to have a valid reason for so doing and had to have acted reasonably overall. The various ways in which the employee was protected are considered later in this chapter.

5.2 Employment law

In general terms, employment law is governed by statute and the rights to which the employee is now entitled are to be found in a considerable number of enactments, the principal one being the Employment Protection (Consolidation) Act 1978. This Act lays down the basic protection afforded to employees and, as its title suggests, consolidates the law up to 1978. It has been varied in a number of ways by subsequent Acts, but should still be the main point of reference when considering the legal aspect of employment. However, heed must also be paid to the provisions of the Disabled Persons (Employment) Act 1944, the Equal Pay Act 1970, the Sex Discrimination Acts of 1975 and 1986, the Employment Protection Act 1975, the Race Relations Act 1976, the Employment Acts of 1980, 1982, 1988, 1989 and 1990 and the Trade Union and Labour Relations (Consolidation) Act 1992. This latter Act consolidates the enactments relating to trade unions, employers' associations, industrial relations and industrial action and came into force in November 1992. It is of considerable importance and will be discussed in greater detail later in this chapter.

5.2.1 The Disabled Persons (Employment) Act 1944

Safety advisers should not forget the existence of this Act which places an obligation upon companies to employ a certain percentage of disabled people. From a safety point of view the conditions of working should be studied carefully to make sure that the disabled employees are not exposed to dangers which, although not a hazard to normal employees, might be to them.

5.2.2 The Equal Pay Act 1970

This Act requires that men and women in the same employment should be treated equally. It should be considered in conjunction with the Sex Discrimination Act 1975 where it is included in full in the First Schedule. It is of particular importance to women, but has caused great difficulty of interpretation of phrases such as 'like work'. The leading case which tries to define a woman's rights in this respect is *Hayward* v. *Cammell Laird Shipbuilders Ltd*[1] which was eventually decided by the House of Lords.

5.2.3 The Employment Protection Act 1975

This Act, among other things, established machinery for promoting the improvement of industrial relations; amended the law relating to workers' rights and otherwise amended the law relating to workers, employers, trade unions and employers' associations; provided for the establishment and operation of a maternity pay fund; provided for the extension of the jurisdiction of Industrial Tribunals; amended the Health and Safety at Work etc. Act 1974 as respects the appointment of safety representatives, health and safety in agriculture, the status of the relevant bodies and the disclosure of information obtained under that Act.

It established the Advisory, Conciliation and Arbitration Service (ACAS) whose aim was to improve industrial relations and who, in the course of this pursuit, published Codes of Practice[2] which established the way in which a good and fair employer should provide for his staff.

5.2.4 The Employment Protection (Consolidation) Act 1978

This Act incorporates those that had preceded it, particularly the Employment Protection Act 1975, and is still the principal Act covering the rights of employees. Section 1 provides that not later than thirteen weeks after beginning an employee's employment the employer shall give him a written statement setting out the terms of his employment. Section 1(2) and s. 2 of the Act stipulate what information should be given to the employee including the scale or rate of pay, when it is paid, any terms and conditions relating to working hours, holidays, sick pay and pensions, any disciplinary rules applying to the employee and identifying the document in which they are contained. If subsequently, during the course of the employment, there are any changes in these terms, the employee must be notified within one month of their taking effect. The employee should also be given a copy of any company rules which affect him in any way, or at least be told where a copy is available.

The Act provides for trade union representatives to have time off to carry out their duties, and similar rights exist for safety representatives appointed by recognised trade unions under the Safety Representatives and Safety Committees Regulations 1977[3].

One of the most important provisions of the 1978 Act was that an employee was granted the right not to be unfairly dismissed, thus giving him much greater protection in his employment than he had enjoyed hitherto. There are various exceptions to this right, one of the most important being that, to achieve the protection, the employee must have been in continuous employment with his employer for a period of 2 years. This also applies with regard to qualification for redundancy payment. Further, the employee, if dismissed, could require his employer to give him written reasons for such dismissal. This is particularly relevant to the employee since there are only certain grounds on which he can be fairly dismissed. These grounds are detailed in section 5.6.1 but it is incumbent upon the employer to show that the dismissal was, firstly, on one of the grounds detailed and, secondly, was reasonable in all the circumstances of the case.

To have acted reasonably the employer must, among other things, have followed his disciplinary procedures. His failure to do so could make what was otherwise a fair dismissal unfair. See *Polkey* v. *Drayton Services*[4].

If an employee feels that he has been unfairly dismissed he is entitled to make a complaint to an Industrial Tribunal. Such application should be made within three months of the date of dismissal unless it was not reasonably practicable to have done so. If the Tribunal finds his dismissal to have been unfair, they can order that he be re-instated or re-engaged or award him compensation for his loss. The object in awarding compensation is to try and restore the employee to the position he was in prior to being unfairly dismissed.

Decisions requiring re-engagement and re-instatement are, to a certain extent, misleading as they do not impose absolute obligation upon the employer to either re-engage or re-instate the employee. Should the employer unreasonably refuse to do so, the employee is entitled to a special award of compensation in addition to whatever he would receive otherwise.

If either party is dissatisfied with a decision of a Tribunal they can appeal on a point of law to, firstly, the Employment Appeal Tribunal (EAT) and thereafter to both the Court of Appeal and the House of Lords if they so desire.

Once a complaint has been made to a Tribunal the Conciliation Officer appointed by ACAS will normally contact the parties to see whether or not a settlement can be reached without the necessity of an actual hearing.

5.2.5 The Employment Act 1980

This Act provided principally as follows:

(1) It gave power to the Secretary of State to provide payments towards expenditure incurred by independent trade unions for secret ballots.
(2) It gave the right to a trade union to ask that ballots should be held on the employer's premises and, provided it was reasonably practical to do so, the employer was bound to comply with the request.
(3) It gave the Secretary of State power to issue various codes of practice containing such practical guidance as he saw fit for the purpose of promoting the improvement of industrial relations.
(4) It gave an employee the right not to have an application for membership of a specified trade union unreasonably refused nor to be unreasonably expelled from a specified trade union.
(5) It gave the employee powers to make a complaint if refused membership of or expelled by the union and to be awarded compensation payable by the union.
(6) It removed from the employer the burden of showing that he had acted reasonably in making a dismissal but required that the Tribunal be satisfied that the dismissal was reasonable and fair in all the circumstances.
(7) It made further provisions regarding the membership of a trade union or an employee's dismissal because of it.
(8) It gave an employee the right to return to work after maternity leave provided she gave notice of her intention to do so. It also gave the employee time off for ante-natal care and dealt with maternity rights generally.
(9) It made provision as to the law relating to picketing and secondary industrial action.

5.2.6 The Employment Act 1982

An Act to provide for compensation out of public funds for certain past cases of dismissal for failure to conform to the Union Membership Agreement; to amend the law relating to workers, employers, trade unions and employers' associations; to make provision with respect to awards by Industrial Tribunals and award by, and the procedure of, the Employment Appeal Tribunal; and for connected purposes.

5.2.7 The Trade Union Act 1984

Part I of this Act makes provision for secret ballots for election to certain trade union offices and sets out requirements to be satisfied in relation to such elections. It also imposes upon a trade union the duty to compile a register of the names and proper addresses of its members and to keep the entries up to date as far as possible. It also lists those trade unions that are exempt from this requirement.

Part II provides for secret ballots among the members before any form of industrial action is taken and makes a union automatically liable for any losses if such a ballot is not held. Part III deals with the political objects and funds of the union.

5.2.8 The Employment Act 1988

This Act continued the trend set by the Trade Union Act 1984 of giving the union back to its members by imposing stricter requirements in respect of balloting and by giving members certain powers to question the way the union was run. It removed all legal protection for the union as an institution and removed its powers to establish and maintain closed shops and to bring pressure on employees to become members. This Act also established a Commission for the Rights of Trade Union Members to assist them in enforcing their rights against the union.

Since a new government came into power in the 1980s the union powers have become much more limited than they were in the early days of the 1970s.

5.2.9 The Employment Act 1989

An Act to amend the Sex Discrimination Act 1975 in pursuance of a Directive of the EC dated 9th February 1976 on the implementation of the principle of equal treatment for men and women as regards access to employment, vocational training and promotion, and working conditions; to repeal or amend prohibitions or requirements relating to the employment of young persons and other categories of employees; to make other amendments to the law relating to employment and training; to repeal s. 1(1)(a) of the Celluloid and Cinematograph Film Act 1922; to dissolve the Training Commission; to make further provisions with respect to industrial training boards; to make provision with respect to the transfer of staff employment in the skills training agencies; and for connected purposes. While a number of its provisions are aimed at preventing discrimination at work, it allows a certain discrimination if, and only if, it concerns the protection of women at work. It also exempts Sikhs from the requirement to wear safety helmets on construction sites.

It further introduced for the first time pre-hearing reviews of proceedings before an Industrial Tribunal.

5.2.10 The Employment Act 1990

The purpose of this Act was to:

- make it unlawful to refuse employment to any employee, or any service of an employment agency, on grounds relating to trade union membership;
- amend the law on industrial action and ballots;
- make further provision regarding the Commission for the Rights of Trade Union Members; and
- confer a power to revise or revoke codes of practice.

This Act tends to take rights from the officials of trade unions and hand them back to members.

5.2.11 The Trade Union and Labour Relations (Consolidation) Act 1992

This extensive Act consolidates the law on trade unions, employers' associations, industrial relations and industrial action. There are over 300 sections and three schedules. It is divided into seven parts. In Part I, s. 1 defines the title 'trade union', s. 2 deals with the status and property of trade unions, s. 3 covers trade union administration, s. 4 outlines the duty of trade unions to hold elections for certain positions, while s. 5 sets out the rights of trade union members. Section 6 imposes restrictions on the use of funds for certain political objects and sets out the duties of an employer to deduct union contributions, s. 7 explains amalgamation procedure, and s. 8 details where financial assistance may be provided.

Part II relates to employers' associations, their status and administration. Part III considers collective bargaining and the codes of practice issued by ACAS[2] within which s. 4 deals with the functions of ACAS and its officers. Part V describes the protection available for acts made in contemplation of furtherance of a trade dispute, the requirement to ballot members before a trade union takes action and the loss of unfair dismissal protection during industrial action.

The remaining Parts cover a number of miscellaneous and general provisions including where the employee or employer dies, allowing Tribunal proceedings to be instituted or continued by a personal representative of the deceased employee or defended by a personal representative of the deceased employer. If there is no personal representative the proceeding may be pursued or defended by such person as the Tribunal may appoint.

5.2.12 The Trade Union Reform and Employment Rights Act 1993 (TURER)

This Act received the Royal Assent on 1 July 1993 and is the most comprehensive piece of employment legislation since the Employment Protection (Consolidation) Act 1978. The main purpose of the Act increases the regulation and supervision of internal Trade Union affairs, imposes further restrictions on industrial action and improves rights for employees in particular circumstances, namely in cases of maternity and in the area of health and safety. This is a very important Act that requires detailed study but can only be dealt with in a summary fashion in this chapter.

It is divided into four parts and has ten schedules. Part I deals with union elections and ballots, financial affairs of unions, rights in relation to union membership and industrial action. This whole part needs to be studied and in particular sections 17 to 22 which deal with industrial action and the new requirements relating to it.

Part II deals with employment rights, principally maternity, the provision of employment particulars and employment protection in health and safety cases. It also deals with dismissal on the ground of assertion of statutory rights, compensation when re-instatement or re-engagement orders have been made, sex discrimination and the transfer of undertaking regulations with particular reference to redundancy consultation procedures.

Regarding maternity, a number of important changes have been made which do not come into effect until October 1994 at the earliest. The most important of these are: firstly, the right to 14 weeks maternity leave for all pregnant employees, irrespective of the length of service, with the right to return to work following the 14 week period, subject to notification requirements set out in the Act; secondly, during maternity leave all employment rights other than remuneration must be maintained, including the provision of a company car if appropriate and the approval of holiday entitlements; and, thirdly, protection from dismissal on the grounds of pregnancy irrespective of the length of service.

Sections 26 and 27 widen the scope of the particulars of employment to which any employee is entitled. Section 28 and schedule 5 provide for protection against dismissal and action falling short of dismissal for health and safety representatives, whether union appointed, employer appointed or volunteer, and other employees in circumstances where health and safety is an issue. An employee will now be able to claim compensation if he suffers as a result of his activities relating to health and safety and, if dismissed, will be able to claim unfair dismissal even if otherwise ineligible, and to receive a special award. Schedule 5 amends the Employment Protection (Consolidation) Act 1978 in its new sections 22(a), (b) and (c) to give effect to these provisions.

Section 75 sets out the terms of the special award which amounts to a week's pay multiplied by 104 or £13 400, whichever is the greater, not to exceed £26 800. This award can be increased where an employer fails to comply with an order for re-engagement or re-instatement. Provision is also made regarding dismissal on the ground of assertion of statutory rights and the right to declaration of invalidity of discriminatory terms and rules. Further it deals with procedures relating to consultation on redundancy and places a far heavier burden on employers to consult than previous Acts. This part of the Act imposes a number of obligations upon employers and gives employees additional rights.

Part III deals with other employment matters including the repeal of Part II of the Wages Act 1986 concerning minimum wages, the constitution and jurisdiction of Tribunals and the general function of ACAS.

Part IV deals with supplementary matters which should be noted.

There are ten schedules which set out in greater detail the new provisions outlined above.

5.2.13 Other relevant Acts

The Social Security and Housing Benefits Act 1982 imposes a duty upon employers to provide for the payment of statutory sick pay among other provisions, while the Wages Act 1986 provides protection for workers in connection with the payment of wages to allow payments by other than 'coin of the realm', i.e. by cheque or bank transfer. It also makes provisions with regard to Wages Councils.

5.3 Discrimination

5.3.1 Sex

A facet in employment law which caused much distress until legislation was made to eliminate it was discrimination against employees on the ground of sex. The lot of women was improved by both the maternity provisions of the Employment Act 1980 and the Equal Pay Act 1970 but the sex discrimination Acts provided them with much greater protection. Ostensibly these apply to men as well as women but from a practical point of view the provisions are designed to protect a woman from being discriminated against either in selection for a job or during the course of it. The relevant statutes are the Sex Discrimination Act 1975 and the Sex Discrimination Act 1986 which also created the Equal Opportunities Commission.

5.3.2 Race

A very similar position existed regarding discrimination on the grounds of race and very similar provisions were introduced to eliminate it. The relevant statute is the Race Relations Act 1976 which created, *inter alia*, the Commission for Racial Equality and gave it powers to conduct formal investigations into complaints, to issue and enforce non-discrimination notices, to take action against discriminatory advertisements and persistent discrimination and to help individuals to bring discrimination claims against their employers.

Both Commissions have produced codes of practice[5,6] which provide guide lines which an employer should follow to comply with the statutory provisions. Although the codes of practice are not fully binding at law, it is at his peril that an employer does not follow them. An Industrial Tribunal hearing an application would weigh the fact that an employer had not done so against him.

5.3.3 The Sex Discrimination Act 1975

Section 1 states that a person discriminates against a woman if in any circumstances relevant to the purposes of any provision of the Act he:

(a) on the ground of her sex, treats her less favourably that he treats or would treat a man, or

(b) applies to her a requirement or condition which applies or would apply equally to a man but:
  (i) which is such that the proportion of women who can comply with it is considerably less than the proportion of men who can comply with it, and
  (ii) which he cannot show to be justifiable, irrespective of the sex of the person to whom it applied, and
  (iii) which is to her detriment because she cannot comply with it.

Section 1(2) provides that if a person treats or would treat a man differently according to a man's marital status, his treatment of a woman is for the purposes of s. 1(1)a to be compared with his treatment of a man having like marital status.

S. 1(1)a constitutes direct discrimination and is easily identifiable. In *James* v. *Eastleigh Borough Council*[7] the House of Lords confirmed that a person was directly discriminated against when 'but for' her sex a woman would not have been treated less favourably than a man in similar circumstances. More difficulties arise under s1(1)b where the discrimination is indirect and less easily discernible.

Activities unlawful under the 1975 Act extend to recruitment, promotion and dismissal, and under the Sex Discrimination Act 1986 to retirement, promotion, transfer, training, demotion and dismissal. The 1986 Act repealed restrictive conditions in extant legislation as it applied to terms and conditions of employment of women relating to night work and overtime but retained restrictions for safety (not allowed to clean machinery) and health (contact with certain chemicals).

However, discrimination is allowed on health grounds concerning pregnancy and maternity. A good example is an EAT decision in *Page* v. *Freighthire Tank Haulage Ltd*[8] where a woman lorry driver who was of child bearing age was prevented from driving a tanker lorry containing chemicals that could be harmful to a woman's ability to bear children.

5.3.4 The Race Relations Act 1976

This Act is couched in almost exactly the same terms as the Sex Discrimination Act 1975 in that it provides in s. 1(1) that a person discriminates against another if in any circumstances relevant to the purposes of any provision he:

(a) on racial grounds treats that person less favourably than he treats or would treat other persons, or
(b) he applies to that person a requirement or condition which he applies or would apply equally to persons not of the same racial groups as that person but:
  (i) which is such that the proportion of persons of the same racial group as that person who can comply with it is considerably smaller than the proportion of persons not of that racial group who can comply with it, and

(ii) which he cannot show to be justifiable irrespective of the colour, race, nationality or ethnic or national origins of the person to whom it is applied, and
(iii) which is to the detriment of that other person because he cannot comply with it.

Although identifying cases of direct discrimination is reasonably straightforward, it is much more difficult when it comes to indirect discrimination.

5.4 Disciplinary procedures

A Tribunal, to find a dismissal fair, must be satisfied, among other things, that the dismissal was reasonable in all the circumstances. In the vast majority of cases this will entail the employer having followed his own disciplinary and grievance procedures to the letter, particularly since the House of Lords decision in *Polkey and Drayton Services*[4]. It is important, indeed prescribed in the ACAS Code of Practice[2], that a company should have formal procedures whose contents should be communicated to each and every employee.

The principal underlying reasons for having disciplinary rules and procedures are to promote fairness and order in the treatment of individuals and in the conduct of industrial relations. They also assist an organisation to operate effectively. Further the procedures ensure that standards are adhered to and provide a point of reference when dealing with alleged failures to observe them. Such procedures should:

(a) be in writing,
(b) specify to whom they apply,
(c) provide for matters to be dealt with quickly,
(d) indicate the disciplinary actions that can be taken,
(e) specify the various levels in the organisation that have the authority to take various forms of disciplinary action, ensuring that the immediate supervisor's powers to dismiss are restricted and require reference to a senior manager,
(f) provide for individuals to be informed of the complaints against them and be given an opportunity to state their case before decisions are reached,
(g) give individuals the right to be accompanied by a trade union representative or by a fellow employee of their choice,
(h) ensure that, except for gross misconduct, no employee is dismissed for a first breach of discipline,
(i) ensure that disciplinary action is not taken until the case has been carefully investigated,
(j) ensure that the individual is given an explanation for any penalty imposed, and
(k) provide a right of appeal and specify the procedure to be followed.

A record should be kept of any disciplinary actions taken against an employee for breach of the rules including lack of capability, conduct, etc., and what disciplinary action was taken and the reasons supporting such action. The disciplinary procedures should be reviewed from time to time to ensure that they comply with the then practices of the employer.

Many of these rules and procedures will incorporate items relevant to safety, health and welfare of the employees in that particular employment. The emphasis placed on particular aspects of safety and health will reflect the degree of risk or hazard faced by the employee in his daily work and what effect failure to follow these rules might have on the employees themselves, the environment or the continuing operation of the business. The onus is on the employer to draw up these rules and he may do this unilaterally but it is more prudent of him to consult the employees or their representative to obtain agreement to and acceptance of the various procedures before they are implemented.

The employer should ensure that, except for gross misconduct, no employee is dismissed for a first breach of discipline. Instead the employer should operate a system of warnings consisting of an oral warning, a first written warning and then a final written warning before dismissal is considered.

5.5 Dismissal

S. 55(2) of the Employment Protection (Consolidation) Act 1978 (EP(C)A) provides as follows: 'Subject to sub-section (3) an employee shall be treated as dismissed by the employer if, but only if:

(a) the contract under which he is employed by the employer is terminated by the employer, whether it is so terminated by notice or without notice, or
(b) where under that contract he is employed for a fixed term and that term expires without being extended under the same contract, or
(c) the employee terminates that contract with or without notice in circumstances such that he is entitled to terminate it without notice by reason of the employer's conduct.'

Finally, s. 55(4) construes 'the effective date of termination' as:

(i) in relation to an employee whose contract of employment is terminated by notice, whether given by the employer or the employee, the date on which that notice expires,
(ii) where the termination is without notice, the date on which the termination takes effect, and
(iii) for an employee who is employed under a contract for a fixed term, where the term expires without being renewed under the same contract, the date on which the terms expires.

The language used in this part of the EP(C)A is reasonably straightforward and explicit except for s. 55(2)c which is quoted above. In those circumstances the employee is entitled to make an application to an Industrial Tribunal complaining that he was 'constructively dismissed'. There was initially considerable doubt in the minds of the Tribunals and the Courts as to what circumstances entitled an employee to terminate his own employment in this way and then complain about it to a Tribunal. The doubt was resolved by the Court of Appeal decision in *Western Excavating (ECC) Ltd* v. *Sharp*[9] where the then Master of the Rolls, Lord Denning, pronounced that to show he had been constructively dismissed an employee must satisfy the Tribunal that his employers were guilty of conduct which was a significant breach going to the root of the contract of employment or which showed that they no longer intended to be bound by one or more of its essential terms.

Examples of such behaviour on the part of the employer include reducing the status of the employee, exhibiting a lack of trust in him, reducing his salary, among others.

5.6 Reasons for dismissal

S. 57(2) of EP(C)A sets out the grounds upon which an employee can be fairly dismissed. To these must be added that of 'any other substantial reason of a kind such as to justify the dismissal of an employee holding the position which that employee held' which is contained in s. 57(1)b.

5.6.1 Fair reasons

Reasons which are prima facie 'fair' are set out in s. 57 of EP(C)A and are:

(1) A reason related to the capability or qualifications of the employee to do the particular job which he or she is employed to do.
(2) A reason regarding the employee's conduct.
(3) The employee's redundancy.
(4) That the employee could not continue in his or her particular role without infringing some particular statute.
(5) Some other substantial reason.

5.6.1.1 Capability

The two main classes of capability, or lack of it, are ill-health and the inability of the employee to carry out his duties in a reasonable and acceptable manner.

5.6.1.1.1 Ill-health

The two main sets of circumstances which justify termination by the employer are, firstly, a single long period off work due to sickness and, secondly, regular short spells of illness which, added together, represent a lengthy period of absence.

5.6.1.1.2 Long-term illness

The leading case which sets out the main principles to support a fair dismissal for long-term illness is *Spencer* v. *Paragon Wallpapers Ltd*[10] in which the employee had been absent sick for approximately two months and the medical opinion was that he would return within another four to six weeks. The EAT held that in such cases the employer must take into account:

(a) the nature of the illness,
(b) the likely length of the continuing absence,
(c) the employer's need for the work to be done, and
(d) the availability of alternative employment for the employee.

Since all four criteria had been met, the dismissal was fair.

Consultation with the employee and investigation of the medical position by the employer are important factors although each case has to be decided on its own facts. In *East Lindsay District Council* v. *Daubny*[11] the EAT stated that unless there were wholly exceptional circumstances the employee should be consulted and the matter discussed with him before his employment was terminated on the ground of ill-health.

5.6.1.1.3 Continuing periodic absences

In stark contrast to the above there have been several EAT cases where the employee has been dismissed for persistent absenteeism due to a succession of short illnesses. In *International Sports Company Ltd* v. *Thomson*[12] an employee was persistently absent for minor ailments that could not later be confirmed by medical examination. After review by the employer of her absence record and being given reasonable warnings she was dismissed. There had been no improvement in her attendance and the dismissal was held to be fair.

Note, where instances of ill-health have been caused by the conditions in which the employees are required to work, as in *Glitz* v. *Watford Electric Co. Ltd*[13], the decision on whether or not termination is fair depends upon whether or not the employer has made reasonable investigations into the cause.

5.6.1.1.4 Lack of skill on the part of the employee

When considering dismissals for what can be generally described as incompetence, it is essential for the employer to use his warning procedures to try and improve the employee's performance. If he dismisses without so doing then, except in very rare cases, the Tribunal will hold the dismissal to be unfair. Such an exception occurred in *Taylor* v. *Alidair Ltd*[14] where the employee, an airline pilot, was dismissed because of the disastrous consequences which could have ensued as a result of his lack of skill on one particular occasion when landing an aircraft. In cases of this type it is extremely important that the employer should investigate fully the facts of the case and consider them prior to making the decision to dismiss – see *Cook* v. *Thomas Linnell & Sons Ltd*[15].

Any warning given to an employee should be constructive and must offer all reasonable assistance to help him in improving his performance and he should be allowed sufficient time to do so.

In cases such as *James* v. *Waltham Holy Cross UDC*[16], the employer is entitled to dismiss an employee without warning where there is little likelihood of the employee improving his performance and his continuing presence is prejudicial to the company's best interests. Such cases are rare.

5.6.1.2 Conduct

Conduct and capability tend to overlap. Misconduct falls into two general categories: firstly, for lesser misconduct where disciplinary procedures should be used to try to prevent a repetition of the conduct complained of and, secondly, where the offence amounts to gross misconduct for which an employee can be summarily dismissed.

Examples of lesser misconduct include an employee who refuses to obey an employer's instructions, is regularly absent, fails to clock on and clock off in the prescribed manner and who commits comparatively minor breaches of the rules and procedures. However, in *Austin* v. *British Aircraft Corporation Ltd*[17] the employer's attitude was considered unreasonable. Mrs Austin and her fellow employees were required to wear eye protection. Mrs Austin already wore glasses and the goggles provided were uncomfortable. However, she persevered for three months but eventually stopped wearing them. She raised the problem with her employers and the matter was put in the hands of the safety officer. Six months later nothing had been done so Mrs Austin resigned. The Tribunal hearing her case concluded that Mrs Austin had been constructively dismissed and was entitled to resign by reason of her employer's conduct.

The same principle applied in *Keys* v. *Shoefayre Ltd*[18] where the owner of a retail shop failed to take proper security precautions to protect his employees who worked in a shop in an area with a high crime rate that had suffered two armed robberies. Here it was held that the employer was in breach of a fundamental term of the contract of employment to take reasonable care and provide a safe system of work and to have reasonably safe premises and that Mrs Key's resignation amounted to unfair constructive dismissal.

A good example of competence and conduct overlapping is the case of *Taylor* v. *Alidair Ltd*[14] mentioned in paragraph 5.6.1.1.4.

Conversely, where the employer is doing as much as can be reasonably expected in the circumstances, it is not considered reasonable for the employee to expect him to do more. In the manufacture of tyres, part of the process emits dust and fumes that reports from America indicated might be carcinogenic. Negotiations resulted in face masks being provided as an interim measure until expensive capital equipment could be obtained which would improve matters, a step that was supported by the HSE. However, an employee in *Lindsay* v. *Dunlop Ltd*[19] was not satisfied with these precautions and delayed removing the tyres from the press until the fumes had dispersed. This seriously affected production and, following discussion with his union, the employer dismissed the employee. The Tribunal held that the dismissal was fair, a decision upheld by the Court of Appeal on the grounds that the employer had taken all reasonable steps in the circumstances.

The second category, gross misconduct, is where the offence is so serious that it merits summary dismissal, the most common of which are: theft, fraud, deliberate falsification of records, fighting, assault on another person, deliberate (wilful) damage to company property, serious incapability through alcohol or being under the influence of illegal drugs, serious negligence causing unacceptable loss, damage or injury and serious act of insubordination.

The case of *British Home Stores* v. *Burchell*[20] sets out clearly the steps an employer must take before dismissing an employee on the ground of gross misconduct. The employer must, firstly, believe in the employee's guilt, secondly have reasonable ground on which to so believe and, thirdly, have carried out reasonable investigation to verify the grounds for sustaining such belief. If he has done so, then the Tribunal must uphold his decision although they would not necessarily have taken the same view themselves.

With regard to criminal offences committed outside employment, the ACAS code holds that these should not be treated as automatic reasons for dismissal and that the main consideration should be whether the offence renders the employee unsuitable for his or her job or unacceptable to other employees. Employees should not be dismissed solely because a charge against them is pending or because they are absent through having been remanded in custody.

5.6.1.3 Redundancy

S. 81(2) of EP(C)A provides: 'For the purpose of this Act an employee who is dismissed shall be taken to be dismissed by reason of redundancy if the dismissal is attributable wholly or mainly to:

(a) the fact that his employer has ceased, or intends to cease, to carry on the business for the purpose of which the employee was employed by him, or has ceased, or intends to cease, to carry on that business in the place where the employee was so employed, or
(b) the fact that the requirements of that business for employees to carry out work of a particular kind, or for employees to carry out work of a particular kind in the place where he was so employed, have ceased or diminished or are expected to cease or diminish'.

S. 59 of EP(C)A states that: 'Where the reason, or principal reason, for the dismissal of an employee was that he was redundant, but it is shown that the circumstance constituting the redundancy applied equally to one or more other employees in the same undertaking who held positions similar to that held by him and who have not been dismissed by the employer, and either:

(a) that the reason, or if more than one the principal reason, for which he was selected for dismissal was an inadmissible reason, or
(b) that he was selected for dismissal in contravention of a customary arrangement or an agreed procedure relating to redundancy and there were no special reasons justifying a departure from that arrangement or procedure in his case,

then for the purposes of this part, the dismissal shall be regarded as being unfair'. An employee is not entitled to challenge the employer's decision to make workers redundant where a redundancy structure exists. He is only able to challenge his own dismissal on the ground of redundancy by alleging that the way in which the redundancies were implemented was not reasonable in all the circumstances – see *Moon* v. *Homeworthy Furniture (Northern) Ltd*[21]. In challenging the fairness of redundancies an employee can either allege that the method of selection was contrary to s. 59 of EP(C)A and thus his dismissal is automatically unfair, or that it was unreasonable.

The case of *Williams* v. *Compair Maxam Ltd*[22] sets out the considerations that a Tribunal should take into account where the employees are represented by an independent union. These are that the employer should:

(1) seek to give as much notice as possible;
(2) consult the union as to the criteria to be applied in selecting the employees to be made redundant;
(3) ensure that such criteria do not depend solely upon the opinion of the person making the selection but can be objectively checked against such things as attendance record, efficiency at the job, experience or length of service;
(4) seek to ensure that the selection is made fairly in accordance with these criteria and consider any representation the union may make; and
(5) see whether, instead of dismissing an employee, he could offer him alternative employment.

Whether or not a union is involved, a sensible employer will follow the above rules.

It has been stressed by the Tribunals that consultation between employer and employee is essential. In the case of *Dyke* v. *Hereford and Worcester County Council*[23] the EAT stated that the importance of consultation could not be over-emphasised. Such consultation should involve both unions and individuals where relevant. Instances will be rare, and must be for very special reasons, for a Tribunal to find a dismissal on the ground of redundancy fair where no consultation has taken place.

There is an onus on the employer to find out whether there is any other available work which he can offer to the employee. In *Vokes Ltd* v. *Bear*[24] the Tribunal made it clear that if the employer was part of a group of many companies then enquiries should be made within the associated companies to find out whether any employment could be offered to the redundant employees. It is sensible on the part of the employer to make the offer of an alternative job, even if it represents a demotion, since it is then up to the employee to decide whether or not to accept it.

5.6.1.4 Statutory disability

S. 57(2)(d) of the 1978 Act states that the employer can fairly dismiss an employee when he can show 'that the employee could not continue to work in the position which he held without contravention (either on his part or that of the employer) of a duty or restriction imposed by or under any enactment'.

One of the most common instances of this is where it is part of an employee's job to drive a motor vehicle and he has been disqualified by the courts from holding a valid licence. The reasonableness test applies here also. Health regulations, particularly in the food industry, are a further example where statutory disability can justify dismissal.

5.6.1.5 Any other substantial reason

Where none of the above reasons for fair dismissal apply an employer can still claim that he dismissed the employee for some other substantial reason. This occurs in several ways, one of the most common of which is the changing needs or reorganisation of a business. The law has always endeavoured to permit an employer to run his business properly recognising that on occasions it may necessitate a reduction in the number employed. The main factors to consider here are:

(i) Was the re-organisation necessary?
(ii) Given that the re-organisation was necessary, was it necessary to change the employee's job or dispense with it altogether to implement the re-organisation?
(iii) Was there sufficient consultation?

In *RS Components Ltd* v. *Irwin*[25] the employers were held to have fairly dismissed four employees who were not prepared to accept a restrictive covenant being added to their terms of employment. In the circumstances of the case this was held to be 'for any other substantial reason'. Finally both imprisonment of the employee and breakdown in the trust and confidence between employer and employee can also constitute grounds for dismissal under the heading of some other substantial reason.

5.7 Summary

The main purpose of industrial relations legislation has been to regulate the relationship between employer and employee and to determine the role and powers of trade union representatives in deciding the terms and conditions under which an employee has to work. It has become practice to include under the wing of 'industrial relations' anything that can affect the way in which an employee has to work, and in this respect safety has an important role to play.

This chapter has shown some of the ways in which decisions and actions taken for safety reasons can materially affect the employee's working conditions and, conversely, the ways in which employment legislation can affect safety issues. For the safety adviser to be able to maximise the benefit of his efforts he must be aware of the wider implications of the recommendations he makes, particularly in the field of working conditions.

The law governing industrial relations is extremely complex and covers much more ground than it has been possible to cover in this chapter, but the relevant Acts have been covered in brief and some of the ways in which their application can affect the employer–employee relationship have been shown.

References

1. Hayward *v.* Cammell Laird Shipbuilders Ltd (1985) IRLR 463
2. Advisory, Conciliation and Arbitration Service, *The ACAS advisory handbook: Discipline at Work*, HMSO, London (1989)
3. Health and Safety Commission, *Safety Representatives and Safety Committee Regulations 1977*, HMSO, London (1977)
4. Polkey *v.* A.E. Drayton Services Ltd (1988) IRLR 503
5. Equal Opportunities Commission, *Code of Practice: Sex Discrimination*, SI 1985 No. 387, HMSO, London (1985)
6. Commission for Racial Equality, *Code of Practice: Race Relations*, SI 1983 No. 1081, HMSO, London (1983)
7. James *v.* Eastleigh Borough Council (1990) IRLR 288
8. Page *v.* Freighthire Tank Haulage Ltd (1980) ICR 299; (1981) IRLR 13
9. Western Excavating (ECC) Ltd *v.* Sharp (1978) IRLR 27
10. Spencer *v.* Paragon Wallpapers Ltd (1976) IRLR 373
11. East Lindsay District Council *v.* Daubny (1977) IRLR 181
12. International Sports Company Ltd *v.* Thomson (1980) IRLR 340
13. Glitz *v.* Watford Electric Co. Ltd (1979) IRLR 89
14. Taylor *v.* Alidair Ltd (1978) IRLR 82
15. Cook *v.* Thomas Linnell & Sons Ltd (1977) IRLR 132
16. James *v.* Waltham Holy Cross UDC (1973) IRLR 202
17. Austin *v.* British Aircraft Corporation Ltd (1978) IRLR 332
18. Keys *v.* Shoefayre Ltd (1978) IRLR 476
19. Lindsay *v.* Dunlop Ltd (1979) IRLR 93
20. British Home Stores *v.* Burchell (1978) IRLR 379
21. Moon *v.* Homeworthy Furniture (Northern) Ltd (1976) IRLR 298
22. Williams *v.* Compair Maxam Ltd (1982) ICR 800
23. Dyke *v.* Hereford and Worcester County Council (1989) ICR 800
24. Vokes Ltd *v.* Bear (1973) IRLR 363
25. RS Components Ltd *v.* Irwin (1973) IRLR 239

Chapter 6

Consumer protection

R. G. Lawson

The expression 'consumer protection', and with it the notion of 'consumer law', is of relatively recent vintage. It first found expression in the Final Report of the Committee on Consumer Protection (Cmnd 1781) 1962[1], which led to the enactment of the Trade Descriptions Act in 1968[2]. This latter can fairly be regarded as the starting point of the modern law of consumer protection.

In more recent years, as the following pages will show, much of the impetus for new consumer legislation has come from Brussels. The enactment in the United Kingdom of the Consumer Protection Act 1987[3] and the Control of Misleading Advertisements Regulations 1988[4] in each case derived from a Council Directive. This is a trend which looks set to continue.

6.1 Fair conditions of contract

It is central to any system of consumer protection that a potential customer is given only truthful and accurate information about the goods and services that he is wanting to buy. Even before Parliament had decided to intervene, the courts had already decided to allow a remedy where a contract had been induced by fraud or misrepresentation. Where a consumer has been duped into entering a contract through deception of the kind practised by some salesmen, he would be given the right to put an end to the contract and claim compensation for any loss which he may have suffered. This development in the courts was eventually confirmed by the Misrepresentation Act 1967[5].

Valuable though these controls were, they applied only to what is called the civil law, i.e. the law which regulates the relations between citizens. Where a consumer had been the victim of fraud or misrepresentation, the initiative lay entirely upon him to take remedial action. It was only with the advent of the Trade Descriptions Act that the criminal law came to the aid of the consumer in such cases.

6.1.1 False trade descriptions

The main feature of the Trade Descriptions Act 1968[2], as its name implies, is to outlaw the use of false trade descriptions. Section 2 of the Act contains a comprehensive and exhaustive list of what constitutes a false trade description for the purposes of the Act. This list can be summarised as saying that any statement about goods which when made can be either true or false is a trade description. This has meant that the statement (made in relation to a bar of chocolate) that it was of 'extra value' was not a trade description: it was such a vague kind of claim that no one could say of it that it was true or false.

While it is true to say that this part of the Act seems to have been used almost exclusively to control some of the more dubious antics of the second-hand car trade (convictions for turning back a mileometer have been particularly common) this is far from being entirely the case. One good example arose in the case of *British Gas Board* v. *Lubbock* (1974)[6]. A gas cooker was advertised as being ignited by a hand-held ignition pack. At the time the advertisement was shown, this was no longer true. The Board was prosecuted and convicted for making a false statement about the composition and the physical characteristics of goods.

Another example is the decision in *Queensway Discount Warehouses* v. *Burke* (1985)[7]. A wall unit was advertised in the national press. It was shown ready assembled. The advertisement was seen by a customer who later went to see the unit in store, where it was on display also ready-assembled. The customer agreed to purchase the unit, but when it was delivered he found that it was in sections and that he had to assemble it himself. The advertisement was held to be a false trade description in that it gave a false description of the composition of the goods. It is also possible under the Act for the description of goods to be accurate but still to give rise to an offence if that description is misleading. In *Dixons Ltd* v. *Barnett* (1988)[8] a telescope bore the clear statement that it magnified up to 455 times. This was true, but in fact the telescope had a maximum useful magnification of 120 times: beyond that, the image became less clear and became no clearer with higher magnification. The shop was convicted because, although the statement as to magnification was true, it was misleading.

An offence is committed under this part of the Act regardless of the absence of any blame on the part of the person making the false trade description. Its falsity is enough to secure the commission of an offence. This makes the offence one known as a 'strict liability' offence.

However, the Act does provide for what is called the 'due diligence' defence. This allows the defendant to escape conviction if he can show that he took all reasonable precautions and exercised all due diligence to avoid commission of the offence. The cases show that this is a very difficult defence to satisfy. In the case of *Hicks* v. *Sullam*[9] bulbs were falsely described as 'safe'. The bulbs had been imported from Taiwan. There were 110 000 in all. None had been sampled to test for their safety and no independent test reports were obtained. The defendant's agent in the Far East had checked the bulbs and had reported no defects. The court ruled that the defence had not been made out[10].

The courts have also been prepared to allow the use of disclaimers to avoid conviction, but have insisted that the disclaimer will be effective only if it is as 'bold, precise and compelling' as the false description it is attempting to disclaim. This is laid down in *Norman* v. *Bennett*[11] where a car dealer sought to disclaim a false mileometer reading with the statement 'speedometer reading not guaranteed'. This was contained in the small print of the contract and was held to be ineffective. In contrast, it was held in *Newham London Borough* v. *Singh*[12] that a disclaimer was effective when it was placed over the mileometer and read 'Trade Descriptions Act 1968. Dealers are often unable to guarantee the mileage of a used car on sale. Please disregard the recorded mileage on this vehicle and accept this as an incorrect reading'.

The penalties for breach of the Act depend on the court the case is brought in. Most cases are brought in magistrates' courts where the penalty is a maximum fine of £5000. More serious cases are brought in the Crown Court where the penalty is a fine of any amount, a maximum sentence of two years, or a combination of both. In addition, the Powers of Criminal Courts Act 1973 empowers the court to award compensation to consumers affected by the breach of the Trade Descriptions Act.

6.1.2 Pricing offences

The regulation of misleading pricing used to be the province of s. 11 of the Trade Descriptions Act and the Price Marking (Bargain Offers) Order 1979[13]. Over the years, however, it was found that the former simply gave an inadequate level of protection, while the latter was so hideously complicated that to many it was incomprehensible. After several attempts at reform, success was finally achieved, in a novel form, with the enactment of the Consumer Protection Act 1987[3]. The Act (whose relevant provisions came into force on 1 March 1989) makes it an offence for a price to be indicated which is misleading as to the price at which any goods, services, accommodation or facilities are available. The novel feature of the Act is that it provided for the adoption by the Secretary of State for Trade and Industry of a code of practice which gave practical guidance to traders on how price indications should be given to avoid the commission of an offence. The code has first to be approved by Parliament. In November 1988, the Secretary of State published the Code of Practice for Traders on Price Indications[14].

It is important to understand that the Code is not binding on traders. A contravention of the Code is expressly said by the Act not of itself to give rise to an offence, but can be used as evidence that an offence in fact has been committed. Similarly, a trader who applies the Code cannot be entirely certain in law that his price indication is not misleading, although such compliance will again be evidence that his pricing is indeed not misleading. In practice, however, it will be very unusual for the presumptions raised by compliance with, or breach of, the Code to be displaced.

The Code itself is in four parts, dealing respectively with the use of price comparisons, with the display of the actual price to the consumer and with price indications which become misleading after they have been given. In Part 1, for instance, the Code states that price comparisons should always state the higher price, as well as the price which is to be charged. On the difficult question of comparisons with recommended prices, the Code advises against the use of initials, unless the initials are 'RRP' used to describe a recommended retail price, or 'man rec price' to describe the manufacturer's recommended price. All other descriptions must be written out clearly and in full. The Code also advises against the use of a recommended price comparison unless the recommended price is not 'significantly higher' than the price at which the product is generally sold. Part 2 deals with the actual display of the price to the consumer. Amongst other things, it recommends that the indication should give the full price the consumer will have to pay. All price indications should include VAT. Businesses, such as hotels and restaurants, which impose a non-optional extra such as a service charge, are recommended by the Code to incorporate the charge within fully inclusive prices wherever practicable; and to display the fact clearly on any price list or priced menu, whether displayed inside or outside (e.g. by using statements like 'all prices include service'). Part 3, which deals with indications which become misleading after they have been given, deals amongst other things with changes in the rate of VAT. Where this happens, the Code recommends that the correct price indication should be made clear to any consumer who orders the product. This must be done before the consumer is committed to buying the product and, if practicable, before the goods are sent to the consumer. Finally, Part 4, which takes account of the sale of new homes, broadly applies the relevant provisions of Parts 1 and 2. The penalty for misleading pricing is a maximum fine not exceeding level 5 in the standard scale where the case is prosecuted in the magistrates' court, currently £5000. More serious cases will be taken to the Crown Court where a fine of any size can be imposed. A trader who has committed an offence will still be entitled to an acquittal if he can show that he took all reasonable steps and exercised all due diligence to avoid committing the particular offence[15].

In one case, a firm of estate agents was fined £800 for a 'grossly misleading' advertisement which claimed that the price of a house had been reduced by £30 000. The advertisement said that the price had been brought down from £194 950, but the prosecution said that the highest asking price had been £185 000[16]. In another case, a trader compared sale prices with higher prices described as 'RRPs'. The court held that a consumer would be misled into thinking that the goods were usually offered by the trader at or near the RRP prices when in fact this was not the case[17].

6.1.2.1 Price indications

As well as the general ban on misleading price claims discussed above, there are also enactments imposing positive duties as to price indications. The Price Indications (Method of Payment) Regulations 1990[18] apply to traders who give indications of price for goods, services, accommodation or facilities and who charge different prices for different methods of payment. The Regulations do not apply to motor fuel.

The Price Marking Order 1991[19] implements an EC directive. It applies to goods sold by retail, and to advertisements for such goods, whether the goods are sold in shops, by mail order or by other methods. The Order prescribes unit pricing for goods sold from bulk or pre-packed in variable quantities, and, from 7 June 1995, for goods sold in pre-established quantity packs. Prices should be the final price, inclusive of VAT and, if other goods or services have to be paid for at the same time, this should be spelled out. Special provision is made for the indication of the price of motor fuel, for the sale of jewellery and precious metals, and for indicating price reductions.

Finally, there are the Price Indications (Bureaux de Change) (No 2) Regulations 1992[20]. These apply to any trader carrying on the business of selling foreign currency in exchange for sterling. Clear and accurate information must be given on the buying and selling rates, terms of business, commissions and other charges. Receipts must be given setting out the terms unless the transaction is made by machine.

6.1.3 False statements as to services

The 1968 Act also makes it an offence knowingly or recklessly to make a false or misleading statement about any services, accommodation or facilities. If the material dealt with in 6.1.1 above has been the scourge of the second-hand car trade, this part of the Act has dealt frequent blows to the market in holiday brochures. One striking case was *R.* v. *Thompson Holidays* (1974)[21]. A holiday brochure contained some admitted inaccuracies. The company was fined £500 when a couple of disappointed holiday-makers returned home and made a complaint. When a second set of holiday-makers also complained, another case was brought and the company was fined £1000.

On this occasion, the defendants took their case to the Court of Appeal, saying that they could not be convicted more than once for the same crime. The court took a different line, taking the point that a new offence arose each time that someone read the brochure and was misled. As the court said, misleading advertising might reach millions of people, especially if it goes out at television peak viewing time. The implication of this case is that there could be as many offences as there are people watching.

The House of Lords has widened the scope of this part of the Act in the case of *Wings Limited* v. *Ellis*[22]. A holiday brochure had been published with a false indication that a hotel was air conditioned. The firm publishing the brochure did not know of the inaccuracy and corrected matters as soon as they found out. However, a particular holiday maker booked a holiday on the strength of an uncorrected brochure. It was held that an offence had been committed even though the firm did not know the statement was false when the brochure was first published, but they did know it was false when it was presented to the particular holiday maker.

The House of Lords also said that an offence could arise under this part of the Act whether or not a false statement had been communicated to a customer, but could occur either when the brochures were posted in bulk, or sent in small batches by post or when the information was passed on by telephone.

There have been a number of cases where gaps have been exposed in the Act. One involved the false claim that a video recorder would be given 'free' on the purchase of a particular car. In fact, the price of the car, or the allowance given for a trade-in, was weighted to cover the cost of the recorder. In another case, an advertiser promised a refund on goods if they could be bought cheaper elsewhere but refused to honour the promise on certain goods. In both cases, the court ruled that the statements made were outside the relevant provisions of the Act[23].

The courts have also ruled, in the case of *R.* v. *Broad*[24] that the Act covers false or misleading statements about services provided in the past. The false statement in this case concerned statements made to various consumers that their washing machines were beyond repair when they had not even been examined, the inference being that they had.

The penalties and the due diligence defence set out in paragraph 6.1.1 apply here also. The defence has, however, less scope for application since, in contrast to a charge of applying a false description to goods, a defendant commits an offence in relation to services, facilities or accommodation only when he knew of the inaccuracy or was reckless.

6.1.4 Truth in lending

On fair conditions it is appropriate to conclude with a reference to the idea that a consumer who seeks to borrow money, or to buy goods on credit, should be given full details as to the cost of the available credit so that he can arrange the best deal available or to suit himself. This aim has been achieved by the Consumer Credit (Advertising) Regulations[25] and the Consumer Credit (Quotation) Regulations[26] both of 1989. Under these Regulations a consumer will be able to see just what the true rate of interest is for the credit he is obtaining.

6.2 A fair quality of goods and services

This really takes us to the essence of consumer protection and indeed to its earliest days. The Sale of Goods Act 1893 was the earliest measure of consumer protection, and though it has now been repealed by the Sale of Goods Act 1979[27] the provisions which were contained in that early Act have been retained and its requirements remain the same.

In essence, the 1979 Act stipulates that in every contract of sale, three obligations are laid by law on the seller. First, he must guarantee that he will pass a good title to the buyer; second, he must provide him with goods which are reasonably (not absolutely) fit for their purpose; and last of all, he must provide the buyer with goods which are of 'merchantable quality'. This last broadly means that the goods must be of a reasonably decent quality having regard to all the circumstances of the sale, such as the price of the goods, the description given to them e.g. second-hand, and the function they are intended to perform.

The Sale of Goods Act, as with the law relating to fraud and misrepresentation, is a part of the civil law. But the criminal law has also intervened to give the authority of the State, and its prosecution agencies, to ensure that the consumer obtains goods of a proper quality. There is, to begin with, the Health and Safety at Work Act 1974[28], section 6 of which obliges those who produce and supply articles for use at work to ensure that when used properly those articles do not present a danger or risk to safety or health. This duty includes the supply of adequate information as to proper use. Substances for use at work are to be regarded in a like manner.

That of course relates to the consumer in his place of work. As for the goods which he buys for use at home, the position is now regulated by Part II of the Consumer Protection Act 1987[3], the provisions of which came into force on 1 October that year. This Act repealed and re-enacted some of the provisions contained in the Consumer Protection Act 1961 and the Consumer Safety Act 1978. The provisions which it re-enacted related to safety regulations, prohibition notices and notices to warn, suspension notices and forfeiture. The nature of safety regulations is self-explanatory. They will lay down standards, usually be reference to British Standards, which specified goods must meet. One example is the Ceramic Ware (Safety) Regulations 1988[29] which impose requirements for ceramic ware with reference to British Standard 6748[30].

Prohibition notices are made by the Secretary of State for Trade and Industry and are directed at specified traders ordering them to cease the supply of specified goods which are considered to be unsafe. A notice to warn is also served by the Secretary of State, and requires the recipient to publish a warning that he has supplied specified unsafe goods. A suspension notice is issued by trading standards departments against a trader who is believed to be in contravention of any 'safety provision'. This is a reference to any safety regulations, prohibition notice or the general safety requirement (which is discussed below). A suspension notice has a maximum duration of six months and, as the name perhaps indicates, bans the supply of the particular product during suspension without the authority of the relevant trading standards department. Such departments can also apply to the magistrates' courts for a forfeiture order where there are grounds for suspecting the breach of a safety provision, including breach of a suspension notice. Normally, goods which are the subject of a forfeiture order have to be destroyed, but the court can order them to be handed over to someone who repairs or reconditions goods of the particular type, or who will deal with them as scrap.

The major innovation of the 1987 Act was the introduction of the general safety requirement.

It is an offence for a trader to supply consumer goods which do not conform to this requirement. There will be a breach of the general safety requirement if goods are not reasonably safe having regard to all the circumstances. This enables local trading standards departments to prosecute for the sale of dangerous goods even where no specific safety standard has been infringed. Two examples of successful prosecutions will suffice. In one case, a company was convicted for selling saucepans, the handles of which were unsafe. A fine of £1500 was imposed. In another case, the prosecution related to travel cots which had rough, sharp edges at the end of tubes forming the supports; a protruding hinge on which a child could fall and damage its eye; and it was possible for a child's finger to be trapped under the hinge mechanism. The trader was fined £1000. Water, food, feeding stuffs and fertiliser are not covered by the general safety requirement, nor is tobacco. The defence of due diligence discussed above also applies where the charge relates to breach of the general safety requirement; and the Act also provides that no offence can arise if the goods are made in accordance with standards approved by the Secretary of State for Trade and Industry. A number of standards have been approved as part of a regular programme of adoption of such standards. To date, these standards have all related to British Standards.

6.2.1 EC product safety

The Council of Ministers has now adopted the Product Safety Directive[31] which is to be implemented in all member states by 29 June 1994.

6.2.1.1 The General Safety Requirement

Article 3 says that: 'Producers shall be obliged to place only safe products on the market'. This part of the directive is headed 'The General Safety Requirement', and both this title and the wording of Article 3 are modelled on the provisions of the Consumer Protection Act 1987. Article 2(b) of the directive which defines a product as 'safe' when it, 'under normal or foreseeable conditions of use, including duration, does not present any risk or only the minimum risk compatible with the product's use, considered as acceptable and consistent with a high level of protection for the safety and health of persons . . .'. These definitions call for little change to UK law.

6.2.1.2 Products within the Directive

The 'products' covered by the directive are stated to be those intended for consumers or likely to be used by consumers, whether new, used or reconditioned. This latter will mean a change to the Consumer Protection Act, since it presently says that a person charged will have a defence if he can show that the goods were not supplied as new goods.

More importantly the Act does contain a number of specific exclusions, none of which is contained in the directive. Of these exclusions, perhaps the most important is tobacco. There are also exclusions for aircraft, motor vehicles, controlled drugs and licensed medical products, but these are subject to control in separate enactments. If tobacco is now to be considered as subject to the general safety requirement, the stage looks set for fresh developments in the round of law suits against tobacco companies. The position with regard to food is discussed below.

6.2.1.3 Enforcement

The directive requires member states to nominate authorities charged with the duty of monitoring compliance with the general safety requirement and for those authorities to have powers to be able to enforce the directive, including the imposition of suitable penalties. At present, the Consumer Protection Act is enforced by local trading standards officers. It is doubtful if these powers can be classed as the imposition of 'penalties': still less do trading standards officers have the power to fine. It almost appears that the directive will oblige the UK to empower enforcement officers, in much the manner of road traffic offences, to impose on-the-spot fines for infringements of the general safety requirement, a major change in our current consumer law.

On top of this, Article 6 of the directive spells out a number of powers which member states will have to have in place. Most of the listed powers are already provided for under current UK legislation. Thus, the Consumer Protection Act empowers enforcement officers to make test purchases, enter premises, take records and so forth, while Customs and Excise can detain goods at ports of entry. These take account of the requirements in Article 6 that the enforcement authorities should be able to make appropriate checks, require information, take samples of product lines and subject them to safety checks. Again, prohibition notices and notices to warn cover those provisions in the Article dealing with the publication of special warnings and prohibiting the supply or marketing of dangerous goods. Similarly, the requirement that member states be empowered to subject the marketing of a product to conditions designed to ensure its safety, and for it to be accompanied by suitable warnings, is covered by s. 11 of the Act which allows for regulations to be made dealing with such and other matters.

6.2.1.4 Product recall

What is new is the final requirement of Article 6 that provisions shall exist for 'the effective and immediate withdrawal of a dangerous product or product batch already on the market . . .'. UK legislation does no more than deal with this at the margins by allowing for prohibition notices, suspension and forfeiture orders. This falls short of providing for withdrawal, though an action in negligence could be sustained if a producer fails to withdraw a product. In *Walton* v. *British Leyland*[32], the defendants were held negligent in not arranging a product recall but in relying instead on notices to dealers. Continuing to supply unsafe goods would also involve a breach of the general safety requirement. Even so, these possibilities fall short of the full-blooded power to order a recall which the directive requires. Recall procedures exist in France under the Consumer Safety Act 1983, in the USA under the Consumer Product Safety Act 1972, and in Japan under the Consumer Product Safety Law 1973 and the Law on the Control of Household Products containing Harmful Substances 1973. In practical terms it is doubtful if the introduction of mandatory product recall procedures will make any difference. Voluntary product recall is already a common occurrence in the UK and elsewhere in the EC, precisely because this is the best way to avoid further criminal and civil actions. Motor vehicle and accessory recalls are also effectively dealt with in the UK through the codes of practice agreed to by the industry and the Department of Transport and operated through DVLC at Swansea.

6.2.1.5 Information exchange

A system of exchange of information throughout the EC on dangerous products was established by a Council Decision adopted on 21 December 1988, on a Community System for the Rapid Exchange of Information on

Dangers Arising from the Use of Consumer Products. In the UK, the work is undertaken by the Consumer Safety Unit, a division of the Department of Trade and Industry.

A much more developed system is created by the Product Safety Directive. When a member state restricts the marketing of a product, or requires its withdrawal, this must be notified to the Commission, except where the measures relate to an 'event' whose effect is limited to within that member state. If the Commission views the measures as justified, it will advise the other member states: if it takes a contrary view, it will advise the state concerned. The member state will then be expected to revoke the particular measures and perhaps compensate the affected party for the disruption of his trade, although nothing to such effect is stated in the directive.

The directive also deals in detail with emergencies and action to be taken at Community level. Where a member state adopts 'emergency measures' to deal with products presenting a 'serious and immediate risk', the Commission must be advised. The Commission itself is in addition empowered to take emergency measures, including ordering restrictions on marketing or withdrawal of the product from the market place, if member states are at odds as to the measures to be adopted to cope with the risk, and if effective action for protecting the consumer can only be undertaken at Community level. A Commission decision is to be effective in the first instance for three months, though it may be extended. Member states are given not more than nine days to put the Commission's orders into effect. The affected party must be informed of the decision and advised of the remedies available to him under the laws of the relevant member state. This information must be provided in advance, unless the urgency of the matter dictates otherwise.

6.2.2 Further UK requirements

In addition, there is the Medicines Act 1968[33]. This imposes strict and necessary controls on the manufacture and supply of medicinal products. In particular, most such products will need a 'product licence', while some products will be available only on prescription.

A valuable codifying measure is the Supply of Goods and Services Act 1982[34], which essentially applies to all transactions which are not contracts of sale or contracts of hire-purchase. This Act covers hiring agreements, agreements where goods are exchanged, and agreements, such as repair or installation contracts, where both goods and services are provided. It did not change the law but served usefully to consolidate the common law that had developed in this and the last century. It provides, in terms based on the Sale of Goods Act, that goods which are the subject of a relevant contract, must be of merchantable quality, be reasonably fit for their purpose and conform to their contract description. Services must be performed with reasonable care and skill. If no price is agreed in advance, then only a reasonable price has to be paid: and if no time is agreed on in the contract for performance, the performance must be within a 'reasonable time'.

The Food Safety Act 1990[35] makes it an offence to sell food not conforming to the 'food safety requirements'. This is more narrowly defined in the general safety requirement discussed above. Food is deemed to breach the food safety requirement if it is unfit for human consumption (which was formerly an offence under the Food Act 1984), if it has been rendered injurious to health by any of the operations specified in the Food Safety Act, or if it is so contaminated that it would not be reasonable to expect it to be used for human consumption. Food is not covered by the general safety requirement laid down in the Consumer Protection Act, but it is within the general safety requirement contained in the Product Safety Directive. This will therefore require an amendment of the law to impose an enhanced duty with regard to the supply of safe food.

6.3 Product liability

Part I of the Consumer Protection Act, the provisions of which came into force on 1 March 1988, implemented the EC Directive on Liability for Defective Products[36]. The Directive, and therefore the Act, created what is called a system of 'strict liability' for defective products, allowing an injured person to sue without the need to prove negligence. It had long been regarded as anomalous that the person who bought defective goods could sue under the Sale of Goods Act for any injury caused without the need to prove negligence, whereas a non-purchaser (e.g. a bystander, a member of the purchaser's family or a person to whom the goods had been given as a gift) could only recover damages if he could prove negligence.

Part I of the Act now provides that damages can be recovered simply on proof that the product is defective, which means that its safety is not such as persons generally are entitled to expect. Liability is placed on the producer, which in this context is stated to include any person who 'own-brands' a product as though he were in fact the producer; and the first importer of the product into the EC. The actual supplier of the defective product can also be made liable, but this can only be where he has been asked by the injured party to name the actual producer and he fails to comply with the request or identify the party who supplied him with the goods within a reasonable time.

The producer of a defective product will not be liable in every case. For instance, he will not be liable if he can show that the goods were not defective when supplied by him. Again, the producer of a defective component has a defence if he can show that he was following the instructions of the producer of the product which was to incorporate the component; or if the defect was due to the design of the end-product.

The most contentious defence contained in the Act is usually called the 'development risks' defence. Under this defence, the producer of a defective product will have a defence if he can show that the 'state of scientific and technical knowledge at the relevant time was not such that a producer of products of the same description might be expected to have discovered the defect if it had existed in his products while they were under his control'. Under the Directive, the Council of Ministers has the right to remove this defence following a review of its operation in the middle of 1995.

The Act applies to damage to property in exactly the same way as it applies to personal injury. However, actions for damage to property cannot be brought unless the claim exceeds £275. If it does exceed that amount, then the whole of the loss can be claimed.

All claims for damage, whether to person or property, must be brought within three years, but no claims can be brought at all once ten years has expired from the time of supply of the defective product.

6.4 Misleading advertising

In enacting the Control of Misleading Advertisements Regulations 1988[4], the United Kingdom adopted Council Directive 84/450[37]. Under the Regulations, the Director General of Fair Trading may seek a court injunction against an advertisement where a complaint has been brought that it is misleading. In deciding whether to apply for a court injunction, he must first consider whether the advertisement in question has been the subject of complaint to 'established means' of dealing with such complaints. This is a reference to such bodies as the Advertising Standards Authority whose British Code of Practice[38] represents the industry's efforts at self-regulation. To date, two injunctions have been obtained. In *Director General of Fair Trading* v. *Tobyward Ltd*[39], an injunction was obtained against the company and its director in respect of certain slimming claims. The claims against which an injunction was granted were: (i) that use of the product could result in permanent weight loss; (ii) that the product carried a guarantee of success; (iii) that the product, or some ingredient, represented a medical breakthrough (except where this could be proven); (iv) that the product placed a lining on the stomach or some other part of the alimentary canal so that the fats that cause weight to be gained cannot enter the bloodstream; (v) that the product would enable users to lose specified amounts of weight in specified periods of time; and (vi) that the product was 100% safe in all circumstances. The injunction also restrained the defendants from publishing any advertisement for a slimming product which was in similar terms or which was likely to convey a similar impression to any of the above claims.

In the other case, an injunction was obtained against an individual who, under a number of names, had been advertising misleading home-worker schemes[40]. The injunction specifies that subsequent advertisements must make clear:

(i) exactly what the home-worker has to do;
(ii) that the work under the scheme involves recruiting third parties;
(iii) that persons responding to the advertisement will be required to pay a registration fee in order to participate in the scheme;
(iv) that participants will be required to pay for their own advertising, postage and other disbursements, in addition to the registration fee; and
(v) that the sums that can be earned under the scheme are outside the direct control of the participants (because they depend on the number of responses obtained from third parties).

In addition, the advertisements are to avoid conveying the false impression of the earning that can be achieved. The terms of the injunction apply to any other advertisement likely to convey a false impression in these respects.

In two other cases, injunctions have been avoided on particular advertising by the giving of undertakings which were satisfactory to the Office of Fair Trading. Under the Regulations, the Director General has no power in relation to advertisements carried on any television, radio or cable service. Where an advertisement is misleading, the Independent Television Commission or the Radio Authority, as appropriate, has no power to seek an injunction; instead it can refuse to transmit it.

6.5 Exclusion clauses

One thing that any system of developed consumer law must seek to ensure is that any breach of the law is met with a swift and inexpensive remedy. At one level, this means that the consumer must not be unfairly deprived of rights which would otherwise be his. At one time, particularly in the field of the sale of goods, the small print of the contract would often contain clauses, usually called 'exclusion clauses', which took away from the consumer the rights given him under such legislation as the Sale of Goods Act. The use, or rather abuse, of such clauses became so widespread that Parliament had to intervene, the law now being consolidated in the Unfair Contract Terms Act 1977[41]. This says that in sales to a consumer, it is impossible for the seller to avoid the obligations which are imposed on him by the Sale of Goods Act: see 6.2 above. Even to try to avoid the obligations is now a criminal offence. In the case of sales to other businesses, exclusion clauses will be effective, but only if they are reasonable. Similar constraints are imposed in relation to contracts where possession or ownership of goods passes, but the contract is not one of sale or hire-purchase.

The same Act also controls the operation of other types of exclusion clause whenever they appear in a consumer contract or in a contract which is on written standard terms. The principle is that three types of clause which might be used in such a contract are valid only if they can be proved to be reasonable; such clauses are those which:

seek to allow a person not to perform the contract;
seek to allow him to provide a performance 'substantially different' from that which was reasonably expected; and
allow the person in breach of contract to be free of all liability to pay damages for his breach.

Suppose that a term in a holiday contract says that a person may have to share a room if the tour operator so decides instead of getting the single room that he has booked. Under the Unfair Contract Terms Act, this type of clause will not be valid unless the tour operator can prove that it is reasonable. The assumption in all cases is that a clause is unreasonable until the contrary is proved.

The EC also proposed to legislate on unfair contract terms in consumer contracts. If a term is unfair, in the sense that it causes a significant imbalance in the parties' rights and obligations to the detriment of the consumer, it will be void and of no effect. Under the proposals, ruling can be obtained as to whether terms drawn up for general use, such as those drawn up by a trade association for its members, are unfair. It is intended, if the proposals are adopted by the EC, that they will come into force throughout the Community not later than the end of 1994.

6.6 Consumer redress

Given that the consumer has not been deprived of his rights of action, he may find, if the other party is not prepared to settle, that his only method of getting justice is to go to court. Up until 1973, this could be a daunting and expensive business. But in that year, a small claims or arbitration procedure was set up which operated through the County Court. Any claim within the County Court jurisdiction (generally up to £25 000) may be referred on application to an arbitration heard by the District or County Court judge or even by an outside arbitrator. Any such arbitration has the effect of a full County Court judgement, though it is usually heard in private by the arbitrator in an informal manner and without the normal rules of court procedure applying. If the sum claimed is within £1000, it will go to arbitration automatically if either party desires. Above that limit, both parties will have to agree before the matter can use this procedure. An important feature of the arbitration system is that the rule as to costs has been considerably modified. The loser of a case normally will be asked to pay only a nominal sum. As a rule, he will not have to pay anything in respect of his opponent's legal fees.

The Consumer Arbitration Agreements Act 1988[42] overcomes a problem sometimes faced by consumers who wished to enforce their rights. This was that they might be faced with a term in the particular contract requiring them to take any disputes to arbitration. The Act provides that such a term cannot generally be enforced against the consumer unless he has given his written agreement after the dispute arose, or unless he has already submitted to arbitration under the agreement. The court may also agree to an application (which would presumably only ever be made by the business seeking to enforce the arbitration clause) that the arbitration clause is effective, but generally will not be able to do so where the claim is within the small claims limit of £1000.

References

1. Final Report of the Committee on Consumer Protection, Cmnd 1781, HMSO, London (1962)
2. Trade Descriptions Act 1968, HMSO, London
3. Consumer Protection Act 1987, HMSO, London
4. The Control of Misleading Advertisements Regulations 1988, HMSO, London
5. Misrepresentation Act 1967, HMSO, London
6. British Gas Board *v.* Lubbock [1974] 1 WLR 37
7. Queensway Discount Warehouses *v.* Burke [1985] BTLR 43
8. Dixons Ltd *v.* Barnett [1988] BTLC 311
9. Hicks *v.* Sullam [1983] 91 MR 122
10. Other illustrative cases include:
    Rotherham Metropolitan Borough *v.* Baysun [1988] *The Times* 27 April;
    P & M Supplies (Essex) Ltd *v.* Devon County Council [1991] unreported
11. Norman *v.* Bennett [1974] 3 All ER 351
12. Newham London Borough *v.* Singh [1987] 152 JP 239

13. The Price Marking (Bargain Offers) Order 1979, HMSO, London
14. The Department of Trade and Industry, *A Code of Practice for Traders on Price Indications*, HMSO, London (1988)
15. *See* the discussion of this defence in paragraph 6.1.1
16. Consumer Law Today, April 1991
17. Enfield LBS *v.* Harveys Furnishing Group plc (Wood Green Crown Court, July 1990) unreported
18. The Price Indications (Method of Payment) Regulations 1990, HMSO, London
19. The Price Marking Order 1991, HMSO, London
20. The Price Indications (Bureaux de Change) (No 2) Regulations 1992, HMSO, London (1992)
21. Regina *v.* Thomson Holidays [1974] 1 All ER 1, 823 CA
22. Wings Ltd *v.* Ellis [1984] 3 All ER 577
23. *See:* Newell *v.* Hicks [1983] 148 JP 308; Dixons *v.* Roberts [1984] 148 JP 513
24. Regina *v.* Broad 24 July 1992, unreported
25. Consumer Credit (Advertisements) Regulations 1989, HMSO, London
26. Consumer Credit (Quotations) Regulations 1989, HMSO, London
27. Sale of Goods Act 1979 (superseding the earlier Act of 1893), HMSO, London
28. The Health and Safety at Work etc. Act 1974, HMSO, London
29. The Ceramic Ware (Safety) Regulations 1988, HMSO, London
30. British Standards Institution, *Specification for the performance of handles and handle assemblies attached to cookware*, BS 6748: 1987, BSI, London
31. EC Product Safety Directive, No. 92/59/EEC, HMSO, London
32. Walton *v.* British Leyland [1980] PLI
33. The Medicines Act 1968, HMSO, London
34. The Supply of Goods and Services Act 1982, HMSO, London
35. The Food Safety Act 1990, HMSO, London
36. EC Directive on Liability for Defective Products, No. 85/374/EEC, HMSO, London (1985)
37. EC Directive on Misleading Advertising, No. 84/450/EEC, HMSO, London (1984)
38. Advertising Standards Authority, British Code of Advertising Practice (8th edn), London (1988)
39. The Director General of Fair Trading *v.* Tobyward Ltd [1988] 2 All ER 266
40. *Consumer Law Today*, January 1992
41. The Unfair Contract Terms Act 1977, HMSO, London
42. The Consumer Arbitration Agreements Act 1988, HMSO, London

Further reading

Lawson, R. G. *Advertising Law in the United Kingdom*, Macdonald & Evans, Plymouth (1978)

Lawson, R. G. *Exclusion Clauses after the Unfair Contract Terms Act*, Oyez, London (1983)

Irving, R. *Outline of the Law of Product Liability and Consumer Protection*, B. Rose Publishers Ltd, Chichester (1980)

Consumer Law Today, Monitor Press, published monthly

Consumer Law Statutes, Monitor Press, Sudbury, Suffolk (1988)

Lawson, R. G. The Supply of Goods and Services Act 1982, Oyez, London (1982)

Chapter 7

Insurance cover and compensation

A. West

7.1 Workmen's compensation and the State insurance scheme

The first Workmen's Compensation Act was passed in 1897 (eventually consolidated in the Workmen's Compensation Act 1925) and, as an alternative to a workman's rights at common law, imposed on the employer an obligation to pay compensation automatically in the event of a workman sustaining an accident in the course of his employment. There was no requirement of fault, the legislation being introduced to provide compensation where the workman was injured in purely accidental circumstances with no blame attaching to anyone and resembled therefore an insurance scheme. The system was operated with recourse to the County Court in the event of any dispute arising and facilitated a cheap and relatively quick payment of compensation. The amount of compensation was expressed as a weekly sum and was based on the average wage earnings during the previous 12 months with the employer whereas at common law if successful in establishing liability a workman was awarded a lump sum by way of damages. The workman did however have to elect between claiming at common law or claiming under the Workmen's Compensation Act.

Following the decision in *Young* v. *Bristol Aeroplane Company Limited* [1944] 2 All ER 293 it became established that a workman was precluded from pursuing a claim at common law even where he did not know of his right to elect if he had in fact accepted weekly payments under the Workmen's Compensation scheme. The Workmen's Compensation insurance policies issued at that time indemnified the insured against his liability to pay compensation under the Workmen's Compensation Act, the Employer's Liability Act 1880 and the Factories Act 1846 or at common law in the event of personal injury to any employee arising out of and in the course of his employment.

The introduction of the State scheme by the National Insurance (Industrial Injuries) Act 1946 can be considered as a compromise between the complete abolition of the common law system with its requirement of proof of fault on the part of the employer and the differing opinions of the type of accident insurance which would be most desirable.

Various types of benefits are available under the State insurance schemes for industrial injuries and are payable in respect of any person who has suffered personal injury caused by an accident arising out of and in the course of his employment or where such person suffers from what is termed a prescribed disease with reference to certain industrial occupations which may give rise to that particular disease. The phrases 'accident' and 'arising' out of and in the course of his employment' have given rise to much dispute over the years since their introduction. An accident has been defined as an 'unlooked for mishap or untoward event which is not expected or designed' and by definition may be distinguished from a process involving for example repetitive movements of the hand or wrist which may give rise to a disability such as tenosynovitis where it is difficult to identify any particular event causing injury as opposed to considering the series of events as a whole forming a process.

There are many cases involving the question whether an act of an employee arises out of and in the course of his employment especially under the State insurance scheme and while these are beyond the scope of this text they may be studied in detail elsewhere[1]. For a comparatively recent decision on the topic illustrating some of the problem areas see *Nancollas* v. *Insurance Officer* and *Ball* v. *Insurance Officer* [1985] 1 All ER 833.

An employee suffering from the effects of an accident at work or from a prescribed disease may be entitled to a range of benefits determined by the current Social Security Act and supporting Regulations. The benefits may include:

(1) *Statutory Sick Pay* – The Social Security and Housing Benefit Act 1982 introduced the concept of statutory sick pay payable by the employer for the first 8 weeks of absence due to injury or sickness. From 6th April 1986 the period of payment has been extended to 28 weeks. Payment, which is made at one of two levels, depends upon average weekly earnings and is subject to taxation.
(2) *Industrial Injuries Disablement Benefit* – Where an employee becomes disabled as a result of an injury at work or as a result of one of the prescribed diseases, then he should qualify for Industrial Injuries Disablement Benefit. The requirements for payment of benefit are broadly loss of physical or mental capacity as a result of an industrial accident or disease. This means some impairment of the power to enjoy a normal life and includes disfigurement even though this causes no bodily handicap. The impairment assessment is expressed as a percentage subject to a maximum of 100% and originally no benefit was paid where the loss of faculty was less than 1%. However since the introduction of Social Security (Industrial Injuries and Diseases) Miscellaneous Provisions Regulations 1986 entitlement to benefit only arises where the degree of disablement arising from the loss of faculty is assessed at 14% or more. But it is possible to receive Disablement Benefit if suffering from one of the prescribed diseases and disability is 1% or more, where aggregated assessments exceed 14%. Anyone suffering from disablement assessment between 14% and 19% will be paid at the 20% rate, payment taking the form of a weekly pension.
(3) *Reduced Earnings Allowance* – This replaces the Special Hardship

Allowance and is payable where as a result of the loss of faculty the claimant is incapable and is likely to remain permanently incapable of following his occupation after the end of the 90 day qualifying period. He must also be incapable of following employment of an equivalent standard. Benefit is payable at all levels of disablement provided the loss of faculty is at least 1%. Whether the work he is capable of doing is considered of an equivalent standard will depend mainly upon the rate of remuneration but the prospects of advancement to higher paid work in the normal occupation can also be taken into account.

(4) *Attendance Allowance/Mobility Allowance* – These benefits are not means tested but relate to medical disability. Attendance Allowance is paid under the National Insurance Scheme. Constant Attendance Allowance is paid under the Industrial Injuries Scheme and it is not possible to get both.

The qualifying criteria for Constant Attendance Allowance is a serious handicap sufficient to require constant care and attention as a result of the effects of an industrial accident or disease.

To qualify for Attendance Allowance a person must be so severely disabled that they require frequent attention throughout the day in connection with bodily functions or continual supervision to avoid danger to themselves or others.

The qualifying feature for the payment of Mobility Allowance is the inability to walk a reasonable distance without distress or risk of damage to health.

The employee is not of course precluded from claiming other benefits where the absence from work does not arise from an industrial accident or prescribed disease. All that is required to be shown is that the claimant is incapable of work and he is so incapable if having regard to his age, education, experience, state of health and other personal factors there is no type of work which he can reasonably be expected to do. These benefits may include:

(a) *Sickness benefit*
This is a contributory benefit paid to those incapable of work where they do not qualify for Statutory Sick Pay or Statutory Sick Pay stops earlier than 28 weeks.

(b) *Invalidity Benefit*
This payment arises where the claimant is still incapable of work after 28 weeks following payment of Statutory Sick Pay or Sickness Benefit.

(c) *Severe Disablement Allowance*
Persons not qualifying for Invalidity Benefit because of insufficient National Insurance Contributions may be entitled to Severe Disablement Allowance if they have been unable to work for at least 28 weeks provided they are assessed at 80% disabled unless aged under 20 when no assessment is necessary. Various disabilities can be aggregated[2].

Section 22 of the Social Security Act 1989 makes provision for the Department of Social Security to collect from those paying compensation for injury or illness, the amount of benefit paid to persons as a result of such injury or illness. Effectively this will entitle the Government to repayment of any State Insurance Scheme payments made to those injured or ill where those persons are entitled to compensation following pursuit of a common law claim.

7.2 Employer's liability insurance

Since the 1st January 1972 it has been compulsory for employers to insure against their liability to pay damages for bodily injury or disease sustained by their employees arising out of and in the course of their employment. This was enacted by section 1(1) of the Employer's Liability (Compulsory Insurance) Act 1969 and failure to comply with the provisions of the Act by an employer renders him guilty of an offence and liable to summary conviction – section 5.

The Act contains a definition of the term 'employee' as including an individual who has entered into or works under a contract of service or apprenticeship with an employer whether by way of manual labour, clerical work or otherwise, whether such contract is expressed or implied or in writing – section 2(1). Certain relatives of the employer are outside the ambit of the Act – section 2(2)(a) as are employees 'not ordinarily resident in Great Britain' – section 2(2)(b).

The contract of insurance incorporates conditions compliance with which is itself a condition precedent to liability under the policy. Accordingly whilst an employer may incur liability to one of his employees, in the event of his failing to comply with a condition of the policy, for example failure to notify the insurer in reasonable time of an occurrence which may give rise to liability under the policy, the insurer may invoke non-compliance with the condition as a reason for refusing to indemnify the employer under the policy. In certain circumstances this could prejudice the injured employee's prospects of recovering damages. The Employer's Liability (Compulsory Insurance) General Regulations 1972 restrict the application of conditions in policies of insurance. The regulations do not however prejudice the rights of the insurer to recover from the policy holder sums which they have been required to pay by reason of application of the regulations. To ensure that employees are aware of the existence of the contract of insurance, sections 5 and 6 of the Regulations deal with the requirement on the insurer to issue a certificate and the subsequent requirement on the employer for its display at his place of business in such a position as to be seen and read by every person employed whose claims may be the subject of indemnity under the policy.

Policy Cover – the basic cover indemnifies the insured against liability at law for damages and claimant's costs and expenses in respect of bodily injury or disease sustained during the period of insurance by any person under a contract of service or apprenticeship with the insured whilst employed in or temporarily outside the territorial limits which are normally Great Britain, Northern Ireland, the Isle of Man or Channel Islands and arising out of and in the course of his employment. In view of the increased use of sub-contract labour and to clarify the position regarding temporary staff and others working for an insured under various schemes and arrangements, the definition of employee has now been extended to include persons supplied to, hired or borrowed by the insured in the course of his business.

The criteria by which 'arising out of and in the course of his employment' is established are different in relation to Employer's Liability insurance and the State insurance scheme, the latter incorporating a broader definition. For an illustration of this aspect see *Vandyke* v. *Fender* [1970] 2 All ER 335 concerning the question of which insurer, motor or employer's liability, should deal with a claim where a company provides a car for its employees to go to or from work and an accident occurs on the road.

A more recent example of these issues is *Smith* v. *Stages* [1989] 1 All ER 833. Two employees were sent by their employers to carry out work at a site some distance from the site at which they had previously been working. They were paid 8 hours pay for the travelling time in addition to the equivalent of the rail fare, although no stipulation was made as to the mode of travel. On returning from the site the vehicle crashed killing the passenger. It was held that the employers were vicariously liable for the negligence of the driver. Both men were acting within the course of their employment when returning to their ordinary residence after completing the temporary work as they were travelling back in the employers' time and were paid wages and not merely a travelling allowance.

With effect from 31 December 1992 and to comply with the Third EC Motor Insurance Directive (90/232/EEC) liability to all passengers must be covered by motor insurance including liability arising out of and in the course of employment. To allow motor insurers time to adjust policy wordings and pricing structures transitional arrangements will apply in the UK until 1 July 1994.

The majority of all Employer's Liability claims emanate from accidents on the 'factory floor' often involving injuries sustained through contact with dangerous moving machinery. The Employer's Liability policy is designed to indemnify the employer against his legal liability to pay damages to employees for injuries sustained in such circumstances. This liability may arise either from the employers' breach of certain statutory duties or from a breach of their common law duties to their employees where the injured person can prove that the breach was causative of the injury. An example often encountered is the employer's duty to guard machinery referred to in sections 12–14 of the Factories Act 1961. The duty to fence machinery securely is an absolute one and the fact that compliance with section 14(1) of the Factories Act may render the machine unusable does not absolve the employers from their duty. This principle is illustrated by *Frost* v. *John Summers and Son Limited* [1955] 1 All ER 870 where a grinding wheel was held to have been a dangerous part of machinery within section 14(1). This decision was instrumental in bringing about special regulations (the Abrasive Wheels Regulations 1970) which enabled this type of machine to be operated without being in breach of the Factories Act 1961. Two other sections of the Factories Act, breaches of which are often alleged by injured workmen, are sections 28 and 29 broadly encompassing the employer's duty to provide a safe place of work and safe means of access thereto; however the duty is qualified by the words 'as far as is reasonably practicable'. The duty is not absolute, for example it may be impracticable to maintain passages in a condition such that there are never any slippery patches particularly after it has been raining – for example see *Davies* v. *De Havilland Aircraft Company Limited* [1950] 2 All ER 582. This qualification on the duty of the employer effectively equates the statutory provisions containing the qualification with the employer's common law duty.

At common law the employer's main duties are:

(1) to take reasonable care that the place of work provided for the employee is safe, for example see *Quintas* v. *The National Smelting Co. Limited* [1961] 1 All ER 630,
(2) to provide sufficient safe and suitable plant, for example see *Kilgullan* v. *W. Cooke and Co. Limited* [1956] 2 All ER 294,
(3) to maintain such equipment, for example see *Henderson* v. *Henry E. Jenkins & Sons* [1969] 3 All ER 756, and
(4) to provide a proper and safe system of work, for example see *General Cleaning Contractors Limited* v. *Christmas* [1952] 2 All ER 1110.

In addition an employer has an obligation to use care in the selection of fellow employees although this duty is less often encountered as a result of the development of the doctrine of vicarious liability whereby the employer will be liable for the negligent acts of his employees whilst acting in the course of their employment.

Any breach of these common law duties resulting in injury to an employee will give rise to liability against which the Employer's Liability policy may indemnify the insured in the event of damages being payable to the injured employee.

An insurer will on behalf of the employer, where applicable, raise a defence to a workman's claim. Various defences are available to him. These include the complete defences of:

(1) *volenti non fit injuria* where the injured person has consented to run the risk,
(2) 'inevitable accident' where despite the exercise of reasonable care by the defendant the accident still occurred,
(3) defences based on the Limitation Acts where the plaintiff fails to bring his action within the prescribed time limit, and
(4) partial defences such as contributory negligence (see later text).

The defence of *volenti non fit injuria* has very limited application since the mere continuance in work that involves risk of injury does not imply acceptance of the risk of injury caused by the employer's negligence and this defence has rarely succeeded in circumstances of an injury to a servant by the negligence of his master. See, for example, *Bowater* v. *Rowley Regis Corporation* [1944] 1 All ER 465.

The onus of proving negligence or breach of statutory duty and that this failure was the cause of the accident rests on the plaintiff except where the facts of any accident are such that the accident would not have occurred without negligence. This is the doctrine of *res ipsa loquitur* whereby the defendant must prove that the accident could have occurred without negligence on his part, for example see *Scott* v. *London Dock Company* [1865] 3 H and C 596. For a more modern approach to this concept and a discussion of the problems involved see *Ward* v. *Tesco Stores* [1976] 1 All ER 219.

In contrast to the Public Liability Insurance policy it is not usual to impose a Limit of Indemnity to the Employer's Liability policy. It also usually includes cover for all costs and expenses incurred with the insurance companies' consent and extends to include the cost of representation of the Insured at proceedings in a Court of Summary Jurisdiction arising out of an alleged breach of statutory duty resulting in bodily injury or disease which may be the subject of indemnity under the policy.

The phrase 'sustained during the period of insurance' is designed to pick up the disease risk even where the symptoms do not become manifest until many years later. Insurers are increasingly finding themselves facing claims relating to events which took place many years ago, a situation brought about because of the relaxation in the time limit for bringing claims, in particular the introduction of the 'disapplying' provisions inserted into the 1939 Limitation Act by the Limitation Act 1975 and consolidated by the Limitation Act 1980. These developments are highlighted in the case of *Buck* v. *English Electric Co. Limited* [1978] 1 All ER 271 where the widow of a man who died of pneumoconiosis was allowed to continue her husband's action for damages for personal injuries against his former employers despite the lapse of some 16 years between the deceased's knowledge of the onset of the disease and proceedings being commenced. An insurer however will only indemnify the insured for that part of the damages relating to the period for which the risk was held and during which there was causative exposure to the process to which the disease is in part attributable.

Claims for damages for noise-induced hearing loss are a prime example of retrospective liability giving rise to substantial difficulties for liability insurers. Deafness was added in 1975 to the list of prescribed industrial diseases under the Social Security (Industrial Injuries) (Prescribed Diseases) Regulations 1975. However the right to benefit was limited to deafness caused by exposure to specific noise producing machinery within the metal manufacturing and shipbuilding industries, also requiring an exposure of 20 years or more within that industry. The qualifying occupations have been extended by subsequent regulations now consolidated within the Social Security (Industrial Injuries) (Prescribed Diseases) Regulations 1980 and more recently amended by the Social Security (Industrial Injuries) (Prescribed Diseases) Amendment No 2 Regulations 1983. The major changes introduced by the regulations are:

(1) The minimum period of work in one of the listed occupations is reduced from 20 to 10 years.
(2) The period within which a claim for benefit must be made is lengthened from one to 5 years.
(3) The list of qualifying occupations is extended and now consists of the following:
    (a) The use of, or work wholly or mainly in the immediate vicinity of pneumatic percussive tools or high speed grinding tools, in cleaning, dressing or finishing of cast metal or of ingots, billets or blooms;
    (b) The use of, or work wholly or mainly in the immediate vicinity of pneumatic percussive tools on metal in the shipbuilding or ship repairing industries;

(c) The use of, or work in the immediate vicinity of pneumatic percussive tools on metal, or for drilling rock in quarries or underground, or in mining coal, for at least an average of 1 hour per working day;
(d) Work wholly or mainly in the immediate vicinity of drop-forging plant (including plant for drop-stamping or drop-hammering) or forging press plant engaged in the shaping of metal;
(e) Work wholly or mainly in rooms or sheds where there are machines engaged in weaving man-made or natural (including mineral) fibres or in the bulking up of fibres in textile manufacturing;
(f) The use of, or work wholly or mainly in the immediate vicinity of machines engaged in cutting, shaping or cleaning metal nails;
(g) The use of, or work wholly or mainly in the immediate vicinity of plasma spray guns engaged in the deposition of metal;
(h) The use of, or work wholly or mainly in the immediate vicinity of any of the following machines engaged in the working of wood or materials composed partly of wood, that is to say: multi-cutter moulding machines, planing machines, automatic and semi-automatic lathes, multiple cross-cut machines, automatic shaping machines, double-end tenoning machines, vertical spindle moulding machines (including high speed routing machines), edge banding machines, and sawing machines with a blade width not less than 75 mm and circular sawing machines in the operation of which the blade is moved towards the material being cut.
(i) The use of chain-saws in forestry.

The first reported case of an employee succeeding in a damages claim against his employer for deafness was *Berry* v. *Stone Manganese Marine Limited* [1972] 1 Lloyds Reports 182 although the law has developed since that case and a more recent insight into the problems of a common law claim in this area can be gained from *McGuiness* v. *Kirkstall Forge Engineering Limited QBD Liverpool* 22nd February 1979 (unreported). In this latter case the defendants were forgemasters and the plaintiff had worked for them for most of his working life operating a stamping press. The judge found that there was virtually no evidence that any employer in noisy industries was taking any steps at all to protect his workmen prior to the late 1950s and it was not until the late 1960s that anyone in the drop-forging industry began to show an interest in protecting workmen. The potential damage which might be caused by impact noise was not fully understood until the early 1970s and the judge concluded that the publication of the Ministry of Labour pamphlet 'Noise and the Worker' marked the point where a reasonably careful employer ought to have become aware that, if his employees were exposed to a high level of noise, their hearing might be at risk and there were perhaps steps which could and should be taken to eliminate or at least reduce the danger.

Following the hearing in 1983 of a series of actions claiming damages for noise induced hearing loss sustained whilst working in the ship building industry, it was established that an employer was not negligent at any given time if he followed a recognised practice which had been followed throughout industry for a substantial period, though that practice may not have been without mishap and at that particular time, the consequences of a particular type of risk were regarded as inevitable. Accordingly, 1963 marked the dividing line between a reasonable policy of following the same line of inaction as other employers in the trade and a failure to take positive action. After the publication of *Noise and the Worker* [3] there was no excuse for ignorance.

These cases also confirmed that claimants are only entitled to recover compensation for the additional detriment to their hearing caused during the period when the employers were in breach of their duty – see *Thompson* v. *Smiths Ship Repairers (North Shields) Limited* (1984) 1 All ER 881.

Some 26 years after the publication of *Noise and the Worker*[3], a comprehensive set of regulations was introduced to control the exposure of workers to the effects of noise – The Noise at Work Regulations 1989. These regulations, effective from 1 January 1990, require employers to eliminate or reduce noise exposures above prescribed levels subject to an overiding requirement to reduce, so far as is reasonably practicable, the exposure to noise of employees. What is reasonably practicable will vary according to the circumstances. An employer is required to weigh the quantum of risk against the money, time and trouble involved in remedying the problem and whilst he is not required to incur such cost as to make his business uncompetitive, the protection of the physical health of his employees must demand a high priority. Where it is not possible to reduce noise below the prescribed level protective equipment must be provided and the employee must wear it.

Technological and medical advances in recent years have increased the awareness of the possible relationships between diseases and working environments including contact with injurious substances and operation of machinery. Attempts are constantly being made to extend fields of potential liability. A comparatively recent example is provided by alleged vibration-induced white finger. In 1980 a man who developed symptoms of vibration-induced white finger after working as a caulker/rivetter for many years failed in his claim for damages for personal injury against the Ministry of Defence as employers since in 1973 when the complaint arose little was known of the condition. See *Joseph* v. *Ministry of Defence* Court of Appeal Judgement 29th February 1980 – *The Times* March 4th 1980. Since that time knowledge of the condition has increased and vibration-induced white finger acquired in certain occupations has been introduced as a prescribed disease by the Social Security (Industrial Injuries) (Prescribed Diseases) Amendment Regulations 1985 with effect from 1st April 1985. There have now been a few successful claims brought by employees against their employers following development of the condition of vibration-induced white finger. In the case of *Heal* v. *Garringtons,* unreported, 26th May 1982, it was held that a workman had been exposed to excessive levels of vibration produced by a dressing tool used on a pedestal grinder.

It is generally accepted that industry should have been aware of the risk of vibration-induced white finger from certain processes involving exposure to vibration-inducing equipment by 1976. This does not necessarily imply that an employer is liable from that date as the courts have shown a willingness to realise that an employer cannot modify processes overnight. In another unreported case – *Shepherd* v. *Firth Brown* 1985 – the judge allowed three years after the date of knowledge for the employers to modify an engineering process to reduce vibration.

7.3 Public liability insurance

In addition to the statutory duty to insure against his legal liabilities to his employees an employer will usually insure against his liability to others. This liability may arise from his occupation of premises, the duty to visitors being governed by the Occupier's Liability Act 1957. In addition to the various statutory controls to eliminate the effects of pollution and environmental hazards under the Control of Pollution Act 1974 and the Health and Safety at Work Act 1974, and more recently the Environment Protection Act 1990, the common law has developed doctrines that impose strict liability for the escape of things likely to do damage should they be allowed to escape. See *Rylands* v. *Fletcher* [1861] 73 All ER Reprints No. 1. The occupier may even owe a duty to trespassers in certain circumstances, at least to act with humane consideration. This concept is of particular relevance to injuries to trespassing children, for example see *British Railways Board* v. *Herrington* [1971] 1 All ER 897. The duty an occupier of premises owes to persons other than visitors is now contained in the Occupier's Liability Act 1984. Public Liability insurance has been developed to indemnify the insured against this type of risk, the insurer providing cover against liability for injury to or illness of third parties (other than employees) and loss of or damage to third party property and including claimants' costs and expenses on the same basis as the Employer's Liability policy. It must be emphasised that for Public Liability policies to operate the occurrence must be accidental in origin, for example damage caused to plaster removed by an electrician to facilitate examination of wiring would not be covered. The injury or damage must also occur during the period of insurance and in connection with the business as defined in the policy although it is normally emphasised that the interpretation embraces the insured's legal liability arising from associated activities such as canteens, sports clubs, works fire service, medical facilities and the like.

The Public Liability policy will exclude liability arising out of the ownership, possession or use by or on behalf of the insured of a mechanically propelled vehicle, vessel or craft, the insurances of which are more properly the province of other policies. With regard to motor vehicles liability is often incurred by an employer where the driver of a vehicle who is acting as a servant or agent of the employer is negligent causing injury for which the employer is vicariously liable. However the insurance against liability in respect of the death of or bodily injury to any person caused by or arising out of the use of a vehicle on a road is compulsory by virtue of the Road Traffic Act 1972, see part VI of the Act – Third Party Liabilities.

It is also not the intention of the insurer to provide cover against the insured's liability for damage to property belonging to or in the custody, possession or control of the insured which is more properly the province of material damage policies although often cover is extended in relation to the personal effects including motor vehicles of employees, but in each case legal liability for such damage must devolve on the employer before the policy cover operates.

One particular area of third party liabilities which has been the subject of recent reform is that relating to the supply of defective products. In July 1985 the EEC adopted a directive on product liability[4] which had to be introduced into the laws of Member States within 3 years. This was achieved in this country following the implementation of the Consumer Protection Act 1987. This Act creates a new civil liability for injury or damage caused wholly or partly by a defect in a product. The existing legal framework, under which a person could bring a claim for damage resulting from defective goods either by means of an action in contract or tort, is retained.

Prior to the Consumer Protection Act a very limited form of strict liability existed in the form of statutory liability in contract arising from the direct supply of defective products. This is defined by the Sale of Goods Act 1893 as amended by the Supply of Goods (Implied Terms) Act 1973 (now consolidated into the Sale of Goods Act 1979) and the Unfair Contract Terms Act 1977.

The eventual consumer who sustains injury or damage may be able to succeed in an action in tort under the principle enunciated in *Donoghue (McAlister)* v. *Stevenson* [1932] All ER Reprints 1, if he is able to prove not only that a product was defective and it was that which caused the injury or damage but also that the defendant has failed in his duty of care. The defendant may raise various defences to the claim including contributory negligence or a defence based on the 'state of the art' whereby he asserts that he exercised all reasonable care in accordance with the present level of technological knowledge. This defence is also available to defendants in relation to claims brought under the Consumer Protection Act.

The Products Liability Insurance policy is designed to cover this type of risk, indemnifying the insured against his liability for bodily injury or illness to persons or loss of or damage to property caused by products sold, supplied or repaired by the insured although damage to the defective product itself is excluded.

7.4 Investigation, negotiation and the quantum of damage

Once a claim has been intimated by an injured person or by a solicitor on his behalf the insurer undertakes a detailed investigation into the circumstances of any accident prior to taking any decision regarding liability. Even before this stage is reached it is incumbent upon the insured to notify the insurer of any accident which may be the subject of indemnity under the policy. Some cases, for example fatal accidents, are serious enough to warrant immediate investigation to obviate the possibility of alteration or destruction of physical evidence and to ensure that the witnesses' evidence is secured before the facts become clouded through the passage of time. In fatal cases it is usual for an insurer to instruct solicitors to represent the insured at the inquest who will then report on the proceedings and where necessary obtain the depositions.

Any investigation will usually combine observation of the scene of any accident including the examination of any machinery or apparatus involved and the taking of detailed statements from witnesses, independent where possible. If litigation is in prospect full proofs of evidence may be obtained and particular regard paid to the demeanour of the individual in relation to the form and manner in which he is likely to reproduce his oral evidence in the court. Where both sides in an action produce technical and expert reports a judge will decide which opinion he is disposed to accept. Where the accident or loss involves complex machinery or systems of work it is usual for both the defendant's and plaintiff's advisers to use the services of consulting engineers. However it must be remembered that the rules of the Supreme Court and the rules of the County Court stipulate that where a party intends to rely on expert evidence, the substance of that evidence must be disclosed to the other parties in the form of a written report. In the case of any other oral evidence he shall serve on all other parties written statements of oral evidence he intends to adduce, both within specified time limits. The question of what constitutes the substance of that evidence was dealt with in *Ollett* v. *Bristol Aerojet Limited* [1979] 3 All ER 544 where the judge confirmed that this phrase embraced both the substance of a factual description of the machine and/or the circumstances of the accident *and* his expert opinion in relation to that accident. Where the employer has himself carried out detailed enquiries into the circumstances of an accident a document containing the results of that enquiry may be discoverable, i.e. it must be revealed to the plaintiff's advisers, if litigation is not the dominant purpose of the raising of the document – see in particular *Waugh* v. *British Railways Board* [1979] 2 All ER 1169.

With the aid of experts the insurer will assess the evidence and decide whether liability will attach to the insured. A condition in the policy stipulates that the insured themselves must make no admission of liability, even impliedly, without the consent of the insurers. Conversely insurers do not admit liability to a third party on behalf of their insured without prior consent. Repudiation of a claim will only be made after careful consideration of all of the evidence because litigation is both costly and uncertain in outcome.

The next stage is to assess the quantum of damage, in property damage cases often with the aid of loss adjustors and in personal injury cases with the assistance of medical experts. A medical examination will be arranged where the nature of the injury is sufficiently serious to warrant this expense and where possible an exchange of medical evidence with the plaintiff's advisers is undertaken. Once medical evidence has been clarified the insurer will commence negotiations with a view to agreement of any amount to be paid in settlement of the claim.

The law of damages is complex and in a state of constant evolution. Consequently a full discussion and analysis is beyond the scope of this text. As a brief summary, damages may be classified in the following way (for a full analysis see McGregor on Damages[5] – and for up-to-date case law see Kemp[6]):

(1) Pecuniary loss

This may be sub-divided into:

(a) Past losses – Included under this heading would be the claimant's net loss of earnings, medical expenses, nursing fees, damage to clothing, cost of repairs to property, all of which must have been reasonably incurred.

For accidents occurring or where a claim for benefit naming a disease was or is made on or after 1 January 1989 for which damages above a specified minimum are paid on or after 3 September 1990 the compensator will deduct all relevant State Benefits from the damages and repay them to the Department of Social Security by virtue of the Social Security Act 1989 and the Social Security (Recoupment) Regulations 1990.

In relation to earnings loss prior to the recoupment Regulations the defendant was entitled to deduct 50% of the benefits payable to the claimant under the State insurance scheme for up to a maximum period of five years from the date of the accident by virtue of s. 2 of the Law Reform (Personal Injuries) Act of 1948. Similar rules still apply to cases where compensation is below the current statutory minimum level for the recoupment process to apply, although the benefits which can be deducted have been redefined by the more recent legislation.

(b) Future losses – The court must attempt to predict the plaintiff's needs and the future costs thereof. If the plaintiff can show that his income will be substantially reduced in the future and this will result directly from the accident then this is a recoverable head of damages. In its simplest form it will be calculated by reference to the plaintiff's future earning capacity in relation to his notional pre-accident earnings and multiplied by the number of years over which the loss will exist, due allowance being made for the contingencies of life – see *Lim Poh Choo* v. *Camden and Islington Area Health Authority* [1979] 2 All ER 910.

(c) Loss of future earning capacity – Where the plaintiff has a disability but has returned to equally remunerative employment, compensation may be payable for the risk of loss of opportunity to earn in the future – see *Moeliker* v. *Reyrolle* WLR 4th February 1977.

(d) Loss of profit – In relation to some aspects of this head of damage see *Spartan Steel and Alloys Ltd* v. *Martin and Company (Contractors) Limited* [1972] 3 All ER 557 and *SCM (UK) Limited* v. *W.J. Whittle and Son Limited* [1970] 2 All ER 417.

(2) Non-pecuniary losses

Compensation for pain, suffering and loss of amenity falls into this category. This is awarded by way of general damages and the courts do not apportion individual amounts to each subdivision but merely make a global award. The potential value of any claim must be assessed by reference to previous awards falling within the same general category making due allowance for any individual distinguishing characteristics.

There are other heads of damages including loss of expectation of life and in particular the statutory entitlement of dependants of the deceased person under the Fatal Accidents Act 1976.

Any award or negotiated settlement should also take into account any reduction in the damages possible by virtue of the Law Reform (Contributory Negligence) Act where the plaintiff suffers damage partly as a result of his own fault. The criterion for the proportion of assessment is the degree to which the plaintiff has departed from the accepted norm as compared to the degree of culpability attached to the defendant. The statute itself refers to a reduction in damages 'to such extent as the court thinks just and equitable having regard to the claimant's share in the responsibility for the damage'. Contributory negligence is not always easy to establish. In particular, momentary inadvertence by an employee where the employer is in flagrant breach of his statutory duty will not suffice to mitigate damages, for example see *Mullard* v. *Ben Line Steamers Limited* [1971] 2 All ER 424. Although contributory negligence can amount to a significant degree of culpability it cannot equate to 100% – see *Pitts* v. *Hunt and Another* [1990] 3 All ER 344.

7.5 General

The role of the insurer extends beyond the mere limitations of indemnifying an employer against his liability for certain injury or damage. Accident prevention is of benefit to both the insurer and the insured because in the final analysis premiums are influenced by the claims cost ratio. The social benefits of accident prevention are of course impossible to measure in terms of the avoidance of personal suffering and financial loss. The insurers employ experienced surveyors whose job embraces risk reduction in a direct sense through their observation of potential hazards during surveys prior to the arrangements of Employer's Liability, Public Liability and Engineering insurances resulting in the making of recommendations to improve the risk to be insured.

References

1. Rideout, R.W., *Principles of Labour Law* – 5th edn, Sweet & Maxwell, London (1989)
2. *Disability Rights Handbook*, 14th edn, April 1989–1990, The Disability Alliance Educational and Research Association, London (1989) (revised annually). See also Department of Social Security leaflet FB2 – *Which Benefit?* from local Department of Social Security office
3. Department of Employment, Health and Safety at Work booklet No. 25, *Noise and the Worker,* HMSO, London (1974) (first published in 1963 – out of print)
4. EEC, Directive on the approximation of the laws, regulations and administrative provisions of the Member States concerning liability for defective products. Directive No. 85/374/EEC, official journal No. 1210/29, Brussels (1985)
5. McGregor, Harvey, *McGregor on Damages,* 15th edn, Sweet & Maxwell, London (1988)
6. Kemp, D.A.M., *Quantum of Damages,* Vol. 2, *Personal Injury Reports,* Sweet & Maxwell, London (1989)

Chapter 8

Civil liability

E. J. Skellett

8.1 The common law and its development

The term 'the common law' means the body of case law of universal, or common, application formed by the judgements of the courts. Each judgement contains the judge's enunciation of the facts, a statement of the law applying to the case and his *ratio decidendi* or legal reasoning for the finding to which he has come. The judgements are recorded in the various series of Law Reports and have thus developed into the body of decided case law which we now have and continue to develop.

The doctrine of precedent whereby an inferior court is bound to follow the judgement of a higher court ensures consistency in the law. Thus an earlier judgement of the Court of Appeal will bind a High Court or County Court judge considering a similar situation and a decision of the House of Lords is binding on all inferior courts although the House itself is free to reappraise its previous judgements.

The common law is not a codified body of law clearly defined in its extent and limits. New law is being made all the time. Judges are asked to adjudicate on sets of circumstances which previously might not have been considered by the courts. Moreover a judge, in applying the established principles of common law to the facts he is considering, might well distinguish that particular case from earlier decided cases. Finally in determining whether in a case there has been compliance with standards such as that of 'reasonable care' the judge will of necessity approach the problem in the light of contemporary knowledge and thinking. Thus what is adjudged reasonable conduct in 1950, say, will not necessarily be adjudged reasonable in 1980. In these ways, judges bring up to date the body of common law and adapt and develop it in accordance with the standards and social principles of the era. Such changes are of course slow and gradual, but the common law is also subject to more drastic and immediate change by Parliament, examples being the Employer's Liability (Defective Equipment) Act 1969 and the Occupier's Liability Acts 1957 and 1984. Although Parliament thus exercises dominance over the common law, the statutes in their turn are interpreted by the judges following legal rules and principles already well established.

8.2 The law of tort

This concerns the legal relationships between parties generally in the everyday course of their affairs, the duties owed one to the other and the legal effect of a wrongful act of one party causing harm to the person, property, reputation or economic interests of another.

The law of tort covers relationships generally, compared with the law of contract which applies where two or more parties have entered into a specific relationship between themselves for a specific purpose.

Three separate branches of the law of tort are trespass, nuisance and negligence, the latter being by far the most important and applying in particular to the field of an employer's liability for accidental injury to his employee.

8.2.1 Trespass

This is the oldest branch of the law of tort. An action for trespass is nowadays generally confined to the intentional invasion of a man's person, land or goods involving for example such civil claims for damages as those resulting from battery, assault, false imprisonment, unlawful entry onto the land of another. In the latter case, apart from legal action, direct action can be taken against the trespasser using reasonable force to regain possession against, for example, squatters or 'sit-in' demonstrators.

8.2.2 Nuisance

There are two forms, private nuisance or public nuisance. An action for private nuisance lies only where there has been interference with the enjoyment of land and is appropriate where an occupier of land has acted in such a way as to harm his neighbour's enjoyment of his land. It need not be a deliberate interference and includes such matters as the emission of smoke, fumes or excessive noise. The interference must be sufficiently significant and must be unreasonable. In deciding if it is, the court will take into account all circumstances including the reason for the alleged nuisance, the locality (e.g. whether rural or industrial), the ordinary use of the land and the impracticability of preventing the nuisance.

The second classification of nuisance, public nuisance, constitutes a criminal offence as well as being an actionable wrong at civil law for which damages may be claimed for any injury or damage caused. Public nuisance relates to acts interfering with the public at large and includes, for example, obstruction of the highway, leaving open a cellar flap or leaving unlit scaffolding abutting onto the highway.

8.2.3 Negligence

A broad definition is careless conduct causing damage or injury to another.

Actions based upon the tort of negligence are far commoner than those based upon other torts. Distinctions are not exclusive. Very often the same facts can found an action both in negligence and nuisance. There are three elements necessary to establish a case in negligence:

(1) that there is a duty of care owed by one party to the other,
(2) that there has been a breach of that duty,
(3) that the breach of duty has resulted in damage.

8.2.3.1 The duty of care

To whom is this owed? In the case of *Donoghue* v. *Stevenson*[1] this was defined as follows:

> 'You must take reasonable care to avoid acts or omissions which you can reasonably foresee would be likely to injure your neighbour'.

Neighbours are defined as:

> 'Persons who are so closely and directly affected by my act that I ought reasonably to have them in contemplation as being so affected when I am directing my mind to the acts or omissions which are called in question'.

There are no hard and fast rules as to who might or might not fall into this category, and this must be examined in each case. In some situations, the public at large may be owed a duty, for example, by a motorist. In others, a duty is more closely defined. An employer owes a duty of care in tort to his employee, a manufacturer to the consumer, a solicitor to his client.

The standard of care owed

This requires an examination of the facts of the particular circumstances. The magnitude of the risk of injury and the gravity of the consequences of an accident must be weighed against the cost and difficulty of obviating the risk. A considered decision has to be made. Even though a risk may not warrant extensive precautions, the particular process, place or person may have features that make these vital. In *Paris* v. *Stepney B.C.*[2] for example, the House of Lords held goggles should have been provided for a one-eyed man doing work where there was a risk of metal particles striking the eye although the risk of this happening was such that for a man with normal sight it could be ignored. The question is put succinctly by Denning L.J. in *Latimer* v. *AEC Ltd*[3]:

> 'It is a matter of balancing the risk against the measures necessary to eliminate it'.

The New Zealand courts give a convenient and simple approach to the issue, in the case of *Fletcher Construction Co. Ltd* v. *Webster*[4]:

(1) What dangers should the defendant, exercising reasonable foresight, have foreseen?
(2) Of what remedies, applying reasonable care and ordinary knowledge, should he have known?

(3) Was the remedy, of which he should have known, for the danger he should have foreseen, one he was entitled to reject as unreasonably expensive or troublesome?

8.2.3.2. Breach of duty

Once the existence of the duty of care which arises from the relationship of the parties concerned and its standard are established, one has to consider whether or not there has been a breach of that duty, and if so consideration can be given to the next question.

8.2.3.3. Res ipsa loquitur

This Latin maxim means literally 'the thing speaks for itself'. In other words the circumstances of the accident giving rise to the action are such as impute negligence on the part of the defendant, being an event which, if the defendant had properly ordered his affairs, would not have happened. If this plea by a plaintiff is accepted by the court then a presumption of negligence is raised against the defendants. In other words, effectively it is for the defendant to prove the absence of fault rather than for the plaintiff to prove fault. The defendant can set aside the presumption against him by:

(1) Proof of reasonable care having been taken.
(2) An alternative explanation for the accident which is equally probable and which does not involve negligence on the part of the defendant.
(3) A complete analysis of the facts, i.e. the defendant laying before the court all the facts of the case and inviting full consideration of liability.

Illustrations of the application of this maxim are such cases as bricks falling from a bridge onto a person walking underneath or cargo falling from a crane onto an innocent passerby, i.e. where one would say that prima facie the accident could not have happened without someone's fault.

8.2.3.4 The resultant damage

The damage must result from the negligent act or omission and be caused by it. In other words it must be a direct consequence. Most cases of injury are straightforward but sometimes unexpected complications arise, as in the case of *Smith* v. *Leech Braine*[5] where a plaintiff was entitled to recover damages for cancer developing from a burn on the lip caused by molten metal. This was a direct result of the burn. However the chain of causation must not be broken – there must not be a *novus actus interveniens*, i.e. an independent and unforeseeable cause intervening. For example in *McKew* v. *Holland and Hannen and Cubitts Ltd*[6] it was held that a workman who had sprained his ankle and later fell down stairs when the ankle gave way, resulting in his breaking his leg, could recover from the original wrongdoer damages for the ankle injury but not for the fractured leg because he himself had been negligent for not holding on to the handrail. His negligence was held to constitute a *novus actus*.

If there are more than one possible causes of an injury, it is for the plaintiff to prove causation – *Wilsher* v. *Essex Health Authority*[7]. However, where a pedestrian was injured by one car then further injured by being thrown into the path of a second, it being impossible to say what proportion of injury was caused by each motorist, it was held that the plaintiff did not have to go so far as to prove the extent of injury caused by each – *Fitzgerald* v. *Lane*[8].

8.3 Occupier's Liability Acts 1957 and 1984

The 1957 Act defines the duty owed by the occupiers of premises to all persons lawfully on the premises in respect of:

> 'Dangers due to the state of the premises or to things done or omitted to be done on them'. Section 1(i).

The liability is not confined to buildings and has been held to include, for example, that of the main contractors retaining general control over a tunnel being constructed – *Bunker* v. *Charles Brand & Son Limited*[9].

Section 2 defines the standard of care, owed by the occupier to the persons lawfully on the premises namely:

> 'A common duty of care to see a visitor will be reasonably safe in using the premises'.

Then by section 2(3) 'A person present in the pursuance of his calling may be expected to appreciate and guard against any special risks ordinarily incidental to it, so far as the occupier leaves him free to do so'. In other words this class of visitor is expected to use his own specialist knowledge.

Under section 2(4) 'A warning or notice does not, in itself, absolve the occupier from liability, unless in all the circumstances it was sufficient to enable the visitor to be reasonably safe'. Whilst the occupier could, under this section, avoid his liability by a suitably worded notice, this is superseded by the Unfair Contract Terms Act 1977, which provides that it is not permissible to exclude liability for death or injury due to negligence, by a contract or by a notice and this applies to a notice under section 2(4) of the Occupier's Liability Act 1957. The 1957 Act made no provision for those outside this category of lawful visitors, i.e. contractors, invitees and licensees. The 1984 Act extended the classes of persons to whom the duty of care is owed to those exercising public and private rights of way, ramblers and trespassers. In the latter case the Act was directed to alleviate the position of the innocent, such as the young child or someone walking blithely unaware he had no right to be there, rather than the deliberate trespasser.

8.4 Supply of Goods

In the normal course of obtaining goods, the purchaser can reasonably expect to be supplied with goods that are fit for the purpose for which he purchased them.

8.4.1 Manufacturers

They owe a duty of care to the consumer of their products independently of any rights the purchaser of their products may have under contract law, against the supplier to them of goods. Thus a consumer may be able to sue both his supplier and the manufacturer.

The leading case is *Donoghue* v. *Stevenson*[1] which established the principle, the House of Lords holding that someone who drank ginger beer from an opaque bottle, given her by a friend, and who became ill from the presence of a snail in the bottle was entitled to damages from the manufacturers if she could prove her case.

The manufacturer's duty is to take reasonable care in manufacture to ensure that the product is without defect and not liable to cause injury.

There is no liability on a manufacturer if there is the opportunity of intermediate examination particularly where this is expected, which it could not be in the case of a sealed opaque bottle. Nor for instance is a manufacturer liable to a workman injured by using defective goods the manufacturer supplied which an employer examines, sees are defective but decides to keep in use albeit only until they can be replaced.

8.4.2 Consumer Protection Act 1987

By s. 2, where damage is caused wholly or partially by a defect in a product, then producers, own-branders, importers and suppliers are liable for that damage.

Anyone damaged by a defective product has a right of action against those from whom they obtained the finished product or those involved in the production process. The Act does not cover liability for economic loss (even though recognised by the common law in *Junior Books Co. Ltd* v. *Veitchi*[10]) or damages below £275 or claims against repairers and second-hand dealers. Liability is non-excludable by contract, notice or otherwise.

The Act specifically makes it a defence that the product was supplied other than by way of the defendant's business, e.g. by gift. It also provides for a 'development risks' defence, i.e. that the defect was not one the defendant was aware of at the time, given the state of the scientific and technical knowledge then prevailing.

8.5 Employer's liability

An overall statement of the duty owed by an employer to his employees is that he must take such care as is reasonable for the safety of his employees. That duty is owed to each and every employee as an individual, taking into account his own weaknesses and strengths, and is owed wherever the employee may be in the course of his employment, on or off the employer's premises. It is a duty which the employer owes personally to the employee and the employer remains responsible for a breach of that duty even if he has delegated the performance of that duty to someone else, for example to a safety consultant who might have a separate liability. The same applies if he has put his employee to work under the order of another party – *McDermid* v. *Nash Dredging and Reclamation Co. Ltd*[11].

The employer can be held liable either directly for breach of his own duties or vicariously. Vicarious liability arises where an employee or an agent of the employer has acted negligently and caused injury to another employee. The employer is legally liable for the wrongful act or omission where it has been performed in his interests. However, he is not liable if the employee acts negligently on a frolic of his own independently of his employment. *Smith* v. *Crossley Bros. Ltd*[12] illustrates this, where, as a joke, two apprentices injected compressed air into the body of a third and the employers were held not liable.

The employer's duty at common law can conveniently be considered under five heads. Obviously each will turn on the particular circumstances involving one or more of these elements and it is impossible to give more than general guidelines. The heads are:

(1) system of work,
(2) place of work,
(3) plant and equipment,
(4) supervision and/or instruction,
(5) care in selection of fellow employees.

8.5.1 System of work

The employer is obliged to set up and operate a safe system of work, and it is a question of fact in every case what is safe. This includes such matters as the co-ordination of activities, the layout and arrangement of the way a job is to be done, the use of a particular method of doing a job. The employer is expected to plan and draw up an original method of operation which is safe and free, so far as possible, from foreseeable cause of injury. Regard will be held to established practice and absence of accident in assessing what is safe, but the court will still examine the practice to decide if it is safe. In *General Cleaning Contractors Ltd* v. *Christmas*[13] Lord Oakley said in his judgement:

> 'the common law demands that employers should take reasonable care to lay down a reasonably safe system of work'.

He continued that workmen even though experienced and competent to lay down a system themselves should not be expected to do so, making their decisions at their workplace where the dangers are obscured by repetition, compared with the employer who performs his duty in the calm atmosphere of a boardroom with the advice of experts.

8.5.2 Place of work

The employer is under a duty at common law to provide a reasonably safe place of work, relating to such matters as the provision of gangways clearly marked and free of obstruction, and the maintenance of floors and staircases. The duty is fulfilled through regular inspection of the workplace and keeping it in a safe state, free of hazard so far as reasonably practicable. It does not extend to protection from abnormal hazards which the employer could not reasonably have foreseen. For example, whilst in conditions of ice and snow, paths must as far as possible be sanded before the normal time for employees to arrive at the premises, if there is a sudden totally unexpected snowfall, the employer is not liable if paths are slippery or obstructed until he has had reasonable opportunity to remedy the situation.

The duty extends to any place at which the employee works whether belonging to his employer or not, but it will depend on the circumstances whether the employer should have inspected them before sending his employees to work there, and perhaps had steps taken to make them safer. For example, no court would suggest the employer of a plumber sent out to work at a private house should first send the foreman or supervisor to inspect the house unless the employer had prior knowledge of some particular feature of the premises which introduced added risk. In most cases involving factory or site accidents the relevant section of the Factories Act 1961 or the Construction (Working Places) Regulations 1966 will be pleaded in addition to the duty at common law.

8.5.3 Plant and equipment

The employer owes a duty to his employee to provide safe and proper plant and equipment which must also be suitable for the purpose to which put.

It is a far-ranging aspect of the employer's duty. In the first place the employer may have failed completely to provide equipment necessary for the safe performance of work, for example mechanical lifting equipment for a load too heavy to be manhandled.

Equipment supplied may be unsuitable for the particular function, or it may be the proper equipment but inadequately maintained or defective.

Consideration will be given in deciding if the employer is liable to the procedure followed for reporting and rectifying defects, routine maintenance, the issue of small items of plant and suchlike.

This aspect is relevant also to the question of whether an employer has provided protective equipment such as gloves, goggles and ear-muffs to reduce or prevent exposure to foreseeable risk of injury.

Where a claim for damages arises out of an accident in a factory, the appropriate sections of the Factories Act 1961 will be relied upon, for example relating to the guarding of machinery, in addition to the duty at common law.

The Employer's Liability (Defective Equipment) Act 1969 discussed later is relevant to this aspect, too.

8.5.4 Supervision and/or instruction

An employer must take such care as is reasonable to ensure adequate and proper supervision over and instruction to his employees. What is reasonable must depend on the circumstances, including the complexity of the work to be done, the technicality of the equipment concerned and the age and experience of the workman. It must be obvious that if a young inexperienced man is set to work on a complicated machine or a complicated task where he can injure himself the employer will be held liable. It must not be thought however that an employer can leave even a senior experienced man to his own devices. Supervision and instructions are a matter of degree but always the courts will impute to the employer a superior knowledge of the dangers and risks in a work system with the consequent duty to supervise and instruct his employees.

8.5.5 Care in selection of fellow employees

This aspect of an employer's duty is of less significance since the employer will be held vicariously liable for the act of an employee who negligently injures another, which was not always so.

It is most relevant to the type of case where an employee indulging in horseplay or fighting has injured another and the man concerned has a history of such activities to the knowledge of the employer, who has taken no steps to dismiss him or prevent a recurrence.

8.6 Employer's Liability (Defective Equipment) Act 1969

Prior to the passing of this Act where a workman sued his employer in respect of injury caused by a defective tool or item of plant supplied by the employer to the employee it was a defence for the employer to prove that he did not know and could not reasonably have known of the defect and that he had exercised reasonable care when he obtained the item concerned, by going to a reputable manufacturer or supplier. This was the rule in *Davie* v. *New Merton Board Mills Limited*[14]. The Act changed the law and imposed liability on the employer where an employee was injured in consequence of a defect in equipment provided by his employer for the purpose of the employer's business, if the defect was attributable wholly or partly to the fault of a third party (whether identified or not). In other words the employer no longer has a defence if he provides defective equipment to his employee which results in injury and the defect was the fault of another party. This does not mean that the employer is without remedy against that other party. He is entitled to bring an action against the supplier in respect of the defective plant, but must be able legally to prove his case against the supplier. It is perhaps unnecessary to add that an employer is liable irrespective of the Act if it can be proved that the defect should have been found by the employer on inspection before being put into use or if an employer had caused or permitted his employee to keep in use defective items of plant.

8.7 Health and Safety at Work etc. Act 1974

Although sections 2–8 of the Act impose general duties on parties including employers, a breach of those duties does not provide grounds for a civil claim. However, section 47(2) of the Act provides that an action can be based on a breach of Regulations made under the Act, unless the Regulation has a specific exclusion.

8.8 Defences to a civil liability claim

The first and obvious defence which may be raised is a denial of liability which may be based on a variety of grounds.

(1) That the duty alleged to have been breached by the defendant was never imposed on him in the first place, for example in an employee's claim against his employer that the plaintiff was not an employee but was working for another company.
(2) That the nature of the duty was different from that pleaded against the defendant.
(3) That the duty owed was complied with and not breached.
(4) That the breach of duty did not lead to the damage.
(5) That the plaintiff was himself guilty of contributory negligence resulting wholly in the damage.

Secondly in the defence it may be pleaded that conduct of the plaintiff, constituting contributory negligence, caused and/or resulted in part in the damage he suffered and that any damages which might be payable to him should be reduced accordingly. By way of example, that he failed to see a hole into which he fell. Obviously such a consideration only comes into play in the event of a finding that the defendant is liable. The court will then assess the respective blameworthiness of the parties to decide whether there are grounds for finding the plaintiff partly to blame and, if contributory negligence is established, the court will determine the amount of damages the plaintiff would receive if he succeeded in full and then discount these by the proportion to which the plaintiff is himself found to blame.

Thirdly there is the situation where the accident is the fault not of the defendant sued but of some other party. If another party is blamed in the defence, the usual result is that they are joined in as a co-defendant by the plaintiff, and he sues both. However, if a defendant considers that if he is liable to the plaintiff, then he in turn is entitled to recover from someone else any damages he has to pay to the plaintiff in which case that person can be joined in the proceedings by the defendant as a third party. An example of the circumstances where there may be third party proceedings is one where an injured workman who has fallen into a hole at the place where he works sues his employer, who then brings in by third party proceedings the contractor who had left the hole unfenced. In such a case, the plaintiff would have to establish that the defendant was liable to him, and in turn the defendant would then have to prove his case against the third party. The third party will be liable only if the defendant is liable to the plaintiff.

Compare this situation with an action where the plaintiff sues more than one defendant, such as, in the example given above, both suing his employer and the contractor direct. The plaintiff might fail against both defendants, succeed against one or the other or succeed against both, the judge apportioning the degree of liability attaching to each defendant. In the case of *Fitzgerald* v. *Lane* the House of Lords held that where there were two or more defendants, the first consideration was whether the plaintiff had proved his case against the defendants, then the question of whether he was himself negligent and his damages should be reduced accordingly, and finally the apportionment of liability between the defendants themselves.

Thus a pedestrian who ran into the road and was hit by a car and then by another and who was held equally to blame with the car drivers, had his damages reduced by 50% on account of his own negligence. The car drivers' 50% share of the blame was then apportioned between them at 25% each.

8.9 Volenti non fit injuria

Where a person has agreed either expressly or by implication to accept the risk of injury, he cannot recover damages for damage caused to him by that risk.

For this defence to succeed the person concerned must have had full knowledge of the nature and extent of the risk to be run and have accepted that risk of his own free will. Such a defence is available only in extremely limited circumstances in an action by an employee against his employer. In the case of *Smith* v. *Baker*[15] it was pleaded against an employee drilling rock in a cutting over whose head a crane lifted stones. The court held that although he knew of the danger and continued at work he had not voluntarily undertaken the risk of injury from a stone falling from the crane and hitting him.

Such a defence does not apply to an action for damages brought by a rescuer deliberately running risks to rescue someone who has been injured by dangers created by another. The case of *Baker* v. *T.E. Hopkins & Sons Limited*[16] confirms the entitlement of a rescuer to damages for injury in respect of that negligence.

8.10 Limitation

The Limitation Act 1980 stipulates that an action founded on tort shall not be brought after six years from the date when the cause of action accrued but an action for damages for personal injuries or death must be commenced within three years. Otherwise the actions are barred by the statute and the defendant can plead this as a defence. The three years start to run from the date of the accident or date of the plaintiff's knowledge if later. If the injured person dies within the three years, the period starts to run again from date of death or of his personal representatives' knowledge. The saving provisions of 'knowledge' are aimed primarily at the industrial disease cases where the accidental exposure almost invariably dates back many years before the effects of that exposure had developed and were known. There is a careful definition of what is meant by 'knowledge' in section 11 of the Act.

The Act also permits an overriding discretion to the court to let in late claims where it is equitable or fair to do so. Furthermore in the case of someone under a disability, e.g. an infant or person of unsound mind, the three years do not start to run until the age of 18 or recovery.

8.11 Assessment of damages

Once liability is established the question for consideration is the amount of damages or compensation to be awarded. The object is to put the injured party as far as possible in the same position as before.

In an action for breach of contract or for debt, this amount will already have been defined in the dealing between the parties and is known legally as a liquidated claim. However in an action for damages in respect of tort, where damage and/or injury has been caused, damages are called unliquidated, i.e. they will have to be calculated and assessed after the event giving rise to the claim.

These damages will comprise special damage and general damages.

8.11.1 Special damage

Special damage consists of items of specific expenditure or loss as a result of the accident, damaged goods or loss of wages during time off work. In actions for personal injury, it consists primarily of the loss of wages and the figure recoverable is the net wage lost by the man after deduction of income tax and national insurance contributions, i.e. the actual amount he would have received in his pocket. He will be awarded both his total loss of wages during total incapacity from work and partial loss if by reason of continuing disability, as a result of the accident, he cannot do his full work or has to change to a lighter job and is thereby earning less. Credit is given for sick pay. There are also offset against the loss of earnings claim, any tax refunds and unemployment benefit if, after having been certified fit to return to work following an accident, a man cannot return to his old job and is unable to get another. Redundancy payments under the Redundancy Payments Act 1965 as amended are deductible if attributable to the injury. In actions for damages for personal injury the Social Security Act 1989 compels, in s. 22, a person making a compensation payment in consequence of an accident to obtain a Certificate of Total Benefit paid by the Department of Social Security and then to deduct from the compensation payment the amount of the benefit, accounting to the D.S.S. for this. There are exempting provisions where the damages amount to less than £2500.

8.11.2 General damages

General damages are the amounts paid to compensate for pain, suffering and loss of amenity resulting from an injury. Whilst there are no set tariffs for particular injuries it will be apparent that the damages are assessed by the judge and lawyers in the light of awards for comparable similar injuries cases. However in calculating the appropriate sum to award, account is taken of such matters as the particular idiosyncrasies of the plaintiff, his age, occupation, hobbies and suchlike. The court also has regard to the effect of inflation on past awards or similar injuries. For example, the loss of a finger would attract higher damages for an employee who in his spare time was a skilled musician; similarly damages for an incapacitating leg injury to a keen and energetic sportsman would be higher than those for someone in a sedentary occupation with no active hobbies.

Where there is partial or complete incapacity for work continuing after the trial general damages also include a capital sum awarded for future loss of wages. A sum will, where appropriate, be awarded too for loss of opportunity on the labour market. This is intended to compensate a man, who has suffered a permanent disability but is in work, for the fact that, if he later loses his job, he will find it less easy to get another job than will someone of equal competence who has no such disability.

Awards of damages generally are once and for all. However there is an exception.

8.11.3 Provisional damages

Section 6 of the Administration of Justice Act 1982 introduced provisional damages for cases where there is a chance that some serious disease or serious deterioration in the plaintiff's condition will accrue at a later date. Appropriate cases include industrial disease claims where there may be a risk of the development of cancer or a malignant tumour in the future. Provisional damages are assessed ignoring that possibility. If it occurs then a further award may be made.

8.12 Fatal accidents

A cause of action in tort, save for defamation by or against a person, survives for the benefit of or to the detriment of the estate under the Law Reform (Miscellaneous Provisions) Act 1934.

On behalf of the estate, loss of earnings to date of death and general damages for pain and suffering during lifetime are claimable, without reference to any loss or gain to the estate resulting from the death.

Under the Fatal Accidents Act 1976 damages for loss of financial support can be claimed by or for the dependants. The definition of dependant is set out in the Act as amended by the Administration of Justice Act 1982 and includes spouse or former spouse, ascendants and descendants as well as adopted children and anyone living with the deceased as spouse, the latter subject to certain conditions.

Damages are calculated by the measure of actual financial loss. Thus the deceased's earnings will be established and the proportion expended on the dependant determined. This will then be multiplied by a number of years purchase to allow for the length of time the deceased would have worked. A deduction will be made for capitalisation.

The Administration of Justice Act 1982 also introduced a claim for 'bereavement damages' under which a fixed sum is payable by way of damages – the amount is currently £7500 – for loss of a spouse and to parents for the loss of a child.

References (cases referred to)

1. Donoghue *v.* Stevenson (1932) AC 562
2. Paris *v.* Stepney Borough Council (1951) AC 367
3. Latimer *v.* AEC Limited (1953) 2 All ER 449
4. Fletcher Construction Co. Ltd *v.* Webster (1948) NZLR 514
5. Smith *v.* Leech Braine & Co. Ltd (1962) 2 WLR 148
6. McKew *v.* Holland and Hannen and Cubitts Ltd (1969) 2 All ER 1621
7. Wilsher *v.* Essex Health Authority (1989) 2 WLR 557
8. Fitzgerald *v.* Lane (1988) 3 WLR 356
9. Bunker *v.* Charles Brand & Son Limited (1969) 2 All ER 59
10. Junior Books Co. Ltd *v.* Veitchi (1983) AC 520
11. McDermid *v.* Nash Dredging and Reclamation Co. Ltd (1987) 3 WLR 212
12. Smith *v.* Crossley Bros. Ltd (1951) 95 Sol. Jo. 655
13. General Cleaning Contractors Ltd *v.* Christmas (1953) AC 180
14. Davie *v.* New Merton Board Mills Ltd (1959) 1 All ER 67
15. Smith *v.* Baker (1891) AC 325
16. Baker *v.* T.E. Hopkins & Sons Ltd (1959) 1 WLR 966

Further reading

Munkman, J., *Employer's Liability at Common Law,* 11th edn, Butterworth, London (1990)

Heuston, R. F. V. and Chambers, R. S., *Salmond on the Law of Torts,* 18th edn, Sweet & Maxwell, London (1981)

Kemp, D., *Damages for personal injury and death,* Oyez Publishing, London (1980)

McGregor, Harvey, *McGregor on Damages,* 15th edn, Sweet and Maxwell, London (1988)

Part II

The management of risk

In every activity there is an element of risk and the successful manager is the one who can look ahead, foresee the risks and eliminate or reduce their effects. Risks are no longer confined to the 'sharp end', the shop floor, but all parts of the organisation have roles to play in reducing or eliminating them. Indeed, the Robens' Committee recognised the vital role of management in engendering the right attitudes to, and developing high standards of, health and safety throughout the organisation.

A number of specialised techniques have been developed to enable risks to be identified, assessed and either avoided or reduced but there are other factors related to the culture of the organisation and the inter-relationship of those who inhabit it that have a significant role to play. An understanding of those techniques and the roles and responsibilities of individuals and groups is a necessary prerequisite for high levels of safety performance.

Chapter 9

Principles of the management of risk

L. Bamber

9.1 Principles of action necessary to prevent accidents

9.1.1 Introduction

The Ministry of Labour and National Service[1] postulated six principles of accident prevention in 1956 that are still valid today. These are:

(1) Accident prevention is an essential part of good management and of good workmanship.
(2) Management and workers must co-operate wholeheartedly in securing freedom from accidents.
(3) Top management must take the lead in organising safety in the works.
(4) There must be a definite and known safety policy in each workplace.
(5) The organisation and resources necessary to carry out the policy must exist.
(6) The best available knowledge and methods must be applied.

It would appear that these principles have only received legislative backing in more recent times – i.e. via the Health and Safety at Work etc. Act 1974 and the Safety Representatives and Safety Committees Regulations 1977.

Before a closer examination of principles of action necessary to prevent accidents is undertaken, there is a need to examine more closely what is meant by an accident.

9.1.2 What is an accident?

To start, consider the following axiomatic statements:

(1) All accidents are incidents.
(2) All incidents are *not* accidents.
(3) All injuries result from accidents.
(4) All accidents do *not* result in injury.

An early definition was propounded by Lord MacNaughton in the case of *Fenton* v. *Thorley & Co. Ltd* (1903) AC 443 where he defined an accident as 'some concrete happening which intervenes or obtrudes itself upon the normal course of employment. It has the ordinary everyday meaning of an unlooked-for mishap or an untoward event, which is not expected or designed by the victim'.

This definition refers to an event occurring to a worker that was an unlooked-for mishap having a degree of unexpectedness about it. However, taking into account the axiomatic statements above, this definition would seem to be somewhat narrow, as it is only concerned with accidents resulting in injury to employees.

From research of some 40 accident definitions from general, legal, medical, scientific and safety literature, it appears that the ideal accident definition should have two distinct sections: a description of the causes, and a description of the effects.

Causes should include: unexpectedness or unplanned events, multi-causality and sequence of events; while the effects should cover: injury, disease, damage, near-miss and loss.

Based on the research, the following definition is suggested: 'an accident is an unexpected, unplanned event in a sequence of events, that occurs through a combination of causes; it results in physical harm (injury or disease) to an individual, damage to property, a near-miss, a loss, or any combination of these effects'.

This definition requires recognition of a wider range of accidents than those resulting in injury.

9.2 Definitions of hazard, risk and danger

The HSE leaflet *Hazard and Risk Explained*[2] presents the definitions of 'hazard' and 'risk' in relation to the COSHH Regulations:

9.2.1 Hazard

The *hazard* presented by a substance is its potential to cause harm. Hazard is associated with degrees of danger, and is quantifiable.

9.2.2 Risk

The *risk* from a substance is the likelihood that it will cause harm in the actual circumstances of use. This will depend on: the hazard presented by the substance; how it is used; how it is controlled; who is exposed . . . to how much . . . for how long. Risk should be thought of in terms of 'chance-taking'. What are the odds – the probability – of an accident occurring? Risk can be taken after careful consideration, or out of ignorance. The result can be fortuitous or disastrous, or anything in between.

The link between '*hazard*' and '*risk*' must be understood. In terms of the COSHH Regulations, poor control can create substantial *risk* even from a substance with low *hazard*. But with proper controls, the risk of being harmed by even the most hazardous substance is greatly reduced.

Also, the Approved Code of Practice[3] for the Management of Health and Safety at Work Regulations 1992 clearly states:

(a) a hazard is something with the potential to cause harm (this can include substances or machines, methods of work and other aspects of work organisation);
(b) risk expresses the likelihood that the harm from a particular hazard is realised;
(c) the extent of the risk covers the population which might be affected by a risk; i.e. the number of people who might be exposed and the consequences for them.

Risk, therefore, reflects both the likelihood that harm will occur and its severity. Hence these factors should be taken into account when undertaking either qualitative or quantitative risk assessment.

Danger can be associated with situations where there is a distinct possibility of:

(a) Interchanges of energy above tolerance levels. Such interchanges can be equated to any form of matter, including animals, vegetables and inert objects. The interchanges in energy can be in the form of physical, chemical, biological or psychological energy.

    An example of an interchange of energy occurs when an employee is trapped by a moving part of a machine, e.g. a power press. If the press is inadequately guarded, and the employee is able to get his hand into the danger area, injury results as the energy interchange is above the tolerance level of his hand.
(b) An organisation's financial well-being being placed at risk because of deficiencies in management; deficiencies in design and/or production capabilities; deficiencies in product quality/conformance; inability to expand and/or change; lack of adequate human resources; poor financial stability; lack of market appreciation/penetration; lack of awareness of cultural/social responsibilities; and failure to meet legal obligations.

    An example of a risk to an organisation's wellbeing is the serving by an enforcement officer (e.g. HSE Inspector) of a prohibition notice when a risk of imminent danger exists. The prohibition notice stops the risky process, machine, department, or – in some cases – the whole business. As it is not possible to insure against the consequential financial loss arising from such a stop notice, the loss must be financed from within the organisation's profitability. If profit margins are tight, then an overall loss situation may result, thus threatening the overall financial viability of the organisation as a whole.

9.3 Risk management

Risk management may be defined as the eradication or minimisation of the adverse effects of the pure risks to which an organisation is exposed.

Pure risks can only result in a loss to the organisation, whereas with speculative risks, either gain or loss may result.

An example of a pure – or static – risk concerns a build-up of combustible material in the corner of a large distribution warehouse. If a source of ignition is present in the vicinity, then the risk of fire spread is greatly enhanced by the build-up of combustible material, thus posing the threat of a large loss of stock caused by fire. There will also be consequential loss resulting from the fire to consider, e.g. loss of profit on goods in stock; loss of market share etc.

An example of a speculative – or dynamic – risk concerns commodity purchasing. A company – speculating to accumulate – buys in quantities of a key raw material at price £x/tonne, as the price is favourable, hoping that a cost saving will be made, as the price is likely to increase in the future. The speculative risk may result in gain or loss, as the price of the raw material could continue to fall, after the purchase price has been agreed with the supplier. Alternatively, the price may rise to £(x + y)/tonne, thus achieving the anticipated saving.

It should be borne in mind that the division between 'pure' and 'speculative' risks is not absolute. For example, the speculative risk of operating a machine without an adequate guard has a number of pure risks associated with it – injury; death; prohibition notices; prosecution; fines; loss of profit etc.

The principles of a risk management programme are: risk identification, risk evaluation and risk control. These three principles have been enshrined in recent health and safety legislation – e.g. COSHH, Noise at Work, Lead and Asbestos – and have brought risk management strategies and legislative compliance together.

Indeed, the Management of Health and Safety at Work Regulations 1992[4] have now ensured that risk identification, evaluation – i.e. assessment – and control become the cornerstone of all health and safety management systems.

9.3.1 Risk management – role and process

The role[5] of risk management in industry and commerce is to:

(1) consider the impact of certain risky events on the performance of the organisation;
(2) devise alternative strategies for controlling these risks and/or their impact on the organisation; and
(3) relate these alternative strategies to the general decision framework used by the organisation.

The process of risk management involves: identification, evaluation and control.

Risk identification may be achieved by a multiplicity of techniques, including physical inspections, management and worker discussions, safety audits, job safety analysis, and Hazop studies. The study of past accidents can also identify areas of high risk.

Risk evaluation (or measurement) may be based on economic, social or legal considerations.

Economic considerations should include the financial impact on the organisation of the uninsured cost of accidents, the effect on insurance premiums, and the overall effect on the profitability of the organisation and the possible loss of production following the issue of Improvement and Prohibition Notices.

Social and humanitarian considerations should include the general wellbeing of employees, the interaction with the general public who either live near the organisation's premises or come into contact with the organisation's operations – e.g. transportation, nuisance noise, effluent discharges etc. – and the consumers of the organisation's products or services, who ultimately keep the organisation in business.

Legal considerations should include possible constraints from compliance with health and safety legislation, codes of practice, guidance notes and accepted standards, plus other relevant legislation concerning fire prevention, pollution, and product liability.

The probability and frequency of each occurrence, and the severity of the outcome – including an estimation of the maximum potential loss – will also need to be incorporated into any meaningful evaluation.

9.3.2 Risk control strategies

Risk control strategies may be classified into four main areas: risk avoidance, risk retention, risk transfer and risk reduction.

(1) Risk avoidance

This strategy involves a conscious decision on the part of the organisation to avoid completely a particular risk by discontinuing the operation producing the risk and it presupposes that the risk has been identified and evaluated.

For example, a decision may be made, subject to employees' agreement, to pay all wages by cheque or credit transfer, thus obviating the need to have large amounts of cash on the premises and the inherent risk of a wages snatch.

Another example of a risk avoidance strategy – from the health and safety field – would be the decision to replace a hazardous chemical by one with less or no risk potential.

(2) Risk retention

The risk is retained in the organisation where any consequent loss is financed by the company. There are two aspects to consider under this heading: risk retention with knowledge, and risk retention without knowledge.

(a) With knowledge

This covers the case where a conscious decision is made to meet any resulting loss from within the organisation's financial resources. Decisions on which risks to retain can only be made once all the risks have been identified and effectively evaluated.

(b) Without knowledge
Risk retention without knowledge usually results from lack of knowledge of the existence of a risk or an omission to insure against it, and this often arises because the risks have not been either identified or fully evaluated.

(3) Risk transfer
Risk transfer refers to the legal assignment of the costs of certain potential losses from one party to another. The most common way of effecting such transfer is by insurance. Under an insurance policy, the insurer (insurance company) undertakes to compensate the insured (organisation) against losses resulting from the occurrence of an event specified in the insurance policy (e.g. fire, accident, etc.).

The introduction of clauses into sales agreements whereby another party accepts responsibility for the costs of a particular loss is an alternative risk transfer strategy. However, it should be noted that the conditions of the agreement may be affected by the Unfair Contract Terms Act 1977 and the interpretation placed on 'reasonableness'.

(4) Risk reduction
The principles of risk reduction rely on the reduction of risk within the organisation by the implementation of a loss control programme, whose basic aim is to protect the company's assets from wastage caused by accidental loss.

The collection of data on as many loss producing accidents as possible provides information on which an effective programme of remedial action can be based. This process will involve the investigation, reporting and recording of accidents that result in either injury or disease to an individual, damage to property, plant, equipment,

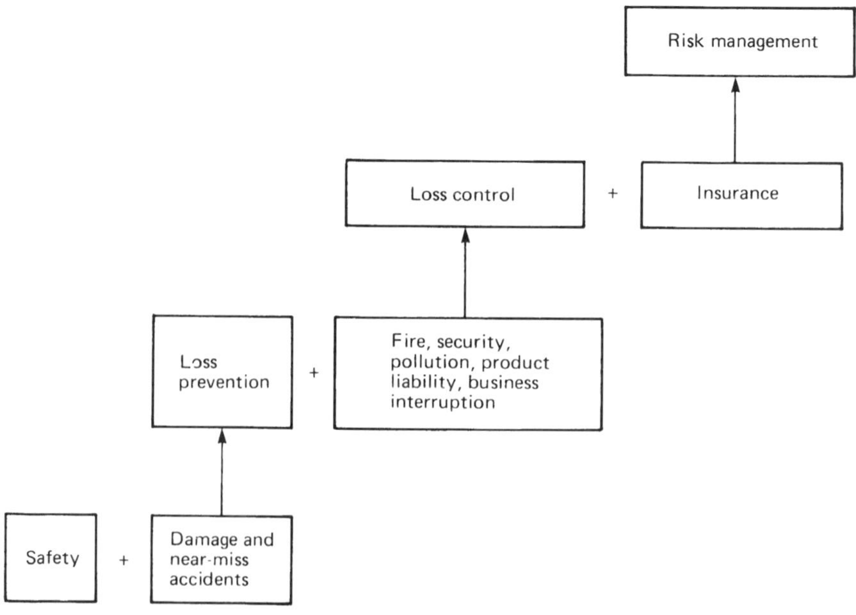

Figure 9.1 Pictogram showing development from safety to risk management

materials, or the product; or those near-misses where although there has been no injury, disease or damage, the risk potential was high.

The second stage of the development towards risk reduction is achieved by bringing together all areas where losses arise from accidents – whether fire, security, pollution, product liability, business interruption etc. – and co-ordinating action with the aim of reducing the loss. This risk reduction strategy is synonymous with loss control.

Loss control (or risk reduction) may be defined as a management system designed to reduce or eliminate all aspects of accidental loss that lead to a wastage of an organisation's assets.

As the emphasis on the economic argument increased, the technique of loss control has become more closely allied to financial matters, and in particular insurance.

The bringing together of insurance (risk transfer) and loss control (risk reduction) was the final stage in the development of the new discipline of risk management.

This logical, progressive development from safety to risk management may be presented in pictogram form as in *Figure 9.1*.

9.4 Loss control

9.4.1 Introduction

Loss control may be defined as a management system designed to reduce or eliminate all aspects of accidental loss that may lead to a wastage of the organisation's assets. Those assets include manpower, materials, machinery, methods, manufactured goods and money. Loss control is based mainly on the economic approach to accident prevention, and loss control management is essentially the application of sound management techniques to the identification, evaluation and economic control of losses within a business. It has been shown in section 9.3.2 that loss control is synonymous with risk reduction, so the practical techniques associated with each stage of the process – i.e. identification, evaluation and control – are closely related.

Bird and Loftus[6] state that loss control management involves the following:

(1) The identification of risk exposure.
(2) The measurement and analysis of exposures.
(3) The determination of exposures that will respond to treatment by existing or available loss control techniques or activities.
(4) The selection of appropriate loss control action based on effectiveness and economic feasibility.
(5) The managing of the loss control programme implementation in the most effective manner subject to economic constraints.

9.4.2 Component parts of a loss control programme

The component parts of a loss control programme may be considered in terms of protecting one or more of the organisation's assets from accidental loss, and will generally include: injury prevention (safety); damage control; fire prevention; security; industrial health and hygiene; pollution; product liability; and business interruption.

Injury prevention is concerned directly with the protection of the manpower asset within an organisation. To a lesser extent, it is indirectly concerned with the protection of the money asset, as a reduction in the number of injuries should result in a reduction in both the insured and the uninsured accident costs. In certain injury prevention programmes, the protection of the manpower asset is extended outside the factory, via the inclusion of off-the-job safety, road safety and home safety.

The following specific areas need to be examined in an injury prevention programme: safety policies; safety training; safety audits; identification of hazards; accident reporting and investigation; safe systems of work; machine and area guarding; housekeeping; personal protective equipment; and legislative compliance.

Damage control is directly concerned with the protection of assets comprising machinery, materials and manufactured goods from accidental loss before they reach the customer. Indirectly, this leads to the protection of the money asset through the elimination of repetitive damage and associated repair/replacement costs. Also, there may be some indirect protection of the manpower asset if damage causes and injury causes are similar. Essentially, damage control is an extension of the injury reporting and prevention programme to encompass also those accidents which result in damage only to plant, property, equipment and materials. The practices and techniques of damage control are considered in depth by Bird and Germain[7], and Bird and Loftus[8].

Incident recall[9,10] is another technique that can be utilised in a damage control programme to gain information about near-miss accidents.

Essentially, the incident recall technique may be used to identify unsafe acts, unsafe conditions, non-compliance with safe systems of work, and near-miss accidents by following a *confidential* interviewing procedure to a stratified random sample of employees. Each interviewee is asked to recall and report verbally any of the above-mentioned situations in which he was involved or has knowledge. Details of near-miss accidents are then obtained to enable remedial action to be taken *before* further similar accidents result in both damage and injury.

Fire prevention may be considered to be a special aspect of damage control in that it protects the machinery, materials and manufactured goods assets. It also protects the manpower asset, since fire can cause injury as well as damage and because fire damage is a very costly item, it is indirectly protecting the money asset as well. The Association of British Insurers has stated that the total cost of fire insurance claims in the UK for 1990 was £1005m, of which £450m was due to arson.

From a practical viewpoint, consideration should be given to aspects of fire prevention techniques, methods of fire control, firefighting and extinguishment, fire protection including fixed equipment (e.g. sprinklers etc.), storage and handling of flammable liquids, fire safety of employees, means of escape, evacuation drills and procedures, explosion potential, handling, storage and use of explosives, electrical installations, disaster contingency planning and the requirements of the Fire Precautions Act 1971 and the Highly Flammable Liquids and Liquefied Petroleum Gases Regulations 1972.

Security protects the materials, methods, manufactured goods and money assets. Its inclusion in a loss control programme is primarily based on economic considerations, as any breaches of security that result in losses of the organisation's assets may not be considered by the organisation to be accidental in nature.

However, certain safety hazards arise because of lack of security: for example, the high risk involved when employees are sent to collect wages; or the potential risk to children and others when a factory or construction site is not physically secure in terms of preventing unauthorised access.

Hence, indirectly, a system of security can improve the overall safety of the organisation, as well as directly protecting the assets mentioned above. Security – in the form of locked doors – can however sometimes conflict with the protection of the manpower asset, especially if the locked doors are the emergency fire exits.

The loss control programme should examine such areas as the physical security of premises; cash collection, handling, and distribution; theft and pilfering; vandalism; storage of valuable and attractive items; sabotage; industrial espionage; and the control of confidential data and methods. Consideration should also be given to defensive techniques, such as stock-taking, accounting/auditing checks, and the use of mechanical/electrical safeguards. Aspects of computer security should also be included in this area, especially in the light of the increasing number of cases being reported that have involved some form of fraud, embezzlement or espionage utilising the computer.

Industrial health and hygiene is concerned with the protection of the manpower asset from the effects of occupational diseases – i.e. long term accidents and other adverse conditions associated with the industrial environment. Indirectly, this also protects the money asset, as an improvement in health and hygiene within a factory should lead to a reduction in the incidence of occupational diseases, and hence a reduction in the associated costs.

Specific areas for consideration include: noise, dusts, gases, vapours, corrosives, toxic materials, radioactive materials, ventilation, heating, lighting, humidity, environmental monitoring, biological monitoring, health checks, general and personal hygiene, health education, counselling, and employment/pre-employment medicals. Compliance with the COSHH Regulations is vitally important to ensure a healthy workforce.

Pollution in all its aspects is concerned not only with the environment within the factory, but also the environment outside and around the factory. Control of air, ground and water pollution protects the manpower asset directly and the money asset indirectly. Adverse publicity resulting from an organisation causing some form of pollution would initially harm the organisation's image, and perhaps harm it economically. Persistent breaches of one or other of the Acts dealing with pollution can ultimately lead to prosecution and fines with further adverse publicity, both at local and national level.

Special attention needs to be paid to the Control of Pollution Act 1974 that deals with the control of noise as a pollutant, in addition to the other areas of air, ground and water pollution. All possible areas and types of pollution should be identified as part of the loss control audit. With the advent of the Environmental Protection Act 1990[11] and BS 7750:1992: Environmental Management Systems[12], the environmental aspects of the loss control programme have increasingly gained in importance in recent years. Indeed, many health and safety practitioners now have the added responsibility to provide advice on environmental matters.

Product liability extends the protection to all consumers of the organisation's products or services, and is therefore primarily concerned with the protection of the money asset. This asset may suffer accidental losses directly because of increased insurance premiums that reflect large compensation payments, or indirectly because of adverse publicity which is detrimental to the organisation's image.

Practical considerations in this area should centre on the development of a product safety strategy that is in keeping with both occupational (section 6, Health and Safety at Work etc. Act 1974) and consumer (Consumer Protection Act 1987) product safety legislation, codes of practice and guidance notes.

Areas to be considered should include: products loss control policy; products loss control committee chaired by senior manager and comprising designers, R and D personnel, manufacturing and production management, safety advisers, quality control, servicing, sales and advertising, and distribution; product safety incorporated at the design and R and D stages; written quality control/assurance procedures i.e. BS 5750[13]; role of sales, marketing, advertising, distribution and servicing personnel in product safety; complaints system; and products recall system.

Business interruption further extends the loss control strategy to take account of the fact that time is money, and, as such, any loss of production or service is detrimental to the overall profitability of the company. Hence, business interruption is primarily concerned with the protection of the money asset. Indirectly, however, it serves to maintain the assets of machinery, materials, manufactured goods and methods.

A programme to prevent business interruption can include planned lubrication, planned preventive maintenance, condition monitoring, statutory inspections, machinery replacement programmes, availability of key spares, identification of key machines, processes, areas, personnel etc. within the organisation, continued supply of raw materials, minimisation of production bottlenecks, and highlighting dependencies on specific items of plant, suppliers, customers, personnel, and/or public utilities (e.g. gas, electricity).

9.4.3 Loss control management in practice

The aforementioned areas – from injury prevention to business interruption – require to be co-ordinated within one senior management function (possibly risk management) in order to ensure a rational and concerted approach to the problem of eliminating or reducing the costly accidental losses that can occur within an organisation. These so-called operational losses inevitably lead to an erosion of profit margins, and also adversely affect the overall performance of the organisation.

This co-ordinated management role is crucial to the success of any loss control programme. The senior manager responsible for programme implementation should have authority to make decisions and take action without the need to seek day-to-day approval for his decisions. He should report on a regular (monthly) basis to the main board on the implementation of the loss control programme within the organisation. Without the backing and commitment of the most senior executives, it is doubtful whether a programme can be successfully introduced.

An effective programme of loss control (risk reduction) not only leads to a more profitable situation, but will also greatly assist legislative compliance, and will result in a reduction in the total number of accidents within the organisation's operations.

9.5 Degrees of hazard

An awareness of the differing degrees of hazard to people will enable appropriate control measures to be developed and implemented.

Immediate physical danger can manifest itself through very short term injury accidents – e.g. hand amputation in a power press; person falling from a height. The result of immediate physical danger – if it goes uncontrolled – will inevitably be *immediate physical injury*. The enforcing agencies use the phrase 'risk of imminent danger' or 'risk of serious personal injury' in connection with the issuing of prohibition notices – a legal control measure designed to reduce the risk of immediate physical danger.

Long-term physical danger is more cumulative or chronic than acute or short-term. Cumulative back strain caused by poor kinetic handling techniques is an example of *long-term physical injury*.

Immediate chemical danger may be caused by strong acids and alkalis being poorly stored and handled, thus leading to a risk of skin contact and corrosive burns – i.e. *immediate chemical injury*.

Long-term chemical danger is again chronic or cumulative – e.g. lead poisoning or exposure to asbestos fibres. The result is some form of occupational disease – i.e. *long-term chemical injury*.

Immediate biological danger may be caused by the presence of contagious diseases or via genetic manipulation. The result is again some form of occupation disease or illness.

Long-term biological danger is usually cumulative in nature, for example, noise-induced occupational deafness.

Immediate psychological danger is linked to short-term trauma – e.g. a disaster at home or work; social problems – domestic illness, etc. This may result in a loss of concentration, abruptness with work colleagues, and other short-term stress-related symptoms.

Long-term psychological danger may be linked to fears connected with fear of failure, unemployment/job security, or lack of career direction and motivation. The symptoms are similar to those described above, but often only become apparent over a longer timescale.

9.6 Accident causation models

9.6.1 Sequence of events – Domino Theory

The 'Domino Theory' attributed to Heinrich[14] is based on the theory that a chain or sequence of events can be listed in chronological order to show the events leading up to an accident:

event a → event b → event c → accident → effect

Each event may have more than one cause, i.e. be multicausal.

Heinrich states that the occurrence of an injury accident invariably results from a completed sequence of factors culminating in the accident itself. He postulates five factors or stages in the accident sequence, with the injury invariably caused by the accident, and the accident in turn the result of the factor that immediately precedes it.

The five factors or stages in the sequence of events are:

(a) ancestry and social environment, leading to
(b) fault of person, constituting the proximate reason for
(c) an unsafe act and/or mechanical hazard, which results in
(d) the accident, which leads to
(e) the injury.

Heinrich likens these five stages to five dominoes standing on edge in a line next to each other, so that when the first domino falls it automatically knocks down its neighbour which in turn knocks down its neighbour and so on. Removal of any one of the first four will break the sequence and so prevent the injury.

In fact, Heinrich suggested that accident prevention should aim to remove or eliminate the middle or third domino, representing the unsafe act, mechanical or physical hazard, thus preventing the accident.

During accident investigations, in addition to asking 'What action has been taken to prevent recurrence?', the investigator needs to be aware of the chain of events leading up to the accident, and tracing it back. Similarly, on safety audits and inspections, when the risk of an accident has been identified, possible event chains should be investigated and action taken to remove potential causes.

9.6.2 An updated domino sequence

Bird and Loftus[15] have extended this theory to reflect the influence of management in the cause and effect of all accidents that result in a wastage of the company's assets. The modified sequence of events becomes:

(a) lack of control by management, permitting
(b) basic causes (personal and job factors), that lead to
(c) immediate causes (substandard practices/conditions/errors), which are the proximate causes of
(d) the accident, which results in
(e) the loss (minor, serious or catastrophic).

This modified sequence can be applied to all accidents, and is fundamental to loss control management.

9.6.3 Multiple causation theory

Multicausality refers to the fact that there may be more than one cause to any accident:

cause a ↘
cause b → accident
cause c ↗

Each of these multicauses is equivalent to the third domino in the Heinrich theory and can represent an unsafe act or condition or situation. Each of these can itself have multicauses and the process during accident investigation of following each branch back to its root is known as 'fault tree analysis'.

The theory of multicausation is that the contributing causes combine together in a random fashion to result in an accident. During accident investigations, there is a need to identify as many of these causes as possible. In reality, the accident model is an amalgam of both the domino and multicausality theories.

Petersen has compared and contrasted both theories and gives an example[16] which illustrates the comparative narrowness of the domino theory in relation to the multicausality theory and concludes that this has severely limited the identification and control of the underlying causes of accidents.

The theory of multicausality has its basis in epidemiology. Gordon[17] points out that accidental injuries could be considered with epidemiological techniques. He believes that if the characteristics of the 'host' (accident victim), of the agent (the injury deliverer), and of the supporting 'environment' could be described in detail, more understanding of accident causes could be achieved than by following the domino technique of looking for a single cause only. Essentially, Gordon's theory is that the accident is the result of a complex and random interaction between the host, the agent and the environment, and cannot be explained by consideration of only one of the three.

9.6.4 Failure modes and effects

This technique involves a sequential analysis and evaluation of the kinds of failures that could happen, and their likely effects, expressed in terms of maximum potential loss.

The technique is used as a predictive model and would form part of an overall risk assessment study.

9.6.5 Fault tree analysis

Fault tree analysis is an analytical technique that is used to trace the chronological progression of factors (events) contributing to the accident situation, and is useful in accident investigation and as a predictive, quantitative model in risk assessment. Again, the principle of multicausality is utilised in this type of analysis. (A fuller treatment on fault tree analysis is given at section 10.6).

9.7 Accident prevention: legal, humanitarian and economic reasons for action

9.7.1 Introduction

In order to get action taken in the field of accident prevention, safety advisers have the three fundamental lines of attack on which to base their strategies for generating and maintaining management activity in this area. These three reasons for accident prevention make use of the legal, humanitarian and economic arguments respectively. An optimum accident prevention strategy for a particular organisation would involve a combination of the three, because they are interrelated and probably reinforce one another.

9.7.2 Legal reasons for accident prevention

The legal argument is based on the statutory requirements of the HSW, FA and other related legislation.

The HSW imposes a general duty on employers to ensure, so far as is reasonably practicable, the health, safety and welfare of all his employees. The term 'reasonably practicable' involves balancing the cost of preventing the accident against the risk of the accident occurring. Thus, economic considerations need also to be taken into account.

The Factories Act lays down more specific statutory requirements which impose a minimum but absolute standard of conduct on the employer.

Any breach of the statutory duties imposed by either of the aforementioned Acts can result in the employer being involved in criminal proceedings. The penalties under the Health and Safety at Work Act include unlimited fines and imprisonment for up to two years, for prosecution on indictment. On average, twenty directors, managers, supervisors, employees are individually prosecuted per annum. No individual has yet been imprisoned, but suspended sentences – of up to one year – have been passed. The maximum fine, on summary conviction, for certain offences is now £20 000, with the maximum for all other offences being £5000.

The safety adviser can therefore reason via the legal argument for accident prevention on the basis that the employer should avoid attracting a prosecution.

The economic argument is also relevant here, because of the fines that may be imposed as a result of statutory breaches, and also because of the impact of Improvement and Prohibition Notices in terms of uninsurable consequential loss arising out of enforced cessation of work.

The image of the company or organisation is also likely to be tarnished as a result of adverse publicity received in connection with any prosecution for breaches of statute. Loss of company image has predominantly economic disadvantages, usually because of the loss of good will or other intangible and invisible company assets, which in turn indirectly leads to a loss of business.

9.7.3 Humanitarian reasons for accident prevention

The humanitarian reason for accident prevention is based on the notion that it is the duty of any man to ensure the general wellbeing of his fellow men. This places an onus on the employer – the common law duty of care – to provide a safe and healthy working environment for all his employees.

An illustration of this occurs in the case of *Wilsons and Clyde Coal Co. Ltd* v. *English*[18], where Lord Wright said that 'the whole course of authority consistently recognises a duty which rests on the employer, and which is personal to the employer, to take reasonable care for the safety of his workmen, whether the employer be an individual, a firm or a company and whether or not the employer takes any share in the conduct of the operations'.

There is some overlap here between common and statute law, as the Health and Safety at Work Act places a general duty on an employer to ensure, so far as is reasonably practicable, the health, safety and welfare of his employees.

The safety adviser is therefore able to argue – via humanitarian reasoning – that it is immoral to have a process or machine which may injure employees, and he can stress the possible outcome of such dangers in terms of pain and suffering.

9.7.4 Economic reasons for accident prevention

The fundamental reason for utilising the economic argument in the promotion of accident prevention is the fact that accidents cost an organisation money. However, in order to press the economic argument, knowledge is needed of the costs to the organisation of all types of accident.

Essentially, there are two types of accident costs – the insured costs, and the uninsured costs.

The insured (or direct) costs are predominantly covered by the Employer's Liability Insurance premium, which to all intents and purposes is the direct accident cost to the majority of organisations.

The uninsured (or indirect, hidden) costs of accidents should also be established. Bamber[19] developed a list of uninsured costs which is considered to be objective, and which will readily be accepted by operational management as being costs associated with accidents:

(1) Safety administration and accident investigation.
(2) Medical and treatment.
(3) Cost of lost time of injured person.
(4) Cost of lost time of other employees.
(5) Cost of replacement labour.
(6) Cost of payments to injured person.
(7) Cost of loss of production and business interruption.
(8) Cost of repair to damaged plant.
(9) Cost of replacement of damaged materials.
(10) Other costs – e.g. photographs, transport, accommodation, wage details, fees, etc.

The above list of costs should be utilised in the calculation of the total accident costs to the organisation, to enable senior management to gauge the relative impact of such costs, by comparing them with other business costs.

The safety adviser is therefore able to reason – via the economic argument – that accident prevention may well be cost-effective. But the organisation is reducing pain and suffering by having an effective system of accident prevention, as well as saving money. Thus, the economic argument gives support to both the legal – via the use of economic sanctions – and the humanitarian arguments. In order to achieve maximum co-operation in any programme of accident prevention, use should be made of an amalgam of all three arguments, i.e. legal, humanitarian and economic. However, from a motivational point of view, it is the economic argument that has the greatest impact with directors and senior management.

References

1. Ministry of Labour and National Service, *Industrial Accident Prevention*, Report of the Industrial Safety Sub-Committee of the National Joint Advisory Council (1956)
2. Health and Safety Executive, *Hazard and Risk Explained – Control of Substances Hazardous to Health Regulations 1988* (COSHH), Leaflet No. IND(G)67(L), HSE Books, Sudbury (1988)
3. Health and Safety Commission, *Management of Health and Safety at Work Regulations 1992: Approved Code of Practice*, 3, HSE Books, Sudbury (1992)
4. *Management of Health and Safety at Work Regulations 1992*, HMSO, London (1992)
5. Carter, R.L. and Doherty, N., *Handbook of Risk Management*, 1.1–06, Kluwer-Harrap, London (1974)
6. Bird, F.E. and Loftus, R.G., *Loss Control Management*, 52, Institute Press, Loganville, Georgia (1976)
7. Bird, F.E. and Germain, G.L., *Damage Control*, American Management Association, New York (1966)
8. Ref. 6, pp. 93–138
9. Ref. 6, pp. 215–246
10. Bamber, L., Incident recall – a (lack of) progress report, *Health and Safety at Work*, **2**, No. 9, 83 (1980)
11. *The Environmental Protection Act 1990*, HMSO, London (1990)

12. British Standards Institution, *BS 7750:1992 Specification for Environmental Management Systems*, BSI, London (1992)
13. BS 5750: Parts 1–6:1987, *Quality systems*, British Standards Institution, London
14. Heinrich, H.W., *Industrial Accident Prevention*, 4th edn, 13–16, McGraw-Hill, New York (1959)
15. Ref. 6, pp. 39–48
16. Petersen, D.C., *Techniques of Safety Management*, 2nd edn, 16–19, McGraw-Hill, Kogakusha, USA (1978)
17. Gordon, J.E., The epidemiology of accidents, *Amer. J. of Public Health*, **39**, 504–515 (1949)
18. Wilsons and Clyde Coal Co. Ltd *v.* English, (1938) AC 57 (HL)
19. Bamber, L., Accident prevention the economic argument, *Occupational Safety and Health*, **9**, No. 6, 18–21 (1979)

Further reading

Heinrich, H. W., Petersen, D. and Roos, N. *Industrial Accident Prevention – A Safety Management Approach*, 5th edn, McGraw-Hill, New York (1980)

DeReamer, R., *Modern Safety Practices*, John Wiley & Sons Inc., New York (1958)

Bird, F. E. and Loftus, R. G., *Loss Control Management*, Institute Press, Loganville, Georgia (1976)

Petersen, D. C., *Techniques of Safety Management*, 2nd edn, McGraw-Hill, Kogakusha, USA (1978)

Hale, A. R. and Hale, M., *A Review of the Industrial Accident Research Literature*, Committee on Safety and Health at Work: Research Paper, HMSO, London (1972)

Crockford, G. N., *An Introduction to Risk Management*, Woodhead-Faulkner, Cambridge (1980)

Carter, R. L. *et al.*, *Handbook of Risk Management*, Kluwer Publishing Ltd, Kingston-upon-Thames (1992)

Chapter 10

Risk management: techniques and practices

L. Bamber

10.1 Risk identification, assessment and control

10.1.1 Introduction

As discussed in section 9.2.2, the risk from a hazard is the likelihood that it will cause harm in the actual circumstances in which it exists.

Essentially, the technique of risk management involves:

(1) identification
(2) assessment
(3) control (elimination or reduction)

Within the workplace, operational management at all levels has a responsibility to identify, evaluate and control risks that are likely to result in injury, damage or loss. Part of these responsibilities should involve implementation of a regular programme of safety inspections of the work areas under their control. These inspections should include physical examinations of the workplace – i.e. the nuts and bolts – and also the systems, procedures, and work methods – i.e. the organisational aspects.

The process of risk management has been briefly outlined in section 9.3.1. The following sections (10.1.2–10.1.4) consider the practical application of the techniques in the workplace.

10.1.2 Risk indentification

Within an organisation, there are several ways by which risks may be identified. These include:

(1) Workplace inspections.
(2) Management/worker discussions.
(3) Independent audits.
(4) Job safety analysis.
(5) Hazard and operability studies.
(6) Accident statistics.

Workplace inspections are undertaken with the aim of identifying risks and promoting remedial action. Many different individuals and groups within an organisation will – at some time – be involved in a workplace inspection: directors, line managers, safety adviser, supervisors and safety representatives. The key aspect is that results of all such inspections should be co-ordinated by one person within the factory, whose responsibility should include (a) monitoring action taken once the risk has been notified, and (b) informing those persons who reported the risk as to what action has been taken.

The vast majority of workplace inspections concentrate on the 'safe place' approach – i.e. the identification of unsafe conditions – to the detriment of the 'safe person' approach – i.e. the identification of unsafe acts.

Heinrich[25] states that only 10% of accidents are caused by unsafe mechanical and physical conditions, whereas 88% of accidents are caused by unsafe acts of persons. (The other 2% are classed as unpreventable, or acts of God!).

Hence for workplace inspections to be beneficial in terms of risk identification and accident prevention, emphasis *must* be placed on the positive safe person approach, using techniques such as:

* safe visiting – talking to people
* catching people doing something right (not wrong)
* positive behavioural reinforcement
* one-to-one training/counselling sessions,

as well as the more traditional safe place approach which tends to be more negative as it evokes fault finding and blame apportionment at all levels within an organisation – i.e. catching people doing something wrong and penalising them for it.

Workplace inspections tend to follow the same format but are given many different names including: safety sampling, safety audits, safety inspections, hazard surveys, etc. Certain of the above are discussed below but all have the same aim – namely risk identification.

Management/worker discussions can also be useful in the identification of risks. Formal discussions take place during meetings of the safety committee with informal discussions occurring during on-the-job contact or in conversations between supervisor and worker. The concept of incident recall[1,2] is an example of management/worker discussion.

In all cases, however, the feedback element is important from a motivational viewpoint. The risk identifier must be kept fully informed of any action taken to prevent injury, damage or loss arising from the risk he has noted.

Independent audits can also be used to identify risks. The term 'independent' here refers to those who are not employees of the organisation, but who – from time to time – undertake either general or specific workplace audits or inspections. Such independent persons may include:

(1) Engineer surveyors – insurance company personnel undertaking statutory inspections of boilers, pressure vessels, lifting tackle etc. They are employed by the organisation as 'competent persons'.

(2) Employers' liability surveyors – insurance company personnel undertaking general health and safety inspections in connection with employers' liability insurance.
(3) Claims investigators – insurance company personnel investigating either accidents in connection with injury or damage claims under insurance policies.
(4) Insurance brokers personnel – risk management or technical consultants undertaking inspections in connection with health and safety, fire, or engineering insurance as part of client servicing.
(5) Outside consultants – undertaking specific investigations on a fee-paying basis. For example, noise or environmental surveys may be commissioned, if the expertise is not available within the organisation. Trade associations may be of assistance in this area.
(6) Health and Safety Executive – factory (and other) inspectors undertaking either general surveys or specific accident investigations.

Again, with all the above there is a need to co-ordinate their independent findings to ensure that action is promptly taken to control any risks identified.

Job safety analysis is another method of risk identification. A fuller discussion of this method is presented below (see 10.2.1).

Hazard and operability studies are useful as a risk identification technique, especially in connection with new designs/processes. The technique was developed in the chemical process industries, and essentially it is a structured, multi-disciplinary brainstorming session involving chemists, engineers, production management, safety advisers, designers etc. critically examining each stage of the design/process by asking a series of 'what if?' questions. The prime aim is to design out risk at the early stages of a new project, rather than have to enter into costly modifications once the process is up and running.

Further information on Hazop studies may be found in the Chemical Industries Association's publication on the subject[3].

Accident statistics will be useful in identifying uncontrolled risks as they will present – if properly analysed from a causal viewpoint – data indicative of where control action should have been taken to prevent recurrence. Ideally, an analysis of *all* injury, damage and near-miss accidents should be undertaken, so that underlying trends may be highlighted and effective control action – both organisational and physical in nature – taken.

10.1.3 Risk assessment

Once a list of risks within a company has been compiled, the impact of each risk on the organisation – assuming no control action has been taken – requires assessment, so that the risks may be put in order of priority in terms of when control action is actually required, i.e. immediate; short term; medium term; long term on the basis of a ranking of the risks relating to their relative impact on the organisation. Such an assessment should take account of legal, humanitarian and economic considerations (as outlined in section 9.7).

The fundamental equation in any risk assessment exercise is:

Risk magnitude = Frequency (how often?) × Consequence (how big?)

In general:

- Low-frequency, low-consequence risks should be retained (i.e. self-financed) within the organisation. Examples include the failure of small electric motors, plate-glass breakages, and possibly motor vehicle damage accidents (via retention of comprehensive aspects of insurance cover).
- Low-frequency, high-consequence risks should be transferred (usually via insurance contracts). Examples include explosions, and environmental impairment.
- High-frequency, low-consequence risks should be reduced via effective loss control management. Examples include minor injury accidents; pilfering; and damage accidents.
- High-frequency, high-consequence risks should (ideally) be avoided by managing them out of the organisation's risks portfolio. If this appears to be an uneconomic (or unpalatable) solution, then adequate insurance – i.e. the risk transfer option – *must* be arranged.

A quantitative method of risk assessment – which takes into account the risk magnitude equation discussed above – considers the frequency (number of times spotted); the maximum potential loss (MPL) – i.e. the severity of the worst possible outcome; and the probability that the risk will actually come to fruition and result in a loss to the organisation.

From this type of quantitative assessment, a list of priorities for risk control can be established, and used as a basis to allocate resources.

A simple risk assessment formula involving frequency, MPL and probability is:

Risk rating = Frequency × (MPL + probability)

In the above formula, frequency (F) is the number of times that a risk has been identified during a safety inspection.

Maximum Potential Loss (MPL) is rated on a 50-point scale where, for example:

multifatality	– 50
single fatality	– 45
total disablement (para/quadraplegic)	– 40
loss of eye	– 35
arm/leg amputation	– 30
hand/foot amputation	– 25
loss of hearing	– 20
broken/fractured limb	– 15
deep laceration	– 10
bruising	– 5
scratch	– 1

Probability (P) is rated on a 50-point scale where, for example:

imminent	– 50
hourly	– 35
daily	– 25
once per week	– 15
once per month	– 10
once per year	– 5
once per five or more years	– 1

Consider an example where the risk to be assessed has been identified once during an inspection. The MPL (worst possible outcome) was considered to be the loss of an eye with the probability of occurrence of once per day.

Thus, for this risk, the rate is:

$$\begin{aligned} RR &= F \times (MPL + P) \\ &= 1 \times (35 + 25) \\ &= 60 \end{aligned}$$

This risk rating figure should then be compared to a previously agreed risk control action guide, such as:

Risk rating	*Urgency of action*
Over 100	Immediate
80–100	Today
60–79	Within 2 days
40–59	Within 4 days
20–39	Within 1 week
10–19	Within 1 month
0–9	Within 3 months

These action scales should be drawn up by individual organisations, taking into account both the human and financial resources available for risk control.

In our example, the risk rating was found to be 60, hence control action to eliminate (or reduce) the risk should be taken within two days.

The above scales, example and action guide serve only to illustrate the principles involved, and – because of resource constraints – may not be generally applicable for practical use in all organisations.

10.1.4 Risk control

The four risk control strategies – avoidance, retention, transfer and reduction – have been discussed in section 9.3.2.

The bulk of the risks identified by regular safety inspections will require some form of risk reduction (or avoidance) through effective loss control management.

The control of risks within an organisation requires careful planning, and its achievement will involve both short-term (temporary) and long-term (permanent) measures.

These measures can be graded thus:

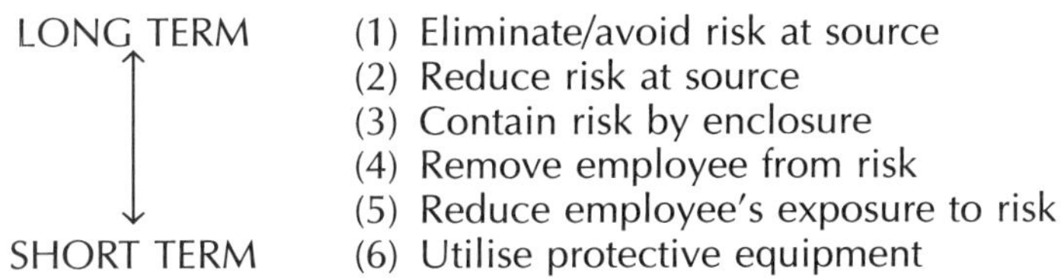

The long-term aim must always be to eliminate the hazard at source, but, whilst attempting to achieve this aim, other short-term actions – for example utilisation of protective equipment – will be necessary. This list indicates an 'order-of' priority for remedial measures for any risk situation.

Various techniques are available to control risks within the workplace.

Mechanical risks may be engineered out of the process, or effectively enclosed by means of fixed guarding. Alternative forms of guarding involve the use of interlocked guards, light-sensitive barriers or pressure-sensitive mats. Trip devices and other forms of emergency stops may also be incorporated.

Risks from the working environment may be controlled by effective ventilation systems, adequate heating and lighting, and the general provision of good working conditions.

Chemical risks may also be controlled by effective ventilation, regular monitoring, substitution of material, change of process, purchasing controls, and the use of protective equipment.

A necessary corollary of risk assessment is the establishment of safe systems of work and training for the work-force to make them aware of the risks in their work areas, and of the methods for the control of such risks.

10.2 Job safety analysis

10.2.1 Job safety analysis – procedure

Job safety analysis (or job hazard analysis) is an accident prevention technique that should be used in conjunction with the development of job safety instructions; safe systems of work; and job safety training.

The technique of job safety analysis (JSA) has evolved from the work study techniques known as method study and work measurement.

The method study engineers' aim is to improve methods of production. In this they use a technique known as the SREDIM principle:

*S*elect (work to be studied);
*R*ecord (how work is done);
*E*xamine (the total situation);
*D*evelop (best method for doing work);
*I*nstall (this method into the company's operations);
*M*aintain (this defined and measured method).

Work measurement is utilised to break the job down into its component parts and, by measuring the quantity of work in each of the component parts, make human effort more effective. From experience standard times have evolved for particular component operations and these enable jobs to be given a 'time'.

Job Safety Analysis Record Chart		
Job title:		Date of job analysis:
Department:		Time of job observation:
Analyst/reviewer:		
Description of job:		
Accident experience:		
Maximum potential loss:		
Legal requirements:		
Relevant codes of practice/guidance notes/advisory publications:		
Sequence of job steps	Risks identified	Precautions advised
Suggested safe system of work:		
Suggested review date:		
Suggested job safety instructions:		
Suggested training programme:		
Signed:		Date:
Department:		Function:

Figure 10.1 Job safety analysis record chart

Job safety analysis uses the SREDIM principle but measures the risk (rather than the work content) in each of the component parts of the job under review. From this detailed examination a safe method for carrying out each stage of the job can be developed.

The basic procedure for job safety analysis is as follows:

(1) Select the job to be analysed. (SELECT)
(2) Break the job down into its component parts in an orderly and chronological sequence of job steps. (RECORD)
(3) Critically observe and examine each component part of the job to determine the risk of accident. (EXAMINE)
(4) Develop control measures to eliminate or reduce the risk of accident. (DEVELOP)
(5) Formulate written and safe systems of work and job safety instructions for the job. (INSTALL)
(6) Review safe systems of work and job safe practices at regular intervals to ensure their utilisation. (MAINTAIN)

From a practical viewpoint, this information can be recorded on a job safety analysis chart of the sort shown in *Figure 10.1*.

This is a typical job safety analysis chart. The detailed format will depend on the process and company and should be adapted to suit.

Criteria to be considered when selecting jobs for analysis will include:

(1) past accident and loss experience;
(2) maximum potential loss;
(3) probability of recurrence;
(4) legal requirements;
(5) the newness of the job; and
(6) the number of employees at risk.

The ultimate aim must be to undertake JSA on *all* jobs within an organisation.

Once the job has been selected, the next stage is to break it down into its component parts or job steps. On average, there will be approximately fifteen job steps; if more than twenty, then the job under study should be sub-divided; if less than ten, then a bigger slice of the job should be analysed. Each job step should be one component part of the total job where something happens to advance by a measurable amount the doing of the work involved. The breakdown should be neither too general nor too specific. An example of such a job breakdown is given below.

Changing a car wheel

Job step	*Risk factor*	*Control action*
1 Put on handbrake	Strain to wrist/arm	Avoid snatching, rapid movement
2 Remove spare from boot and check tyre pressure	Strain to back	Use kinetic handling techniques

Job step	*Risk factor*	*Control action*
3 Remove hub cap	Strain; abrasion to hand	Ensure correct lever used
4 Ensure jack is suitable and is located on firm ground	Vehicle slipping. Jack sinking into ground	Check jack
5 Ensure jacking point is sound	Vehicle collapse	Consider secondary means of support
6 Jack up car part-way, but not so that the wheels leave the ground	Strain; bumping hands on jack/car	Avoid snatching, rapid movements
7 Loosen wheel-nuts	Hands slipping – bruised knuckles. Strain	Ensure spanner brace in good order. Avoid snatching, rapid movements. Use gloves
8 Jack up car fully in accordance with manufacturer's advice	Strain; bumping hands on jack/car	Avoid snatching, rapid movements
9 Remove wheel	Strain to back. Dropping onto feet	Use kinetic handling techniques. Use gloves (if available) to improve grip
10 Fit spare	Strain to back	Use kinetic handling techniques
11 Tighten wheel-nuts	Hand slipping – bruised knuckles. Strain	Use gloves. Avoid snatching, rapid movements
12 Lower car	Strain; bumping hands on jack/car	Avoid snatching, rapid movements
13 Remove jack and store in boot, together with removed wheel	Strain to back	Use kinetic handling techniques
14 Retighten wheel-nuts	Hand slipping – bruised knuckles	Use gloves. Avoid snatching, rapid movements
15 Replace hub cap	Abrasion to hand	Use gloves
16 Ensure wheel is secure, prior to driving off	–	Check wheel and area around car

From the above, it may be seen that each job step has been systematically analysed for its component risk factor. For each identified risk factor a control action has been developed.

The third column – Control action – becomes the Job Safety Instructions, and forms the basis of the written safe system of work.

10.2.2 Job safety instructions

Once the individual job has been analysed, as described above, a written safe system of work should be produced.

The purpose of job safety instructions is to communicate the safe system of work to employees. For each job step, there is a corresponding control action designed to reduce or eliminate the risk factor associated with the job step. This becomes the job safety instruction which spells out the safe (and efficient) method of undertaking that specific job step.

Such job safety instructions should be utilised in as much job safety training both formal (in the classroom) and informal (on the job contact sessions) as possible. All managers and supervisors concerned should be fully knowledgeable and aware of the job safety instructions and safe systems of work that are relevant to the areas under their control.

From a practical viewpoint, job safety instructions should be listed on cards which should be (a) posted in the area in which the job is to be undertaken; (b) issued on an individual basis to all relevant employees; and (c) referred to and explained in all related training sessions.

10.2.3 Safe systems of work

Safe systems of work are fundamental to accident prevention and should: (a) fully document the hazards, precautions and safe working methods, (b) include job training, and (c) be referred to in the 'Arrangements' section (part 3) of the Safety Policy.

Where safe systems of work are used, consideration should be given in their preparation and implementation to the following:

(1) Safe design.
(2) Safe installation.
(3) Safe premises and plant.
(4) Safe tools and equipment.
(5) Correct use of plant, tools and equipment (via training and supervision).
(6) Effective planned maintenance of plant and equipment.
(7) Proper working environment ensuring adequate lighting, heating and ventilation.
(8) Trained and competent employees.
(9) Adequate and competent supervision.
(10) Enforcement of safety policy and rules.
(11) Additional protection for vulnerable employees.
(12) Formalised issue and proper utilisaton of all necessary protective equipment and clothing.

(13) Continued emphasis on adherence to the agreed safe method of work by all employees at *all* levels.
(14) Regular (at least annually) reviews of all written systems of work to ensure:
  (a) compliance with current legislation,
  (b) systems are still workable in practice,
  (c) plant modifications are taken account of,
  (d) substituted materials are allowed for,
  (e) new work methods are incorporated into the system,
  (f) advances in technology are exploited,
  (g) proper precautions are taken in the light of accident experience, and
  (h) continued involvement in, and awareness of the importance of, written safe systems of work.
(15) Regular feedback to all concerned – possibly by safety committees and job contact training sessions – following any changes in existing safe systems of work.

The above 15 points give a basic framework for developing and maintaining safe systems of work.

10.3 System safety

10.3.1 Principles of system safety

A necessary prerequisite in connection with the study of system safety is a working knowledge of the principles of safe systems of work and job safety analysis. Also an appreciation of how hazard and operability studies[3] can be used will be of assistance.

System safety techniques have primarily emanated from the aviation and aerospace industries, where the overriding concern is for the complete system to work as it has been designed to, so that no one becomes injured as a result of malfunction.

Therefore, system safety techniques may be applied in order to eliminate any machinery malfunctions or mistakes in design that could have serious consequences. Thus, there is a need to analyse critically the complete system in order to anticipate risks, and estimate the maximum potential loss associated with such risks, should they not be effectively controlled.

The principles of system safety are founded on pre-planning and organisation of action designed to conserve all resources associated with the system under review.

According to Bird and Loftus[4], the stages associated with system safety are as follows:

(1) The pre-accident identification of potential hazards.
(2) The timely incorporation of effective safety-related design and operational specification, provisions, and criteria.
(3) The early evaluation of design and procedures for compliance with applicable safety requirements and criteria.
(4) The continued surveillance over all safety aspects throughout the total life-span – including disposal – of the system.

System safety may therefore be seen to be an ordered monitoring programme of the system from a safety viewpoint.

It may be seen that the system safety approach is very closely allied to the risk management approach. Indeed, the logical progression of system safety management techniques has been incorporated into many risk management processes, and also to other linked disciplines such as total quality management and environmental management systems.

10.3.2 The system

The system under review is the sum total of all component parts working together within a given environment to achieve a given purpose or mission within a given time over a given life-span.

The elements or component parts within a system will include manpower, materials, machinery and methods.

Each system will have a series of phases, which follow a chronological pattern; the sum total of which will equate to the overall life-span of the system. These phases are: conceptual phase, design and engineering phase, operational phase, and disposal phase:

(1) The conceptual phase considers the basic purpose of the system and formulates the preliminary designs and methods of operation. It is at this stage that hazard and operability studies should be undertaken.
(2) The design and engineering phase develops the basic idea from the conceptual phase, and augments them to enable translation into practical equipment and procedures. This phase should include testing and analysis of the various components to ensure compliance with various system specifications. It is at this stage that job safety analysis should be undertaken.
(3) The operational phase involves the bringing together of the various components – i.e. manpower, materials, machinery, methods – in order to achieve the purpose of the system. From a practical viewpoint, it is at this stage that safe systems of work should be developed and communicated.
(4) The disposal phase begins when machinery and manpower are no longer needed to achieve the purpose of the system. All components must be effectively disposed of, transferred, re-allocated or placed into storage.

10.3.3 Method analysis

There are many methods of analysis in use in systems safety including:

(1) Hazard and Operability Study[3]
This analytical method has been discussed above.
(2) Technique of Operations Review[5]
This analytical technique or tracing system directs system designers and managers to examine the underlying and contributory factors that combine together to cause a failure of the system. It is associated with the theory of multicausality of accidents.

(3) Gross Hazard Analysis
This analysis is done early in the design stage, and would be a part of a 'Hazop' (hazard and operability) study. It is the initial step in the system safety analysis, and it considers the total system.
(4) Classification of Risks
This analysis involves the identification and evaluation of risks by type and impact (i.e. maximum potential loss) on the company. A further analysis – Risk Ranking – may then be undertaken.
(5) Risk Ranking
A rank ordering of the identified and evaluated risks is drawn up, ranging from the most critical down to the least critical. This then enables priorities to be set, and resources to be allocated.
(6) Failure Modes and Effects
The kinds of failures that could happen are examined, and their effects – in terms of maximum potential loss – are evaluated. Again this analysis would form part of an overall Hazop study.
(7) Fault-Tree Analysis
Fault-tree analysis is an analytical technique that is used to trace the chronological progression of factors contributing to the accident situation, and is useful not only for system safety, but also in accident investigation. Again, the principle of multicausality is utilised in this type of analysis.

10.4 Systems theory and design

The word 'system' is defined in the Oxford Dictionary as 'a whole composed of parts in orderly arrangement according to some scheme or plan'. In present day parlance, we tend to think of 'systems' as connected with computers. However, the word is used in a wider sense in Operational Research to imply the building of conceptual and mathematical models to simulate problems and provide quantitative or qualitative information to executives who have to control operations, e.g. a maintenance system, a system governing purchase and use of protective clothing, a training system etc.

In this chapter only an outline can be given of the concepts underlying systems theory and the theory will be presented mainly as an aid to clear thinking. The mathematical techniques associated with quantifying it can be found in textbooks of operational research.

The essential components of a systems model are goals or objectives, inputs, outputs, interactions between constituent parts of the system (e.g. storage, decision making, processing, etc.) and feedback.

The stages in establishing and using a systems model are:

(1) Define the problems clearly.
(2) Build a systems diagram (including values).
(3) Evaluate and test the system using already solved problems to check that the model gives the correct answer.
(4) Use the model on new problems.

If we take as example the provision of cost-effective machinery guards, we might produce a diagram such as *Figure 10.2* to indicate some of the factors affecting the process.

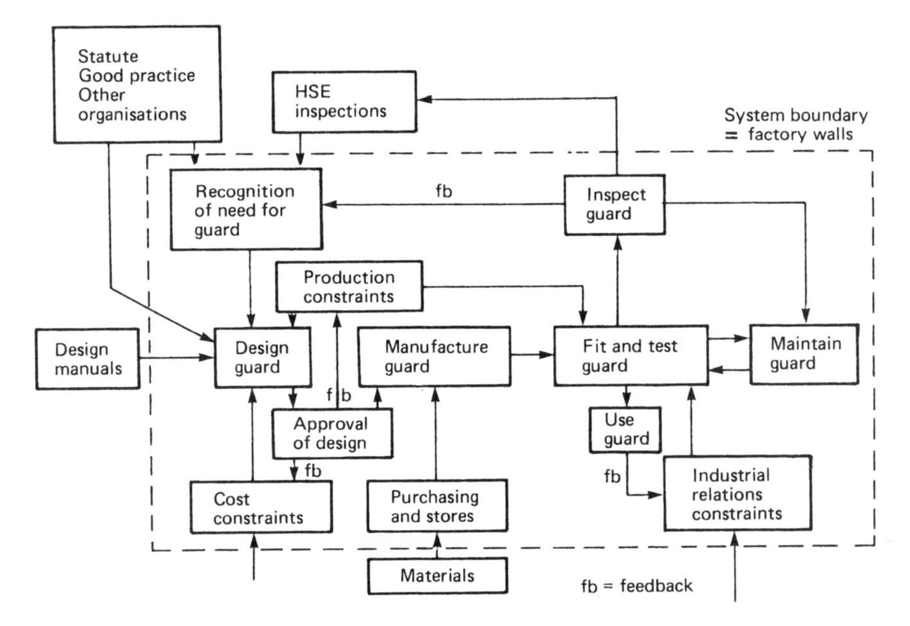

Figure 10.2 Systems diagram of the provision of a guard

Such a conceptual type of model shows not only the sequence of events taking place, but further highlights feedback (fb on model) which informs management whether or not legal requirements are being satisfied. Besides the legal and technical considerations the diagram shows that the new guard could upset previously agreed incentive earnings and lead to conflict between management and unions, which in turn may lead to work stoppage and delays. The aim of the organisation (the system within which the subsystem is embedded) is to satisfy its customers and this can only be achieved by consistent output both in terms of quantity and quality. It can be seen that the fitting of a relatively insignificant machine guard can affect wider areas of the company's operations. Systems diagrams can direct the attention of those who are responsible for the effective running of a company to possible interactions, between either individuals or groups, inside or outside the company which could lead to conflicts and hence work disruption, long before it actually occurs, thus enabling suitable provisions to be made.

It is a useful exercise to consider how the safety adviser fits into the above system and which activities he should be involved in.

The example shows a system boundary drawn to correspond to the boundary of the organisation. The fact that there are inputs and outputs across this boundary indicates that the system is an open one. Closed systems have no transactions across the system boundary. Consider a simple example, which compares these two main types. A steam engine's speed is controlled by a valve which controls the supply of steam. If the valve is adjusted by an attendant (an outside agent) the system is open, whereas if the valve is controlled by a governor responsive to the engine speed, the system is a closed one.

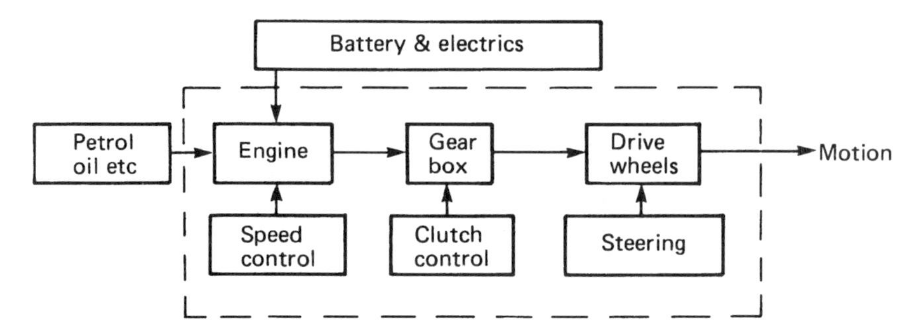

Figure 10.3 System diagram of a car

The system boundary could be drawn at various levels – e.g. in the guarding example it could be drawn at the level of the department in which the machine is located, the works, the company, or the country (in the last case there might be inputs across the boundary (frontier) of materials, designs or EC regulations which would still make it an open system). For a full systems analysis and model it is usually necessary to produce a hierarchy of diagrams showing the total system, main sub-systems and sub-sub-systems etc.

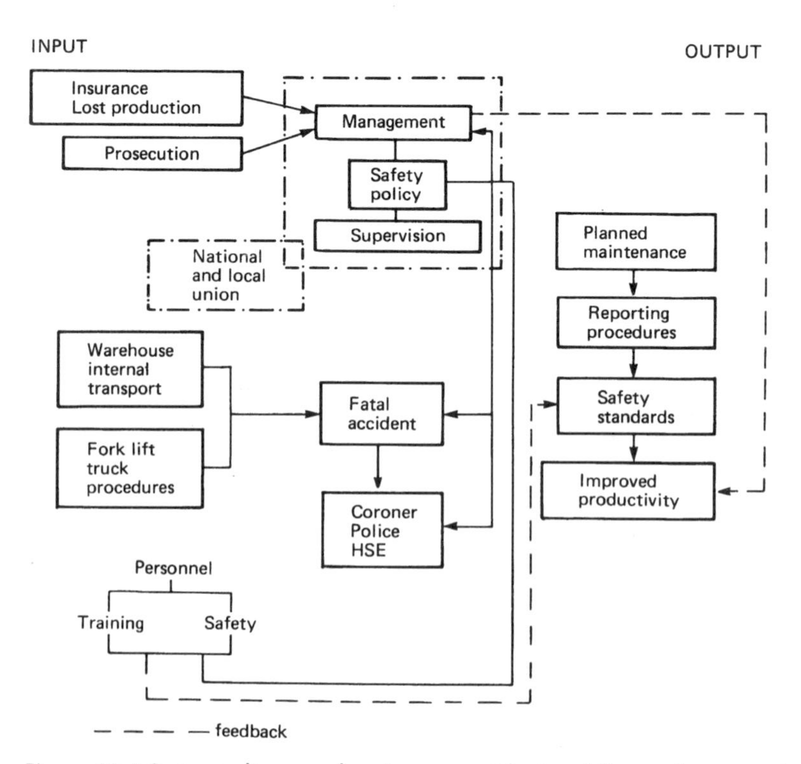

Figure 10.4 Systems diagram showing an accident and the environment in which it occurred

System diagrams sometimes only contain the hardware or technical elements as in *Figure 10.3* of a simple diagram of a car.

This is a very incomplete system diagram as it leaves out human control. A complete model or sociotechnical system should include both technical and human aspects (both desired and undesired, e.g. vandalism, sabotage, etc.). Only in this way can the mode of operation or breakdown of the whole system be investigated and if necessary redesigned. A useful exercise is to add the human element to the car system above, or to devise a complete sociotechnical systems diagram for a company.

Accidents can be modelled as breakdowns of systems. The individual as a system set out in Chapter 13 is one example. Another which illustrates a fatal forklift truck accident in a warehouse is given in *Figure 10.4*.

10.5 System safety engineering

System safety engineering has been defined[6] as an element of systems engineering involving the application of scientific and engineering principles for the timely identification of hazards and initiation of those actions necessary to prevent or control hazards within the system.

It draws upon professional knowledge and specialised skills in the mathematical, physical and related scientific disciplines, together with the principles and methods of engineering design and analysis to specify, predict, and evaluate the safety of the system.

An examination of system safety engineering methodology requires the consideration of two basic and interrelated aspects, namely system safety management and system safety analyses.

System safety management provides the framework wherein the findings and recommendations resulting from the application of system safety analysis techniques can be effectively reviewed and implemented.

System safety analyses employ the three basic elements of identification, evaluation, and communication to facilitate the establishment of cause. System safety analyses provide the loss identification, evaluation and communication factors and interactions within a given system which could cause inadvertent injury, death or material damage during any phase or activity associated with the given system's life-cycle.

Examples of system safety analyses include: routine hazard spotting; job safety analysis; hazard and operability studies; design safety analysis; fault-tree analysis; and simulation exercises using a computer.

10.6 Fault-tree analysis

10.6.1 Introduction

Fault-tree analysis is a technique that may be utilised to trace back through the chronological progression of causes and effects that have contributed to a particular event, whether it be an accident (industrial safety) or failure (system safety).

The fault-tree is a logic diagram based on the principle of multicausality that traces all the branches of events that could contribute to an accident or failure.

10.6.2 Methodology

In constructing a fault-tree to assist in cause analysis, firstly the event – the accident or failure – must be identified. Secondly, *all* the proximate causes (contributory factors) must be investigated and identified. Thirdly, each proximate cause (i.e. each branch of a contributory factor) must be traced back to identify and establish all the conceivable ways in which each might have occurred. Each contributing factor or cause thus identified is then studied further to determine how it could possibly have happened, and so on, until the beginning or source of the chain of events has been highlighted for each branch of the fault-tree.

Certain standard symbols are used in the construction of a fault-tree; some of the more common are:

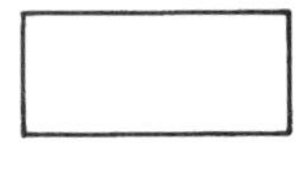

The rectangle identifies an event or contributory factor that results from a combination of contributory factors through the logic gates.

The 'AND' logic gate describes the logical operations whereby the co-existence of *all* input contributory factors are required to produce the output event or contributory factor.

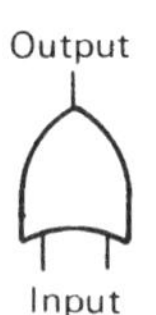

The 'OR' logic gate defines the situation whereby the output event or contributory factor will exist if *any* one of the input contributory factors is present.

Examples of fault-tree analyses are presented in Bird and Loftus[7] and Petersen[8].

By tracing back in this way the causes of accidents during accident investigation, a clearer and more objective assessment may be made of *all* the contributory factors, and hence more effective preventive action may be taken to ensure that there is no recurrence.

10.7 Probabilistic risk assessments

A probabilistic risk assessment consists of the following stages:

(a) the identification of undesired events and the mechanisms by which they occur (WHAT IF?),
(b) the likelihood – probability – that these undesired events may occur (HOW OFTEN?),

(c) the consequences of such an event once it occurs (HOW BIG?),
(d) a calculated judgement as to the significance of (b) and (c) (SO WHAT?) which may or may not lead to,
(e) the taking of control action.

Stage (a) – Identification of undesired events
Primarily, this stage involves the use of HAZOP studies (see section 10.1.2). Hazops are usually carried out at various stages of a system's cycle:

- conceptual design stage
- detailed design stage
- operational stage

Stage (b) – Likelihood of undesired events
This stage involves an estimation of the probability of whether the undesired event is likely to occur or not.

Probabilities can be established mathematically, based upon the probable failure rates of individual components within the system. Data on individual components may be obtained from manufacturers' reliability statistics or quality assurance information. Specific failure rate data for individual items can also be obtained from reliability data banks such as that operated by the United Kingdom Atomic Energy Authority's (UKAEA) System Reliability Service.

Aspects such as maintenance schedules, condition monitoring, replacement criteria and human reliability/failure should also be taken into account.

An ideal technique for summating these individual probabilities to obtain the overall probability of the event occurring is Fault-Tree Analysis (see section 10.6), which is in essence a logic diagram with the event at the top of the tree.

Stage (c) – Consequences of undesired event occurring
Initially a hazard analysis (HAZAN) is undertaken so as to ascertain the magnitude of the potential problem and its potential for harm to the people, plant, process and the public.

A subsequent risk analysis will then go on to examine the actual consequences – worst possible case considerations – and express them in quantifiable terms. This then enables Stage (d) to be performed.

Stage (d) – Is the risk of the event occurring significant?
The output from Stage (c) may be expressed in the form of individual risk or of societal risk. Individual risk is the probability of death to an individual within a year (e.g. 1 in 10^4 per year). Societal risk is the probability of death to a group of people – either employees or members of the general public – within a year (e.g. a risk of 500 or more deaths of 10^{-8} per year).

Societal risks are usually given as fatal accident frequency rates (FAFRs).

The fatal accident frequency rate is defined[9] as the number of fatal injury accidents in a group of 1000 in a working lifetime (10^8 hours).

In making judgements to enable decisions concerning control action (Stage (e)) to be made, use is made of (published) risk criteria. These criteria are expressed in the form of numerical risk targets and they provide a yardstick for decision-makers against which to judge the significance of estimated risks. Generally two forms of risk criteria – as indicated above – are used:

- employee risks (on site)
- public risks (off site)

Stage (e) – Taking control action
If the calculated risk criteria figure is above the agreed accepted (published) figure, then control action is necessary. The amount by which the calculated figure is higher than the agreed figure is useful in setting priorities, i.e. the greater the difference, the higher is the priority for control action.

If the calculated risk factor is below the accepted figure, then the safety provisions of the system may be considered to be adequate, and hence no further control action is required.

For employees in the UK chemical industry the FAFR is approximately 3.5. An acceptable target would obviously be below this, at approximately half, i.e. 1.7.

For the general public, FAFRs are rare. However, it has been suggested[10] that from an individual risk viewpoint as involuntary risks expose members of the public to a risk of death of about 10^{-7} per person per year, then industrial activities should not increase this figure. Hence, a risk criteria of less than 10^{-7} is acceptable.

10.8 Health and safety in design and planning

10.8.1 Introduction

The consideration of health and safety aspects at the design and planning stages of new projects, buildings, plant and processes is vitally important in order to ensure that health and safety are built in, rather than bolted on.

It is therefore essential that all engineers, designers, and architects receive education and training in such matters, so as to ensure that relevant legislative and technical factors appertaining to health and safety are taken into account at the design and planning stage of all new projects.

Certain risk identification (e.g. Hazop) and risk evaluation techniques (e.g. Hazan, fault-tree analysis) that may prove useful in this regard have been discussed above.

10.8.2 Project design

It is imperative that assessment and control of all new projects take health and safety aspects into account at the earliest – and at all – stages of a project's development.

The project originator should ensure that the project is appraised from a health and safety viewpoint, and it should not be allowed to proceed until it has been approved by a safety adviser. Ideally the safety adviser should be involved at all stages of the project's design and planning, so that specialist guidance and advice may be incorporated as necessary.

The risks associated with new projects may include: use of hazardous substances; insufficient product data; faulty electrical equipment; poor access/egress; poor ergonomics; noisy equipment; poorly guarded machinery; imported equipment/materials; lack of risk assessment; lack of training/awareness on behalf of management, supervision and employees; poor environmental control; inadequate emergency procedures; inadequate maintenance considerations; poor construction methods; little or no consideration of waste disposal/demolition.

The whole life cycle of the project – from inception to ultimate disposal – should be considered at the design stage.

10.8.3 Project design: health and safety action plan

When health and safety is considered at the design stage, the following action plan should ensure that risks are designed and engineered out of the system *before* they are able to cause injury, disease, damage or loss:

- Ensure advice on health and safety is made available to the project team/originator.
- Ensure a Hazop-type brainstorming meeting of key personnel associated with the development of the project is held to identify risks and establish control actions. (The list of risks in section 10.8.2 may serve as a checklist.)
- Ensure that suitable written safe systems of work are prepared and communicated to all concerned – i.e. develop the 'software' to go with the 'hardware'.
- Ensure that all aspects of the project comply with relevant legislative and technical standards.
- Ensure that all personnel concerned with the project receive necessary health and safety training.
- Ensure that suitable emergency procedures are developed.
- Ensure that the project commissioning procedure involves approval by the safety and health adviser at all stages of the project's development.

10.8.4 Project commissioning

Once the project has been approved, practical aspects of supply, installation, commissioning, use and ultimately disposal follow on. As stated in section 10.8.2, health and safety should already have been considered at the design stage of the project.

From a legislative viewpoint, the supply, installation and commissioning aspects are covered by section 6 of the Health and Safety at Work etc. Act, 1974, which requires manufacturers, suppliers, installers etc. of articles and substances to ensure that they are safe and without risks to health when set, used, cleaned, maintained, stored, transported and disposed of.

In order to ensure safe commissioning of new plant and equipment, a three-part plant acceptance system should be utilised:

Part one – provisional safety certificate (for test purposes)
– this enables only design/engineering personnel to undertake testing, once approved by safety adviser and project originator.

Part two – trial production run
– this enables production employees to become familiar with the new equipment under the supervision of the project originator/safety adviser and enables any previously unforeseen risks to be engineered out at the man/machine interface.

Part three – final certification
– once all testing and production trials have been satisfactorily completed, the plant/equipment is handed over to production management.

By involving a multidisciplinary team – including a safety adviser – at all stages of the design, planning and commissioning process, the risk of having to provide additional – and more costly – safeguards *after* the plant is in full use is minimised.

10.9 Safety and quality assurance

10.9.1 Introducton

Established quality assurance procedures provide a sound basis for the development of systems for health and safety management[11]. The introduction of the HSE's publication *Successful Health and Safety Management* clearly states[19] that many of the features of effective health and safety management are indistinguishable from the sound management practices advocated by proponents of quality and business excellence.

10.9.2 Quality systems

The British Standard on quality systems[12] – BS 5750 – is the UK National Standard which tells suppliers and manufacturers what is required of a quality-orientated system from a practical viewpoint. It identifies basic disciplines and specifies procedures and criteria to ensure customer requirements.

Within the context of BS 5750, quality means that the product is fit for the purpose for which it has been purchased, and has been designed and constructed to satisfy the customer's needs.

The Standard sets out how an effective and economic quality system can be established, documented and maintained.

The Standard considers that an effective quality system should comprise: management responsibility; quality system principles; quality system audits; quality/cost considerations; raw material quality control; inspection and testing; control of non-conforming product; handling, storage, packaging and delivery; after sales service; quality documentation and records; personnel training; product safety and liability; and statistical data/analyses.

10.9.3 Quality and safety

Although the Standard does not explicitly refer to 'people safety', there are obvious parallels to be drawn between the quality systems approach and health and safety management.

Indeed, the management systems described in BS 5750 are as applicable to health and safety management as they are to product liability risk management[11]. BS 5750 is concerned with the achievement of quality, which is measurable against specific criteria. It lays down systems which demonstrate achievement against these specified criteria.

One of the benefits of an effective quality system is to minimise the risk of product liability claims and losses.

In the case of product liability risk management, the specified criteria of performance are:

- Compliance with section 6 of the Health and Safety at Work etc. Act, 1974 (as modified by the Consumer Protection Act, 1987).
- Compliance with all other relevant statutory provisions, especially the Management of Health and Safety at Work Regulations 1992.
- The ability to adhere to all product contract conditions.
- The minimisation of defective products.
- The maximisation of health and safety benefits to the consumer.

This parallels very closely the perceived criteria of an effective health and safety management system, namely:

- Compliance with all aspects of the Health and Safety at Work etc. Act, 1974.
- Compliance with all other relevant statutory provisions – e.g. Factories Act, 1961.
- The ability to adhere to the common law duty of care, and relevant aspects of employment contract conditions.
- The minimisation of risks likely to cause injury or disease.
- The maximisation of health and safety benefits to employees, third parties, and the general public.

From the above, it may be seen that the application of quality systems to the management of health and safety at work has distinct benefits, especially when consideration is given to the tremendous overlap between the two subject areas. Overlap examples include: policies; systems and procedures; standards; documentation – records; training (including record keeping); statistical analyses – causal, numerical; accident/complaint investigations; audits/inspections (internal and external); and the taking of remedial control action.

Hence effective quality systems management will greatly enhance the management of health and safety, and will lead to an overall improvement in the level of safety performance.

10.10 Use of data on accidents

In addition to the use of qualitative and quantitative accident data when identifying and evaluating risk (see sections 10.1.2 and 10.1.3), there is a need to consider the relative occurrence of different accident types, in order that effective accident control measures may be implemented throughout an organisation.

Accidents, whether they result in injury, damage, disease or loss, need to be controlled. Similarly those that have no end result – i.e. the near misses – should be considered for control action.

To enable an accident control system to be developed, it is necessary that *all* accidents are reported, recorded, investigated and analysed, so that after remedial action has been decided, plans can be drawn up to prevent a recurrence. The most important question to be asked in any accident investigation is: 'What action has been taken to prevent a recurrence?'

The collection of accident data on a much broader base to facilitate the planning of control action has been undertaken by a number of researchers and one of the most widely applied accident ratios is that propounded by Bird[13] in 1969.

The Bird study is generally depicted in triangular form:

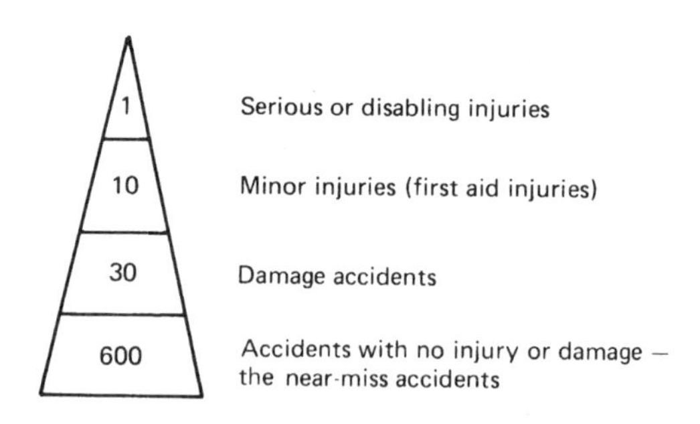

From this it can be seen that the majority are either 'near misses' or damage only accidents, so that any accident control programme, for its greatest effect, must concentrate much of its attention on these two areas. Related techniques of damage control and incident recall are dealt with in sections 9.4.2 and 10.12 and also by Carter[14].

Accident data are only one form of risk information which assists in the identification, assessment and control of risks. Other forms of risk information include: reliability data; epidemiological studies; mortality/morbidity data; vulnerability analysis; results of audits/inspections; insurance claims data; and – probably the best – personal experience. Sources of risk information are listed in the Further Reading section of this chapter (see especially the *Handbook of Risk Management* by Carter *et al.*).

10.11 Maintenance systems and planned maintenance

10.11.1 Maintenance – risk management aspects

From a risk management viewpoint, health and safety aspects concerning all maintenance operatives should be considered and planned for at the design stage of a new project or process.

Aspects such as: safe access; operations at unguarded machinery; breaking into pressurised systems; blanking off; inerting; hot work; installation; setting up; inching; dismantling; and demolition, should have been taken into account.

Effective maintenance systems that take into account health and safety aspects serve not only to increase the lifetime of key plant and equipment, but also ensure safer and more cost efficient operation.

Indeed, a planned maintenance system is essential to the continued safe operation of a range of plant and equipment, as well as meeting both general and specific legislative requirements – e.g. statutory inspections.

Plant and equipment requiring such regular inspection and planned maintenance include: boilers; pressure vessels – air receivers; power presses; cranes; lifting equipment – chains, pulleys; ventilation systems; pressurised systems; access equipment – ladders; scaffolding; electrical equipment – portable electric tools; office equipment; emergency eye-wash points and showers; fire alarm systems and fire points; fire extinguishers; fixed firefighting appliances/systems; stairways, walkways, gangways; in addition to all key items of production plant and equipment.

10.11.2 Planned maintenance

It is generally accepted that a system which only allows for the maintenance/repair of plant on a crisis or breakdown basis is not cost-efficient, especially in the longer term.

Hence any form of planned maintenance – whether it be planned lubrication at one end of the scale, or sophisticated techniques such as condition monitoring at the other end – will improve both safety and plant integrity/reliability.

A system of planned maintenance may be introduced progressively, commencing with limited maintenance routines only on key items of plant and equipment. This may then be extended to a wider range of plant, utilising more complex routines, until such time as a 100% planned maintenance system has been implemented. As with quality systems, maintenance records and histories are essential to ensure smooth operation.

When a total (100%) system of planned maintenance has been implemented, then, by definition, any breakdown that occurs must be accidental in nature. Hence such accidental breakdowns – i.e. damage accidents – should be reported and investigated (see section 10.12 for fuller discussion).

10.12 Damage control

10.12.1 Introduction

The technique of damage control involves the systematic reporting, investigation, costing and control of damage accidents within an organisation.

10.12.2 Reporting

The reporting system for damage accidents within an organisation should, as far as possible, be the same as that for injury accidents. The fundamental differences between injury and damage accidents are that in the former, the victim (the injured person) starts the reporting chain by advising his supervisor or first-aider of his injury.

However, in the case of damage accidents, the victim is an inanimate object, physically incapable of reporting the accident. Therefore, there is a need to develop a system by which accidental damage can be highlighted and clearly identified.

It would appear that the situations likely to give rise to accidents can be divided into three main categories:

(i) Fair wear and tear – if we define this as the failure of an item in service after a lifetime of proper use, then, in theory, such a failure will be eliminated by a total (100%) system of planned maintenance and condition monitoring. In such a situation any failure will be an accident, i.e. unexpected, unplanned.
(ii) Malicious damage – this will need to be identified and dealt with separately.
(iii) Accident damage – it is this final category that we are dealing with in this section.

There will, of course, be many occasions in practice where it will be necessary to determine whether damage arises from (i) or (iii) above and the opinion of all relevant staff will need to be sought.

10.12.3 Investigation

In order to ensure that prompt action is taken to prevent the damage accident recurring, a detailed investigation should be undertaken. This should aim to establish the sequence of events, and to identify the combination of causes of the damage accident.

It should be remembered that the prime aim of any accident investigation is prevention, rather than blame. As with injury accidents, the investigating team should generally comprise the relevant supervisor and/or manager, together with the safety adviser and safety representative (if appointed).

To aid the investigation it is suggested that a detailed accident investigation report form is utilised.

10.12.4 Costing

The prime aim in costing damage accidents is to obtain an objective measure of severity. This will in turn assist in the setting of priorities and hence the allocation of resources and the overall planning of the damage control programme.

Information regarding the cost of all accidents – both injury and damage – may also be used to motivate management action in accident prevention.

10.12.5 Control

During the investigation the question 'What action should be taken to prevent a recurrence?' should be clearly answered. This should ensure that the correct remedial action is promptly taken.

The findings of the investigation should be clearly communicated to the manager responsible for instigating the control action.

To gain the full value from a control system, it is important that findings be communicated to all management within the organisation who may be faced with similar problems. This will greatly assist in the elimination of the causes of potential accidents before either injury or damage results.

10.13 Cost-effectiveness of risk management

10.13.1 Accident costing

The three fundamental arguments that may be employed to promote action in the field of risk management concern the legal, humanitarian, and economic aspects (see section 9.7).

For the vast majority of organisations, there will be few or no quantitative data relating to the economic facets, and particularly to the costs associated with accidents. In order to be able to demonstrate cost-effectiveness (or cost benefit), there is therefore a need to be able to quantify the cost of all losses associated with accidents.

An investigation of the costs of occupational accidents for the UK as a whole[15] has considered the costs under two headings: resource costs covering lost output, damage to plant, medical treatment and administrative costs, and subjective costs relating to pain and suffering of the victim and his family.

	Resource costs £	*Subjective costs* £	*Total costs* £
Fatality	71 500	38 000	109 500
Serious injury	1 800	2 750	4 550
Temporary disabling diseases	1 600	2 750	4 350
Minor injuries	300	150	450

However figures relating to the national situation are often not pertinent to individual factories and departments. Some figures obtained in 1992 for the uninsured costs of three types of accident[16] were:

Lost time injury	£1672
Non-lost time injury	£ 25
Damage	£ 112

The above figures – in isolation – do not have much impact, but must be related to the total costs of accidents in an individual factory or organisation. When these accident costs are compared with other business costs, such as production, sales and distribution, a true indication of the drain on financial resources can be appreciated. The following hypothetical case is presented as an example:

Consider Factory Z in 1992; the data given below apply:

Number of lost time injury accidents	=	30
Number of non-lost time injury accidents	=	750
Number of damage accidents (estimated)	=	780
Employers' liability premium (£1.10% wages)	=	£90 000
Number of employees	=	1000

When the average uninsured accident cost figures are applied to the above data, the following result:

Uninsured cost of lost time injury accidents			
	30 × £1672	=	£50 160
Uninsured cost of non-lost time injury accidents			
	750 × £25	=	£18 750
Uninsured cost of damage accident	780 × £112	=	£87 360
Total uninsured accident cost		=	£156 270

Adding to this the Employer's Liability insurance premium (£90 000), the total accident cost becomes £246 270, or approximately £246 per employee per year.

Of particular concern is the impact of such costs on the overall profitability of the organisation. Although the total cost of accidents within a company may be relatively small, it can amount to approximately 2% of the annual running costs, and represents a direct drain on profits.

The table below illustrates the sales necessary to cover these accident costs.

Accident costs £	*If your organisation profit margin is:* 1%	3%	5%
1 000	100 000	33 000	20 000
10 000	1 000 000	330 000	200 000
100 000	10 000 000	3 300 000	2 000 000

Any reduction of these costs that may be made through a cost-effective risk management programme will lead both to a safer and more profitable organisation.

The use of cost figures presented in this way will be more meaningful to managers and executives and is likely to stimulate their motivation to reduce the number of accidents.

An alternative – more specific – method of relating accident and wastage costs to costs of production is presented in the following example:

On a construction site, the number of facing bricks lying around was estimated by random counting to be 1300, valued at £455.

The cost in terms of profit on turnover (of 6.3%) was:

$$\frac{\text{Cost} \times 100}{\text{Profit on turnover}} = \frac{455 \times 100}{6.3\%} = £7222$$

i.e. a turnover of £7222 was required to pay for the bricks – i.e. to break even.

The cost expressed as a percentage of the total contract price (of £1 650 000) was:

$$\frac{\text{Turnover to pay for bricks} \times 100}{\text{Contract price}} = \frac{£7220 \times 100}{£1\ 650\ 000} = 0.43\%$$

Such exercises may be undertaken for all types of accident and wastage situations, to enable the costs involved to be judged in relative – rather than absolute – terms.

10.13.2 Cost benefit analysis

Cost benefit analysis techniques have been developed in recent years, as decisions concerning risk management have been made on a cost versus risk basis – i.e. so far as is reasonably practicable.

Indeed, all proposed legislation has to pass the cost/benefit test at the consultative stage, before being allowed to pass on to the statute books. Although, in the majority of cases, the benefits of proposed legislation are not accurately quantified, nevertheless the qualitative benefits are listed, and these may be seen to outweigh the costs. Further discussion on the cost/benefit of specific proposed legislation may be found in relevant Health and Safety Commission Consultative Documents which contain a review of the costs and benefits associated with proposed changes in legal requirements.

To undertake a cost benefit analysis, answers to the following questions are required:

- What costs are involved to reduce or eliminate the risk?
- What degree of capital expenditure is required?
- What ongoing costs will be involved, e.g. regular maintenance, training?
- What will the benefits be?
- What is the pay-back period?
- Is there any other more cost-effective method of reducing the risk?

The cost factors associated with poor risk management have been discussed in sections 9.7 and 10.13.1 and include both insured and uninsured elements.

The benefit factors should initially be listed, and should always be quantified, where possible, so that the pay-back period can be established. Some benefits are easier to quantify than others.

Benefits of effective risk management include:

- few claims resulting in lower insurance premiums,
- less absenteeism,
- fewer injury and damage accidents,
- better levels of health,
- higher productivity/efficiency,
- better utilisation of plant and equipment,
- higher morale and motivation of employees,
- reduction in cost factors.

The costs are then balanced against the benefits (both qualitative and quantitative) and then an objective decision may be made on whether to allocate resources to the project or not. This will usually be based on the length of the pay-back period. Most health and safety projects will generally have a pay-back period of between three and five years – i.e. medium-term rather than short-term.

10.14 Performance evaluation and appraisal

10.14.1 Introduction

In the vast majority of company health and safety policies the health and safety responsibilities of line and functional management are clearly laid down, together with – in some cases – the mechanism by which the fulfilment of these responsibilities will be monitored.

Indeed, the Accident Prevention Advisory Unit (APAU) have produced three excellent publications[17–19] which provide additional guidance and discussion on the aspects of policy management, implementation and monitoring.

10.14.2 Financial accountability and motivational theory

However, the most effective way to fix accountability for health, safety and indeed risk management responsibilities is by financial accountability of directors and managers.

This is borne out when consideration is given to the use of the legal, humanitarian and economic arguments for health and safety at work (section 9.7).

Maslow[20] related his theory of motivation to human needs. He suggested five sets of goals which are usually depicted as a progression or hierarchy:

Self actualisation (Achievement; Doing a good job)
↑
Esteem (Status; Approval)
↑
Love (Social)
↑
Safety (Security)
↑
Physiological (Sustenance)

The logic is that once the lower needs are well satisfied, the individual is motivated to attain satisfaction at the next higher level, and so on up the hierarchy.

From a health and safety viewpoint, therefore, the humanitarian argument tends to operate in terms of doing a good job, caring for people, and being well thought of and accepted, i.e. the higher end of the hierarchy: esteem/ self actualisation.

The legal argument tends to motivate via the safety (security) need – the real (if remote) threat to physical security – and the esteem (approval) need of others. This motivation is in the middle of the hierarchy.

However the economic argument derives its motivational impact from the fact that the goals involved tend to be generally lower in the hierarchy – i.e. the safety (security) need and, in certain cases, the physiological need. The security need involves the manager's performance in his job – the ability to effectively carry out his responsibilities whilst keeping within budgetary constraints. Hence adverse performance could result in a decrease in or loss of the next salary increment or merit rise, as part of the performance appraisal exercise, thus directly threatening the security needs. In times of economic recession, this failure to achieve satisfactory performance could even affect the manager's position in the company and his ability to maintain employment, thus threatening the physiological needs.

Hence it would appear that financial accountability is the key.

10.14.3 Use of accident costs

However, the present system of accounting for health, safety, accident prevention and risk management operating in most of industry does not attempt to make line management financially accountable for accidents and uninsured losses, and very little use is made of economic arguments in stimulating management interest in risk management. Any arguments put forward rely mainly on legal and humanitarian considerations which, in some instances, fail to convince management that there is a need for risk management, at least beyond compliance with statutory duties.

One method that may prove useful involves the use of budgetary control which would introduce economic accountability into the field of accident prevention. Such measures would involve the re-organisation of the existing accounting procedures in most companies in order to overcome the lack of accountability for accident prevention and risk management.

When an accident occurs within a factory department, the cost of the accident usually is absorbed into the running costs of the factory as a whole, and will not be itemised on the departmental balance sheet. Nor will many of the uninsured costs be paid for from the departmental manager's budget. Furthermore, the insured cost – i.e. insurance premiums, such as employer's liability – will generally be paid from a central fund, usually administered by Head Office.

However, when the company safety adviser or factory inspector recommends safety measures such as guarding for machinery, the cost is usually charged against the departmental manager's budget, though it is very unlikely that it would be itemised as an accident prevention cost.

Thus under accounting systems currently employed in many companies, accident costs are not charged to departmental managers' budgets, whereas accident prevention is. Hence the departmental manager has no economic motivation to undertake any accident prevention; rather the reverse.

A positive economic motivating factor for encouraging accident prevention may be introduced by interchanging the budgetary system. For each accident – injury or damage – that occurs within a department, a charge is made against that department. Any accident prevention expenditure that is required within the department is financed from a central fund subject to approval by the risk manager or safety adviser. The result – as far as the departmental manager is concerned – is that it is costly to have accidents but not to prevent them.

Thus line management become accountable for the accidents occurring within their areas of control. At the end of the financial year, when the budgets are drawn up, a realistic allowance for accidents will be set within the budget as a target for the manager to achieve. Failure to achieve the agreed target would adversely reflect on performance and should be taken into account at job performance appraisal interviews. This allowance will form an integral part of the management plan as with budgetary control in other business areas.

However, the number – and hence the cost – of accidents budgeted for should be less than the previous year so that reduction of accidents becomes part of the management plan. The re-organised system would bring about the necessary economic accountability and would make full use of the knowledge and data obtained in establishing what accidents were costing the company in financial terms.

Once the costs of accidents have been established, the re-organised budgetary system can be implemented. The charges to be made against the departmental manager's budget can then be calculated and allocated on a monthly basis. The departmental manager might receive a monthly report giving information on the costs of accidents and accident prevention expenditure. This would enable him to plan any action necessary to maintain or improve the level of safety within his department. Also it would facilitate decision-making in connection with the allocation of scarce resources. Any deficiencies in the current programme would be highlighted in cost terms rather than by a frequency rate – a measure of safety increasingly questioned by safety advisers.

On their own, the legal and humanitarian arguments for risk management may not be sufficient to achieve a reduction in accidents and other losses. The addition of economic accountability – through accident costing – should greatly assist in reducing losses arising from accidents and ill-health at work.

10.15 Loss control profiling

General aspects of loss control are discussed in section 9.4. Loss control profiling is one of the major evaluation and control techniques associated with loss control management. The technique of profiling has formed the basis for a number of proprietary auditing systems such as International Safety Rating System (ISRS), Complete Health and Safety Evaluation (CHASE) and Coursafe.

Between 1968 and 1971, Bird[21,22] designed a loss control profile to quantify management's efforts in this area. He considered thirty areas of management activity that are connected either directly or indirectly with the reduction of loss.

These thirty areas are:

(1) Management involvement and policy making.
(2) Professional competence of loss control manager.
(3) Technical experience of loss control manager.
(4) Aptitude and talents of loss control manager.
(5) In-depth accident investigations.
(6) Plant and facility inspection.
(7) Laws, policies, standards.
(8) Management group meetings.
(9) Safety committee meetings.
(10) General promotion through the use of posters, banners, signs.
(11) Personal protection.
(12) Supervisory training.
(13) Employee training.
(14) Selection and employment procedures.
(15) Reference library.
(16) Occupational health and hygiene.
(17) Fire prevention and loss control.
(18) Damage control.
(19) Personal communications.
(20) Job safety analysis.
(21) Job safety observations.
(22) Records and statistics.
(23) Emergency care and first aid.
(24) Product liability.
(25) Off-the-job safety including on the road and at home.
(26) Incident recall and analysis.
(27) Transport including managers and salesmen driving cars.
(28) Security.
(29) Ergonomic applications.
(30) Pollution and disaster control.

Each of the thirty areas needs to be evaluated to pinpoint where action is necessary to improve the organisation's control of losses. The evaluation should be undertaken by trained personnel using the technique of asking a series of questions related to each of the thirty areas. Up to five hundred questions may be required to cover the thirty areas.

Thus, the first stage in loss control profiling is to develop the list of questions that relate specifically to the organisation under review.

The answer to each question is rated on a scale ranging from 0% for a bad to 100% for a good response. An evaluation of each of the thirty areas is then calculated by taking the average percentage of those answers relating to a particular area.

A percentage figure of 25% or less indicates those areas where immediate action needs to be taken. A percentage figure between 25% and 50% indicates those areas where there is a need for improvement within the near future. A percentage figure between 50% and 75% indicates those areas in which an acceptable level has been achieved, but in which there is still room for improvement. A percentage figure of 75% or more indicates those areas where the organisation is operating at optimum performance, but which have to be monitored to ensure that this performance is maintained.

The results of such evaluations can be presented graphically in the form of a horizontal bar chart where each subject area is shown on a separate line. Those areas giving cause for concern, i.e. the short lines, are immediately highlighted.

Fletcher and Douglas[23] and Fletcher[24] developed Bird's original ideas on profiling and formulated their own detailed evaluation questionnaire in which the answer to each question was rated on a six-point scale, ranging from 'fully implemented and fully effective' (score 5) to 'nothing done to date' (score 0). The scores of each question in a subject area are then summated, and the value is expressed as a percentage of the maximum attainable score. Loss control profiles are then constructed and utilised in a similar manner to that described above.

Once the losses – both actual and potential – have been evaluated, and a loss control profile developed, then – and only then – can a definite action programme of loss control be planned and implemented.

This would be based on assessing the deficiencies highlighted by the loss control evaluation and profile, then initiating a programme of work to make good those deficiencies.

Annual profiles may be undertaken to assess progress made, and also to ensure that all areas under review are maintained at an acceptable level.

References

1. Bird, F.E., and O'Shell, H.E., Incident recall, *National Safety News,* **100,** No. 4, 58–60 (1969)
2. Bamber, L., Incident recall – a (lack of) progress report, *Health and Safety at Work,* **2,** No. 9, 83 (1980)
3. Chemical Industries Association Ltd, *A Guide to Hazard and Operability Studies,* Chemical Industries Association Ltd, London (1977)
4. Bird, F.E. and Loftus, R.G., *Loss Control Management,* 464, Institute Press, Loganville, Georgia (1976)
5. Ref. 4, pp. 165–171
6. Ref. 4, p. 474
7. Ref. 4, p. 493 et seq.
8. Petersen, D.C., *Techniques of Safety Management,* 2nd edn, p. 174 et seq., McGraw-Hill, Kogakusha, USA (1978)

9. Kletz, T.A., *Hazop and Hazan: notes on the identification and assessment of hazards,* Institution of Chemical Engineers, Rugby (1983)
10. Kletz, T.A., Hazard analysis – its application to risks to the public at large (Part 1), *Occupational Safety & Health,* **7,** 10 (1977)
11. Industrial Relation Services, A systems approach to health and safety management, *Health & Safety Information Bulletin* No. 168, 5–6, Industrial Relations Services, London (1989)
12. BS 5750: 1987, Quality systems, British Standards Institution, Milton Keynes
13. Bird, F. E, *Management Guide to Loss Control,* 17, Institute Press, Atlanta, Georgia, (1974)
14. Carter, R.L., The use of non-injury accidents in risk identification, 4.6-01–4.6-05, *Handbook of Risk Management,* Kluwer Publishing Ltd, Kingston-upon-Thames (1992)
15. Morgan, P. and Davies, N., Cost of occupational accidents and diseases in GB, *Employment Gazette,* 477–485, HMSO (Nov. 1981)
16. Ref. 14, pp. 6.4-01–6.4-12
17. Health and Safety Executive, *Managing Safety, Occasional Paper Series No. OP3,* HSE Books, Sudbury (1981)
18. Health and Safety Executive, *Monitoring Safety, Occasional Paper Series No. OP9,* HSE Books, Sudbury (1985)
19. Health and Safety Executive, Accident Prevention Advisory Unit, publication No. HS(G)65, *Successful Health and Safety Management,* 1, HSE Books, Sudbury (1991)
20. Maslow, A.H., A theory of human motivation, *Psychological Review,* **50**, 370 – 396 (1943)
21. Ref. 13, pp. 151–165
22. Ref. 4, pp. 185–197
23. Fletcher, J.A. and Douglas, H.M., *Total Loss Control,* 113–154, Associated Business Programmes, London (1971)
24. Fletcher, J.A., *The Industrial Environment – Total Loss Control,* 18–122, National Profile Ltd, Willowdale, Ontario, (1972)

Further reading

Diekemper and Spartz, A quantitative and qualitative measurement of industrial safety activities, *J. Amer. Soc. Safety Engrs,* **15,** No. 12, 12–19 (1970)

Fine, W. T., Mathematical evaluation for controlling; hazards, *J. Safety Research,* **3,** No. 4, 57 –166 (1971)

Chemical Industries Association Ltd and Chemical Industry Safety and Health Council, *A Guide to Hazard and Operability Studies,* Chemical Industries Association Ltd, London (1977)

Petersen, D. C., *Techniques of Safety Management,* 2nd edn, McGraw-Hill, Kogakusha, USA (1978)

Bird, F. E. and Loftus, R. G., *Loss Control Management,* Institute Press, Loganville, Georgia (1976)

Heinrich, H. W., Petersen, D. and Roos, N., *Industrial Accident Prevention – A Safety Management Approach,* 5th Edn, McGraw-Hill, New York (1980)

Dewis, M. *et al. Product Liability,* Heinemann, London, (1980)

Carter, R. L. *et al., Handbook of Risk Management,* Kluwer Publishing Ltd, Kingston-upon-Thames (1992)

Health and Safety Executive, Report – Canvey: an investigation of potential hazards from operations in the Canvey Island/Thurrock area, HSE Books, Sudbury (1978)

Health and Safety Executive, Advisory Committee on the Safety of Nuclear Installations (ACSNI), ACSNI Study Group on Human Factors, *2nd Report – Human Assessment: A Critical Overview,* HSE Books, Sudbury (1991)

Health and Safety Executive, Accident Prevention Advisory Unit, Publication No. HS(G)65: *Successful Health and Safety Management,* HSE Books, Sudbury (1991)

Chemical Industries Association, *Guidance on Safety, Occupational Health and Environmental Protection Auditing,* Chemical Industries Association, London (1991)

Health and Safety Commission, *Management of Health and Safety at Work, Approved Code of Practice: Management of Health and Safety at Work Regulations 1992,* HSE Books, Sudbury (1992)

Chapter 11

Accident investigation and reporting

E. W. Adrian

11.1 Introduction

In addition to the requirements of UK laws for reporting certain injuries and specified dangerous occurrences, it is customary, because of claims at common law, for the majority of serious injuries sustained in the occupational environment to be investigated. Caring employers would, in any event, have a policy requiring as many accidents as possible, whether causing serious injuries or not, to be investigated and reported. Ideally, every accident, dangerous occurrence or 'near-miss situation' should be properly investigated. It must be borne in mind that the difference between a minor injury and a serious injury is often a matter of chance. The purpose of an investigation is to establish all the facts relating to the incident, to draw conclusions from the facts and make recommendations to prevent recurrence. The right attitude towards accident investigation is that of establishing the cause rather than apportioning blame. Joint investigations with safety representatives, or making all the facts available to them, is one way of helping. With the right of trade unions to appoint safety representatives, joint investigations have become more common.

To develop the right attitude to accident prevention, it is sometimes useful, from a training point of view, to learn from previous incidents, either by way of group role play or by having a discussion leader to guide the direction of the investigation exercise. Details of the accident can be illustrated with slides, photographs or sketches if available. By questioning and drawing out the facts of the original investigation, individuals can learn in greater depth than they would otherwise do had they merely read about it or listened to a lecture on accident investigation.

11.2 Statutory requirements

The requirements for the reporting of accidents in all employment, with the exception of 7 particular work situations, are defined in The Reporting of Injuries, Diseases and Dangerous Occurrences Regulations 1985[1]. While these regulations include the reporting of certain diseases and ill-health, the

2.5 The content of the induction should include as a minimum those items detailed on the induction checklist contained within Appendix A. and a copy of the "Brief summary of legal duties" (Appendix B).The checklist should be customised to suit the particular requirements of the location and person being inducted.

2.6 The induction programme should recognise the special needs of employees who may have sensory, physical or other learning difficulties.

2.7 Induction should preferably be done on the first day of work. If, for operational reasons, or due to the short duration of the work, this is not possible, the person or persons as a minimum should be provided with the information identified by an asterisk on the checklist (Appendix A). Induction training should be completed no later than ten working days from commencement of work.

2.8 Formal records should be kept of all induction sessions by the local Line Manager. If local systems are not in place, the induction checklist can be utilised for this purpose.

2.9 Where contractors are engaged to carry out works in Cafcass premises, information must be supplied on the following arrangements: Fire and evacuation, First aid, Incident reporting & any specific workplace hazards that may affect the contractor. Induction of Contractors and Consultants should be carried out by, or arranged by, the person responsible for overseeing/supervising the particular activity/project.

3 REFERENCES

Reference should be made to the following documents:
The Management of Health and Safety at Work Regulations 1999 Regulation 8 (information to employees).

4 DEFINITIONS

4.1 **Health and Safety Induction:** This is the process of providing essential health and safety information to all categories of worker who work either as direct or indirect labour for CAFCASS.

4.2 **Local Induction:** This should be induction for a local office area or building.

Cafcass Health and Safety Procedure

Title: Local Health & Safety Induction			
Document Number:	Cafcass – SP01	Revision 2	Date of Issue: 6/2/03

APPENDIX A – Induction Checklist

HEALTH AND SAFETY INDUCTION (new and transferred staff)		
Name of person being inducted		
Name of person conducting induction		
Start date		
The completed induction sheet to be signed off by the inducted person and their line manager and a copy kept in the local office		
ITEM	**DETAILS**	**COMPLETED Initial and date**
1. General obligations and responsibilities	Refer to attached summary of legal obligations.	
2. Safety Management System	Refer to CAFCASS health and safety management system, which can be located on the Intranet site in particular the Policy and Procedures Document. Provide a copy of the H&S General Policy Statement.	
* 3. Line Manager	Provide details of the line manager	
* 4. Safety Representative	Provide information and contact details for the accredited trade union health and safety representative.	
* 5. First Aid	Show the location of first aid facilities together with name and contact details of relevant first aid personnel.	
* 6. Emergency Evacuation	Explain the local emergency evacuation procedure and where the written information can be found. Provide name and contact details of the local Fire Marshals.	

* 7. Accident and Incident Reporting	Refer to CAFCASS health and safety management procedure HS 08 and explain.	
8.Workplace hazards	List and discuss the main hazards that the individual may encounter in the course of their activities, both location and task related. Provide details of any local policy and procedures	
9. Occupational health provision	Provide details of the OH provision and how it can be accessed for advice and counselling etc.	
10. Personal alarm systems	Provide details of the operation and procedures covering the local office personal alarm system.	
11.Home visiting and lone working arrangements	Explain the local procedure and risk assessments covering both home visiting and lone working. Emphasise the importance of following the control measures	
Induction complete (signatures) and date	……………………… Date……… …	Line manager
	……………………… Date…….	Inducted person

Cafcass Health and Safety Procedure

Title: Local Health & Safety Induction			
Document Number:	Cafcass – SP01	Revision 2	Date of Issue: 6/2/03

This procedure is written and issued in accordance with Cafcass' Safety Management System Framework and Protocol.
The responsibility for upkeep and amendment of this procedure rests with the corporate Health & Safety Advisor. All requests for modification should be made to Cafcass Headquarters.

CONTENTS

1. Policy

CAFCASS is committed to ensuring that any person who is called upon to do work for the organisation is provided with sufficient health and safety information to enable them to do their work safely and in a controlled manner.
The purpose of this procedure is to alert managers to the need for local induction and suggest what the minimum content of such an induction programme might be.

2. Procedure

2.1 All new employees or employees transferred to other jobs will be given local health and safety induction training.

2.2.1 All persons who are not direct employees i.e. temporary staff, consultants, contractors etc. will also be provided with an appropriate degree of induction training

2.3 The induction should be tailored to the local area and should be flexible enough to meet the individual needs of various categories of worker.

2.4 It is the responsibility of the Regional Business Manager to ensure health and safety induction training is carried out for all new and transferred staff. Inductions can be provided by anyone who is deemed competent to communicate the necessary information. In most cases induction should be seen as a "line managers" responsibility

onus placed on the general practitioner doctor to report the diseases listed in the Social Security (Industrial Injuries) (Prescribed Diseases) Regulations 1980[2] remains. The accident reporting requirements are considered in detail below:

11.2.1 Reporting of Injuries, Diseases and Dangerous Occurrences Regulations 1985

Made under section 15 of the HSW Act, the Regulations[1] relate to all accidents occurring in any employment in the UK and lay down the reporting and recording of details that must be made. They supersede the reporting requirements of existing legislation (listed in schedule 7) with the exception of certain industries (listed in schedule 6) and cover the whole range of employment situations.

The regulations are concerned with the reporting to the enforcing authorities deaths at work, certain types of injuries, diseases and dangerous occurrence incidents known to have a serious injury potential. Included in the types of injury to be reported are those lesser injuries that result in an inability to work for more than 3 days.

Guidance on the interpretation of these regulations together with an explanation of their application is given in an HSE booklet[3].

11.2.1.1 What type of incident must be reported?

Regulation 3 defines those injuries and illnesses arising out of work that must be reported forthwith.

(1) *Fatal accidents* Fatal accidents have to be reported if they arise out of or in connection with work whether the person who dies is employed or not. For example, if a scaffolding which has been erected near a public highway collapsed and killed a member of the public who was passing, or if a child trespassing on a construction site fell into a hole and was suffocated by falling earth, both would now be reportable direct to the HSE by the employer who had control over the plant or area where the death occurred. There is a further duty to provide a written report of an accident which has resulted in death within one year of the date of the accident although it may not have been previously reported or have been reported as an injury, illness or dangerous occurrence.

(2) *Accidents causing injury or illness* The second type of accident that has to be reported is one where a person suffers a major injury at work or certain types of illness requiring medical attention. The injuries and illnesses are defined in regulation 3 as:

(a) fracture of skull, spine or pelvis;
(b) fracture of any bone in the arm, wrist, leg or ankle but not the hand or foot;
(c) amputation of a hand, finger, thumb, foot or toe;
(d) the loss of sight of an eye;

(e) injury or loss of consciousness requiring immediate medical attention resulting from:
  (i) electric shock
  (ii) lack of oxygen
  (iii) decompression
  (iv) absorption of any substance
  (v) exposure to pathogens or infected material;
(f) other injury requiring immediate admission to hospital for more than 24 hours;
(g) any other injury resulting in an inability to work for more than 3 days – including days which may not have been work days.

The extent of the injury may not be apparent at the time of the accident or immediately afterwards; or the injured person may not be admitted immediately to hospital. Once, however, one of the above injuries or illnesses is confirmed or more than 24 hours is spent in hospital, then a reportable incident has been identified and the enforcing authority must be notified.

These reporting requirements relate to all employment except in a limited number of industries with special reporting arrangements, such as railways, the merchant navy, nuclear installations, the use of poisonous substances in agriculture and the use of radioactive substances.

(3) *Diseases* A requirement incorporated in regulation 4 is for the reporting of certain diseases contracted as a result of work. The list of diseases is contained in schedule 2 to the Regulations which gives not only the diseases but the occupation to be followed if the condition is to be reportable.
The diseases are split into 5 groups:

(i) poisoning covering 15 substances and their compounds regardless of the type of work being carried out;
(ii) skin diseases resulting from contact with:
  (a) chromic acid or chromium compounds
  (b) mineral oil, tar, pitch or arsenic
  (c) ionising radiations;
(iii) lung diseases covering the following conditions:
  (a) occupational asthma
  (b) extrinsic alveolitis including Farmer's lung
  (c) pneumoconiosis
  (d) byssinosis
  (e) lung cancer, mesothelioma and asbestosis
  (f) cancer of the bronchus or lung;
(iv) infections mainly from animals or handling human tissue including:
  (a) leptospirosis
  (b) hepatitis
  (c) tuberculosis
  (d) any illness caused by a pathogen
  (e) anthrax;
(v) other conditions such as:
  (a) malignant diseases of the bones from working with ionising radiations

(b) cataracts from exposure to electro-magnetic radiations
(c) decompression sickness or barotrauma
(d) cancer of the nasal cavity, liver or urinary tract
(e) vibration-induced white finger.

(4) *Gas incidents* There is a requirement which places on the supplier of gas, whether through a fixed pipe or by refillable containers, a duty to report immediately to the enforcing authority any incident resulting from the use of their product that causes death or any of the injuries listed above. Particularly referred to is poisoning due to incomplete combustion of gas and inadequate ventilation.

(5) *Dangerous occurrences* These are listed in schedule 1 to the Regulations and cover 17 types of incident in any employment with additional incidents listed for mines, quarries and railways.

Details of any of these incidents must be reported to the enforcing authority (see *Table 11.1*) as quickly as possible (i.e. by telephone) with a written confirmation to follow within 7 days using the prescribed form F2508[4] (see *Figure 11.1*). Responsibility for reporting is placed on:

(i) in industry and commerce – the employer;
(ii) in a mine – the manager;
(iii) in a quarry, closed tip or pipeline – the owner;
(iv) in a road vehicle conveying a dangerous substance – the operator;
(v) in any other case – the person in control of the premises where the accident happened.

For those injured while working away from base they must inform their employer, or get someone else to do so, in order that he can report to his local enforcing authority.

It is recognised that the responsible person cannot report an accident if he is not aware of it (regulation 11) but it is his responsibility to institute

Table 11.1. List of enforcing authorities to whom accident reports should be sent

Premises by main activity	*To whom reportable*
1 Factories and factory offices	Factory inspector
2 Mines and quarries	Mine and quarry inspector
3 Farms (and associated activities), horticultural premises, and forestries	Agricultural inspector
4 Civil engineering and construction sites	Factory inspector
5 Non-statutory railways	Inspecting officer of railways
6 Shops, offices catering services, launderettes	Local authority
7 Hospitals, research and development services, water supply postal services and telecommunications, entertainment and recreational services, local government services, educational services and road conveyance	Factory inspector

Health and Safety Executive
Health and Safety at Work etc Act 1974
Reporting of Injuries, Diseases and Dangerous Occurrences Regulations 1985

Spaces below are for office use only

Report of an injury or dangerous occurrence

- Full notes to help you complete this form are attached.
- This form is to be used to make a report to the enforcing authority under the requirements of Regulations 3 or 6.
- Completing and signing this form does not constitute an admission of liability of any kind, either by the person making the report or any other person.
- If more than one person was injured as a result of an accident, please complete a separate form for each person.

A Subject of report *(tick appropriate box or boxes) – see note 2*

Fatality ☐ 1 · Specified major injury or condition ☐ 2 · "Over three day" injury ☐ 3 · Dangerous occurrence ☐ 4 · Flammable gas incident (fatality or major injury or condition) ☐ 5 · Dangerous gas fitting ☐ 6

B Person or organisation making report (ie person obliged to report under the Regulations) – *see note 3*

Name and address –

Post code –

Name and telephone no. of person to contact –

Nature of trade, business or undertaking –

If in construction industry, state the total number of your employees –

and indicate the role of your company on site (*tick box*) –

Main site contractor ☐ 7 · Sub contractor ☐ 8 · Other ☐ 9

If in farming, are you reporting an injury to a member of your family? (*tick box*) Yes ☐ No ☐

C Date, time and place of accident, dangerous occurrence or flammable gas incident – *see note 4*

Date ☐☐ 19☐ (*day month year*) Time –

Give the name and address if different from above –

Where on the premises or site – ENV

and

Normal activity carried on there

Complete the following sections D, E, F & H if you have ticked boxes, 1, 2, 3 or 5 in Section A. Otherwise go straight to Sections G and H.

D The injured person – *see note 5*

Full name and address –

Age ☐ Sex ☐ (M or F) Status (*tick box*) – Employee ☐ 10 · Self employed ☐ 11 · Trainee (YTS) ☐ 12 · Trainee (other) ☐ 13 · Any other person ☐ 14

Trade, occupation or job title –

Nature of injury or condition and the part of the body affected –

F2508 (rev 1/86)

continued overleaf

Figure 11.1(a) Statutory form for reporting injuries and dangerous occurrences. Reproduced by permission of Controller of HMSO. (Source: HSE)

E Kind of accident - *see note 6*

Indicate what kind of accident led to the injury or condition (*tick one box*) –

Contact with moving machinery or material being machined ☐ 1	Injured whilst handling lifting or carrying ☐ 5	Trapped by something collapsing or overturning ☐ 8	Exposure to an explosion ☐ 12
Struck by moving, including flying or falling, object ☐ 2	Slip, trip or fall on same level ☐ 6	Drowning or asphyxiation ☐ 9	Contact with electricity or an electrical discharge ☐ 13
Struck by moving vehicle ☐ 3	Fall from a height* ☐ 7	Exposure to or contact with a harmful substance ☐ 10	Injured by an animal ☐ 14
Struck against something fixed or stationary ☐ 4	*Distance through which person fell ☐ (metres)	Exposure to fire ☐ 11	Other kind of accident (give details in Section H) ☐ 15

Spaces below are for office use only.

F Agent(s) involved – *see note 7*

Indicate which, if any, of the categories of agent or factor below were involved (*tick one or more of the boxes*) –

Machinery/equipment for lifting and conveying ☐ 1	Process plant, pipework or bulk storage ☐ 5	Live animal ☐ 9	Ladder or scaffolding ☐ 13
Portable power or hand tools ☐ 2	Any material, substance or product being handled, used or stored. ☐ 6	Moveable container or package of any kind ☐ 10	Construction formwork, shuttering and falsework ☐ 14
Any vehicle or associated equipment/ machinery ☐ 3	Gas, vapour, dust, fume or oxygen deficient atmosphere ☐ 7	Floor, ground, stairs or any working surface ☐ 11	Electricity supply cable, wiring, apparatus or equipment ☐ 15
Other machinery ☐ 4	Pathogen or infected material ☐ 8	Building, engineering structure or excavation/underground working ☐ 12	Entertainment or sporting facilities or equipment ☐ 16
			Any other agent ☐ 17

Describe briefly the agents or factors you have indicated –

☐

G Dangerous occurrence or dangerous gas fitting – *see notes 8 and 9*

Reference number of dangerous occurrence ☐ Reference number of dangerous gas fitting ☐

H Account of accident, dangerous occurrence or flammable gas incident - *see note 10*

Describe what happened and how. In the case of an accident state what the injured person was doing at the time –

☐

Signature of person making report ☐ Date ☐

Figure 11.1(a) (continued)

Health and Safety Executive
Health and Safety at Work etc Act 1974
Reporting of Injuries, Diseases and Dangerous Occurrences Regulations 1985

For HSE use

Report of a case of disease

- This form is to be used to make a report to the enforcing authority under the requirements of Regulation 5.
- Completing and signing this form does not constitute an admission of liability of any kind, either by the person making the report or any other person.

A Person or organisation making report
(ie person obliged to report under the Regulations)

Name and address

Post code

Name of person to contact for further inquiry

Tel. No.

Nature of trade, business or undertaking

B Details of the person affected

Surname ____ Forenames ____

Date of birth ☐☐ ☐☐ 1 9 ☐☐ Sex (M or F) ☐
day *month* *year*

Occupation ____

Please indicate whether Employee ☐
(tick box)
Other person ☐

If not an employee, what is the ill person's status?
(eg self-employed or trainee)

F2508A (1/86)

Figure 11.1(b) Statutory form for reporting a case of disease. Reproduced by permission of Controller of HMSO. (Source: HSE)

arrangements or working rules that ensure he is informed of any injuries to his employees or those under his control.

The HSW requires employers and the self-employed to conduct their undertakings in such a way that people other than their own employees are not exposed to risks to their health and safety. However where a non-employee or member of the public receives fatal or major injuries as a result of a work activity, this fact must be reported to the HSE in the same way as for employees. Accidents to members of the public which arise out of work activities may occur in many circumstances. For example:

(a) In the street or other areas where the public go about their ordinary business, and are injured e.g. by the collapse of scaffolding or by falling into an excavation.
(b) In shops or business premises, e.g. by falling on moving stairways, or being struck by the collapse of shop fittings, etc.
(c) At building sites and enclosed factory premises, injuries to children or others trespassing on the premises.

However, for those accidents that have not caused major injury but which occasion more than three days incapacity for work, the injured person may enter a claim for industrial injury benefit with the DSS who will send the employer a form BI 76 (*Figure 11.2*). Normally benefit is paid only in respect of absences of more than three days' duration but a claim following an absence of shorter duration may also be entered for an injury which in the long term could attract a degree of disability benefit.

Certification for absences of up to five days is by the individual himself with claims for social security benefit being made on a 'New Sickness Benefit Claim Form' – form SC1 – obtainable from the DSS (see *Figure 11.3*). For absences beyond the first five days, a medical certificate is required. Under statutory sick pay (SSP) arrangements payment of benefit for the first twenty eight weeks of absence is by the employer who is entitled to deduct the appropriate sums from the moneys he pays to the DSS in respect of his and his employees' contributions under the Social Security Act 1975.

(6) *Written records* A written record (regulation 7) must be kept by the responsible person of all reportable injuries and dangerous occurrences that occur in connection with the work or premises under his control. The details to be kept are listed in schedule 3 of the Regulations and have been contained in the form F2508[4]. However, company forms or reports that contain this information are suitable alternatives.

There is also a requirement to record incidents of disease. The details to be recorded are given in part II of schedule 3 namely:

(i) date of diagnosis of the disease;
(ii) occupation of the person affected;
(iii) name or nature of the disease.

The name of the patient is not required.

These obligations to keep records extend to include the self-employed even though there may be no duty on him to report accidents. These records must be kept readily available and be retained for at least three years. If the

Department of Social Security

If you get in touch with us tell us this reference number

If you ring ask for extension

Date

Please reply by .... / ...... / .....

....

....

....

....

Industrial Injuries Benefits — we need some information

Name .... Works number ....

National Insurance Number

Type of job .... Works department ....

Date of accident .... / .... / .... Time .... am/pm

Place ....

Details of Incapacity/Injury ....

The person named above has made a claim under the industrial injuries provisions due to an accident at work. To enable the claim to be decided will you please answer the questions on the next 2 pages.

As no benefit can be paid until the claim has been decided please return the form as soon as you can, and by the date at the top of this page at the latest. Use the envelope we have sent you. It does not need a stamp.

The information you give will normally be used only by the authorities who decide the claim, but details may have to be disclosed to the claimant and discussed at a hearing.

If at some future date you want a copy of the information you give on this form we will send it to you on request.

Form BI76

Figure 11.2 First two pages of form BI 76 for confirmation of injury circumstances. Reproduced by permission of Controller of HMSO. (Source: DSS)

Please answer all questions and tick the boxes that apply

Reply

1 After investigation, are you satisfied an incident occurred on the date, time and place given on page 1?

If **No,** please explain

a date Yes ☐ No ☐ ....

b time Yes ☐ No ☐ ....

c place Yes ☐ No ☐ ....

2a On the date of the incident was the claimant employed by you in the job given on page 1? (You should state if the claimant was a trainee on a Manpower Services Commission Scheme.)

Yes ☐ No ☐ If **No** please explain

2b Was the claimant a sub-contractor, partner, your wife or husband?

Yes ☐ No ☐ If **Yes** state which

3a On the day of the incident what hours was the claimant expected to work?

from .... to ....

3b What hours did the claimant actually work?

from .... to ....

---

4 At the time of the incident, what was the claimant doing?

---

5 What was the incident and how did it happen? (If the claimant fell or something fell onto the claimant, what was the height of the fall?)

6a At the time of the incident was the claimant authorised for the purposes of work to be where he/she was?

Yes ☐ No ☐ If **No** please explain

6b Was the claimant authorised to do what he/she was doing?

Yes ☐ No ☐ If **No** please explain

Self-certificate
And Sickness and Invalidity Benefit claim form **SC1**

1 Yourself
PLEASE USE BLOCK LETTERS
If you cannot fill this form in yourself ask someone else to do so and to sign it for you
Surname Mr Mrs Miss Ms
First names
Present address
Postcode
Date of birth
National Insurance number
CLOCK/STAFF/WORKS NUMBER

2 Details of sickness — Give details of your sickness. Words like 'illness' or 'unwell' are not enough
Please say briefly why you are unfit for work
• Is your sickness due to an accident which happened while you were working for any employer?
Tick one box YES ☐ NO ☐
• Is your sickness due to a prescribed industrial disease caused by conditions at work while you were working for an employer?
Tick one box YES ☐ NO ☐
NOTE: This does not apply if the accident or prescribed disease occurred while you were self employed
FORM SC1

3 Period of sickness
If you do not fill in this section or if you fill it in wrongly, any payment of benefit due to you may be delayed.
Everyone to fill in — Date you became unfit for work — day / month / year 19
Do not complete if you are unemployed — Date you last worked — day / month / year 19
Time you finished work — time am/pm
Everyone to fill in — Do you expect to be unfit for work for more than 7 days? YES ☐ NO ☐ (Tick one box)
If you ticked YES go to part 5
Night shift workers only — When did your last shift begin? — time am/pm — day / month / year 19

4 Returning to work
Last day of sickness before starting or seeking work — day / month / year 19
Date you intend to start or seek work — day / month / year 19
Night shift workers only — Shift will begin at time am/pm and end next day at time am/pm

5 If you are claiming Sickness or Invalidity Benefit
Go to part 6 'Your work' — do not sign below
If you are using this form for Statutory Sick Pay
Stop here. Sign below, and send this form to your employer
Signature Date
Remember if you are sick for a second week, your employer may want a sick note

6 Your work
Are you?
Tick one or more boxes — Employed ☐ Self-employed ☐ Unemployed ☐ Other ☐
What is your usual job?

Figure 11.3 Sickness benefit claim form. Reproduced by permission of Controller of HMSO. (Source: DSS)

enforcing authorities require it, copies of these records must also be sent to them.

These records can also be used by the employer to work out over a period of time the number of accidents which have occurred in his premises; and establish whether accident figures are improving, and whether there are any particular trends in causation or location. This information helps employers to make cost-effective decisions about preventative action.

11.2.2 Social Security Act 1975

This Act deals with, *inter alia*, the benefits paid to workpeople who are unable to work due to disability, illness or injury. In the last circumstances regulations[5] under this Act require an employee who suffers personal injury by accident at work to inform his employer immediately (either orally or in writing) or as soon as possible, after the accident occurred.

For his part, the employer must take reasonable steps to investigate the circumstances of every such accident notified to him and, if there is any discovered discrepancy between his findings and the information provided by the employee, must record that discrepancy for possible future reference.

An employer, who normally employs 10 or more employees on or about the same premises must maintain an Accident Book[6] in which can be recorded, if the injured employee wishes it, the appropriate particulars of any accident causing personal injury to him. The Accident Book must be kept, when filled, for at least three years after the date of the last entry in the book. It must also be available for examination on demand by an accredited officer of the Department of Social Security.

11.2.2.1 Industrial diseases

It has long been recognised that certain industrial processes can cause diseases and there is a duty under section 82 of FA 1961 on doctors who diagnose patients suffering poisoning due to compounds of lead, phosphorus, arsenic and mercury to report the fact to the HSE. In addition under the Social Security (Industrial Injuries) (Prescribed Diseases) Regulations 1980[2], there are over 50 prescribed occupational diseases, which, if they occur or develop whilst claimants are in prescribed occupations, will qualify claimants to receive Industrial Injury Benefits.

There are a number of conditions which apply in such cases. Those diseases or injuries will be presumed to have resulted from prescribed occupations, unless there is evidence to the contrary, provided an individual claimant has been employed in a particular occupation for a minimum period (ranging from one month in most cases, to 20 years in the case of occupational deafness).

The Department of Social Security which is responsible for the payments of claims under the Social Security Act 1975 will send employers a Form B177 for confirmation of employment and details of the work carried out by employees who are making claims for industrial disease benefits.

11.3 Learning of the event

While the legal obligation is placed on the employer to report accidents he can in fact only report them if he is aware of them, so it is incumbent upon him to include in his works rules a procedure to be followed for informing him that an accident has happened. Any such procedure must depend upon the size of the organisation, the larger the organisation the more formal the procedure. This could range from the manager of a small firm himself giving the first aid treatment and recording it in either his own log or the Accident Book[6] to attendance in a large firm or site at a fully staffed medical centre which issues formal notes of treatment for the injured person to take back to his supervisor.

Where the injury is sufficiently serious to warrant attendance at hospital, if the local supervisor is not already aware of the accident, then he should be contacted by telephone so that he can start his investigation into the cause. Similarly the safety adviser should also be informed so that he can follow the matter up.

With serious accidents, details of the injuries may have to be obtained from the hospital, although there may be some reluctance for them to give this information to other than relatives.

In the event of a fatality, in addition to the deceased's family, the police, as coroner's officers, and HSE should be informed immediately. The employer's insurance company will also wish to know as soon as possible so they can make financial provisions and also keep a watching brief on investigations and any legal actions.

With notifiable dangerous occurrences and damage only accidents, the incident should be reported to the supervisor in whose area of control the damage occurred. In the case of notifiable dangerous occurrences he should contact the person responsible for reporting the matter to the enforcing

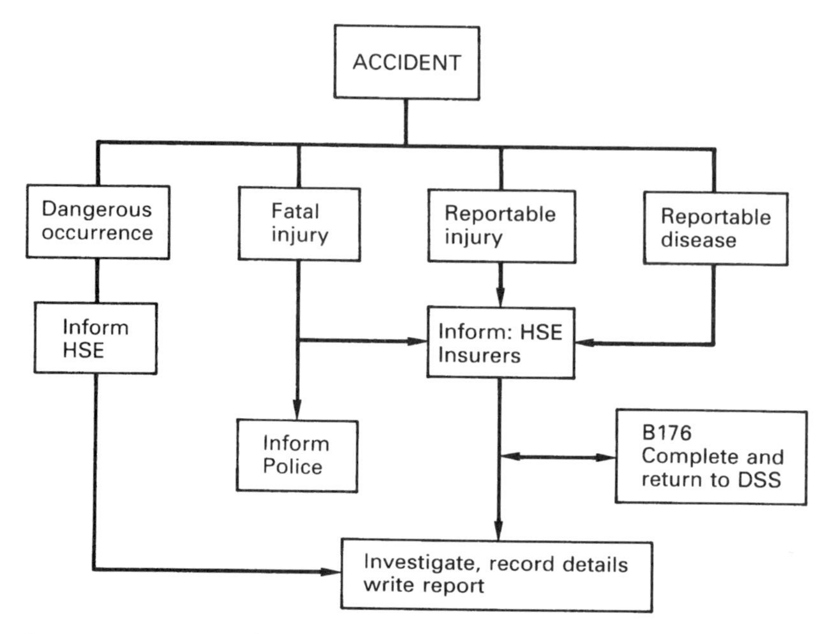

Figure 11.4 Diagram of action to be taken on learning of an accident

authority. Other damage only accidents should be fully investigated, either by the supervisor or using the laid down procedures, so that the cause can be established and remedial action implemented. Again the company's insurers may need to be informed if the plant is covered by an insurance policy.

Injuries occurring while an employee is working on another company's premises should be reported to his employer either by the injured person himself or by the company on whose premises the injury occurred. For accidents on the road to drivers, while these would normally come under the Road Traffic Act 1972, the driver if he can should inform his employer, otherwise the police would do so.

A summary of action to be taken on hearing of an accident is shown diagrammatically in *Figure 11.4*.

11.4 Arriving at the scene

In any accident the first and most immediate action should be to succour the injured and take action to minimise the possible effects of his injuries. First aiders should be given all assistance possible and their advice on handling the patient followed. In the case of serious injuries a check should be made that an ambulance and medical assistance have been called. When they arrive the paramedics or doctor will take charge of treating the patient. The patient should not be moved but made as comfortable as possible.

Where he is trapped in machinery efforts should be made to dismantle the machine so that he can be released.

Once the patient has been removed and taken away for treatment the site should be isolated and nothing disturbed. An investigation should start immediately. In the event of a fatal accident, the deceased should be identified and relatives informed. There could be some religious problems in the case of Indian, Pakistani, Jewish or Roman Catholic employees, etc., and their wishes in respect of, e.g. the local priest etc. should be complied with. The police should be informed and the site, including property of the deceased, tools etc., should be left undisturbed until they arrive. The police investigation will be concerned with the cause of death rather than the cause of the accident. The HSE should also be informed and, if the death was due to gassing or poisoning, a doctor from EMAS will be involved.

11.5 Investigation

There should be a laid down procedure for investigating accidents with, ideally, the supervisor of the section where it occurred carrying out an immediate examination. If the seriousness of the accident warrants it, a further examination by a qualified specialist, such as a safety adviser, engineer etc., should be carried out. How the findings of these examinations are recorded will depend on the sophistication of the firm's organisation, but could be a pre-printed form or as a manuscript report or a combination of the two.

Any investigation, to obtain the most accurate evidence possible, should be undertaken immediately after the accident. It should be concerned with obtaining facts to establish the cause of the accident and not with finding out who was to blame. Wherever possible the accident should be discussed at an early stage with the injured person to get his version of the events. If the injured person is either sent home or to hospital this may well have to wait till he can be visited.

Witnesses should be interviewed as soon as possible and in taking note of their evidence care must be exercised to differentiate between what are measurable facts and what is hearsay or opinion. They should be interviewed one at a time although, if they wish it, they may be accompanied by a representative of their union. The object of the investigation, to establish the cause of the accident and not apportion blame, should be explained to them. Any discrepancies between witnesses' evidence should be checked.

Where a safety representative requests it, a joint investigation should be carried out. A list of typical information required from an investigation is given in section 8 of this chapter. It is often helpful when providing information on an accident to supplement the written word by sketches, drawings, photographs and makers' manuals.

11.6 Investigation report

Where the investigation of an accident or damage incident requires a separate written report care must be taken 'to organise and plan material for report writing and . . . to present the report in such a way that the key points are easily established and the conclusions easily drawn'[7].

The material for the written report would be the same as that gathered during the investigation but the form of its presentation is left to the author. Its format should follow a logical sequence which could be:

Title – to say what it is about.
List of Contents – but only if the report is long enough to warrant it.
Summary – so that senior managers can quickly get the gist of the report content and findings.
Introduction – to give background to the report.
Findings – recording the information gathered during the investigation.
Conclusions – drawn up from the findings.
Recommendations.
Appendices – may be necessary for supporting material, such as tables, sketches, photographs etc.

Whether the report is dictated or written in longhand it should be typed in draft and its contents checked critically to ensure that the points that need to be brought out have been correctly emphasised and that the conclusions and recommendations say what was intended to be said. Before issue, the report, either on its title page or the last page should be signed and dated.

11.7 Accident investigation checklist

The following list can be used as an *aide-mémoire* for those matters that should be considered when undertaking an accident investigation. It is not exhaustive, but covers the range of essential items likely to need consideration.

Objectives	Discover facts – Prevent recurrence.
	Legal duty to inquire into reported accidents.
	National Insurance (Industrial Injuries) Act.
	Accident Book – BI 510A – Form BI 76.
	Notification to HSE – Form 2508.
	Employer's liability insurance – Third party insurance.
	Evidence for possible civil actions for damages.
Who Investigates	Supervisor.
	Safety adviser.
	Union representative (joint consultation).
	Formal inquiry.
When	Early as possible.
	Effect of delay – lost evidence.
Course of Action	Disaster plan.
	Camera, tape measure.
	Emergency telephone numbers, test equipment, etc.
	Early visit and questioning injured person – effect of shock.
APPROACH	
Faults of Persons	Dangerous occurrence – Major injuries or damage control.
	Strains and sprains – minor injuries.
	Distinguish between injury and accident.
Technique	Interview one witness at a time.
	Explain reason – to discover cause.
	Check knowledge of the witness.

Real Evidence	(Productions). Broken machinery. Sketches and photographs. Models. Expert examination – expert may give opinion.
DANGEROUS OCCURRENCES	
	Types notifiable to HSE. Others of interest to Company (Risk Manager).
Report Writing	State type of machine. What was injured person doing? What went wrong? Why? Recommendations to prevent recurrence.

11.8 Typical information required after a lost-time accident

(1) Name of owner or occupier.
(2) Department or section or name of contractor and/or employer's address.
(3) Type of process or industry, e.g. engineering, cabinet-making, paper-making, civil engineering etc.
(4) Exact place of accident – sketch map or photograph if a serious accident. A photograph especially if conditions likely to change or plant be moved.
(5) (a) Name of injured person and clock number.
(b) Occupation.
(c) Age – married or single – sex.
(d) Home address.
(e) How long continuously employed by the Company?
(f) How long continuously employed in present job?
(g) Registered disabled.
(h) Names of witnesses and clock numbers.
(j) Addresses of witnesses.
(k) Name and address of foreman/chargehand directly in charge of job.

} Some detail often obtainable from personnel records

(6) (a) What was he doing at the time of the accident? If different from 5(b) state reasons. (The safety adviser will try to assess whether training and/or instructions given were adequate, and especially if a contributory cause was a hearing loss, failing eyesight, etc.) In the event of an accident due to slippery surface, enquiries should be made as to the type of footwear worn at that time, and the type of floors.
(b) Was the job he was doing authorised?
(7) Full nature of injury, where treated, own doctor's name or name of hospital.

(8) (a) Date and time of accident.
(b) Date and time he actually ceased work due to accident if different from above.
(c) What time was he supposed to work on the day of the accident? i.e. From .... to ....
(9) Supply sketch if necessary to explain report. Give measurements, weights, heights, angles, condition of ground etc. – anything you consider may have a bearing on the accident.
(10) If due to machinery – state: type, name, what used for, i.e. process. State name of part causing injury. If it is guardable – was it already guarded? Was guard removed or not adequate? Check statements as to how it occurred. Check on previous accidents on the machine etc. or any other relevant history. Check drawings from drawing office or area engineer – maker's handbook, etc.
(11) Explain probable cause of accident giving reasons.
(12) Outline action taken to prevent recurrence.

11.9 Legal status of accident reports

Following an investigation into an accident a number of reports have to be prepared or forms completed that have different roles and that have different standing at law.

In law, accident reports fall into two broad categories, discoverable or privileged. Discoverable documents are those which one party who possesses them must disclose to the other party if requested to do so. Typical documents within this category would be the statutory forms F2508 and F2508A. At the other end of the scale are those documents that are private to the party holding them. Typical of these would be the insurance claim form made out by the employer to obtain indemnity under his insurance policies following an accident, and the form completed by trade union members requesting their union to initiate a claim on their behalf. These are private between the employer and his insurer and the union and its members and are given the privilege of immunity from discovery.

Between these two extremes there are a range of documents of which a written accident report is typical. The status of such documents was determined by the judgement *Waugh* v. *British Railways Board*[8] where it was held that, if the prime purpose of the investigation and report was to discover the cause of the accident, then the document was discoverable. For a document to be accorded professional privilege it would be necessary to prove that 'the dominant purpose for which it was prepared was that of submitting it to a legal adviser for advice and use in litigation'. The report of an insurance surveyor into an accident would be privileged since its dominant purpose is to obtain information for use in possible litigation. Other documents used in the day-to-day business of an employer such as memoranda, works requests, daily reports etc. are also normally discoverable, but medical records of an identifiable individual are confidential and are not normally disclosed except with the agreement of the individual concerned or on the direction of the courts.

References

1. Health and Safety Executive, *The Reporting of Injuries, Diseases and Dangerous Occurrences Regulations 1985*, HMSO, London (1985).
2. Department of Health and Social Security, *Social Security (Industrial Injuries) (Prescribed Diseases) Regulations 1980*, HMSO, London (1980)
3. HSE, Booklet no. HS(R) 23, *A guide to The Reporting of Injuries, Diseases and Dangerous Occurrences Regulations 1985*, HSE Books, Sudbury (1986)
4. HSE, Report of an injury or dangerous occurrence, Report Form F2508 (Rev 1/86), HSE Books, Sudbury (1986)
5. DSS, *Social Security (Claims and Payments) Regulations 1975*, HMSO, London (1975).
6. DSS, *Accident Book*, Form BI 510A, HMSO, London (1972)
7. Vidal-Hall, J., *A Guide to Report Writing*, Industrial Society, London (1972)
8. Waugh *v.* British Railways Board [1980] AC 521: [1979] 2 All ER 1169

Further reading

Gowers, Sir E., *The Complete Plain Words*, Pelican Books, London (1969)
Cooper, B. M., *Writing Technical Reports*, Penguin Books, London (1964)
Roget, J. L. and Roget, S. R., *Thesaurus*, Penguin Books, London (1960)

Chapter 12

Records and statistics

Dr A. J. Boyle

12.1 Introduction

There is a popular belief that statistics can be made to show whatever the statistician wants them to show. This is nearly true, it *is* easy to lie with statistics, but it is equally easy to lie with English, and no one suggests that this is a reason for dispensing with English! The difference arises because people are used to and understand verbal arguments so that if what they hear or read appears false, they challenge it and continue to ask relevant questions until they are satisfied. The same does not apply with statistics which the majority of people do not understand, so that when presented with statistical arguments, they either accept them without being able to question them, or they reject the whole subject out of hand. This is unfortunate when it happens with safety advisers since statistical techniques, properly applied, can be useful to them in their work of preventing accidents. Some of these techniques and their application in quantifying certain common safety factors are described below.

The preparation of statistical data requires two quite distinct skills, mathematics and arithmetic. These two skills are often confused but, in general, mathematics is concerned with the relationships between numbers or values which are shown through the use of symbols, as is the case with algebra. Arithmetic deals with the numbers themselves and involves the skills we use every day when working out the correct change or how long it will take to complete a journey.

The mathematics required in statistics are very limited and are mainly confined to notation, that is, mathematical symbols used to summarise what has to be done. (It is much quicker to write 'Σx' than it is to write 'add up all the x's' and to write 'x' rather than 'the number of accidents sustained by each person in the company'). However, should difficulty be experienced with either mathematics or arithmetic one of the elementary texts, available from most large bookshops, should be consulted.

12.2 Statistical analysis

The aim of most statistical analysis is to provide the answer to a question. However, unless the question is formulated in an appropriate way, the answer may be misleading. The diagram in *Figure 12.1* shows a typical

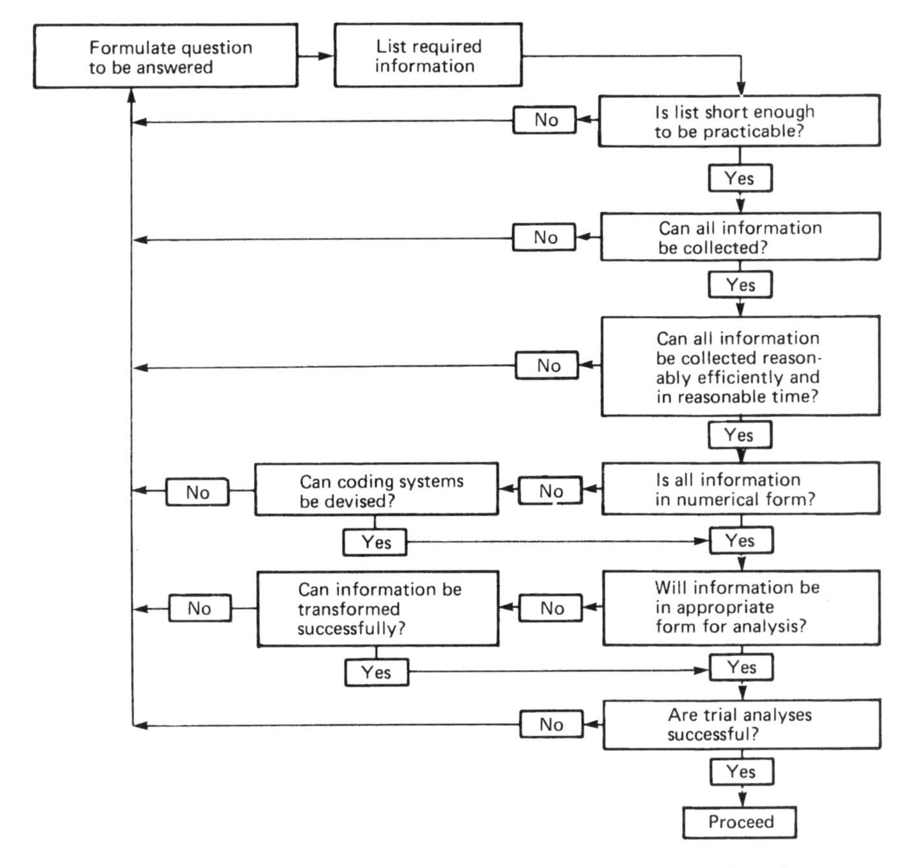

Figure 12.1 Diagram of steps in designing a project using statistical analysis

procedure for checking that the question to be asked is formulated appropriately.

It can be seen from *Figure 12.1* that there are a number of factors which might lead to a reformulation of the question being asked. This process of reformulation is an important one since, to a large extent, it is what ensures that relevant questions are asked and that the answers obtained are realistic. The use of the procedure in *Figure 12.1* is illustrated below using the question 'How much do accidents cost my company?'

Having formulated this question, the next stage is to list the information required. Accidents have associated with them a wide range of costs and as many as possible should be listed so that as accurate an overall cost as possible can be arrived at. However, it is easy to attempt to list too much information, so a check needs to be carried out to ensure that what is asked for is realistic and practicable. This may require reformulation of the original question to exclude, for example, the costs associated with near misses.

Next it is necessary to determine whether the required information can be collected bearing in mind the number of accidents involved. This is particularly important in the present case since not all accidents are reported, especially those involving minor injuries and damage. This latter point may necessitate reformulation of the question to ensure that it reflects accurately the information that can be collected. For example, 'How much do major injury accidents cost my company?'

When considering whether the required information can be collected reasonably efficiently and in a reasonable period of time, further practical problems may be met. A variety of techniques are available for collecting data including the searching of records, observation, measurement, questionnaires and interviews with individuals. The appropriate technique will depend on the type of information required. For example, while it may be possible by examining accident reports to assess the time lost by injured persons, it may require interviews to determine how much working time was lost by colleagues and supervisors as a direct result of particular accidents.

It is also necessary to consider the time it will take to collect the data, both in terms of the calendar period, as when gathering information on say the next fifty major accidents, and also in terms of the man-days required by the collector to assemble the required data. Should either of these times prove prohibitively long, then the original question can be reformulated to, for example, 'How much did major injury accidents cost my company last month?'

Next it is necessary to find out if all the information gathered can be put in numerical form. Typical costs which are difficult to quantify include pain and suffering and loss of customers' goodwill due to late deliveries.

Having assembled the information, it is necessary to assess if it is in a suitable form for analysis. For example, it may be that surgery costs are well documented and broken down into staff costs, material costs, overheads etc., but it may nevertheless not be possible to decide what part of these costs should be apportioned to the treatment of a particular accidental injury.

The final stage in the setting up of an analysis is to carry out a trial check of the proposed analyses, either using a preliminary sample of the collected information or information invented for the purpose. This enables an estimate to be made of the time the complete analyses will take, there being little point in collecting so much information that the analysis takes too long to be useful. An additional benefit from a trial run is that it often shows up problems which were not identified earlier, and enables final refinements to be made to the formulation of the questions before the full analysis is commenced.

Having established that the question to be asked is relevant and can be answered, the range of statistical techniques available needs to be considered.

12.3 Descriptive statistics

Descriptive statistics is the term used for those statistical techniques which convert numerical data into a more 'user friendly' form. Because they are the simplest of the statistical techniques they are dealt with first. However, in practice they would be the last step in the analysis procedure.

The use of descriptive statistics is illustrated by giving an extract from a hypothetical annual report, then presenting the pertinent parts of the information using descriptive statistics.

From the annual report:

'Concerning our activities in the safety field, as in past years we have reviewed the four main areas of company activity, Production, Dispatch, Maintenance and Offices, looking at the number of lost-time accidents and the average number employed. The numbers of accidents were 53 for Production, 16 for Dispatch, 26 for Maintenance and 12 for Offices. The average numbers employed over the year were Production 117, Dispatch 11, Maintenance 37 and Offices 83. An accident rate was calculated for each department by dividing the number of accidents in a department by the average number of people employed. The resulting accident rates were 0.45 for Production, 1.45 for Dispatch, 0.7 for Maintenance and 0.14 for the Offices. The comparable rates for last year were Production 0.41, Dispatch 1.21, Maintenance 0.72 and Offices 0.14'.

There is a lot of information in this extract and presenting it in tabular form, as in *Table 12.1*, makes it much easier for the reader to assimilate. It also has benefits for the person preparing it since it helps to order his thoughts and any items missed will show as obvious gaps.

In addition to Tables, numerical information can be presented diagrammatically using, for example, pie charts, pictograms and histograms (or bar charts), all of which are illustrated below using the information from Table 12.1.

Pie charts use a circle (the 'pie') to represent the total quantity or number of the feature being considered, and the radial subdivisions of the circle (the 'slices') represent parts of this whole. In the example there were 107 lost-time accidents in 1992 which is represented by the whole 'pie', and the proportion of the total accidents which occurred in each of the four departments is illustrated by the appropriate size of slice of the 'pie'. This is shown in *Figure 12.2* both as a plain and isometric presentation.

Pictograms (or pictographs) show pictorial representations using the feature being considered and indicate the quantity by either relative sizes or relative numbers of drawings. In *Figure 12.3*, a simple pictogram is used to show the relative numbers of people employed in each department.

Table 12.1. Lost-time accidents and average numbers employed in four departments in 1992 and comparison of 1991 and 1992 accident rates

Deparment	*No. of lost-time accidents (1979)*	*Average no. employed (1979)*	*Accident incidence rate (X/Y)*	
	X	*Y*	*1992*	*1991*
Production	53	117	0.45	0.41
Dispatch	16	11	1.45	1.21
Maintenance	26	37	0.7	0.72
Offices	12	83	0.14	0.14

Production
Maintenance
Offices
Dispatch
(a) Plain diagram
Production
Maintenance
Offices
Dispatch
(b) Isometric diagram

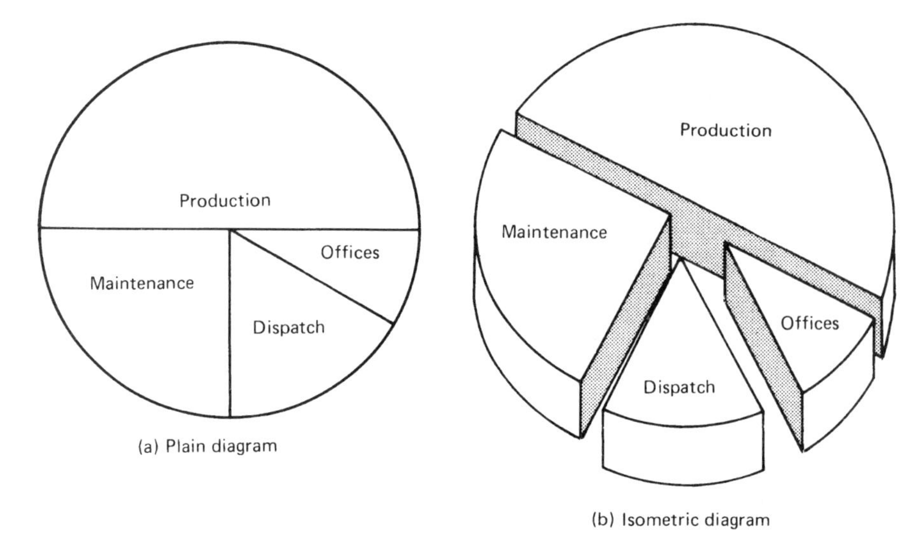

Figure 12.2 The distribution of 107 lost-time accidents in 1992 in four departments (a) plain diagram; (b) isometric diagram

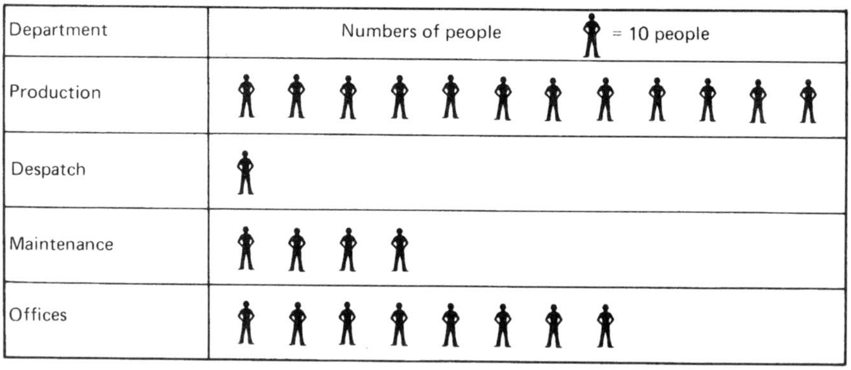

Department
Numbers of people
= 10 people
Production
Despatch
Maintenance
Offices

Figure 12.3 Pictogram of average number employed

Presenting two related pieces of information diagrammatically can be achieved by the use of compound diagrams. In the example, it would be possible to combine *Figures 12.2* and *12.3* by drawing the appropriate numbers of stick men into the correct slice of the 'pie'. In general, information can be more readily assimilated when presented in an appropriate diagrammatic form as illustrated in *Figure 12.4*[2].

The final type of descriptive statistic described in this section is the histogram or bar chart. *Figure 12.5* is a histogram which shows a comparison between accident rates in 1991 and 1992. It will be noted that in this histogram the scale is a value obtained from a prior calculation (i.e. accidents per 100 employees); it could, in other circumstances, equally well be a straightforward number such as the number of accidents. In practice it

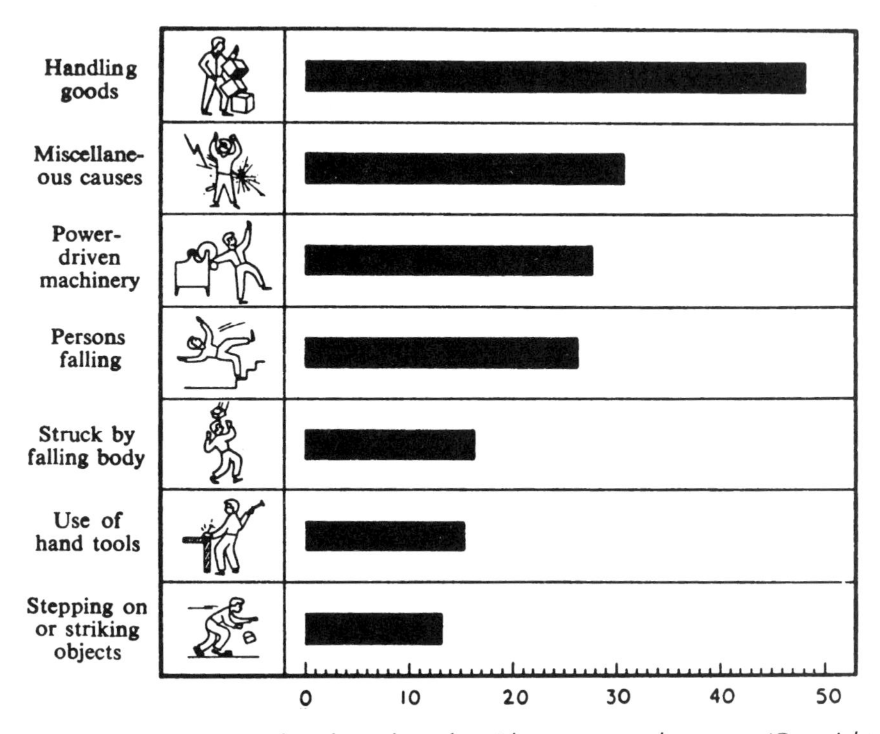

Figure 12.4 Pictogram of total number of accidents per year by cause. (Copyright 1961, International Labour Office, Geneva[2])

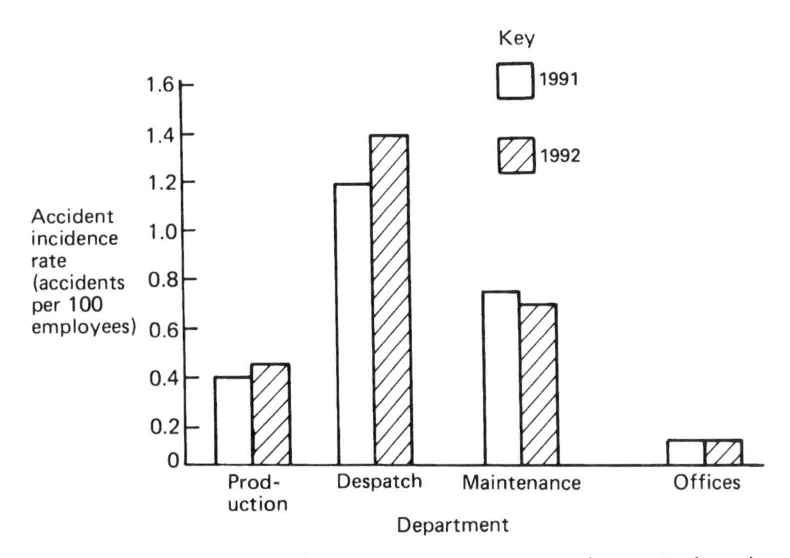

Figure 12.5 Comparison of accident rates in 1991 and 1992 in four departments

is possible to use any scale that is relevant to the analysis being displayed, perhaps those most often used being frequency rate and percentage. For displaying certain types of data it is useful for the scale to be horizontal (as in production programmes), and these are normally referred to as bar charts.

One further example of descriptive statistics is the cumulative frequency graph which is described later in the chapter.

Further information on the production and use of tables and diagrams is given in Moroney[3] and Huff[4].

12.4 Summary statistics

In this section, two topics will be introduced: general summary statistics which can be used for most types of numerical information and specialised accident summary statistics.

12.4.1 General summary statistics

It is possible to summarise a large set of numbers by using a few numbers, just as it is possible to summarise a long report in a few words. In the case of numbers, however, there are laid down procedures for producing such summaries so that legitimate comparisons can be made with other summaries. Three such summaries are considered: range, central tendency and variance. Calculations to obtain these will be illustrated, using as

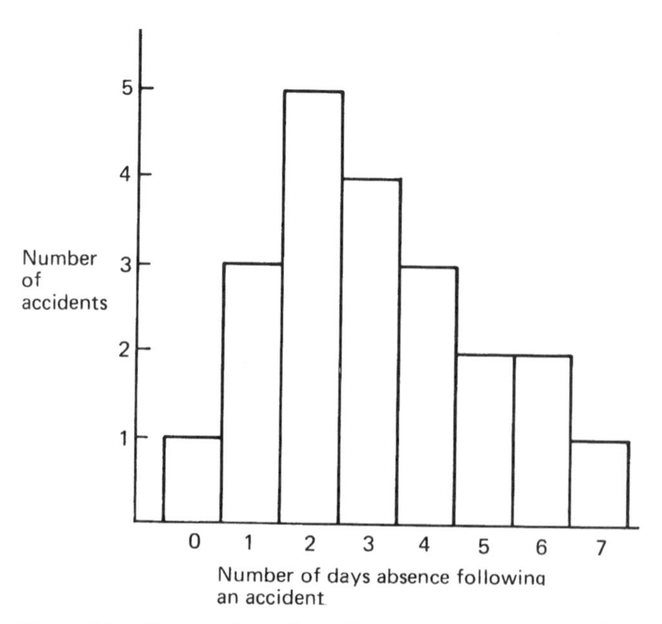

Figure 12.6 The number of accidents occasioning specified number of days absence from work

examples the information in *Figure 12.6* which, for 21 accidents, shows the number that resulted in absences of 0, 1, 2, 3 etc. working days.

12.4.1.1 The range

This summary statistic gives the spread of the numbers in the set and is calculated simply by subtracting the lowest number in the set from the highest number. For the days lost data in *Figure 12.6*, therefore the range is

$$7 - 0 = 7$$

12.4.1.2 Central tendency

As its name suggests, this type of summary statistic is used to indicate the mid-point of a set of numbers. Usually it is referred to as the 'average' but this term can be used for a number of different types of measure of central tendency and it is not normally used by statisticians. Instead, they use terms which describe specific measures of central tendency, and three of these measures, the arithmetic mean, the median and the mode, are described below.

(1) *The arithmetic* mean (usually referred to as just the mean) is the measure which is most commonly understood by the term 'average'. To find the mean days lost per accident add up the total number of days lost and divide by the number of accidents (including the one where no days were lost). The appropriate calculations are set out in *Table 12.2.*

(2) *The median.* In the example, if the accidents are arranged in ascending or descending order of days lost, then the median value is the number of

Table 12.2. Steps in calculating the arithmetic mean

Days lost per accident x	*Number of accidents causing days lost* *f (frequency)*	*Man days lost* $f \times x$ *(or fx)*
0	1	0
1	3	3
2	5	10
3	4	12
4	3	12
5	2	10
6	2	12
7	1	7
Total	21	66

Mean number of days lost $\bar{x}$ =

$$\frac{\text{Total of } fx}{\text{Total of } f} = \frac{\Sigma fx}{\Sigma f} = \frac{66}{21} = 3.1$$

Note that the mathematical notation for the mean is $\bar{x}$ (read as 'x bar')

days lost as a result of the middle accident of the sequence. Since there are 21 accidents in the example, the middle accident is number 11. Arranging the accidents in increasing order by days lost thus:

accident 1 lost no days
accidents 2–4 lost 1 day
accidents 5–9 lost 2 days
accidents 10–13 lost 3 days

The middle accident, no. 11, lost 3 days; therefore, the median number of days lost is 3.

It does not matter that the 10th and 12th accidents also resulted in 3 lost days. The median will be the same if the calculation started from the highest number of days lost. Thus the median is the middle point (or central tendency) of a set of numbers in the sense that there are as many numbers greater than the median as there are less.

Obtaining the median is slightly more complicated when there is an even number of accidents. For example, had there been 22 accidents, there would not have been a 'middle' accident. In this case, the median is the mean of the measures immediately above and below the mid-point. If a 22nd accident is added to *Figure 12.6* that resulted in 30 days away from work, then the mid-point will be between the 11th and 12th accidents, each of which caused 3 lost days. The median, the mean of these two, is

$$\frac{3 + 3}{2} = 3.$$

It is possible to calculate the middle points necessary to obtain the median as follows:

(a) if there are an odd number of measurements, add 1 to the number of measurements and divide by 2. Thus for 21 accidents the middle one will be

$$\frac{21 + 1}{2} = 11$$

(b) If there is an even number of measurements, divide by 2 to get the lower one of the required pair and add 1 to this number to get the higher one. Thus for 22 accidents, the lower one will be 11 and the upper one 12.

(3) *The mode* (or modal value) is simply the most frequently occurring value in a set of numbers. In *Figure 12.6*, the number of lost days that occurred most frequently, as seen by the tallest column, was 2, so the mode for days lost is 2.

The particular measure of central tendency used will depend upon the circumstances since each has its strengths and weaknesses. The 'mean' is the most 'sensitive' measure and is therefore affected by a single extreme event. If a 22nd accident occasioning 30 lost days is added to the *Table 12.2* data the mean would rise from 3.1 to 4.4. The median, on the other hand, may be the more appropriate measure when fractional values are not realistic, as with accidents, while the mode, because it represents the most frequently occurring value, is the value most likely to occur in the future.

Further details of measures of central tendency can be found in Moroney[5].

12.4.1.3 Variance and standard deviation

Summarising sets of numbers using one or more measures of central tendency can be adequate in some circumstances, but in others it is not. The problem is illustrated in *Figure 12.7* where three distributions of days lost are shown, each with the same mean ($\bar{x} = 2$) and range ($r = 4$).

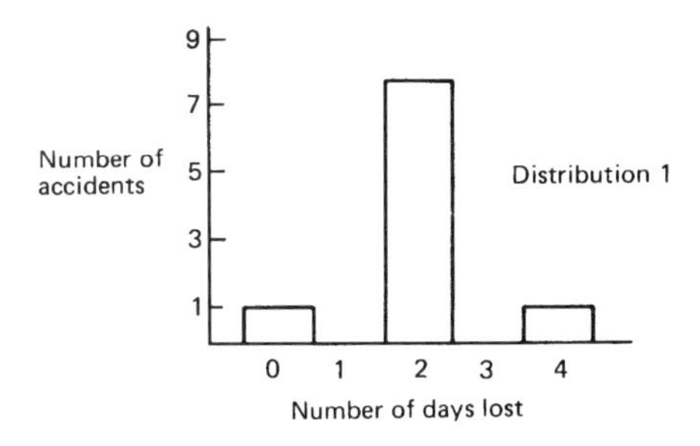

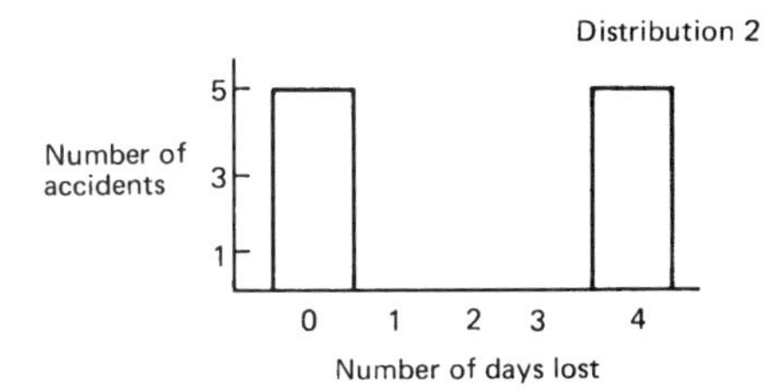

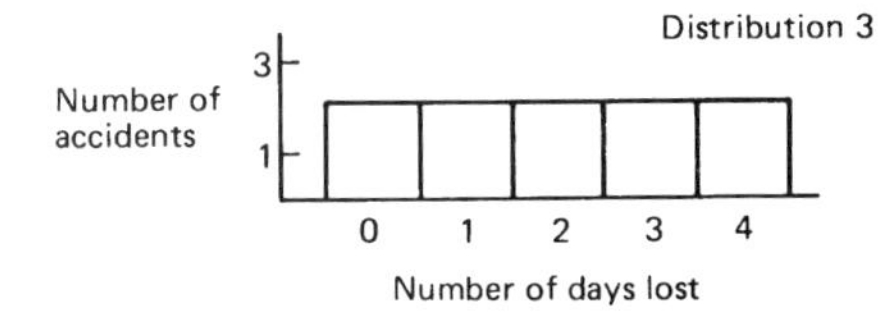

Figure 12.7 Three different distributions of number of days lost, each having the same mean and range

Thus, some measure other than mean and range is required if the differences between the three distributions are to be summarised. Such a measure is the *variance*. The steps to calculate the variance are:

(1) Calculate the mean of the measurements as described above (*Table 12.2*).
(2) Subtract the mean from each of the measurements represented. (Note that for measurements greater than the mean, these subtractions will result in positive numbers but for measurements less than the mean the results will be negative numbers.) The results of these subtractions are called the differences (d).
(3) Square each difference (i.e. multiply it by itself so that $d \times d = d^2$). This eliminates all minus signs since for example $-4 \times -4 = 16$.

(4) Calculate the mean of the squared differences. This mean is the variance.

Another measure, the *standard deviation,* is frequently used and this is calculated by:

(5) Taking the root of the variance (i.e. standard deviation = $\sqrt{\text{variance}}$). This is done to remove the effect of step (3).

The calculations to obtain these values are given in *Table 12.3* where it will be seen that the calculated values for the variance and standard

Table 12.3. Calculation of variance and standard deviation for the three distributions from Figure 12.7.

No. of days lost per accident (x)	*No. of accidents (f)*	*Total days lost = no. of accidents × days lost (fx)*	*Difference = mean – no. of days lost (d)*	*Difference squared (d^2)*	*Difference squared × no. of accidents (fd^2)*
Distribution 1					
0	1	0	2	4	4
1	–	–	–	–	–
2	8	16	0	0	0
3	–	–	–	–	–
4	1	4	–2	4	4
	$n = \Sigma f = 10$	$\Sigma fx = 20$			$\Sigma fd^2 = 8$
mean $\bar{x} = \frac{\Sigma fx}{n} = \frac{20}{10} = 2$					variance $= \frac{\Sigma fd^d}{n} = \frac{8}{10} = 0.8$
					standard deviation $= \sqrt{\text{variance}} = 0.9$
Distribution 2					
0	5	0	2	4	20
1	–	–	–	–	–
2	–	–	–	–	–
3	–	–	–	–	–
4	5	20	–2	4	20
	$n = \Sigma f = 10$	$\Sigma fx = 20$			$\Sigma fd^2 = 40$
mean $\bar{x} = \frac{\Sigma fx}{n} = \frac{20}{10} = 2$					variance $= \frac{\Sigma fd^d}{n} = \frac{40}{10} = 4.0$
					standard deviation $= \sqrt{\text{variance}} = 2.0$
Distribution 3					
0	2	0	2	4	8
1	2	2	1	1	2
2	2	4	0	0	0
3	2	6	–1	1	2
4	2	8	–2	4	8
	$n = \Sigma f = 10$	$\Sigma fx = 20$			$\Sigma fd^2 = 20$
mean $\bar{x} = \frac{\Sigma fx}{n} = \frac{20}{10} = 2$					variance $= \frac{\Sigma fd^d}{n} = \frac{20}{10} = 2.0$
					standard deviation $= \sqrt{\text{variance}} = 1.4$

deviation reflect accurately the 'spread' of measurements. That is the more measurements there are close to the mean, the smaller are the variance and standard deviation. Further information on the calculation and use of the mean and standard deviation can be found in Moroney[6].

It is possible to represent the variance of a distribution diagrammatically on a cumulative frequency diagram. To do this, cumulative frequencies are calculated in the way illustrated in *Table 12.4* for the three distributions of days lost used in *Figure 12.7*. The cumulative frequencies so calculated are then plotted on a diagram as shown in *Figure 12.8* where it can be seen that

Table 12.4. Cumulative frequencies for the three distributions shown in Figure 12.7

Distribution		*Number of days lost*				
		0	*1*	*2*	*3*	*4*
1	Frequency	1	–	8	–	1
	Cumulative frequency	1	1	9	9	10
2	Frequency	5	–	–	–	5
	Cumulative frequency	5	5	5	5	10
3	Frequency	2	2	2	2	2
	Cumulative frequency	2	4	6	8	10

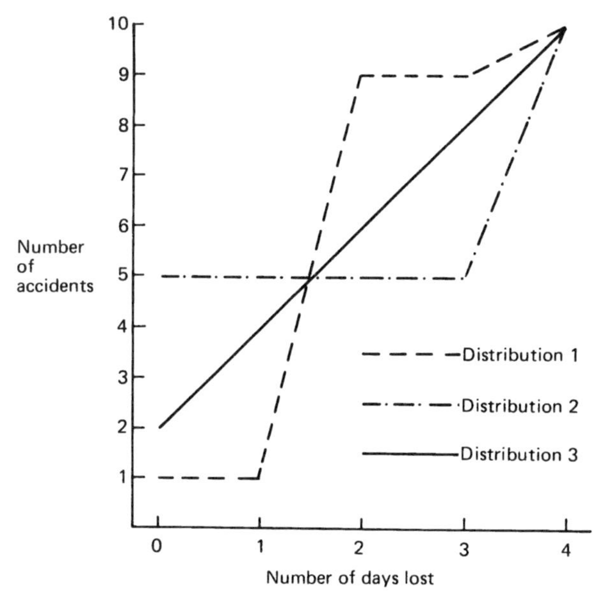

Figure 12.8 Cumulative frequencies for the three distributions of days lost in Table 12.4

when the distribution is even, as in distribution 3, the cumulative frequency shows as a straight line.

Uneven distributions, like 1 and 2, diverge from this straight line and the extent of the divergence gives a diagrammatic representation of the variance.

Using the general summary statistics of central tendency, range and variance, it is possible to give, in a very few numbers, a clear idea of the information you are summarising.

12.4.2 Specialised accident statistics

The specialised accident statistics to be dealt with in this section are sometimes referred to as accident performance or injury statistics and they are frequently used as a means of monitoring the effectiveness of accident prevention strategies. A number of summary statistics are used in presenting these data and the International Labour Office[7] recommends the following:

(1) Frequency rate

$$= \frac{\text{Total number of accidents}}{\text{Total number of man hours worked}} \times 1\,000\,000$$

i.e. accidents per 1 000 000 hours worked.

(Note, however, that in the United Kingdom the multiplier used is frequently 100 000 and when comparing frequency rates allowance must be made for the different multiplier.)

(2) Incident rate

$$= \frac{\text{total number of accidents}}{\text{number of persons employed}} \times 1000$$

i.e. accidents per 1000 employees.

(Note, however, that when considering fewer than 1000 employees a multiplier of 100 is occasionally used and again adjustments must be made when making direct comparisons with ILO figures.)

(3) Severity rate

$$= \frac{\text{total number of days lost}}{\text{total number of man-hours worked}} \times 1000$$

i.e. the average number of days lost per 1000 hours worked.

Other ratios that are useful summary statistics include:

(4) Mean duration rate

$$= \frac{\text{total number of days lost}}{\text{total number of accidents}}$$

i.e. average number of days lost per accident.

(5) Duration rate

$$= \frac{\text{number of man-hours worked}}{\text{total number of accidents}}$$

i.e. average number of man-hours between accidents.

Unfortunately, the data used to obtain these rates are not always consistent so that when interpreting figures supplied from other sources it is essential to have information on:

(a) The definition of an accident being used. Typically only accidents involving injury are used, with different organisations adopting different degrees of severity of injury for classification as accident.
(b) The multiplier being used varies and can range from 1 through 100 to 1 000 000. It is inappropriate to compare figures having different multipliers.
(c) The employees and the nature of their work vary to such an extent that it is not appropriate to make comparisons. For example, comparing frequency rates of shop-floor and office workers, or incidence rates where the proportion of part-time staff varies widely.
(d) The period over which the data were collected can materially affect the rates causing them to vary widely. Examples include seasonal variations, production cycles, plant shutdowns, weather conditions and the economic situation. This makes it necessary to know the particular time to which the rates refer.

It is essential when presenting any of these rates to quote the equation used and identify the way in which all four points are covered.

Injury accidents have been used in the examples in this chapter because these are the data most commonly encountered in the safety world. It should be remembered, however, that a number of other types of incident represent losses for an organisation and can be dealt with in similar ways to those used for injury accidents. Examples of these other types of incident include: damage accidents, sickness absence, machine or plant failures, defective products identified in-house and customers' complaints about product faults.

The statistical techniques dealt with in this chapter can be generalised to all of these categories of incident so long as care is taken to be clear about the definitions being used. For example, making it clear that rates are 'injury accidents' or 'damage accidents'

12.5 Probability and its application

Most people have some intuitive idea of probability in that they know, for example, that a tossed coin has a 50% chance of coming down 'heads'. Probability as a field of study is an attempt to formalise these intuitive ideas so that calculations can be carried out using them.

To do this, statisticians adopt the convention that the whole range of probabilities can be expressed as decimal fractions between 0 and 1. Thus if a thing is certain not to happen, it is said to have a probability of 0; if it is certain that it will happen, then it has a probability of 1. (Note that this corresponds to expressing probabilities as percentages since 0% probability means a thing will not happen, while 100% probability means that it is inevitable.) There are many different ways in which statisticians use probability but for the present purposes it is necessary to consider only the more basic ideas and these are described below.

12.5.1 The different types of probability

There are three types of probability which are likely to be encountered; *a priori* probability, empirical probability and conditional probability.

A priori probabilities are those for which exact probabilities can be worked out or deduced before the event. They are usually restricted to games. For example, when tossing a coin, it is known beforehand that the probability of its coming down heads equals the probability of its coming down tails, and since these are the only two alternatives, each probability = $\frac{1}{2}$ = 0.5. Similarly, when throwing a dice there are six possible outcomes so that the probability of any chosen number = $\frac{1}{6}$ = 0.17.

Empirical probabilities, on the other hand, are those probabilities which can only be worked out on the basis of past experience. They are extensively used by insurance companies to determine the risk associated with a particular cover and hence the size of insurance premiums. The probability figure is arrived at by determining how many times an event has happened and then dividing this by the number of times the event could have happened. For example, in working out life insurance premiums for men aged 50 to 55, the number of insured men in this group who have died over a specified period is determined (say it is 100) and this number is divided by the total number of insured men in this age group, including those who have died. If this total number was, say 10 000, then the probability of a man in this age group dying in an equivalent period of time, other things being equal, is

$$\frac{100}{10\ 000} = 0.01.$$

Empirical probabilities are also used in working out certain accident probabilities. For example, if there have been ten accidents during one hundred shifts, then it is possible to express the probability of an accident on a given shift as

$$\frac{10}{100} = 0.1.$$

However, great care should be taken when using empirical probabilities in this context since it is theoretically possible to have 101 accidents in 100 shifts which would give a probability of

$$\frac{101}{100} = 1.01.$$

and, as was mentioned above, true probabilities cannot exceed 1.0.

In the examples used above, the events described were *mutually exclusive*, that is, the fact that a coin comes down heads on one throw has no influence on what it will come down on the second throw, one man aged 50–55 dying has no influence on the death of others of the same age, and an accident on one shift does not make accidents on other shifts more or less likely. However, there are events which do influence each other, and where this is the case, the third type of probability, conditional probability, applies. These probabilities express, in numerical terms, the chance of event A occurring, given that event B, which will influence it, may or may not happen.

12.5.2 The two laws of probability

There are two very simple rules or laws used in the calculation of probability, known respectively as the law of addition and the law of multiplication.

The law of addition says that if an event can happen in a number of different ways, then the probability of the event happening is obtained by adding up the probabilities of each of the separate ways. Thus in calculating the probability of getting an even number with one roll of a dice, it is known that the probability of each number is $\frac{1}{6}$ (since there are six possible numbers), and the event wanted (an even number) can occur in three possible ways (for numbers 2, 4 or 6). So by the law of addition, the probability is obtained by adding the separate probabilities. In this case it will be $\frac{1}{6} + \frac{1}{6} + \frac{1}{6} = \frac{1}{2}$; therefore probability $P = 0.5$.

The law of multiplication says that if event A can only happen when a number of other events have also happened, then the probability of event A is obtained by multiplying together the probabilities of the other related events. For example, for two throws of a dice, what is the probability of getting a total of 12? It is known that the probability of getting a 6 on any one throw is $\frac{1}{6}$ and in order to get a total of twelve, two 6s are needed. Therefore the probability of getting 12 is $\frac{1}{6} \times \frac{1}{6} = \frac{1}{36}$, i.e. $P = 0.028$.

Although the laws of probability are simple ones they have a wide range of applications, and are referred to in Moroney[8] (a brief introduction), and in Huff[9].

12.5.3 Statistical significance

From the safety adviser's point of view, one particular application of probability is important, namely the idea of statistical significance. This idea is necessary because in most things in life there is a chance or random element which has to be taken into account. For example, if there were fifty minor injuries in a factory in one month, it would be reasonable to expect (assuming no changes), that there would be *around* fifty, though not necessarily exactly fifty, the next month.

It is possible to work out, given fifty accidents in the first month, the exact probabilities of any number of accidents in the second month. When this is done it is found that the further the number of accidents in the second month is from fifty the smaller is its probability of occurrence, so that, for example, 48 accidents is less likely than 49, and 52 less likely than 51. Once these probabilities have been calculated[10] it is possible to decide on a probability level below which the variation in accident rate is considered not to be due to chance. Usually the level chosen is $P = 0.05$. Thus if the number of accidents in the second month is less likely than a probability level of less than $P = 0.05$, the variation is said to be statistically significant, that is, it is assumed to be not due to chance variation, and warrants investigation of the causes.

12.5.4 Monitoring accident rates

Shipp[11] has applied the idea of statistical significance outlined above to produce two useful techniques for monitoring accident rates. The first uses monthly injury frequency rates and provides a system for determining whether variations in monthly frequency rates are statistically significant. The second technique assesses the statistical significance of trends in injury rates. Both techniques make use of diagrams for the presentation of their results. Worked examples of both these techniques and diagrams of the results are given below.

Variations in monthly frequency rate In a particular year the statistical significance of variations in injury rates from month to month is assessed against the injury rate for the whole of the previous year. This annual injury frequency rate (IFR) is:

$$\frac{\text{total number of injuries in the year}}{\text{total number of man-hours worked in year}} \times 100\,000$$

Suppose that in a particular company there were, last year, 68 injuries and that 2 000 000 hours were worked. This would give

$$\text{IFR} = \frac{68 \times 100\,000}{2\,000\,000} = 3.4$$

Further supposing that the monthly injuries, hours worked and monthly IFRs for the first 6 months of the following year were as shown in the first three rows of *Table 12.5*, it would be possible to draw in these monthly IFRs on a chart of the type shown in *Figure 12.9*. Comparison of each of the monthly IFRs with the previous year's IFR shows that overall the monthly IFRs are higher. However, there is some fluctuation and, by referring to the tables provided by Shipp it is possible, on the basis of the previous year's results to draw onto the chart the 'upper' and 'lower' lines shown in *Figure 12.9*. The area between these lines represents the differences from the previous year's IFR which have a probability of 0.05, i.e. are random differences. IFRs outside these lines, having a probability of less than 0.05, are statistically significant and require investigation.

Trends in IFR over a year Although the chart in *Figure 12.9* suggests that there is a tendency for IFRs to increase over the first six months of this year, it is not possible to determine from this Figure whether or not the trend is

Table 12.5. Calculations required for Shipp's techniques and for three-month moving means

Shipp's techniques	*JAN*	*FEB*	*MAR*	*APR*	*MAY*	*JUNE*
Number of injuries	5	3	7	5	6	7
Number of hours worked	110 000	90 000	120 000	115 000	130 000	85 000
Monthly IFR	4.5	3.3	5.8	5.2	4.6	8.2
Cumulative injuries	5	8	15	20	26	33
Cumulative hours worked	110 000	200 000	320 000	435 000	565 000	650 000
Cumulative IFR	4.5	4.0	4.7	4.6	4.6	5.1
3-month moving mean	–	–	JAN–MAR	FEB–APR	MAR–MAY	APR–JUNE
3-month	–	–	15	15	18	18
3-month hours worked	–	–	320 000	325 000	365 000	330 000
3-month moving mean IFR	–	–	4.7	4.6	4.9	5.5

significant. In order to do this, Shipp's second technique is required which is based on the IFR for the previous year but instead of using this number to look up the upper and lower lines in the form shown in *Figure 12.9*, it is used in conjunction with a second set of tables to produce the 'curved' upper and lower lines shown in *Figure 12.10*. The monthly IFRs are charted on *Figure 12.10*, not directly as was the case in *Figure 12.9*, but in the form of 'cumulative IFRs'. These have to be calculated each month and the necessary calculations are shown in *Table 12.5*.

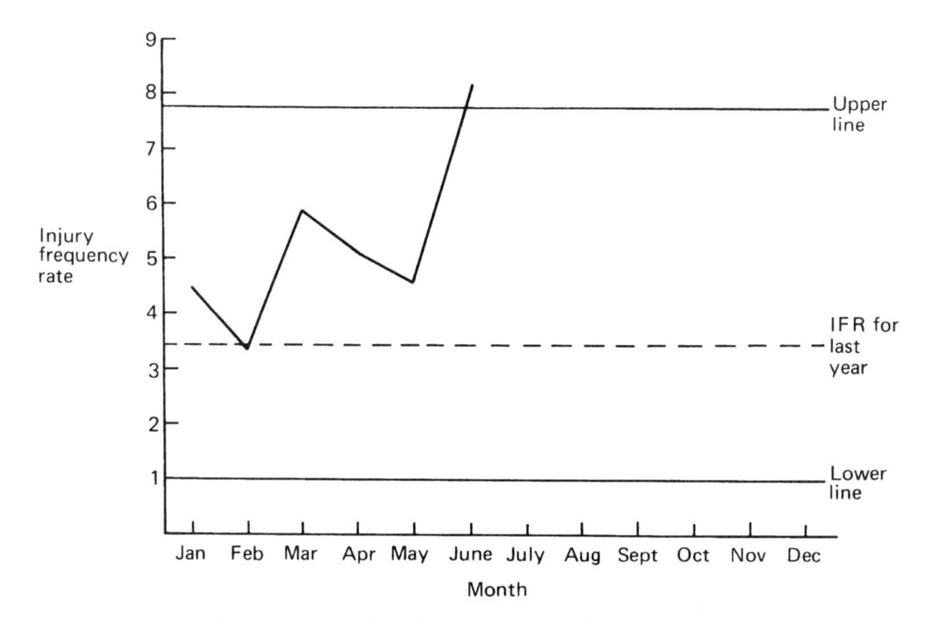

Figure 12.9 Month by month injury frequency rate chart

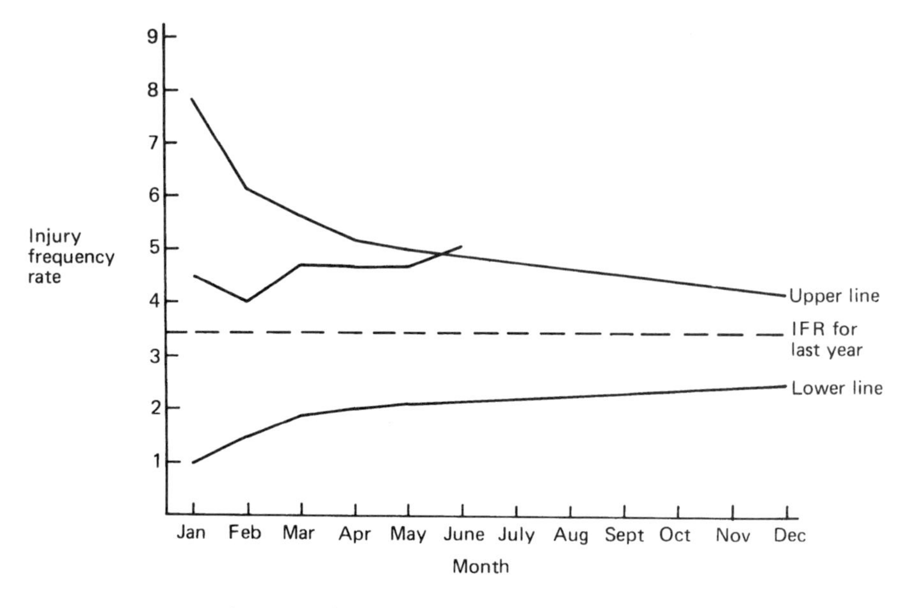

Figure 12.10 Cumulative total injury frequency rate

The calculated values for each month's cumulative IFR are plotted as shown in *Figure 12.10*. These plots show that the IFR is increasing in the current year but it is only in June that the trend becomes significant since it is only at this point that the upper line is crossed. As with *Figure 12.9* the upper line is the 5% ($P = 0.05$) significance level for an increasing IFR and the lower line is the 5% ($P = 0.05$) significance level for a decreasing IFR.

Further information on Shipp's techniques and copies of the relevant tables are given in his report[11].

12.5.5 Moving mean in time

However, one disadvantage of Shipp's report is that the tables provided require a minimum of 10 000 hours worked per month. This may make these tables unsuitable for some organisations which require a simpler method of showing injury trends diagrammatically.[12] One method is known as a 'moving mean in time' and the necessary calculations are shown in the last three rows of *Table 12.5*. The method is as follows:

(1) Choose an appropriate period of time over which to calculate the mean. In the example a three month period has been chosen.
(2) Calculate the mean IFR for the first three months (in the example these are January to March).
(3) Drop the first of the months used (January) and add the next month (April) and work out a new moving mean (i.e. for February to April).
(4) Repeat this process for each three month period available.

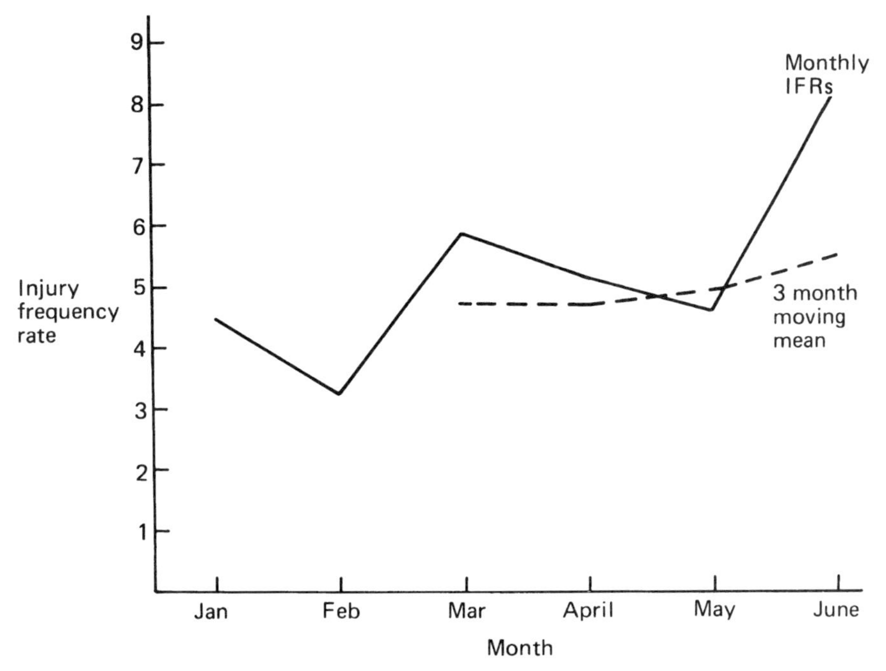

Figure 12.11 Smoothing effect of 3 month moving mean

These moving means can then be plotted, as in *Figure 12.11* which illustrates clearly the smoothing effect of the three month moving mean IFRs compared with the individual monthly IFRs. In some industries it may be more convenient to collect injury statistics in three monthly units and calculate the moving mean for 4 units, i.e. 12 months. This then smooths out any variations due to such factors as weather, holidays, annual wage negotiations, etc.

12.5.6 The Poisson distribution

One of the common types of information available to safety advisers is the number of accidents sustained by employees in a factory or department. Statistically, this information is referred to as the distribution of accidents over the group, and it is usual to find that most of the group have no accidents, some have only one, while certain individuals have four, five or more. Where a high number is found, the obvious action is to investigate the reasons for them, but to save effort it is useful to have a method for determining whether or not an observed distribution of accidents can be explained by chance variation. The relevant method makes use of Poisson's distribution[10] which enables the expected distribution of the number of accidents occurring by chance in a given group to be calculated. Investigation of the causes of accidents among high accident rate individuals will be worthwhile only where there is significant difference between the expected and the observed distribution.

Suppose that in a department of 100 people there were 100 accidents over a one year period. We would not expect each person to have had one accident, nor would we expect one person to have had 100. The Poisson distribution is used to calculate what could be expected *if the accidents occurred randomly among the 100 people in the department.*

The steps in the calculation are as follows:

(1) Work out the mean number of accidents per person.

$$x = \frac{100 \text{ (accidents)}}{100 \text{ (people)}} = 1$$

(2) Look up in tables, or obtain from a calculator, the value of e^{-x}. This e is a mathematical constant (like π) and there is no need to know why it is used.

In our example $e^{-x} = e^{-1} = 0.368$

(3) Work out the probability of 0 accidents, 1 accident, 2 accidents, 3 accidents and so on. This is done by working out successive terms in the equation for the Poisson distribution, which is

$$e^{-x}\left(1 + x + \frac{x^2}{2!} + \frac{x^3}{3!} + \frac{x^4}{4!} \ldots\ldots \text{etc.}\right)$$

2!, 3! and 4! are mathematical shorthand for two factorial, three factorial and four factorial. What they are in practice is

$2! = 2 \times 1 = 2$
$3! = 3 \times 2 \times 1 = 6$
$4! = 4 \times 3 \times 2 \times 1 = 24$

Table 12.6. Steps in the calculation of the Poisson distribution

Number of accidents	e^{-x}	*Term*	*Calculation*	*Probability*	*Number of people*
0	0.368	1	1 × 0.368	0.3683	37
1	0.368	x	1 × 0.368	0.368	37
2	0.368	$\frac{x^2}{2!}$	(1/2) × 0.368	0.184	18
3	0.368	$\frac{x^3}{3!}$	(1/6) × 0.368	0.063	6
4	0.368	$\frac{x^4}{4!}$	(1/24) × 0.368	0.015	2
			Totals	0.998	100

The successive terms within the brackets in the Poisson distribution, when multiplied by e^{-1x} give the probabilities of 0, 1, 2, 3, and so on, accidents. Note that the terms within the brackets are a simple sequence. For example, the term for 7 accidents is simply

$$\frac{x^7}{7!}$$

Table 12.6 shows how these calculations are made and how the probabilities are converted to the numbers of people expected to have a given number of accidents by multiplying the probabilities by the number of people in the sample (100 in the example).

From *Table 12.6* it can be seen that if the accidents were distributed randomly you would expect two people to have four accidents each. If a greater number of people had four accidents each, or one or more people had 5, 6 or 7 accidents further investigation would be needed. The appropriate next step would be to calculate whether the expected distribution that has been calculated is *significantly* different from the distribution observed, using the Chi-squared test[15].

12.5.7 Correlation

This is the statistical term used to describe the fact that certain measurements are related in such a way that they increase and decrease together. This may be the case with, for example, number of hours worked and number of accidents. Where this occurs and both measures increase or decrease together they are said to be positively correlated. Negative correlation is also found and this occurs when, as one measure increases, the other decreases as could be the case with job experience and frequency of accidents. Calculating the correlation between two measures can be useful since, if there is a strong correlation between them, the extent to which changes in one measure will be associated with changes in the other can be defined. For example if a high correlation is found between ambient temperature and accident rates, it is possible to predict accident rates at any given ambient temperature.

The first step in investigating a possible correlation between two measures is to prepare a *scattergram*. This is a graph showing each pair of related readings or measurements as one dot. If there is a 'cloud' of dots which is circular or has no particular shape, then there is no correlation. If, however, the 'cloud' is narrow and slopes from bottom left to top right, then the correlation is positive. A 'cloud' sloping from top left to bottom right indicates negative correlation.

Typical scattergrams for no correlation, positive correlation and negative correlation are illustrated in *Figure 12.12*. It is also possible to calculate a numerical value for correlations without drawing a scattergram. The appropriate measure for correlations involving accident data is Spearman's Rank Correlation Coefficient.[13]

Statistical test of correlation can only indicate the degree to which two variables are correlated; they cannot give any indication of which causes which, or whether both variables are caused by some other unstudied

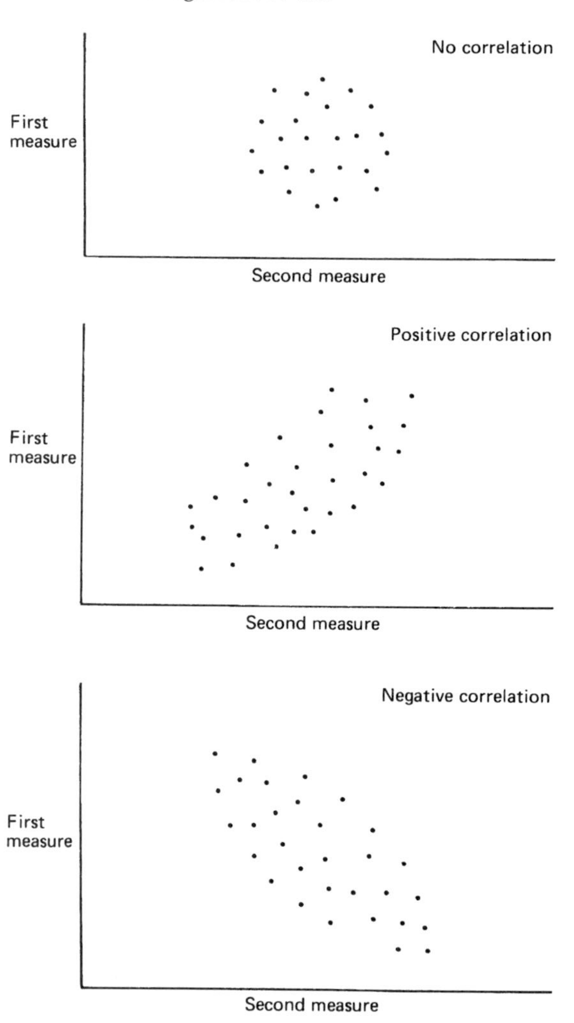

Figure 12.12 Typical scattergrams illustrating no correlation, positive correlation and negative correlation

variable. An example comes from Powell *et al.*[1] in their study of shop floor accidents. They found that there was a correlation between number of injuries recorded against individual at the first aid centre, and a measure of extroversion. It might have been tempting to assume that the results proved that extroverts had more accidents. In fact, a further study showed that extroverts were more likely to report their injuries to the first aid room and, hence, the causal link was between personality and reporting of injury rather than between personality and the occurrence of injury. To infer cause, therefore, it is necessary both to have found a correlation between the factors and to have investigated a plausible causal mechanism to link them together.

12.5.8 Confidence limits

You may have seen scattergrams or points plotted on other types of graph in the form shown in *Figure 12.13* which illustrates the IFRs for a company over the years 1988 to 1992. In *Figure 12.13* the points have been plotted and joined to form a graph in the usual way. However, additional calculations have been made of the confidence the author has in the accuracy of each measure. The lines above and below the plotted points indicate diagrammatically the likely range of error associated with that particular point. The shorter the line, the more confidence the author has in the accuracy of the measure. In *Figure 12.13* the confidence limit lines indicate that, for whatever reason, the author has less confidence in the measures for 1988 and 1990. See Moroney[14] for further information on confidence limits.

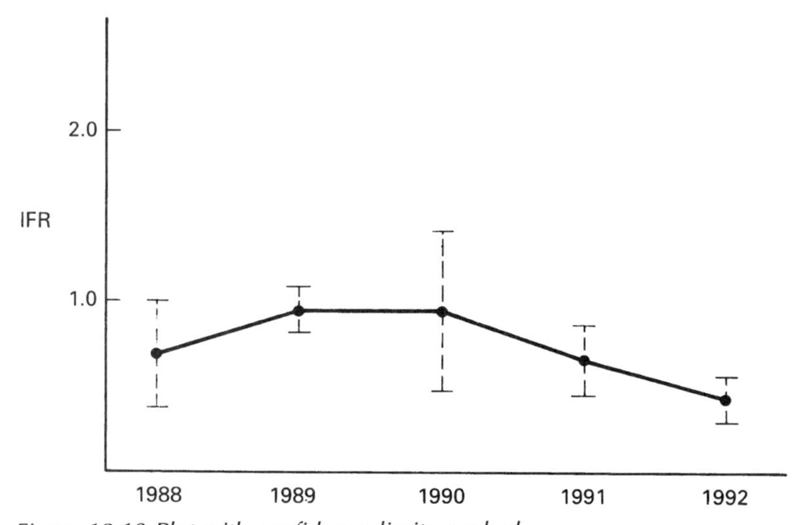

Figure 12.13 Plot with confidence limits marked

12.6 Drawing conclusions

All statistical operations are based on assumptions about the data used and the samples on which the calculations were carried out. This was illustrated earlier in the calculation of different types of accident rates. Only if these assumptions are valid will the statistical operations be valid. For example, comparing two frequency rates for two companies is valid only if the assumption that they used the same definition of an accident is valid.

When presenting conclusions, it is essential to describe the assumptions that have been made. When checking other people's conclusions, their assumptions should be identified and their validity checked. It should be noted that much of this work has little to do with mathematical or arithmetical operations, it has much to do with clear, analytical thinking.

Remember that statistical methods never allow conclusions to be drawn with complete certainty, they simply quantify the uncertainty and enable more rational argument.

12.7 The use of computers

There is a wide range of computer software available for health and safety applications, as well as statistical packages designed for general use. The main software categories, so far as health and safety is concerned, are described below.

12.7.1 Accident recording and analysis

There are a number of accident recording and analysis packages which provide facilities to record basic data and most will satisfy the data recording requirements of RIDDOR. Some of the packages provide a number of additional features:

- automatic printing of the form F2508,
- automatic report generation enabling periodic reports to be prepared automatically,
- records based round incidents so that near misses can be recorded and analysed in the same detail as injury accidents,
- multiple records so that injuries and property and vehicle damage arising from the same incident can be recorded in detail,
- linked incident costing so that reports can automatically include details of the costs involved, and
- automatic summaries of incident records and graphical displays of these summaries.

12.7.2 Information on chemicals and substances

A number of organisations maintain data bases on chemicals and substances which they make available in one or more of the following ways:

- On-line computing using own computer to contact the publisher's computer to obtain and read off the information wanted.
- CD–ROM discs like conventional CDs but holding words rather than music. They require a special CD–ROM (Compact Disk–Read Only Memory) player for the computer and special software. The information from the disk is displayed on the computer screen.
- Floppy disks which hold the information on the same type of disk used in normal computer operation. The information is read off using the special software supplied.

Some packages link information on chemicals and substances with a COSHH data base (see below) and this can save a great deal of retyping. Other packages allow a selection of information from the records supplied and print it out as a tailored substance data sheet.

12.7.3 Data bases for COSHH

The recording requirements for COSHH are extensive in range (substances, operations, personnel, training, LEV, etc.) and in the time for which they have to be kept (up to 30 years). There are also requirements for confidentiality of parts of the medical records, complex links between records for the analyses required by the HSE and the ability to produce 'anonymous' summaries. There are COSHH data base packages that meet all these requirements and also provide facilities for other types of recording, for example, noise exposures.

12.7.4 Audit recording and analysis

This is a relatively new field for microcomputer software which stems from the increasing pressure on organisations to audit their health, safety and environmental management performance. The most convenient way to do this is by using a set of audit questions, often referred to as an audit protocol. These can be purchased as an 'off the shelf' system or be developed in-house.

Organisations undertaking safety, health and environmental audits soon find that they have a data handling problem stemming from the need to record and analyse the responses to hundreds of questions, from a number of locations, together with the associated auditor's notes and recommendations for action. To help with these data handling problems a number of software suppliers have produced audit recording and analysis packages. These are available either with, or independent of, sets of audit questions.

Other features of these packages include:

- The ability to edit the audit questions and add audit questions as new risks are identified.
- The ability to add guidance for the auditors to specific questions, including details of any relevant company standards.
- Facilities for displaying two or more sets of audit results on the screen at the same time so that trends over time can be examined for the same audit area and comparisons made with performances of different audit areas.
- The automatic summarising of results from two or more audit areas so that a measure of performance by region, division or function, depending on the organisation, can be obtained.
- Graphical display of audit summaries.
- Automatic generation of reports on audits, including action plans.
- Diary facilities to assist in managing an audit schedule and keeping track of recommendations for remedial action.

12.7.5 The general benefits of software

Software is only of benefit if it enables a job to be done more quickly or to a higher standard than would otherwise be achieved without it. It is usually quite easy to see how software will speed up a task, once the setting up and learning phases are over. However, raising standards is more difficult. If what the software will do is not known, it will not be possible to assess its value in raising standards of performance. Individual software packages are aimed at raising standards in particular areas, such as the recording requirements under COSHH and RIDDOR. However, most good software will also raise standards in three general areas:

(a) *Expertise.* A good software package will encapsulate the expertise of the authors and, if it has been thoroughly piloted, reflect general 'best practice'. This expertise ranges from the fairly straightforward, for example prompting to record all the relevant information for an accident report form, to the very detailed, for example the large number of specific questions in a health and safety or environmental audit protocol. Buying good software can save a lot of 'wheel reinventing'.
(b) *Up to date information.* Most software is revised periodically to take into account changes in legislation, lessons learned from existing users and other relevant changes such as the annual re-issue of EH40. Signing up for an appropriate package is one way of keeping up to date in a particular area.
(c) *Improved management.* The maxim 'If you can't measure it, you can't manage it' is being applied more and more in the management of environmental and health and safety issues. Measurements, and the analysis of the results, are time consuming tasks if they are done manually, which is probably why they have not been carried out in the past. The range of software now available reduces the time required for analysis and allows the maxim to be met.

12.8 Other sources of statistical information

There is a wide variety of sources of information available which can be classified by the medium used. The main ones are: paper-based material such as books and periodicals; computer-based materials such as on-line data bases and CD–ROM; and other media including videos, film and microfiche.

Paper-based sources of particular relevance to the safety adviser include the Annual Reports of Her Majesty's Chief Inspector of Factories, government reports on population statistics and trends including sickness rates and disease incidence, and a large range of publications from insurance companies and industry trade associations.

Computer-based sources include on-line and CD–ROM data bases containing abstracts from health and safety journals, information on hazardous chemicals and details of relevant legislation.

The other media are relevant in terms of general health and safety information, but little statistical information is, at present, available from them.

The sorts of information listed above are useful only if they are stored in a systematic manner and there are suitable mechanisms for the retrieval of the relevant data. Similar considerations apply when the data come, not from an outside source, but from within the organisation. For example, accident data (both injury and damage), sick absence and product complaints are all internally generated data. It is necessary to set up systematic methods for collecting, storing and retrieving these data if the best use is to be made of them.

Data and information handling (information technology (IT) as it is now known) are beyond the scope of this chapter but see Waring, Kendall and Boyle[16] for a detailed description of the techniques involved and a more complete list of sources of available information.

12.9 Coda

Some of the ways in which statistics can be used have been described, but it has only been possible to give detailed descriptions of a limited number of the available techniques. However, detailed information is available in the references cited. The techniques described require no more than elementary arithmetic and should enable the safety adviser to make a start on preparing simple statistical data and displaying it in a way readers will find acceptable.

References

1. Powell, P.I., Hale, M., Martin, P. and Simon, M., *2000 Accidents*, National Institute of Industrial Psychology, London (1971)
2. International Labour Office, *Accident Prevention*, 8th impression, 30, ILO, Geneva, 1976
3. Moroney, M.J., *Facts from Figures*, 19–33, Penguin Books Ltd, London (1980)
4. Huff, D., *How to Lie with Statistics*, 58–71, Penguin Books Ltd, London (1975)
5. Ref. 3, pp. 34–54
6. Ref. 3, pp. 56–65
7. International Labour Office, *Encyclopaedia of Occupational Health and Safety*, 3rd edn, 34, ILO, Geneva (1983)
8. Ref. 3, pp. 4–17
9. Huff, D., *How to Take a Chance*, 9–129, Penguin Books Ltd, London (1978)
10. Ref. 3, pp. 96–107
11. Shipp, P.J., *The Presentation and Use of Injury Data*, British Iron and Steel Research Association, London, open report (no date)
12. Siegel, S.S., *Non-parametric statistics for the behavioural sciences*, **42–47**, McGraw-Hill, New York (1956)
13. Ref. 12, pp. 202–213
14. Ref. 3, pp. 238–245
15. Ref. 3, pp. 249–270
16. Waring, A.E., Kendall, S.V. and Boyle, A.J., *Safety Information and Information Technology*, Portsmouth Polytechnic (1986)

Further reading

Boyle, A. J., 'Accident Information Data', *The Safety Practitioner*, **1,** No 1, 12, (1983)

Boyle, A. J., 'Accurate Analysis of Accident Information', *The Safety Practitioner*, **1,** No 2, 10, (1983)

Boyle, A. J., 'Accident Data Analysis with Microcomputers – An Update', *The Safety Practitioner*, **4,** No 11, 11 (1986)

Chapman, M. and Mahon, B., *Plain Figures*, HMSO, London (1986)

Chapter 13

The individual

Professor A. R. Hale

13.1 What is behavioural science?

Behavioural science has three main aims: to describe, to explain and to predict human behaviour. Systematic description is the essential foundation of this, as of any other scientific subject. The safety adviser is interested particularly in behaviour at work and more particularly in the behaviour of people in situations which endanger their health or safety. Even describing behaviour in such situations is not easy; many people describe what they expected to see rather than what they actually did see. Explaining behaviour requires theories about why it happens; we need to get to this level of understanding in order to understand behaviour in accidents and to decide how to design hardware and organisations which will be useable by people and complement their skills. What we really want to do is predict in detail how decisions on selection, training, design and management will influence the way people will behave in the future. This is a very severe test of theories about individual behaviour, and psychology is often not far enough advanced as a science to withstand such scrutiny. In this chapter the aim is to describe in broad terms what is known and can be used to help us understand and guide human behaviour.

A human individual is far more complex than any machine and when individuals are placed together in groups and organisations the interactions between them add many times to the complexity which needs to be understood. Individuals are also extremely adaptable. They change their behaviour as they learn and if they know that they are being observed. Also, each individual is to an extent unique because of their unique experience. Because of all this the behavioural scientists' task can be seen to be daunting indeed. It is also difficult to measure all the factors which may affect an individual's behaviour at any one moment.

The explanations and predictions of behavioural science therefore have wider margins of error than those which can be offered by engineers or doctors. Statements made about behaviour will usually be qualified by words such as 'probably' or 'in general'. Individual exceptions to the predictions will always occur.

Because of its limitations behavioural science is dismissed by some as being no more than common sense dressed up in fancy language. All individuals must have some ability to explain and predict the behaviour of themselves and others, or they would not be able to function effectively in the world. However, most individuals' explanations and predictions are often proved wrong. Behavioural science used in a systematic and rigorous way can always improve on unaided 'common sense'.

Behavioural science commonly works by developing models of particular aspects of human behaviour. These models are inevitably simplifications of real life, in order to make it comprehensible. The models are frequently analogies drawn from other branches of knowledge. They represent the brain as a telephone exchange or a computer, the eye as a camera, etc. Different behavioural scientists use different analogies. This, to some extent, explains why there sometimes appear to be parallel and incompatible theories about the same aspect of human behaviour. Analogies are powerful and useful, but they have limitations which must always be acknowledged. They can never be perfect descriptions of the way that an individual functions, and will be useful only within their limits. In the sections which follow some of the models will be described and used to explain particular aspects of behaviour. Readers are urged to use them, but with care.

13.2 The relevance of behavioural science to health and safety

Here are some of the questions which behavioural science can help to answer:

What sort of hazards will people spot easily, and which will they miss?
At what times of day and on what sorts of job will people be least likely to notice hazards?
Can you predict what sorts of people will have accidents in particular circumstances?
Why do people ignore safety rules or fail to use protective equipment and what changes can be made in the rules of equipment to make it more likely that people will use them?
If people understand how things harm them, will it make them take more care?
What knowledge do people need to cope with emergencies?
What extra dangers arise when people work in teams?
How do company payment, incentive and promotion schemes affect people's behaviour in the face of danger?
How can training help to make people take care?
Can you frighten people into being safe?
When are committees better than individuals at solving health and safety problems?

The list of questions can go on almost indefinitely. Before studying behavioural science it is a valuable exercise to draw up a list of questions relevant to your own work place. See how many of the questions you have answered, or reformulated at the end of your study.

13.3 The human being as a system

A common model used in behavioural science, and in the biological and engineering sciences, is the 'systems' model. Systems are defined as organised entities which are separated by distinct boundaries from the environment in which they operate. They import things across those boundaries, such as energy and information; they transform those inputs inside the system, and export some form of output back across the boundaries. Open systems are entities which have goals or objectives which they pursue by organising and regulating their internal activity and their interchange with their environment. They use the feedback from the environment to check constantly whether they are getting nearer to or further away from their objectives. *Figure 13.1* shows a generalised system model.

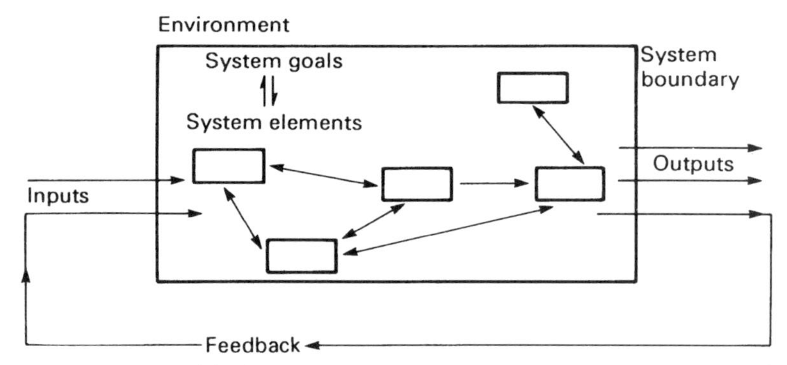

Figure 13.1 Simplified system model. (Adapted from Hale and Glendon[1])

Such models can be applied to a single cell in the body, to the individual as a whole, to a group of individuals who are working together, and to an organisation such as a company.

Figure 13.2 considers the human being as a system for taking in and processing information. Accidents and ill-health are conceived as damage which occurs to the system when one or other part of the system fails. The human factor causes of accidents can be classified according to which part of the system failed. The following sections discuss the various stages in the model set out at *Figure 13.2*.

13.4 Facets of human behaviour

13.4.1 Goals, objectives and motivation

Any understanding of human behaviour must start with an attempt to describe the goals and objectives of the human system. Individuals have many goals. Some such as the acquisition of food and drink are innate. Others are acquired, sometimes as means of achieving the innate goals, and

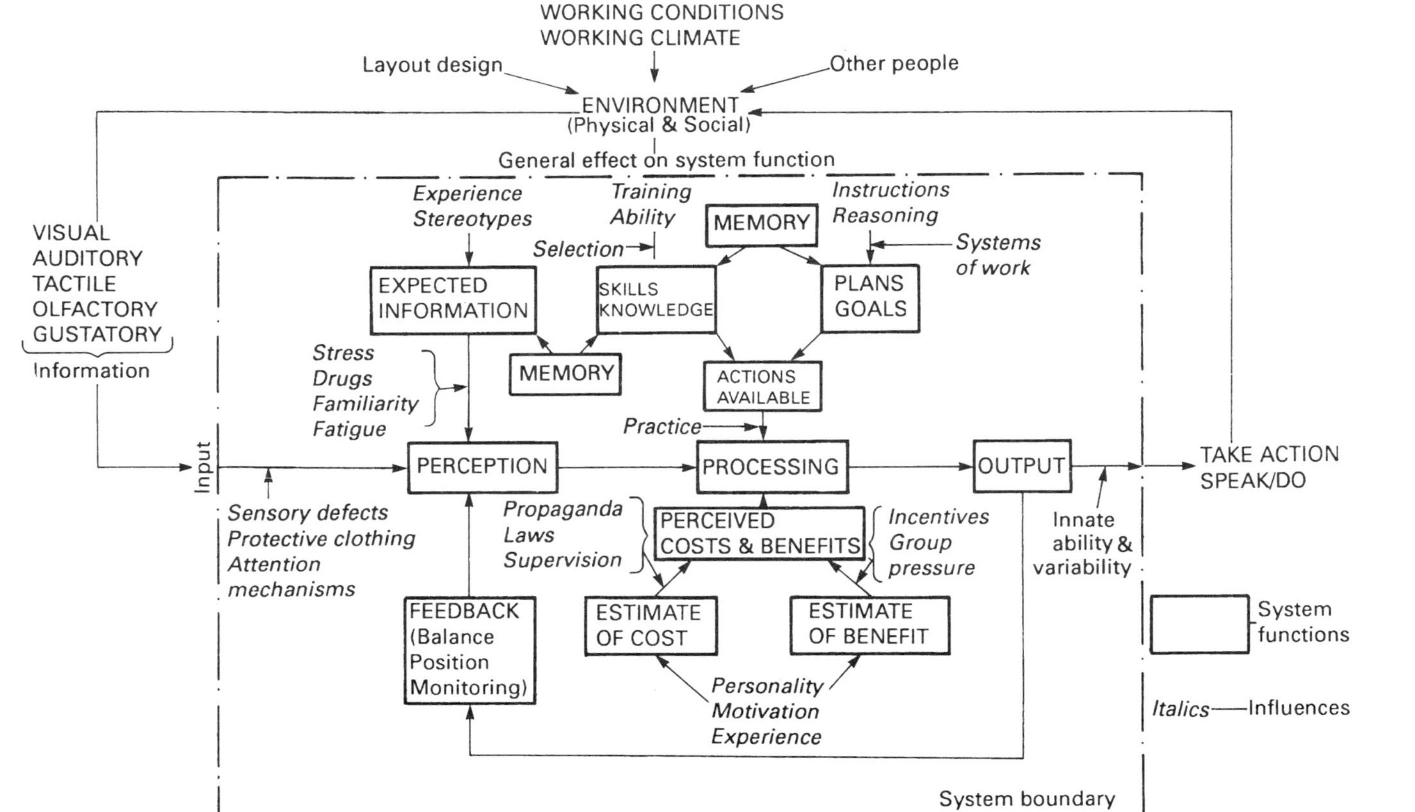

Figure 13.2 Systems model of human behaviour. (Adapted from Hale and Hale[2])

sometimes as ends in themselves, for example the acquisition of money, attainment of promotion, purchase of a house, etc. Some are short term, e.g. food at dinner time; others are much longer term, e.g. earning enough for retirement. In some cases the short and long term goals may be in conflict; e.g. a person may fail to check equipment before starting work in order to satisfy the short term goal of getting the job over as fast as possible, as a result jeopardising the long term goal of preserving his own health and safety.

Not all goals are consciously pursued, either because the person may not want to admit even to himself that he is pursuing a particular goal, or because the goal is so basic that it has been built into the person's behaviour and no longer requires any conscious thought.

An individual's goals can be conceived of as vying with each other to see which one will control the system from moment to moment. People will therefore show to some extent different and sometimes contradictory behaviour from day to day, and certainly from year to year, depending which goal is uppermost at the time. However, people will also show consistency in their behaviour, since the power of each of their goals to capture control of the system will change only slowly over the sort of time periods which concern those interested in behaviour at work.

Many theorists have written about motivation, particularly motivation at work. They have emphasised different aspects at different times. The following is a brief historical survey of the main currents of theory.

(1) *F.W. Taylor and Economic Man.* Taylor divided people into two groups: potential managers who were competent at and enjoyed planning, organising and monitoring work, and the majority of the workforce who did not like those activities but preferred to have simple tasks set out for them. Taylor considered that, once work had been rationally organised by the former and the latter had been trained to carry it out, money was the main motive force to get more work out of them. His ideas of Scientific Management[3] encouraged the development of division of labour and the flow line process, work study and the concentration on training, selection and study of the optimum conditions for work.

(2) *Elton Mayo[4] and Social Man.* Studies in the 1930s at the Hawthorne works of the Western Electric Company in Chicago which set out to discover optimum working conditions led to the realisation that people were not automata operated by money, but that they worked within social norms of a fair day's work for a fair day's pay. It showed that they were responsive to social pressure from their peers, and to interest shown in them by the company. This led to a new emphasis on the role of the supervisor as group leader, rather than as autocrat, and also to a greater emphasis on building group morale.

(3) *Self Actualising Man, Maslow[5].* Maslow looked at the motivation of people who were successful and satisfied with their work. He found that there was always an important element of achievement, self-esteem and personal growth in their descriptions of their behaviour. He put forward his theory of the hierarchy of needs (*Figure 13.3*) to express this concept of growth. He postulated that the homeostatic needs had to be satisfied before the growth needs would emerge. Although this hierarchy has not

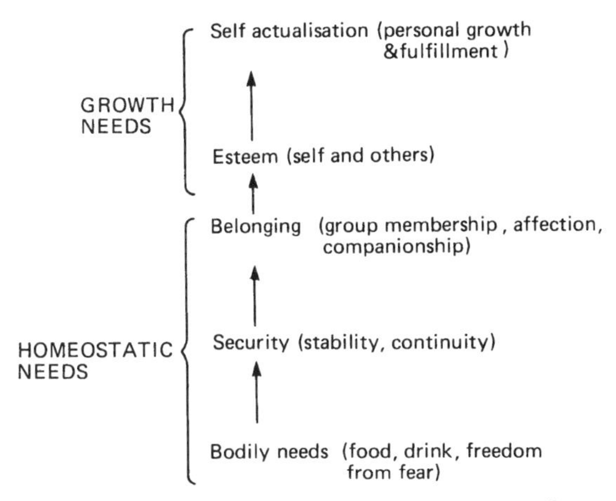

Figure 13.3 Hierarchy of needs. (After Maslow[5])

been subjected to rigorous scientific confirmation, it is broadly borne out by research studies.

Other theorists who have studied achievement motivation are Atkinson[6] and McClelland[7] who studied the motivation of entrepreneurs compared with government employees, showing a clear difference in the importance they accorded to commercial risk taking and to success.

(4) *Complex Man.* Modern motivation theory tries to incorporate what is valuable from all of the earlier theories, and recognises that there are individual differences in the strengths of different motivations both between individuals and over time in the same individual. As far as possible, incentives need to be matched to the individual and the situation (the job of person-oriented man management). It is also recognised that the human system is more complex than many early theories postulated, and that expectations play a strong part in motivation[8]. In other words the force of a motivator is dependent on the sum of the value of the reward and the expectancy that a particular behaviour will lead to the reward. If someone perceives that it will take a great deal of effort to gain any increase in reward, or that the reward does not appear to be dependent upon how much effort is actually put in, their behaviour will not be influenced by that reward.

The unique combination of goals and behaviour which represents each individual's adaptation to the environment in which he finds himself is one definition which is given to the word personality (see section 13.7.1.4). Thus, those who habitually place a high value on their need for acceptance by people around them are called gregarious or friendly, whereas those people who habitually subordinate their need for approval by others to their goal of achieving high status in the organisation, we call ambitious.

It may be thought axiomatic that the preservation of the self (i.e. of safety and health) would be one of the basic goals of all individuals. Clearly this is not a goal of all people at all times, as the statistics of suicides must indicate.

Some psychologists[9] have argued that there are deep seated unconscious desires in some people at some times to damage themselves, which constitute a major factor in accident causation. However, research carried out among normal populations who are not receiving psychiatric treatment indicates that there is little, if any, ground to support this view[10]. It is perhaps best to assume that preservation of health and safety is indeed a basic goal of the normal human system, and that most failures to achieve that goal are because individuals do not perceive that their safety is immediately threatened, and so other goals which the individual has are given priority over the one of self-preservation. If risks are perceived to be small and gains great, the individuals are willing to trade off a slight increase in risk for a bigger short term gain in speed or comfort. In exceptional circumstances, such as in saving a child from drowning, personal safety may be put at considerable risk, but this does not often happen otherwise.[11]

13.4.2 System inputs

Information gets into the human system through the sense organs. These have traditionally been considered as five in number – sight, hearing, touch, smell and taste – but touch can be sub-divided into the senses of pain, pressure and temperature, and in addition there is the 'proprioceptive' or 'kinaesthetic' sense which transmits information from the muscles and joints to the brain, informing it about their position in space, and their orientation one to another. Closely related to this is the sense of balance.

Hazards which are not perceptible to the senses will not be noticed unless suitable alarms are triggered by them or warnings given of them. Examples are odourless, colourless gases such as methane, X-rays, or ultrasonics, or hazards in the dark. The canary falling off its perch in the mine because of its greater sensitivity to methane was an early example of a warning device, subsequently superseded by the colour change in a safety lamp flame and now the methanometer.

If any of the senses are defective the necessary information may not arrive at the brain at all, or may be so distorted as to be unrecognisable. Some sensory defects are set out in *Table 13.1*. Sensory defects can also be

Table 13.1. Some sensory defects

Sense	*Natural and 'imposed' sensory defects*
Sight	Colour blindness, astigmatism, long and short-sightedness, monocular vision, cataracts, vision distortion by goggles and face screens
Hearing	Obstructed ear canal, perforated ear drum, middle ear damage, catarrh, ear plugs or muffs altering the sound reaching the ear
Taste and smell	Lack of sensitivity, genetic limitations, catarrh, breathing apparatus screening out smells
Touch senses	Severed nerves, genetic defects, lack of sensitivity through gloves and aprons
Balance	Ménière's disease, alcohol consumption, rapid motion, etc.

'imposed' by some of the equipment or clothing provided to protect people against exposure to danger, e.g. safety goggles, gloves or ear defenders.

The sense organs themselves have a severely limited capacity for receiving and transmitting information to the brain. The environment around us always contains far more information than they can accept and transmit. There are two types of mechanism which the individual uses to overcome this potential overload of information.

(1) Switching or attention mechanisms.
(2) Expectancy.

13.4.3 Switching or attention mechanisms

The brain classifies information by its source and type. It is capable of selecting on a number of parameters those stimuli that it will allow through a filter into the system. The setting of the filters on each sensory mode is partly conscious and partly unconscious.

The main visual attention mechanism is the direction of gaze which ensures that the stimulus from the object being looked at is directed to the most sensitive part of the retina (fovea) where it can be analysed in detail. The rest of the field of view is relegated to the less sensitive parts of the retina. In normal activity this centre of focus is shifted constantly in a search pattern which ranges over the field of view until an object of interest is picked out. The visual sense can be tuned to seek out a particular facet such as a defect in a machine or component, provided that we know in advance what characteristics to tune it to. This ability is known as 'perceptual set'.

In the hearing sense there is a similar ability that enables us to tune into the various characteristics of sounds such as direction, pitch, intensity, or timbre. People can therefore 'tune' themselves to pay attention to strange sounds coming from a particular part of a machine, while ignoring all others. It may take several occasions of looking at something before the filter is confident enough that this is what is being sought and allows it through. So we can sometimes look but not see. As well as being variable from time to time perceptual set shows longer term settings which produce differences between individuals because of their interests and their experience; safety advisers notice hazards because they are interested in them and used to finding them; motor cycle addicts spot a Bonneville in a crowded street where others would not even notice that there was a motor bike.

Inputs which do not vary at all are usually not particularly useful to the system, e.g. a constant noise or smell, a clock ticking, the sensation of clothes rubbing on the skin. The filter alters over time to exclude such constant stimuli from consciousness. As soon as they change, however, e.g. the clock stops, the filter lets through this information and we notice it.

These selective attention mechanisms are extremely efficient and invaluable in many tasks. But any mechanism which is selective carries with it the penalty that information which does not conform to the characteristics selected by the filter will not get through to the brain, however important that information is. People can be concentrating so hard on one task that they are unaware of other information. Hence someone can fall down a hole because they were staring at some activity going on in the opposite direction. Presetting the filter can also lead to false alarms; searching a list for the name Jones we can sometimes be fooled by James. The cost of a rapid response is an increase in errors.

13.4.4 Expectancy

Perception is strongly influenced by a source which is located entirely inside the brain. This is expectancy, or the model of the real world which has been built up from experience over an individual's lifetime. This can form a very 'real' alternative to direct input from the world itself. Every one living in an industrial society knows what a motor car looks like and can conjure up a mental picture of one comparatively easily. This means that when confronted with a particular car in the real world there is no need to take in all of the details which are already on file in the brain. The person can concentrate upon only those characteristics which differentiate this car from the 'standard' car of his mental picture, e.g. its colour, or make, or its driver. Again there is a cost. In the UK we are so used to expecting the driver to be sitting on the right of the car that we may not see that this particular car is from abroad and that the person on the left is the one driving.

Machines, processes, people, and whole situations are stored in the brain and can be recalled at will, like files from the hard disc of a computer. This cuts down enormously on the amount of information about any scene which an individual needs to take in in order to perceive and understand it. But where the real world differs from the expectation, problems occur. This is most often the case with situations which go against population stereotypes, for example a machine on which moving a lever downwards turns it off. Other population stereotypes are red for danger and stop, clockwise turns the volume up or shuts the valve. These examples are very widely shared, but in other cases stereotypes for one population may contradict those held by others, e.g. you turn the light on by putting the switch down in Britain but by pushing it up in the USA and parts of Europe. Designs which do not match expectation can trap people into making errors.

Many of the classic illusions seem to be caused by misplaced expectations. For example, if you are sitting in a train at a station and another train alongside moves away, you can experience a sensation that you are moving because that is what you expect. Similarly, sitting in the cinema you believe the voices come from the characters on the screen, not from the loudspeakers at the side.

We tend to perceive pattern, regularity and constancy in the world when it is not really there because we fill out any imperfections with our thoughts, so failing to notice flaws and other irregularities.

In some circumstances these false expectations may result in little more than annoyance and delay. In other cases it may be a prelude to physical damage or injury. For example, a machine operator may reach rapidly towards his pile of components without looking and gash his hand on the sharp edge of one of them which has fallen off the pile and is nearer than he expected; the truck driver may drive rapidly through the doors which are reserved for trucks without looking or sounding a warning because no one is supposed to be there, only to find that someone is using the truck doors as pedestrian access.

The reliance upon expectation is an essential mechanism in skilled operation and many tasks would take a great deal longer to carry out if this was not so. Thought therefore needs to be given to ways in which reality can be made to fit people's models, rather than vice versa. Standardisation of machine controls, layout of work places, colour coding and symbols, etc. are all designed to achieve this result, as are codes of rules such as the Highway Code or Plant Operating Procedures. But standardisation has a hidden snag. The more standard things normally are, the more likely are exceptions to trap someone into an error. So, standards and rules must be enforced 100% to avoid this danger.

Any circumstances which are unclear or ambiguous (e.g. fog, poor lighting) or where an individual is under pressure of time, is distracted or worried, or fatigued, will encourage expectancy errors. In extreme cases individuals may even perceive and believe in what are in fact hallucinations.

13.4.5 Storage

The storage facility of the human system is the memory. The memory is divided into two different types of storage, a long-term, large capacity store which requires some time for access, and a short-term working storage which is of very small capacity and rapidly decays, but can be tapped extremely rapidly.

The short-term memory is extremely susceptible to interference from other activities. It is used as a working store to remember where one has got to in a sequence of events, for example in isolating a piece of equipment for maintenance purposes. It also stores small bits of information between one stage of a process and another, such as the telephone number of a company between looking it up in the directory and dialling the number.

Long-term memory contains an abundant store of information which is organised in some form of classification. Any new information is perceived in terms of these categories (closely related to expectations) and may be forced into the classification system even when it may not fit exactly. In the process it can become distorted. This process probably also results in specific memories blurring into each other, with the result that the wrong memory may be retrieved from the store when it is demanded.

People are not able to retrieve at any one occasion all the things which they have stored in their memory. There are always things which they know, but cannot recall, and which 'pop out' of store at some later stage. They are there but we have forgotten where we put them. This sort of limitation can frequently be overcome by recalling the circumstances in which the original memory was stored, or by approaching it via memories which we know were associated with it. Unavailability of memories may be crucial in emergency situations where speed of action is essential. A technique for overcoming unavailability is to recall and reuse the memories (knowledge and skills) at regular intervals. Refresher courses, emergency drills, and practice sessions all perform this function. However, one unwanted side effect of constant recall and reuse of memories is that they may undergo significant but slow change. When the memory is unpleasant, or shows the individual in a bad light, it is extremely likely that distortion will occur at each recall and it will be these that will be remembered rather than the original story. Testimony following an accident is notoriously subject to such distortion. People can quite genuinely remember doing what they should have done (the rule) rather than the slip they actually made.

13.4.6 Processing

To use information it must be processed. This may be done 'on-line' or 'off-line'.

13.4.6.1 On-line processing. Routines and skills

This is the moment-to-moment decision making about what action to take next in order to cope with and respond to the environment around the individual. This processing has to be funnelled through a narrow capacity channel which can only handle small numbers of items at one time. This limited capacity can be used to best effect by grouping actions together as packages or habits which can be set in motion as one, rather than as separate actions. Such habits form the basic structure of many repetitive skills, for example signing one's name, loading a component into a machine, changing gear in a car. Such grouping of activities does, however, carry with it the penalty that, once initiated, the sequence of actions is difficult to stop until it has run its course. Monitoring is turned down low during the routine. This can result in injury, for example if someone steps off a loading platform at the point where the steps always used to be, without remembering that recently they have been moved. The packaging of actions in these chunks also places greater premium on correct learning in the first place, since it becomes very difficult to insert any new actions into them at a later stage (see section 13.7.4.1).

13.4.6.2 Off-line processing. Decision making and intelligence

This is the facility whereby people can simulate in their mind the results of different possible courses of action before they make any decision about which course to choose. This skill is an immensely valuable one because it allows some courses of action to be rejected without ever trying them on account of the unpleasant consequences which we correctly predict. However, as a skill it depends upon knowledge of how factors interact and the ability to manipulate many factors together in the mind. This in turn is related to intelligence, and to the amount of practice in using the skill.

The ability to learn, to manipulate concepts in the head and to solve problems is one way of defining intelligence or cognitive efficiency. Psychological research is full of conflicting views of exactly how intelligence should be defined, whether it is largely innate or can be modified significantly by environment, and how it should be measured. This conflict arises from the complexity of obtaining evidence with which to support or disprove the various theories. Intelligence testing was first carried out by child psychologists – notably Binet in the first years of this century and developed through its massive use by the American army in the first world war. Binet[12] invented the term IQ (intelligence quotient) to describe his finding that, in children, greater intelligence seemed to result in a child being able to do certain tasks sooner than the average child.

$$\text{Intelligence quotient, IQ} = \frac{\text{mental age} \times 100}{\text{chronological age}}$$

Therefore by definition an average IQ is 100. Most people who gain a university degree or equivalent professional qualification have an IQ of more than 120, while an IQ of 70 or less will make all but simple routine work beyond the person's intellectual capacity.

Most theories of intelligence agree that it can be subdivided into special aptitudes e.g. verbal, numerical, spatial, manual, mechanical, musical. These represent the sort of problems that a given individual is best at solving in the head. Tests can be found for all these aptitudes. However, they need to be selected, administered and interpreted by experts if they are to be valid and useful.

Errors in off-line processing usually relate to attempts to simplify the decision so that it can be handled in the head. Such simplifications involve ignoring the effect of some factors, rejecting out of hand some courses of action without considering them, and limiting the degree to which the consequences of a course of action are thought through before making a decision. All these limitations, which may be unconsciously imposed, can result in unsafe decisions being made.

Individuals are also not entirely logical in the way in which they make decisions. The value assigned to particular outcomes, such as amount of effort saved, money earned, approval obtained or withheld by colleagues and superiors, etc. is a subjective one. It will be influenced by the personality of the individual making the decision, and by experience of the way in which previous decisions have turned out. Small probabilities are consistently poorly assessed and people rely upon such illogical factors as 'luck', which they believe they can influence, rather than accepting that some things such as roulette wheels or dice are entirely random machines.

The rate and efficiency of mental processing are also limited by the level of arousal of the brain. At low arousal levels performance is poor, rising with increased arousal to an optimum and then falling with further increases. This change in arousal corresponds roughly to a movement from drowsiness, through optimum coordinated performance, to the confused activity resulting from over-anxiety and panic.

13.4.7 Output

Once the decision to act has been made, the remaining limitations on the human system are those of its capacity to act, e.g. its speed, strength, versatility. Humans differ from machine systems in that their actions are not carbon copies of each other, even when the individual is carrying out the same task again and again. The objective may be unchanging, but the system adapts itself to small changes in body position, etc. to carry out a different sequence of muscular actions each time to achieve that same result. This use of constantly changing combinations of muscles in coordination is an essential means for the body to avoid fatiguing any particular muscle combination.

In all human actions there is also a trade-off between the speed of an action and its accuracy. Speed can be improved by reducing the amount of monitoring which the brain carries out during the course of the action, but only at the cost of reducing the accuracy.

13.4.8 Effects of the environment and degradation of performance

All of the above limitations have been described in isolation from the effects of the external environment. General environmental conditions such as noise, glare and lighting level, dust and fumes, social environment, etc. will influence the factors which have been described above. For example, noise and high temperature both have an effect on the arousal level. Noise increases it, heat decreases it, and both have an effect on the accuracy of detection of information and the speed of processing it. These physical environmental factors are dealt with in the chapters on Occupational Health and Hygiene. The effects of fatigue and the social environment upon individual behaviour are dealt with here.

The performance of the human system is only at an optimum within certain environmental limits. As part of the price of its sensitivity and flexibility the human system is susceptible to the influence of a very large range of factors which can affect its performance. Unlike machines, human beings show a slow and often subtle degradation of performance over a wide range of environmental conditions, but arrive at a total breakdown only comparatively rarely. This means that individuals can maintain some sort of functioning long after they have passed the peak of their performance, but it also blurs the point at which they should stop in order to avoid errors. Regulations and good practice on working hours for coach drivers, hospital doctors and others must wrestle with the problem of making these black and white decisions at some point of the continuum of shades of grey.

Performance is degraded under the following types of situations:

(1) Working for too lengthy a period, which produces fatigue.
(2) Working at times of day when body mechanisms are not functioning efficiently, i.e. the diurnal rhythm is disturbed.
(3) Loss of motivation to perform.
(4) Lack of stimulation resulting in lowered arousal.
(5) Working under conditions of conflict, threat, both physical and psychological, or conditions which threaten the body's homeostatic or coping mechanism and cause stress.

13.4.8.1 Length of working periods and fatigue

The 1833 Factory Commission[13] said in its report that the then existing hours of work of children led to 'permanent deterioration of the physical constitution and the production of various diseases often wholly irremediable, and exclusion from the means of obtaining education, elementary and moral, and of profiting by those means by reason of excessive fatigue'. This early quotation underlines the fact that the length and distribution of working hours have social and physical effects, both of which must be borne in mind when considering permitted or agreed working hours. Long periods of working produce:

(1) Muscular fatigue resulting from overloading of individual muscle groups, either through static loading to maintain a posture, or through awkward or repetitive dynamic loading. In addition to limiting working hours, the cure to this problem lies in design of work places to minimise static working load and to allow for the utilisation of the most efficient muscle groups, and the opportunity to rest muscle groups by shifting posture.
(2) General mental fatigue characterised by an increase in the length and variability of reaction time, especially for decision making. This leads to an increase in errors and a tendency to neglect peripheral aspects of tasks such as checking routines. These effects can be demonstrated in most tasks after periods of uninterrupted working of between 10 and 50 minutes, depending upon the task load. Rest pauses of one or two minutes interposed when the performance begins to fall from its optimum level are sufficient to restore functioning to its former level. If performance is allowed to carry on without a break until more obvious signs of degradation have appeared, then proportionately longer rest pauses are required for complete recovery. If the task is machine paced, or strong motivation from pressure of work or incentive bonus schemes prevents natural breaks, artificial breaks should be introduced in order to maintain performance at an optimum level.

    If the working day exceeds about ten hours including overtime there is a highly significant reduction in output, and increased risk of accidents on many tasks. Depending upon the intensity of work, the type of work load and any loads imposed by activity outside working hours this effect may occur earlier. It occurs with mental work as much as with manual and physical work[10].

13.4.8.2 Distribution of working hours[14]

Around 10% of the working population spend some time working on a night shift, and a larger proportion who are on shift work, work outside the period 7.00 a.m. to 1.00 p.m.

Body systems follow a cyclical variation in activity which is linked to the 24 hour light-dark cycle. This rhythm is known as the diurnal or circadian rhythm. *Figure 13.4* shows the diurnal rhythm for performance on simple mental tasks. The rhythm for other activities such as motor skills, or tasks involving short-term memory may be out of phase with this, resulting in performance on different tasks being best at different times of day. The difference between performance at the peak and the trough of the curve is of the order of 10%, which is as significant as the degradation in performance caused by a blood alcohol level at the legal limit or by approximately two hours' loss of sleep on the previous night.

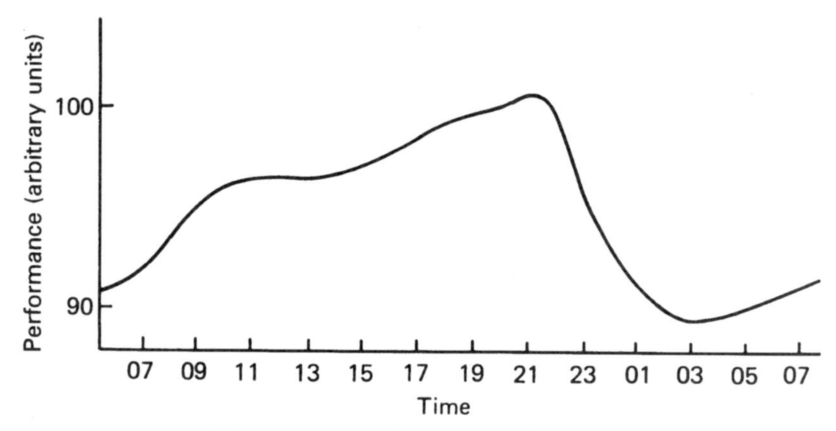

Figure 13.4 Diurnal rhythm for simple mental tasks

The rhythms are keyed into the sleeping/waking cycles as well as the light/dark cycle. These normally work together, but if people work at night, rhythms are thrown into some disarray and take time to begin to adjust. Adjustment begins to be apparent after 2–3 days, and goes on increasing up to a period of about 14 days provided that the person continues both to live and work on a night-time schedule, and does not return to day-time living over a weekend. Even after two weeks the curves have not fully reversed, but have flattened out.

Night workers have the additional disadvantage that they are trying to sleep when the rest of the world is awake and making a noise. Hence their sleep is far more disturbed than that of day workers. Night and evening workers also suffer from a major disruption to social life. This can result in conflicts and stress generated within the family. Studies of night workers show that they tend to have a higher incidence of gastro-intestinal disease such as ulcers, and nervous disorders.

There is no clear evidence that the physical and health effects on women and young persons are greater than on men. The original reasons for the introduction of the ban on night work for protected persons were as much for its supposed moral dangers as for its health effects.

Many firms cope with the problems of shift and night work by adopting rotating shift patterns. Research on the physical effects of rotation show that the traditional British shift pattern of a weekly rotation period is the worst possible compromise, since it results in a constant semi-adaptation and de-adaptation of the diurnal rhythms. Regimes of permanent night work or very rapidly rotating shifts with no more than two nights on night work are far better from these points of view. However, social factors often make the weekly rotation pattern more acceptable to those who have to work shifts.

13.4.8.3 Stress

To the psychologist stress is the result of an internal or external disturbance which threatens the balance of the system. Causes of stress, named stressors, can be divided into two main groups:

(1) Environmental stressors such as extremes of temperature, noise, vibration, bodily injury, hunger, etc. all have both physiological effects as well as psychological effects.
(2) Psychological stressors such as threat, the inability to achieve valued goals, conflict, physical or social isolation, intense periods of mental activity or excitement, or dramatic changes in life style, e.g. retirement, redundancy, marriage.

There is extremely wide variation in the reaction of different individuals to the same stressor. In the case of physiological stressors this variation is largely a case of differences in physical or physiological tolerance. In the case of psychological stressors there is the important intervening variable of the perception of the stressor and the degree to which it is seen as a threat to valued goals. Individuals have different abilities to deploy coping responses when faced with a stressor which will not go away. *Figure 13.5* summarises these factors.

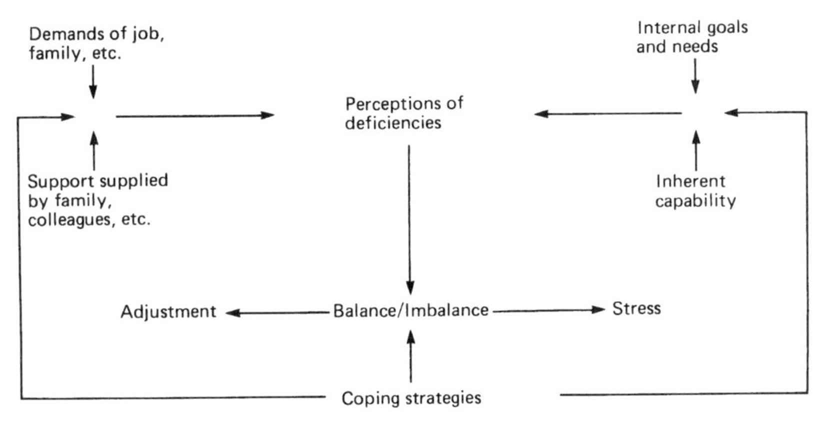

Figure 13.5 Stress components. (Modified from Mackay and Cox[15, 16])

The physical response of the body to stressors is to increase the activity of the sympathetic nervous system, and to increase the secretion of the hormone adrenalin. This mobilises the body in the primitive condition for fight or flight by increasing muscle tension and heart rate, diverting the blood from inessential activities such as digestion to the muscles and the brain, mobilising stored food reserves, etc. However in modern industrial society physical activity is often not an appropriate response to stressors such as impending examinations, or an angry boss. These physical preparations if maintained for long periods of time can result in harmful side effects such as gastric ulcers, high blood pressure and arteriosclerosis.

The psychological symptoms of stress are disturbed concentration, impaired memory, impaired decision making, tension and aggression, sleep disturbance, and, in severe cases, mood change.

There are a number of coping strategies which can be adopted by individuals. They can withdraw from the source of the stress either physically by leaving their job or going absent, or psychologically by lowering their ambitions, e.g. ceasing to fight for promotion when promotion prospects are blocked. Companies can remove some of the unnecessary demands of the work place, e.g. by work restructuring or by providing greater support through discussion groups, meetings, or counselling services. Finally, there has been some success in bolstering the individual's own resources for countering stress through the teaching of relaxation techniques, through counselling and psychotherapy. A coping strategy open to the employing organisation is to identify people who would appear to be susceptible to stress and to redeploy them into jobs where the demands are low. Some activities such as drinking, smoking or use of tranquillisers which start as coping strategies can end up as health problems in their own right if they result in addiction.

13.5 Types of error

The description of behaviour given up to now has blurred a distinction which is vital in understanding the types of error which people make. This distinction has arisen from the work of Reason[17] and Rasmussen[18]. They distinguish three levels of behaviour which show an increasing level of conscious control:

(1) Skill-based behaviour in which people carry out routines on 'automatic pilot' with built-in checking loops.
(2) Rule-based behaviour in which people select those routines, at a more or less conscious level, out of a very large inventory of possible routines built up over many years of experience.
(3) Knowledge-based behaviour where people have to cope with situations which are new to them and for which they have no routines. This is a fully conscious process of interaction with the situation to solve a problem.

As a working principle we try to delegate control of behaviour to the most routine level at any given time. Only when we pick up signals that the more routine level is not coping with do we switch over to the next level (see *Figure 13.6*). This provides an efficient use of the limited resources of attention which we have at our disposal, and allows us to a limited extent to do two things at once. The crucial feature in achieving error-free operation is to ensure that the right level of operation is used at the right time. It can be just as disastrous to operate at too high a level of conscious control as at too routine a level.

Each level of functioning has its own characteristic error types, which are described briefly in the following sections.

13.5.1 Skills and routines

All routines consist of a number of steps which have been highly practised and slotted together into a smooth chain, where completion of one step automatically triggers the next. Routine dangers which are constantly or

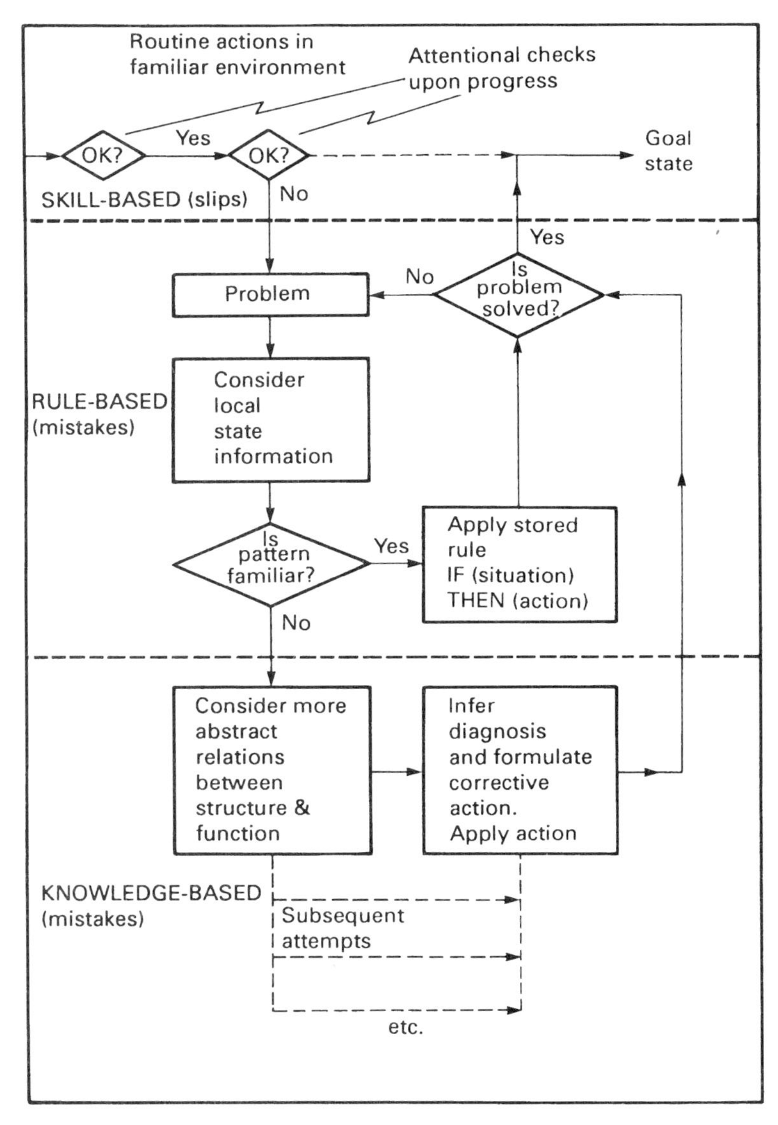

Figure 13.6 Dynamics of generic error-modelling system (GEMS). (From Reason[17])

frequently present in any situation are (or should be) kept under control by building the necessary checks and controls into the routines as they are learned. The checks still require a certain amount of attention and the comparatively small number of errors which occur typically at this level of functioning are ones where that attention is disturbed in some way.

(1) If two routines have identical steps for part of their sequence, it is possible to slip from one to the other without noticing. This nearly always occurs from the less frequently to the more frequently used

routine; for example arriving at your normal workplace rather than turning off at a particular point to go to an early meeting in another building. Almost always these slips occur when the person is busy thinking about other things (e.g. making plans, worrying about something, under stress).

(2) If someone is interrupted half way through a routine they may return to the routine at the wrong point and miss out a step (e.g. a routine check) or carry out an action twice (e.g. switching off the instrument they have just switched on because both actions involve pushing in the same button).

(3) The final problem at this level is that routines are dynamic chains of behaviour and not static ones. There is a constant tendency to streamline them and to drop steps which appear unnecessary. The most vulnerable steps are the routine checks for very infrequent problems in very reliable systems. (e.g. checking the oil level in a new car engine).

Many of these errors occur because the boundary between skill-based and rule-based activity has not been correctly respected.

These sorts of error will be immediately obvious in many cases because the next step in the routine will not be possible; the danger comes when the routine can proceed apparently with no problem and things only go wrong much later. The cure for the errors does not lie in trying to make people carry out their routines with more conscious attention. This will take too long and so be too inefficient, and will be subject over a short time to the erosion of the monitoring steps. It lies to a great extent with the designer of the routines (and so of the apparatus or system) to ensure that routines with different purposes are very different so that unintended slipping from one to the other is avoided. Where this is not possible extra feedback signals can be built in to warn that the wrong path has been entered by mistake (see 13.6.1.1).

The second line of defence is to train people thoroughly so that the correct steps are built into the system, and then to organise supervision and monitoring (by the people themselves, their work or reference group, and supervisors or safety staff) so that the steps do not get eroded.

13.5.2 Rules and diagnosis

When the routine checks indicate that all is not well, or when a choice is needed between two or more possible routines, people must switch to the rule level. Choice of a routine implies categorisation of the situation as 'A' or 'B' and choice of routine X which belongs to A or Y which belongs to B. This is a process of pattern recognition. This is analogous to computer programme rules of the form IF . . . . , THEN . . . .

The errors which people make at this level are linked to a built-in bias in decision making. We all have the tendency to formulate hypotheses about the situation which faces us on the basis of what has happened most often before. We then seek evidence to confirm that diagnosis rather than doing what the scientific method bids us and seeking to disprove the hypothesis. This means that people tend to think they are facing well-known problems until they get unequivocal evidence to the contrary. The Three Mile Island was a classic case where operators persisted with a false diagnosis for several hours in the face of contradictory evidence until a person coming on shift (and so without the perceptual set coming from having made the initial diagnosis) detected the incompatibility between the symptoms and diagnosis.

The solution lies in aiding people to make diagnoses more critically (e.g. checking critical decisions with a colleague or supervisor before implementing them).

13.5.3 Knowledge and problem solving

When people are facing situations they have no personal rules for, they must switch to the fully interactive problem solving stage, when they have to rely upon their background knowledge of the system and principles on which it works to derive a new rule to cope with the situation. There are meta-rules for problem solving which can be taught (see section 13.7.4.3). Besides these there is the creativity and intelligence of the individual and the thoroughness of their training in the principles underlying the machine or system. Errors at this level can be traced to:

(1) Inadequate understanding of these principles (inadequate mental models).
(2) Inadequate time to explore the problem thoroughly enough.
(3) The tendency to shift back to rule-based operation too soon and to be satisfied with a solution without checking out the full ramifications it has for the system.

The first two are typical errors of novices, the last more of the expert. Experts are by definition the most capable of functioning at this level, but also the people who need to do so least often, because they have learned to reduce most problems to rules. They may also become less willing to accept that there are situations which do not fit their rules. Almost all experts overestimate their own expertise.

13.6 Individual behaviour in the face of danger

Hale and Glendon[1] combined the insights of the two major models presented so far (*Figures 13.2* and *13.6*) with other sources[19,20] into a model of individual behaviour in the face of danger (*Figure 13.7*). Their model allows us to discuss a number of practical issues in which we are trying to influence human behaviour, e.g. through task design and training.

Danger is always present in the work situation (as in all other situations). The task of the individual is to keep it under control; to avoid errors which provoke an increase in danger and to detect and avoid or recover from danger increasing from other reasons. Much of this activity occurs at the skill- and rule-based levels by more or less routine reaction to warning signals. Only occasionally do people in their normal work situations need actively to contemplate danger. However these occasions are vitally important when they do occur. Examples of such activities are:

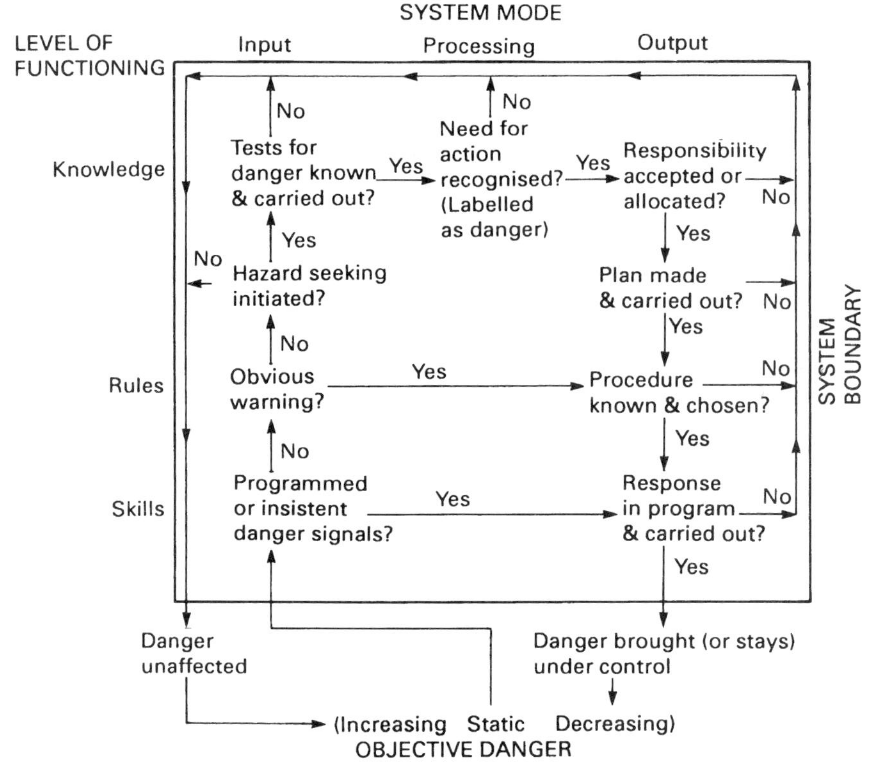

Figure 13.7 Behaviour in the face of danger model[1]

(1) Designers making decisions about machine or workplace design, plant layout, work procedures, etc. They need to predict the actions of the people who will use these products and the hazards which will arise in use.
(2) Operators, safety committees, safety advisers and factory inspectors carrying out hazard inspections, safety audits and surveys, who need to seek out hazards.
(3) Policy makers in industry and government deciding whether a level of risk associated with a technology or plant location is to be accepted. Members of the public assessing whether that policy decision is acceptable to them.
(4) Planners designing emergency plans for reacting to disasters.

Such decisions and activities are all largely carried out at the knowledge-based level and the borders with the rule-based level.

13.6.1 Hazard detection

This is important in three situations:

(1) Detection by operators in routine tasks of deviations from normal which could lead to harm. This is largely skill-based.

(2) During formal or informal inspection in the ordinary work environment. This is largely rule-based.
(3) Prediction of danger at the design and planning stage of a system. This is largely knowledge-based.

Surprisingly little scientific study has been made of how hazard recognition operates in any of these situations and what alerts people to the presence of danger. What follows is a summary of the available information[1].

13.6.1.1 Routine hazard detection and warnings

Implicit in the structure of skills is a built-in monitoring loop which detects deviations from the programme at the skill level. Deviations from the plan at the rule level are less automatically picked up and depend on good warnings, upon building in physical or procedural blocks which do not allow you to proceed further after such a mistake, or upon intervention by others.

Designers should improve upon the availability of information about deviations and consciously build into their designs feedback about the actions which the individual has just taken, and their consequences. Examples are displays on telephones which show the number you have just keyed in, a click or bleep when a key is pressed hard enough to enter an instruction on a keyboard, commands echoed on a visual display as they are entered on a keyboard, and the use of tick boxes on a checklist to indicate the stage in the check reached.

The following common sense criteria can be used for the design and placing of warnings. They should:

- be present only when and where needed,
- be clearly understandable and stand out from the background in order to attract attention,
- be durable,
- contain clear and realistic instructions about action, and
- preferably indicate what would happen if the warning is not heeded.

Warnings should preferably not be present when the hazard is absent, otherwise people will soon learn that it is not necessarily dangerous in that area. They will then look for further confirmatory evidence that something really is a problem before taking preventive action. The philosophy of 'if in doubt put up a warning sign' is counterproductive unless an organisation is prepared to go to great lengths in enforcing it even in the face of the patent lack of need for the precautions at some times.

If an alarm goes off and there proves to have been no danger, there will be a small, but perhaps significant loss of confidence in it. If false alarms exceed true ones, the first hypothesis an individual will have when a new alarm goes off is that it is a false one. Tong[21] reports that less than twenty per cent of people believe that a fire alarm bell going off is a sign that there really is a fire. The rest interpret the bell in the absence of other evidence as either a test, a faulty alarm or a joke. The recognition of the presence of fire is therefore often delayed, and the first reaction to warnings of a fire is often to approach the area where it is, in order to find out more, rather than to go the other way.

Because warnings must be understood quickly and sometimes under conditions of stress, there can never be too much attention to their ease of comprehension. The language used must be consistent, whether verbal or visual (e.g. a black triangle round a yellow sign always meaning a warning, a red circle with a bar a prohibition). The language must be taught. Written warnings must take account of the reading age of the intended audience and the proportion of illiterates or foreign nationals with a poor understanding of the language. A word such as 'inflammable' should, for example, be avoided because many misunderstand it to mean 'not flammable' (by analogy with 'inappropriate' or 'incomprehensible').

13.6.1.2 Inspection and hazard seeking

During workplace safety inspections the people concerned are already alerted to the possibility that there are hazards present, and are actively seeking them. But to seek is not always to find. Untrained inspectors characteristically miss hazards which have one or more of the following features:

(1) Not detectable by the unaided eye, but requiring active looking in, behind or under things, rattling guards or asking questions about the bag of white powder in the corner.
(2) Transient, e.g. most unsafe behaviour which can only be discovered by asking questions and using the imagination.
(3) Latent; i.e. contingent upon other events, such as a breakdown, a fire, or work having to be done by artificial light.

Inspection must be an active and creative search process, of developing hypotheses about how the system might go wrong. It requires the allocation of time and mental resources. Checklists can help to make the search systematic and to avoid forgetting things, but they should not be allowed to become a substitute for active thinking.

13.6.1.3 Predicting danger. Techniques for human reliability assessment

Prediction at the design stage is an extension of the problem of inspections, made more difficult because there may be no comparable system in existence from which to learn. Imagination and creativity are therefore relevant attributes for the risk analyst in addition to both plant knowledge and expertise in behavioural sciences.

There are large individual differences in how good people are at imagining the creative misuse that operators will make of their systems. Those who are good are known as 'divergent' thinkers. There is some evidence that people who gravitate towards the sciences, mathematics and engineering are more 'convergent' in their thinking, and tend to be more bound by experience and convention, than those who choose social sciences and the arts. They may therefore be less able to anticipate the more unusual combination of events which could lead to harm. This suggests the need for teamwork in design and hazard prediction.

Risk assessment techniques such as HAZOP, design reviews and fault and event trees are systematic methods to guide and record the process of creative thinking. They are also often used to quantify the chance of failure. That step is only legitimate if we are certain that all models of failure have been identified. That is particularly difficult with human-initiated failures because people can act (and fail) in so many more ways than hardware elements. Human reliability assessment techniques should be seen within the framework of the following list of steps[22,23].

(1) Familiarisation;
 Plant visit
 System analysis review
 Information gathering
(2) Qualitative analysis:
 Task analysis and talk-through
 Performance objectives definition
 Performance situation specification
 Modelling of human performance
 Identification of potential human errors
(3) Quantitative analysis:
 Determination of basic error probabilities
 Identification of performance shaping factors and dependencies
 Allowance for recovery factors
 Calculation of final human error contribution
(4) Incorporation in Probabilistic Risk Assessment:
 Sensitivity analysis
 Combination with hard- and software failure probabilities

A systematic task analysis is essential for a good prediction of human error. The data for such an analysis come partly from logical analysis of what should happen and partly from observation of what does happen. Such techniques are dealt with in chapter 10.

Task analysis forms the basis for the use of techniques for error prediction, which are currently in an early stage of development. They all depend upon one or other sort of checklist. For example each sub-task can be subjected to the following standard list of questions to specify what would happen if these types of error occurred and how such an error could happen (cf. HAZOP):

- sub-task omitted
- sub-task incorrectly timed:
  - too soon
  - too late
  - wrong order
- inadequate performance of sub-task:
  - input signals misinterpreted (misdiagnosis)
  - skills not adequate
  - tools/equipment not correctly chosen
  - procedure not correct
  - inappropriate stop point
    - too soon
    - too late
    - not accurate enough
    - quality too low
- routine confusable with other routines

Other checklists[24] have been produced linked to the models of Reason[17] and Rasmussen[18] or derived from ergonomics[25].

Kirwan[26] carried out an evaluation of five techniques of this sort to see which gave the best results when used by safety assessors on four representative tasks. No one technique came out as best in all circumstances. It is a question of horses for courses. Whatever the technique used it must distinguish clearly between:

- The external error mode, which is the effect the human error has on the system (e.g. fails to turn switch, misreads dial, etc.). This is important in linking human error analysis to other parts of the risk assessment.
- The error cause, which is the psychological description of why the error is made. This relates to prevention measures which can be taken.
- The system factors which affect performance on a range of tasks (the Performance Shaping Factors).
- The possibility of error recovery because the initial error is recognised.

Human error quantification is one of the most bitterly disputed areas of risk analysis. Risk analysts starting from the reasonably successful experience of quantifying hardware reliability try to treat human reliability in the same way, so that it can be integrated into their fault tree and event tree analyses. Psychologists doubt the possibility of doing this because:

- Human behaviour is far more complex than component behaviour and can fail in many different ways in different circumstances. The data collection problem is therefore enormous.
- Human failure rate is dependent upon the way in which each individual interprets the task and adapts it or themselves to prevailing circumstances.
- Inter-individual differences in error rate are large (typically a factor 3). Intra-individual differences can be considerable over time because of learning processes, working situation, fatigue etc.
- The initial error rate of humans is high, but they detect a great majority of their own errors and failures and correct them.

Kirwan[26] reviews eight techniques for human reliability quantification. Again his conclusion is that no one method stands out above all others and that there is room for much work to develop better methods.

13.6.2 Knowledge of causal networks

Hazard detection has been shown above to be dependent on the mental models people have of the way in which events happen and systems develop. If these mental models are incomplete or wrong they can lead to inappropriate behaviour in the face of hazards. Interview studies[1] show that such problems frequently occur, particularly in relation to occupational disease hazards. Examples of such significant inaccuracies are:

- Men sawing asbestos cement sheets who said that they only wore their face masks when they could see asbestos particles in the air. (Yet it is the microscopic, invisible particles which are the most dangerous because they are in the size range which can penetrate to the lung.)

- Wearers of ear defenders who fail to incorporate the notion of time weighted average exposure in their concept of what constitutes dangerous noise. Hence they fail to realise that 'just taking the ear-muffs off for a few moments to let the ears breathe' in high noise areas can negate much of their protective effect.
- Misconceptions about the link between posture and musculo-skeletal damage such as considering postures as 'relaxed' and therefore good when they showed the body and notably the spine in a slumped position (which in fact puts extra load on the back muscles to stabilise the spine in that position).

Such misconceptions must be put right by training.

13.6.3 Reactions to perceived risk

The reactions of different groups to risks they perceive have been the subject of much research in the past two decades[1,27,28]. Much of it has concentrated on decisions about siting hazardous plants or developing technologies such as nuclear power. A main focus has been the question of 'acceptability of risk'. This term has led to much confusion because it implies that people are, or should be content, or even actively happy with a particular risk level. The word 'accepted' or 'tolerated'[29] gives a better assessment of the situation, since it carries with it an idea that the opportunity to do something about the hazard is a relevant factor in any decision. There is also overwhelming evidence that people do not consider the risk attached to an activity or technology in isolation from the benefits to be gained from it[30]. Therefore no absolute 'acceptable' level for a wide range of different hazards can be meaningful, since the benefits which go with them will vary widely.

A clear distinction has emerged from the research between threats to personal safety, threats to health and threats to societal safety[31]. The factors which people use in assessing each of them appear to weigh differently, and this is likely to be related to the sort of action which people perceive they can take against the differing threats. For example, moving house or changing jobs will solve the threat to an individual's safety from a particular chemical plant, but will do nothing for the threat to societal safety from that same plant.

Despite these differences there appear to be common factors which people use to make assessments of danger and to apply a label to a situation indicating that something must be done about it. What varies between types of hazard and between responses to different questions is the weighting given to the different factors.

The research has used two basic approaches. Either to ask people directly what they think about hazards and how they react to them, or to consider actual behaviour in respect of different hazards. The first is called expressed preference research, the second revealed preference.

One clear result of expressed preference research is that people use a more sophisticated assessment process in judging risk decisions than just considering probability of harm. They also consider a wide range of other factors, which can be grouped under the following headings:

(1) Whether the victim has a real choice to enter the danger or not, or to leave it once exposed.
(2) Whether the potential for harm in the situation is under the control of the potential victim or another person, or outside any human control.
(3) The foreseeability of the danger.
(4) The vividness, and severity of the consequences.

13.6.3.1 Choice to enter and leave danger

Those who choose to engage voluntarily in activities like skiing which they know to be dangerous seem to be considered to know what they are doing, to realise the nature of the hazards and to have accepted their own responsibility to control them. The problem of accidents or disease is then seen as their affair, and they are presumed to be in control of the hazard. On the other hand if there is no choice about exposure to the danger, e.g. in having a nuclear plant built near your village, far higher demands on the level of safety are made. The situation is, however, seldom clear cut. Can, for example, the choice of a person to take a job on a construction site in an area of high unemployment be called a voluntary acceptance of risks associated with that job? Hazards frequently come as part of a package with other costs and benefits. In the early years of the industrial revolution workers were deemed to have accepted voluntarily the hazards of the job that they accepted. Therefore they were deemed liable for their own accidents. Now both society's view and the law have changed. The employee is not considered to accept occupational hazards voluntarily, unless there is talk of some gross deviation from normal carefulness.

The demand for increased safety levels is also stronger if the risks and benefits are not equitably shared and one group profits from the risk exposure of another.

There is some evidence that dangerous activities which are voluntarily chosen are positively valued partly because of their finite element of danger. Mountaineers choose to attempt climbs of increasing difficulty as their skill increases, finding the old ones tame. There is an element here of testing the degree of control which one has over a situation to check that it is real. The element of apparent loss of control is one of the attractions of fairground rides such as the 'wall of death'. But the fascination seems to go further than this. Greater danger, such as in war or time of disaster, is associated in the minds of survivors with greater group friendliness, shared feelings, sense of purpose and competence which makes that danger in retrospect positively valued, or at least willingly accepted.

13.6.3.2 Controllability

The largest element here seems to be the feeling of personal control. Those who believe themselves knowledgeable about and in control of a dangerous situation, even where the magnitude of the consequences is potentially great, show little fear or concern about it. This also applies to attitudes towards the safety of others. Thus construction site supervisors may consider[32] that the site hazards are under the control of skilled craftsmen and not personally concern themselves with them, even when they see that

that control is not being fully exercised. Similarly workers in a plant are much less concerned about the hazards from it than those who live nearby but do not work there.

If the assessment of personal control is such an important factor in evaluating hazards it is very important that the assessment is accurate, and that people do not believe they are in control when they are not. But there is ample proof that people can have illusions of great control where none or less exists. Svenson[33] quotes a number of examples from the field of driving. For example between 75 and 90% of drivers believe themselves to be better than average when it comes to driving safely; only 50% can be right. Similarly 88% of trainees in cardiopulmonary resuscitation felt confident after an interval of several months to perform it, while only 1% actually performed adequately. Experts always tend to be overconfident in their expertise. This is particularly dangerous when specific knowledge, for example of a theoretical nature, about a process is taken to mean control over the whole activity involving that process. This can be a serious source of overconfidence in skilled personnel such as research chemists or toolroom personnel, most of whose accidents in fact come from the everyday hazards of the machinery or the laboratory which have little to do with their speciality.

When people have no personal control they may place their trust in others to keep the situation safe. Again it is a question of whether the assessment is accurate and whether the trust is justifiably placed. The work of Vlek & Stallen[30] suggests that one of the clusters of beliefs characterising those who oppose large nuclear, transport and chemical plant developments is personal insecurity and lack of trust in those controlling the technology. The situation is made worse by the spectacle of experts disagreeing violently with each other about the safety issues of the developments in question.

In the field of health promotion the concept of control has also been shown to be important. One of the main thresholds to be crossed before people will act to change their own behaviour is to admit that they personally are susceptible to the health threat, for example, from smoking, alcohol, drugs, or heart disease, i.e. that they have lost control. The other side of this coin is the need to believe in the efficacy of the preventive action before it will be adopted. This means believing that the proposed action would restore the lost control; that giving up smoking would reduce the risk of cancer and heart disease, that wearing the protective earmuffs would reduce the hearing loss and so on. The opportunity to prove for oneself the effectiveness of protective devices is therefore important in persuading people to wear them.

13.6.3.3 Foreseeability

Foreseeability is a word familiar from the case law of the English legal system relating to health and safety. It has been defined by judges with reference to what the 'reasonable man' would expect to happen given access to the current state of knowledge at the time of making a decision. It is used to draw a dividing line between situations in which people should have taken action to prevent an accident, and those for which it is not reasonable to hold them liable.

At an individual level this also affects the assessment of risk with hazard detection being limited by what is foreseeable or foreseen. But if people cannot foresee exactly what may happen, but suspect that it may still go wrong, they will be afraid. If this feeling goes hand in hand with the belief that there could be severe consequences and that the person is powerless to do anything, the reaction may be extreme. Evidence that a new and unknown technology like genetic engineering is not as much under control as previously thought would therefore have a profound effect on people's beliefs, shifting them rapidly from indifference to strong opposition. This is approximately the effect which Chernobyl had on nuclear power.

13.6.3.4 Vividness, dreadfulness and severity

The most recent accident or tragedy weighs heavily in the minds of people when they are asked about priorities for prevention, but this may fade rapidly. On a more permanent basis, people have reasonably consistent ratings of how nasty particular types of injury or disease are[34]. For example cancer is greatly feared, an eye is worth more than a leg and some injuries such as quadriplegia and brain damage are consistently rated as worse than death.

An important element in memorability is 'kill size', the number of people who either actually do, or potentially could, get killed in an incident.

13.6.3.5 Conclusion

With such a complex of factors determining the reaction of both individuals and society to risk it is not surprising that no simple scale such as Fatal Accident Frequency Rate can capture its essence. Managers and planners may wish to reduce decision making to a tidy consideration of probability times cost of harm (usually deaths). They may even wish to label as irrational any opposition to this definition of risk. But this is no more than one powerful group putting an emotive label on something they seek to oppose. A better approach is to treat the factors for what they are, namely the basic elements which must be influenced if we wish to change behaviour. If you want someone to use a safety device, you must convince them that the danger is foreseeable, unpleasant and avoidable, that the safety device is effective and that they can choose how to use it.

13.6.4 Assessment of probability

For a normal person probability is not a concept that comes naturally. This is an idea that will be readily accepted by anyone who has tried to learn the fundamentals of statistics. Most normal people have little need for accurate probability judgements and little practice in making them. Individuals normally only rate whether they think things will remain under control. Despite this, it is surprising how good the correlation is between measured probability of particular types of accident and subjective assessments by the general population. The major bias is that the subjective scale is compressed and foreshortened in relation to the objective, typically spanning only three orders of magnitude instead of six. Very rare risks are treated as non-existent, slightly less rare risks may be overestimated and common ones underestimated. Some hazards are raised in the order of probability; typically those which receive media coverage.

The framing of questions about probability and of statements about risk can strongly affect people's responses to them. It is more effective as an argument to get people to be vaccinated to tell them that a vaccine offers total protection against one strain of disease that accounts for half of a given sickness, than to tell them that the vaccine offers 50% protection against the sickness; the presence of the word 'total' in the message gives the illusion of certainty. Framing information about road accidents in terms of the probability of accidents over a lifetime (probability of death c. 0.01 – and of disabling injury c. 0.33) is much more effective in getting people to wear seat belts than quoting the probability over one trip (probability of death c. 1 in 3.5×10^{-6}, and of disabling injury c. 1 in 10^{-5}).

13.6.5 Responsibility for action

Even if people can see a danger and appreciate the need for action, they may not act because they think they cannot or should not. This may be because it is seen as someone else's job or responsibility. This attitude is found among supervisors who are not willing to tell skilled workers how to avoid risks in their job[32]. Social pressures determining what is or is not acceptable behaviour may discourage people from warning others because of the fear of being told to mind one's own business, or of being thought to be interfering.

The major factor at this stage will be the way in which people view the courses of action open to themselves and others to influence the danger. Again the crucial role of the mental models of cause and effect are clear. If I believe as a supervisor that accidents to my staff are caused by their own carelessness and lack of attention to rules I will only think in terms of selection, training and discipline as actions. If I believe that the machine design is such that nobody can be expected to concentrate 100% of the time to avoid injury, I shall give attention to redesign or guarding as well. Biases in the way people allocate responsibility for accidents or prevention are therefore of vital importance and should be the subject of training. Such biases can be summarised as follows[1]:

(1) When people are looking at their own future behaviour they think that they can exercise more control than is usually the case. Hence they accept great (even too great) responsibility to act to control the situation and to prevent future accidents.
(2) When people personally suffer an accident, they are inclined to attribute it too much to the force of external circumstances rather than to personal responsibility.
(3) When people observe others' behaviour they grossly underestimate the effect that the situation has on determining it; hence they overestimate the control that others have over what they do, and blame them unfairly for their accidents. This can lead to a reluctance to intervene in situations to warn, instruct or help people.

These biases arise when there is some ambiguity in a situation which allows for more than one interpretation. Such occasions are most frequent in rapidly changing situations, and when people are trying with hindsight to reconstruct an accident of which they may have been a witness or about which they have merely heard reports. Putting the three biases together goes some way towards explaining the inactivity in accident prevention in a number of situations. Designers must consider hazards to others (the users). They tend to overestimate users' ability to look after themselves, and so underestimate the need to build in safeguards. Supervisors place the onus for avoiding accidents on the victims and not on themselves; while their bosses think that it is the supervisors' job and so shuffle off their own responsibility for the climate of rules and priorities which they create. If managers and designers sit round a table with workers from the shop floor who are talking about their own accidents (and so are subject to the first two biases), a great measure of agreement is possible. Both sides will agree that the main onus lies on the worker to prevent accidents. However, this agreement does not lead to any action, as both sides will also tend to believe that everything is under control and nothing needs to be done.

13.6.6 Action plans

The major problem under this heading is whether the necessary actions have been learned and are available when needed. The latter point is particularly crucial in emergencies, such as evacuation, plant shut-down and first aid treatment. Considerable investment in refresher training, sometimes on simulators, is needed to keep rarely used skills available enough to be deployed accurately when they are wanted.

13.7 Change

Underlying the above discussion of human characteristics, limitations and differences has been the notion of change. The human system is in dynamic equilibrium with its physical, social and cultural environment, constantly adjusting itself to changes in that environment while it pursues its goals and objectives. The environment changes as society and technology change and as family responsibility and job type and location change. At the same time the individual changes, grows up, matures and grows old, learns new skills, forgets old knowledge and acquires new goals. Change is intrinsic in the human condition. It is a mistake to think of solutions to problems in health and safety in terms of changing a person from one stable state to another. The problem is better seen as one of prodding and guiding behaviour along one path out of the many possible ones and trying to stabilise it in a new path which will maintain its essential characteristics while it carries on adapting to other changes in the world. An appropriate analogy for change is steering a sailing ship on a turbulent ocean rather than the action of picking up a piece of plasticine, remoulding it and setting it down again.

In the following sections the various processes of change are briefly presented. First the 'natural' processes of growing up and change are sketched, which lead to individual differences in personality, motivation, knowledge and skill. Then the process of learning is outlined, which must form the basis for the systematic acquisition of experience in training courses.

13.7.1 Growth, change and individual differences

Individuals differ because of three influences: genetic or inherited characteristics, learned or environmental characteristics, and situational factors operating at a particular instant.

13.7.1.1 Genetic factors

Many of the physical characteristics of individuals are strongly influenced by genetic factors, for example hair and eye colour, height, physical build and body dimension, although the last three are also influenced by environmental factors such as nutrition. When it comes to factors such as intelligence or personality there is far greater argument about the importance of genetic factors in determining the final measured characteristic in an individual. This debate is important wherever we are trying to change behaviour through education or by changing social or company policy, since what is genetically determined is likely to be more or less unchangeable.

13.7.1.2 Environmental or learned characteristics

Environment forms the most important influence on the personality and social characteristics of individuals. This category is often sub-divided according to which stage of development or aspect of the environment is responsible for the influence.

The newspaper headlines of recent years are eloquent witnesses to some of the more dramatic effects occurring between conception and birth which can produce individual differences. Drugs such as thalidomide have massive deforming effects, as do diseases such as German measles contracted during critical phases of pregnancy. Alcohol and cigarette consumption in pregnancy have also been shown to have less severe but widespread effects on the foetus.

The opportunities provided by the family, and the goals and the motivations which are learned because of the behaviour which is rewarded during childhood have a great effect on the attitudes, skills and abilities shown in maturity. During later childhood and adolescence school, teachers and peer groups take over from the immediate family as the major influence on the development of personality and attitudes. The rate at which personality and attitudes change slows down once the age of 20 is reached, but the working environments and the social groups within which time is spent still have a formative and changing influence. Individuals tend to seek out social groups which suit their personality and attitudes, but when there is a mismatch the individual will be changed by the group to a greater or lesser extent depending upon the importance that the individual attaches to acceptance by the group.

Some of the factors which have been mentioned are often drawn together and labelled as cultural factors if they are influences which are shared by a defined group of people, whether in a particular country, region, social class, age, or occupational group. Thus certain attitudes towards risk taking may be shared by members of one adolescent culture, which are markedly different from those of other age groups.

13.7.1.3 Situational factors

Besides the innate and learned factors which an individual carries around all the time, behaviour at any one moment will be influenced by situational factors. For example, people slow down when a police car drives past, get excited when their team scores a goal or conceal their political views to avoid arguing with their prospective father-in-law. Situational factors can sometimes make people behave in ways which are untypical of them, or of which they disapprove.

13.7.1.4 Personality and attitudes

Since each individual will have been subject to a unique mixture of all of the factors set out above, the result is that no two individuals will be entirely alike in the combination of characteristics which make up their behaviour. No two people will perceive the world in quite the same way. No two individuals will react in quite the same way to the same circumstances confronting them. To predict with certainty how any one individual will behave in a particular set of circumstances would require a complete knowledge of all the factors which had gone to make up that person. However, the position is not entirely hopeless since there is enough common ground in individual responses to most circumstances to make predictions worthwhile. That common ground within one person is labelled personality; where it is common ground between people in a group we label it norms or group attitudes.

The study of personality is an area of psychology which has spawned many parallel and conflicting theories. One style of theory tries to explain where personality comes from and classifies people into 'types' or groups based on differences in personality development; other theories merely classify the end result and measure existing differences (trait theories).

Freudian psychology defines personality as developing through phases of coming to terms with the conflicts between the instinctual and social parts of a person and in learning to express and control sexual drives. It classifies people according to the degree to which their full development and resolution of these conflicts is incomplete. The insights of Freudian analysis have been enormously influential in many fields, but apart from enriching language and the arts, their importance lies mainly in understanding mental pathology.

Other developmental theories stress social influences more (Laing), or concentrate on the cognitive development of the means of making sense of, and classifying events and its influence on personality (Kelly).

Cattell's trait theory is taken as an example of a descriptive theory. From extensive research based upon the responses to questionnaires on their

Table 13.2. Cattell[35] personality factors

1. Reserved, detached, critical	Outgoing, warm-hearted
2. Less intelligent, concrete thinking	More intelligent, abstract thinking
3. Affected by feelings, easily upset	Emotionally stable, faces reality
4. Humble, mild, accommodating	Assertive, aggressive, stubborn
5. Sober, prudent, serious	Happy-go-lucky, impulsive, lively
6. Expedient, disregards rules	Conscientious, persevering
7. Shy, restrained, timid	Venturesome, socially bold
8. Tough-minded, self-reliant	Tender-minded, clinging
9. Trusting, adaptable	Suspicious, self-opinionated
10. Practical, careful	Imaginative
11. Forthright, natural	Shrewd, calculating
12. Self-assured, confident	Apprehensive, self-reproaching
13. Conservative	Experimenting, liberal
14. Group-dependent	Self-sufficient
15. Undisciplined, in self-conflict	Controlled, socially precise
16. Relaxed, tranquil	Tense, frustrated

beliefs and preferences by many thousands of individuals, Cattell has produced a list of 16 personality factors (see *Table 13.2*). The factors are envisaged as 16 dimensions on which an individual's position can be plotted to produce a profile which describes that unique individual. Since someone can score from 1 to 10 on each scale, these 16 scales provide 10^{16} unique character combinations or personalities, which is more than the total number of human beings who have ever walked the earth since Homo sapiens evolved.

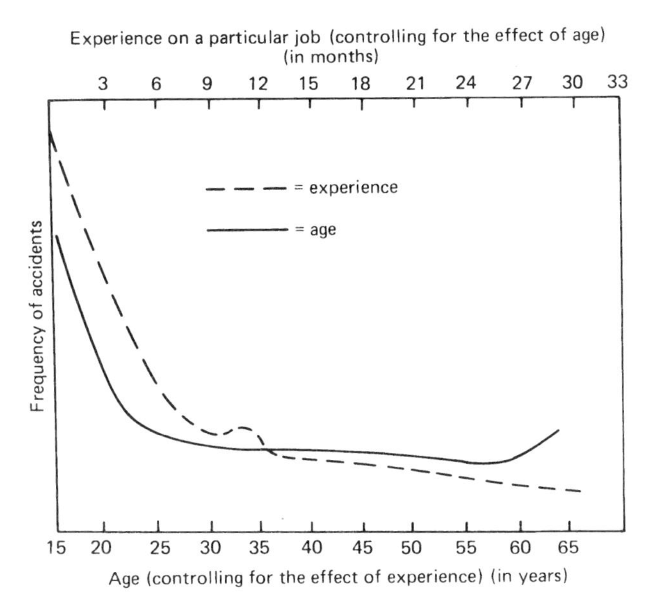

Figure 13.8 Distribution of accidents by age and experience. (Derived from Hale and Hale[10])

13.7.1.5 Individual differences in accident susceptibility

Research into individual differences in accident rate (accident proneness) has a long and complex history littered with mistaken conclusions from invalid methodology and poor experimentation[1,10]. It set out to discover whether there were stable differences in accident susceptibility when individuals were subjected to equal hazards (both in type and length of exposure). It was established very early in the research that both age and experience were correlated with differences in accident susceptibility (see *Figure 13.8*). The exact shape of the graphs will vary from job to job.

Job-related experience appears to be that most relevant to accident rate (Powell *et al.*[36]) although the effects of number of years in industry and of number of hours worked on a specific task (where the job involves a number of tasks) can also be demonstrated.

The relationship of physical and anthropometric differences to accident susceptibility has also been shown in many specific tasks. For example, colour blindness can be a danger where hazard perception depends on colour discrimination; extremes of height, reach and slimness of arms, wrists or fingers can result in individuals being able to reach into danger areas around or through guards; susceptibility to epilepsy, bronchitis or eczema can be problems on jobs involving moving machinery, dust and oils respectively; etc. Research on sex and ethnic differences in accident liability often shows apparent differences in gross accident rates, but these almost always turn out on closer examination to be differences in risk exposure (i.e. immigrant workers and men tend to be found more often in the dirtier and more dangerous jobs).

Research on the relationship of other factors to accident susceptibility has produced few clear-cut results; personality factors, intelligence coordination and attention skills and many other characteristics have been studied but the correlations produced have usually been low and have been specific to the job or task studied. Accident proneness as an explanation for accidents or a basis for a safety policy is therefore very unprofitable and only helps to reinforce a blame culture instead of a problem-solving one.

13.7.1.6 Attitude

Attitude is sometimes defined as 'a tendency to behave in a particular way in a certain situation'. Underlying this definition is one of the thorniest problems in psychology, the consistency between what people say they believe or will do and what they actually do. As with personality many theories abound in this area. In their theory Fishbein and Ajzen[37] define:

Attitude. Attraction to or repulsion from an object, person or situation. Evaluation, e.g. liking children, favouring trades unions etc.
Belief. Information about an object, person or situation (true or false) linking an attribute to it, e.g. that guards are a hindrance.
Behavioural intention. A person's belief about what he will do if a given situation arises in the future, e.g. that he will use his safety belt when driving on a motorway.
Behaviour. Actual overt action, e.g. telling the interviewer that you will wear your seat belt; actually doing so.

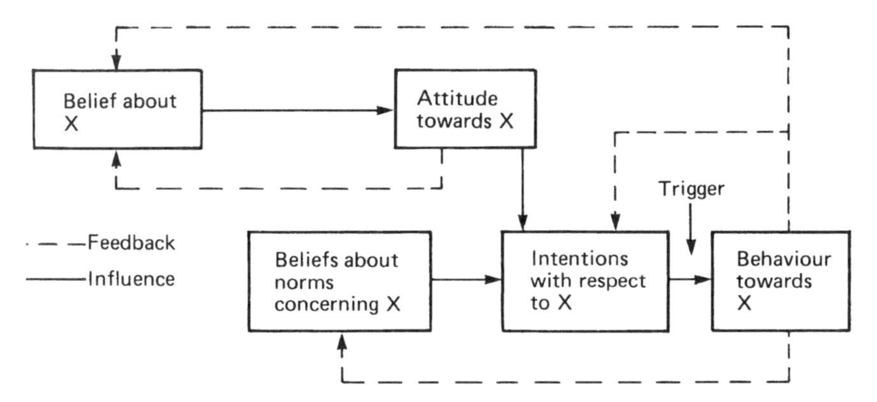

Figure 13.9 System diagram showing links between attitudes and behaviour

All these are linked as shown in *Figure 13.9.*

Thus, for example, someone may believe that breathing apparatus is uncomfortable and dislike it (which may feedback to beliefs by making that person hypercritical of the comfort of any new apparatus) to the extent that the person develops resistance to wearing it, but knowing that it is a company rule (norm) will hurriedly put it on (behaviour) when the safety adviser walks by (trigger). If this happens many times he may find it is not so bad after all and there will be feedback which will change the beliefs.

13.7.2 Stimuli for change and resistance to change

A stimulus for the individual (or indeed for the organisation) to change will be the perception that his or her adaptation to the environment is no longer as close as desired. If the failure in adaptation is not perceived there will be no acceptance by the individual that change is needed. In that case attempts to impose change will be met with the sort of resistance which is characterised by the remark 'We have done it this way for 50 years and it's been OK; why change now?' This conservatism in attitudes and beliefs seems to become more marked with advancing age, perhaps because there is more past experience to call on for support in rejecting the need for change.

Social groups are frequently bastions against change. If a number of people can be found to share the view that things are all right as they are, they reinforce each other's view of the world and unite to resist change.

Those wishing to promote change have the task of convincing individuals, groups and organisations that change is needed because the old adaptation is no longer appropriate. This process is least difficult when something dramatic such as an accident demonstrates that old methods were not safe, or there is a new law, a new machine, a new boss or a new job, all of which self-evidently require change.

It is the slow changes in environment and individuals which often go undetected, and so unresponded to. Examples are changes in technology, affluence and life expectancy or in attitudes to work which occur as the result of growing up or of growing old. If these changes are to be brought to the notice of individuals or organisations it is often necessary to dramatise them in order to get through the conservatism or blinkered vision which is failing to see them.

13.7.3 Methods of change

Change can be brought about by:

(1) Changing the physical or social environment of the individual so that the old behaviour does not bring the rewards it did, or new behaviour brings greater rewards.
(2) Providing new information to an individual which shows that the existing adaptation is not as close as the individual thought it was, or as it could be.
(3) Changing the goals and objectives which an individual is seeking.

13.7.3.1 Changing the environment

Under this heading come modifications to physical work design and layout, new machinery and work methods, changes in allocation of jobs to people, introduction of new incentive schemes so that different behaviour is rewarded and punished (e.g. payment schemes, promotion systems, direct safety incentives or simply what behaviour elicits praise from the boss), changes to safety rules, company policy or legal standards.

In all cases the change must be brought to the notice of individuals, and conformity to the new rules or situation must be appropriately rewarded and persistent non-conformity punished. Provided that the person, group or organisation introducing the change is perceived to have sufficient authority (or legitimate right) to make the change, or is sufficiently powerful in wielding reward and punishment, or has great enough charisma or expertise, the change will be accepted and adapted to.

13.7.3.2 Giving information on maladaptation

Under this heading fall the provision of information about danger, provision of training on more effective ways of conforming with any of the changes outlined in section 13.7.3.1 above, or pointing out the way in which behaviour in one area conflicts with strongly held beliefs in another area. Another approach is to improve communication skills so that people are more open to influence (e.g. sensitivity training). The success of all these endeavours will depend upon the credibility of the source of information and the ability of that person or organisation to organise and put over information.

13.7.3.3 Changing long-term goals and objectives

Under this heading come education, media and advertising campaigns to build and change 'images' and also long-term changes in law and culture, etc. These methods of change are usually very long-term, and are often poorly understood. They operate by training people to look at, question, and so develop their own goals. They also expose people to different opportunities and chances for achievement, and present them with examples of what are labelled 'acceptable' and 'unacceptable' behaviour for people to copy.

13.7.4 Learning and training

This section deals with the psychological principles of learning. A number of authors have classified learning processes into a hierarchy of levels, building from the simplest stimulus response learning to the most complex processes of research and problem solving. Three levels of learning are particularly relevant:

(1) Stimulus response learning (Skill).
(2) Concept and rule learning (Rule).
(3) Problem solving (Knowledge).

13.7.4.1 Stimulus response learning

This is the building block which is particularly important in all 'habits' and repeated invariant sequences of behaviour. The sequences are built up by a process called conditioning. This has three essential elements:

(1) Evoking the correct response when the stimulus is presented; by trial and error, explanation or demonstration.
(2) Rewarding correct responses: the rewards can include praise for correct performance, administering punishment for incorrect performance and the use of monetary rewards. It is also possible to play upon the motives of achievement and interest by feeding back information of how well the task is being learnt or how near to the objective the learner has reached.
(3) Practice of the correct sequence which establishes the response more and more firmly until it occurs without conscious effort.

Once sequences of action of this sort have become established it is extremely difficult to add to them or subtract from them. It is therefore important that in safety all the necessary steps are built into the sequence at the learning stage. For example, if the response of donning protective goggles is built into the sequence of setting up and starting an abrasive wheel, it will become an automatic part of that habit. If some responses are not built in at the time of first learning, then either the behaviour sequence must be broken down and relearned with them in place, or, as a poor second best, individuals must be trained consciously to break into the chain at the appropriate point in order to carry out the missing response action.

The sequences of actions will become less automatic if they are not used regularly. Behaviour in response to infrequent emergencies will therefore not be available unless it is practised in the interim.

13.7.4.2 Concept and rule learning

A fundamental human characteristic is the tendency to think about and classify objects, experiences or ideas into mental categories or 'concepts', which share things in common.

Concepts are built up from experience and each person's set of concepts will then differ slightly. Thus, for example, the concept 'dangerous' to one individual may contain hundreds of items which include any situation which is new or strange (a person we might label 'nervous'); another's might contain very few items and leave out some which should be there, such as noisy discotheques or bottles of weed killer kept in the larder.

Acquiring concepts depends on amassing examples which have elements in common until the individual arrives at a tentative definition of the boundaries of the group of objects or ideas. This process can be abbreviated by defining the concept for the person. It must then be consolidated by providing examples which fall inside it, and outside it, gradually reducing the difference between these positive and negative examples until the boundaries become clearly defined. People continue even then to test out and modify the boundaries of the concept.

Concept learning is appropriate whenever someone is expected to recognise a new stimulus as belonging to a particular category, so that they can respond to it even though they have never met it before. It is also necessary whenever someone is expected to follow a rule. For example the rule 'all hazards must be reported to a responsible person' requires that individuals should learn a correct concept of what is a hazard, and a correct concept of who are responsible persons, as well as an adequate idea of what reported means (a casual mention, a formal verbal report, or a formal written statement).

13.7.4.3 Problem solving

Where someone is faced with a situation which they have never met before, and cannot clearly place it into an existing concept category, they are faced with the need to produce a solution new to them. Problem solving is a creative process which relies on the basic building blocks described above and consists of re-ordering them and re-interpreting them.

Learning to solve problems can be aided by teaching the steps in systematic problem solving:

(1) To recognise the problem area and define it in broad terms.
(2) To explore all the possibilities for solving the problem.
(3) To analyse all the facts available to determine whether or not subsequent problems will follow from the solution suggested.
(4) To choose the best possible solution and implement a plan of action for introducing it.

Some techniques for creative thinking (step 2) can also be taught, e.g. brainstorming. A problem is presented to a group of people who are encouraged to throw out ideas, sparked off by each other, no matter how wild. These are recorded by the group leader on a board for all to see. It is important that all ideas be put forward without either interruption or criticism because cross-fertilisation and the building up on ideas produced by others will generate many new ideas. After a suitable period, the ideas noted can be discussed in open forum as part of the evaluation stage to obtain the best solution to the problem from the large pool of ideas so formed.

13.7.5 Communication

Learning and attitude change depend on the motivation to change and on good communication of the message about what change is needed. Earlier sections have dealt in passing with aspects of communication: warnings, information about hazards and prevention measures, checklists and training courses. In this section the basic rules for the process of communication are covered.

Communication is the process whereby one person makes his ideas, feelings and knowledge known to others and learns in exchange about theirs. It is therefore a two-way process which depends crucially on both clear sending and receipt of the message (*Figure 13.10*).

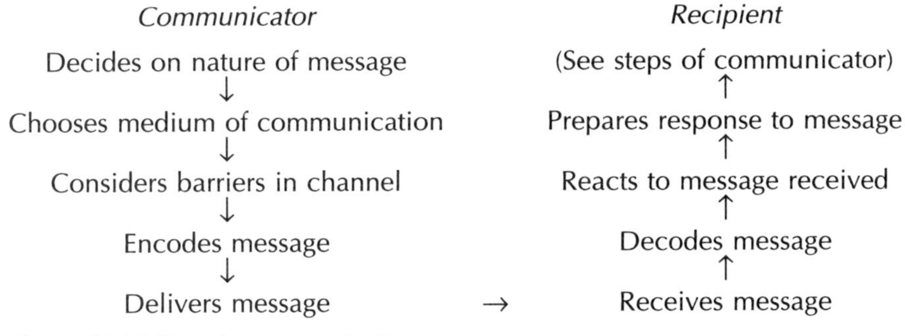

Figure 13.10 Steps in communication

The first essential is to know with whom you are communicating (directors, line managers, accountants, the workforce, the factory inspectorate). Then the precise objectives of the message must be planned (what change is wanted? what are the precise obstacles? must the message succeed with more than one different group?). The message must be coded in terms appropriate for the audience. It must latch onto their way of thinking, priorities and language. It must not use jargon they do not know. To do all of this it is necessary to think about the subject of the communication from the point of view of the receiver and to plan the message so that it leads from that point, covers the disadvantages of the change from that viewpoint and how they will be overcome and ends by spelling out the advantages of the change.

On the basis of this planning the medium can be chosen:

(1) *Face to face communication* has the advantage that it allows feedback and adjustment of the message based on the response. It is also friendlier and less formal. However it is less easy to control because it is

interactive. It is important to remember that it consists of two elements, verbal and non-verbal. The words seem dominant and must indeed be chosen appropriately and put over clearly, but the non-verbal clues can either reinforce or destroy their effect. The tone of voice can indicate boredom, the stance friendliness, hand movements nervousness. The very different effect of messages over the telephone and face-to-face demonstrate this point. It is excellent practice for communicators to listen to themselves on tape and watch themselves on video to see and correct these elements of their style.

(2) *Written communication* allows for much more complex messages to be sent and understood because they can be reread and carefully weighed. It also forms a permanent record for future reference, both for and against the sender.

(3) *Visual communication* allows for very rapid transmission of the relations between things in one glance. It can therefore have great power and emotional impact, but is less easily controllable. There is just as much a language of pictures which must be learned by both parties to the communication, as there is a language of words.

13.8 Methods of change and control

Finally two topics will be briefly reviewed which have been behind much of the discussion in this chapter:

- design as an influence on behaviour
- the response to safety rules.

13.8.1 Design

Ergonomists have been saying for a generation that design is the biggest influence on use. The operator or user has too often in the past been saddled with the impossible task of recovering the inadequacies of the designer. Machines which operate in ways which do not fit expectations, routines which are easily confusible, and tasks which have been left to the operator only because technology cannot yet take them over, are all examples of accidents planted like time-bombs in the system. Users are to be congratulated that they manage for so much of the time to operate safely despite them. The increase in feedback from users to designers and the heavier emphasis on product liability which have occurred in the last decade (e.g. in the EC Product Safety Directive) are welcome signs that this tolerance is reaching its limit.

Designers should also be humble enough to realise that their designs are not eternal and that operators have a need to modify or adjust their workplace and not to have to work within rigid constraints which are not perfectly attuned to operating conditions. Therefore, enough information and training must be provided to the operators to enable them to recognise the possibilities for modification and prevent them from falling into unsuspected traps; in addition, predictable modifications which will lead to danger (like removing guards or defeating safety interlocks) should be made as difficult as possible.

Designers should not think of people as automata. There are differences both between and within individuals. People will never be as consistent in their response as hardware components. Therefore designs must be error tolerant and make error recovery as easy as possible. Designers must predict foreseeable misuse and error and cater for it. This is now a principle enshrined in European machinery standards[38]. Nor should designers respond to human error with an unthinking push to automate the individual out of the system as much as possible. That is a recipe for creating residual monitoring tasks which are boring and unsatisfying. It will also result in the loss of skill and insight to such an extent that the operator cannot intervene effectively when the hardware fails.

Finally, designers should not have unrealistically high hopes of their hardware solutions. People will always adapt to system changes by altering their behaviour, sometimes trading off increased safety margins against other gains; e.g. straightening out roads with dangerous bends results in an increase in traffic speed; more reliable hardware results in less spontaneous checks of its functioning. This behaviour, sometimes called risk compensation[1], should be anticipated by the designer, who should design against such trade-offs.

13.8.2 Safety rules

We should be suspicious of anyone who claims that safety is merely a matter of laying down and enforcing rules. It can never do any harm to define clearly and as exhaustively as possible how the system should operate to overcome all known hazards. But enforcement of such detailed rules is difficult to achieve. This approach takes away all individual freedom and control over the work. It will only work where danger is very evident and it can be guaranteed that application of the rules will always result in safety. Even then it will work only with difficulty if following the rules is also not the easiest and most obvious way of doing the job.

The following extract from a study of rules[39] is typical:
50 railway workers were asked about safety rules governing work on and near railway tracks;

- 80% considered that the rules were mainly concerned with pinning blame.
- 79% thought there were too many rules (12% too few).
- 77% found the rules conflicting.
- 95% thought that work could not be finished on time if the rules were all followed.
- 85% found it hard to find what they wanted in rule book.
- 70% found the rules too complex and hard to read.
- 71% thought there was too little motivation to follow rules.
- Not one could remember ever having referred to the rules in a practical work situation.

Rules are subject to exceptions and to erosion. Safety manuals and safety laws tend to be full of complex specifications with many 'if . . . , then . . .'

clauses which are perfect if followed, but which are too complex to remember. Execution of all the checks to see which sub-clause applies in any one case would often take too long in practice. Such rule books only serve to assuage the conscience of the rule makers. After an accident they can establish exactly who should have done what and so who was to blame. The existence of such a complex edifice of rules is a signal that the system is inwardly sick and in urgent need of redesign to incorporate behavioural rules into either training or hardware design. Ideally, design should precipitate the right action, and articulated written rules are only necessary where the way someone would expect to have to operate in a given situation is not in fact correct.

This conflict between establishing rules and leaving the flexibility to cope with exceptions and with changes can be seen at all levels in safety. It is reflected in the arguments about rigid central specification in laws and standards in contrast with enabling frameworks with objectives and freedom for each company to comply in the way it wishes. It can be seen at the level of the company where operating managers are keen to reduce problems as fast as possible to fixed rules in order to be able to get on with production. Safety departments have a task here to act as the protagonists of continual revolution in the firm. Safety rules need to be written with the involvement of those who must follow them. They also need to be updated at regular intervals using the critical experience of those same people.

References

1. Hale, A.R. and Glendon, A.I., *Individual Behaviour in the Control of Danger*, Elsevier, Amsterdam (1987)
2. Hale, A.R. and Hale, M., Accidents in perspective, *Occupational Psychology*, **44,** 115–121 (1970)
3. Taylor, F.W., *Principles of Scientific Management*, Harper & Row, New York (1911)
4. Mayo, Elton, *The Social Problems of an Industrial Civilisation*, Routledge & Kegan Paul Ltd, London (1952)
5. Maslow, A.H., *Motivation and Personality.* Harper, New York (1954)
6. Atkinson, J.W., Motivational determinants of risk-taking behaviour, *Psychological Review*, **64,** 359–372 (1957)
7. McClelland, D., Risk-taking in children with high and low need for achievement. In Atkinson, J.W. (Ed.), *Motives in Fantasy, Action and Society*, van Nostrand (1958)
8. Porter, L.W., Lawler, E.E. and Hackman, J.R., *Behaviour in Organisations*, McGraw-Hill, Kogushawa, Tokyo (1975)
9. Adler, A., The psychology of repeated accidents in industry, *American Journal of Psychiatry*, **98,** 99–101 (1941)
10. Hale, A.R. and Hale, M., A review of industrial accident research literature, Committee on Safety and Health at Work, *Research Paper*, HMSO, London (1972)
11. Wagenaar, W.A. and Groeneweg, J., Accidents at sea: multiple causes and impossible consequences, *International Journal of Man-machine Studies*, **27,** 1–90 (1987)
12. Binet, A. and Simon, T., The development of intelligence in the child, *Année Psychologique*, **14,** 1–90 (1908)
13. Report of the Commissioners appointed to collect information in the manufacturing districts relative to the employment of children in factories. *Parliamentary papers* – three reports, 1833/34
14. Waterhouse, J.M., Minors, D.S. and Scott, A.R., Circadian rhythms, intercontinental travel and shiftwork. In Ward Gardiner, A. (Ed.), *Current Approaches to Occupational Health*, **3,** Wright, Bristol (1987)
15. Mackay, C. and Cox, T., *A Transactional Approach to Occupational Stress*, Department of Psychology, University of Nottingham (1976)

16. Cox, T., *Stress*, Macmillan Press, London (1978)
17. Reason, J.T., A framework for classifying errors. In Rasmussen, J., Leplat, J. and Duncan, K. (Eds.), *New Technology and Human Error*, Wiley, New York (1986)
18. Rasmussen, J., What can be learned from human error reports. In Duncan, K., Gruneberg, M.M. and Wallis, D.J. (Eds.), *Changes in Working Life*, Wiley, Chichester (1980)
19. Hale, A.R. and Pérusse, M., Perceptions of danger – a prerequisite to safe decisions, *Proceedings of the Institution of Chemical Engineers*, Rugby (1978)
20. Surry, J., *Industrial Accident Research*, Department of Industrial Engineering, University of Toronto (1969)
21. Tong, D., The application of behavioural research to improve fire safety, *Proc. Ann. Conf. Aston Health and Safety Society*, Birmingham (1983)
22. *Procedures Guide for Probabilistic Risk Assessment*, U.S. Nuclear Regulatory Commission Report NUREG/CR 2000 (1980)
23. Swain, A.D. and Guttman, H.E., Handbook of human reliability analysis with emphasis on nuclear power applications, *US Nuclear Regulatory Commission Report NUREG/CR 1278*, Sandia Laboratories, Albuquerque, New Mexico (1983)
24. Tettelaar, H.C., de Vries, K.L.M. and Phaf, R.H., *General Risk Assessment Procedure (GRASP)*: Incorporating Human Error into Risk Analysis. Report R-88/28, Centre for Safety Research, University of Leiden (1988)
25. Feggetter, A.J., A method for investigating human factors aspects of aircraft accidents and incidents, *Ergonomics*, **11,** 1065–1075 (1982)
26. Kirwan, B., Human reliability assessment. In Wilson, J.R. and Corlett, N. (Eds.), *Evaluating Human Work: an Ergonomics Methodology*, Taylor and Francis, London (1989)
27. Royal Society, *Risk Assessment*; a Study Group Report, London (1983)
28. Lowrance, W., *Of Acceptable Risk: Science and Determination of Safety*, W. Kaufman, Los Altos, California (1976)
29. Health and Safety Executive, *The Tolerability of Risk from Nuclear Power Stations*, HSE Books, Sudbury (1988)
30. Vlek, C. and Stallen, P-J., Judging risks and benefits in the small and in the large, *Organisational Behaviour and Human Performance*, **28,** 235–271 (1981)
31. Starr, C., Social benefit versus technological risk, *Science*, **16,** 1232–1238 (1969)
32. Abeytunga, P.K., *The role of the first line supervisor in construction safety: the potential for training*, PhD Thesis, University of Aston in Birmingham (1978)
33. Svenson, O., Risks of road transportation in a psychological perspective, *Accident Analysis and Prevention*, **10,** 267–280 (1978)
34. Green, C.H. and Brown, R.A., *The perception of, and attitudes towards, risk: Preliminary report: E2, Measures of safety*, Research Unit, School of Architecture, Duncan of Jordanstone College of Art, University of Dundee (1976)
35. Cattell, R.B., *The Scientific Analysis of Personality*, Penguin Books, London (1965)
36. Powell, P.I., Hale, M., Martin, P. and Simon, M., *2000 Accidents*, National Institute of Industrial Psychology, London (1971)
37. Fishbein, M. and Ajzen, I., *Belief, Attitude, Intention and Behaviour – an introduction to Theory and Research*, Addison-Wesley, Mass. (1975)
38. Centre Européen de Normalisation, *Safety of Machinery – Basic concepts – General principles for design. Part I: Basic terminology and methodology* (EN 292-1:1991). *Part II: Technical principles and specifications* (EN 292-2:1991). BS EN 292-1 and BS EN292-2 respectively, BSI, London (1991)
39. Hale, A.R., Safety rules OK? Possibilities and limitations in behavioural strategies, *Journal of Occupational Accidents*, **12,** 3–20 (1990)

Further reading

The primary text for further reading is:

Hale, A. R. and Glendon, A. I., *Individual Behaviour in the Control of Danger*, Elsevier, Amsterdam (1987). This chapter is in great part a summary of the material covered there in great detail. It also contains detailed references for still deeper reading.

Many of the following texts have an overlapping coverage of subject matter. The reader should therefore select from among them. The brief notes attached will guide that choice.

Canter, D., *Fires and Human Behaviour,* Wiley, Chichester (1980). A good review of work on the specific topic of reactions to fire.

Coleman, J. C., *Introductory Psychology,* Routledge & Kegan Paul, London (1977). Written with medical and nursing students in mind. The individual chapters are by different experts. Covers almost the full range of the subjects in these chapters.

Cohen, J. and Clark, J. H., *Medicine, Mind and Man,* W. H. Freeman & Co., Reading (1979). A parallel text to Coleman also written for students of health sciences.

Hoyos, C. G. and Zimolong, B., *Occupational Safety and Accident Prevention: Behavioural Strategies and Methods,* Elsevier, Amsterdam (1988). A parallel text to Hale and Glendon written somewhat more from the viewpoint of safety management.

Maier, N. R. F., *Psychology in Industrial Organisations,* 4th edn, Houghton Mifflin, Atlanta (1973). Written for industrial managers and supervisors. The examples are therefore good and relevant though with an American bias.

McCormick, E. J., *Human Factors Engineering,* 3rd edn, McGraw-Hill, New York (1970). A very thorough and complete source book for ergonomics and human factors in design.

Hale, A. R. and Hale, M., *A Review of Industrial Accident Research Literature,* Committee on Safety and Health at Work: Research Paper, HMSO, London (1972). A brief review of the literature up to 1972 on human factors in accident causation. Valuable source of further references.

Powell, P. I., Hale, M., Martin, P. and Simon, M., *2000 Accidents,* National Institute of Industrial Psychology, London (1971). Summary report of a four year field study of accident causes. Good overview of the priorities in the field.

Rasmussen, J., Duncan, K. and Leplat, J. (Eds.), *New Technology and Human Error,* Wiley, Chichester (1987). A very valuable book of readings of both theory and practice in human error assessment and control.

Reason, J., *Human Error,* Cambridge University Press (1989). An excellent book setting out the theories of a very influential researcher.

Stammers, R. B. and Patrick, J., *The Psychology of Training,* Methuen Essential Psychology E3, London (1975). Short text covering the main psychological approaches and insights into the subject.

Vroom, V. H. and Deci, E. L. (Eds.), *Management and Motivation,* Modern Management Readings, Penguin Books, London (1970). Fairly advanced text but an excellent summary of the many theories that exist in the area.

Readers wishing to keep up to date with research on this topic will find research and review articles in scientific journals such as *Safety Science* and *Applied Ergonomics.* Road traffic safety papers are to be found in the *Journal of Safety Research* and in *Accident Analysis and Prevention.*

Chapter 14

Organisation for safety

J. R. Ridley

14.1 Introduction

For any organisation to function effectively and successfully it is necessary that those who constitute that organisation understand the goals of the enterprise and identify with them, know where they fit into the executive structure and are competent and confident in the work they have to do. This applies to all the many facets of the enterprise's activities, whether production, financial, administrative or safety.

There needs to be an understanding of the influences, both internal and from outside, that bear on the success of the component parts and the organisation as a whole, an appreciation of the conflicts that can arise, their causes and techniques to defuse the situation.

Similarly there needs to be an understanding and recognition of the informal structures and relationships that occur within the more formal imposed organisation structure, informal arrangements that very often 'oil the wheels' of industry and keep the organisation running against all apparent odds. While a firm structure is necessary to ensure a consistent direction of the efforts of the enterprise, that structure must not be so rigid that it cannot adjust to changes in the trading, economic, legislative and other facets of the operating environment over which the enterprise has no control. This chapter looks at the various relationships and organisations that occur within an enterprise and which can materially affect its success not only in serving its customers and the community but also the satisfaction it gives to, and the safety it provides for, the people it employs.

14.2 Structure and functions of an organisation

Three centuries ago John Donne observed that 'No man is an island' and crystallised one of the basic tenets of society in that man has a basic need to be involved with and to relate to others of his kind. As societies develop they evolve structures and rules to regulate what the society does, how it does it and for whose benefit. Eventually many of those functions become highly specialised so that within the society specific groups emerge to play a limited but necessary part in the continuing and healthy development of the parent body. In modern society and commerce many of these functions have become stereotyped into quite rigid and often exclusive sub-organisations that form discrete limbs existing almost in isolation in (and sometimes in spite of) the parent to which they belong. This can prove disastrous unless there are good communications and interpersonal relationships with other members and parts of the organisation.

Many organisations have formal structures, determined rationally by the senior executives and often displayed as charts or 'organograms' showing the functions which are considered essential for the effective and smooth running of the organisation. Equally important but very rarely committed to paper are the informal working groups established socially by members within the organisation.

While the overt aim of an organisation may be to serve its customers and in so doing make a profit, there is an often unrecognised aim to ensure perpetuation of itself. This can be seen in the tendency of large companies to have special departments charged with investigating areas into which the company can diversify should it become necessary.

14.3 Formal organisation structure

When setting up an enterprise the owner will determine how many workpeople he considers necessary to enable him to achieve the objects of output and profit he has set himself. He will also know what sort of work they need to do and he will organise them into groups of different sizes and expertise to that end. With time considerable degrees of specialisation are developed and the different groups establish formal relationships.

Those who belong to a company need to know how the company is structured and where they fit into it.

For an enterprise to succeed it needs to have some sort of organisation and the most common is hierarchical with authority flowing through definite channels from top management to the workpeople. A typical hierarchical organisation chart is shown in *Figure 14.1* and for it to be effective certain positions must be vested with power to exercise their authority to direct and control the activities of the organisation. How that power is exercised will be determined by the culture of the organisation, whether it is authoritarian or bureaucratic.

14.3.1 Power

Power has been described as the capacity to influence others to do that which they would not have done voluntarily and is defined by C. Wright Mills[1] as 'the capacity to make and carry out decisions even if other people resist'. Power can derive from authority vested in a person by the organisation, the control of a desired product, such as money, or by virtue of special knowledge or expertise.

Kaplan[2] says that there are at least three dimensions of power which are of practical importance: weight, domain and scope. He describes 'weight' as

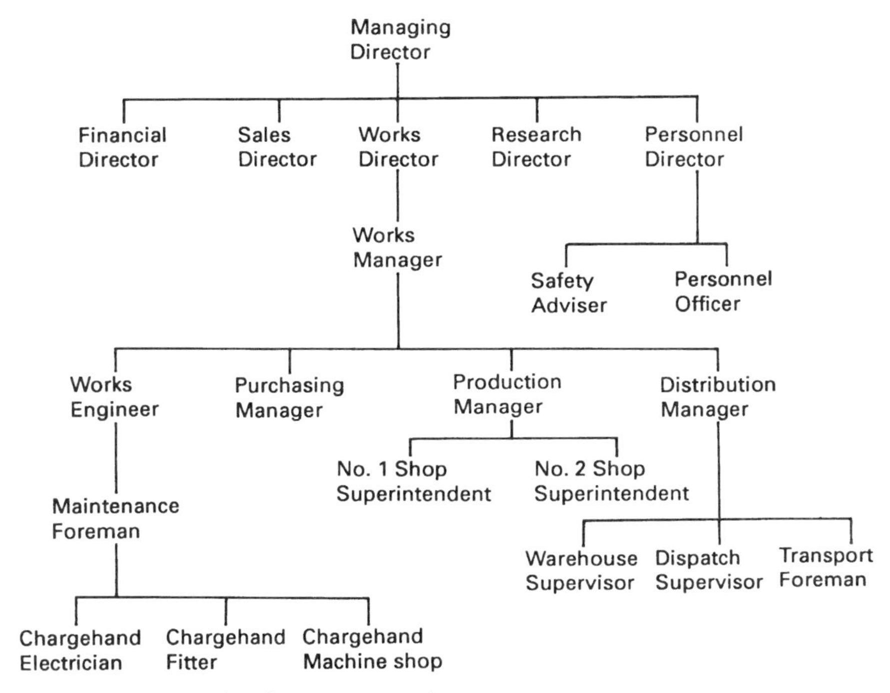

Figure 14.1 Hierarchical organisation chart

the ability of an individual to affect the probability that another individual will act in a certain way under certain circumstances, 'domain' as the span or number of individuals or groups influenced and 'scope' as the range of outcomes over which power can be exercised.

A safety adviser in a factory may have the power to cause a manager to abandon a particular production method or process immediately on his advice, or at the other extreme so little power that his advice is completely ignored. The scope of his power may extend to authorising production procedures within the factory but not extend to influencing basic safety matters that affect employees in their leisure pursuits, thus his domain encompasses the factory premise but does not extend outside it.

The power referred to above is 'power over' – the exercising of control over the actions of others, but it is arguable whether this is as effective use of power as 'power with' – the concept of two people pulling in the same direction exerting greater force than one person trying to drive another.

14.3.2 Authority

A simple definition of authority is 'legitimate power' and Max Weber[3] saw authority and its legitimacy as central to the question of organisational structure. He maintained that authority exists when instructions are obeyed in the belief that they are legitimate, in other words that they are justified and that obedience is the appropriate response. Weber classified authority into three kinds:

- rational legal authority based upon rules and procedures and with bureaucracy as its purest form,
- charismatical authority where the personality of the individual predominates, and
- traditional authority founded on respect for custom and practice.

These three authority types are not mutually exclusive but can appear in any organisation at any hierarchical level.

Authority often has at its base the rules and regulations that the organisation has drawn up to regulate its affairs, but for the authority to be effective relies on a number of presumptions:

- that the rules exist and are accepted by those they are aimed at,
- that the rules are relevant to the particular circumstances,
- that the rules will be obeyed by every member of the organisation including those exercising the authority, and
- that the authority is vested in the office and not the individual.

A question that has frequently been debated but never resolved is the relationship between authority and responsibility. Which should come first? If authority is vested in an individual does responsibility follow or does the holding of responsibility grant the taking of authority? Common sense suggests a balance has to be struck if internal conflict is to be avoided and an organisation is to operate at its most effective.

14.3.3 Bureaucracy

This term has come to be used to describe what are felt to be the worst features of contemporary organisation and conjures up visions of over-regulation, inflexible procedures, 'red tape', disinterest in the customer and accountability to a 'faceless' committee. However, Weber considers that 'bureaucracy has a crucial role in our society as the central element in any kind of large scale administration' but in its most rational form depends upon rules, procedures and authority to achieve its control. He suggests it has the following characteristics:

- specialisation between positions,
- hierarchy of authority,
- a system of rules even extending to the recruitment of new members,
- impersonality, and
- written records of administrative acts, decisions and rules.

A bureaucratic organisation can be thought of as one that aims to maximise its efficiency in administration.

Claims that a bureaucratic organisation offered benefits from cost reduction, precision, impersonality, inflexibility, etc., may owe more to the informal staff relationships and practices than to the organisation itself. However it must be recognised that elements of bureaucratic organisation can probably be found in parts of most medium and large organisations.

14.3.4 Informal organisation structure

Within any formal organisation will be found whole networks of informal organisations based on personal relationships, social needs, personal allegiances and often a desire to be helpful in by-passing red tape. These organisations are rarely committed to paper, they come into existence to serve a perceived need of those involved and may remain for many years or can disappear when the need is satisfied.

An informal organisation may use the formal structure but create unofficial lines of communication to circumvent failures and weaknesses in individuals that make up the formal organisation in attempts to achieve the enterprise's objectives. This is shown in *Figure 14.2* which maps those actual relationships occurring in the structure shown in *Figure 14.1* where the Production Manager has been running affairs for some time because of the indifferent health of the Works Manager. The Production Manager has gathered a group of key personnel who recognise him as 'next-in-line' after the Works Manager and acknowledge his authority. However, the Works Engineer, who happens to be the Managing Director's son-in-law, does not acknowledge the Production Manager's authority and endeavours to exert his own authority by resisting the Production Manager. As a result the Works Engineer is often by-passed when maintenance work needs to be done and indeed the maintenance foreman refuses to take work from him since he, the foreman, only acknowledges the Production Manager as his superior.

The informal structure serves to improve communication and to develop non-official roles. It has an important part to play in the resolution of conflicts between roles and positions. Questions raised in the formal

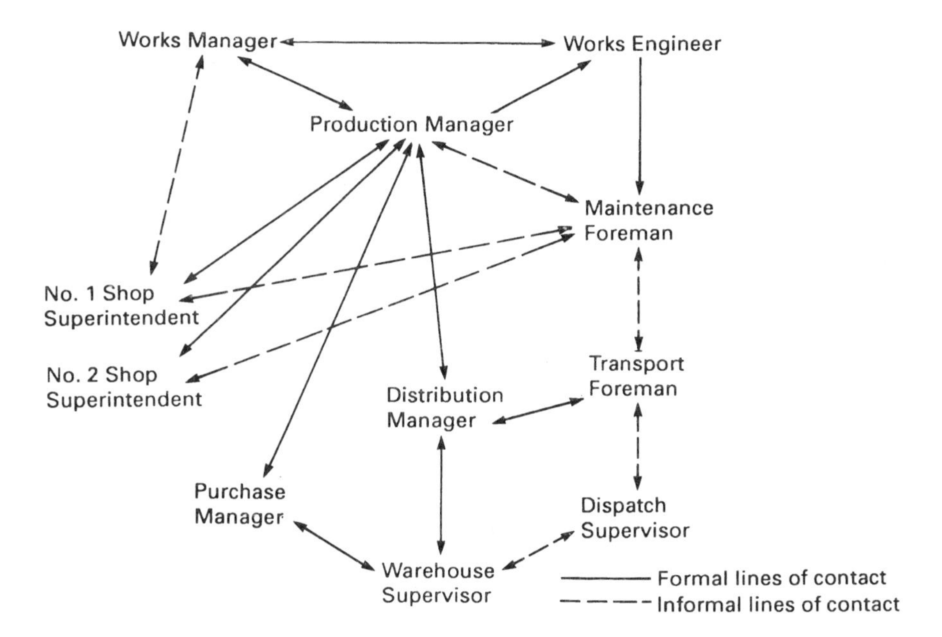

Figure 14.2 Diagram of workplace contacts

organisation often elicit answers that rely on the informal structure. Questions such as 'what's the best way to get this information?' or 'who can get this done?' produce answers that cut right across the formal structure – but get the job done.

Occasionally there is tacit recognition of the informal structure by the formal since the informal level allows for practical interpretation of rules and procedures that can place restrictions on achieving enterprise targets. This flexible interpretation of the formal rules may be to the common organisational good, but may also be detrimental where they concern safe working practices. The fact of these informal interpretations can be seen in the dramatic effect of a 'work-to-rule' in a formalised rule-based organisation.

14.3.5 Work groups

Work groups may be defined as collections of individuals interacting with each other in the pursuance of a common work related task or goal and who, for this purpose, are dependent upon each other. An important characteristic of groups is the existence of standards of expected behaviour or norms which are informally binding upon members. The benefits of belonging to a group stem from the social experience of working together, having a group identity, the sharing of common goals and from the social approval given by other group members to those who meet the group norms. There may be a presumption that the goals of the work group are the same as those of management, but this is not necessarily the case.

Work groups as a special kind of social phenomenon have been studied by many observers. The best known of these studies, the Hawthorne Experiment[4], which started in the early 1920s and ran for 12 years, initially monitored work and behaviour responses to varying physical conditions. Later in the Bank Room study, the experiments investigated work norms where Mayo and his fellow workers found that the act of monitoring workers' behaviour influenced that behaviour (the Hawthorne Effect), and that the social environment within the groups must be considered equally with the physical conditions as factors governing production rates. Mayo's work stimulated the birth of the human relations movement as opposed to the mechanistic scientific management school of thought of Taylor and Gilbreth.

The size of a work group may be determined by the tasks to be undertaken but other factors can be its role in the formal organisation, the social atmosphere and the geographical spread of the work. Work groups may extend their activities to beyond the work place when there are strong social ties and common social interests between the members.

Within a work group there may be two levels, the 'primary' group whose relations are personal and informal and the larger 'secondary' group which is established within the formal structure and where the role relationships predominate. Primary groups tend to be small in size dividing into smaller units as they grow in size so maintaining the personal contact and satisfying the social needs of its members.

Membership of a work group carries with it obligations in the form of pressure to conform as well as the benefit of rewards. The need to remain within a work group stimulates a desire to conform to group norms that can be so strong as to change individual attitudes and beliefs. The group itself may determine its own boundaries and membership often through an informally recognised leader or leaders. Where the work group is management-organised there is a risk of the inclusion of an informally unacceptable member – a sort of bad apple in the barrel – that can disrupt the whole group.

Consultation with informal work groups by management in order to gain acceptance of changing work patterns has become a recognised work practice. As such, work groups have been trained in accident prevention techniques in the hope of stimulating group as well as the individual action. A typical example of this practice is the establishment of Quality Circles. Changes devised or approved by the group have been found to have more ready acceptance among members. Formal recognition of the autonomy of the work groups has been advocated in Sweden in an attempt to improve job satisfaction and output, but there is some evidence that this technique is not totally successful since it relies on management and group goals being identical. An autonomous group may seek to redefine its goals in ways which may not be to the benefit of the organisation, and are unacceptable to the management. The need to maintain a balance of interests within the organisation may be the most significant constraint on the development of work groups within it.

14.4 Organisational control

Control in any organisation is necessary to ensure the steady progress of the enterprise towards its goals, but the controls used need not be restrictive. By the same token they do not have to be controls imposed by the management, which are the formal controls imposed on the organisation, but can be voluntarily set up by the workforce to assist them to meet what they see as their targets. If the workforce targets coincide with those of the enterprise then a co-operative and productive environment is developed; on the other hand if the targets do not coincide or even clash, then the environment becomes restrictive and often acrimonious. These self-imposed controls are the informal controls that mirror the informal organisation. Finally there are the external controls exerted from outside the organisation by bodies such as the Health and Safety Executive, the Trade Unions, the employers' associations and indeed by the local community.

14.4.1 Formal controls

The most obvious of these is the authoritarian control imposed by the employer as part of his planning strategy to achieve the goals of the enterprise. He plans for certain sales targets which demand a certain output from the factory and the easiest way of ensuring those targets are achieved is by placing tight controls on production rates, material usage, working hours and delivery dates. In theory fine, but to be effective the workforce must be prepared to accept them and there is no guarantee that this will occur.

Controls are also exercised through the supervisor whose job is to so organise the workforce that they meet the required targets. A good supervisor will take the workforce along with him, get their co-operation and create a happy productive workplace. On the other hand a supervisor who misuses the authority vested in him by the company and behaves in a dictatorial manner with the workforce can create an atmosphere of acrimony where energies are expended, not in achieving targets but in defending needless entrenched positions. This latter example has all the makings of a situation which if allowed to develop can blow up out of all proportion and become a major industrial relations confrontation. To be effective, supervisors must supervise sensibly.

The sensible exercise of discipline can engender respect and through it an acceptance of the discipline as a means to carry the work forward. But where discipline is used it must be seen to be used fairly and with moderation to match the degree of control that the workforce perceive as being necessary. Unnecessary or harsh discipline will be counter-productive and will antagonise the workforce. Discipline has a more ready acceptance where there is the power of law to back it up as in some safety situations.

Joint consultation, as a technique where the employees have an opportunity to comment on and influence the way the work is carried out before the work is commenced, can bring benefits of higher production and also a more ready acceptance of controls. A development of joint consultation that moves towards a distribution of industrial power is *participation* where representatives of the workforce are involved in the decision-making process, an arrangement that is intended to bring a benefit in the form of greater workforce acceptance of the targets and controls being imposed. However it seems to have been overlooked that to participate in a decision means a sharing of the responsibility for the outcome of that decision, for better or for worse, and there is evidence of a reluctance by trade unions to accept such a commitment.

Further formal controls can occur through careful job design both in the functional sense and in the mechanical sense. A functional job design, or job description, determines the limits within which a person can operate and can prevent deviations outside that field into areas for which the individual has not been trained and is not qualified. It also sets boundaries that are reasonable and realistic to an individual and within which he can develop more freely. In the mechanical sense the job design relates to the physical layout of the work area and places physical restraints on the extent to which the worker needs to move in the course of doing his job. In this way control can be kept on his movements.

The fact of evolving or developing controls, no matter how comprehensive, relevant or acceptable they might be, is a waste of time unless the substance of what is required can be communicated to those who have to exercise the control and to those on whom the controls are to be imposed. Knowing what is required and accepting the reasons behind it are prerequisites for accepting the control. However the clarity of the communication brings no guarantee of its acceptance; that will depend on how the recipients regard it. Certainly good communications at the right time can do much to increase acceptability of decisions and controls and will often reduce any potential conflict that could arise. Communications must not be the means of by-passing intermediate levels to reach another particular level or stratum in the organisation. Too often in the past, in discussions with union representatives, whether on safety or other matters, senior managers have left out the supervisors and then wondered why joint agreements were not implemented smoothly. For communications to be effective each level in the train must be involved and allowed to participate within the limits permitted by their position in the hierarchy. If conflicts do arise between work groups and management, communications can play a vital role in its resolution but will not be the solution itself.

There are schools of thought that believe that the greatest influence or control can only be exercised through that feature of employment that affects each worker personally – the pay packet. While there is no doubt that wages do play a big part in attitudes to work, there is also evidence that it is not necessarily the biggest nor the most effective influence. It is one of the spectrum of factors that go to make up the whole panoply of industrial relations and working together.

14.4.2 Informal controls

Under the formal controls proposed and implemented by management is an undercurrent of real but unwritten controls that centre on the informal organisation. They stem from the primary group where norms of behaviour are established that group members are required to comply with. These norms can relate to production matters and if they coincide with the company aims can be to the great benefit of both the company and the employee. However if these two aims are at variance it can be to the detriment of both.

The urge to comply with these informal controls can be much stronger than the need to comply with the formal controls because of the great pressure exerted, firstly by the wish to comply with the group norm and secondly by the degree of pressure a peer group can put on a member. The level or extent of the informal control will often bear no relationship to the formal needs of the organisation but will reflect the social needs of the informal group.

14.5 Conflict

Within any organisation, people have their own ideas about priorities for themselves and for the organisation which not infrequently conflict with the 'official view'. Many of the individuals in an organisation are likely to be subject to conflicting demands upon their time, energy and their principles not only in their work where they may play a number of roles but also in their private lives. Conflicts can arise within and between individuals, groups, departments and organisations.

A side effect of conflict which is attracting increasing attention is *stress* which can occur whenever an individual is put in a position of having to attack or defend. As stress builds up so equanimity is eroded and the propensity to argue, disagree or openly oppose grows with the risk of an escalation of potential conflict.

14.5.1 Sources of conflict

Conflicts arise whenever there are differences between individuals or groups and other individuals or groups and it can be between those at the same level or at different levels. At the individual level, there can be a reaction to not being consulted about a matter that materially affects the individual or resentment when the reason for working in a particular way is not understood or has not been explained.

Often the cause of the conflict is either obscure or not appreciated by those taking entrenched positions such as occurs in the case where a union official insists on representing a group of members with whom he has previously had little contact and without fully investigating the reason for the conflict, antagonising not only the employer but often the members he purports to represent. Similarly, where the supervisor is not given the support by management in resolving a relatively minor difference on the shop floor but where management (often the Personnel Manager) insists on handling the affair without fully appreciating the points at issue and finishes up with a full-blown failure-to-agree and a major industrial relations problem. This can discredit the supervisor and antagonise the workforce, both of which militate towards further conflicts.

Conflicts can stem from the different ways in which individuals or groups believe that affairs should be run. This can be seen in the broad differences between political parties over the allocation of national resources, in the differing views on how a social club should be organised and, within an organisation, the different views on whether, for example, promotion should be on the basis of merit or seniority. Again, employees may be disenchanted with the way their tasks are organised because the planned way does not match their natural way of working.

Perhaps the more frequent, but less disrupting, conflicts are those between one individual and another. However, these can escalate where there are strong allegiance ties with other individuals who rally to support the contesting parties.

Refusal by an individual to conform to group standards of thought or behaviour can result in pressure to do so or isolation – 'being sent to Coventry'. Between groups, demarcation of jobs, pay scale and differentials and the threat of redundancy with the competition for diminishing job opportunities are fruitful sources of inter-group conflicts.

Demands for more say in corporate decision making at their roots pose questions over the use of authority and power by management and individuals.

Conflicts can arise between organisations which compete for shares in a fluctuating market where the creation and the removal of jobs is at stake. Also organisations that exercise control over others in the way they perform their tasks can have important results in the workplace. Typical of the latter is the effect of those who enforce statutory regulations where unnecessarily expensive safety controls and procedures can be insisted upon with consequent adverse effect both on operator earnings and on the profitability of an operation.

Conflict sources may be inter-personal – a clash of personality or frustration of an ambition, based on fact – overtly bad production planning or unjust exercise of authority, or philosophical – involving a clash of beliefs or aims.

14.5.2 Conflict and professional standards

A special case of conflict arises when the demands of a position clash with the ethics of a profession. This is a situation that is faced by many of the professions – accountants, doctors, solicitors, etc. and safety advisers. In all these cases there is a pressure resulting from the requirements of legislation but, in the safety adviser's case, he has the added pressure of being in direct contact with, or indeed often reporting to, the person who is creating the conflict. Thus he is placed in the position of having to decide between job security and professional ethics. Should he resign if his advice is consistently subjugated for less safe but more productive methods, should he continue to provide advice to the best of his ability in the knowledge that it will only be partially acted upon with consequent increased risk of injury to an employee and at worst a possible breach of legislation, or is there another solution that will satisfy the demands of production and at the same time ensure the necessary high standard of safety . . . How a safety adviser resolves this sort of problem will depend on his role in the organisation, his interpersonal skills and the culture of the organisation.

14.5.3 The resolution of conflict

Much attention has been paid to ways of minimising conflict within and between the parties to an industrial enterprise. Three general philosophies emerge, none of which is a panacea in the sense that none offers the perfect solution, because conflict is both inevitable and functional in the dynamics of any enterprise.

The *human relations* philosophy has developed from the work of Mayo and others, who believed that individuals should be encouraged to align with the goals of the enterprise by improving leadership skills and communication on the part of management – thus engendering a better appreciation of the way the enterprise is moving and becoming more identified with the goal. There can be an added benefit of increased respect for the manager with consequent reduced resistance to his instructions.

Institutionalisation based on the theory that procedures can be developed to deal with competitive conflict and to achieve compromises. Collective bargaining is a practical result but this is not always successful. The use of cartels in the open market is another example which achieves short-term results until exposed to the buying public. The implication of the institutional approach is that all conflict can be resolved if the right procedure can be found, but this is an over-simplification of reality.

Social engineering involves the alteration of tasks and roles, and changes in the methods of acquiring wealth and power in organisations and society. Progress in this area is a long-term goal of the trade unions who see benefits in job redesign and more involvement in corporate affairs. While it has application in some areas, this approach does not provide a simple solution to all forms of conflict.

14.6 Organisational techniques

In a working community there are a number of techniques available that, used in various combinations, will assist in ensuring the achievement of the enterprise goals through enlisting the co-operation of the workforce, as individuals and as groups. The techniques work on the premise of involving the work people to the greatest extent consistent with maintaining discipline and control.

14.6.1 Communications

Good communications are an essential part of ensuring that not only the goals to be achieved are made known and understood by all involved in the process, but the timing, the reasons and the methods are fully understood so they are followed in producing the required quality of product or efficiency of service.

Communications should be clear and in terms that can be understood by the recipient so that decisions can be clearly conveyed, confusion is prevented and any procedures properly followed. This is particularly apposite where safety issues arise and the health and safety of employees is reliant on following specific work methods.

Care must be taken to ensure that communications do not leap-frog levels in an organisation. Those most often left out or by-passed by communications are the supervisors who have a critical role to play. To get the maximum benefit from communications, all relevant levels in the organisation should be included whether written or verbal means are used. Good communications will reduce conflict and improve workplace relations but they are not the only means nor are they a complete solution.

14.6.2 Involvement and participation

Modern industrial and commercial practices encourage the involvement of the work people in those aspects of the work which directly affect them. It encourages them to put forward suggestions on working methods, equipment, materials, safety, etc., so that their ideas and opinions can be taken account of when the final decision is made. While the decision is ultimately made by a manager or designer, if it incorporates the suggestions received from the work people, there is a higher chance of acceptability of that decision. A typical example could involve the design of a guard. If the ideas of the people who work the machine are incorporated into the guard it is much more likely to be accepted and used than if it just appears on a Monday morning.

A projection of this practice is that the workers should be party to, and participate in, the decision-making process. The trade union interest seems to centre on having a larger say in the running of companies but they are reluctant to accept responsibility for the results of those decisions. A recent EC directive[5] includes a requirement for Member States to provide 'balanced participation' for the workers in the organising of the work. An interpretation of the phrase 'balanced participation' has not been given, but it is likely that local interpretations will restrict its application to matters that are of immediate concern to the workplace health and safety. In the UK, this requirement is largely met by compliance with s. 2(6) of HSW and the safety representatives regulations[8].

14.6.3 Work and job design

All too often in the past, the job or the machine has arrived without prior warning or consultation and this has caused considerable resentment. Plant layouts have been decided without reference to those who have to operate them. There is a double benefit in discussing with and involving the work people in projects at an early stage, whether it is for a new process, new machine or changes in the material. Firstly, those who have carried out the operations know the snags and operating requirements and can put forward suggestions to overcome problems before they arise, and secondly, if they have been involved at this early stage and their ideas used, the new plant or process will be much more readily accepted.

The application of ergonomic principles to the design of new plant and equipment is a regulatory requirement[11] for achieving conformity with the EC Machinery Directive[12]. This is a necessary step before any new plant or equipment can be put on the market or used in the EC. Employing ergonomic techniques and practices can improve the physical aspects of the work and should be used to complement existing proven safety rules and practices taking account of ideas put forward by the work people who will have to use the items.

14.6.4 Collective bargaining

In the early days of the industrial revolution, each worker agreed his wages with the employer, a situation that led, in times of unemployment, to a great vulnerability of the individual worker and the gross exploitation of him. However, with the growth of recognition of the trade unions, they took on the role of bargaining with the employer, on behalf of the workers, on such matters as working conditions, rates of pay, holiday entitlement, etc. Initially negotiations were with individual employers but eventually the trade unions grew big enough to represent whole industries in discussions with employers' federations and so established a system known as collective bargaining.

Collective bargaining, usually carried out by experienced negotiators on both sides, could be acrimonious with each side trying to out-manoeuvre the other using rates of pay, hours of work, conditions of workers' employment, health and safety, etc. as point-scoring bargaining items.

Unfortunately, collective bargaining was often the only point of contact between employer and employee on matters of conditions of employment and, not infrequently, on matters of health and safety also. However, many organisations who relied on national or industry-wide collective bargaining to determine rates of pay, etc. set up formal in-factory groups of volunteer members to discuss health and safety matters. These safety committees proved an effective force in raising standards of health and safety and their role is now recognised in legislation. This has effectively removed health and safety from the collective bargaining arena and placed it firmly in the local joint consultation arena.

The Safety Representatives and Safety Committees Regulations 1977[8] give to recognised trade unions the right to appoint safety representatives to act as spokesmen on health and safety matters with the employers for the members they represent. It is not a bargaining right but is aimed at developing an atmosphere of co-operation to improve standards of health and safety in the workplace and so benefit both employer and employee.

14.6.5 Payment and reward systems

The overt object of working is to obtain a wage to be able to live, be able to enjoy one's leisure and to take a place in society. Once the individual has joined an organisation and begins to make a contribution to it, then payment of some kind, which by no means consists solely of wages, is made in return. Because wages, or money, can be measured and counted it has become the means used in bargaining and as the means to stimulate to greater effort.

Methods for determining the wages paid to the workpeople can range from those based on the number of items produced (piece work), which brings in its train pressure to cut corners to increase the amount produced, to straight day work where a guaranteed day's wage is paid regardless of the amount of work done. Within this range are a multiplicity of payment schemes developed over the years to suit particular ways of working, local attitudes towards how payments should be calculated and the culture of the different organisations. A trend which is developing, particularly with the growth of service industries, is to give all workers staff status, i.e. pay a guaranteed weekly or monthly salary and rely on integrity and goodwill to achieve the required production levels. Day work and staff systems have the advantage from a safety point of view that they do not put pressure on the individual to cut corners to achieve higher output.

With fixed wage systems, the motivation to keep high levels of output needs to be boosted using non wage (and hence untaxed) benefits or 'perks'. The types of benefit offered can range from simple cash bonuses at the end of each accounting period provided certain targets or output criteria have been met to prize draws in the successful department with large prizes such as overseas holidays. Similar schemes have been tried to boost health and safety standards and a number of commercial schemes are available that work on the principle of a graded award for meeting or maintaining a target safety performance, the awards being either cash or kind. While these schemes give good results, they are expensive to run and experience has shown that when the scheme is terminated safety performance reverts to pre-scheme levels.

14.6.6 Quality assurance

The major aim of every company must be to design, make and put on the market its products or services in the most economical manner possible, so ensuring profitability of its operations and enhancing its chances of survival. However, unless the product or service meets the standard specified in the contract, or expected by the customer, the viability of the organisation will be brought into question. Products made to a published standard carry some guarantee of quality, but increasingly customers are demanding reliability as well as quality.

For many years contracts for the supply of defence equipment have demanded compliance with standards at every stage of manufacture in a requirement that has become known as quality assurance (QA). The scheme is explained in BS 4891[6], and BS 5750[7] specifies requirements for three basic levels of system for the assurance of quality of materials (stock in trade) and services. In essence, QA demands comprehensive planning from the outset of a project with detailed specifications for each stage of manufacture, monitoring to check that the specifications are being met with recording and reporting so that deviations can be corrected.

QA cannot be implemented overnight and extensive training is necessary for all those involved to develop the right attitude required to ensure maintenance of the system and achievement of the specified quality. An essential feature of successful QA schemes is the inculcation of the right attitudes in all involved.

Similar standards of excellence should be the aim in health and safety since the attitudes and procedures are identical, namely: clear specification, audit and report back, monitor to identify and correct deviations in materials, behaviour and systems, sample checks and training. Where QA schemes exist or are being implemented, not only should the quality of the product improve but so too should the standards of safety and health.

14.7 Safety organisations

In the UK there are a number of safety organisations all of whom are working towards the achievement of high standards of health and safety in the work place. These organisations range from the enforcement authorities backed by the power of the law to the voluntary bodies who encourage and support local and individual employment efforts to raise health and safety standards.

14.7.1 The Health and Safety Commission and Executive

Set up under the HSW, the Health and Safety Commission (HSC) consists of a chairman and nine members, three nominated by the CBI, three by the TUC, two by the local authorities and one independent person who has yet to be appointed.

The HSC is responsible to the government, through the Minister, for the administration of HSW, initiating investigations into major accidents and research into health and safety matters and for encouraging employers and employees to attain high standards of health and safety in all places of work.

Enforcement of safety laws and the carrying out of research, investigations, etc., rests with the Health and Safety Executive (HSE). The Executive consists of three members, the Director General and two Deputy Director

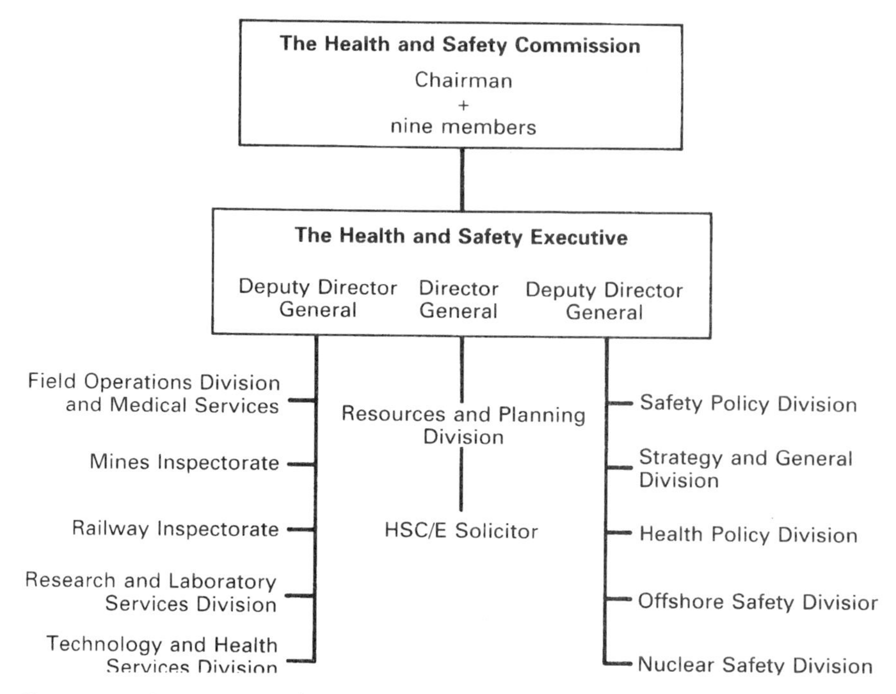

Figure 14.3 Organisation chart for the Health and Safety Commission and Executive (at July 1993)

Generals, who are supported by a Senior Managers Board comprising the Directors of the various divisions and the Chief Inspectors of the various inspectorates. (see *Figure 14.3*).

Specialist inspectorates, such as the Nuclear, Mines, Railway, Agriculture, Explosives and Offshore, carry out the work of enforcement of health and safety laws in their own particular areas. Responsibility for the enforcement of fire matters is delegated to the Fire Authority and for offices and shops to the Local Authorities where the duty falls on the Environmental Health Officers.

For enforcement purposes, the country is divided into 21 areas with each area having the responsibility for developing a specialism in a particular industry or service through groups known as National Industry Groups (NIGs). In addition to providing specialist information on their particular industries to other enforcement areas, the NIGs provide the secretariat for the Industry Advisory Committees (IACs) set up to advise the HSC on matters relating to health and safety in those industries. Technical advice is available to each area through the seven regional offices of the Field Operations Division (FOD) with their professionally qualified staff.

14.7.2 Employers' organisations

Since the Industrial Revolution, employers have met together to consider their joint interests and to form pressure groups to represent their case to government. While most industries have their association or federation of employer members, much of their representation to government is now through the Confederation of British Industries (CBI). The CBI also co-ordinates policies and opinions on all facets of business life both nationally and internationally. It represents the employers in discussions with the employees' organisation, the Trades Union Congress (TUC) and supplies its members with data on a range of political and technical subjects. It has an active Environmental, Health and Safety Group.

Many of the major employers have their own trade organisations that make representations to government in their own right on matters of specific concern to the industry. Many also belong to international organisations of like industries.

14.7.3 Employees' organisations

Unlike the employers, the employee organisations tend to be made up of members of similar trades or skills rather than from particular industries. This has its reason in the historic development of the trade unions to protect their members' employment by restricting entry into a trade and so maintain a demand for those skills. In the early part of the 20th century national negotiations on behalf of groups of unions in an industry had become the practice but this is now being eroded by employment laws and a recognition that it may not always be for the benefit of all members to do so.

The majority of trade unions are affiliated to the TUC which acts as their collective spokesman to the employers and to government. Like the CBI, the TUC has an active Health and Safety Department which publishes advisory booklets on health and safety issues for its members. Most individual trade unions also have their own health and safety sections offering specialist advice to their members.

14.7.4 Voluntary safety bodies

These exist, mainly as registered charities, independent of government, industry or employees' organisations. Of the three major voluntary organisations, the *Royal Society for the Prevention of Accidents* (RoSPA) is the longest established and the only one to hold a Royal Charter. It is concerned with safety in its widest sense and has sections covering leisure activities, water safety, agriculture, road safety, schools and young persons and occupational safety. Affiliated to the Occupational Safety Section are Industrial Safety Groups throughout the UK. RoSPA has its own training school and runs courses on a wide range of safety subjects.

The *British Safety Council* concentrates on occupational safety matters and those environmental and health issues that are major national concerns. It provides a wide range of training together with an advisory service on safety matters.

The Institution of Occupational Safety and Health (IOSH) is the safety adviser's professional body which lays down the academic and experience qualifications necessary for membership. Academic qualifications can be obtained through the various universities that run specialist degree courses in occupational safety and health or through the National Examination Board in Occupational Safety and Health (NEBOSH) which provides a syllabus for and examines in the appropriate health and safety subjects to the academic standard required for membership of IOSH. IOSH provides professional services to its members and represents their interests in discussions with government and other bodies.

14.8 Workplace safety organisation

How safety is organised in any workplace will depend on many factors, perhaps the most important of which is the attitude of the Board and particularly its Managing Director, and the extent to which they are committed to achieving high standards of health and safety throughout the organisation.

14.8.1 Responsibility for safety

When looking at the health and safety responsibilities of the different levels in an organisation, cognisance must be taken of the role played by each in the normal day-to-day running of the operation. Health and safety responsibility must be matched to the authority – both line and financial – vested in an individual. Thus the Managing Director should not be expected to be responsible for ensuring that guards are kept in place but he is responsible for setting the tone of the company's attitude to health and safety, exhorting those who report to him to improve standards and ensuring that, where a case has been made for it, financial resources are allocated to health and safety projects and that those funds are properly applied. The MD should be seen to be committed to high standards of health and safety and one overt way he can demonstrate his commitment is by his knowledge and advocacy of health and safety matters when talking to supervisors and operators during his regular tours of the shopfloor.

Similarly, the Works Manager is responsible for ensuring that the allocated resources are properly utilised, authorising expenditure within the limits of his authority, making a case to the Board where expenditure for a safety item is outside his authority and overseeing his managers in the manner in which they meet their health and safety responsibilities.

Responsibility for ensuring guards are kept in position must rest with the operator, with the supervisor responsible for ensuring that the operator continues to follow the rules.

14.8.2 Organising for safety

To indicate the sort of arrangements and procedures that promote high standards of health and safety in an organisation, a hypothetical example is considered below of a Production Department based on the organisation shown in *Figure 14.1*.

14.8.2.1 Safety arrangements

Headed by the Works Manager, the department is dedicated to the manufacture of the product, its distribution and the maintenance of the production plant. It is located in the factory premises and while it works closely with other departments enjoys a large degree of autonomy. Safety is recognised as an inherent part of the day-to-day responsibilities. The factory has its own safety policy and recognises the major trade union.

Because safety is regarded as an essential part of the work, it is written into the job descriptions of all in the management team, from supervisors upwards. All employees received induction training when they joined the company and this included safety matters. Safety was also included in their subsequent skills and, where relevant, supervisor training.

The Works Manager is anxious that his factory achieves and maintains a high standard of health and safety, so in addition to supporting the safety adviser, he has established a safety committee and includes in his regular Works Committee meetings, which are attended by the trade union representatives, a report from the safety committee.

14.8.2.2 The safety adviser

A safety adviser has been appointed by the Managing Director and he reports, for administrative purposes only, to the Personnel Director. However, for all matters to do with health and safety he has the right of direct access to all managers and directors, including the MD.

The safety adviser's brief is very broad – 'to do all things necessary to ensure that the company's safety performance compares favourably with the best in the industry'. He is given access to all departments at all times to carry out inspections and checks and to monitor that agreed safety procedures are being followed. He is popular with the managers because whenever he finds a fault or matter that requires correction, he raises it with the supervisor with immediate responsibility for the area and advises him on the action required. The only times he has gone further up the line of authority have been when supervisors have failed to take the necessary action. He sees his role as an adviser to all levels of management, and deals with them in a way that reinforces their executive authority by leaving them to make the final decision on what action to take.

Both managers and workpeople respect him for his impartiality on safety matters and like the fact that he consults them before making recommendations. He has not, in fact, had to have recourse to his right to approach the Managing Director because he has always managed to persuade the foreman or supervisor concerned to take action. He is adamant that his job is not to by-pass the authority line but to use it.

There is a procedure for reporting injuries and damage causing accidents which he co-ordinates with the reports being initiated by the local supervisor for the area where the incident occurred. Only if the matter is serious does the safety adviser carry out his own investigation.

Other facets of his job include providing the safety input in skills, supervisory and management training and overseeing fire prevention.

14.8.2.3 Safety representatives

Properly appointed by the union in writing, they are recognised by the works manager and encouraged to participate in safety activities such as safety inspections. They are not particularly active because the other, voluntary, members of the committee are enthusiastic. However, they do insist on exercising their rights under the safety representative regulations of being members of the safety committee[8,9].

14.8.2.4 Safety committee

The works manager is keen on this committee but insists that it works within a set of procedural rules that were agreed at one of its earliest meetings. He chairs the meeting on the grounds that if a decision is needed to initiate an action arising from a recommendation of the committee, he has the authority to make that decision. Alternating the chair with a trade union representative was tried but proved less successful so it was agreed that he remain as the permanent chairman.

Apart from the safety representatives, there are representatives from the work people of each department, the supervisors and the heads of sections. The nurse also attends as does the works doctor when a meeting coincides with his periodic visit. There are more representatives from the workforce than from management. The committee agreed that the safety adviser should not be minute secretary since he should be an active participant. The minutes are taken by the works manager's secretary. Meetings are held every two months.

Business follows an agenda circulated well in advance of the meeting with the major items being the safety adviser's report and a review of progress on outstanding safety matters. One rule that the chairman adheres to is that no matter of complaint may be brought to the committee until the supervisor involved has had time to deal with it and then only if the action taken, or not taken, had resulted in the matter not being resolved. Also general welfare and production matters are excluded since they belong more correctly to the Works Committee.

The final item on the agenda is a review by the chairman of future orders and work load, any special projects, new machinery or plant changes that are proposed.

14.8.2.5 Visitors

Because of the product they make, the company receives a great number of visitors. To ensure their safety, all visitors are booked in by Reception and given a visitor's badge and a card which gives instructions in what to do in the event of an emergency. Once inside the factory they are the responsibility of the person they are visiting until booked out on departure.

14.8.2.6 Contractors

There are always changes being made to the building, plant layout, etc. and there is a constant flow of contractors of one sort or another. Since they can be working in areas where employees are working, a system of strict control of contractors has been implemented. This starts at the enquiry stage which outlines the conditions the contractor must meet and includes a section on health and safety. At the order stage the detailed health and safety requirements are spelt out – the contractor must:

- comply with all statutory requirements,
- obey the safety rules of the company,
- provide all plant necessary to complete the job,
- may only use factory services with written permission,
- restrict himself and his workmen to agreed areas,
- hold third party insurance cover for a minimum of £1m in respect of:
  - damage to the company's property and goods
  - injury to any company's employees caused by his negligence, and
- produce a copy of his safety policy.

A supervising officer is appointed by the company to be the point of contact for the contractor and to ensure that the contractor meets his obligations, both contractual and legislative. The supervising officer may be the safety adviser, works engineer, project engineer or other person qualified to do the job. At the completion of the contract, and before the contractor is paid, his work and the area in which he has been working are inspected to ensure that he has met his contractual obligations and has left the site clean and tidy. Useful guidance on the control of contractors is given in an IOSH publication[10].

14.8.2.7 Results

The safety performance of the department has improved over the last two years and compares with the best in the company. There is a broad awareness of health and safety matters, the work area is much tidier, morale is high and productivity has risen.

14.9 Summary

This chapter has considered, briefly, some of the behavioural and practical matters that can arise in the course of an enterprise's operations particularly in respect of their endeavours to establish a working environment where health and safety are regarded as natural and inherent parts of every operation and process. It has referred to a 'typical' organisation, but no organisation is typical since no two are the same, each having its own culture, practices and people. What has been discussed has to be adapted and used to suit particular local needs.

References

1. Mills, C.W., *The Sociological Imagination*, Oxford University Press (1959)
2. Kaplan, A., Power in perspective, *in* Kahn, R.L. and Boulding, E. (Eds.), *Power and Conflicts in Organisations*, Tavistock, London (1964)
3. Weber, M., *The Theory of Social and Economic Organisation*, Free Press, New York (1964)

4. Mayo, E., *The Social Problems of an Industrial Civilisation*, Routledge, London (1949) Reprinted in *Organisation Theory*, (Ed. Pugh), Penguin Modern Management Texts, 'Hawthorne and the Western Electric Company', London (1971)
5. Commission of the European Community, *Council Directive of the 12 June 1989 on the introduction of measures to encourage improvements in the safety and health of workers at work*, Official Journal of the European Communities No. L 183, Office for Official Publications of the European Communities, Luxembourg, and HMSO, London (1989)
6. British Standards Institution, BS 4891, A guide to quality assurance, BSI, London (1981)
7. BS 5750: 1981 – Quality systems
   Part 1: Specification for design, manufacture and installation
   Part 2: Specification for manufacture and installation
   Part 3: Specification for final inspection and report
   Parts 4, 5 and 6: Guides to Parts 1, 2 and 3 respectively.
   British Standards Institution, London
8. Health and Safety Executive, *Safety Representatives and Safety Committees Regulations*, (SI 1977 No. 500), HMSO, London (1977)
9. Health and Safety Executive, Approved Code of Practice: Safety Representatives and Safety Committees, HSE Books, Sudbury (1988)
10. Institution of Occupational Safety and Health, *Safety with contractors in the motor industry*, IOSH, Leicester (1987)
11. The Department of Trade and Industry, *The Supply of Machinery (Safety) Regulations 1992*, schedule 3 para. 1.1.2(d), HMSO, London (1992)
12. Commission of the European Community, *Council Directive of the 14th June 1989 on the approximation of the laws of the Member States relating to Machinery, No. 89/392/EEC*, Official Journal of the European Communities No. L183, 9–32, Office of the Official Publications of the European Communities, Luxembourg and HMSO, London (1989)

Further reading

Roethlisberger, F. J. and Dickson, W. J., *Management and the Worker*, Harvard University Press (1939)

McGregor, D., *The Human Side of Enterprise*, McGraw-Hill, New York (1960)

Likert, R., *New Patterns of Management*, McGraw-Hill, New York (1961)

Argyris, C., Interpersonal Competence and Organisational Effectiveness, Tavistock, London (1962)

Burns, T. and Stalker, G. M., *The Management of Innovation*, 2nd edn, Tavistock, London (1968)

Lupton, T., *On the Shop Floor*, Pergamon, Oxford (1963)

Lupton, T., *Management and the Social Sciences*, Penguin, London (1971)

Goulder, A., *Patterns of Industrial Bureaucracy*, Free Press, Glencoe, Illinois (1954)

Health and Safety Executive, Occasional Papers Series:
  OP 3: Managing safety: a review of the role of management in occupational health and safety (1981)
  OP 9: Monitoring safety: an outline report on occupational safety and health by the Accident Prevention Advisory Unit (1985)
  HSE Books, Sudbury

Health and Safety Executive, *Health and Safety series booklets*:
  HS(G)48: Human factors in industrial safety, HSE Books, Sudbury (1989)
  HS(G)65: Successful health and safety management, HSE Books, Sudbury (1991)

Health and Safety Executive, *Management of Health and Safety at Work Regulations 1992; an Approved Code of Practice*, No. L21, HSE Books, Sudbury (1992)

Chapter 15

Employer's obligations for safety

J. R. Ridley

As safety legislation was being developed in the mid and late 19th century, the onus for preventing accidents and looking after oneself rested with the individual worker. Gradually it became apparent that the worker had little control over the conditions under which he worked and often was not trained in anything other than the immediate skills needed to operate the machinery. The person who had the control and could influence working conditions was the employer and gradually the pendulum of responsibility for safety swung from the employee to the employer. By the late 1960s the pendulum had swung so far that virtually all responsibility for ensuring safe working had been removed from the worker. The pendulum is swinging back and new laws are recognising that both employer and employee have responsibilities for developing and maintaining high standards of occupational health and safety.

However, the major responsibility remains with the employer who provides the place of work and machinery and still controls what work is done and, to a large extent, how it is done. This chapter looks at some of those responsibilities which have extended beyond statutory and common law duties for health and safety *per se* and now encompass health and safety issues that arise as adjuncts to employment and discrimination laws.

Useful guidance on management techniques that contribute to high standards of health and safety is contained in an HSE publication[1] which considers five major areas: policies, organisation, planning, measuring performance and auditing.

15.1 Duties of care

Evolving in the civil courts, the employer's duty of care to each of his employees can conveniently be considered under five headings. These require that the employer must provide:

(1) Safe systems of work.
(2) A safe place of work.
(3) Plant and machinery that is safe to use.
(4) Competent supervision and/or suitable training.
(5) Care in the selection of fellow employees.

These common law duties, which are considered in more detail in Chapter 8, are now incorporated into statutory law under S. 2(2) of HSW which has extended the employer's obligations to the provision of training, instruction and information in sufficient detail to enable the employee to understand the hazards faced and be familiar with the techniques for avoiding them.

These obligations have been further extended by the requirements of the Management of Health and Safety at Work Regulations 1992[2] (MHSWR) under which employers must carry out assessments of all operations and processes that present a potential risk to the health and safety of employees. Additionally, they must include in working procedures the proper organisation to meet health and safety responsibilities and ensure that they have available to them adequate and competent assistance on health and safety matters

15.2 Use, handling and storage

HSW requires the employer to make arrangements for the safe use, handling and storage of substances and equipment. The Manual Handling Operations Regulations 1992[3] (MHOPR) advocates the elimination of manual handling, but where this cannot be achieved requires the risks from handling to be assessed and precautionary measures taken to prevent injury.

The detailed requirements for the use, handling and storage of certain substances are contained in the relevant regulations such as the Control of Lead at Work Regulations[4], HFL[5] and COSHH[6]. Safe storage of substances relates not only to the physical arrangements such as racking etc., but also to possible chemical reactions between adjacently stored incompatible substances. A number of serious warehouse fires reinforce this point[7,8].

The Supply of Machinery (Safety) Regulations 1992[9] (Machinery Regulations) lays down that new plant and equipment must be designed so that it is stable when in use and being moved, or arrangements made to ensure it will be stable. Adequate and suitable handling equipment must be provided to enable plant and equipment to be moved safely.

15.3 Transport

The transport of substances, whether within the factory premises or on the public highway, must be in a safe manner with the container of the substance labelled to identify its contents – for certain hazardous substances the label should carry standard warning signs. The duties of the employer, as consignor or employer of the driver, are much more onerous once the load leaves the factory premises and goes on the public highway. Very strict requirements are imposed by road traffic laws and, in the case of certain chemicals, by Regulations[10] supported by an Approved Code of Practice[11]. If an accident occurs the HSW is likely to be invoked only if the driver (an employee in the course of his employment) is injured or the substances are not properly labelled. Any other enforcement action would be taken by the police under road traffic or highway laws but could involve the employer as owner of the vehicle.

15.4 Safe access and egress

The occupier or employer should provide safe means of access to and egress from all the places where employees and others have need to resort to in the course of their work. The Workplace (Health, Safety and Welfare) Regulations 1992[12] (WHSWR), which apply to all new or modified workplaces, and will apply to all workplaces from 1996, place general duties on the employer or occupier in that respect. These regulations are more specific than earlier legislation regarding the condition of access routes and the organisation of traffic routes.

15.5 Environment

The environment of a workplace must be such that it does not put the health of the workpeople at risk. WHSWR lays down the requirements to be met within the workplace and covers such conditions as keeping the workplace clean, properly decorated, ventilated, well lit, suitably heated and that adequate welfare facilities such as first aid, toilets, washrooms, changing and rest rooms are provided.

Noxious fumes must not be emitted into the atmosphere and the disposal of waste, particularly waste containing dangerous chemicals, is strictly controlled by regulations[16].

15.6 Safety policy

One of the major recommendations of the Robens Report[17] was that all employers should develop and publish a statement setting out their intentions with regard to protecting the health and safety of their employees. This is now enshrined in s2(3) of HSW and all employers, except those employing less than five persons, are required to have a written statement of their safety policy. However, in preparing this statement of safety policy sight should not be lost of the prime aim of the company which is to make a profit through the manufacture and sale of the products it makes or the service it provides to its clients. The safety policy should state that the intention is to do so without putting employees, or others, at risk of their health or physical well-being.

Copies of the policy should be 'brought to the notice of employees' either by displaying it on notice boards if this is the normally accepted means of general communication, or by giving a copy to each employee. For new starts, their induction training should include discussion of the company's safety policy when copies can be issued to them.

Supporting the policy should be information on the organisation that exists to implement the policy and the facilities or arrangements that are in being for achieving the policy's intent. These two points are dealt with below.

The fact of drawing up a safety policy document is not sufficient to ensure that the company will be a safe place to work or that the employees will not suffer injury. The successful implementation of the policy depends to a large extent for its success on the degree of the employer's commitment and on his attitude towards safety. Without the commitment and involvement of the employer the policy is not likely to be effective.

15.7 Organisation

For an organisation to be effective it must have some sort of structure or framework into which the functions and functionaries who make up that organisation fit. A diagram of an example of an organisation is shown in *Figure 14.1* from which the inter-relationship between the different office holders can be readily appreciated and understood. But this is not sufficient to ensure that the organisation will work and it rests with the manager to provide the motivation and to ensure its continued smooth running. His role extends to co-ordinating the various activities so that they work as a team and all contribute towards the success of the enterprise.

He must prepare the plan of action, allocate jobs and determine what roles and functions each of the members is required to fulfill. The functions are shown on the diagram as mechanistic parts of an organisation and these can be expanded by written explanations or job descriptions. What is not so easily defined are the actual roles each individual plays since these can extend beyond the bald function as outlined. For example, functionally, the supervisor is responsible for allocating work, ensuring that the materials are available and that the required number of the product are produced at the right time. However, his role is considerably more complex. He has to deal with the inter-personal problems that arise, handle technical queries to do with the materials that have been supplied, ensure that the safety rules are followed and cope with the effects of absenteeism on production targets.

Each individual in the organisation is faced with informal or fringe problems not directly associated with the main thrust of his job, but problems that are inevitably part of the job. It is part of the manager's responsibility to ensure that each part of the organisation he sets up is capable of meeting the demands put on it in its efforts to achieve the company's operational goals.

Into this organisation must be fitted the health and safety role, whether full-time appointment, part-time or split responsibility or as a hired-in service. Whichever arrangement is employed, the person or body providing the advice must be competent in health and safety matters – this is a requirement of MHSWR.

15.8 Arrangements

Attached to the safety policy document should be a note that outlines the arrangements that exist within the company for carrying forward its efforts to develop and maintain high levels of health and safety. Included in this note could be reference to any rules that relate to health and safety matters. Typical of the arrangements described in the notes could be:

- The role and function of the safety committee.
- Recognition of safety representatives.
- The need for personal involvement.
- Provisions for safety training.
- Existing safety standards and rules.
- The issue and use of protective clothing.
- Control of contractors and visitors.
- Safety in the use of fork lift trucks.
- Handling of chemicals.
- Noise and hearing protection.
- Testing and control of the environment.
- Permit-to-work and locking-off procedures.
- etc.

15.9 Risk assessments

Much of the success in achieving high standards of health and safety lies in removing or reducing the risks faced by employees before an accident happens. The value of this has been recognised and requirements are contained in legislation for employers to carry out risk assessments for all operations.

In many cases, risk assessments are carried out subjectively, almost subconsciously, in the normal course of work, and indeed of living. More conscious identification of hazards can be based subjectively on experience or analytically using one or other of the techniques described in 10.2–10.7.

Risk is defined in a harmonised standard[13] as a combination of severity, frequency and duration of exposure and of the probability of an event occurring. Putting a numerical value to the risk, using one of the methods outlined in either 10.1.3 or in an IOSH publication[14], enables priorities to be assigned to further investigative work for reducing the risk or providing protective measures.

MHSWR requires every employer to carry out risk assessments and, for those employing more than five people, to record the findings of the assessment. The degree of detail recorded must be determined by the risk faced, but it can range from a simple statement of the date, location, hazard viewed, assessed level of risk and precautionary measures, if any, to be taken through to a detailed analysis of risk using one of the techniques listed in chapter 10 and the harmonised standard[13]. Risk assessments should be reviewed and updated whenever changes occur to the work situation that make the earlier assessment unsound.

15.10 Monitoring safety performance[15]

The key elements in achieving and maintaining high standards of health and safety are a commitment by management, effective training, systems of work that are safe, regular monitoring and a committed work-force. In manufacturing, regular checks on production performance enable a manager to assess whether he is likely to meet the goals that have been set and to identify those areas where there is slippage so that he can initiate corrective action. In much the same way, regular checks on safety matters can highlight problem areas and enable action to be taken before an accident occurs. These safety checks can be either *reactive* or *proactive*.

Reactive checks are those that are made post-incident in response to reports which highlight something that is not according to plan. The most common such checks in health and safety are the accident statistics used to measure accident performance. These record the failures (accidents) after they have occurred and, while they can give an indication of where corrective action needs to be taken, the price of an accident has already been paid. There is also a delay between the incident, or series of incidents, that have attracted attention, and the initiation of corrective action. Unless care is taken in the way in which an investigation into an accident is carried out, obvious causes, such as lack of guards etc. especially when there is the possibility of future litigation, can divert attention from the underlying, but not necessarily obvious, cause. Accident statistics are relatively easy to collect but have a limited role and must be seen as one of a number of techniques that can be employed in the identification, elimination and control of hazards.

The *proactive* approach, on the other hand, endeavours to identify, evaluate and control hazards and risks before they develop to the stage of causing an accident. The technique is inherent in the process of risk management (Chapters 9 and 10). One of the more common methods employed to identify potential hazards before they manifest themselves is the safety inspection, or survey, carried out by the safety adviser, manager and safety representative, where the work area is inspected for any hazards which are noted. After the inspection the identified hazards are assessed and a plan of action to put them right is formulated. Variations on this theme include:

- safety tours where a pre-agreed route is followed looking for all hazards,
- safety sampling which involves inspecting the whole work area but looking only for particular types of hazard – i.e. the sample, and
- safety audits which endeavour to obtain a numerical measure of the number of examples of each hazard found.

These techniques identify hazards before accidents occur and enable preventative measures to be taken. If the inspections or surveys are repeated on a regular basis and compared with earlier results, an indication of improving or worsening safety performances can be obtained without suffering the trauma of an accident.

Other schemes for the proactive assessment of safety performance include CHASE[18] which concentrates on assessing the individual manager's awareness, attitudes and knowledge of health and safety matters on the premise that if the manager knows what is required he will ensure that it is either provided or achieved. In another scheme, ISRS[19], the survey is carried out by trained assessors who look at a number of operating facets to determine management and operator safety performance and attitudes.

15.11 Training

A key element in achieving and maintaining high levels of safety is knowledge of the hazards, their effects and the techniques to avoid or ameliorate those effects. With this knowledge comes the confidence to deal with the hazards but it is important that the knowledge acquired is relevant and is technically correct. Much harm can be done when using incorrect information in the misguided belief of its veracity. The responsibility for ensuring that the knowledge employees of all levels have about the hazards they may meet is correct and with a sound technical base lies with the employer. The provision of that knowledge through training, instruction and information constitutes a major contribution towards high safety performance.

In providing the training, account must be taken of the level of employee involved. At a senior manager level with high qualifications, the information can be of a fairly technical nature whereas at operator level the information needs to be more basic and relate to the identification of the substance in terms meaningful to the operator, recognition of the hazard, its likely effect on an individual and the action necessary to avoid ill-effects. How, and to what degree, the training is given will depend on the local circumstances, practices, types of labour and seriousness of the hazard faced.

15.12 Use of information

The flow of information is the life blood of an organisation; if it dries up the organisation dies, if it becomes a flood the organisation drowns. Controlling the flow of information is important for the health of the enterprise and ensuring that the right information gets to the right people is a major feature of successful and efficient organisations. This applies in safety as much as in production, sales, finance and administration. As new laws are published, new techniques become known and new rules are agreed, the early dissemination of information about them, possibly accompanied by an explanation of how they affect the organisation, enables an early response and adjustment of working practices to be made.

The speedy availability of information is of vital importance should an emergency arise where the emergency services may require to be appraised of the latest situation and may require details of the plant, its layout and key equipment. The employer should not wait for an emergency to arise but should plan beforehand and liaise with the emergency services to ensure that all the information they might require is readily available.

It is not only the emergency services that will require up-to-the-minute information. If an incident occurs that could affect the neighbours or the community, the media will want a statement or summary of the current position for their readers and viewers. This can be a serious distraction to a manager deeply involved in a major crisis. A spokesman should be nominated who knows the plant and process and who can be briefed to deal with the media.

Pressure can also come from the public and community for information on proposals for future developments, particularly if they are of a sensitive nature either politically, environmentally or could be seen to have an adverse effect on the immediate neighbourhood. This information needs to be readily available and issued without delay if credibility as a caring employer and factory owner is to be maintained. Any delay could be interpreted as having something to hide and seriously prejudice future relations with the community. The use of a spokesman previously briefed can do much to ensure prompt response to media enquiries and public concern.

15.13 Special risks with machinery

Where the operation to be performed by or on a piece of equipment, such as adjusting a moving machine or the emergency servicing to enable a machine to complete a particular operation, is likely to put the operator at particular risk to his health and safety, special precautions must be taken. These include ensuring that only fully trained, competent and authorised persons carry out the work and that all appropriate safety measures are in place to prevent or reduce the extent of possible injury. In addition to proper training and instruction, these measures can include the use of special safety overalls with no loose or flapping parts and someone standing by the emergency stop button at all times when the work is being done. The Provision and Use of Work Equipment Regulations 1992[20] (PUWER) do not detail the precautions to be taken but rely on the employer to take precautions most suited to the circumstances of the operation.

15.14 Systems of work

Reference is made in s. 2(2)(a) of HSW to 'systems of work . . . that are safe and without risk to health' and similar phrases are used in codes of practice and guidance booklets. It is a general exhortation that wherever systems of work are prepared they should be safe, and the onus for ensuring this is placed with the employer. No responsible employer should adopt a system of work that is not safe otherwise he is actively supporting a breach of the law.

What is frequently as important as the safety of the system is that it should be properly communicated to those who should be using it. For lower-risk activities, verbal communication may be adequate, but where the hazard or risk is high and the safety of the operator depends on the system, the system needs to be understood without ambiguity. In such cases the details of the system should be in writing.

The use of safe systems of work raises the question of whether it is sufficient to rely on systems of work as a strategy for accident prevention. To do so means, in essence, that reliance is put on the behaviour of the operator as the means of preventing injury – the 'safe person' concept. Because it is impossible to control human behaviour under every circumstance the concept of a safe person is not acceptable as a primary means of protecting an employee. Thus, safe systems of work should be seen as a back-up to more positive physical, mechanical, electrical, etc., techniques to keep an employee from being injured.

15.15 Permits-to-work

In certain operations, such as work on live electrical equipment or entry into confined spaces, where in spite of the precautions being taken the operator is still likely to be exposed to a high level of risk, the employer must institute very strict controls on the circumstances under which the work can be performed, the parts of the plant to be worked on and the people who carry out the work through the use of systems of permits-to-work[21,22]. Before any work can commence, instructions in writing have to be issued detailing the work to be carried out, identifying the specific pieces of plant to be worked on and any special precautions to be taken. These must be signed by a competent person who is required to certify that the particular plant is safe to work on. When the work has been completed, the person in charge of the work must sign the permit saying the work has been completed and the plant is safe to return to production. The competent person must then check that the plant is indeed safe before restoring energy and signing it off for return to production.

A typical form of permit-to-work is given in section 28.10.

15.16 Locking off

In maintenance operations where it is necessary for work to be undertaken within a machine or where entry is necessary to a vessel into which chemicals are piped, special precautions need to be taken to ensure that the machine cannot be started or the chemicals flow into the vessel. This is achieved by 'locking off' the source of supply, in the case of the machine by locking the switch on the incoming supply in the OFF position or in the case of the vessel by locking the various supply valves in the SHUT position. Locking off must be by padlock with the key retained by the operator carrying out the work. A strict rule should be enforced that the lock can be removed only by the person who applied it. This rule should only be varied in exceptional circumstances and then only on the written authority of a senior manager. Proper and adequate training will be required to ensure the effective functioning of the locking off arrangements which should be monitored regularly. Guidance on this technique is given in an HSE publication[23].

15.17 Emergency services

Even in the best run organisations situations arise that involve the emergency services and could affect the local community. The extent of the emergency and its effect on the community will depend on the materials and processes being used. Under Regulations[24] that apply to certain industries there is a statutory requirement to prepare emergency plans and inform the local community about the processes being carried out. For those industries, their position is clear and they know what they have to do. However for the great majority of employers there is no such direction and they are left to their own devices and decisions. Unless particularly hazardous materials are being used, the most likely emergency for the average employer is either a bomb threat or a fire.

For bomb threats, the police have issued extensive advice which if followed reduces the risk to employees but does cause disruption of production. This advice should be made known to all employees who should also be trained in what they have to do.

Fire presents a more permanent and on-going risk with the potential to cause greater financial and employment loss. The requirement to take precautions against fire is part of health and safety law which is broadly aimed at protecting life and limb but not the business. An onus is put on the employer to train his employees in what to do in the event of fire. Should a fire occur, the best protection for the business is to put the fire out and so minimise the damage to buildings, plant and machinery. First aid fire fighting, which endeavours to limit the spread of the fire, can be undertaken voluntarily by employees but they will need to be given special training in the use of portable fire-fighting equipment. Arrangements may need to be made to make a payment to them to ensure that they are covered by Employer's Liability insurance when fire fighting (which may be outside the job they were employed to do).

The main fire fighting will be carried out by the fire brigade who will need to get to the seat of the fire as quickly as possible and begin tackling it. To help them in this the employer should have immediately available plans of the workplace and should encourage the fire brigade to visit the site to become familiar with it. In addition he should ensure that there is always someone on duty who knows the site and can direct the brigade and other emergency services when they arrive.

15.18 Plant and machinery

There is an obligation on employers to ensure that the plant and machinery they provide for use at work is safe for the purpose to which it is put. All new machinery put on the market in the European Community must meet the essential safety requirements (ESR's) contained in the EC directive on machinery[25] and repeated in the Supply of Machinery (Safety) Regulations 1992[9] (SMSR). Evidence of compliance is by way of the EC mark applied to the machine by the manufacturer who is required to provide a Certificate of Conformity supported by a Technical File giving details of how conformity with the ESR's was achieved. However, it is incumbent on the purchaser to check that the documentation provided relates to the machine being bought and that the machine carries the EC mark.

Secondhand machinery, whether modified, refurbished or not, is required to comply with lesser requirements contained in PUWER. However existing machinery in current use on 1st January 1993 has until 1st January 1997 to comply with the standards demanded by PUWER. In these latter cases, no documentary evidence of conformity is required; it rests with the employer to ensure that the machines he is using meet the requirements of PUWER.

Hired plant, if first hired out after 1st January 1993, must conform with SMSR and carry the EC mark whereas hired plant first used before 1st January 1993 should comply with PUWER where the onus for compliance rests with the person from whom the equipment is hired. However, when hiring plant care needs to be exercised by the person who is going to use it to check the wording of the lease or hire agreement for specific conditions that may transfer the compliance responsibility to him. When hiring lifting plant and plant incorporating pressure vessels, the current certificate of examination should also be checked.

15.19 Designing out hazards

There is an obligation on employers to provide plant that is safe. Bought out machinery should already be adequately guarded, but for custom designed and built plant, such as process plant, machining stations, robot assemblies, etc., the onus rests firmly with the user to ensure that the plant is safe when put to work.

The most economically beneficial time to consider the safeguarding of plant is at the early design stage when the guards can be included at relatively small cost and can be incorporated as an inherent part of the plant (this also makes them more acceptable to the operators). Delaying consideration of the necessary safeguards until the plant is commissioned increases the cost enormously and the guards will always look what they are, add-ons.

Consideration of safe guards at the early design stage offers an opportunity to review the overall safe operation of the plant when a number of questions can be asked:

- Are there hazards associated with the materials to be used? If so, are there other technically acceptable but safer alternatives?
- Are the proposed operating methods safe or can they be made safer? (Consultation with the operators of existing plant may bring some useful comments.)
- Are all the dangerous parts, or access to them, adequately guarded?
- Has adequate and proper access been allowed for operation, setting, adjustment and maintenance?
- Does the layout of the plant take account of existing practices?
- Have ergonomic principles and behavioural techniques been considered in the layout of the operating controls and work stations?
- etc.

At this stage specialists such as the chemist, production engineer, safety adviser and representative of the operators should be involved so that the designer can draw on their collective experiences and knowledge.

By building in safety from the design stage, much time and cost can be saved when the plant is commissioned and the employer will be able to put the plant to work confident that it meets or is better than statutory safety requirements.

15.20 Welfare facilities

The earliest health and safety laws were concerned with what we now consider to be welfare matters but the obligations they imposed remain and have been joined by many more. Essentially aimed at ensuring that workpeople do not have their health put at risk as a result of the conditions under which they are required to work, the obligations contained in WHSWR cover matters such as temperature in the workplace, lighting, heating, toilets, washing facilities, changing rooms, rest rooms, first aid, etc. The present trend is to provide facilities that go beyond health and hygiene requirements into the realm of improving the quality of working life.

15.21 Personal protective equipment

There are many situations at work where, in spite of the precautions taken, it is not possible to eliminate entirely exposure to hazardous substances, physical hazards or potentially dangerous equipment. In these circumstances the final protection must be personal protective equipment (PPE). However, in providing PPE it is necessary to ensure, *inter alia*, that it is suitable to protect against the particular hazard, does not interfere with other protective equipment, allows the wearer to do his job and is sufficiently robust to withstand reasonable wear and tear. The provision and use of this equipment is covered by Regulations[26] which place obligations on the employer to train and instruct his employees in its proper use and maintenance. It is not sufficient just to provide the equipment but employees should be allowed to select from a range of suitable equipment that which suits them best. Failure to allow this choice has led to employees leaving the employment and successfully appealing to a Tribunal on the grounds of constructive dismissal[27].

15.22 Employment law

Under these complex and constantly changing laws matters relating to health and safety may not be concerned with the protection provided but with the manner of its provision or lack of provision. The spirit of these laws is that the employer should treat his employees in a manner which is reasonable in all the circumstances. This applies particularly when contemplating disciplinary action for a breach of a health and safety rule when suitable warnings must be given and the employee allowed an opportunity to mend his ways. The detailed procedure that should be followed is covered in chapter 5. Successful appeals have been made to a Tribunal on the grounds of both unfair and constructive dismissal where the employer, not having followed the appropriate procedure, was called to justify his actions.

15.23 Discrimination laws

Under the legislation on sex and racial discrimination a body of Tribunal case opinions is building up that puts further obligations on employers. In the health and safety area, the restrictions that existed until recently prevented the employment of women in certain jobs and on night work. These restrictions have been removed except where a woman's childbearing ability is put at risk[28]. The moral issues surrounding the employment of men and women together on night work no longer carry any weight and the exclusion of women from jobs hitherto considered suitable only for men can lead to complaints of sexual discrimination to Industrial Tribunals.

Similarly, employment in any occupation and prospects for promotion cannot be refused on grounds of colour or ethnic origin and an employer can be required by a Tribunal to justify why he appointed, say, a white person in preference to a coloured person for a particular job. Certain jobs, such as handling food, have general rules of hygiene that apply to all employees. Where the customs of an ethnic group clash with those hygiene requirements, the employer is justified in excluding members of that group from that particular job[29,30].

15.24 Summary

This chapter has endeavoured to outline some of the main obligations placed on employers by health and safety laws, by other laws where health and safety is a point at issue and by decisions made by Tribunals and the Courts in common law. The number of statutes and the volume of case law continue to grow with the result that the body of law involving health and safety issues is becoming more extensive and more complex.

References

1. Health and Safety Executive, *Health and Safety: Guidance Booklet No. HS(G)65, Successful health and safety management*, HSE Books, Sudbury (1991)
2. Health and Safety Executive, Management of Health and Safety at Work Regulations 1992, HMSO, London (1992)
3. Health and Safety Executive, Manual Handling Operations Regulation 1992, HMSO, London (1992)
4. Health and Safety Executive, The Control of Lead at Work Regulations 1980, (SI 1980 No. 1248), HMSO, London (1980)
5. Health and Safety Executive, The Highly Flammable Liquids and Liquefied Petroleum Gases Regulations 1972 (SI 1972 No. 917), HMSO, London
6. Health and Safety Executive, The Control of Substances Hazardous to Health Regulations 1988 (SI 1988 No. 1790), HMSO, London
7. Health and Safety Executive, *Investigation Report: Fire and Explosions at B and R Hauliers, Salford, 25 September 1982* (ISBN 0 11 883702 8), HSE Books, Sudbury (1983)
8. Health and Safety Executive, *Investigation Report: Fire and Explosions at Cory's Warehouse, Toller Road, Ipswich*, 14 October 1982 (ISBN 0 11 883785 0), HSE Books, Sudbury (1985)
9. The Department of Trade and Industry, The Supply of Machinery (Safety) Regulations 1992, (SI 1992, No. 3073), HMSO, London (1992)
10. Health and Safety Executive, The Road Traffic (Carriage of Dangerous Substances in Packages, etc.) Regulations 1986, HMSO, London (1986)
11. Health and Safety Executive, *Approved Code of Practice COP19 Classification and labelling of dangerous substances for conveyance by road in tankers, tank containers and packages*, HSE Books, Sudbury (1990)
12. Health and Safety Executive, Workplace (Health, Safety and Welfare) Regulations 1992, HMSO, London (1992)
13. British Standards Institution, *BS EN1050, Safety of Machinery – Risk Assessment*, British Standards Institution, London (1993)
14. Holt, A. StJ., and Andrews, H., *Principles of Health and Safety at Work*, 2nd edn, The Institution of Occupational Safety and Health, Leicester, 84 (1993)

15. Health and Safety Executive, *Occasional Paper No. OP9: Monitoring Safety, (ISBN 0 11 883783 4)*, HSE Books, Sudbury (1985)
16. Health and Safety Executive, The Control of Pollution (Special Waste) Regulations 1980, HMSO, London
17. Report of the Committee on Safety and Health at Work, 1970–72 (Robens Report), Cmnd 5034, HMSO, London (1972)
18. HASTAM, *The Complete Health and Safety Evaluation Manual (CHASE)*, Health and Safety Technology and Management Ltd., Birmingham (1989)
19. International Loss Control Institute, *International Safety Rating System (ISRS)*, ILCI, Loganville (1978)
20. Health and Safety Executive, Provision and Use of Work Equipment Regulations 1992, HMSO, London (1992)
21. Health and Safety Executive, *General Series Guidance Note: No. GS5, Entry into confined spaces*, (ISBN 0 11 883067 8), HSE Books, Sudbury (1977)
22. Health and Safety Executive, *Oil Industry Advisory Committee Report, Guide to the principles and operation of permit-to-work procedures as applied in the UK petroleum industry*, (ISBN 0 11 883885 7), HSE Books, Sudbury (1986)
23. Health and Safety Executive, *Legislation Booklet No. L22: Guidance on the Provision and Use of Work Equipment Regulations 1992*, HSE Books, Sudbury (1992)
24. Health and Safety Executive, The Control of Industrial Major Accident Hazards Regulations 1984, HMSO, London (1984)
25. Commission of the European Communities, *Council Directive of 14 June 1989 on the approximation of the laws of Member States relating to machinery*, No 89/392/EEC, Office for Official Publications of the European Communities, Luxembourg, also HMSO, London (1989)
26. Health and Safety Executive, Personal Protective Equipment at Work Regulations 1992, HMSO, London (1992)
27. British Aircraft Corporation Ltd. *v.* Austin (1978) IRLR 332
28. Page *v.* Freighthire Tank Haulage Ltd., (1980) ICR 299; (1981) IRLR 13
29. Panesar *v.* The Nestlé Co. Ltd., (1980) ICR 144; (1980) ICR 144; (1980) IRLR 64; CA
30. Singh *v.* Rowntree Mackintosh Ltd., (1979) ICR 554; (1979) IRLR 199

Part III

Occupational health and hygiene

In his work, the safety adviser may be called upon to recommend measures to overcome health problems that have been identified by the doctor or nurse. Part of his duties may include the identification of processes and substances that are known to give rise to health risks and advising on the procedures to be followed for their safe use.

The advice he can give will be more pertinent if the safety adviser has an understanding of the nature of the substance and the manner in which it affects the functioning of the human body.

This Part explains the functions of the major organs of the body, considers the characteristics and hazards of a range of commonly used substances and processes and discusses the techniques that can be employed to reduce the effects of those risks on the health and well-being of the workpeople.

Chapter 16

The structure and functions of the human body

Dr T. Coates

16.1 Introduction

Occupational health is that branch of medicine concerned with health problems caused by or manifest at work. Some health problems, although not caused by the job, may be aggravated by it.

A knowledge of the structure and functioning of the organs and tissues of the body is of value in the understanding of occupational illness and injury.

Some substances are particularly liable to damage certain organs; e.g. hydrocarbon solvents may affect the liver, cadmium may damage the lungs or kidneys and mercury may affect the brain.

A brief description of anatomy and physiology is given below and more details may be obtained from textbooks on the subject.

16.2 History

Although many hazards of work were well recognised in ancient times, very little was done to prevent occupational disease. Mining was a dangerous unpleasant occupation performed by slaves. The latter were expendable and the frightful conditions in which they worked may have been a deterrent to slaves on the surface!

By the second century A.D. some miners were using bladders to protect themselves from dust inhalation. (Apart from armour and shields this is probably the first example of protective clothing worn at work.)

Little is known about occupational diseases in the dark ages but by the sixteenth century there was extensive mining for metals in central Europe and several accounts of associated diseases. The year 1556 saw the publication of a work of 12 books on metal mining by a mining engineer and doctor called Agricola. The latter part of book VI was devoted to the diseases of miners. Agricola advised the use of loose veils worn over the face to protect the miner against dust and ventilating machines to purify the air.

Eleven years later another doctor with an interest in mining published a work on diseases of mining and smelting. Paracelsus was physician to an Austrian town and local metallurgist. He used several metals including lead, mercury, iron and copper to treat diseases. He described the signs and symptoms of mercury poisoning and recommended the use of mercury in treating syphilis. When challenged that some of his drugs were poisonous he replied 'All things are poisons, for there is nothing without poisonous qualities. It is only the dose which make things a poison'.

In 1700 a book on trade diseases was published by an Italian physician by the name of Bernardino Ramazzini. He based the book on personal observations in the workshops of Modena where he was professor of medicine and on the writings of earlier doctors. Ramazzini was the first person to advise that physicians should ask specifically about the patient's occupation when diagnosing illness.

The development of the factory system saw the rapid movement of people from the country to the towns with consequent disruption of family life. Large numbers of workers and their families housed near the factories resulted in overcrowding, poor housing and poor sanitation. At work, people suffered appalling injury and disease and worked very long hours until eventually the pressures of humanitarians such as the Earl of Shaftesbury promoted legislation which improved working conditions and reduced the hours of employment of workers in factories, mines and elsewhere.

During this time Charles Turner Thackrah, a doctor from Leeds, wrote a book about occupational diseases in his native city which was the first such work to be published in this country. But this was 1832 and his work raised little interest, but did influence the House of Commons on future factory legislation.

The Factories Act of 1833 saw the appointment of Factory Inspectors and the need for doctors to certify that a child appeared to be at least nine years of age before being employed in textile mills. When birth certification was introduced in 1837 the assessment of children's ages became unnecessary. In 1844, the Factory Inspectors appointed Certifying Surgeons and by 1855 they were required to investigate industrial accidents and to certify that young persons were not incapacitated by disease or bodily infirmity.

By the mid nineteenth century the Registrar General had amassed a great deal of statistical information about occupational disease. Dr E.H. Greenhow of St Thomas' Hospital showed from these figures that much of the chest disease in certain areas of the country was due to the inhalation of dust and fumes at work.

In 1895, poisoning by lead, phosphorus and arsenic and cases of anthrax became notifiable to the Factory Inspectorate. Certifying surgeons examined workers in match factories, lead paint works, trinitrobenzene explosive factories and india-rubber factories using the vulcanising process which involved carbon bisulphide. The widespread occurrence of 'phossy-jaw' in phosphorus workers and lead poisoning gained much publicity and provoked the appointment in 1898 of Dr Thomas Legge as the first Medical Inspector of Factories. Legge devoted the next 30 years to investigating and preventing occupational disease. His book 'Industrial Maladies' was published posthumously in 1934.

By 1948, the Certifying Surgeons had become 'Appointed Factory Doctors' and numbered over 1800. They examined young people under the age of 18, investigated patients suffering from notifiable diseases and carried out periodic medical examinations on people employed in specific dangerous trades. The Appointed Factory Doctor system was replaced by the Employment Medical Advisory Service in 1972. This service, the nucleus of which was formed by the medical branch of the factory inspectorate, gives advice to employers, employees, trade unions and others on medical matters related to work.

Occupational Health Services in private industry were slow to develop and although there are rare instances of medical services at work even before the industrial revolution the first Workman's Compensation Act of 1897 was the first real stimulus which provoked employers to seek medical advice in their factories. At that time, the main reason for such appointments was to help protect the firm against claims for compensation. Exposure to hazards in munitions factories in World War I initiated many new medical and nursing appointments and increased the number of trained first aiders. Although the depression of the 1920s reversed the trend, interest returned in the 1930s and 1935 saw the founding of the Association of Industrial Medical Officers with some 20 members. This organisation grew into the Society of Occupational Medicine with a current membership of almost 2000 doctors.

A new surge of growth in Occupational Health Services occurred in World War II. The large factories were required to have their own doctors. After the war medical services grew but slowly. Many larger industries developed comprehensive medical services with X-ray, laboratory and other facilities. Some smaller factories shared medical services with their neighbours in schemes set up by the Nuffield Foundation.

In 1978, the Royal College of Physicians of London established a Faculty of Occupational Medicine as an academic centre for the subject. The Faculty has established criteria for the training and examination of specialists in the field and has a membership of over 1700.

Meantime, occupational health nursing has developed as an important aspect of health at work. Many factories with no occupational health physician employ one or more occupational health nurses. The first such nurse was employed by Colemans of Norwich in 1877. The Royal College of Nursing has recently formed a Society of Occupational Health Nursing for members employed in industry and commerce and provides training courses for those engaged in this branch of nursing. The House of Lords produced a report on Occupational Health and Hygiene Services in 1984. The report recommended development of group services which would benefit the smaller companies and suggested a Government-financed fund administered by HSE to initiate such services.

16.3 The functions of an occupational health department

These fall into clinical and advisory categories.

Health assessments

(1) Pre-employment and other medical examinations, e.g. employees returning from sickness or those changing jobs.
(2) Examination of people exposed to specific occupational hazards.

(3) Treatment of conditions on behalf of the hospital or general practitioner. This may include physiotherapy or rehabilitation for which purposes a physiotherapist may be employed.
(4) Emergency treatment of illness or injury occurring at work.
(5) Immunological services, e.g. vaccination of overseas travellers, tetanus prevention, influenza prevention. Hospital workers require protection against hepatitis and tuberculosis.

Advisory services
(1) The study and prevention of occupational disease.
(2) Advice on problems of medical legislation and codes of practice.
(3) Advice on medical aspects of new processes and plant.
(4) The study of sickness absence.
(5) Advice on the reduction or prevention of common non-occupational diseases such as alcoholism and the effects of smoking.
(6) Training first-aiders.
(7) Advice to employees prior to retirement.
(8) The preparation of contingency plans for major disasters at the place of work.

Nurses have an important part to play in these activities and much of the clinical treatment of patients is in their hands. Nurses may be State Registered (S.R.N.) or Registered General Nurse (R.G.N.) with 3 years' training or State Enrolled (S.E.N.) with 2 years' practical training. Full-time and part-time training courses in occupational health nursing are run by the Royal College of Nursing at various centres. The R.G.N. may obtain a diploma in occupational health nursing after an examination. The S.E.N. may take part in one of the courses which will help her carry out her duties in this field of nursing. As most nurses in industry and commerce lack full-time medical advice the need for formal training in the subject is very clear.

16.4 Overseas developments

Not all EC countries have introduced legislation on occupational health. In France, for example, there is no law requiring treatment services but pre-employment medical examinations are mandatory.

Holland and Belgium require medical services in companies of over a specified size. In Germany, doctors trained in occupational health must be employed in factories as must safety advisers, and in a number of other European countries the major concerns have occupational health services.

Some countries use the factory as the site for a medical centre which provides clinical services for workers and their families as well as making available similar medical facilities to those provided by many factory medical departments in this country.

In the USA the National Institute for Occupational Safety and Health (N.I.O.S.H.) determines standards of occupational health and safety at work and organises training and research facilities. The Occupational Safety and Health Act 1970 applies to workers in industry, agriculture and construction sites and requires that employers must provide a place of work free from hazards likely to cause death or serious harm to employees[1].

16.5 Risks to health at work

The main hazards are of three kinds, physical, chemical and biological, although occupational psychological factors may also cause illness.

16.5.1 Physical hazards

Noise, vibration, light, heat, cold, ultraviolet and infrared rays, ionising radiations.

16.5.2 Chemical hazards

These are liable to occur as a result of exposure to any of a wide range of chemicals.

Ill-effects may arise at once or a considerable period of time may elapse before signs and symptoms of disease are noticed. By this time the effects are often permanent.

16.5.3 Biological hazards

These may occur in workers using bacteria, viruses or plants or in animal handlers and workers dealing with meat and other foods. Diseases produced range from infective hepatitis in hospital workers (virus infection) to ringworm in farm labourers (fungus infection).

16.5.4 Stress

This may be caused by work or may present problems in the time spent at work. Work related stresses may be due to difficulties in coping with the amount of work (quantitative stress) or the nature of the job (qualitative stress).

16.6 Occupational hygiene

In 1959, the American Industrial Hygiene Association defined occupational hygiene as 'that science and art devoted to the recognition, evaluation and control of the environmental factors or stresses arising in or from the workplace which may cause sickness, impaired health and well-being or significant discomfort and inefficiency among workers or among citizens of the community'[2].

The British Occupational Hygiene Society was founded in 1953 'to provide a forum in which specialist experience from many different but related fields could be exchanged and made available to the growing number of occupational hygienists at both national and international level, and to encourage discussion with other managerial and technical professions'. The Society holds frequent conferences and meetings.

The British Examining Board in Occupational Hygiene was set up by the British Occupational Hygiene Society in 1968 to examine candidates to well defined professional standards.

The first stage in the practice of good occupational hygiene is to recognise the potential or manifest hazard. This may result from an inspection of the process in question or may be suggested by symptoms and signs of disease in the operatives. Ideally the potential risk should be considered at the planning stage before plant is installed.

The next stage is to quantify the extent of the hazard. Measurements of physical and chemical factors and their duration must be related to levels of acceptability and the likelihood of injury or disease arising if the hazard is allowed to continue. These measurements often involve the use of sophisticated measuring devices which must be calibrated and used very carefully in order to produce meaningful results.

For smaller firms or small isolated units in larger organisations, the person who carries out a limited range of tests may benefit from attendance at a short course leading to a Preliminary Certificate in Occupational Hygiene at a College of Further Education. Full-time specialists in Occupational Hygiene will require a professional qualification in the subject. The introduction in 1989 of the Control of Substances Hazardous to Health Regulations 1988 has increased the role of occupational hygienists, both full-time and part-time.

Having assessed the dangers of the process, the final stage is to decide how best to control the hazard. This may require some radical modification of plant design, special monitoring devices which will warn of increasing danger or the need for protective devices to be used by plant operators.

In deciding on appropriate ways of dealing with such problems, the occupational hygienist will often require the co-operation and understanding of the occupational physician, nurse, safety adviser, personnel officer and line management in order to achieve his aims.

The degree of involvement and co-operation of advisers in the medical, nursing, engineering and safety fields will vary from one problem to another. Failure to achieve adequate health and safety measures may be due to lack of understanding or co-operation between advisers in the various disciplines or failure to influence line management.

The appointment of safety representatives under the Health and Safety at Work Act 1974 has focussed attention on the part played by all who are involved in occupational health, safety and hygiene. Advice on prevention and safety measures is less likely to be ignored but more likely to be challenged than in the past. It is vital that the adviser's opinions can withstand challenge and are seen to be fair and unbiased.

16.7 First aid

The first aid requirements to be met by employers were revised in 1981 and were consolidated in the Health and Safety (First Aid) Regulations 1981[3]. These Regulations apply to all employments, except those prescribed in

Regulation 7, and have endeavoured to match the first aid facilities with the needs of particular employments.

First aid is defined as treatment to preserve life and minimise the consequence of injury until help from a medical practitioner or nurse arrives. It also includes the treatment of minor injuries which may otherwise receive no treatment or would not require treatment by a medical practitioner. The Regulation is supported by an approved Code of Practice and a Guidance Note which lay down in broad terms the first aid facilities to be provided according to the number of people employed, covering the number of first aiders, the provision of first aid rooms and the content of first aid boxes which are related to the size of the working population served. Recommended contents are shown in *Table 16.1*.

Table 16.1. Contents of first aid boxes

Item	*Number of employees*				
	1–5	*6–10*	*11–50*	*51–100*	*101–150*
Guidance card	1	1	1	1	1
Individually wrapped sterile adhesive dressings	10	20	40	40	40
Steril eye pads, with attachment	1	2	4	6	8
Triangular bandages	1	2	4	6	8
Sterile coverings for serious wounds (where applicable)	1	2	4	6	8
Safety pins	6	6	12	12	12
Medium sized sterile unmedicated dressings	3	6	8	10	12
Large sterile unmedicated dressings	1	2	4	6	10
Extra large sterile unmedicated dressings	1	2	4	6	8

The Regulations, Code of Practice and Guidance Notes are contained in an HSE publication[4,5].

Those who can give first aid treatment at work have to be 'suitable persons' and have been trained by organisations or employers approved by the HSE. Where there is no trained 'suitable person', employers are required to appoint a person to take charge in the event of a serious injury or major illness.

16.8 Basic human anatomy and physiology

Anatomy is the study of the structure of the body. Physiology is concerned with its function. Although the various organs which make up the body can be studied individually it is important to remember that these organs do not function independently but are interrelated so that if one part of the body is not functioning properly it may upset the health of the body as a whole.

An organ like the stomach or the brain contains structures within it such as arteries, nerves and other specialised components. These components, which contain cells of a similar kind, are referred to as tissues. So we have nervous tissue, arterial tissue, muscular tissue and so on making up specialised organs which have a specific function. (The stomach, for example, is concerned with the first stage in the digestive process.)

The cells which make up the tissues and organs are so small that they are invisible to the naked eye. Under the microscope a cell consists of a mass of jelly-like material called protoplasm held together by a surrounding membrane. The shape and function of the cell vary according to the tissue of which it is composed and depend upon the job which the cell is required to do. For example, the nerve cell has long fibres capable of conducting electrical impulses while some cells in the stomach wall produce hydrochloric acid. Cells in the thyroid gland produce a chemical which influences other body cells. With these varying roles it is not surprising that cells differ from one another in appearance.

Although the human body is composed of many million cells, the work of each one is controlled so as to serve the body as a whole. If this co-ordination is lost, some cells can grow rapidly relative to others and the result may be disastrous. This sort of cell behaviour occurs in cancer when a group of cells may grow rapidly invading adjacent tissues.

Each cell is a sort of miniature chemical factory. It takes in food and converts it into energy to perform work. The sort of work carried out depends on the type of cell, e.g. locomotion (muscle cells), oxygen transport (trachea, lungs, blood vessels, red blood cells). Energy is also needed to repair the wear and tear of body cells. We refer to the chemical processes which convert food into energy as metabolism.

16.8.1 Foodstuffs

The energy needed to perform work and to maintain body temperature is provided by oxygen and various foods. Any diet which maintains life must contain six basic ingredients in a digestible form. These are as follows:

(1) Proteins.
(2) Carbohydrates.
(3) Fats.
(4) Salts.
(5) Water.
(6) Vitamins.

Proteins are composed of large complicated molecules which contain atoms of carbon, hydrogen, oxygen, nitrogen and often sulphur. They are made up of simpler substances called amino-acids which form the basic structures of body cells. Foods such as meat, milk, beans and peas contain protein. This is broken down in the digestive process into its constituent amino-acids which are then re-aligned in a new pattern to make human proteins.

As the name suggests, *carbohydrates* are composed of carbon, hydrogen and oxygen. Sugars and starches are common examples of carbohydrate. *Fats* are used as reserve foodstuffs and insulate the body thus protecting it against cold. Carbohydrates and fats are a ready source of heat energy.

Various *salts* including those of sodium, iron and phosphorus are obtained from food such as milk and green vegetables. They are needed for the formation of bone, blood and other tissues.

Water is a vital constituent of all cells and a regular intake is essential to maintain life. Small quantities of a range of chemicals known as *vitamins* are also needed for healthy existence. The absence of a vitamin leads to a deficiency disease such as scurvy which occurs when vitamin C is absent from the diet. It is available in fresh vegetables, oranges and lemons. Vitamin D is formed by the action of ultraviolet light on a chemical in the skin (7-dehydrocholesterol), and is present in milk and cod liver oil. Absence of this vitamin may lead to rickets.

As well as containing the substances listed above an adequate diet must provide enough calories to satisfy metabolic requirements. This will depend on the physical demands of the person's work and hobbies as well as on his stature.

16.8.2 Digestion

When food is taken into the body much of it is in a form which cannot be used directly by the tissues as its chemical structure is too complicated. The larger molecules of food therefore need to be broken down into simpler molecules. This process takes place by chemical action and occurs in the digestive tract (*Figure 16.1*) which is a long tube of varying dimensions which starts at the mouth from where the food passes to the gullet, the stomach, the small intestine and finally the large intestine.

Alcohol is absorbed in the stomach and water is absorbed in the large intestine but the majority of energy-containing foods are absorbed in the small intestine. The products of digestion pass through the walls of the digestive tract into blood vessels and thence to the liver. This is a very large organ situated in the upper right side of the belly cavity below the diaphragm. It is made up of many units of cells arranged around blood vessels in a circular fashion. Sugar from the digestive tract is taken to the liver where it is changed into a chemical called glycogen which is stored in the liver cells. Proteins are broken down in the liver and form urea as a waste product which is then excreted via the kidneys and is the main chemical constituent of urine.

Old red blood corpuscles are removed from circulation by the liver which retains iron from them for later use. The liver is also responsible for the manufacture of bile which assists digestion and which is stored in the gall bladder adjacent to the liver.

Many poisons are dealt with by the liver which attempts to render the poison less toxic (detoxification) before it is excreted. Sometimes the poison damages or destroys liver cells but fortunately the liver has such a large reserve of cells that a great deal of damage is necessary to affect its function adversely. Liver damage may result from certain types of industrial poisons as well as from excessive consumption of alcohol.

16.8.3 Excretion

Just as a motor car needs to get rid of exhaust gases so the human body has to dispose of waste materials left over from metabolic chemical reactions. Special organs are involved in the process of excretion. Water and urea are

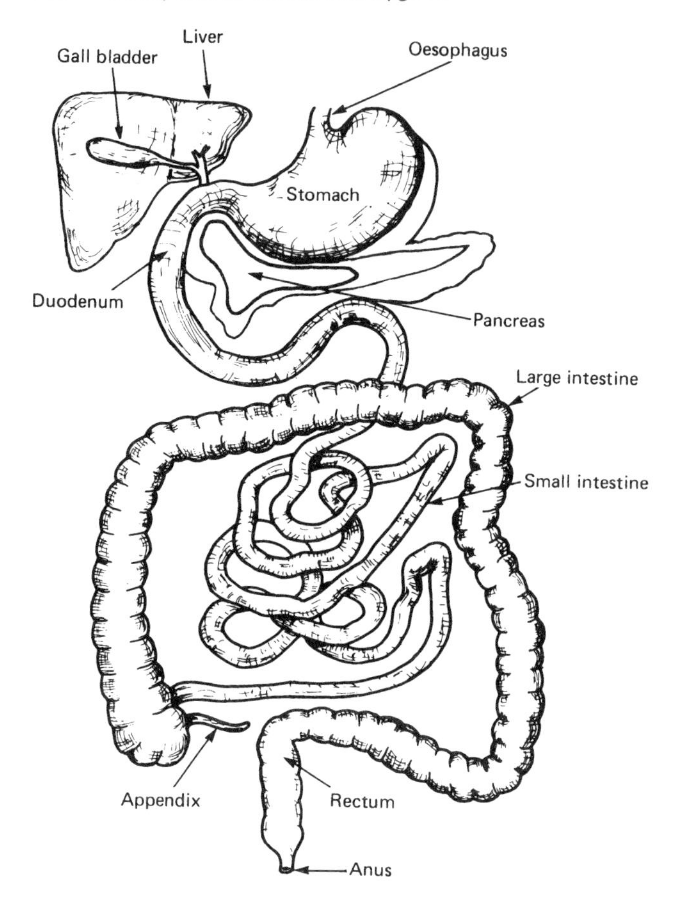

Figure 16.1 Diagram of digestive system

disposed of by the kidneys, although some water is lost via the skin. Solid waste leaves the body through the bowel after water has been extracted in the large intestine.

The voiding mechanisms are of paramount importance when the body is affected by poisonous materials and may be by vomiting, diarrhoea or by being excreted in the urine. Sometimes the passage of a poison through the body may leave a trail of destruction in its wake and result in permanent liver or kidney damage.

16.8.4 The respiratory system

The foodstuffs absorbed from the digestive tract are converted into energy. In order to produce this energy the body cells require oxygen just as a motor car engine needs oxygen in order to function. This converting process generates carbon dioxide which, in large quantities, is poisonous and must be got rid of.

During the process of respiration oxygen is transferred from the air to the body cells and carbon dioxide is disposed of in exhaled air. Because body cells function at different rates their oxygen requirements vary from one tissue to another. If brain cells are starved of oxygen for more than four minutes there is little prospect of recovery of intellectual function. Other body cells can do without oxygen for longer periods of time.

A great deal of oxygen is needed to provide the requirements of all the body cells. The apparatus which fulfills this requirement is the respiratory tract (*Figure 16.2*) which is made up of the nose and mouth, the throat, the larynx (voice box) and trachea (windpipe), the bronchi and the lungs. The respiratory tract is lined by a wet shiny membrane which contains mucous secreting cells that keep the walls moist. Other cells are fringed with little hairs or cilia which by moving in one direction can evict towards the mouth particles of dust which have entered the airways. Sometimes the quantity of material which has entered the airway is greater than the cilia can cope with. The lungs then eject collections of particles by the mechanism of

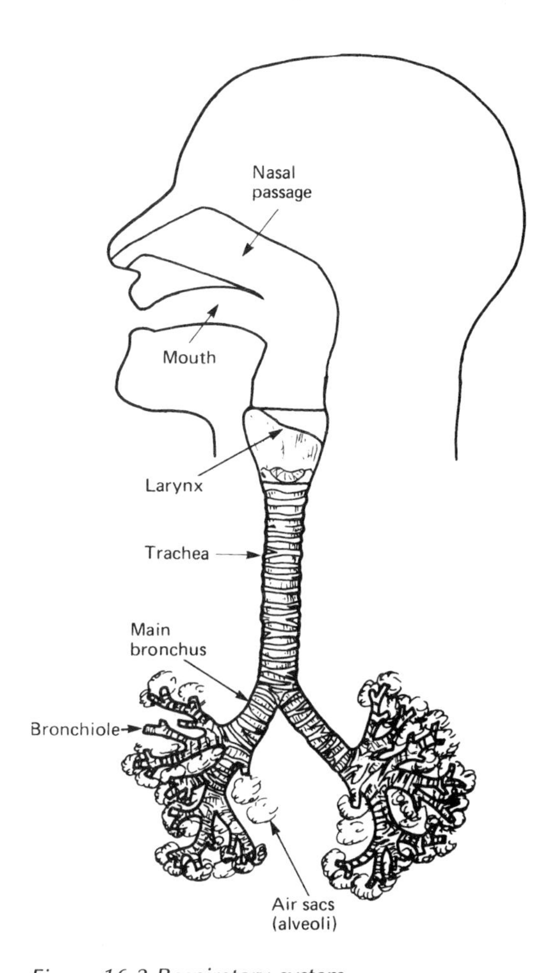

Figure 16.2 Respiratory system

coughing. Primary filtration of air entering the respiratory tract occurs in the nose but many of the smaller particles enter the air passages where some hit the walls of the bronchi and are rejected by the cilia but others go on to reach the lungs.

The respiratory tract is rather like an inverted tree. The windpipe is the trunk and the two main bronchi (one for each lung) the main branches. Each main bronchus then sub-divides into smaller branches (bronchioles). All these tubes are held open by rings of cartilage which prevent their collapse when subject to suction when breathing in. The smallest tubes end in air sacs or alveoli which have very thin walls. These walls allow oxygen to pass into the small blood vessels with which they are surrounded and carbon dioxide to pass in the opposite direction from the blood into the air sac.

The blood vessels lining the air sacs carry the oxygen-containing blood to the left side of the heart from whence it is pumped via the arteries to all parts of the body. When the tissues have used the oxygen to carry out the metabolic processes described earlier carbon dioxide is produced which like oxygen is dissolved in the blood. The carbon dioxide is carried back to the lungs via the veins and the right side of the heart.

The lungs are surrounded by a tough layer of smooth membrane called the pleura which if it becomes inflamed gives rise to a condition known as pleurisy. Blue asbestos fibres can irritate the pleura to produce a type of cancer called a mesothelioma.

The lungs are located in the chest cavity in a space limited by the ribs, breastbone, backbone and diaphragm (*Figure 16.3*). The latter is a dome-shaped sheet of muscle which separates the chest and belly cavities. When the diaphragm moves downward the dome shape is flattened and at the same time the ribs move upwards, thus increasing the volume of the chest

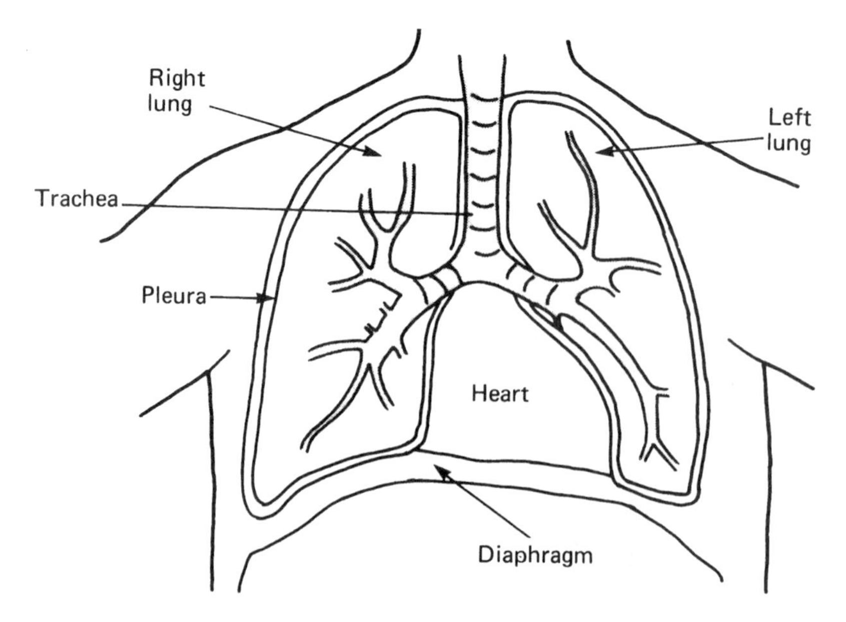

Figure 16.3 Diagram showing lungs and heart within chest

cavity creating a suction which draws air into the lungs and then into the air sacs (about 20% of air is oxygen). This movement is called inhalation. When the diaphragm expands the chest cavity contracts and the elastic recoil of the lungs forces air in the opposite direction (exhalation).

If respiratory movement ceases (e.g. due to electric shock damaging the breathing mechanism) artificial respiration is needed at once. This may be carried out by breathing into the casualty's mouth (mouth to mouth resuscitation) or by exerting pressure on the casualty's chest, thereby forcing air out of the lungs and allowing the recoil of the chest wall to draw air into the lungs.

Divers and caisson workers may suffer from pain in the joints if they return to the surface too quickly after working in deep water or under elevated air pressure. In these conditions, air contained in the body tissues, which was dissolved under high pressure, is released to form air bubbles (mostly nitrogen gas) in the joints and elsewhere to produce unpleasant symptoms. 'The bends' is the name of the illness resulting. Mild cases affect the elbows, shoulders, ankles and knees. As the illness develops, the pain increases in intensity and the affected joint becomes swollen. Serious cases of the bends may affect the brain and/or the spinal cord. In cases of brain damage the patient may suffer visual problems, headaches, loss of balance and speech disturbances. Spinal cord damage may cause paralysis of the limbs, loss of sensation, pins and needles and pain in the shoulders and/or hips. The problem is obviated by reducing the rate of change of pressure to which the workers are subjected to a level at which the bubbles of gas do not form.

16.8.5 The circulatory system

This consists of the heart, the arteries, the veins and the smaller blood vessels which permeate all tissues of the body (*Figure 16.4*). The heart is a muscular pump divided into left and right sides. The left side is larger and stronger than the right since it has the bigger job to do in pumping to all parts of the body via the arteries oxygen-containing blood which it has received from the lungs. The blood then takes carbon dioxide from the tissues and carries it back to the lungs through the veins via the right side of the heart.

Each side of the heart (*Figure 16.5*) has two chambers, an auricle or intake chamber and a ventricle or delivery chamber which are separated by valves which ensure that the blood travels in one direction only from auricle to ventricle. A further set of valves ensures that the blood being ejected cannot flow back into the ventricles each time the heart contracts.

The muscles of the heart are less dependent on the brain than the muscles which cause body movement. Cutting the nerves to a muscle in the leg, for example, will stop that muscle working. Cutting the nerves to the heart will not stop the heart since it is not entirely controlled by nervous impulses coming from outside the organ as is the case with voluntary muscle. The heart has a dual nerve supply. One nerve supply increases the heart rate, the other reduces it. The rate of the heart beat, which in the healthy adult at rest is about 70 beats per minute, may be slowed by stimulating the vagus nerve and increased by stimulating the sympathetic nerves. The latter are activated when people are afraid so that more blood is sent to the muscles and the person is keyed up for action.

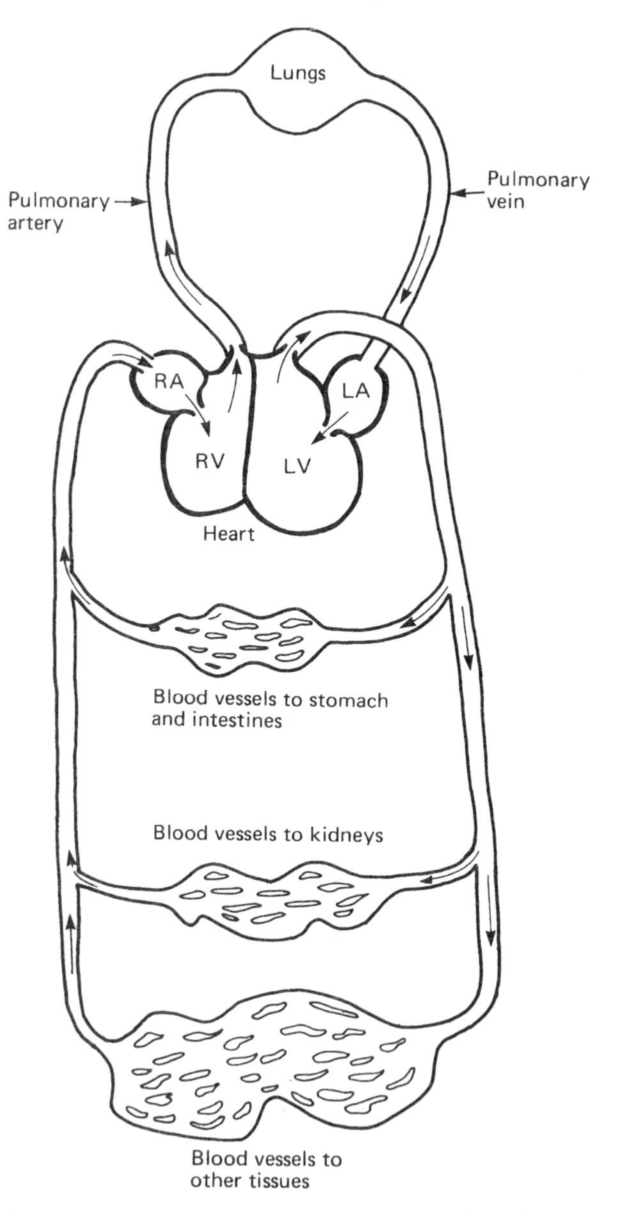

Figure 16.4 Diagram showing circulation of blood (RA – right auricle; LA – left auricle. RV – right ventricle; LV – left ventricle.)

Electric shock may affect the heart rate either by stimulating the nerves to the heart or by interfering with the conduction of electricity to the heart muscle. This may stop the heart or change the rhythm thus affecting its efficiency as a pump. Fibrillation of the heart muscle may occur as a result.

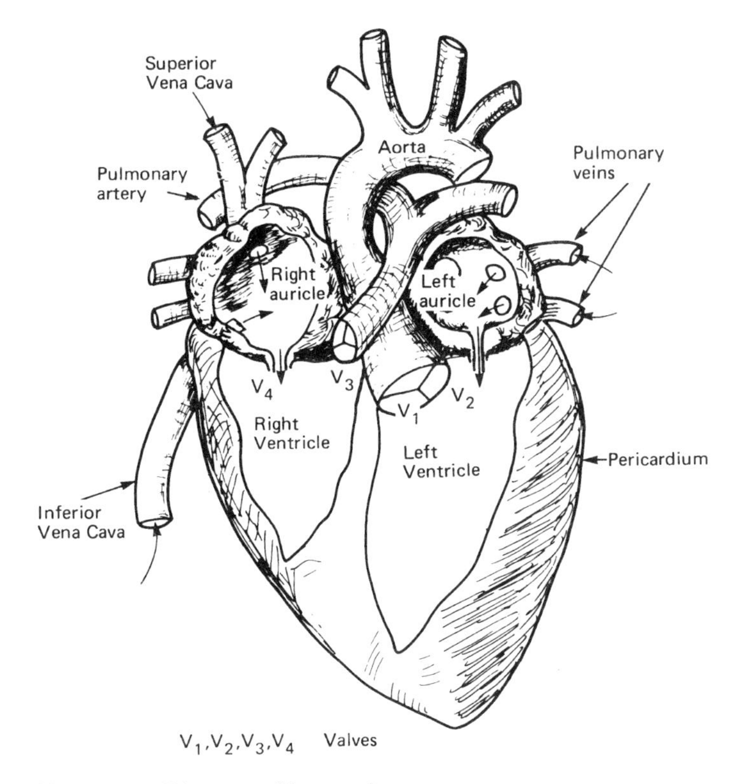

Figure 16.5 Diagram of human heart

In addition to its oxygen-carrying capacity, the blood also carries nutrients to the tissues and removes waste products. It enables heat generated to be dissipated and its white cells defend the body when attacked by viruses and bacteria. The blood accounts for about one thirteenth of body weight and in the average adult amounts to about 5 litres (12 pints).

It is composed of a straw-coloured fluid (plasma) and cells of two different colours – red and white. The red blood cells which account for the colour of blood are made up of minute circular discs which contain red pigment (haemoglobin) with which the oxygen and carbon dioxide transported in the blood combine temporarily. Haemoglobin also combines very readily with carbon dioxide.

The white blood cells are somewhat larger than the red blood cells but fewer in number. There is only one white cell for every 500 red cells. Several kinds of white cell exist which are mobilised when the body is infected by germs and viruses and attempt to destroy them.

16.8.6 Muscles

One of the characteristics of all animals is movement. In man, movement is brought about by the contraction of muscles. There are three varieties of human muscle (*Figure 16.6*). Voluntary muscles which can be moved at will are made up of many fibres which are covered by transverse stripes when seen under the microscope. The muscle is often attached to bones (hence the term skeletal muscle) by connective tissue fibres which collectively form a tendon. When the muscle contracts, the bones to which it is attached are drawn together producing movement of one part of the body relative to another part.

Involuntary muscle does not have the characteristic striped appearance of voluntary muscle. It exists in the digestive tract, the walls of the blood vessels and in the respiratory and genito-urinary apparatus. Involuntary muscle is controlled automatically by the autonomic nervous system.

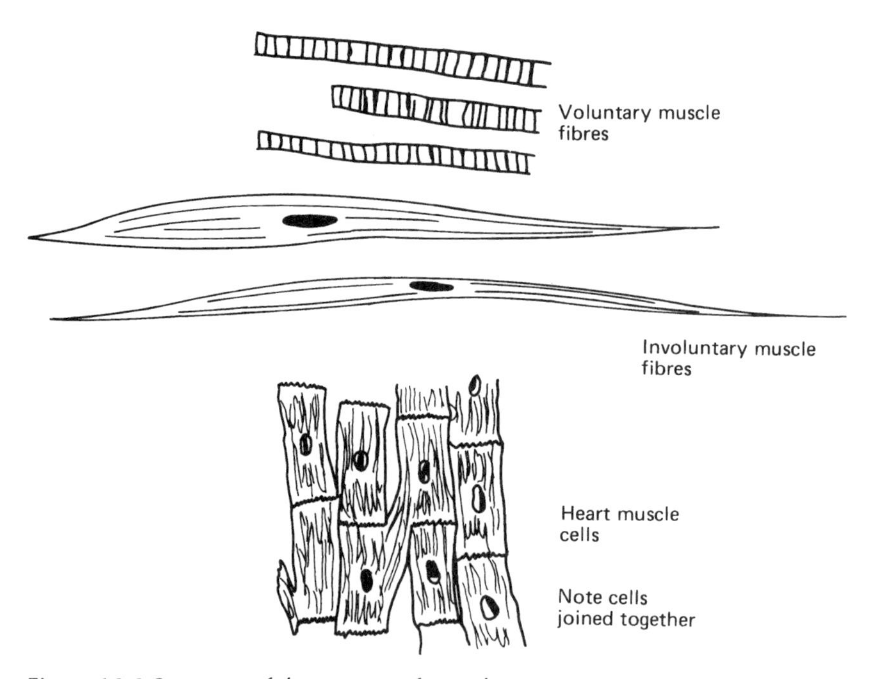

Figure 16.6 Structure of three types of muscle

The third type of muscle is that found in the heart. This shows some striated fibres under the microscope but is not under voluntary control and is not entirely dependent on its nerve supply. A 'pacemaker' within the heart muscles produces, at a rate appropriate to the body's need for oxygen, electrical impulses which cause the heart muscles to contract thus producing heart beats.

16.8.7 Central nervous system

This is made up of the brain and spinal cord. The brain is a highly developed mass of nerve cells at the upper end of the spinal cord. The largest part of the brain is taken up by the two cerebral hemispheres. These receive sensory messages from various parts of the body and originate the nerve impulses which produce voluntary movements. The layers of grey tissue (known as the cerebral cortex) overlying the cerebral hemispheres are covered in folds. This tissue is concerned with the intellectual function of the individual. The various parts of the cortex of the brain are associated with specific activities. For example, there are centres concerned with speech, hearing, vision, skin sensation and muscle movement.

The cerebellum is the part of the brain concerned with balance and complicated movements. Part of the base of the brain is involved with emotional behaviour (the hypothalamus).

The portion of the brain nearest the spinal cord contains centres which control the rate of respiration and heart beat.

16.8.8 The special senses

These specialised organs measure environmental factors such as light and noise and facilitate communication with other human beings.

The nerve cells within these organs pass their messages to the brain which then interprets them and determines appropriate action.

16.8.9 The eye

The eyeball (*Figure 16.7*) is a globe 25 mm (1 inch) in diameter which is made up of a transparent medium (the vitreous) through which light is focused by a lens onto a light sensitive layer (the retina). The front of the globe (the cornea) is transparent. After light rays have entered the eye they pass through a fluid (the aqueous humour) and the lens which changes in

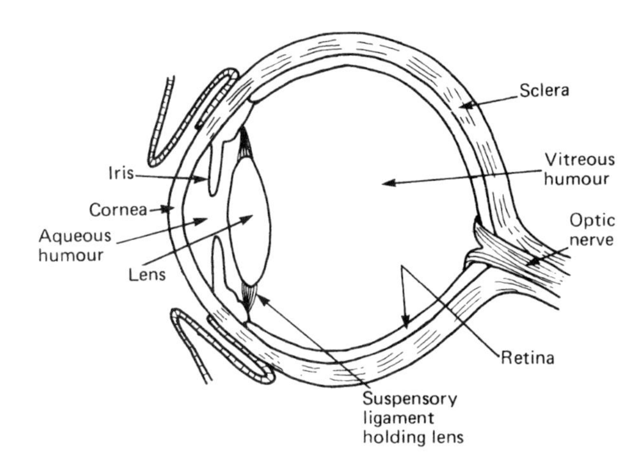

Figure 16.7 Diagram of section through eyeball

shape in order to focus on to the retina light rays from objects at varying distances.

If the path of light rays is interrupted, e.g. by a foreign body or by an opacity of the lens (cataract), the vision may be distorted and/or diminished.

Light produces changes in the cells of the retina which transmit electrical impulses to the visual cortex at the back of the brain. The movements of the eyeball are controlled by six muscles in each eye which are carefully synchronised with those of the other eye. Imbalance of these muscles may give rise to double vision or a squint. Temporary muscle imbalance may result from exposure to toxic materials or alcohol.

Burns of the eye may result from chemical splashes and exposure to ultraviolet radiation as can occur with electric welding. It is vital that chemical burns be treated at once by irrigation with copious quantities of running water.

16.8.10 The ear

This organ is concerned with hearing and with orientating our position in space. The portion concerned with hearing consists of three parts (*Figure 16.8*). Sound travels through the ear canal to the ear-drum which is a

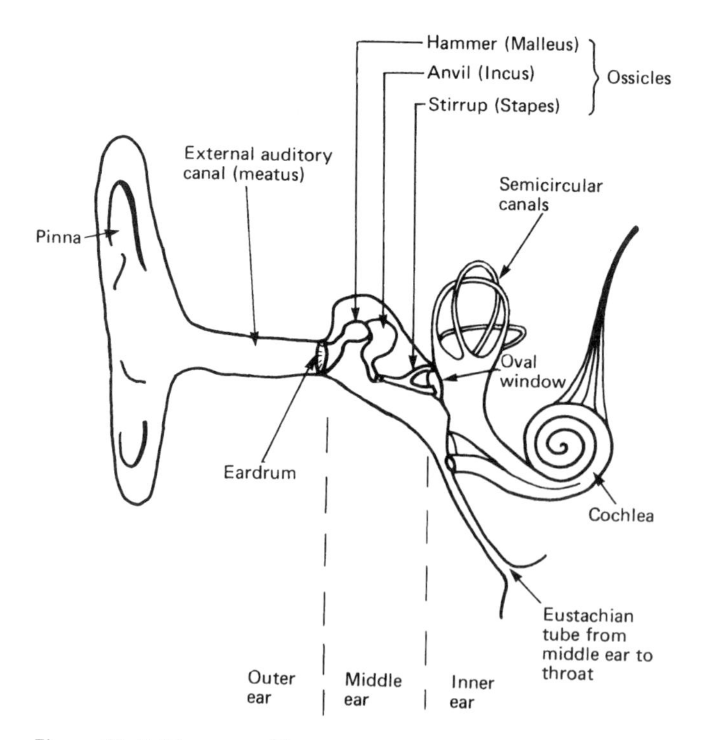

Figure 16.8 Diagram of human ear

membrane stretched across the canal and separating it from the middle ear. Vibrations of the ear-drum are produced by the sound waves passing along the ear canal.

These vibrations are transmitted through the middle ear by three tiny bones known as the 'ossicles', being the hammer (malleus), anvil (incus) and stirrup (stapes). The hammer bone is fixed to the ear-drum and the stirrup to another membrane (the oval window) which separates the middle and inner parts of the ear. The section of the inner ear which receives sound waves is shaped like a snail's shell (the cochlea) and contains strands of tissue under varying tensions. These strands or hairs vibrate in response to sound waves of particular frequency which have entered the inner ear from the bones of the middle ear and produce nerve impulses in the auditory nerve which are then transmitted to the cortex of the brain. It is at this point that the signals are received as sound of a certain pitch, intensity and quality.

Various factors may interfere with the transmission of sound impulses. Normally, the pressure on either side of the ear-drum is equal but when a difference occurs, as with airline passengers who fly when suffering from a cold, temporary hearing impairment can be experienced. Infection of the middle ear may occur and this may result in thickening and scarring of the ear-drum. Some unfortunate people suffer an inherited form of deafness in which the ossicles develop damage and are unable to transmit sound.

The inner ear is a very sensitive part of the hearing mechanism and may be damaged by prolonged loud noise. Usually, the frequencies around 4000 Hertz (cycles/second) are first affected but the damage can extend to other frequencies as well as becoming more pronounced. In addition to this, deafness is associated with the ageing process and is more noticeable in males (presbyacusis).

A balance mechanism is also situated in the inner ear. This is composed of three semicircular canals which are at right angles to each other. Inside each canal are specialised nerve endings.

Moving the body into an unbalanced position stimulates the nerve endings in one of the canals in each ear and results in an urge to return the body to a normal balanced posture.

16.8.11 Smell and taste

The lining of the inside of the nose contains special cells which are capable of detecting chemicals in the air. Nerve fibres pass from these cells into the skull and connect with the brain.

The sense of smell may be an important safety factor. A cold may diminish or remove the facility. Hydrogen sulphide gas smells of bad eggs. Continuing exposure to increasing concentrations of this gas saturates the nerves concerned with smell so that the person exposed to this substance may be unable to smell it even if the concentration increases further.

The sense of taste originates when chemical stimulation of the taste buds occurs. These are collections of cells concentrated in certain areas of the tongue. The sides, tip and rear third of the organ have the most taste buds. The back of the tongue most readily detects bitterness and the tip sweetness.

16.8.12 Hormones

These are chemicals which act as messengers provoking action in some distant part of the body. They are produced by various hormone or endocrine glands. For example, the thyroid gland is a gland situated in the front of the neck which produces the chemical thyroxine. Too much thyroxine produces a rapid pulse and an over-active jumpy person. Too little thyroxine may result in a slowing of the pulse and too slow a rate of metabolism with the face becoming swollen and the skin dry and aged; the hair becomes coarse and falls out (myxoedema).

The suprarenal glands are two small glands situated above the kidneys. They produce a number of hormones including adrenaline and cortisone. Adrenaline is released in conditions causing fear or anger and makes the muscles in the artery walls contract. This increases the blood pressure, and consequently the supply of oxygen to the muscles, so that an animal is ready to meet a confrontation by either 'fight or flight'. This is not always an appropriate reaction for human beings in present day stressful situations where they cannot fight or run away.

Cortisone is released from the adrenal cortex at times of stress. It delays physical fatigue by increasing the ability of muscles to contract and has a euphoric effect on the brain which may give added confidence at a stressful time.

Stress is an engineering term describing the force applied to an object and the resulting deformity is referred to as strain. It has become customary to refer to the result of applying such force or pressure on human beings as 'stress'.

The effects of long-term stresses on the human body and mind are not clearly understood. Certain so called 'stress diseases' such as asthma, duodenal ulcer and coronary heart disease may be aggravated at times of stress but a direct cause/effect relationship is difficult to prove.

Nevertheless, people in stressful situations, whether caused by work or by non-occupational factors, may be more liable to accidents.

Another hormone-producing organ, the pancreas, has two important functions. Its secretions flow into the digestive tract where the gland's products are involved in the digestion of carbohydrates. A different secretion which passes straight into the blood stream is a chemical called insulin. Without insulin the body is unable to use available carbohydrates as a source of energy and has to obtain energy from the breakdown of body tissues. The person whose pancreas is unable to make enough insulin for his needs is diabetic. Diabetes is a condition which can be treated by replacing the insulin deficiency and by careful dietary control.

The sex glands also produce hormone secretions which determine the growth of beard hair and the deep voice of the male and breast development in the female. Other endocrine glands include the parathyroid glands which are concerned with calcium metabolism and the pituitary gland which controls the other hormone glands. The pituitary gland situated at the base of the brain has two lobes: the front one controls growth in children while the rear lobe secretions cause contraction of the muscles of the womb and increase the output of urine. The part it plays in regulating the other endocrine glands has been referred to as 'direction of the endocrine orchestra'.

16.8.13 The skin

This is the largest organ in the body (*Figure 16.9*) and performs a variety of functions. Its most obvious purpose is a protective one. The superficial layers of cells keep out chemicals and germs as well as acting as a physical barrier. If the physical pressures on certain cells of skin are considerable, the tissues may be much thickened, for example on the soles of the feet.

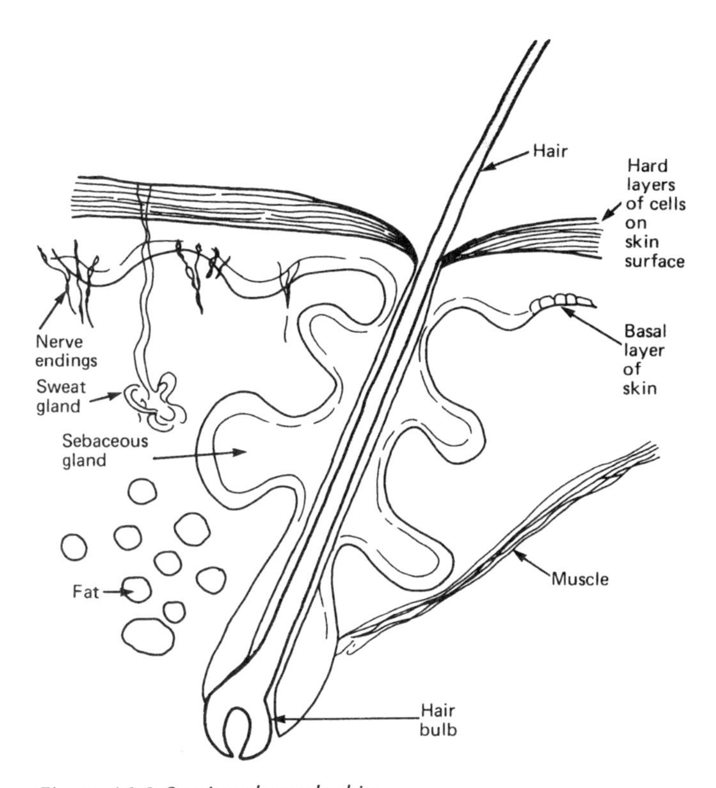

Figure 16.9 Section through skin

Four different kinds of sensation may be appreciated through the skin, namely heat, cold, touch and pain.

The skin is not uniformly sensitive to all these stimuli, some parts being more sensitive than others; for example, the tips of the fingers are more sensitive than the back of the hands. Apart from our awareness of touch and temperature on the surface of the skin, we are also aware of the position of our arms and legs even when our eyes are closed. This is because special nerve endings from the muscles return nerve impulses to the brain which is provided with information on the relative positions of various parts of the body. Water and various salts are lost through the skin as sweat. This may be important to people working in very hot environments such as in deep mines who may sweat profusely and as a result of the loss of salt suffer severe cramps in the muscles.

The human being is able to keep his body temperature fairly constant whatever the range of ambient temperature. This is because heat loss from the skin may be increased by sweating, followed by evaporation of the sweat and increase in the size of the blood vessels in the skin which encourages more heat to reach the skin surface.

In a cold environment the blood vessels in the skin contract and the muscles under the skin produce the phenomenon of shivering in an attempt to generate heat.

The skin also secretes sebum which is a waxy material produced by special glands. This may help protect the skin surface from attack by foreign substances. Exposure to various chemicals including solvents may remove such secretions and predispose the skin to attack by germs.

16.9 Cancer and other problems of cell growth

Cells reproduce in order to replace other cells which are continually being lost, e.g. on the surface of the skin by wear and tear or by damage such as that caused in a wound.

Sometimes, cells do not develop as nature intended and a variety of abnormal cells may be formed, some of which may endanger life. Some examples of maldevelopment are given below:

Aplasia The tissues may fail to develop. This happened in the unfortunate children whose mothers had taken the drug thalidomide during their pregnancies. In some of these babies, the limbs were only partly developed (hypoplasia).

Hyperplasia The organ or tissue has an increased number of cells. When the cells increase in size this is termed hypertrophy. The latter condition may occur in the muscles of a leg when the other has been amputated and the remaining limb has to work harder.

Metaplasia This process involves changes in the types of cell present in a tissue often from a complicated cell to a more simple one. Long lasting irritation from chemicals may bring about this change which is often seen in the lining of the air passages of smokers.

Mutagenesis This occurs when any abnormal changes which may have arisen in a group of cells are passed on to the next generation as the original genetic material within the cell has been damaged.

Teratogenesis This refers to congenital abnormalities which occur in the new-born as a result of damage to the genetic material or to the developing embryo.

Neoplasia This is the process in which there is a mass of new tissue (a tumour) which is made up of abnormal cells which cannot stop growing. Malignant tumours are more likely to occur in tissues which are frequently producing new cells.

The chemical DNA which carries the coding mechanism which determines the structure and function of new cells is liable to be damaged by some chemicals. New malformed cells may form and grow in an uncontrolled manner producing more damaged cells of the same type. Malignant tumours consist of undifferentiated cells which grow into surrounding tissues. Benign tumours are slower to develop and their cells look more like normal cells.

Cancer cells may spread and 'seed' themselves elsewhere in the body. From these new sites further secondary cancerous growths may develop, hastening the destructive process which may well end in the victim's death. Neoplasms may spread via the blood stream or along lymph vessels.

References

1. U.S. Public Law 91–596. The Occupational Safety and Health Act 1970
2. Gardner, A. Ward (Ed.), *Current Approaches to Occupational Medicine*, 205, John Wright & Sons, Bristol (1979)
3. Health and Safety Executive, *The Health and Safety (First Aid) Regulations 1981* (SI 1981 No. 917), HMSO, London (1981)
4. Health and Safety Executive, *First Aid at Work*, Health and Safety Series Booklet No. HS(R)11, HSE Books, Sudbury (1981)
5. Health and Safety Commission, *Approved Code of Practice: First Aid at Work, Health and Safety (First Aid) Regulations 1981 and Guidance*, No. COP 42, HSE Books, Sudbury (1990)

Further reading

International Labour Office, *Encyclopaedia of Occupational Health and Safety*, ILO, Geneva (1983)

Ffrench, G., *Occupational Health*, Medical and Technical Publishing Company, Lancaster (1973)

Gauvain, Suzette (Ed.), *Occupational Health: A Guide to Sources of Information*, Heinemann, London (1974)

Gardner, A. W. (Ed.), *Current Approaches to Occupational Medicine*, John Wright & Sons, Bristol (1979)

Gardner, A. W. (Ed)., *Current Approaches to Occupational Health – 2*, John Wright & Sons, Bristol (1982)

Gardner, A. W. and Taylor, P., *Health at Work*, Associated Business Programmes, London (1975)

Hunter, D., *The Diseases of Occupations*, 6th edn, English Universities Press, London (1978)

Kinnersley, P., *The Hazards of Work: How to Fight Them*, Pluto Press, London (1974)

Muir, G. D. (Ed.), *Hazards in the Chemical Laboratory*, 4th edn, Royal Institute of Chemistry, London (1986)

Sax, N. I. (Ed.), *Dangerous Properties of Industrial Materials*, 6th edn, Reinhold Book Corporation, New York (1984)

Schilling, R. S. F. (Ed.), *Occupational Health Practice*, 2nd edn, Butterworths, London (1981)

Ashton, I. and Gill, F. S., *Monitoring for Health Hazards at Work*, 2nd edn, Blackwell Scientific Publications, Oxford (1991)

Croner's Handbook of Occupational Hygiene, loose leaf publication with regular up-dates. Croner Publications Ltd, Kingston-upon-Thames

Harrington, J. M. and Gill, F. S., *Occupational Health Pocket Consultant*, Blackwell Scientific Publications, Oxford (1992)

Health and Safety Executive, *Guidance notes: Environmental Series*, (an on-going series of guidance notes on environmental hygiene), HSE Books, Sudbury

Health and Safety Executive, *Methods for the determination of hazardous substances*, (an on-going series of notes dealing with the measurement and determination of concentrations of hazardous substances in the work place environment), HSE Books, Sudbury

Howard, J. K. and Tyrer, F. H., *Textbook of Occupational Medicine*, Churchill Livingstone, Edinburgh (1987)

Raffles, P. A. B., Adams, P., Baxter, P. S. and Lee, W. R. (Eds), *Hunter's Diseases of Occupation*, 8th edn, Hodder & Stoughton, London (1992)

Chapter 17

Occupational diseases

Dr A. R. L. Clark

17.1 Introduction

Large companies may employ a number of specialists in the field of health, hygiene and safety who co-operate as a team and pool their particular expertise, but in small firms the safety adviser is often the only local source of advice on these matters, other expertise being brought in when considered necessary.

The task of safeguarding the health of persons at work is a formidable one especially during periods of rapid technological and organisational change. Thousands of chemical substances are used by industry and in commerce but only about 800 of those in common use have been recognised as presenting a risk to the health of workpeople.

This chapter provides a brief introduction to some of the more important diseases and materials that cause them. Conveniently, these fall into four major areas covering illnesses and diseases due to:

chemical agents	(sections 17.2–17.10)
physical agents	(sections 17.11–17.15)
biological agents	(section 17.16)
psycho-social causes	(section 17.17).

In addition, section 17.18 deals with target organs.

17.2 Toxicology

The toxicity of a substance is its potential to cause harm by reaction with body tissues. Measures of toxicity include lethal dose (LD 50) and lethal concentration (LC 50). The LD 50 is the single dose of a substance which when administered to a batch of animals under test kills 50% of them. It is measured in terms of mg of the substance per kg of body weight. Typical degrees of LD 50 toxicity[1] are:

extremely toxic	1 mg or less
highly toxic	1–50 mg
moderately toxic	50–500 mg

slightly toxic	0.5–5 g
practically non-toxic	5–15 g
relatively harmless	15 g or more

In a similar way, the LC 50 refers to an inhaled substance and is the concentration which kills 50% in a stated time. The Ames test is a useful test for carcinogenic substances which measures the ability of a substance to damage genetic material (DNA) in special strains of bacteria.

The toxic response to a given substance depends on the dose, and this may be demonstrated in an animal's tissues, physiology, biochemistry and behaviour. However, with all animal testing there is difficulty in extrapolating these results to man, especially when the animals are tested in conditions to which man is not exposed.

17.2.1 Portals of entry

Occupational poisons gain entry to the body via the lungs, skin and sometimes the gut. Absorption of a poison depends on its physical state, particle size and solubility. Of the substances entering the lung some may be exhaled, coughed up and swallowed, attacked by scavenger cells and remain in the lung or enter the lymphatics. Soluble particles may be absorbed into the blood stream. The skin is protective unless abraded when soluble substances can penetrate to the dermis, as they may also do via the hair follicles, sweat and sebaceous glands, and then be absorbed into the blood stream.

In pregnancy, harmful substances in the mother's body may cross the placenta to affect the unborn baby.

17.2.2 Effects

Effects may be acute, i.e. of rapid onset and short duration; or chronic, i.e. of gradual onset and prolonged. They may be local, occurring at the site of contact only, or general following absorption. Toxic substances may disturb normal cell function, damage cell membranes, interfere with enzyme and immune systems and RNA and DNA activity. Pathological response may be irritant, corrosive, toxic, fibrotic, allergic, asphyxiant, narcotic, anaesthetic and neoplastic.

17.2.3 Metabolism

Most substances absorbed will be carried by the blood stream to the liver where they may be rendered less harmful by a change in their chemical composition. However, some may be made more toxic, e.g. naphthylamine which is responsible for bladder cancer and tetra-ethyl lead which is converted into the tri-ethyl form and is toxic to the central nervous system.

17.2.4 Excretion

The body eliminates harmful substances in the urine, lungs and less commonly the skin. Some are also excreted in the faeces and milk. The time taken to reduce the concentration of a substance in the body to 50% is known as its biological half-life. A knowledge of bio-chemical changes and excretory routes of products is important in the design of screening tests.

17.2.5 Factors influencing toxicity

A number of factors are important when considering the toxic effect of a substance on the body. These include:

(1) The inherent potential of a substance to cause harm.
(2) Its ease of body contact and entry: particle size and solubility.
(3) Dose received (concentration and time of exposure).
(4) Metabolism in the body (bio-transformation) and its half-life.
(5) Susceptibility of the individual which depends on a number of factors:
  (a) Body weight; the same dose of a substance is more damaging to the smaller person.
  (b) The extremes of age in the working population are more prone to skin damage.
  (c) Fair skinned persons are more liable than the dark skinned to chemically induced dermatitis, and to radiation induced skin cancer.
  (d) Physical and physiological differences between the sexes may cause a variation in toxic response.
  (e) Failure to reach a health standard set for a particular job may expose the individual to greater risk.
  (f) The level of training, information, supervision and protection provided.

17.2.6 A no-adverse-effect level

In the setting of Occupational Exposure Standards (OES)[2], which are health based exposure concentration limits, it is necessary to establish, with reasonable certainty, the airborne concentrations which will not result in ill-health even if inhaled day after day. This concentration may be derived from the no-adverse-effect level in animal species which is arrived at after careful epidemiological and toxicological tests. This level relates to the average work person but will not apply to individuals who, under certain circumstances, are susceptible and, therefore, at special risk[3].

17.3 Diseases of the skin

17.3.1 Non-infective dermatitis

The term 'dermatitis' simply means an inflammation of the skin. When the condition is due to contact with a substance at work it is called 'occupational' or 'industrial' dermatitis. It is a common cause of occupational disease but the number of cases is declining owing to improved work conditions.

The skin has two layers, the outer layer called the 'epidermis' and the inner the 'dermis'. The epidermis has a protective function. It consists of densely packed flat cells, thicker in some areas, like the palms of the hands, which are more subject to injury. It is covered by a moist film known as an 'acid mantle', made up of secretions from sweat and sebaceous glands, that helps to protect from acids, alkalis and also excessive water, and to some degree from heat and friction by preventing the skin from drying out. The natural grease of the skin can be removed by solvents. In the deeper layer of the epidermis are pigment cells which produce the 'tan' following exposure to sunlight and protect the body from ultraviolet radiation.

Some persons are more susceptible to skin damage than others, particularly the young, those with soft, sweaty skin, the fair complexioned and those with poor personal hygiene. Occupational dermatitis can affect any part of the body, but the hands, wrists and forearms are most commonly involved. Damage to the skin may follow exposure to chemical and biological substances as well as physical agents. Dermatitis is of two kinds: irritative and sensitising – the former is four times more common. Chemicals which cause irritative dermatitis include acids, cement, solvents, some metals and their salts. Their effect on the skin depends on the concentration and duration of exposure, and will affect most people in contact with them. At first the response may be minor, but it worsens with repeated contact.

Sensitisers, on the other hand, do not cause dermatitis until the individual has first become sensitised by them. This involves an allergic response in the blood initially, the dermatitis following on subsequent exposure. Once sensitisation has occurred a small dose may be sufficient to cause a rash. Sensitisers include chrome-salts, nickel, cobalt, plastics made of epoxy, formaldehyde, urea or phenolic resins, rubber additives, some woods and plants[4]. Some substances act as both irritant and sensitiser, e.g. chrome, nickel, turpentine and mercury compounds.

17.3.1.1 Symptoms

The onset of dermatitis may be unnoticed, especially as it usually clears up when away from work, i.e. at weekends and holidays. On return to work and further exposure the condition recurs, worsening with each subsequent contact. The skin, at first rough and raw, may itch, become cracked and sore, prompting the individual to seek medical advice. The rash may be diffuse, as with eczema, or pimple-like as with acne – the former following exposure to irritative and sensitising agents, and the latter from exposure to mineral oils, pitch and chlorinated hydrocarbons. Patch testing, in which a dilute quantity of chemical is applied to the skin under a plaster and left for several hours to see if a reaction develops, is useful only for determining allergic response to chemicals but requires specialist interpretation.

17.3.1.2 Protective measures

Persons with dermatitis or sensitive skin may need to be excluded from certain kinds of work. Good personal hygiene is essential and barrier creams may be helpful. Protective clothing should be considered.

17.3.2 Cancer of the scrotum

The first occupational skin cancer was reported by Percivall Pott in 1775, among chimney sweeps. In those days children were apprenticed to master sweeps to climb inside and clean chimneys; their skin became ingrained and their clothes impregnated with soot, and as they seldom washed or changed their clothes the skin was constantly irritated. From puberty onwards a 'soot wart' might appear on the scrotum and develop into a cancer. In 1820, Dr Paris wrote of the influence of arsenical fumes affecting those engaged in copper smelting in Cornwall and Wales, giving rise to a cancerous disease of the scrotum similar to that affecting sweeps[5]. From 1870 a number of substances in a variety of industries were found to cause scrotal cancer – shale oil in those engaged in oil refining and cotton mule spinning; pitch and tar in those making briquettes from pitch-containing coal dust; mineral oil used by engineers and gunsmiths, and paraffin in refinery workers. Others at risk include creosote-timber picklers and anthracene chemical workers, and also sheep-dippers using arsenic[6].

Commonly, workers' clothes become begrimed with the offending substance, making close contact with the scrotum, the wrinkled skin of which favours the harbouring of the carcinogen.

The cancer begins as a wart, which enlarges and hardens, then breaks down into an ulcer with spread of malignant cells to neighbouring glands and other parts of the body.

The carcinogen might be an arsenic compound, but, more often, a polycyclic or aromatic hydrocarbon of the benzpyrene or benzanthracene type. Skin cancers, in general, have been attributed to sunlight, ionising radiation, hydrocarbons and arsenic compounds.

17.3.2.1 Prevention

The use of non-carcinogen oils: carcinogens can be removed from mineral oil by washing with sulphuric acid or solvents. Workers should be educated to avoid contact as much as possible. The use of splash guards on machinery, protective clothing, avoidance of an oily rag placed in a pocket, which could spread oil through the clothes to the scrotum. To wash the hands before toilet and to have a daily bath. Clothes should be kept reasonably clean and a laundry service provided, so that overalls can be changed once or twice a week as need requires. Workers should not wear their dirty overalls after duty, but be encouraged to change into their home clothes. Workers also should be medically examined prior to employment and periodically to ensure that their skin is clear, and be encouraged to report to the doctor any doubtful 'wart' that might appear.

17.3.3 Coal tar and pitch

The destructive distillation of coal yields a variety of products, depending on the temperature at which distillation takes place, e.g.

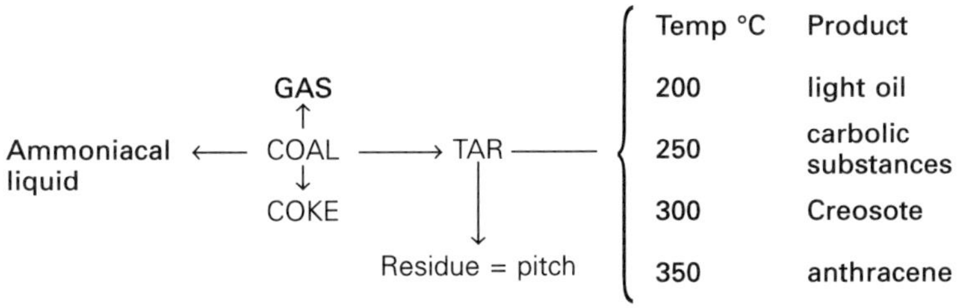

GAS ↑ Ammoniacal liquid ⟵ COAL ⟶ TAR ↓ COKE

TAR ↓ Residue = pitch

Temp °C	Product
200	light oil
250	carbolic substances
300	Creosote
350	anthracene

Distillation at high temperatures results in aromatic polycyclic hydrocarbons retained in the pitch which are harmful to health. Pitch is used in many industries: briquetting of coal, roofing materials, waterproofing of wood, manufacture of electrodes, impermeable paper, optical lenses, dyestuffs and paints.

17.3.3.1 Symptoms

Exposure of a worker to pitch dust or vapour may harm the skin by causing irritation, tumour or dermatitis. Irritation is the earliest and commonest reaction, occurring after a few days or weeks of exposure and affecting the face and neck. There is complaint of itching or burning, aggravated by cold, wind or sunlight (Pitch Smarts). Usually it clears up soon after exposure ceases. Benign tumours or warts occur on exposed areas of skin, chiefly the face, eyelids, behind the ears, the neck, arms and, occasionally, on the scrotum and thighs. Their recurrence is related to duration and degree of exposure to pitch[7]. Many regress spontaneously, especially those appearing early, but some undergo malignant change, particularly those appearing in the older age groups. They need to be removed and examined under the microscope, i.e. biopsied, to check for any malignant change. A variety of other skin conditions may occur such as darkening and thickening of the skin, acne, blackheads, cysts and boils, pitch burns and scarring. There is also a risk of damage to the cornea.

17.3.3.2 Prevention

Pitch dust and vapour must be avoided by transporting the raw material in a liquid or granular state and enclosing the process as far as possible. Workers require clean protective clothing for head, neck and forearms and eye protection should be worn[8]. Employees ought to be warned of the risk, and advised to report any skin disease which develops. Good personal hygiene is essential, and adequate wash and shower facilities need to be provided. Barrier creams applied before work are helpful. Those susceptible to warts should be excluded from further exposure, and each worker needs to be medically examined regularly to detect possible skin disorders.

17.4 Diseases of the respiratory system

17.4.1 Pneumoconiosis

The term pneumoconiosis means 'dust in the lung', but medically refers to the reaction of the lung to the presence of dust[9].

17.4.1.1 Body defence to inhalation of dust

During inspiration particles of dust in the air larger than 10 μm in diameter are filtered off by the nasal hairs. Others, which enter through the mouth, are deposited in the upper respiratory tract. Particles between 5 μm and 10 μm tend to settle in the mucus covering the bronchi and bronchioles and are then wafted upward by tiny hairs (ciliary escalator) towards the throat. They are then coughed or spat out, though some may be swallowed. Particles less than 5 μm in diameter are more likely to reach the lung tissue. However, fibres (e.g. asbestos) which predispose to disease have a length to diameter ratio of at least 3:1 with a diameter of 3 μm or less; the longer the fibre the more damaging it may be.

17.4.1.2 Respirable dust

Respirable dust is that dust in the air which on inhalation may be retained by the lungs. The amount of dust deposited depends on the duration of exposure, the concentration of dust in the respired air, the volume of air inhaled per minute and the nature of the breathing. Slow, deep respirations are likely to deposit more dust than rapid, shallow breathing. Dust in the lung causes a tissue reaction, which varies in nature and site according to the type of dust.

17.4.1.3 Causes of pneumoconiosis

(a) *Benign* The inhalation of some metal dusts, such as iron, tin and barium, results in very little structural change in the lungs and, therefore, few symptoms. The tissue reaction, nevertheless, is detectable on X-ray as a profusion of tiny opacities.
(b) *Symptomatic* The most important causes include coal dust, silica and asbestos. Symptoms of cough and breathlessness develop usually after many years of exposure, but only in the later stages of disease.

Beryllium dust causes acute and chronic symptoms. Early features are breathlessness, cough with bloody sputum and chest pain. Recovery follows removal from exposure, but a chronic state can develop insidiously with cough, breathlessness and loss of weight.

Organic dusts, such as mouldy hay, when inhaled cause a disease known as extrinsic allergic alveolitis with 'flu-like symptoms; cough and difficulty in breathing occur within a few hours of exposure. Repeated exposure leads to further lung damage and chronic breathlessness.

Talc is a white powder consisting of hydrous magnesium silicate. Although some talc presents little risk to health, commercial grades may contain asbestos and quartz and provoke pneumoconiosis and lung cancer.

Cobalt combined with tungsten carbide forms a hard metal used for the cutting tips of machine tools and drills. Inhalation of the dust may give rise to fibrosis of the lungs causing cough, wheezing and shortness of breath.

Man-made mineral fibres irritate the skin, eyes and upper respiratory tract. A maximum exposure limit has been set based on the risk of lung cancer because a 'no-adverse-effect' level cannot be established with reasonable certainty[3].

17.4.1.4 Diagnosis of pneumoconiosis

This depends on:

(1) A complete occupational history of all jobs.
(2) A characteristic appearance on the chest X-ray. There is an international grading system which is used to assess radiologically the extent of the disease.
(3) A clinical examination.
(4) Lung function tests.
(5) In some cases involving organic dust, specific blood tests.

17.4.2 Silicosis

Silicosis: the commonest form of pneumoconiosis is due to the inhalation of free silica.

Free silica (SiO_2) or crystalline silica occurs in three common forms in industry: quartz, tridymite and cristobalite. A cryptocrystalline variety occurs in which the 'free silica' is bound to an amorphous silica (non-crystalline). It includes tripolite, flint and chert. Diatomite is the most common form of amorphous silica capable of producing lung disease. Some of these forms can be altered by heat to the more dangerous crystalline varieties, such as tridymite and cristobalite.

e.g.

Quartz
Cryptocrystalline
Amorphous

} → ⟶ tridymite → → cristobalite (800°C+ ⟶ directly to cristobalite)

17.4.2.1 Lung reaction

Industrial exposure occurs in mining, quarrying, stone cutting, sand blasting, some foundries, boiler scaling, in the manufacture of glass and ceramics and, for diatomite, in the manufacture of fluid filters. Particles of free silica less than 5 μm in diameter when inhaled are likely to enter the lungs and there become engulfed by scavenging cells (macrophages) in the walls of the tiniest bronchioles. The macrophages themselves are destroyed and liberate a fluid causing a localised fibrous nodule which obliterates the air sacs. The nodules are scattered mainly in the upper halves of the lungs. They gradually enlarge to form a compressed mass of nodules. Sometimes a single large mass of tissue may occur, known as progressive massive fibrosis. If much of the lung is affected the remaining healthy tissue is likely to become over-distended during inhalation.

17.4.2.2 Symptoms

There are no symptoms in the early stage. Later the initial complaint is of a dry morning cough. Next occurs some breathlessness, at first noticeable on exercise but, as destruction of lung tissue proceeds, breathlessness worsens until it is present at rest. The interval between exposure and the onset of symptoms varies from a few months in some susceptible individuals to, more usually, many years, depending on the concentration of respirable free silica and the exposure time at work. Silicosis is the one form of pneumoconiosis which predisposes to tuberculosis, when additional symptoms of fever, loss of weight and bloody sputum may occur. In the presence of gross lung destruction the blood circulation from the heart to the lung may be embarrassed and result in heart failure.

17.4.2.3 Diagnosis

This depends on a history of exposure and, in the early stages, a chest X-ray showing tiny radio opaque nodules and, later, a history of cough and breathlessness and sounds in the chest detectable with a stethoscope. Lung function tests may be helpful, but usually not until the late stages.

17.4.2.4 Medical surveillance

Where exposure to free silica is a recognised hazard, a pre-employment medical is advised, which should enquire into previous history of dust exposure, of respiratory symptoms, with examination of the chest, lung function testing and a chest X-ray. The medical should be repeated periodically as circumstances demand.

17.4.2.5 Prevention

Reduction of the dust to the lowest level practicable and where necessary by the provision of personal respiratory protective equipment.

17.4.3 Asbestosis

There are three important types of asbestos, blue (crocidolite), brown (amosite) and white (chrysotile). Asbestosis is a reaction of the lung to the presence of asbestos fibres which, having reached the bronchioles and air sacs, cause a fibrous thickening in a network distribution, mainly in the lower parts of the lung[10]. There follows a loss of elasticity in the lung tissue, (relative to the concentration of fibres inhaled and the duration of exposure) resulting in breathing difficulty.

Among those at risk are persons engaged in milling the ore, the manufacture of asbestos products, lagging, asbestos spraying, building, demolition, and laundering of asbestos workers' overalls.

Symptoms develop slowly after a period of exposure which varies from a few to many years. In some cases exposure may have begun so long ago that it cannot be recalled. Breathlessness occurs first and progresses as the lung loses its elasticity. There may be little or no cough and chest pain seldom occurs. The individual becomes weak and distressed on effort and, eventually, even at rest. Unless periodic medicals are introduced the diagnosis will not be made until symptoms appear. Early diagnosis is essential in order to prevent further exposure and an exacerbation of the condition. Asbestosis predisposes to cancer of the bronchus, a risk increased by cigarette smoking. The chest should be X-rayed every two years and special lung function tests are helpful. Diagnosis depends on history of exposure, chest X-ray, lung function testing, symptoms and physical signs.

17.4.4 Mesothelioma

Mesothelioma is a malignant tumour of the lining of the lung (pleura) or abdomen (peritoneum). The abdominal form is less common. The disease is significantly related to exposure to asbestos, especially the blue and brown varieties. However, in some 10–15% of cases there is no such history of exposure[10]. Those at risk are miners, manufacturers of asbestos, builders and demolition workers, and even residents in the neighbourhood of blue asbestos working. While the exposure time may have been only minimal, there is no safe threshold of dose below which there is no risk of asbestos related disease. The onset of the disease is delayed, often by some 20 to 50 years. It affects men and women, but the 'attack rate' of the tumour in the exposed population is only about 5%.

17.4.4.1 Symptoms

The lung variety of tumour is more common. Symptoms begin with a gradual onset of breathlessness, particularly noticeable on effort, and due to the growth of tumour and fluid compressing the lung. There may occur pain on one side of the chest, with tenderness, cough and fever. More obvious is a rapid loss of weight and weakness. A chest X-ray reveals an opacity on one side of the chest suggestive of the tumour. The symptoms of the abdominal form also develop slowly, beginning with a swelling, loss of weight, impaired appetite and weakness. Death usually follows within two years of making the diagnosis.

17.4.5 Other dust causes of lung cancer

These include: chromate, in the manufacture of chromate from the ore; nickel compounds in the refining of nickel; benzpyrenes in coke-oven work; uranium and radon; and arsenic compounds in mining.

17.4.6 Bronchial asthma

Bronchial asthma is defined as 'breathlessness which is due to narrowing of the small airways' and it is reversible. There are many occupational causes of which fourteen have been prescribed as resulting in industrial diseases:

(1) Isocyanates.
(2) Platinum salts.
(3) Epoxy resin curing agents.
(4) Colophony fumes.
(5) Proteolytic enzymes.

(6) Animals and insects in laboratories.
(7) Flour and grain dust.
(8) Antibiotic manufacture.
(9) Cimetidine used in manufacturing cimetidine tablets.
(10) Hard wood dusts of cedar, oak and mahogany.
(11) Ispaghula used in the manufacture of laxatives.
(12) Caster bean dust.
(13) Ipecacuanha used in the manufacture of tablets.
(14) Azodicarbonamide used in plastics.

Other asthma-like diseases are found.

Byssinosis occurs in workers in the cotton processing industry who may develop tightness of the chest on Mondays which decreases as the week progresses. However, with continuing exposure to cotton dust they are affected for more days of the week. Steam treatment of the raw cotton can prevent chest symptoms from this material.

An allergic lung reaction also occurs after exposure to spores on sugar cane (*bagassosis*). The sugar cane spores can be killed by spraying with propionic acid.

17.4.7 Extrinsic allergic alveolitis (farmer's lung)

A disorder due to inhalation of organic dust and characterised by chest tightness, fever and the presence of specific antibodies in the blood.

17.5 Diseases from metals

17.5.1 Lead

Lead (Pb) is a relatively common metal, mined chiefly as the sulphide (galena) in many countries – USA, Australia, USSR, Canada and Mexico[11]. In this country we use about 330 000 tonnes of lead annually, much of which comes from recycled scrap.

Lead has a great variety of uses, e.g. (percentages approximated from annual production figures issued by World Bureau of Metal Statistics, London):

Electric batteries	27%
Electric cables	17%
Sheet, pipe, tubes	16%
Anti-knock in petrol	11%
Solder and alloys	9%
Pottery, plastics, glass, paint	4%
Miscellaneous	15%

Lead, as a fume or dust hazard, is therefore met in many industries. The pure metal melts at 327°C and begins to fume at 500°C, but the presence of impurities alters these properties and may form a slag on its surface and thereby reduce fuming, except at higher temperatures. Particle size and solubility are important factors governing the absorption of lead via the lungs. In the gut, however, solubility differences of ingested compounds are of less significance. Amongst lead miners lead poisoning does not occur due to the insolubility of the sulphide ore.

17.5.1.1 Inorganic lead

Inorganic lead can enter the body by inhalation or ingestion[12]. Up to about 50% of that inhaled is absorbed and only about 10% of that ingested. It is then transported in the blood stream and deposited in all tissues, but about 90% of it is stored in the bone. It is a cumulative poison; excretion is slow and occurs mainly in the urine and faeces.

Symptoms Early features are vague and include fatigue, loss of appetite, and metallic taste in the mouth. Constipation is the commonest complaint and is sometimes associated with abdominal pain. This may be so severe as to mimic an acute abdominal emergency. Classically, a blue line appears along the margin between the teeth and gums, but this usually occurs only in the presence of infected teeth and is indicative of lead exposure rather than poisoning.

Lead interferes with the normal formation of haemoglobin, causing anaemia, but the diagnosis of excessive absorption should be made before anaemia appears. The same interfering mechanism causes abnormal products to appear in the urine, e.g. amino laevulinic acid (ALA) which is a useful indicator of excessive lead absorption or poisoning.

Paralysis, though rare nowadays, can occur as wrist or foot drop due to the effect of lead on nerve conduction. It may begin with a weakness in the fingers and wrists, which is a useful early sign.

Lead is transported in the blood and can cross the placental barrier in pregnant women and affect the unborn child. Abortion was common in women employed in lead industries during the nineteenth century and was believed to be due to excessive lead absorption. The brain can also be affected, a condition known as encephalopathy, causing abnormal behaviour, convulsions, coma and death. Children are much more susceptible than adults.

Because of the excretion of lead in the urine, kidney damage is a likely long-term effect.

17.5.1.2 Organic lead

Tetra-ethyl and tetra-methyl lead are the most important organic forms used in industry, especially in petrol to improve the octane rating. These substances can be absorbed via the lungs and the skin. In the liver they are changed respectively to tri-ethyl and tri-methyl lead, which are much more toxic. They have a particular predilection for the brain and cause psychiatric disturbance, headache, vomiting, dizziness, mania and coma. Excretion occurs mainly via the urine. The blood is less affected than with inorganic lead.

17.5.1.3 Biological monitoring

For lead workers periodic medical examination is a statutory requirement. Blood samples should be taken as required for haemoglobin and lead. Lead level in normal blood is about 20 μg/100 ml but for lead workers can be 40–60 μg/100 ml. The acceptable upper limit of blood lead concentration in adults if 70 μg/100 ml except men who have worked in lead for many years. For women of child-bearing age the limit is 40 μg/100 ml.

A useful indicator of excessive lead effect is the presence of zinc protoporphyrin (ZPP). It can be measured from a small quantity of blood obtained by finger-prick. For confirmatory evidence of excessive lead absorption or poisoning, urine estimation of amino laevulinic acid are helpful. Inorganic lead is best monitored by blood sampling and organic lead by urine sampling.

17.5.2 Mercury

Mercury (Hg) occurs naturally as the sulphide in the ore known as cinnabar, and also in the metallic form quicksilver. It is mined chiefly in Spain, but also in Italy, Russia, USA and elsewhere. The ore is not particularly hazardous to miners, as the sulphide is insoluble. Risk is greater in other industries, such as in the manufacture of sodium hydroxide and chlorine, electrical and scientific instruments, fungicides, explosives, paints and in dentistry.

17.5.2.1 Symptoms

Acute mercury poisoning is rare but can occur following the inhalation of quicksilver – it being very volatile at room temperature. There is particular risk should spillage occur in an enclosed space. About 80% of that inhaled can be absorbed[13], and a few hours later there occurs cough, tight chest, breathlessness and fever. Symptoms last a week or so, dependent upon degree of exposure, but its effects are reversible. Acute poisoning may also occur by ingestion of soluble salts, such as mercuric chloride which has a corrosive action on the bowel, causing bloody diarrhoea.

Ingestion of metallic mercury is not generally toxic as it is not absorbed.

Chronic poisoning is the more usual presentation, following absorption by lung or gut of soluble mercury salts. Symptoms develop almost imperceptibly, usually beginning with a metallic taste in the mouth and sore gums. Later tremor of the hands and facial muscles develops; gums may bleed and teeth loosen. Personality changes of shyness and anxiety, inability to concentrate, impaired memory, depression and hallucinations may occur. As excretion is mainly via the urine, the kidney is subject to damage.

Organic mercury can be absorbed via the lung, gut and skin, and also cause chronic poisoning. There are two varieties: aryl and alkyl, and they have different effects on the body. The aryl variety, of which phenyl mercury is an example, has a similar metabolic pathway to inorganic mercury and has a similar clinical effect.

Alkyl mercury is much more dangerous – methyl mercury is an example. It causes irreversible damage to the brain, resulting in a constriction of visual fields, disturbance of speech, deafness and inco-ordination of movement. Most of it (90%) is excreted without change, slowly in the faeces.

All forms of mercury may give rise to dermatitis. Mercury can cross the placental barrier and affect the unborn child of exposed mothers.

17.5.2.2 Medical surveillance

Those at risk should be medically examined periodically and attention paid to the mouth, tremor of the hands (a writing test is useful), personality, and for those exposed to methyl mercury, vision, hearing and co-ordination. The urine should be checked for protein and mercury excretion. Mercury does not normally occur in urine, but may be detected in some persons with no apparent occupational exposure. In organic exposure, owing to the different metabolic pathway from that of inorganic, the urine concentration does not correlate with body levels. The upper limit which requires further investigation is for inorganic mercury 1000 nmol/litre and, for organic mercury, 150 nmol/litre[14].

17.5.3 Metal fume fever

Inhalation of the fume of some metal oxides such as zinc, copper, iron, magnesium and cadmium causes an influenza-like disease. Similar effects may follow the inhalation of polytetrafluoroethylene (ptfe) fumes. Usually there is recovery within one or two days. Zinc fume fever is probably a very common disease, the diagnosis of which is often missed because of the short duration of the illness. Cadmium fume inhalation can be much more serious.

17.5.4 Chromium

Chromium (Cr) is a silvery hard metal used in alloys and refractories. Chrome salts are used in dyeing, photography, pigment manufacture and cements. Electroplating tanks contain solutions of chromic acid which forms a mist during the electrolysis process.

Chromates and dichromates used in cement manufacture and chromium plating may cause skin irritation or ulceration and chrome ulcers in the skin of the hands or in the inside of the nose where the ulcer may penetrate the cartilage of the nasal septum.

17.5.5 Arsenic (As)

Inorganic arsenic compounds cause irritation of the skin and may produce skin cancer. It is used in alloys to increase hardness of metals, especially with copper and lead.

17.5.6 Arsine (arseniuretted hydrogen – AsH_3)

Arsine is a gas which arises accidentally in many metal working industries. It damages the red blood cells, releasing the red pigment haemoglobin from them. This may cause jaundice, anaemia and the urine may appear red due to the presence of haemoglobin pigment. Poisoning by arsine can result in rapid death. Organic arsenic compounds have been used as war gases, and can produce severe and immediate blistering of the skin and severe lung irritation (pulmonary oedema).

17.5.7 Manganese (Mn) and compounds

This is used to make manganese alloy steels, dry batteries and potassium permanganate which is an oxidising agent and a disinfectant. Poisoning is rare and follows inhalation of the dust causing acute irritation of the lungs and affects the brain leading to impaired control of the limbs rather like Parkinson's disease.

17.5.8 Nickel (Ni) and nickel carbonyl ($Ni(Co)_4$)

Nickel is a hard blue-white metal used in electroplating and in a range of alloys. Nickel salts (green) cause skin sensitivity (nickel itch). Nickel carbonyl (a colourless gas) causes headache, vomiting and later pulmonary oedema.

17.5.9 Cadmium (Cd)

This metal is used in alloys, rust prevention, solders and pigments. A fume may be released during smelting, alloy manufacture or when rust-proofed metals are heated, e.g. in welding cadmium-plated metals, which produces irritation of the eyes, nose and throat. With continued exposure tightness of the chest, shortness of breath and coughing may increase and can lead to more severe lung damage which may be fatal.

Long-term damage by smaller quantities of dust or fumes may lead to loss of elasticity of the lungs. Cadmium may cause kidney damage and while it has been suggested that lung cancer may occur after cadmium exposure this has not been proved in man.

17.5.10 Vanadium (V)

This material occurs as vanadium ore but is also found in petroleum oil. It is also used to make alloy steels and as a catalyst in many chemical reactions. Exposure to the metal occurs when oil-fired boilers are cleaned and manifests itself in eye irritation, shortness of breath, chest pain and cough. The tongue becomes greenish-black in colour. Severe cases may develop broncho-pneumonia. Removal from contact with the dust usually leads to rapid recovery.

17.6 Pesticides

17.6.1 Insecticides

Various organo-phosphorus compounds are used; two of the commonest are demeton-S-methyl and chlorpyrifos. Poisoning causes headaches, nausea and blurred vision. Further symptoms include muscle twitching, cramps in the belly muscles, severe sweating and respiratory difficulties. Extreme exposure may lead to death. All these effects are due to interference with a chemical enzyme called cholinesterase which is concerned with the passage of nerve impulses. The level of this enzyme in the worker's blood can be measured and if it falls below a certain value the worker must be removed from contact with the chemical until his blood returns to normal. The appropriate protective clothing must be worn at all times when working with these materials.

17.6.2 Herbicides

Commonly used as a weedkiller (e.g. paraquat). Ingestion may result in damage to the liver, kidneys and lung. There is no antidote and death occurs in about half the cases.

17.7 Solvents

A solvent is a liquid that has the power to dissolve a substance: water is a common example[15]. In industry organic liquids are often used as solvents, and these are mainly hydrocarbons used as degreasing agents and in the manufacture of paints and plastics.

Examples of solvents (classification after Matheson[16])

Hydrocarbons	
(i) Aromatic	Benzene; toluene; styrene
(ii) Aliphatic	Paraffin; white spirit
Aliphatic alcohols	Methyl alcohol; ethyl alcohol
Aliphatic ketones	Methyl-ethyl-ketone
Aliphatic ethers	Diethyl ether
Aliphatic esters	Ethyl acetate
Aliphatic chlorinated	Trichloroethylene; carbon tetrachloride
Non-hydrocarbons	Carbon disulphide

17.7.1 General properties

All organic solvents are volatile and have a vapour density greater than one, i.e. their vapours are heavier than air and will therefore settle at floor level; this is important to note when considering ventilation. With the exception of the chlorinated hydrocarbons they tend to be flammable and explosive and in the liquid form most have specific gravities of less than one so will float on water. In the event of a fire, attempt should not be made to extinguish with water, as the solvent will float away and the fire be spread. The chlorinated solvents, being neither flammable nor explosive but heavier than water, have been used as fire extinguishants.

17.7.2 Toxic effects

Solvents vary widely in their toxicological properties. In common they cause dermatitis by removing the natural grease from the skin, and narcosis by acting on the central nervous system; additionally some can damage the peripheral nerves, the liver and kidneys and interfere with blood formation and cardiac rhythm. Chlorinated solvents can decompose if exposed to a naked flame to produce acidic fumes (hydrochloric acid and small amounts of phosgene) which are harmful to the lungs. Any harmful effect is related to the amount of solvent absorbed.

Skin penetration varies with the solvent, hence in the list of Occupational Exposure Limits[2] some are designated 'skin', but other factors include surface area exposed and the thickness of the skin, e.g. less may be absorbed via the palms than the forearms while the scrotum is especially liable to absorption[17].

Absorption is also related to the breathing pattern, activity, obesity and addiction. Because of this individual variation, the amount taken up by the body is a more important estimate of potential harm than the concentration to which the body is exposed. Body uptake correlates well with blood concentration and to less extent with quantities excreted in the urine[18].

However, periodic urine testing of excreted solvent or its metabolite is a more convenient means of biological monitoring.

17.7.3 Trichloroethylene

Structural formula:

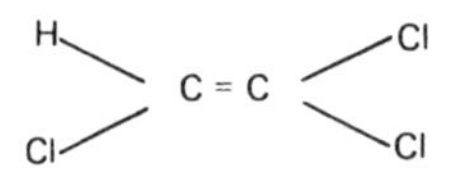

Other names: Tri, Trike, Trilene.
Properties: Non-flammable
Vapour density 4.54
Specific gravity 1.45
Boiling point 87°C
MEL 100 ppm 8 hour TWA

Exposure to naked flames or red-hot surfaces can cause it to dissociate into hydrogen chloride, possibly with small amounts of phosgene or chlorine.

Use Its main use is as a solvent especially in the degreasing of metals. It has also been used as an anaesthetic. *Figure 17.1* shows a single compartment vapour type plant used for cleaning by solvents.

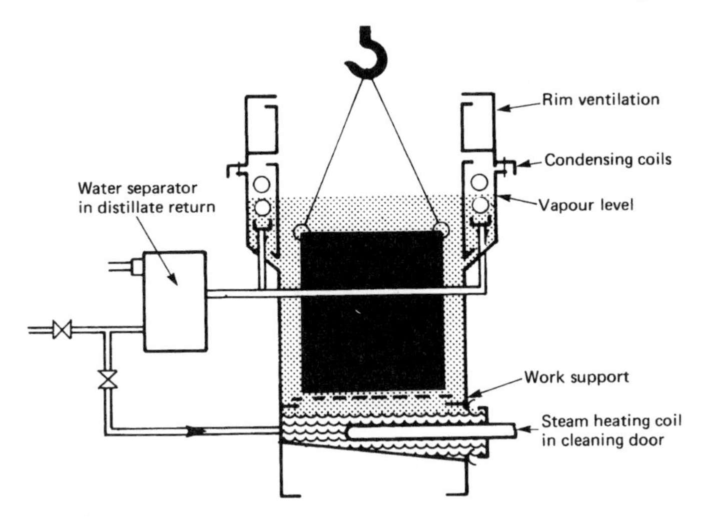

Figure 17.1 Cleaning by solvents: single compartment vapour type plant. (Courtesy ICI, PLC, Mond Division)

Metabolism Its main route into the body is via the lungs, where it is rapidly absorbed. Some is excreted into the expired air, while the remainder is converted to trichloroacetic acid and passed in the urine. It is usually cleared quickly from the body, but daily exposure may tend to its cumulation. The estimation of trichloroacetic acid in the urine is a useful test for checking excessive exposure and its concentration should not exceed 100 mg/litre urine, standardised to a specific gravity of 1.016. Samples should be collected at the end of a shift towards the end of a working week.

17.7.3.1 Harmful effects

Acute Trichloroethylene is a powerful anaesthetic and can be dangerous in confined spaces. Early features include headache, dizziness, and lack of concentration and eventually unconsciousness. Its vapour may cause irritation of the eye and the skin can be blistered by the liquid.

Chronic The main problem from repeated exposure is a dermatitis of the hands, due to the solvent's action in removing the normal grease of the skin which then becomes rough, red, raw, and cracks – a condition known as eczema. Some people become addicted to trichloroethylene, usually by repeated 'sniffing' of the vapour, or even drinking the fluid, and then display abnormal behaviour known as 'tri-mania'. Rare cases of sudden cardiac arrest have been reported in situations of gross short-term over-exposure. After long-term exposure there have been a number of individual case reports of liver damage, and recently, following animal tests in the USA, it has been under suspicion as a carcinogen.

17.7.3.2 Prevention

Employees should be made aware of the risks. Local exhaust ventilation around the lips of vapour degreasing tanks is necessary, and in confined spaces good general ventilation is essential. In work areas, atmospheric monitoring is recommended to ensure that exposure is kept to a minimum and certainly below the Maximum Exposure Limit (MEL) of 100 ppm 8 hour TWA. Body absorption can be monitored by a urine sample taken at the end of a shift near the end of a working week and analysed for trichloroacetic acid. Those being tested must refrain from drinking alcohol as it inhibits excretion.

17.7.4 Carbon tetrachloride

Structural formula:

```
Cl     Cl
  \   /
    C
  /   \
Cl     Cl
```

Properties: Non-flammable
Vapour density 1.5
Specific gravity 1.6
Boiling point 76.8°C
OES–TWA 2 ppm (skin)

Use Its main use is in the manufacture of chlorofluorocarbons, also aerosols and refrigerants. It has been used in fire extinguishers and grain fumigation. Its use in dry cleaning has declined because of its toxicity.

Metabolism Carbon tetrachloride is absorbed into the blood mainly via the lungs, but also via the skin and gut. Some is excreted in the expired air and the remainder in the urine, but in altered form.

17.7.4.1 Harmful effects

Acute In common with other solvents it has a narcotic effect, with features varying from headache and drowsiness to coma and death. If taken by mouth it can cause abdominal pain, diarrhoea and vomiting. Acute over-exposure can result in liver and kidney damage.

Chronic Carbon tetrachloride can also cause damage to the kidneys and liver; in the long term it is more toxic than trichloroethylene. An early sign of kidney damage may be detected by urine examination for protein and cells. Liver damage may be indicated early by special tests or later by the appearance of jaundice. It is also under suspicion as a carcinogen.

17.7.5 Other common solvents

17.7.5.1 Benzene (C_6H_6)

OES–TWA 5 ppm. This excellent solvent is seldom used today because of its toxic effects. It may be inhaled or absorbed via the skin and is readily absorbed by fatty tissues. A large proportion of benzene which enters the body is stored in the bone marrow which it may damage, causing anaemia or more rarely leukaemia. Benzene is altered chemically in the body and then excreted in the urine. For exposures about the MEL, blood benzene is a useful measurement. For lower exposures, breath benzene is suitable. Urinary excretion as a 'phenol' is no longer recommended.

17.7.5.2 Toluene ($C_6H_5CH_3$) (methylbenzene) and xylene ($C_6H_4(CH_3)_2$)

Toluene, OES–TWA 100 ppm and xylene, OES–TWA 1000 ppm are frequently used solvents which have toxic effects common to other solvents. They produce narcosis and can damage the liver and kidneys. Blood or breath toluene is suitable for monitoring; for xylene, urine is tested for methyl hippuric acid.

17.7.5.3 Tetrachloroethylene ($CCl_2.CCl_2$) (perchloroethylene)

OES–TWA 50 ppm. This solvent is a narcotic and may cause liver damage. Like trichloroethylene it may break down to release phosgene when exposed to naked flames or red-hot surfaces. Monitor using blood sample taken towards the end of the working week.

17.7.5.4 Trichloroethane ($CH_3.CCl_3$) (methyl chloroform)

MEL–TWA 350 ppm. This was regarded as one of the safest solvents but in the last decade there have been several deaths in young people working with this substance. This substance is to be withdrawn as a result of EC Regulation 594/91[19].

17.7.5.5 Carbon disulphide (CS_2)

MEL-8 hour TWA 10 ppm. Carbon disulphide is an inorganic solvent used mainly in the manufacture of viscose rayon fibres. It is absorbed through the lungs and skin and is a multi-system poison affecting the brain, peripheral nerves and the heart. Monitoring is of urinary metabolites.

17.8 Gassing

17.8.1 Gassing accidents

In the UK those gassing accidents that are reported annually occur in the following approximate order of frequency:

(1) Carbon monoxide
(2) Chlorine
(3) Hydrochloric acid
(4) Trichloroethylene
(5) Sulphur dioxide
(6) Ammonia
(7) Hydrogen sulphide
(8) Phosgene
(9) Carbon dioxide
(10) Nitrous fumes
(11) Phosphorus oxychloride
(12) Carbon tetrachloride

Chlorine, sulphur dioxide and ammonia are highly soluble gases and will irritate the upper respiratory tract first while nitrous fumes and phosgene, being less soluble gases, will pass further down the respiratory tract to irritate the lung tissue.

Asphyxia caused by gassing falls into two broad categories:

Simple; in which oxygen in the lungs is replaced by another gas such as carbon dioxide, nitrogen or methane.

Toxic; in which there is a metabolic interference with the oxygen taken up by the body. This occurs with gases such as carbon monoxide, hydrogen sulphide and hydrogen cyanide.

17.8.2 Chlorine and hydrochloric acid (HCl)

These highly irritant gases may affect the air passages and lungs causing bronchitis and difficulties in breathing due to fluid in the lungs (pulmonary oedema).

17.8.3 Carbon monoxide (CO)

This colourless odourless gas may be found wherever incomplete combustion occurs such as in motor vehicle exhausts, furnaces, steelworks and domestic boilers.

Inhalation results in a rapid rise in CO concentration in the blood within the first hour and a much slower rise thereafter. The gas is more readily absorbed by the blood's red cells to the exclusion of oxygen and so impairs the supply of oxygen to vital organs, particularly the heart.

The effects of the gas are shown in the following table:

Exposure ppm	*probable concentration of CO in blood after* 1 hr *exposure (carboxy haemoglobin)*	*Effect*
200	20%	Headache, flushed appearance, breathlessness
400	40%	Dizziness
500	50%	Collapse
600	60%	Unconsciousness

17.8.4 Hydrogen sulphide (H_2S)

Occurs in sewers, oil refineries and chemical processes. Its odour of rotten eggs can be detected at concentrations of 0.3 ppm but increasing the concentrations of exposure impairs the sense of smell. Even at low concentrations the gas irritates the eyes. Higher concentrations irritate the lungs causing pulmonary oedema (although the onset may be delayed), headache, dizziness, convulsions and unconsciousness.

17.8.5 Carbon dioxide (CO_2)

This occurs in bakeries, breweries etc. and is a result of fermentation. The gas is heavier than air. Low concentrations of CO_2 increase the rate of breathing but higher levels depress respiration causing rapid unconsciousness and even death.

17.8.6 Sulphur dioxide (SO_2)

OES-8 hour TWA 2 ppm. A colourless irritant gas with a pungent smell which causes bronchitis and pulmonary oedema.

17.8.7 Nitrous fumes (commonest form NO_2)

Pungent brown fumes which cause lung irritation after a delay of a few hours. Occurs in explosions and blasting, silo storage and diesel engine exhaust.

17.8.8 Phosgene ($COCl_2$)

Arises from burning chlorinated hydrocarbons, e.g. trichloroethylene. Effects similar to nitrous oxide.

17.8.9 Ammonia (NH_3)

OES-8 hour TWA 25 ppm. Ammonia has a corrosive action that will burn the skin, severely irritate or burn the cornea, cause bronchitis and pulmonary oedema.

17.9 Oxygen deficiency

Normal respiration requires:

(1) An adequate concentration and partial pressure of oxygen in the inspired air.

(2) A clear airway to the lungs.
(3) Transfer of oxygen in the air sacs to the blood.
(4) The transport of oxygen by the red cells to the tissues.

Normal oxygen requirements depend on body size, activity and fitness, and interruption of the supply can occur through failure at any of the above indicated levels. Fresh air contains approximately 21% oxygen, 79% nitrogen, 0.03% carbon dioxide. Although inspired air contains 21% oxygen, that in the air sacs has only 14% which at sea level exerts sufficient partial pressure to cross the lung-blood barrier.

At altitudes above sea level the percentage of oxygen in air is unaltered, but because the barometric pressure is less, the partial pressure of oxygen drops accordingly and makes breathing more difficult. At sea level barometric pressure equals 760 mm Hg, therefore oxygen partial pressure equals $760 \times 21/100 = 160$ mm Hg[20].

In the air sacs, however, there is vapour pressure present. It equals 47 mm Hg irrespective of altitude and diminishes the effective partial pressure which the oxygen would otherwise exert. For example, in the air sacs oxygen partial pressure at sea level equals $(760 - 47) \times 14/100 = 100$ mm Hg. In confined spaces the oxygen concentration can fall by several means. It can be displaced by another gas, e.g. a simple asphyxiant such as carbon dioxide. In a disused and ill ventilated coal mine the oxygen present could be used up in oxidising the coal, resulting in a condition known as 'black damp'. Combustion requires oxygen, so that in a confined space a flame will burn up the oxygen present. Similarly, oxygen can be 'combusted' by ordinary respiration of persons working in the space. Canister type respirators should not be worn in a confined space, because of the danger of a depletion of oxygen in the atmosphere; instead full breathing apparatus should be used.

The presence of disease can also embarrass breathing, as during an attack of bronchial asthma, or in pneumoconiosis, when transfer of oxygen across the lungs is impeded. In anaemia the red cell's capacity for carrying oxygen is diminished, and in heart disease the blood may be inadequately pumped around the body. A similar effect is found with carbon monoxide poisoning, in which the normal uptake of oxygen by the red cells is prevented. Each of these mechanisms results in an inadequate oxygen supply to the tissues, a condition known as anoxia.

17.9.1 Oxygen requirement

The 'average man' of 70 kg body weight requires 0.3 litres of oxygen per minute at rest, but considerably more with activity[21].

Degree of work	*Oxygen requirement (l/min)*
Rest	0.3
Light	0.3–1
Moderate	1–1.5
Heavy	1.5–2
Very heavy	2–6

17.9.2 Response to oxygen deficiency

At oxygen concentration of 21–18%, the fit body tolerates exercise well. Below 18% the response depends upon the severity of work undertaken. Between 18 and 17% the body will probably not be adversely affected, unless the work undertaken is heavy, when there is likely to develop oxygen insufficiency which could lead to unconsciousness. Between 17 and 16% heavy work is not possible. Light activity will result in an increase in pulse and respiration rate in order to improve oxygen supply to the tissues.

In an environment in which the oxygen concentration is diminished it is the rate of its decline which influences body response. A sudden reduction in which the partial pressure of oxygen is inadequate for it to cross the lung-blood barrier, as might occur when the oxygen supply to an aviator at very high altitude is dramatically cut off, results in convulsions and unconsciousness within a minute and, unless promptly relieved, death. A gradual reduction in oxygen concentration may be unnoticed by the victim, there being at first a feeling of well-being and over-confidence. Then mistakes in thinking and action may occur until, at a level of 10% or lower, unconsciousness follows and, possibly, death. Should the oxygen level be restored and the individual recover, the incident might not be recalled and there could be a repetition of the mistakes as before[22]. Recovery may be complete, or there may be residual headache and weakness for some hours. The most sensitive tissues are the brain, heart and retina, which are liable to sustain damage.

17.10 Occupational cancer

Cancer is a disorder of cell growth. It begins as a rapid proliferation of cells to form the primary tumour (neoplasm) which is either benign or malignant. If benign it remains localised, but may produce effects by pressure on neighbouring tissue. A malignant tumour invades and destroys surrounding tissue and spreads via lymph and blood streams to distant body parts (metastasis) such as the lung, liver, bone or kidney (secondary tumours). The patient becomes weak, anaemic and looses weight (cachexia). Pneumonia is the commonest form of death. The incidence of cancer increases with age and is responsible for 24% of all deaths.

Cancer is caused either by the inheritance of an abnormal gene, or exposure to an environmental agent acting either directly or indirectly on the cell genes.

Of all cancers, less than 8% are occupational and due to chemical and physical agents (see *Table 17.1*). Occupational cancers tend to occur after a long latent period of some 10–40 years and at an earlier age than spontaneous cancers.

Some carcinogens act together (synergistically); an example is found in asbestos workers who smoke and are 10 times more likely to develop cancer of the bronchus than those who do not.

Identification of occupational cancer often depends in the first place on the observation of a cluster of cases, as occurred with cancer of the scrotum in chimney sweeps in 1755, skin cancer in arsenic workers in 1822 and

Table 17.1. Table of some causes of occupational cancer in man

Agent	*Body site affected*	*Typical occupation*
Sunlight	Skin	Farmers and seamen
Asbestos	Lung, pleura, peritoneum	Demolition workers, miners
2-napthylamine	Bladder	Dye manufacture, rubber workers
Polycyclic aromatic hydrocarbons	Skin, lung	Coal gas manufacture, workers exposed to tar
Hard wood dust	Nasal sinuses	Furniture manufacture
Leather dust	Nasal sinuses	Leather workers
Vinyl chloride monomer	Liver	PVC manufacture
Chromium fume	Lung	Chromate manufacture
Ionising radiations	Skin and bone marrow	Radiologists and radiographers

cancer of the liver in PVC manufacture in 1930. Following observation of cases it is necessary to establish the potential link between cause and effect. This requires either a cohort or case control epidemiological study.

A cohort study compares people exposed to the suspect cause with those not exposed to determine if more people develop the disease among the exposed than the non-exposed. Such studies are expensive and time-consuming.

A case control study compares those with the disease with those without to determine if the suspect cause occurs more frequently among those with the disease than those without. It is quicker and cheaper than a cohort study.

These studies rely on the existence of good records of hygiene, exposure and health and also of lifestyle factors such as smoking which is the commonest single cause of respiratory cancer. The results of these studies may be confirmed where the individuals have continued to be exposed to the risk.

Once a risk is identified, animal studies are used to predict the carcinogenicity of substances, but they can take two to three years, are expensive and caution is needed when extrapolating the results to man. An alternate is the Ames Test in which a strain of bacteria, such as *Salmonella typhimurium*, is mixed with rat liver enzyme and the suspect chemical then incubated for two days. The carcinogenicity of the substance is indicated by the number of mutants induced. The Ames Test is sensitive, quick and cheap.

The classification of carcinogens is based on internationally agreed epidemiological and animal studies[23] and are:

Group 1 Carcinogenic to humans.
Group 2a Probably carcinogenic to humans with sufficient evidence from animal studies.

Group 2b Possibly carcinogenic to humans but absence of sufficient evidence from animal tests.
Group 3 Not classifiable as to its carcinogenicity to humans.
Group 4 No evidence of carcinogenicity in humans or animals.

Many chemical substances have been assigned the risk phrase 'R-45; May cause cancer' and these are listed in Appendix 9 of Occupational Exposure Limits[2].

Although the total number of deaths from cancer in this country is rising there is no evidence that the increase is due to the effect of industrial chemicals. The two most important factors leading to this increase appear to be the ever increasing number of lung cancer deaths due to smoking and fewer deaths from other causes such as infection thus putting more people at risk of developing cancer who otherwise would have died from other causes[24].

17.10.1 Angiosarcoma

Angiosarcoma is a rare 'cancer' of the liver, known to be associated with vinyl chloride monomer and, more rarely, with thorium dioxide. Much more commonly, angiosarcoma has occurred without a recognised association with any chemical. Vinyl chloride monomer (VCM) can be polymerised to form polyvinyl chloride (PVC) and was first discovered in Germany in the 1930s[25]. In 1966 VCM was known to cause bone disease, affecting the hands of Belgian autoclave workers employed in the manufacture of PVC. When, in 1971, the chemical was given to animals to reproduce the bone disease, it was found instead to have carcinogenic properties.

17.10.2 Vinyl chloride monomer (VCM) ($CH_2=CH$)

This gas is polymerised when heated under pressure (i.e. molecules of the gas are joined together in long chains) to form polyvinyl chloride (PVC). Although the explosive dangers of the gas have long been recognised, it was not until 1974 that three cases in American factory workers who were making PVC from VCM indicated that it could cause a rare liver tumour, angiosarcoma. Symptoms include abdominal pain, impaired appetite, loss of weight, distention of abdomen, jaundice and death. A Code of Practice[26] gives useful guidance on the control of this substance in the work environment.

17.11 Physical agents

In recent years there has been an increasing recognition of the harm that physical agents can do to the health of people at work. Injuries from this source now account for two-thirds of the new successful claims for industrial disease compensation.

17.11.1 Hand–arm vibration syndrome (HAVS)

HAVS follows from exposure to vibrations in the range 2–1500 Hz which causes narrowing in the blood vessels of the hand, damage to the nerves and muscle fibres and to bones and joint[27] evidenced by pain and stiffness in the joints of the upper arm. The impaired circulation of blood to the fingers leads to a condition known as *vibration white finger* (VWF). The most damaging frequency range is 5–350 Hz.

17.11.1.1. Vibration white finger

There is a latent period from first exposure to the onset of blanching which can vary from one to several years depending on the magnitude and frequency of the vibration and the length of exposure. Early symptoms include numbness, tingling of the index, middle and ring finger tips; coldness, pain and loss of sensation may follow. Later, there may be loss of finger dexterity (e.g. picking up objects and fastening buttons) and impairment of grip. Eventually the finger tips become ulcerated and gangrenous. The vascular and nervous effects may develop independently but usually occur concurrently. Disability is graded in accordance with the Stockholm scale (see *Table 17.2*)

17.12 Ionising radiations

Ionising radiations are so called because they produce 'ions' in irradiated body tissue. They also produce 'free radicals' which are parts of the molecule, electrically neutral but very active.

Table 17.2. Stockholm scale for the classification of the hand–arm vibration syndrome

Stage	*Grade*	*Description*
1. Vascular component		
1	Mild	Occasional blanching attacks affecting tips of one or more fingers
2	Moderate	Occasional attacks distal and middle phalanges of one or more fingers
3	Severe	Frequent attacks affecting all phalanges of most fingers
4	Very severe	As in 3 with trophic skin changes (tips)
2. Sensorineural component		
0_{SN}	–	Vibration exposed. No symptoms
1_{SN}	–	Intermittent or persistent numbness with or without tingling
2_{SN}	–	As in 1_{SN} with reduced sensory perception
3_{SN}	–	As in 2_{SN} with reduced tactile discrimination and manipulative dexterity

The staging is made separately for each hand.

The biological consequences of radiation depend on several factors:

(1) The nature of the radiation – some radiations being more damaging than others. Alpha particles are not harmful until they enter the body by inhalation, ingestion or via a wound. Beta particles can penetrate the skin to about 1 cm and cause a burn. X-rays, gamma rays and neutrons can pass right through the body and cause damage on the way.
(2) The dose and duration of exposure.
(3) The sensitivity of the tissue.
(4) The extent of the radiation.
(5) Whether it is external or internal.

17.12.1 Sensitivity of tissue

Tissues vary in their sensitivity to radiation, the most sensitive being the lymphocytes of the blood: they respond to excess radiation by a drop in their number within a couple of days. Next are the gonads, followed by blood cells formed in the bone marrow, the cells lining the bowel, and then the skin. Least sensitive are lung, liver and kidney cells.

17.12.2 Extent of radiation

Localised radiation is generally less immediately serious than whole body radiation for the same total dose.

17.12.3 Localised external radiation effects

Exposure to a small area of the body may result in redness of the skin, or even a blister, which either heals or ulcerates. The hands are very susceptible to localised radiation, the fingers becoming swollen and tender and, if the blood vessels are affected, gangrene could develop: the nails may become ridged and brittle. Exposure to the eyes in a dose of about 2 Sievert may lead to cataract after a lapse of about two years. Exposure to the gonads can cause mutation and loss of fertility.

Visible injury or loss of tissue function is called 'non-stochastic' or 'deterministic', while injury manifest as a neoplasm, being of a random statistical nature, is called 'stochastic'.

17.12.4 Whole body external radiation effects

Dose		*Effect*
rem	Sv	
Up to 25	Up to 0.25	Probably none. Lymphocyte count might fall in two days. Sperms and chromosomes may be damaged.
25–100	0.25–1.00	Damage more likely. Drop in total white cell count.
100–200	1.00–2.00	Nausea, vomiting, diarrhoea.
200–500	2.00–5.00	Above effects plus increasing mortality.
500–1000	5.00–10.00	Rapid onset of above symptoms, shock and coma.

17.12.5 Acute radiation syndrome

A dose of some 2 Sievert or more to the whole body may give rise to an 'acute radiation syndrome'. The response, depending on the intensity of the dose, begins with vomiting and diarrhoea within a few hours. By the second or third day there is an improvement, but the blood count falls. By the fifth day there is a return of symptoms, with fever and infection.

17.12.6 Internal radiation

These effects depend upon the nature of the radioactive material, its route of entry and concentration in a particular tissue, and due mainly to α or β particles. Lung cancer has been observed in miners following inhalation of radon, and severe anaemia and bone tumour following ingestion of radium in luminising dial painters.

17.12.7 Long-term effects

These may take several years to develop. Cancer of the skin or other organs has a peak incidence about seven years after exposure. The blood can be affected in two ways, either by leukaemia, which is a cancer of the white cells or, less commonly, by a severe anaemia in which the bone marrow fails to produce red cells. Chronic ulceration, loss of hair, cataracts, loss of fingertips, diminished fertility, and mutations may also occur. The maximum permitted doses are indicated in *Table 17.3.*

17.12.8 Medical examinations

A pre-employment medical is required for employees likely to receive a dose of ionising radiation exceeding 3/10 of the relevant dose limit. The examination will include a test of blood.

Table 17.3. Radiation dose limit

Body part	*Dose limit per calendar year mSv*
Whole body	50.0
Individual organs and tissues	500.0
Lens of eye	150.0
Woman of childbearing age	13 per 3 months
Pregnant woman	10 for period of pregnancy
Trainees under 18 years old:	
whole body	15
individual organs and tissues	150
lens of the eye	45

A certificate issued by the examining Employment Medical Adviser or factory doctor will be valid for one year.

17.12.9 Principles of control

The following simple precautions should be adopted to reduce to a minimum hazards from the use of radioactive materials:

(1) Employ the smallest possible source of radiation.
(2) Ensure the greatest distance between source and person.
(3) Provide adequate shielding between source and person.
(4) Reduce exposure time to a minimum.
(5) Practise good personal hygiene where there is risk of absorption of radioactive material.
(6) Personal sampling by use of (a) film badge and/or (b) thermal luminescent dose meter.

17.13 Noise-induced hearing loss

17.13.1 Mechanism of hearing

What we perceive as sound is a series of compressions and rarefactions transmitted by some vibrating source and propagated in waves through the air[28]. The compressions and rarefactions impinge on the ear drum (tympanic membrane) causing it to vibrate and transfer the movements through three small bones in the middle ear to the fluid of the inner ear. There they are received by rows of hairs (in the organ of corti), which vary in their response to different frequencies of sound, and are then transmitted to the brain and interpreted as sound.

17.13.2 Sensitivity of the ear

The ear can interpret frequencies between 20 and 20 000 Hz approximately. Frequencies below (infrasonic) and above (ultrasonic) this range are not heard. The range of frequency for speech is between 400 and 4000 Hz.

17.13.3 Definition and effects

Noise is commonly defined as unwanted sound. The definition is dependent on individual interpretation and may or may not include the recognition that some sounds produce harmful effects. Some 'sounds' cause annoyance, fright, or stress; others may interfere with communication. Loud sounds can cause deafness. 'Noise'-induced deafness is of two kinds: temporary and permanent.

17.13.4 Temporary deafness

Exposure to noise levels of about 90 dBA for even a few minutes may induce a temporary threshold shift (change of the threshold at which sound can just be heard), lasting from seconds to hours, and which can be detected by audiometry. Temporary threshold shift (TTS) may be accompanied by 'noises' in the ears (tinnitus) and may be a warning sign of susceptibility to permanent threshold shift (PTS) which is an irreversible deafness.

17.13.5 Permanent deafness

The onset of permanent deafness may be sudden, as with very loud explosive noises, or it may be gradual. A gradual onset of deafness is more usual in industry and may be imperceptible until familiar sounds are lost, or there is difficulty in comprehending speech. The consonants of speech are the first to be missed: f, p, t, s and k. These are of high frequency compared with the vowel sounds, which are of low frequency. Speech can still be heard, but without the consonants it is unintelligible. There is a risk too that a person exposed to excessive noise may believe himself to be adjusting to it when, in fact, partial deafness has already developed.

17.13.6 Limit of noise exposure

As noise effects are cumulative, the noise emission levels should be below 85 dBA. If this is not possible they should be reduced to the lowest level possible and suitable hearing protection provided. 10 years exposure at 90 dBA ($L_{EP.d}$) can be expected to result in a 50 dB hearing loss in 50% of the exposed population.

If the noise energy is doubled, then it is increased by 3 dBA and requires a halving of the exposure time. e.g.[28]

dBA	*Hours of exposure*
90	8
93	4
96	2
99	1
102	½
105	¼

The above table is helpful provided the noise level remains constant. For variable noise exposure, however, the daily personal noise exposure ($L_{EP.d}$) must be calculated.

Individuals exposed to 85 dBA must be offered hearing protection, but at 90 dBA or more hearing protection must be provided and worn.

17.13.7 The audiogram

An audiogram (*Figure 17.2*) is a measure, over a range of frequencies, of the threshold of hearing at which sound can just be detected. Early deafness occurs in the frequency range 2–6 kHz and is shown typically as a dip in the

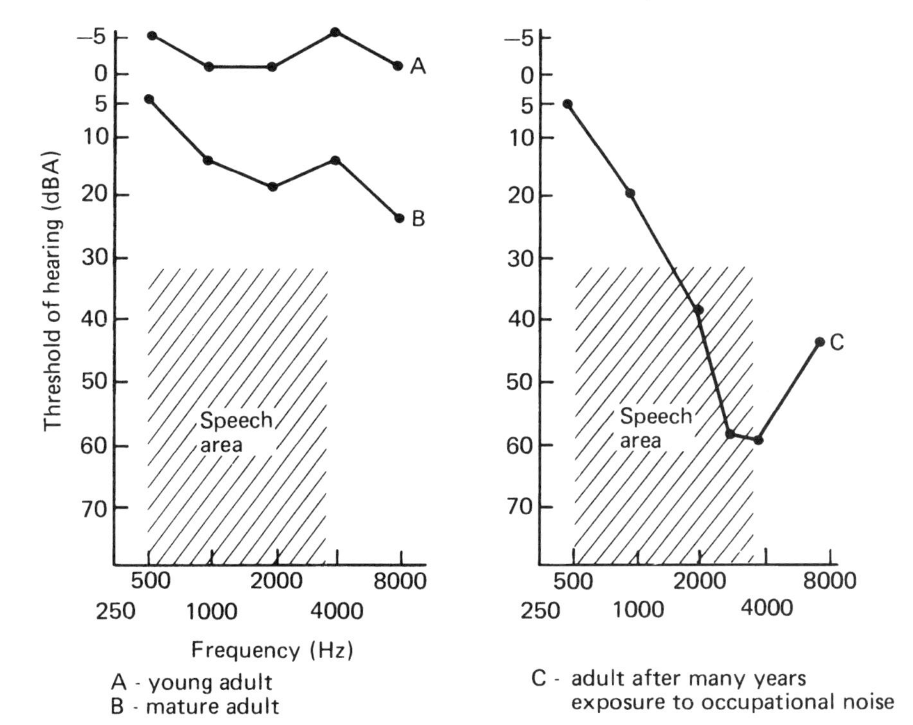

Figure 17.2 Audiograms

audiogram at 4 kHz. The depth of the dip depends on the degree of hearing damage and, as this worsens, so the loss of hearing widens to include neighbouring frequencies. The advantages of an audiogram are that it:

(1) provides a base line for future comparison;
(2) is helpful in job placement; and
(3) can be used to detect early changes in hearing and in the diagnosis of noise-induced deafness.

17.13.8 Occupational deafness

Disablement benefit may be awarded if deafness follows from:

(1) employment in a prescribed occupation for at least 10 years, or an incident at work, and
(2) the hearing loss is 50 dB averaged over the frequencies 1, 2 and 3 kilohertz in each ear.

17.14 Working in heat

Normally the human body maintains its core temperature within the range 36°–37.4°C by balancing its heat gains and losses. Maintaining an employee's health in a hot environment requires the control of air temperature and humidity, body activities, type of clothing, exposure time and ability to sweat. To sweat freely the individual must be fit, acclimatised to the heat with sufficient water intake to ensure a urine output of about 2½ pints per day. When the air temperature reaches 35°C plus, the loss of body heat is by sweating only, but this may be difficult when humidity reaches 80% or more.

Body reactions to overheating are:

An increase in pulse rate. The rate should fall by 10 beats/minute on cessation of exposure.
Muscle cramp due to insufficient salt intake.
Exhaustion with the individual feeling unwell and perhaps confused.
Fainting and dizziness with pallor and sweating.
Heat stroke is the most serious with the body temperature very high, the skin dry and flushed.
Dehydration due to insufficient fluid intake. Prolonged dehydration may lead to the formation of stones in the kidney.

Following first aid care, the patient needs to be referred to a doctor.

17.15 Work related upper limb disorders (WRULD)[29]

WRULD covers a number of conditions variously known as telegraphist's, writer's or twister's cramp and tenosynovitis, all of which became prescribed diseases in 1948. Other common conditions include carpal tunnel syndrome, tennis and golfer's elbow.

The condition arises from frequent forceful repetitive arm movement. Early symptoms include aches and pain in the hands, wrists, forearm, elbows and shoulders with tenderness over the affected tendons and muscles. Following rest there is a quick recovery. If similar work is resumed too soon, there is likely to be a worsening to the second phase when, in addition to the symptoms, there appears a redness, swelling and marked limitation of movement. A longer period of rest is then required with possible splinting of the limb and injections of cortisone. If not treated in time the condition can become extremely disabling and may require surgical intervention.

17.15.1 Prevention

Identify those jobs involving frequent prolonged rapid forceful movements, forceful gripping and twisting movements of the hand and arm, where the wrist is angled towards the little finger, the arm held above shoulder height or uncomfortably away from the body, and those where repetitive pushing, pulling and lifting are necessary.

Ensure hand tools are designed with good mechanical advantage and have a comfortable grip, are suitable for those who use them and that cutting edges are kept sharp.

Those involved in the work should be warned of the risks and trained in the correct use of the tools. Rest periods and work rotation should be introduced and piece work avoided.

Where the condition is suspected, complaints should be monitored and checks made of first aid records and absence certification. Susceptible persons should be examined by a doctor before further exposure.

17.16 Diseases due to micro-organisms

Micro-organisms include a variety of minute organisms such as viruses, bacteria, fungi and protozoa that can only be seen with the aid of a microscope. Micro-organisms gain entry to the body through the lungs, gut or breaks in the skin. If their virulence overcomes the body's defence, disease may result. The term pathogen covers all micro-organisms which cause disease. Diseases of animals transmitted to man are known collectively as zoonoses.

Micro-organisms account for about 10% of successful new occupational disease claims. Typical examples are:

Organism	*Disease*
Viruses:	hepatitis A and B, AIDS, orf
bacteria:	anthrax, legionella, leptospirosis, tuberculosis, tetanus, ornithosis, Q fever, dysentery
fungi:	Farmer's lung, ringworm, athlete's foot
protozoa:	malaria, amoebiasis
nematodes:	hookworm

17.16.1 Hepatitis

Hepatitis is an inflammatory condition of the liver. When occurring at work it is usually caused by infections or toxic substances such as alcohol and organic solvents. Most commonly the cause is a virus of which there are three main kinds.

17.16.1.1 Infective hepatitis or 'Hepatitis A'

The virus is transmitted from infected stools to the mouth. After 2–6 weeks there occurs fever, nausea, abdominal pain and jaundice. Recovery usually occurs in 1–2 weeks and recurrence is rare.

Precautions to be taken include good personal hygiene, washing hands after the toilet and before handling food. There is a vaccine which gives long-term protection.

17.16.1.2 Serum hepatitis or 'Hepatitis B'

The virus is transmitted in infected blood or serum, especially among drug addicts who share needles; there is also a risk in renal dialysis units. The disease manifests itself after 2–6 months with symptoms similar to Hepatitis A but the effects are more prolonged and damaging.

Precautions that should be taken include the screening of donor blood for the presence of antigen and the non-reuse of needles and syringes. People at special risk are those who come into contact with blood or blood products and they should be immunised with Hepatitis B vaccine.

17.16.1.3 Hepatitis C

This is transmitted by the same route as Hepatitis B and poses an occupational hazard to a similar group of workers. There is no vaccine available.

17.16.2 Leptospirosis

Leptospirosis, sometimes known as Weil's disease, is a form of hepatitis caused by bacteria from the urine of infected rats or dogs which enter by a break in the skin or by ingestion. A week later a 'flu-like disease occurs sometimes with jaundice. The liver and kidneys may be severely damaged with a 20% mortality. Those most at risk include workers in abattoirs, sewers, mines, tunnels, canals and veterinary workers. They should be informed of the risk and issued with a Weil's disease warning card to be presented to their doctor.

17.16.3 Legionnaire's disease

In 1976 nearly 200 American Legionnaires attending a convention at a hotel in Philadelphia collapsed with a 'flu-like disease, some with pneumonia, and there were 29 deaths. The disease was later attributed to bacteria, named *Legionella,* of which there are several types differing in pathogenicity.

Legionella bacteria occur in soils, rivers and streams, but the recent dramatic appearance of the disease in hotels, hospitals and industry is related to modern building design which allows water in air-conditioning and water systems to stagnate. At temperatures of 20–50°C bacterial growth is encouraged, which if released as a spray and inhaled, leads to pneumonia. Middle-aged smokers are most vulnerable. The disease has a 15% mortality. Diagnosis is confirmed by the presence of antibodies in the blood.

Water systems and bath shower heads should be cleaned and chlorinated periodically and a record maintained.

Pontiac fever is a less serious non-pneumonic form of the disease.

17.16.4 Anthrax

Anthrax is a highly infectious disease of ruminants: goats, cattle, sheep and horses, and is due to a bacillus. Man can be affected by direct contact with the animal, or indirectly by contact with the animal products. The disease is rare amongst animals in Britain, but it can be introduced into the country by infected materials, such as hides and skin, hair and wool, dried bones and bone-meal, hooves and horn. At risk particularly are those engaged in tanning, wool sorting, manufacture of brushes, bone-meal, fertiliser and glue. Also at risk are dockers and agricultural workers.

17.16.4.1 Symptoms

The disease involves the skin in about 95% of cases – the bacillus entering through an abrasion, commonly on the arm. In 2–5 days there appears a red-brown spot or pimple, which becomes a black ulcer surrounded by tiny blisters and inflamed tissue. It is usually painless, but the individual feels unwell with fever, headache, sickness and swollen glands, usually under the arm or in the groin if the leg is infected. Should the bacilli be inhaled, there follows a severe pneumonia with cough and blood-stained sputum. The mortality rate is high. Abdominal infection following ingestion is very rare. Fortunately, the disease responds to an antibiotic like penicillin if given early.

17.16.4.2 Prevention

Employees in the risk industries should be informed of the danger and carry with them an HSE card MS(B)3. All cuts must be treated and covered with a dressing while at work. Attention must be given to personal hygiene, washing hands, arms and face before meals and at end of shift. Protection can be provided by immunisation, which requires three injections at three-week intervals, a fourth six months later and then annually. Protective gloves should be worn wherever possible.

17.16.5 Humidifier fever

Humidifier fever is a 'flu-like condition which follows the inhalation of a variety of organisms such as amoeba, bacilli and fungi that grow in humidifying systems. Symptoms of cough, limb pain and fever occur within a few hours of starting work. The disease is usually short term with recovery by the next day, but symptoms are likely to recur on returning to work after a few days off. It is sometimes known as Monday Morning Fever.

Water systems need to be cleaned and chlorinated periodically and a record of this maintained.

17.16.6 Tuberculosis

The incidence of tuberculosis is increasing in some communities owing to resistance to the drugs used in its treatment and to lowered resistance in AIDs patients. Infection is by inhalation of bacteria.

Mainly a disease of the lungs, its symptoms are a persistent cough, bloody sputum, night sweating and loss of weight. Sometimes no symptoms occur and the disease is first discovered on chest X-ray. Those most at risk are medical, veterinary and mortuary staff.

17.16.7 Other diseases of micro-organisms

A number of other diseases that have to be reported include:

Shigella flexna dysentery: caused by a bacillus, transmitted by faecal–oral route where there is poor hygiene and overcrowding.

Ornithosis: Bacterial disease of birds transmitted by inhalation of dried faeces causing a 'flu-like disease with cough.

Q fever: A bacterial infection spread from cows and sheep to farm workers who inhale infected dust causing a 'flu-like illness.

Orf: A viral disease of sheep and goats transmitted to farm workers causing a pustular ulcer usually on the hands.

Farmer's lung: An asthma-like complaint resulting from the inhalation of mould from rotten hay.

17.17 Psycho-social disorders

This group is probably the largest group of occupational diseases. It stems from the complex interaction of individual, social and work factors and is responsible for a great amount of sickness absence.

17.17.1 Stress

Stress is a reaction of the body to external stimuli which ranges from the apparently normal to the overtly ill health. It varies with the individual personality but is one of the commonest occupational diseases.

The initial response is physiological and shows as an increase in pulse, blood pressure and respiratory rates. Although the body adjusts, persisting stimuli cause fatigue and the display of signs of 'over stress' with sweating, anxiety, tremors and dry mouth. There is difficulty in relaxing, appetite is lost, sleep disturbed and a loss of concentration. Some may eventually become depressed, aggressive and try to avoid the cause through absenteeism or the use of alcohol or drugs. Other diseases may appear affecting the skin, peptic ulcer and coronary heart disease.

Studies have identified two personality groups: type A people who are competitive, impatient achievers and who are at greatest risk from the severe effects of stress, whereas type B people are easy-going, patient and less susceptible to pressure. Causes of stress may be considered under a number of headings:

The person – lack of physical and mental fitness to do the job; inadequate training or skill for the particular job; poor reward and prospects; financial difficulties; fear of redundancy; lack of security in the job; home and family problems; long commuting distances.

Work demand – long hours; shift work; too fast or too slow a pace; boring repetitive work; isolation; no scope for initiative or responsibility.

Environment – noise; heat; humidity; fumes; dust; poor ventilation; diminished oxygen; confined space; heights; poor house-keeping; bad ergonomic design.

Organisation – poor industrial relations, welfare services and communications; inconsiderate supervision; remote management.

17.18 Target organs

Target organs are those body parts which sustain some adverse effect when exposed to or contaminated by harmful substances or agents.

Many of the body target organs have been referred to in the text and the table below gives a summary of them with causes.

Body part	*Condition*	*Cause*
HANDS/ARMS	Vibration white finger:	using vibrating tools.
	Tenosynovitis:	repetitive pulling and twisting actions with forceful movements.
	Dermatitis:	exposure to irritants.
LUNGS	Fibrosis:	coal, silica, asbestos, dust.
	Allergy:	isocyanates, flour and grain dust, colophony, epoxy resins, hard wood dusts, moulds.
	Irritation:	nitrous fumes, phosgene, chlorine, hydrogen sulphide, sulphur dioxide, ammonia.
	Infection:	*legionella*, tuberculosis.
	Cancer:	asbestos, radon, nickel.
SKIN	Dermatitis:	solvents, acids/alkalis, mercury, chrome, nickel, arsenic, mineral oils, wood, plants, resins, heat
	Cancer:	aromatic polycyclic hydrocarbons, arsenic, uv light, ionising radiations.

Body part		*Condition*	*Cause*
HEAD:	EARS	Deafness:	noise
	EYES	Cataracts:	ionising radiations, uv light, heat, acids/alkalis.
	NOSE	Ulceration:	chrome.
	TEETH	Loosening:	mercury.
		Erosion:	sulphuric acid.
		Mottling:	fluorides.
		Discolouration:	vanadium, iodine, bromine.
BRAIN		Narcosis:	chlorinated solvents.
		Encephalopathy:	mercury, lead, manganese, carbon disulphide, carbon monoxide.
PERIPHERAL NERVES		Neuritis:	lead, mercury, carbon disulphide, tetrachloroethane, trichloroethylene, organo-phosphorus compounds.
HEART		Cardio-vascular disease:	lead, ionising radiations, organic solvents, arsine, vinyl chloride, carbon disulphide.
LIVER		Hepatitis:	organic solvents, viral hepatitis, leptospirosis, arsenic, manganese, beryllium.
		Cancer:	Hepatitis B and C, vinyl chloride monomer.
KIDNEY		Disease:	halogenated solvents, infections, lead, mercury, cadmium, micro-organisms.
BLADDER		Cancer:	2-naphthylamine.
BONE		Osteolysis:	vinyl chloride monomer, vibration, pressure.

References

1. Harrington, J.M. and Gill, F.S., *Occupational Health Pocket Consultant,* Blackwell Scientific Publications, Oxford (1992)
2. Health and Safety Executive, *Guidance Note EH40, Occupational Exposure Limits,* HSE Books, Sudbury (latest issue)
3. Health and Safety Executive, Guidance Note, Environmental Health Series No. EH 64 *Occupational exposure limits: criteria document summaries.* Synopses of the data used in setting occupational exposure limits, HSE Books, Sudbury (1992)
4. Fregert, S., *Manual of Contact Dermatitis,* A. Munksguard (1974)
5. Bishop, C. and Kipling, M.D., Dr J. Ayston Paris and Cancer of the Scrotum, 'Honour the Physician with the Honour due unto Him', *J. Soc. Occup. Med.,* **28,** 3–5 (1978)
6. Hunter, D., *The Diseases of Occupations,* 8th edn, Hodder & Stoughton, London (1992)
7. Hodgeson, G.A. and Whiteley, H.J., Distribution of pitch warts – personal susceptibility to pitch, *Brit. J. Ind. Med.,* **27,** 160–166 (1970)

8. Ref. 7, p. 20
9. Parkes, W.R., *Occupational Lung Disorders,* (3rd Edn), Butterworths, London (1990)
10. Ref. 9, p. 231
11. Alexander, W.S. and Street, A., *Metals in the Service of Man,* 6th edn, 30, Penguin Books, London (1976)
12. Waldron, H.A., Health care of people at work – workers exposed to lead, inorganic lead, *J. Soc. Occup. Med.,* **28,** 27–32 (1978)
13. Clarkson, T.W., *Mercury Poisoning Clinical Chemistry and Chemical Toxicology of Metals,* 189–204, Elsevier, Amsterdam (1977)
14. Health and Safety Executive, Mercury – medical surveillance, Guidance Notes MS 12, HSE Books, Sudbury (1978)
15. Uvaroy, E.B., Chapman, D.R. and Isaacs, A., *Dictionary of Science,* 5th edn, Penguin Books, London (1976)
16. Matheson, D., *Occupational Health and Safety,* 2085, ILO, Geneva (1983)
17. Bird, M.G., Industrial solvents: some factors affecting their passage into and through the skin, *Annals of Occupational Hygiene,* **24,** No. 2 (1981)
18. Gompertz, D., Solvents: the relationship between biological monitoring stratagem and metabolic handling. A review, *Annals of Occupational Hygiene,* **23,** No. 4 (1980)
19. Commission of the European Communities, *Regulation No. 594/91 on banning the use of methyl chloroform,* EC Publications Department, Luxembourg and HMSO, London (1991)
20. Green, J.H., *An Introduction to Human Physiology,* 78–79, 3rd edn, Oxford University Press (1974)
21. Lamphier, E.H., *The Physiology and Medicine of Diving,* 59–60, Bailliere, Tindall & Cassell, London (1969)
22. Miles, S. and Mackay, D.E., *Underwater Medicine,* 4th edn, 107–108, Adlard Coles Ltd, St. Albans (1976)
23. International Agency for Research on Cancer, *Monograph on the evaluation of carcinogenic risks to humans,* **46,** World Health Organisation, Geneva (1989)
24. Editorial: What proportion of cancers are related to occupation?, *Lancet,* 1238, Dec. 9 (1978)
25. Gauvain, S., Vinyl chloride, *Proc. Royal Soc. Med.,* 69 (1976)
26. Health and Safety Executive, Approved Code of Practice No. COP 31, *Control of vinyl chloride at work,* HSE Books, Sudbury (1988)
27. Royal College of Physicians, *Hand Transmitted Vibrations,* 2 vols., Royal College of Physicians, London (1993)
28. Department of Employment, *Noise and the Worker,* HMSO, London (1971)
29. Health and Safety Executive, Health and Safety Guidance Booklet No. HS(G)60, *Work related upper limb disorders: a guide to prevention,* HSE Books, Sudbury (1990)

Further reading

National Radiological Protection Board, *Living with Radiation,* 4th edn, National Radiological Protection Board, Didcot (1989)

British Medical Association, *The BMA Guide to Living with Risk,* Penguin Books, London (1990)

Olsen, J., Merletti, F., Snashall, D. and Vuylsteek, K., *Searching for Causes of Work related Diseases,* Oxford Medical Publications, Oxford (1991)

Rose, G. and Barker, D. J. P., *Epidemiology for the Uninitiated,* British Medical Association, London (1979)

Chapter 18

Occupational hygiene

C. Hartley

Occupational hygiene is defined by the British Occupational Hygiene Society as: 'the applied science concerned with the identification, measurement, appraisal of risk, and control to acceptable standards, of physical, chemical and biological factors arising in or from the workplace which may affect the health or well-being of those at work or in the community'.

It is thus primarily concerned with the identification of health hazards and the assessment of risks with the crucial purpose of preventing or controlling those risks to tolerable levels. This relates both to the people within workplaces and those who might be affected in the surrounding local environment.

Occupational hygiene deals not only with overt threats to health but also in a positive sense with the achievement of optimal 'comfort conditions' for workers, i.e. the reduction of discomfort factors which may cause irritation, loss of concentration, impaired work efficiency and general decreased quality of life.

The American Industrial Hygiene Association in its corresponding definition begins: 'Industrial hygiene is that science and *art* devoted to the recognition, evaluation and control . . .' [author's emphasis] indicating that although much of occupational and industrial hygiene is underpinned by proven scientific theory, a considerable amount relies on 'rule of thumb'; thus in the practical application of occupational hygiene, judgemental and other skills developed by the experienced practitioner are important.

18.1 Recognition

People at work encounter four basic classes of environmental stress, examples of which are given in *Table 18.1*.

18.2 Evaluation

When the potential for harm of a particular hazard in the workplace has been identified, it is necessary to assess the consequent risk. Some common environmental measurement techniques together with their interpretation as they relate to accepted standards are reviewed.

Table 18.1. Environmental stress

Environmental stress	*Example*
Chemical	Exposure of the worker to dusts, vapours, fumes, gases, mists etc. 100 000 chemicals are believed to be in common use in the UK at present
Physical	Noise, vibration, heat, light, ionising radiation, pressure, ultraviolet light etc.
Biological	Insects, mites, yeasts, hormones, bacteria, viruses, proteolytic enzymes
Ergonomic	Man-machine interaction, e.g. body position in relation to task on machine

18.2.1 Environmental measurement techniques

18.2.1.1 Grab sampling

Stain detector tubes are used for measuring airborne concentrations of gases and vapours. Several proprietary types are available which operate on a common principle. A sealed glass tube is packed with a particular chemical which reacts with the air contaminant. The tube seal is broken, a hand pump attached, and a standard volume of contaminated air is drawn through the tube (*Figure 18.1*). The packed chemical undergoes a colour change which passes along the tube in the direction of airflow. The tube is calibrated so that the extent of colour change indicates the concentration of contaminant sampled (*Figure 18.2*).

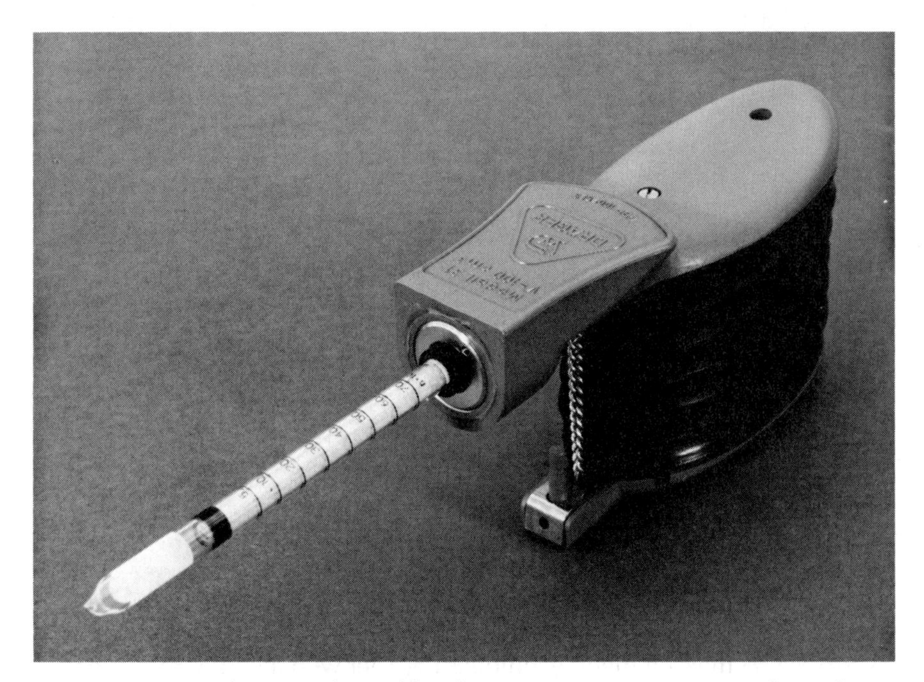

Figure 18.1 Stain detector tube and hand pump. (Courtesy Draeger Safety Ltd)

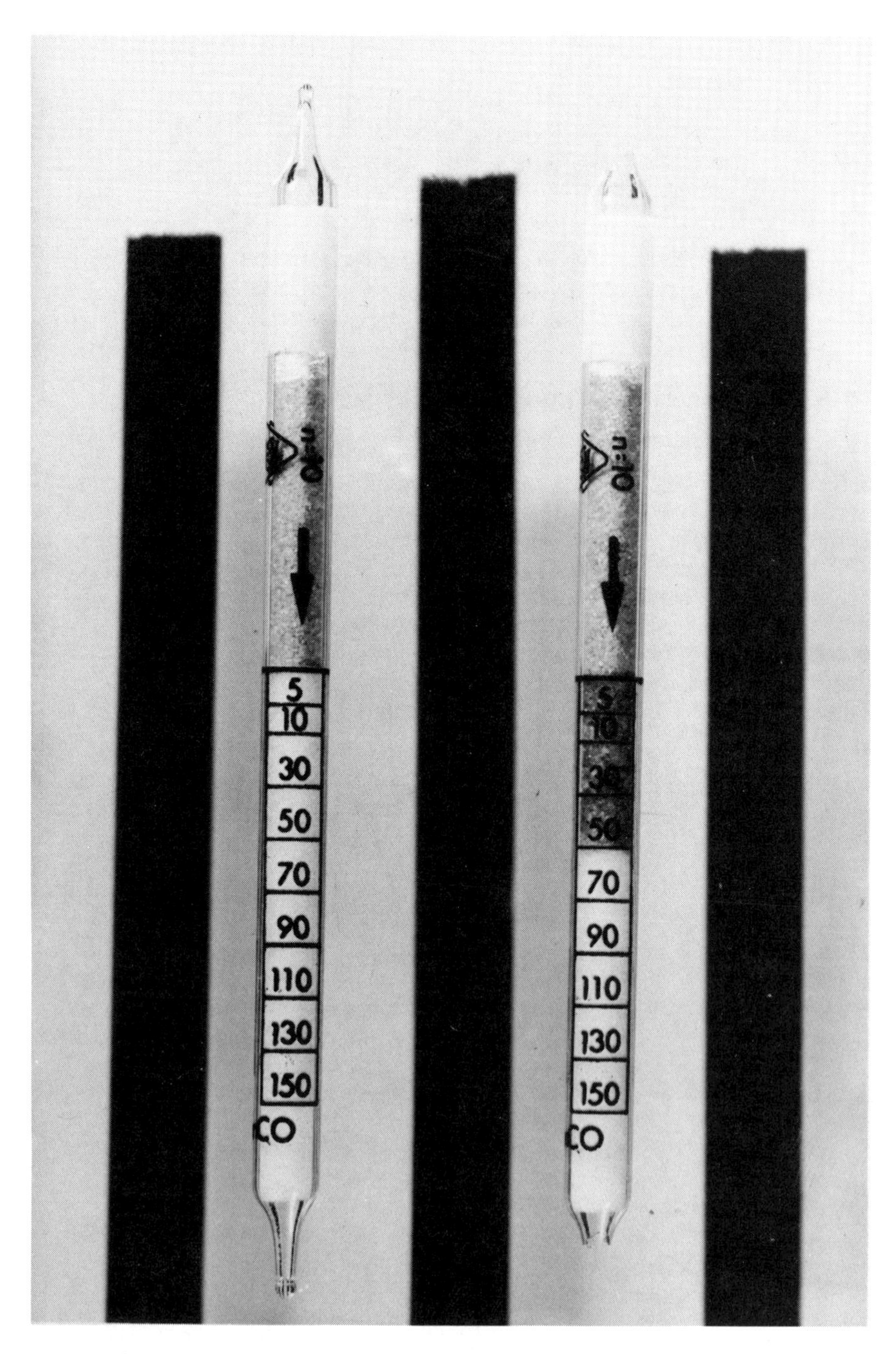

Figure 18.2 Stain detector tubes, illustrating the principle of detection. (Courtesy Draeger Safety Ltd)

The hand pump must be kept in good repair and re-calibrated at intervals to check that it is drawing the standard volume of air and care taken to ensure that a good seal is obtained between the pump and tube.

This method of measurement has several advantages:

(1) It is a quick, simple and versatile technique.
(2) Stain detector tubes are available for a wide range of chemical contaminants.
(3) Measurement results are provided instantaneously.
(4) It is a relatively economical method of measurement.

However it is important to be aware of the limitations of stain tubes:

(1) The result obtained relates to the concentration of contaminant at the tube inlet at the precise moment the air is drawn in. Short-term stain tubes do not measure individual worker exposure.
(2) Variations in contaminant levels throughout the work period or work cycle are difficult to monitor by this technique.
(3) Cross-sensitivity may be a problem since other chemicals will sometimes interfere with a stain tube reaction. For example, the presence of xylene will interfere with stain tubes calibrated for toluene. Manufacturer's handbooks give information on the known cross-sensitivities of their products. This point emphasises the need to consider the whole range of chemicals used in a process rather than just the major ones.
(4) Stain tubes are not re-usable.
(5) Random errors associated with this technique can range up to ±25% depending on tube type.

Using a planned sampling strategy rather than an occasional tube will give a better picture of toxic contaminant levels and manufacturers will give guidance about this. However since tubes are not re-usable this method could be more costly than some of the long-term sampling techniques.

Some basic practical guidance with regard to toxic substance monitoring is given in an HSE publication[1].

18.2.1.2 Long-term sampling

This involves sampling air for several hours or even over the whole workshift. Results give average levels of contaminant across the sample period. The air sampling may be carried out in the worker's breathing zone (Personal Sampling) or at a selected point or points in the workplace (Static or Area Sampling). Long-term stain detector tubes are available for this purpose and are connected to a pump which draws air through the tube at a predetermined constant rate. At the end of the sampling period the tube is examined as above to give the amount of contaminant absorbed, from which the average level of contamination can be calculated.

For a few substances, direct indicating diffusion tubes are available and these devices may be used for up to 8 hours. In these cases, because the contaminant is collected by diffusion, no sampling pump is required – with obvious benefits.

Often more accurate methods are required in the assessment of worker exposure and charcoal tube sampling is commonly used. Air is drawn by an

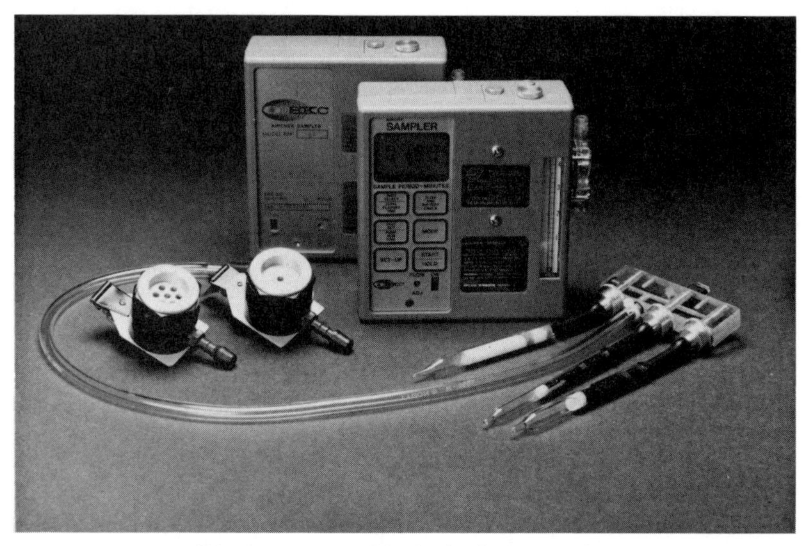

Figure 18.3 Sampling pumps suitable for both dust and vapour monitoring with collection heads. (Courtesy SKC Ltd)

attached portable low-flow pump through a tube containing absorbent charcoal (*Figure 18.3*). The chemical contaminant is trapped in the tube by adsorption on the charcoal granules. After sampling the tube is sealed and taken to the laboratory, where the contaminant is removed (desorbed) by chemical or physical means, followed by quantitative analysis, often using gas chromatographic techniques to determine the weight of contaminant adsorbed. When related to the known total air volume sampled (from pump flow rate and sampling time) the airborne concentration of the chemical over the sample period can be calculated.

Diffusive monitors or badges containing various absorbents are now becoming much more widely used. These are easy to use since they do not require a sampling pump with the result that sampling can be carried out by suitably trained non-specialists.

Long-term sampling methods are reliable, versatile and accurate, being widely used by occupational hygienists in checking compliance with hygiene standards.

18.2.1.3 Direct monitoring of gases and vapours

A wide range of instruments is used in the detection of gases, vapours and dusts. These devices make a quantitative analysis giving a direct read-out of contaminant level on a meter, chart recorder or other display equipment.

For example, a portable infrared gas analyser allows direct monitoring of gases and vapours in the workplace (*Figure 18.4*). The principle utilised is that many gases and vapours will absorb infrared radiation and under standard conditions the amount of absorption is directly proportional to the concentration of the chemical contaminant. This instrument takes a sample

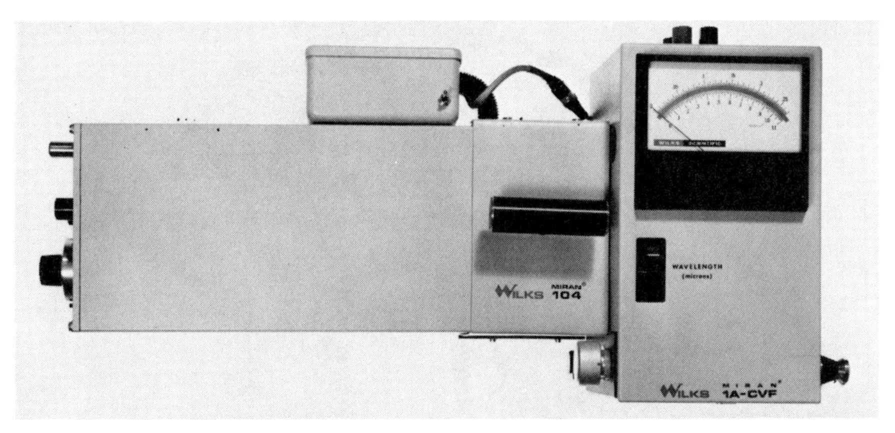

Figure 18.4 Portable infrared gas analyser. (Courtesy Foxboro Analytical Ltd)

of air, detects the extent of interruption of an infrared beam and gives a read-out of the result. Before use the instrument must be set up and calibrated for the particular chemical to be measured. Because so many substances absorb infrared radiation this is an extremely versatile instrument.

There is a wide variety of commercially available direct monitoring instruments, which are based on different physical principles of detection. Many of the physical principles involved and the range of instruments available are considered by Nader[2].

Direct monitoring is particularly useful where there is a need to have immediate readings of contaminant levels, for example in the case of fast acting chemicals. Also periods of peak concentration during the work cycle or work shift can be detected and this may be useful in determining a control strategy.

18.2.1.4 Oxygen analysers

Deficiency of oxygen in the atmosphere of confined spaces is often experienced in industry, for example inside large fuel storage tanks when empty. Before such places may be entered to carry out inspections or maintenance work a check must be made on the oxygen content of the atmosphere throughout the vessel. Normal air contains approximately 21% oxygen and when this is reduced to 16% or below people experience dizziness, increased heartbeat and headaches. Such atmospheres should only be entered when wearing air supplied breathing apparatus.

Portable analysers are available which measure the concentration of oxygen in the air by the depolarisation produced at a sensitive electrode mounted in the instrument (*Figure 18.5*). Several different devices are available which vary in sensitivity, reliability and ease of maintenance but they must all be checked and carefully calibrated to the manufacturer's instructions before use. Long extension probes may be attached which allow remote inspection of confined spaces.

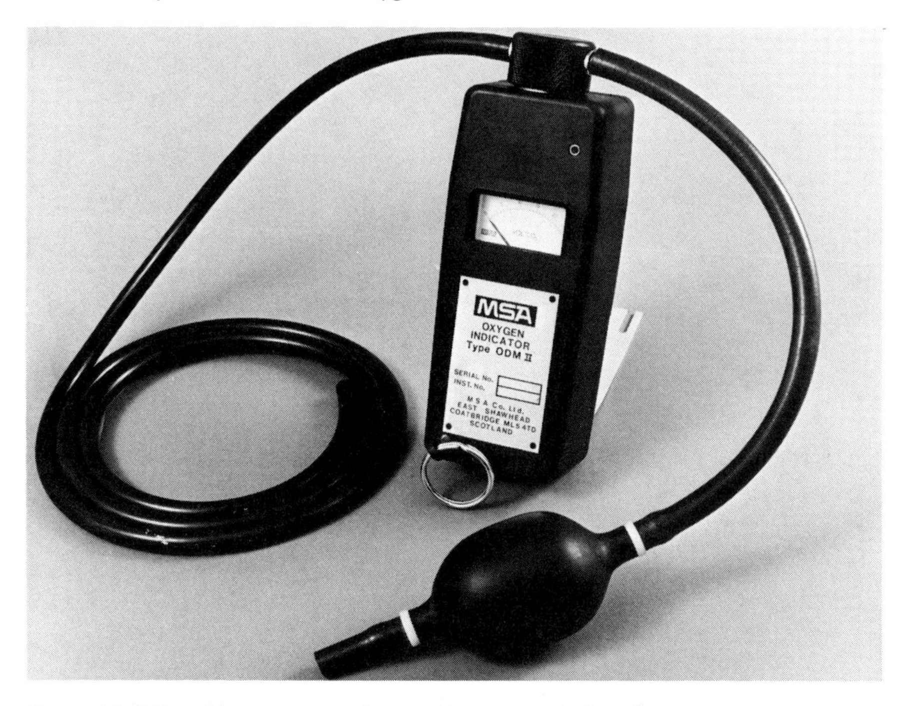

Figure 18.5 Portable oxygen analyser. (Courtesy M.S.A. Ltd)

18.2.1.5 Hygrometry

Hygrometers are instruments used in the measurement of the water vapour content of air, i.e. the humidity. The comfort and efficiency of people depends on their ability to lose heat to the environment so that body temperature may remain constant. In conditions of high temperature and humidity this heat loss cannot occur fast enough. Thus measurement of humidity and its control are important in ensuring the thermal comfort of people at work. There are also some industrial processes, e.g. textile manufacture, whose success depends on a controlled humidity in the mill.

Humidity is generally expressed as relative humidity and quoted as a percentage. It is the ratio at a given atmospheric temperature and pressure of the mass of water in a given volume of air to the mass if the air had been saturated with water. (See *Table 18.2.*)

18.2.1.6 Wet and dry bulb hygrometer

This instrument consists of two normal thermometers, one of which has its bulb exposed to the air while the other has its bulb surrounded by a wick connected to a water reservoir. Evaporation of moisture from the wick to the surrounding air causes the 'wet bulb' thermometer to show a lower reading than the corresponding 'dry bulb' thermometer. The difference between the wet and dry bulb temperature related to the dry bulb temperature defines the hygrometric state of the atmosphere. Tables have been produced, from large

Table 18.2. Relative humidity table (for whirling hygrometer). (Courtesy F. Darton & Co. Ltd).

DEPRESSION OF WET BULB

DRY BULB READINGS °C	·25	·5	·75	1·0	1·25	1·5	1·75	2·0	2·25	2·5	2·75	3·0	3·25	3·5	3·75	4·0	4·25	4·5	4·75	5·0	5·25	5·5	5·75	6·0	6·25	6·5	6·75	7·0	7·25	7·5	7·75	8·0	8·25	8·5	8·75	9·0	9·25	9·5	9·75	10·0	10·25	10·5	10·75	11·0
—5	94	88	83	77	71	66	60	54	49	43	38	32	29	26	23	20																												
4	95	89	84	78	73	67	62	57	51	46	41	36	31	27	24	22	20																											
3	95	89	84	79	74	69	64	59	54	49	44	39	34	29	25	23	21	19																										
2	95	90	85	80	75	70	66	61	56	52	47	42	37	33	28	24	22	19	17																									
1	96	91	86	81	76	72	67	63	58	54	49	45	40	36	32	27	24	21	19	17																								
0	96	91	86	82	78	73	69	65	60	56	52	48	43	39	35	31	27	24	20	18	16																							
+1	96	91	87	83	79	75	71	66	62	58	54	50	46	42	38	34	30	26	22	19	17	15																						
2	96	92	88	84	80	76	72	68	64	60	56	52	49	45	41	37	33	30	26	22	20	17	15																					
3	96	92	88	84	81	77	73	69	66	62	58	54	51	47	44	40	36	33	30	26	23	20	17	15																				
4	96	92	89	85	82	78	74	70	67	63	60	56	53	49	46	42	39	35	32	29	26	23	20	17	15																			
5	97	93	89	86	83	79	75	72	69	65	62	58	55	51	48	45	42	38	35	32	28	24	21	19	17	15	12	10																
6	97	93	90	86	83	79	76	73	70	66	63	60	57	53	50	47	44	41	37	35	32	29	25	23	20	18	16	12	10															
7	97	93	90	87	84	80	77	74	71	67	64	61	59	55	53	49	46	43	40	37	35	31	28	26	23	20	17	14	12	10														
8	97	94	91	87	85	81	78	75	72	69	66	63	60	57	54	51	48	45	42	40	37	34	31	29	26	23	20	18	16	13	11													
9	97	94	91	88	85	82	79	76	73	70	67	64	62	58	56	53	50	47	44	42	39	36	33	31	29	26	23	21	18	16	13	11												
10	97	94	91	88	86	82	79	76	74	71	68	65	63	60	58	54	52	49	46	44	41	39	36	34	31	29	26	24	22	19	16	14	12	10										
11	98	94	91	88	86	83	80	77	75	72	69	66	64	61	58	56	54	51	48	46	43	41	38	36	33	31	28	26	24	22	19	17	15	12	10									
12	98	94	91	89	87	83	80	78	75	73	70	68	65	62	59	57	55	53	50	48	45	43	40	38	35	33	31	29	26	24	22	20	18	16	14	11								
13	98	95	91	89	87	84	81	79	76	74	71	69	66	64	61	59	56	54	51	49	47	45	42	40	37	36	33	31	29	27	24	23	21	18	16	14	12							
14	98	95	92	90	87	84	81	79	77	74	72	70	67	65	62	60	57	56	53	51	49	46	44	42	39	38	35	33	31	29	27	25	23	21	19	17	15	12						
15	98	95	92	90	88	85	82	80	77	75	72	71	68	66	63	61	59	58	54	52	50	48	46	44	41	40	38	36	34	32	29	27	25	23	21	20	18	16	14	12				
16	98	95	92	90	88	85	83	81	78	76	73	71	69	67	65	63	61	58	56	54	52	50	47	46	43	42	39	37	36	34	32	30	28	26	24	22	20	18	16	15	13	11	9	8
17	98	95	92	90	88	86	83	81	78	77	74	72	70	68	66	64	62	59	57	55	53	51	49	47	45	43	41	39	38	36	33	32	30	28	26	24	22	21	19	17	16	14	12	11
18	98	95	93	91	89	86	83	82	79	77	75	73	71	69	67	65	63	61	58	56	54	53	50	49	47	45	43	41	39	37	35	34	32	30	28	27	25	23	21	20	18	16	14	13
19	98	95	93	91	89	86	84	82	80	78	76	74	72	70	67	65	64	62	59	58	56	54	52	50	48	46	44	43	41	39	37	36	34	32	30	29	27	25	23	22	20	19	17	16

DEPRESSION OF WET BULB

DRY BULB READINGS

°C	·5	1·0	1·5	2·0	2·5	3·0	3·5	4·0	4·5	5·0	5·5	6·0	6·5	7·0	7·5	8·0	8·5	9·0	9·5	10·0	10·5	11·0	11·5	12·0	12·5	13·0	13·5	14·0	14·5	15·0	15·5	16·0	16·5	17·0	17·5	18·0	18·5	19·0	19·5	20·0
+20	96	91	87	83	78	74	70	66	63	59	55	51	48	44	41	37	34	30	27	24	21	18	15	12	9	6	3													
21	96	91	87	83	79	75	71	67	64	60	56	52	49	46	42	39	36	32	29	26	23	20	17	14	11	8	6	3												
22	96	92	88	83	80	76	72	68	64	61	57	54	50	47	44	40	37	34	31	28	25	22	19	16	13	11	8	5	3											
23	96	92	88	84	80	76	72	69	65	62	58	55	51	48	45	42	39	36	33	30	27	24	21	18	16	13	10	8	5	3										
24	96	92	88	84	80	77	73	69	66	63	59	56	53	49	46	43	40	37	34	31	28	26	23	20	18	15	12	10	7	5	3									
25	96	92	88	84	81	77	74	70	67	63	60	57	54	50	47	44	41	38	36	33	30	27	25	22	19	17	14	12	10	7	5	3								
26	96	92	88	85	81	78	74	71	67	64	61	58	55	52	48	46	43	40	37	34	32	29	26	24	21	19	16	14	12	9	7	5	3							
27	96	92	89	85	81	78	75	71	68	65	62	59	55	52	50	47	44	41	38	36	33	30	28	25	23	21	18	16	14	11	9	7	5	3						
28	96	93	89	85	82	78	75	72	68	65	62	59	56	53	50	48	45	42	40	37	34	32	29	27	25	22	20	18	15	13	11	9	7	5	3					
29	96	93	89	86	82	79	76	72	69	66	63	60	57	54	51	49	46	43	41	38	36	33	31	28	26	24	21	19	17	15	13	11	9	7	5	3				
30	97	93	89	86	83	79	76	73	70	67	64	61	58	55	52	50	47	44	42	39	37	34	32	30	27	25	23	21	19	17	15	13	11	9	7	5	3			
31	97	93	89	86	83	79	77	73	70	67	64	61	59	55	53	50	48	45	43	40	38	35	33	31	29	27	24	22	20	18	16	14	12	10	8	7	5	3		
32	97	93	89	86	83	80	77	74	71	68	65	62	60	56	54	51	49	46	44	41	39	37	34	32	30	28	26	24	22	20	18	16	14	12	10	8	7	5	3	
33	97	93	90	86	84	80	77	74	71	68	66	63	60	58	55	52	50	47	45	42	40	38	36	33	31	29	27	25	23	21	19	17	15	14	12	10	8	7	5	3
34	97	93	90	87	84	80	77	74	71	68	66	64	61	58	56	53	51	47	46	42	41	38	37	34	32	30	28	26	24	22	21	19	17	15	13	12	10	8	7	5
35	97	93	90	87	85	81	78	75	72	69	66	64	61	59	56	54	51	49	47	44	42	40	38	36	33	31	29	27	26	24	22	20	18	16	15	13	11	10	8	7
36	97	93	90	87	85	81	78	75	72	69	67	64	61	59	57	55	52	50	48	45	43	41	39	37	34	32	30	29	27	25	23	21	20	18	16	14	13	11	10	8
37	98	93	91	87	86	82	79	76	73	70	67	65	62	60	58	55	53	51	48	46	44	42	40	38	35	34	32	30	28	26	24	22	21	19	17	16	14	13	11	10
38	98	93	91	88	86	82	79	76	73	70	67	65	62	60	58	56	53	51	48	46	44	42	40	39	37	35	33	31	29	27	25	24	22	20	19	17	16	14	13	11
39	98	94	91	88	86	82	79	76	74	71	68	66	63	61	58	57	54	52	50	48	46	44	42	40	37	35	33	31	30	28	26	25	23	21	20	18	17	15	14	12
40	98	94	91	88	86	82	79	77	74	71	68	66	63	61	59	57	54	52	50	48	46	44	42	40	38	36	34	31	29	28	27	26	24	22	21	20	18	17	15	14
41	98	94	91	88	86	83	79	77	74	72	69	67	64	62	59	57	55	53	51	49	47	45	43	41	38	36	34	32	30	29	28	26	25	23	22	20	19	18	16	15
42	98	94	92	88	86	83	80	78	75	72	69	67	64	62	60	58	56	54	51	49	47	45	44	42	39	37	35	33	31	29	28	27	26	24	22	21	20	19	18	16
43	98	94	92	88	86	83	80	78	75	73	69	68	65	63	60	58	56	54	52	50	48	46	44	43	40	38	36	34	32	30	29	27	27	25	23	22	21	20	19	17
44	98	94	92	88	87	83	80	78	75	73	70	68	65	63	61	59	57	55	53	51	49	47	45	44	41	39	37	35	33	30	29	28	27	25	24	22	22	20	19	18
45	98	94	92	89	87	83	80	78	75	73	70	68	65	64	62	59	57	55	53	51	49	47	45	44	41	39	37	35	33	31	30	28	27	25	24	23	22	21	20	19
46	98	94	92	89	87	84	80	78	75	73	70	68	66	64	62	60	57	56	54	52	50	48	46	44	42	39	37	35	34	31	30	28	27	26	25	24	23	22	21	20
47	98	94	92	89	87	84	81	79	76	74	71	69	66	65	62	60	58	56	54	52	50	49	47	45	42	40	38	36	34	32	31	29	28	27	26	25	24	23	22	21
48	98	94	92	89	87	84	81	79	76	74	71	69	66	65	62	61	58	57	55	53	51	49	47	45	42	40	38	36	34	32	31	30	29	28	27	26	25	24	23	22
49	98	94	92	89	87	84	81	79	76	74	71	70	67	66	63	61	58	57	55	53	51	49	47	45	42	40	38	36	35	34	32	31	30	29	28	27	26	25	24	23
50	99	94	92	89	87	84	81	79	76	74	71	70	67	66	63	62	59	58	56	54	52	50	48	46	43	41	39	37	36	35	34	32	31	30	29	28	27	26	25	24

numbers of observations, which give the relative humidity corresponding to likely wet and dry bulb temperature combinations.

The Mason hygrometer is mounted in a static position so that readings can be taken whenever required, whilst the 'whirling hygrometer' needs to be rotated rapidly for short periods to obtain a reading.

Instruments are now available which will give a direct reading of relative humidity. In them, air is drawn across two precision matched temperature sensors; one measures dry air (dry bulb) and the other is fitted with a moistened wick (wet bulb). Signals from both sensors are processed electronically to give a direct digital read-out of relative humidity.

18.2.2 Interpretation of results

18.2.2.1 Hygiene standards

When measurements of airborne contamination levels or other parameters have been made it is necessary to interpret results against a *standard*. This interpretation will form the basis for the control strategy. In considering the exposure of workers to chemicals two broad options may be presented:

(1) A zero exposure policy.
(2) Permit certain levels of exposure.

To achieve zero exposure to all workplace chemicals is an impossible objective in the light of present day industrial processes. However this approach has been adopted in some countries for proven human carcinogens.

Since no-exposure as a general policy is not possible, hygiene limits have been introduced in an attempt to quantify 'safe' permissible levels of exposure. These are applied to the workplace environment and attempt to reconcile the industrial use of a wide range of materials with a level of protection of the health of exposed workers.

The setting of a hygiene limit is a two-stage process – it involves firstly the collection and evaluation of scientific data and secondly the decision-making process by a committee which also has to take into account socio-economic and political factors.

18.2.2.2 Threshold Limit Values

The Threshold Limit Value (TLV) system is that developed and published in the USA by the American Conference of Governmental and Industrial Hygienists (ACGIH), which in spite of its title is a non-governmental organisation. It is a scientific society similar to the British Occupational Hygiene Society.

The preface to 1989 ACGIH List states:

> 'Threshold Limit Values refer to airborne concentrations of substances and represent conditions under which it is believed that nearly all workers may be repeatedly exposed day after day without adverse effect'[3].

Because of individual variation in susceptibility some workers will suffer effects ranging from discomfort to sensitisation to chemicals and occupational disease at exposure levels well below the TLV. The basis for TLVs is intended to be reasonable freedom from irritation, narcosis, nuisance or impairment of health for the majority of workers. Reference to TLVs relates to these US based standards.

18.2.2.3 UK exposure limits

In the past the HSE has incorporated the TLV system in a Guidance Note[4] which was published annually. However, in the mid-1980s the HSE first published a list of British exposure limits to chemical substances[5] which, in their present form, are part of the requirements of the Control of Substances Hazardous to Health Regulations 1988 (COSHH)[6,7]. These standards are used for determining adequate control of exposure (by inhalation) to hazardous substances.

When applied, the limits should not be used as an index of relative hazard and toxicity, nor should they be used as the dividing line between 'safe' and 'dangerous' concentrations. The list is not comprehensive and the absence of a substance does not indicate that it is safe.

The list of exposure limits is divided into two basic parts – namely Maximum Exposure Limits (MELs) and Occupational Exposure Standards (OESs). For substances that have been given MELs, exposure should be reduced as far as reasonably practicable, and in any case should not be exceeded. With respect to OESs, it will be sufficient to ensure that the level of exposure is reduced to the OES level. This latter requirement reflects the standard demanded for compliance with the COSHH Regulations but it should not obscure the desirable aim of reducing all exposure as far as reasonably practicable.

In order to guide the decision as to which type of standard should be assigned to an individual substance, the HSE has published its *Indicative Criteria*. With respect to OES values these are:

the ability to identify the concentration (with reasonable certainty) at which there is no indication of injurious effects on people, with repeated daily exposure;
the OES can reasonably be complied with; and
reasonably foreseeable over-exposures are unlikely to produce serious short- or long-term effects on health.

An MEL value may be set for a substance which is unable to satisfy the above OES criteria and which may present serious short- and/or long-term risks to man. In some cases where a substance has been assigned an OES value, a numerically higher MEL figure may be assigned because socio-economic factors require that substance's use in certain processes.

With two exceptions, levels embodied in these values relate to personal exposure via the inhalation route, i.e. monitoring is carried out in the person's breathing zone. The list will be reprinted annually, with a list of proposed changes, together with notification of those standards which are priorities for review in the reasonably near future.

18.2.2.3.1 Maximum exposure limits (MEL)

This is the maximum concentration of an airborne substance, averaged over a reference period, to which employees may be exposed by inhalation under any circumstances. Details of the legal requirements are contained in Schedule 1 of the COSHH Regulations, as amended, and the relevant substances are listed in the first part of the HSE's Guidance Note EH 40. Currently there are 38 substances listed but this is reviewed each year. A few selected examples are: hardwood dust, rubber fume, cadmium and compounds and trichloroethylene.

18.2.2.3.2 Occupational exposure standards (OES)

This is the concentration of an airborne substance, again averaged over a reference period, at which '... according to current knowledge, it is believed that there is no evidence that it is likely to be injurious to employees if they are exposed by inhalation, day after day at that concentration ...'[5]. However, current knowledge of the health effects of some chemicals is 'often limited'[5]. This is particularly the case with respect to long-term health effects on humans.

While control of exposure to the OES level satisfies minimally the requirements of COSHH Regulations, it should not discourage the application of good hygiene principles in reducing the concentration levels still further, especially in view of the limited scientific data available in respect of many of these chemicals. It would be prudent for employers to aim for concentrations of 25–50% of these levels.

18.2.2.3.3 Long-term and short-term exposure limits

Two types of exposure limit are listed with the aim of protecting against both short-term effects, such as irritation of the skin, eyes and lungs, narcosis etc., and long-term health effects. Both MEL and OES values are given as time weighted averages (TWA), i.e. the exposure concentrations measured are averaged with time over 8 hours to protect against long-term effects and over 10 minutes for protection against short-term effects.

In both the British and American systems concentrations are given in parts per million (ppm), i.e. parts of vapour or gas by volume per million parts of contaminated air, and also in milligrams of substance per cubic metre of air (mg/m^3).

18.2.2.3.4 Time weighted average concentrations (TWA)

The limits refer to the maximum exposure concentration when averaged over a 10-minute period or an 8 hour day. The time weighted average value (Cm) may be obtained from the following formula:

$$Cm = \frac{(C_1 \times t_1) + (C_2 \times t_2) + \ldots (C_n \times t_n)}{t_1 + t_2 + \ldots\ldots t_n}$$

where C_1, C_2 = concentrations measured during respective sampling periods;

t_1, t_2 = duration of sampling periods.

A simple example is where the person working an 8 hour day was exposed for 4 hours at 20 ppm vapour and then for 4 hours at 10 ppm.

$$Cm = \frac{(20 \times 4) + (10 \times 4)}{4 + 4}$$

This gives an 8 hour TWA of 15 ppm.

18.2.2.3.5 Mixtures

Most of the listed exposure limits refer to single substances or closely related groups, e.g. cadmium and compounds, isocyanates, etc. A few exposure limits refer to complex mixtures or compounds, e.g. white spirit, rubber fume. However, exposure in workplaces is often to a mixture of substances and such combinations may, by their nature, increase the hazard. Mixed exposure requires assessment with regard to possible health effects, which should take into account other factors such as the primary target organs of the major contaminants and possible interaction between the latter substances.

General guidance on mixed exposures is given in EH40[5] together with a rule-of-thumb formula which may be used where there is reason to believe that the effects of the constituents of a mixture are *additive*.

$$\text{Exposure ratio} = \frac{C_1}{L_1} + \frac{C_2}{L_2} + \frac{C_3}{L_3} + \ldots\ldots$$

where C_1, C_2 = time weighted average concentrations of constituents;
L_1, L_2 = corresponding exposure limits.

The use of this formula is only applicable where the additive substances have been assigned OESs. If the ratio is greater than 1 then the limit for the mixture has been exceeded. If one of the substances has been assigned an MEL then the additive effect should be taken into account when deciding to what extent it is reasonably practicable further to reduce exposure.

Example

If air contains 50 ppm acetaldehyde (OES = 100 ppm) and 150 ppm sec-butyl acetate (OES = 200 ppm), applying the formula:

$$\text{Additive ratio} = \frac{50}{100} + \frac{150}{200} = 1.25$$

The threshold limit is therefore exceeded.

This is a relatively crude formula and would not be applicable to a situation where two or more chemicals enhance each other's effects as is the case with synergistic reactions.

18.2.3 Physical factors

Physical factors such as heat, ultraviolet light, high humidity, abnormal pressure etc. place added environmental stress on the body and are likely to increase the toxic effect of a substance. Most standards have been set at a level to encompass moderate deviations from the normal environment. However, for gross variations, e.g. heavy manual work where respiration rate is greatly increased, continuous activity at elevated temperatures or excessive overtime, judgement must be exercised in the interpretation of permissible levels.

18.2.4 Skin absorption

Some substances have the designation 'Sk' and this refers to the potential contribution to overall exposure of absorption through the skin. In this case airborne contamination alone will not indicate total exposure to the chemical and the 'Sk' designation is intended to draw attention to the need to prevent percutaneous absorption. In the application of the assigned exposure limit it is assumed that additional exposure of the skin is prevented.

18.2.5 Sensitisation

Similarly, in the list of exposure standards[5], the designation 'Sen' is assigned to selected substances to indicate that their potential for causing sensitisation reactions has been recognised. Such substances may cause respiratory sensitisation on inhalation, for example allergic asthma, or skin effects where contact occurs, for example allergic contact dermatitis. Although not all exposed persons will become sensitised when exposed to such substances, those that do will develop ill-health effects on subsequent exposure at extremely low concentrations. Once a person has become sensitised to a substance, the occupational exposure limits are not relevant for indicating 'safe' working concentrations.

18.2.6 Carcinogens

There are differing views as to whether carcinogens should be assigned exposure limits, with one body of opinion advocating that the only safe level for a substance that can cause cancer is zero. An Approved Code of Practice[7] for the Control of Carcinogenic Substances, published under the COSHH Regulations, lists 9 single or closely related groups of substances together with materials that have been assigned the risk phrase 'R 45 . . . may cause cancer' or 'R49 . . . may cause cancer by inhalation' in the Classification, Packaging and Labelling of Dangerous Substances Regulations 1984[8]. Recently these groups of substances have been included in *Schedule 10* of the amended COSHH Regulations[6].

However, there is no clear definition of a carcinogenic substance in an occupational context, where the criteria are stated, that has been used in deriving the above selection of materials, except in the general sense that they are: '*. . . specific substances with which a cancer hazard is associated*'.

18.2.7 Derivation of Threshold Limit Values

Ideally, hygiene standards should be derived from the quantitative relations between the contaminant and its effects, i.e. *X* ppm of substances causes *Y* amount of harm. However such relations are very difficult to establish in humans. The problems involved have been considered in some detail by Atherley[9]. Attempts have been made to relate human disease patterns to industrial experience, but unfortunately sufficient data do not exist. The effects of harmful agents have been studied by various methods. These include chemical analogy, which assumes that similar chemicals have biologically similar effects; and short-term testing, which may involve bacteria, animal exposure experiments and human epidemiology.

The major criteria which have been used to develop the TLV list are the effect of a substance on an organ or organ system (49%), irritation (40%), and thirdly to a lesser extent narcosis (5%) and odour (2%).

An ACGIH publication[10] summarises toxicological information on substances for which TLVs have been adopted and shows that for some substances the hazards are clear, whereas for others there is very little information on human danger. This inherent uncertainty is not reflected in the bland listing of adopted values.

In recent years, the HSE has published summaries[11] of the information used in setting its MEL and OES values. Such information may be useful in assessing the applicability of a standard to a particular workplace situation.

18.2.8 Variation in international standards

Hygiene standards vary from country to country depending upon the interpretation of scientific data and the philosophy of the regulations.

International hygiene standards for trichloroethylene illustrate this variation.

	mg/m^3	ppm
+ Australia	535	100
+ UK	535	100
+ USA (ACGIH)	267	50
+ Sweden	105	20
++ Hungary	53	10

\+ Time Weighted Average.
++ Maximum Allowable Concentration.

In some countries great emphasis is put on neurophysiological changes in experimental animals as well as behavioural effects in human beings. It should be noted that although low levels may be embodied in national regulations this does not mean they are achieved in practice. In the past this has been acknowledged by former Soviet representatives. In the European Community context, there are moves to harmonise exposure limits but there are also simultaneous needs to harmonise compliance strategies.

18.2.9 Changes in hygiene limits

With new scientific evidence and changing attitudes hygiene limits are constantly being revised. A startling example is provided by vinyl chloride monomer (VCM) which occurs in the manufacture of PVC plastics. The acute effects of VCM were identified in the 1930s as being primarily 'narcosis'. To prevent such effects during industrial use a TLV of 500 ppm was set in 1962. After further research VCM was identified as affecting the liver, bones and kidneys and the adopted value (TLV) was lowered to 200 ppm in 1971. In 1974 some American chemical workers died of a rare liver cancer (angiosarcoma) which was traced to exposure to VCM, with the result that in 1978 the adopted value (ACGIH) was dropped to 5 ppm. Hence the adopted TLV for vinyl chloride monomer has been reduced a hundredfold in under 20 years.

A section in the TLV list formally notes chemicals for which a change in the standard is intended.

Similarly, the HSE in its Guidance Note[5] publishes a list of substances where the OES is new, has been changed, or where it is intended to assign an MEL value. There is also a list of substances for which the occupational exposure limits are to be reviewed.

18.3 Control measures

When, in a workplace, a hazard has been identified and the risk to health assessed, an appropriate prevention or control strategy is then required. The general control strategy should include consideration of:

Specification.
Substitution.
Segregation.
Local exhaust ventilation.
General dilution ventilation.
Good housekeeping and personal hygiene.
Reduced time exposure.
Personal protection.

These control options should be complemented and underpinned by adequate administrative arrangements which should include the provision for regular re-assessment of risks.

18.3.1 Specification

The design of a new plant or process is the ideal stage to incorporate hazard prevention and control features, e.g. limiting the quantities of toxic materials handled, the provision of remote handling facilities, utilising noise control features in the design and layout of new machinery etc. Including safety features at the design stage will be much less costly than having to add them later. This emphasises the need for safety advisers to be involved at the earliest stage of developments.

18.3.2 Substitution

This involves the substitution of materials or operations in a process by safer alternatives. A toxic material may be replaced by another less harmful substance or in another context something less flammable. An example is the widespread replacement of carbon tetrachloride by other solvents such as dichloromethane and 1,1,1-trichloroethane. In turn this substance is to be phased out because of its ozone depleting characteristics and industry is looking for suitable substitutes.

Care needs to be exercised in the selection of 'safer' substitutes since they may be considered safer simply because there is less information available about their hazards.

Alternatively the process itself may be changed to improve working conditions with a possible benefit of increased efficiency as well.

Arc welding has been widely introduced to replace rivetting and subsequently noise levels have been reduced. Again such alterations may introduce new hazards (e.g. welding fumes) and the risks from these must be similarly assessed with the implementation of adequate controls.

18.3.3 Segregation

If a substance or process cannot be eliminated, another strategy is to enclose it completely to prevent the spread of contamination. This may be by means of a physical barrier, e.g. an acoustic booth surrounding a noisy machine or handling toxic substances in a glove box. Relocation of a process to an isolated section of the plant is another possibility that reduces the number exposed to the hazard. A particular process may be segregated in time, e.g. operated at night, when fewer people are likely to be exposed. However, in the latter case, such workers are already subjected to the additional stress of night shift working and generally function less efficiently, a point that should be borne in mind by the occupational hygienist when considering alternatives.

18.3.4 Local extract ventilation

Where it is not practicable to enclose the process totally, other steps must be taken to contain contaminants. This can be achieved by removing vapours, gases, dusts and fumes etc. by means of a local extract ventilation system. Such a system traps the contaminant close to its source and removes it so that nearby workers are not exposed to harmful concentrations.

Local extract ventilation systems have four major parts:

(1) Hoods – collection point for gathering the contaminated air into the system.
(2) Ducting – to transport the extracted air to the air purifying device or the outside atmosphere.
(3) Air purifying device – such as charcoal filters are often used to remove organic chemical contaminants.
(4) Fan – provides the means for moving air through the system.

There are several different types of local extract ventilation systems and adherence to sound design principles is necessary to achieve effective removal of contaminants. Ventilation systems must also be adequately maintained to ensure that they are operating to design specifications. This subject is dealt with in greater detail in Chapter 21.

18.3.5 Dilution ventilation

Sometimes it is not possible to extract the contaminant close to its source of origin and dilution ventilation may be used under the following circumstances where there is:

(1) Small quantity of contaminant.
(2) Uniform evolution.
(3) Low toxicity material.

Dilution ventilation utilises natural convection through open doors, windows, roof ventilators etc. or assisted ventilation by roof fans or blowers which draw or blow in fresh air to dilute the contaminant. With both of these systems the problem of providing make-up air at the proper temperature, especially during the winter months, has to be considered.

18.3.6 Personal hygiene and good housekeeping

Both have an important role in the protection of the health of people at work. Laid down procedures are necessary for preventing the spread of contamination, for example the immediate clean-up of spillages, safe disposal of waste and the regular cleaning of work stations.

Dust exposures can often be greatly reduced by the application of water or other suitable liquid close to the source of the dust. Thorough wetting of dust on floors before sweeping will also reduce dust levels.

Adequate washing and eating facilities should be provided with instruction for workers on the hygiene measures they should take to prevent the spread of contamination. The use of lead at work is a case where this is particularly important.

Recently enacted wide ranging regulations[12,13] dealing with workplace safety and welfare, formally require that workplaces are kept 'sufficiently clean' and waste materials are kept under control.

18.3.7 Reduced time exposure

Reducing the time of exposure to an environmental stress is a control strategy which has been used. The dose of contaminant received by a person is generally related to the level of stress and the length of time the person is exposed. A noise standard for maximum exposure of people at work of 90 dBA over an 8-hour work day has been used for several years and is now contained in the Noise at Work Regulations 1989[14] as the 'second action level'. Equivalent doses of noise energy are 93 dBA for 4 hours, 96 dBA for 2 hours, etc. (The dBA scale is logarithmic.) Such limiting of hours has been used as a control strategy but does not take into account the possibly harmful effect of dose rate, e.g. very high noise levels over a very short time even though followed by a long period of relatively low levels.

18.3.8 Personal protection

Making the workplace safe should be the first consideration but if it is not possible to reduce danger sufficiently by the methods outlined above the worker may need to be protected from the environment by the use of personal protective equipment which may be broadly divided as follows:

(1) Hearing protection.
(2) Respiratory protection.
(3) Eye and face protection.
(4) Protective clothing.
(5) Skin protection.

Personal protective devices have a serious limitation in that they do nothing to attenuate the hazard at source, so that if they fail and it is not noticed the wearer's protection is reduced and the risk he faces increases correspondingly.

Making the workplace safe is preferable to relying on personal protection; however, this regard for personal protection as a last line of defence should not obscure the need for the provision of competent people to select equipment and administer the personal protection scheme once the decision to use this control strategy has been taken. Personal protection is not an easy option and it is important that the correct protection is given for a particular hazard, e.g. ear muffs/plugs prescribed after octave band measurements of the noise source.

Else[15] outlines three key elements of information required for a personal protection scheme:

(i) nature of the hazard,
(ii) performance data of personal protective equipment, and
(iii) standard representing adequate control of the risk.

18.3.8.1 Nature of the danger

The hazards need to be identified and the risks assessed; for example, in the case of air contaminants the nature of the substance(s) present and the estimated exposure concentration, or, with noise, measurement of sound levels and frequency characteristics.

18.3.8.2 Performance data of personal protective equipment

Data are required about the ability of equipment to protect against a particular danger. This will often be provided by manufacturers who carry out tests under controlled conditions which are often specified in national or international standards. In the UK the British Standards Institution is the predominant standards setting body. For example, the British Standard BS 2091[16] gives general guidance about maximum service life of respirator cartridges and canisters at maximum permitted concentration of contaminant gas or vapour (i.e. 0.1% and 1% v/v for cartridges and canisters respectively). For approval canister types are tested against an 'appropriate' gas whilst cartridges are tested against trichloroethylene under specified conditions of relative humidity and air flow rate. Similarly the method used to determine the noise attenuation of hearing protectors at different frequencies (octave bands) throughout the audible range is specified in BS 5108.[17]

18.3.8.3 Standards representing adequate control of the risk

For some risks such as exposure to potent carcinogens or protection of eyes against flying metal splinters the only acceptable level is zero. The informed use of hygiene limits, bearing in mind their limitations, would be pertinent when considering acceptable levels of air contaminants.

A competent person would need these three types of information to decide whether the personal protective equipment could *in theory* provide adequate protection against a particular hazard.

Once theoretically adequate personal protective equipment has been selected the following factors need to be considered:

(1) Fit. — Good fit of equipment to the person is required to ensure maximum protection.

(2) Period of use. — The maximum degree of protection will not be achieved unless the equipment is worn all the time the wearer is at risk.

(3) Comfort. — Equipment that is comfortable is more likely to be worn. If possible the user should be given a choice of alternatives which are compatible with other protective equipment.

(4) Maintenance. — To continue providing the optimum level of protection the equipment must be routinely checked, cleaned, and maintained.

(5) Training. — Training should be given to all those who use protective equipment and to their supervisors. This should include information about what the equipment will protect against and its limitations.

(6) Interference. — Some eye protectors and helmets may interfere with the peripheral visual field. Masks and breathing apparatus interfere with olfactory senses.

(7) Management commitment. — This is essential to the success of personal protection schemes.

Appropriate practice should ensure *effective* personal protection schemes are based on the requirements of regulations and codes of practice[18,19]

18.3.8.4 Hearing protection

There are two major types of hearing protectors:

(1) Ear-plugs – inserted in the ear canal.
(2) Ear-muffs – covering the external ear.

Disposable ear-plugs are made from glass down, plastic-coated glass down and polyurethane foam, while re-usable ear plugs are made from semi-rigid plastic or rubber. Re-usable ear-plugs need to be washed frequently.

Ear-muffs consist of rigid cups to cover the ears, held in position by a sprung head band. The cups have acoustic seals of polyurethane foam or a liquid-filled annular sac.

Hearing protectors should be chosen to reduce the noise level at the wearer's ear to at least below 85 dB(A) and ideally to around 80 dB(A). With particularly high ambient noise levels this should not be done from simple A-weighted measurements of the noise level, because sound reduction will depend upon its frequency spectrum. Octave band analysis will provide the necessary information to be matched against the overall sound attenuation of different hearing protectors which is claimed by the manufacturers in their test data.

18.3.8.5 Respiratory protective equipment

This may be broadly divided into two types in the manner shown in *Figure 18.6.*

(1) Respirators – purify the air by drawing it through a filter which removes most of the contamination.
(2) Breathing apparatus – supplies clean air from an uncontaminated source.

18.3.8.5.1. Respirators

There are five basic types of respirator:

(1) Filtering Facepiece Respirator.
The facepiece covers the whole of the nose and mouth and is made of filtering material which removes respirable size particles. (These should not be confused with nuisance dust masks which simply remove larger particles.)
(2) Half Mask Respirator.
A rubber or plastic facepiece that covers the nose and mouth and has replaceable filter cartridges.
(3) Full Face Respirator.
A rubber or plastic facepiece that covers the eyes, nose and mouth and has replaceable filter canisters.
(4) Powered Air Purifying Respirator.
Air is drawn through a filter and then blown into a half mask or full facepiece at a slight positive pressure to prevent inward leakage of contaminated air.
(5) Powered Visor Respirator.
The fan and filters are mounted in a helmet and the purified air is blown down behind a protective visor past the wearer's face.

Filters are available for protection against harmful dusts and fibres, and also for removing gases and vapours. It is important that respirators are never used in oxygen-deficient atmospheres.

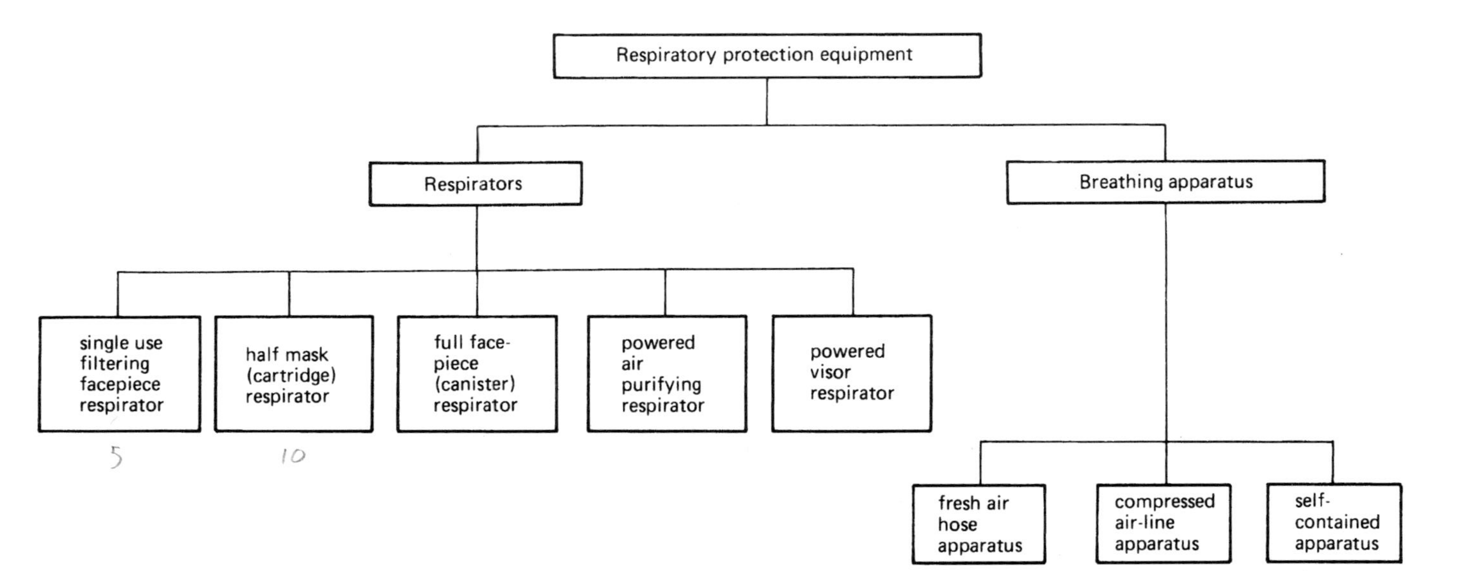

Figure 18.6 Types of respiratory protection equipment

18.3.8.5.2 Breathing apparatus

(1) Fresh Air Hose Apparatus.
Air is brought from an uncontaminated area by the breathing action of the wearer or by a bellows or blower arrangement.
(2) Compressed Air Line Apparatus.
Air is brought to the wearer through a flexible hose attached to a compressed air line. Filters are mounted in the line to remove nitrogen oxides and it is advisable to use a special compressor with this equipment. The compressor airline is connected via pressure-reducing valves to half-masks, full facepieces or hoods.
(3) Self-contained Breathing Apparatus.
A cylinder attached to a harness and carried on the wearer's back provides air or oxygen to a special mask. This equipment is commonly used for rescue purposes.

The British Standard BS 4275[20] gives guidance on the selection, use and maintenance of respiratory protective equipment. The nominal protection factor (npf) measures the theoretical capability of respiratory protection and is used in the selection of equipment.

$$\text{npf} = \frac{\text{concentration of contaminant in the atmosphere}}{\text{concentration of contaminant in the facepiece}}$$

BS 4275 lists npf values for different devices.

18.3.8.5.3 Eye protection

After a survey of potential eye hazards the most appropriate type of eye protection should be selected. Safety spectacles may be adequate for relatively low energy projectiles, e.g. metal swarf, but for dust, goggles would be more appropriate. For people involved in gas/arc welding or using lasers, special filtering lenses would be required.

18.3.8.5.4 Protective clothing

Well designed and properly worn, protective clothing will provide a reasonable barrier against skin irritants. A wide range of gloves, sleeves, impervious aprons, overalls etc. is currently available. The factors listed above should be considered when the selection of this equipment is being made. For example, when selecting gloves for handling solvents knowledge of glove material is required:

Neoprene gloves – adequate protection against common oils, aliphatic hydrocarbons; not recommended for aromatic hydrocarbons, ketones, chlorinated hydrocarbons.
Polyvinyl alcohol gloves – protect against aromatic and chlorinated hydrocarbons.

For protective clothing to achieve its objective it needs to be regularly cleaned or laundered and replaced when damaged.

18.3.8.5.5 Skin protection

Where protective clothing is impracticable, due to the proximity of machinery or unacceptable restriction of the ability to manipulate, a barrier cream may be the preferred alternative. Skin protection preparations can be divided into the following three groups:

(1) Water-miscible – protects against organic solvents, mineral oils and greases, but not metal-working oils mixed with water.
(2) Water-repellent – protects against aqueous solutions, acids, alkalis, salts, oils and cooling agents that contain water.
(3) Special group – cannot be assigned to a group by their composition. Formulated for specific application.

Skin protection creams should be applied before starting work and at suitable intervals during the day.

However, these preparations are only of limited usefulness as they are rapidly removed by rubbing action and care must be taken in their selection since, with some solvents, increased skin penetration can occur. The application of a moisturising cream which replenishes skin oil is beneficial after work.

18.4 Summary

The overall strategic approach is summarised in *Figure 18.7*. Although this approach to hazard identification, risk assessment and control has long been established in occupational hygiene, detailed supporting legislation has

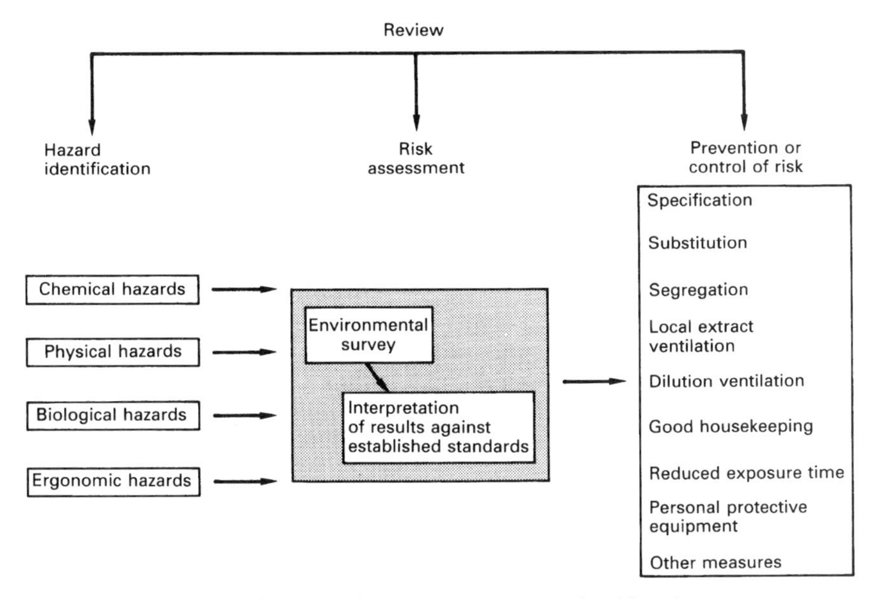

Figure 18.7 Diagram of strategy for protection against health risks

been restricted to selected health hazards, e.g. substances hazardous to health, noise, lead, etc. However, new regulations[21] require this same approach to all hazards thus reinforcing hygiene practice.

References

1. Health and Safety Executive, *Guidance Note EH42*, Monitoring Strategies for Toxic Substances, HSE Books, Sudbury (1989)
2. Nader, J.S., *Air Sampling Instruments for Evaluation of Atmospheric Contaminants*, 7th edn, American Conference of Governmental Industrial Hygienists, Cincinnati, Ohio (1989)
3. American Conference of Governmental Industrial Hygienists, *Threshold Limit Values and Biological Exposure Indices for 1992–1993*, Cincinatti, Ohio (1992)
4. Health and Safety Executive, *Guidance Note EH 15/80, Threshold Limit Values*, HSE Books, Sudbury (1980)
5. Health and Safety Executive, *Guidance Note EH40, Occupational Exposure Limits*, HSE Books, Sudbury (This is up-dated annually)
6. H.M. Government, The Control of Substances Hazardous to Health Regulations 1988, HMSO, London (1988)
7. Health and Safety Commission, *Control of Substances Hazardous to Health and Control of Carcinogenic Substances, Approved Codes of Practice*, HSE Books, Sudbury (1992)
8. H.M. Government, The Classification, Packaging and Labelling of Dangerous Substances Regulations 1984, HMSO, London (1984)
9. Atherley, G.R.C., *Occupational Health and Safety Concepts*, Applied Science, London (1978)
10. American Conference of Governmental Industrial Hygienists, *Documentation for Threshold Limit Values*, ACGIH, Cincinnati, Ohio (1971) Updated
11. Health and Safety Executive, *Guidance Notes: Environmental Health Series No. EH 64, Occupational Exposure Limits: criteria document summaries.* Synopses of the data used in setting Occupational Exposure Limits, HSE Books, Sudbury (1992)
12. H.M. Government, Workplace (Health, Safety and Welfare) Regulations 1992, HMSO, London (1992)
13. Health and Safety Commission, *Legislation publication No. L24 Approved Code of Practice: Workplace (Health, Safety and Welfare) Regulations 1992*, HSE Books, Sudbury (1992)
14. H.M. Government, The Noise at Work Regulations 1989, HMSO, London (1989)
15. Else, D., *Occupational Health Practice*, (Ed. R.S.F. Schilling), 2nd edn, Ch. 21, Butterworth, London (1981)
16. BS 2091:1969, Respirators for protection against harmful dusts, gases and scheduled agricultural chemicals, British Standards Institution, London
17. BS 5108:1991, Method of measurement of attenuation of hearing protectors at threshold, British Standards Institution, London
18. H.M. Government, Personal Protective Equipment at Work Regulations 1992, HMSO, London (1992)
19. Health and Safety Executive, *Legislation publication No. L25 Personal Protective Equipment at Work Regulations 1992, Guidance on the Regulations*, HSE Books, Sudbury (1992)
20. BS 4275:1974, Recommendations for the selection, use and maintenance of respiratory protective equipment, British Standards Institution, London
21. H.M. Government, Management of Health and Safety at Work Regulations 1992, HMSO, London (1992)

Chapter 19

Radiation

Dr A. D. Wrixon

19.1 Introduction

Radiation is emitted by a wide variety of sources and appliances used in industry, medicine and research. It is also a natural part of the environment. The purpose of this chapter is to give the reader a broad view of the nature of radiation, its biological effects, and the precautions to be taken against it.

19.2 Structure of matter[1–3]

All matter consists of elements, for example, hydrogen, oxygen, iron. The basic unit of any element is the atom, which cannot be further subdivided by chemical means. The atom itself is an arrangement of three types of particles.

(1) Protons. These have unit mass and carry a positive electrical charge.
(2) Neutrons. These also have unit mass but carry no charge.
(3) Electrons. These have a mass about 2000 times less than that of protons and neutrons and carry a negative charge.

Protons and neutrons make up the central part of the nucleus of the atom; their internal structure is not relevant here. The electrons take up orbits around the nucleus and, in an electrically neutral atom, the number of electrons equals the number of protons. The element itself is defined by the number of protons in the nucleus. For a given element, however, the number of neutrons can vary to form different isotopes of that element. A particular isotope of an element is referred to as a nuclide. A nuclide is identified by the name of the element and its mass, for example, carbon-14. There are 90 naturally-occurring elements; additional elements, such as plutonium and americium, have been created by man, for example, in nuclear reactors.

If the number of electrons does not equal the number of protons, the atom has a net positive or negative charge and is said to be ionised. Thus if a neutral atom loses an electron, a positively-charged ion will result. The process of losing or gaining electrons is called ionisation and occurs in the course of many chemical and physical processes.

19.3 Radioactivity[1–3]

Some nuclides are unstable and spontaneously change into other nuclides, emitting energy in the form of radiation, either particulate (e.g. α and β particles) or electromagnetic (e.g. γ-rays). This property is called radioactivity, and the nuclide showing it is said to be radioactive. Most nuclides occurring in nature are stable, but some are radioactive, for example, all the isotopes of uranium and thorium. Many other radioactive nuclides (or radionuclides) have been produced artificially, such as strontium-90, caesium-137 and the isotopes of the man-made elements, plutonium and americium.

19.4 Ionising radiation[4]

The radiation emitted during radioactive decay can cause the material through which it passes to become ionised and it is therefore called ionising radiation. X-rays are another type of ionising radiation. Ionisation can result in chemical changes which can lead to alterations in living cells and eventually, perhaps, to manifest biological effects.

The ionising radiations encountered in industry are principally α, β, γ and X-rays, bremsstrahlung and neutrons. Persons can be irradiated by sources outside the body (external irradiation) or from radionuclides deposited within the body (internal irradiation). External irradiation is of interest when the radiation is sufficiently penetrating to reach the basal layer of the epidermis (i.e. the living cells of the skin). Internal irradiation arises following the intake of radioactive material by ingestion, by inhalation or by absorption through the skin or open wounds.

The α particle consists of two protons and two neutrons. It is therefore heavy and doubly charged. Alpha radiation has a very short range and is stopped by a few centimetres of air, a sheet of paper, or the outer dead layer of the skin. Outside the body, it does not, therefore, present a hazard. However, α-emitting radionuclides inside the body are of concern because α particles lose their energy to tissue in very short distances causing relatively intense local ionisation.

The β particle has mass and charge equal in magnitude to an electron. Its range in tissue is strongly dependent on its energy. A β particle with energy below about 0.07 MeV would not penetrate the outer dead layer of the skin, but one with an energy of 2.5 MeV would penetrate soft tissue to a depth of about 1.25 cm. Energy is expressed here in units of electron volt (eV), which is a measure of the energy gained by an electron in passing through a potential difference of one volt. Multiples of the electron volt are commonly used; MeV stands for Mega electron Volts (1 MeV = 1 000 000 V). As β particles are slowed down in matter, bremsstrahlung (a type of X-radiation) is produced, which will penetrate to greater distances. Thus a β-radiation source outside the body may have more penetrating radiation associated with it than is immediately apparent from the energy of the β radiation. Beta-emitting radionuclides inside the body are also of concern, but the total ionisation caused by β particles is less intense than that caused by α particles.

Gamma-rays, X-rays and bremsstrahlung are all electromagnetic radiations similar in nature to ordinary light except that they are of much higher frequencies and energies. They differ from each other in the way in which they are produced. Gamma-radiation is emitted in radioactive decay. The most widely known source of X-rays is in certain electrical equipment in which electrons are made to bombard a metal target in an evacuated tube. Bremsstrahlung is produced by the slowing down of β particles; its energy depends on the energy of the original β particles. The penetrating power of electromagnetic radiation depends on its energy and the nature of the matter through which it passes; with sufficient energy it can pass right through a human body. Sources of these radiations outside the body can therefore cause harm. With X-ray equipment, the radiation ceases when the machine is switched off. Gamma-ray sources, however, emit radiation all the time.

Neutrons are emitted during certain nuclear processes, for example nuclear fission, in which a heavy nucleus splits into two fragments. Neutrons, being uncharged and therefore not affected by the electric fields around atoms, have great penetrating power, and sources of neutrons outside the body can cause harm. Neutrons produce ionisation indirectly. When a high-energy neutron strikes a nucleus in the material through which it passes, some of its energy is transferred to the nucleus which then recoils. Being electrically charged and slow-moving the recoiling nucleus creates dense ionisation over a short distance.

19.5 Biological effects of ionising radiation[4–6]

Information on the biological effects of ionising radiation comes from animal experiments and from studies of groups of people exposed to relatively high levels of radiation. The best-known groups are the workers in the luminising industry early this century who used to point their brushes with the lips and so ingest radioactivity; the survivors of the atomic bombs dropped on Japan, and patients who have undergone radiotherapy. Evidence of biological effects is also available from studies of certain miners who inhaled elevated levels of the natural radioactive gas, radon, and its radioactive decay products.

The basic unit of tissue is the cell. Each cell has a nucleus, which may be regarded as its control centre. Deoxyribonucleic acid (DNA) is the essential component of the cell's genetic information and makes up the chromosomes which are contained in the nucleus. Although the ways in which radiation damages cells are not fully understood, many involve changes to DNA. There are two modes of action. A DNA molecule may become ionised, resulting directly in chemical change, or it may be chemically altered by reaction with agents produced as a result of the ionisation of other cell constituents. The chemical change may ultimately mean that the cell is prevented from further division and can therefore be regarded as dead.

Very high doses of radiation can kill large numbers of cells. If the whole body is exposed, death may occur within a matter of weeks: an instantaneous absorbed dose of 5 gray or more would probably be lethal (the unit gray is defined below). If a small area of the body is briefly exposed to a very high dose, death may not occur, but there may be other early effects: an instantaneous absorbed dose of 5 gray or more to the skin would probably cause erythema (reddening) in a week or so, and a similar dose to the testes or ovaries might cause sterility. If the same doses are received in a protracted fashion, there may be no early signs of injury. The effect of very high doses of radiation delivered acutely is used in radiotherapy to destroy malignant tissue.

Low doses or high doses received in a protracted fashion may lead to damage at a later stage. With reproductive cells, the harm is expressed in the irradiated person's offspring (genetic defects), and may vary from unobservable through mildly detrimental to severely disabling. So far, however, no genetic defects directly attributable to radiation exposure have been unequivocally observed in human beings. With other cells, cancer induction may result. There is always a delay of some years, or even decades, between irradiation and the appearance of a cancer.

It is assumed that within the range of exposure conditions usually encountered in radiation work, the risks of cancer and hereditary damage increase in direct proportion to the radiation dose. It is also assumed that there is no exposure level that is entirely without risk. Thus, for example, the mortality risk factor for all cancers from uniform radiation of the whole body is now estimated to be 1 in 25 per sievert (see below for definition) for a working population, aged 20 to 64 years, averaged over both sexes[5]. In scientific notation, this is given as 4×10^{-2} per sievert.

19.6 Quantities and units

All new legislation in force after 1986 is required by the Units of Measurement Regulations 1980 to be in SI units. Only the SI system of units is described in full here, although the relationships between the old and new units are given.

The *activity* of an amount of a radionuclide is given by the rate at which spontaneous decays occur in it. Activity is expressed in a unit called the becquerel, Bq. A Bq corresponds to one spontaneous decay per second. Multiples of the becquerel are frequently used such as the megabecquerel, MBq (a million becquerels).

The *absorbed dose* is the mean energy imparted by ionising radiation to the mass of matter in a volume element. It is expressed in a unit called the gray, Gy. A Gy corresponds to a joule per kilogram.

Biological damage does not depend solely on the absorbed dose. For example, one Gy of α radiation to tissue is more harmful than one Gy of β radiation. In radiological protection, it has been found convenient to introduce a further quantity that correlates better with the potential harm that might be caused by radiation exposure. This quantity, called the *equivalent dose,* is the absorbed dose averaged over a tissue or organ multiplied by the relevant radiation weighting factor. The radiation weighting factor for γ radiation, X-rays and β particles is set at 1. For α particles, the factor is 20. Equivalent dose is expressed in a unit called the sievert, Sv. Submultiples of the sievert are frequently used such as the millisievert, mSv (a thousandth of a sievert) and the microsievert, μSv (a millionth of a sievert).

The risks of malignancy, fatal or non-fatal, per sievert are not the same for all body tissues. The risk of hereditary damage only arises through irradiation of the reproductive organs. It is therefore appropriate to define a further quantity, derived from the equivalent dose, to indicate the combination of different doses to several tissues in a way that is likely to correlate well with the total detriment due to malignancy and hereditary damage. This quantity, derived for the fractional contribution each tissue makes to the total detriment, is called the *effective dose*. This is defined as the sum of the equivalent doses to the exposed organs and tissues weighted by the appropriate tissue weighting factor. This quantity is also expressed in sieverts.

It should be noted that the above quantities, equivalent dose and effective dose, are those defined in the latest recommendations[5] of the International Commission on Radiological Protection (ICRP). They have yet to be adopted into UK regulations, which currently use the old quantities, dose equivalent and effective dose equivalent. The differences between the old and new quantities are beyond the scope of this chapter but their relationships are summarised in *Table 19.1*.

Table 19.1. Relationship between SI units and old units

Quantity	*New named unit and symbol*	*In other SI units*	*Old unit and symbol*	*Conversion factor*
Absorbed dose	gray (Gy)	$J\,kg^{-1}$	rad (rad)	1 Gy = 100 rad
Dose equivalent	sievert (Sv)	$J\,kg^{-1}$	rem (rem)	1 Sv = 100 rem
Activity	becquerel (Bq)	s^{-1}	curie (Ci)	$1\ Bq = 2.7 \times 10^{-11}$ Ci

19.7 Basic principles of radiological protection

Throughout the world, protection standards have, in general, been based for many years on the recommendations of the ICRP. The primary aim of radiological protection as expressed by ICRP[5] is to provide an appropriate standard of protection for man without unduly limiting the beneficial practices giving rise to radiation exposure. For this, ICRP has introduced a basic framework for protection that is intended to prevent those effects that occur only above relatively high levels of dose (e.g. erythema) and to ensure that all reasonable steps are taken to reduce the risks of cancer and hereditary damage. The system of radiological protection by ICRP[5] for proposed and continuing practices is based on the following general principles:

(a) No practice involving exposure to radiation should be adopted unless it produces sufficient benefit to the exposed individuals or to society to offset the radiation detriment it causes. (The justification of a practice).
(b) In relation to any particular source within a practice, the magnitude of individual doses, the number of people exposed, and the likelihood of incurring exposure where these are not certain to be received should be kept as low as is reasonably achievable, economic and social factors being taken into account. This procedure should be constrained by restrictions on the doses to individuals (dose constraints), or risks to individuals in the case of potential exposure (risk constraints), so as to limit the inequity likely to result from the inherent economic and social judgements. (The optimisation of protection).
(c) The exposure of individuals resulting from the combination of all the relevant practices should be subject to dose limits, or to some control of risk in the case of potential exposures. These are aimed at ensuring that no individual is exposed to radiation risks from these practices that are judged to be unacceptable in any normal circumstances. Not all sources are susceptible to control by action at the source and it is necessary to specify the sources to be included as relevant before selecting a dose limit. (Individual dose and risk limits).

The ordering of these recommendations is deliberate; the ICRP limits are to be regarded as backstops and not as levels that can be worked up to.

For workers, the effective dose limit recommended by ICRP is 20 mSv per year averaged over defined periods of 5 years with no more than 50 mSv in any single year, the equivalent dose limit for the lens of the eye is 150 mSv in a year and that for the skin, hands and feet is 500 mSv in a year.

For comparison, the principal effective dose limit for members of the public is 1 mSv in a year. However, it is permissible to use a subsidiary dose limit of 5 mSv in a year for some years, provided that the average annual effective dose over 5 years does not exceed 1 mSv per year. The equivalent dose limits for the skin and lens of the eye are 50 mSv and 15 mSv per year respectively.

In the application of the dose limits for both workers and the public, no account should be taken of the exposures received by patients undergoing radiological examination or treatment and those received from normal levels of natural radiation.

It should be noted that these limits are in some cases more restrictive than those currently given in UK regulations (see section 19.8.1).

19.7.1 Protection against external radiation[4,6]

Protection against exposure from external radiation is achieved through the application of three principles: shielding, distance or time. In practice judicious use is made of all three. Shielding involves the placing of some material between the source and the person to absorb the radiation partially or completely. Plastics are useful materials for shielding β radiation because they produce very little bremsstrahlung. For γ and X-radiation a large mass of material is required; lead and concrete are commonly used.

Radiation from a point source falls off with the square of the distance and through absorption by the intervening air. Remote handling is one way of putting distance between the source and the person (for example, tweezers may be used when handling β-emitting sources).

19.7.2 Protection against internal radiation[4–7]

Protection against exposure from internal radiation is achieved by preventing the intake of radioactive material through ingestion, inhalation and absorption through skin and skin breaks. Eating, drinking and smoking should not be carried out in areas where unsealed radioactive materials are used. The degree of containment necessary depends on the quantity and type of material being handled: it may range from simple drip trays through fume cupboards to complete enclosures such as glove boxes. Surgical

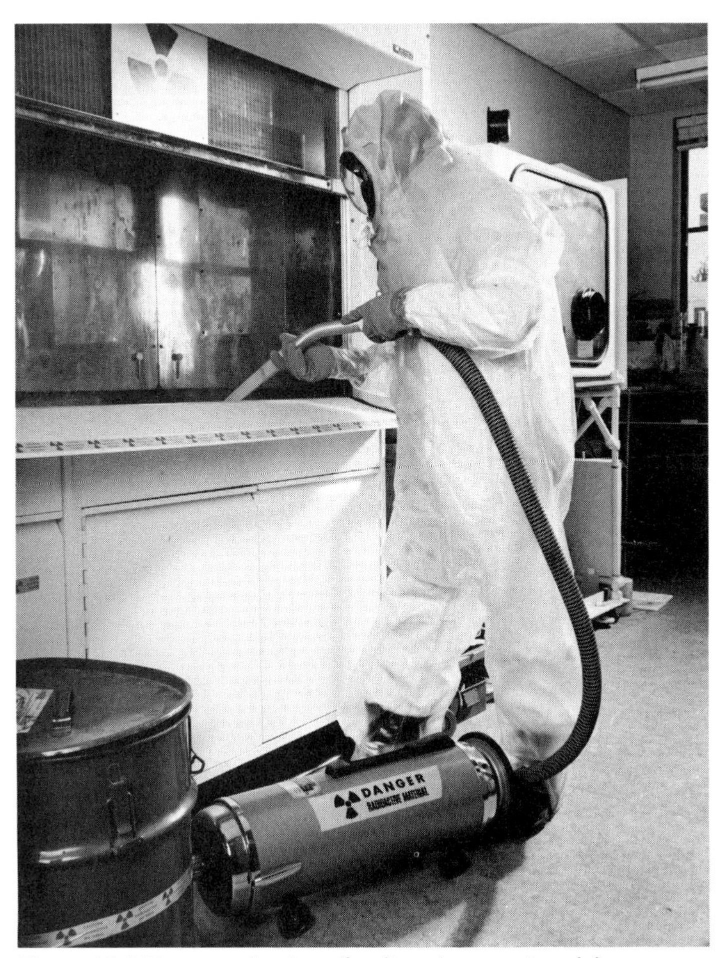

Figure 19.1 Decontamination of radioactive area in a laboratory

gloves, laboratory coats and overshoes are frequently worn. A high standard of cleanliness is required to prevent the spread of radioactive contamination and great care is necessary in dealing with accidental spills (*Figure 19.1*). Anyone working with unsealed radioactive material should wash and monitor his hands on leaving the working area; this is particularly important before meals are taken. Cuts and wounds should be treated immediately and no-one should work with unsealed radioactive substances unless breaks in the skin are protected to prevent the entry of radioactive material.

The radiation dose received through the intake of radioactive material depends on the mode of intake, the quantity involved, the organs in which the material becomes deposited, the rate at which it is eliminated (by radioactive decay and excretion) and the radiations emitted.

19.7.3 Radiation monitoring

The main objectives of monitoring are to evaluate occupational exposures, to demonstrate compliance with standards and regulatory requirements and to provide data needed for adequate control. For the latter, monitoring can serve the following functions:

(1) detection and evaluation of the principal sources of exposure,
(2) evaluation of the effectiveness of radiation control measures and equipment,
(3) detecting of unusual and unexpected situations involving radiation exposures,
(4) evaluation of the impact of changes in operational procedures, and
(5) provision of data on which the effect of future operations on radiation exposure can be predicted so that the appropriate controls can be devised beforehand and instituted.

The most appropriate means of assessing a worker's exposure to external radiations is through personal monitoring involving the wearing of a 'badge' containing radiation sensitive material, in particular, a thermoluminescent chip or powder or a small piece of film (*Figure 19.2*). Doses from the intake of airborne contamination can be assessed through the use of air samplers either worn by the person or set up at appropriate points in the workplace. Radioactive material within the body can be determined by excreta or whole body monitoring, depending on the particular radionuclide involved.

The appropriate detector to be used to monitor the workplace environment depends on the type and energy of the radiation involved and whether the hazard arises from external radiation or surface or air contamination. Most survey instruments can be divided into two groups:

(a) Dose rate meters

These measure the radiation in units of dose rate and normally contain an ionisation chamber or Gieger-Müller tube. They are usually used to monitor β, γ and X-radiation fields. Special instruments are used for measuring neutron radiation dose rates.

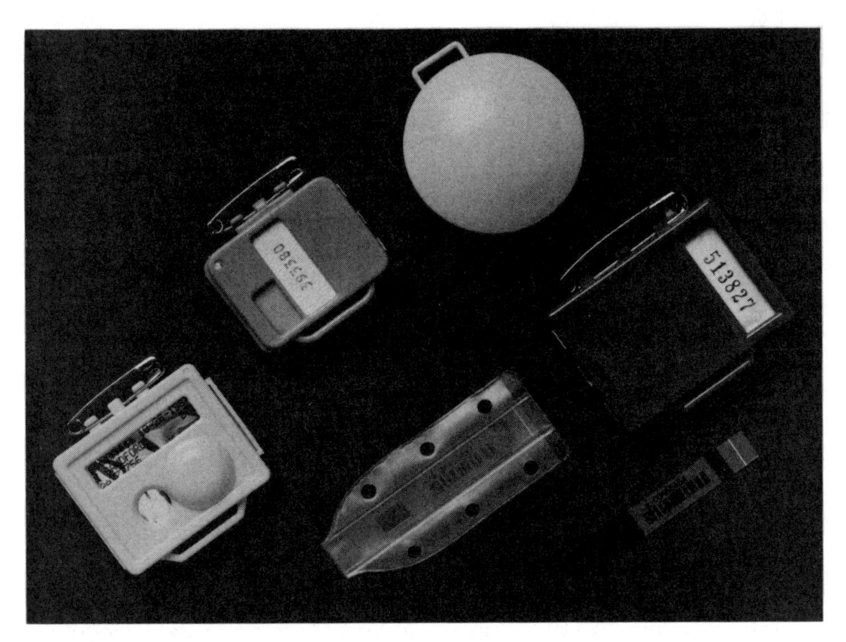

Figure 19.2 Devices for monitoring the exposure of workers to various types of radiation. (Courtesy NRPB)

(b) Contamination monitors
These measure the surface activity of radioactive contamination in counts per unit time. They normally contain a Geiger-Müller tube or scintillation counter. For α contamination, the detector normally employed would be a scintillation counter. The efficiency depends on the particular radionuclide being measured and the instrument should be calibrated for each radionuclide of interest.

The selection and use of monitoring instruments may be complex and should be discussed with a Radiation Protection Adviser (see below) or other suitable expert.

19.8 Legal requirements

The principal legislation in the UK affecting the use of ionising radiations in industry is summarised briefly below. However, readers should consult the appropriate documents for full details.

19.8.1 The Ionising Radiations Regulations 1985

These regulations, which were made under the Health and Safety at Work etc. Act, 1974, came fully into effect on 1st January 1986. They apply to all work with ionising radiation rather than just work in a factory. They take account of the recommendations of ICRP and are in conformity with a Council Directive of the European Communities which lays down basic safety standards for the health protection of the general public and workers against the dangers of ionising radiation. Details of acceptable methods of meeting the requirements of the regulations are given in the supporting Approved Code of Practice[8]. The following is a summary of some of the main requirements of the Regulations.

The dose limits for employees over the age of 18 years are those recommended by ICRP prior to 1990[9]. Lower limits apply to trainees under the age of 18 years. Special restrictions apply to the rate at which women of reproductive capacity can be exposed and to the exposure of pregnant women during the declared term of pregnancy. The limits for any other person are 5 mSv in a year for the effective dose and 50 mSv in a year for the dose to individual organs or tissues other than the lens of the eye for which the value is 15 mSv in a year. The main requirement however is for employers to 'take all necessary steps to restrict so far as reasonably practicable the extent to which his employees and other persons are exposed to ionising radiation', in keeping with the emphasis of ICRP. For practical purposes, this requirement is backed up by action levels. If the effective dose equivalent to an employee exceeds 15 mSv in a year, the employer is required to make an investigation to determine whether it is reasonably practicable to take further steps to reduce exposure. If the effective dose exceeds 30 mSv in a calendar quarter, the dosimetry service is required to send details of the exposure to the Health and Safety Executive (HSE).

Part 4 of the Approved Code of Practice[8] requires employers to make arrangements with their dosimetry service to be informed whenever one of their employees reaches a cumulative effective dose equivalent of 75 mSv within any five consecutive calendar years. Reaching such a cumulative dose should trigger an additional investigation centred around the relevant employee. This is intended to determine what action may be necessary in the light of the risk to the individual's health from the actual and projected radiation exposure.

To facilitate the control of doses to persons, the Regulations specify criteria for designating areas as controlled or supervised areas. The underlying basis of designation is to define areas where doses may exceed 3/10 or 1/10, respectively, of the annual dose limit for employees. Areas are designated on the basis of dose rate, total activity being handled, air activity concentration and surface contamination levels.

Employers are required to 'designate as classified workers those of his employees who are likely to receive a dose of ionising radiation which exceeds three-tenths of any relevant dose limit'. Only employees aged 18 years or over who have been certified as fit to be designated as a classified person can be so designated. Employees or other persons are only permitted to enter a controlled area if they are classified or enter under a 'written system of work'. If a written system of work is used the employer must be able to justify non-classification of the workers involved.

The Radiation Protection Adviser (RPA) is a key figure in the Regulations. An RPA is to be appointed if any employee is exposed to an instantaneous dose rate above 7.5 μSvh^{-1} or there is a controlled area which is entered. His function is to advise the employer 'as to the observance of these Regulations and as to other health and safety matters in connection with ionising radiation'. He should, for example, be consulted about restricting the exposure of workers, the identification of controlled and supervised areas, dosimetry and monitoring, the drawing up of written systems of work and local rules, the investigation of abnormally high exposures and over-exposures and training. The potential appointment of an RPA should be notified to HSE one month before he takes up his role. The notification should include details of qualifications and experience and a description of the scope of the advice that the RPA will be required to give.

In relation to employees who are designated as classified persons, the Regulations require employers to ensure that assessments are made of all significant doses. For this purpose, the employer is to make suitable arrangements with an approved dosimetry service (ADS). The ADS will also keep the dose record for the employer. The purpose of the approval system is to ensure as far as possible that the doses are assessed on the basis of accepted national standards.

The Regulations also specify requirements for the medical surveillance of employees and the maintenance of individual records of medical findings and assessed doses. The general requirement to keep doses as low as reasonably practicable is strengthened by the inclusion of a basic requirement to control the source of ionising radiation and by subsequent specific requirements to provide appropriate safety devices, warning signals, handling tools, etc, to leak test radioactive sources, to provide protective equipment and clothing and test them, to monitor radiation and contamination levels (see *Figure 19.3*), to store radioactive substances safely, to design, construct and maintain buildings, fittings and equipment so as to minimise contamination, and to make contingency arrangements for dealing with foreseeable but unintended incidents. In addition, undertakings holding

Figure 19.3 Checking contamination levels after a fire

large quantities of radioactive substances will need to make a survey of potential hazards and prepare a report, a copy of which should be sent to HSE prior to commencing work.

There are also requirements for employers to notify HSE of work with ionising radiation, over-exposures and certain accidents and losses of radioactive material. The provision of information on potential hazards and appropriate training are also required. In addition, there are requirements to formulate written local rules and to provide supervision of work involving ionising radiation. This last requirement will normally necessitate the appointment by management of a radiation protection supervisor (RPS) whose responsibilities should be clearly defined.

The RPS should not be confused with the RPA. While the latter may be an outside consultant, the RPS plays a supervising role in assisting the employer to comply with the Regulations and should therefore be an employee directly involved with the work with ionising radiations, preferably in a line management position that will allow him to exercise close supervision to ensure that the work is done in accordance with the local rules, though he need not be present all the time. The RPS should therefore be conversant with the Regulations and local rules, command sufficient respect to allow him to exercise his supervisory role and understand the necessary precautions to be taken in the work that is being done.

It should be noted that the EC directive on basic safety standards referred to above is being revised in the light of the new ICRP recommendations[5]. It is expected that, once made, the revised directive will necessitate revision of the Ionising Radiations Regulations 1985.

19.8.2 The Radioactive Substances Act 1960

The main purpose of this Act is to ensure effective control over radioactive wastes. Under the Act those who keep or use radioactive materials on premises used for the purposes of an undertaking (trade, business, profession etc.) are required to register with Her Majesty's Inspectorate of Pollution, Her Majesty's Industrial Pollution Inspectorate of the Scottish Office or the Department of the Environment for Northern Ireland, according to the region, or to be exempted from registration. Conditions may be attached to registrations and exemptions, and these are made with regard to the amount and character of the radioactive waste likely to arise.

Furthermore, no person may dispose of or accumulate radioactive waste unless he is authorised by the appropriate Department or is exempt. Whenever possible local disposal of radioactive waste should be used but with many industrial sources, such as those used in gauges and radiography, disposal should be by transfer to British Nuclear Fuels plc at Drigg in Cumbria.

A number of generally applicable exemption orders have been made under the Act for those situations where control would not be warranted. The orders cover such things as substances of low activity, luminous articles, electronic valves, smoke detectors and some uses of uranium and thorium. The orders should be consulted for details of the conditions under which exemption is granted. The orders are currently under review.

Readers are also referred to a useful guide to the administration of the Act[10].

19.8.3 Transport Regulations

Protection of both transport workers and the public is required when radioactive substances are transported outside work premises. The Regulations and conditions governing transport in the UK and internationally follow those specified by the International Atomic Energy Agency. The latest version of the Agency's regulations is listed in reference 11. The particular regulations that apply depend on the means of transport to be used. Those that apply to the transport of radioactive materials by road are given in reference 12. These regulations are under review and are likely to be replaced in the near future by regulations made under the Radioactive Material (Road Transport) Act 1991. Guidance on the requirements of the regulations is given in a Code of Practice[13]. Requirements for sending radioactive materials by post are specified in the Post Office Guide.

A full review of the UK regulation of the transport of radioactive materials is given in reference 14.

19.9 National Radiological Protection Board

The National Radiological Protection Board (NRPB) was created by the Radiological Protection Act 1970. The Government's purpose in proposing the legislation was to establish a national point of authoritative reference in radiological protection. The NRPB's principal duties are to advance the acquisition of knowledge about the protection of mankind from radiation hazards and to provide information and advice to those with responsibilities in radiological protection. The NRPB also provides technical services to organisations concerned with radiation hazards, and training in radiological protection. It is empowered to charge for its services. Its headquarters are at Chilton and it has centres at Glasgow, Leeds and Chilton for the provision of advice and services. The services provided include: radiation protection adviser (RPA), reviews of design, monitoring of premises, personal monitoring, record keeping, instrument tests, testing of materials and equipment, leakage tests on sealed sources and assistance in the event of incidents and accidents. The Board runs scheduled and custom-designed training courses.

19.10 Incidents and emergencies[4,7]

In any radiological incident or emergency, the main aim must be to minimise exposures and the spread of contamination. Pre-planning against possible incidents is essential and suitable first-aid facilities should be provided. Where significant quantities of radioactive substances are to be kept, procedures for dealing with fires should be discussed in advance with the local fire service.

Spills should be dealt with immediately and appropriate monitoring of the person and of surfaces should be carried out. Anyone who cuts or wounds himself when working with unsealed radioactive material must obtain first-aid treatment and medical advice. This is particularly important as contamination can be readily taken into the bloodstream through cuts. If a radioactive source is lost immediate steps must be taken to locate it and, if it is not accounted for, Her Majesty's Inspectorate of Pollution (or appropriate Regional Department) and the HSE must be notified.

The National Arrangements for Incidents involving Radioactivity (NAIR) enables police to obtain expert advice on dealing with incidents (for example, transport accidents) that may involve radiation exposure of the public and for which no other pre-arranged contingency plans exist or, for some reason, those plans have failed to function. A source of radiological advice and assistance exists in each police administrative area – hospital physicists and health physicists from the nuclear industry, government and similar establishments. The scheme is co-ordinated by the National Radiological Protection Board at Chilton from whom further details are obtainable.

19.11 Non-ionising radiation

There are several forms of non-ionising electromagnetic radiation that may be encountered in industry[15,16]. They differ from γ and X-rays in that they are of longer wavelength (lower energy) and do not cause ionisation in

Figure 19.4 Characterisation of electric fields from a radio frequency PVC welding machine

matter. They are ultraviolet (a few tens of nanometres (nm) to 400 nm wavelength), visible (400 to 700 nm) and infrared (700 nm to 1 mm) radiations in the optical region, and microwave and radiofrequency radiations and electric and magnetic fields. The ability of radiation within one of these defined regions to produce injury may depend strongly on the wavelength. *Figure 19.4* illustrates the monitoring for non-ionising radiation around a microwave oven.

19.11.1 Optical radiation

Ultraviolet radiation is used for a wide variety of purposes such as killing bacteria, creating fluorescence effects and curing inks[17]. It is produced in arc welding or plasma torch operations and is emitted by the sun. Short wavelength ultraviolet radiation of wavelength approximately less than 240 nm is strongly absorbed by oxygen in the air to produce ozone which is a chemical hazard. The OES for ozone is 0.1 ppm. Even below this level it may cause smarting of the eye and discomfort in the nose and throat. It has a characteristic smell.

Ultraviolet radiation does not penetrate beyond the skin and is substantially absorbed in the cornea and lens of the eye. The human organs at risk are therefore the skin and the eyes. The immediate effects are erythema (as in sunburn) and photokeratitis (arc eye, snow blindness). Long-term effects are premature skin ageing and skin cancer, and possibly cataracts. No cases of skin cancer due to occupational exposure to artificial sources of ultraviolet radiation have been identified, but a causal link between skin cancer and exposure to solar ultraviolet radiation is now accepted, particularly for those with white skin. Some chemicals such as coal tar can considerably enhance the ability of ultraviolet radiation to produce damage.

Wherever possible, ultraviolet radiation should be contained[17,18]. If visual observation of any process is required, this should be through special observation ports transparent to light but adequately opaque to ultraviolet radiation. Where the removal of covers could result in accidental injurious exposures, interlocks should be fitted which either cut the power supply or shutter the source. Protection is also achieved by increasing the distance between source and person, covering the skin and protecting the eyes with goggles, spectacles or face shields.

Intense sources of visible light such as arc lamps and electric welding units and, of course, the sun can cause thermal and photochemical damage to the eye; they can also produce burns in the skin. Adequate protection is normally achieved by keeping exposures below discomfort levels.

Infrared radiation is emitted when matter is heated. The principal biological effects of exposure can be felt immediately as heating of the skin and the cornea. Long-term exposure can cause cataracts. Protection is achieved by shielding the source and through the use of personal protective equipment especially eye wear.

The intensity of laser sources in the ultraviolet, visible and infrared regions can be orders of magnitude higher than that of other optical sources. Because of their very low beam divergence some lasers are capable of delivering large amounts of power to a distant target. Of particular importance is the injury that can be caused to the eye, such as retinal burns and cornea damage. Protection is achieved through appropriate design of equipment and through the use of a combination of the following: enclosure of the device, safety interlocks, shutters, warning signs, eye and skin protection and adequate operator training. It is also necessary to guard against stray reflections.

19.11.2 Electric and magnetic fields

Time varying electric and magnetic fields are emitted by numerous devices in the transmission and distribution of electricity and in the use of electrical and electronic equipment. Sources of exposure include power lines, induction heaters, broadcast and telecommunication systems, microwave ovens and radar[19,20].

The restriction on extremely low frequency (ELF) fields such as those generated at power distribution frequencies are based on the avoidance of induced electric currents affecting central nervous system function. At radiofrequencies which include microwaves, the fields can penetrate the body and cause heating; restrictions on exposure are designed to prevent adverse responses to increased heat load and elevated body temperatures.

In addition to the direct effects of the interaction of the electric and magnetic fields with people there is also the possibility of people touching metallic objects in the field. The indirect effect of such contact may be shock or burn. Whilst there is no good evidence of harm to people from exposure to electromagnetic fields at existing environmental levels, NRPB has developed advice that is coherent in respect of the way the guideline exposure levels vary with frequency for both electric and magnetic fields[21]. The advice takes into account both direct and indirect effects.

The NRPB guidelines are based on the study of human populations. The NRPB's view is that epidemiological studies are not sufficient to form the basis for restricting human exposure to electromagnetic fields. Nevertheless, the guidelines are supported by the findings of two reports that deal with general health[22] and the risk of cancer[20].

At frequencies between 10 Hz and 1 kHz, a cautious estimate of the threshold current density to adversely affect the central nervous system is 10 mA m^{-2}. Progressively larger current densities are necessary at frequencies above and below this range.

At frequencies above 100 kHz, the restrictions on heat load are averaged over the whole body or small masses of tissue to avoid the adverse effect of localised heating. The NRPB has produced guidance based on a restriction of 0.4 W kg^{-1} averaged over the whole body mass. This value is considered to incorporate a sufficient margin such that it should not be necessary to account for environmental factors and work loads for healthy people.

Localised exposures are restricted to 10 W kg^{-1} in the head and trunk and 20 W kg^{-1} in the limbs, the averaging masses being 10 g and 100 g according to target tissue. Further details of the basic restrictions on exposure and derived reference levels of external electric and magnetic field strength are given in the NRPB document[21].

References

Atomic structure and radioactivity

1. Evans, R.D., *The Atomic Nucleus*, McGraw-Hill, New York (1955)
2. Royal Commission on Environmental Pollution, 6th Report, *Nuclear Power and the Environment, Cmnd. 6618, HMSO London (1976)*
3. Burchman, W.E., *Elements of Nuclear Physics*, Longman, London (1979)

Ionising radiation

4. Bennellick E.J., Ionising radiation. In *Industrial Safety Handbook*, (Ed. W. Handley), 2nd edn, McGraw-Hill, London (1977)
5. ICRP, 1990 Recommendations of the International Commission on Radiological Protection, ICRP Publication 60, Pergamon Press, Oxford. *Ann. ICRP*, **21,** No. 1–3 (1991)
6. Hall, G.J., *Radiation and Life*, Pergamon Press, Oxford (1984)
7. Martin, A. and Harbison, S.A., *An Introduction to Radiation Protection*, 3rd edn, Chapman and Hall London (1986)
8. Health and Safety Commission, *Approved Code of Practice No. COP 16, Protection of Persons against Ionising Radiations arising from any work activity*, HMSO, London (1985)
9. ICRP, Recommendations of the International Commission on Radiological Protection (adopted January 17, 1977), ICRP Publication **26**, Pergamon Press, Oxford. *Ann. ICRP*, **1**, No. 3 (1977)
10. Department of the Environment, *Radioactive Substances Act 1960, A guide to the administration of the Act*, HMSO, London (1982)
11. International Atomic Energy Authority, *IAEA Safety Series Publication No. 6, Regulations for the Safe Transport of Radioactive Material*, 1985 edition as amended in 1990, IAEA, Vienna (1990)
12. Department of the Environment, *The Radioactive Substances (Carriage by Road) (Great Britain) Regulations 1974*, HMSO, London (1974) *The Radioactive Substances (Carriage by Road) (Great Britain) (Amendment) Regulations 1985*, HMSO, London (1985)
13. Department of the Environment, *Code of Practice for the Carriage of Radioactive Material by Road*, HMSO, London (1989)
14. Advisory Committee on the Safe Transport of Radioactive Materials. *The UK Regulations of the Transport of Radioactive Materials: Quality Assurance and Compliance Assurance*, HMSO, London (1988)

Non-ionising radiation (general)

15. McHenry, C.R., Evaluation of exposure to non-ionising radiation. In *Patty's Industrial Hygiene and Toxicology, Vol. III, Theory and Rationale of Industrial Hygiene Practice* (Eds. L.V. Cralley and L.J. Cralley), John Wiley & Sons, New York (1979)
16. Sliney, D.H., Non-ionising radiation. In *Industrial Environmental Health* (Eds. L.V. Cralley *et al.*), Academic Press, London (1972)

Ultraviolet radiation

17. McKinlay, A.F., Harlen, F. and Whillock, M.J., *Hazards of Optical Radiations. A Guide to Sources, Uses and Safety*, Adam Hilger, Bristol (1988)
18. Health and Safety Executive, *Guidance Notes, Medical Series No. MS 15, Welding*, HMSO, London (1978)

Radiofrequency radiation

19. Allen, S.G. and Harlen, F., *NRPB report No. R144, Sources of Exposure to Radiofrequency and Microwave Radiations in the UK*, NRPB, Chilton (1983)
20. National Radiological Protection Board, *Electromagnetic Fields and the Risk of Cancer, Report of an Advisory Group on Non-ionising Radiation*, **3,** No. 1, NRPB (1992)
21. Dennis, A.J. Muirhead, C.R. and Ennis, J.R., NRPB report R241, *Human Health and Exposure to Electromagnetic Radiation*, NRPB, Chilton (1992)
22. National Radiological Protection Board, *Guidance as to Restrictions on Exposure to Static and Time Varying Electromagnetic Fields and Radiation*, NRPB (to be published)

Chapter 20

Noise and vibration

R. W. Smith

The first four sections of this chapter explain what noise is, how it is defined and the theory and practice behind the measurement of noise levels. The rest outlines the way the ear works and the damage that can occur to cause noise-induced hearing loss. Some of the problems created by vibrations are considered. Reference is made to the guidelines, recommendations and legislation that exist and which are aimed at limiting the harmful effects of noise in the workplace, and the nuisance effect on the community.

20.1 What is sound?

A vibrating plate will cause corresponding vibrations or pressure fluctuations in the surrounding air, which would then be transmitted through to the receiver. For example, when an alternating electrical signal is fed into a loudspeaker, the cone vibrates causing the air in contact with it to vibrate in sympathy, and a sound wave is produced. These pressure waves are

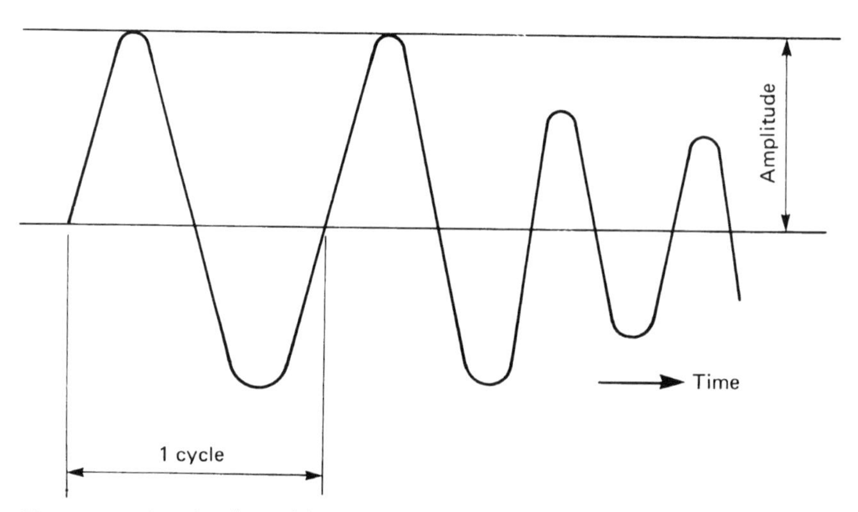

Figure 20.1 Amplitude and frequency

transmitted through the air at a finite speed. This is easily demonstrated by observing the time interval between a flash of lightning and hearing a clap of thunder. The velocity of sound in air at normal temperature and pressure is approximately 342 metres per second (1122 feet per minute), at 20°C. Increasing the temperature of air increases the velocity.

These pressure fluctuations or vibrations have two characteristics: firstly the amplitude of the vibration, and secondly the frequency – both are illustrated in *Figure 20.1*.

20.1.1 Amplitude

The amplitude of a sound wave determines loudness, although the two are not directly related, as will be explained later. Typically these pressure amplitudes are very small. For the average human being the audible range is from the threshold of hearing at 20 μPa up to 200 pascals (Pa) where the pressure becomes painful. This is a ratio of 1 to 10^6. The intensity of noise is proportional to the pressure squared hence the range of intensity covers a ratio of 1 to 10^{12}.

With such a range it becomes more convenient to express the intensity of pressure amplitude on a logarithmic base. The intensity level is proportional to the square of the pressure, thus the sound pressure level (Lp) can be defined as:

$$Lp = 10 \log_{10} (P_1/P_0)^2 \quad (1)$$

where the sound pressure level (SPL) is expressed in decibels (dB), P_1 equals the pressure amplitude of the sound and P_0 is the reference pressure 20 μPa. All logarithmic calculations are to the base 10. Typical examples of sound pressure levels for a variety of environments are shown in *Figure 20.2*.

Note that, since the decibel is based on a logarithmic scale, two noise levels cannot be added arithmetically. Hence, the resultant L_{Pr} from adding sources L_{P1}, L_{P2} etc. is obtained thus:

$$L_{Pr} = 10 \log_{10} \left[\left(\frac{P_1}{P_0} \right)^2 + \left(\frac{P_2}{P_0} \right)^2 + \cdots \right] \quad (2)$$

For two equal sources $L_{P1} = L_{P2}$

$$\therefore\ L_{Pr} = 10 \log_{10} \left[\left(\frac{P_1}{P_0} \right)^2 \times 2 \right] \quad (3)$$

$$= 10 \log_{10} \left(\frac{P_1}{P_0} \right)^2 + 10 \log_{10} 2 \quad (4)$$

$$= L_P + 3.1 \text{ (dB)} \quad (5)$$

Thus for all practical purposes doubling the sound intensity increases the sound pressure level by 3 dB. For example:

$$90\,\text{dB} + 90\,\text{dB} = 93\,\text{dB} \quad (6)$$

Sound pressure level in dB	Environment
140	
	Threshold of pain
130	
	Pneumatic drill
120	
	Loud motor car horn (dist. 1 m)
110	
	Rush hour traffic freely flowing
100	
	Inside tube train
90	
	Inside motor bus
80	
	Average traffic on street corner
70	
	Conversational speech
60	
	Typical business office
50	
	Living room, surburban area
40	
	Library
30	
	Bedroom at night
20	
	Broadcasting studio
10	
	Threshold of hearing
0	

Figure 20.2 Typical sound pressure levels

Similarly, 103 dB + 90 dB = 103.2 dB, showing that where the effect of a small source is to be added to a large source, the resultant noise level would be relatively unchanged.

Of course, the converse is true, and this is important when considering the control of noise from a number of different sources, since treatment of a minor source may not result in any change in overall levels.

20.1.2 Frequency

The rate at which these pressure fluctuations take place is called frequency and is measured in hertz (Hz) or cycles/s. The human ear is normally capable of hearing over a range from 20 Hz to 16 000 Hz (16 kHz) although this range can be considerably reduced at the high frequency end for older people and for those suffering from hearing impairment.

20.2 Other terms commonly found in acoustics

20.2.1 Loudness

The human ear does not respond equally to all frequencies. To obtain the same subjective loudness at low frequencies as at higher frequencies requires a larger physical amplitude (greater L_P), since the ear is less sensitive at low frequencies. These curves of equal loudness are illustrated in *Figure 20.3*. The curves of equal loudness are defined in dB phon, and are obtained experimentally using pure tones to create the same sensation of loudness at different frequencies. From these curves it can be seen that the difference between physical amplitudes at different frequencies required to produce the same loudness curve reduces as the loudness increases.

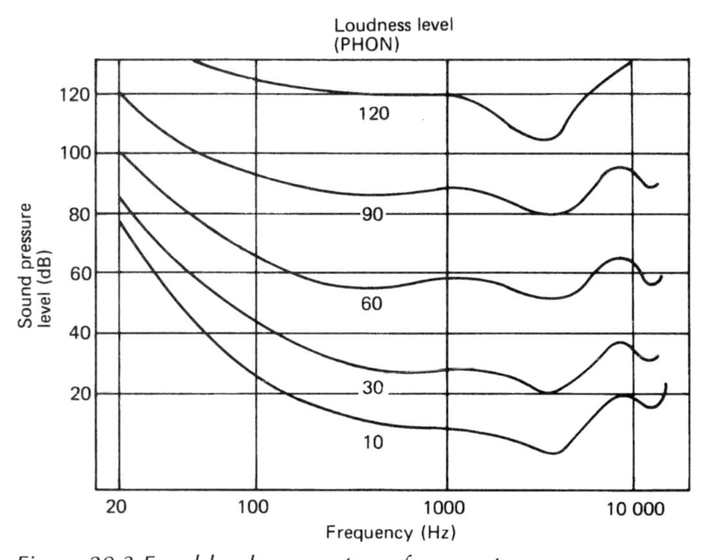

Figure 20.3 Equal loudness contours for pure tones

As an approximation, an increase in sound pressure level of 8–10 dB corresponds to a subjective doubling of loudness. There are a number of different ways of calculating the loudness, and the loudness level in phon should be referenced to the method used.

It is apparent that the subjective response to noise is extremely complex and these complexities should always be borne in mind when dealing with individual people. A noise or noise level which is acceptable to one individual may not be to another.

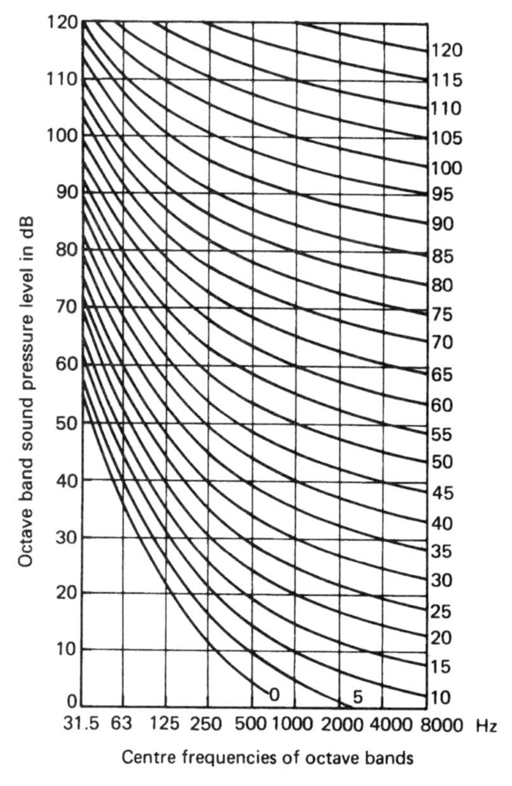

Figure 20.4 Noise rating curves

20.2.2 Noise rating curves

There are a standard series of Noise Rating (NR) Curves (*Figure 20.4*) which are stylised forms of the loudness response curves. These NR curves are often used as a criterion for noise control, and as such are internationally accepted. Other criteria may also be encountered such as NC curves (Noise Criteria).

20.2.3 Octave bands

The previous two sections have not attempted to define frequency in terms of a bandwidth. The generally accepted bandwidth used within the field of noise control are octave bands, that is a range or band of frequencies with the upper frequency limit f_u equal to twice the lower limit f_l. Each octave band is defined by the centre frequency f_{ob} where

$$f_{ob} = (f_u \times f_l)^{1/2} = (2f_l \times f_l)^{1/2} = 1.414\ f_l \qquad (7)$$

The commonly used octave bands are illustrated in *Table 20.1*.

Other bandwidths may be encountered in analysis and noise control work.

Table 20.1. Octave bands

Octave band centre frequency (f): Hz	*Octave band range: Hz*
31.5	22–44
63	44–88
125	88–177
250	177–354
500	354–707
1000	707–1414
2000	1414–2828
4000	2828–5657
8000	5657–11814

20.2.4 The decibel

The methods of assessing sound and developing noise criteria are complex and there are two approaches that may be used to obtain a quantitative measure of subjective noise. Either measure the physical characteristics of the sound over the frequency range on a meter and correct for the response of the ear, or alternatively, devise a measurement system which gives an output of loudness rather than amplitude. The first method can be undertaken but is rather long-winded 'by hand'. It can be approximated in instrumentation by feeding the sound pressure level into an electronic weighting network which automatically corrects the octave band levels and adds them together to give an approximation of the subjective level. The unit of measurement used is the decibel (dB).

The shapes of the weighting curves most commonly used are shown in *Figure 20.5* and the sound pressure level curve that follows the hearing characteristic of the ear is termed the A weighted curve. This curve gives the most widely used unit (dBA) for quantifying a noise level. Other curves on

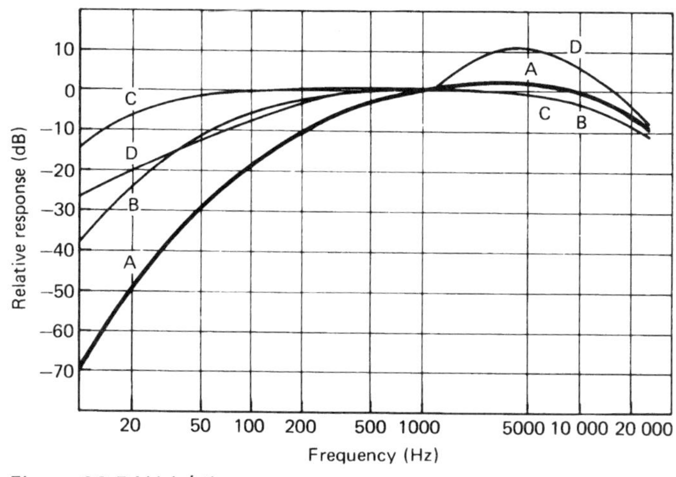

Figure 20.5 Weighting curves

the weighting network are the B, C and D. Curves B and C are not commonly used and D is sometimes used for aircraft noise.

20.2.5 Sound power level

The sound pressure level (L_P) must always be defined at a specific point in relation to the noise source. It is dependent on the location relative to the source, and the environment around the source and the receiver. The Sound Power Level (L_W) is a measure of the total acoustic energy radiated and is independent of the environment around the source. It is defined as:

$$L_w \text{ (dB)} = 10 \log_{10} \frac{W}{120} \qquad (8)$$

where W is the power in watts. In many cases it is calculated from sound pressure levels by the approximate formula:

$$L_W = L_P + 10 \log_{10} A \qquad (\text{m}^2) \qquad (9)$$

or $$L_P + 10 \log_{10} A - 10 \qquad (\text{ft}^2) \qquad (10)$$

where A = the surface area of measurement.

The concept of sound power level is best illustrated by the following analogy.

Suppose that the temperature measured at 1 metre from 1 kW electric fire is the same as that measured 1 metre from a large steam boiler. Although the temperatures are the same, the boiler would be radiating more heat because it is radiating over a larger surface area. Similarly, the sound pressure level 1 metre from a large industrial cooling tower may be the same as that 1 metre from a small pump, but the acoustic power radiated by the cooling tower would be very much greater.

The sound power level is used to calculate the reverberant and pressure level and community noise.

20.3 Transmission of sound

The simplest concept in the transmission of noise is the inverse square law. For example, if the distance from the source is doubled, the reduction in noise level would be:

$$10 \log_{10} \left(\frac{2}{1}\right)^2 = 10 \log_{10} 4 = 6 \text{ dB} \qquad (11)$$

Where the sound power level is known the resulting sound pressure level at a distance is given by:

$$L_P = L_W - 10 \log 2\pi r^2, \text{ where } r \text{ is in metres} \qquad (12)$$

Where distances exceed 200–300 metres other absorption factors should be taken into account. These include atmospheric attenuation and 'ground' absorption effects. Atmospheric attenuation (molecular absorption) is proportional to distance, but 'ground' absorption is dependent on the intervening terrain between the source and receiver.

20.4 The sound level meter

The majority of noise control work undertaken by the safety adviser will involve the measurement and possibly the analysis of noise. It is therefore important that the use and the limitations of sound pressure level measurements and sound level meters are understood. Errors in measurement technique or interpretation could lead to costly mistakes or over-specifying in remedial measures.

20.4.1 The instrument

There are many sound level meters on the market, but all work in a similar manner. The basic hand-held set (*Figure 20.6*) consists of a microphone, an

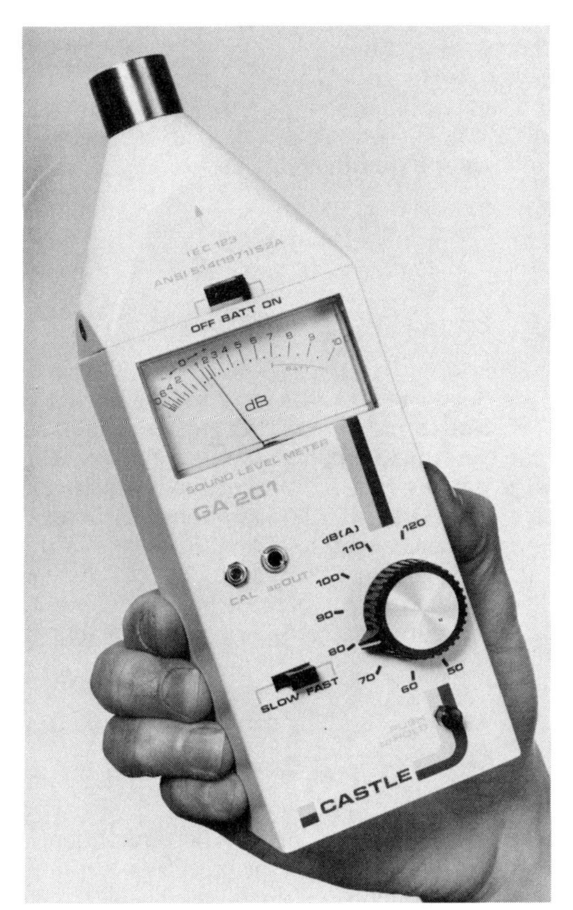

Figure 20.6 Industrial hand-held sound level meter. (Courtesy General Acoustics Ltd)

amplifier with a weighting network and a read-out device in the form of a meter or digital presentation. The microphone converts the fluctuating sound pressure into a voltage which is amplified and weighted (A, B or Linear etc.). The electrical signal then drives a meter or digital read-out. In many instruments an octave or ⅓ octave band filter is incorporated, to enable a frequency analysis to be performed. The signal would then by-pass the weighting network and be fed into the read-out device via the frequency filters.

The difference between the two available grades of sound level meter, precision and industrial, is effectively the degree of accuracy of the measurements particularly at high and low frequency. Meters should comply with British Standards[1] or with IEC[2] recommendations.

20.4.2 Use of a sound level meter

For the operation of sound level meters of different types the safety adviser should refer to the manufacturer's instruction book. The prime requirement for any instrument for noise measurement is that it should not be more sophisticated than necessary and it should be easy to use and calibrate. A typical sound level meter for use by the safety adviser should have the facility for measuring dBA and octave band sound pressure levels. More sophisticated meters have facilities for measuring equivalent noise level (L_{eq}, $L_{EP.d}$) (see section 20.6) and undertaking statistical analysis.

20.5 The ear

20.5.1 Mechanism of hearing

Before appreciating how deafness occurs the manner in which the ear works should be understood. Briefly, the sound pressure waves enter the auditory canal (*Figure 20.7*) and cause the eardrum (tympanic membrane) to move in sympathy with the pressure fluctuations. Three minute bones called the hammer, anvil and stirrup (the ossicles) transmit the vibrations via the oval window to the fluid in the inner ear and hence to the minute hairs in the cochlea. These hairs are connected to nerve cells which respond to give the sensation of hearing. Damage to, or deterioration of these hairs, whether by exposure to high noise levels, explosions or ageing will result in loss of hearing.

20.5.1.1 Audiogram

The performance of the ear is evaluated by taking an audiogram. Audiograms are normally performed in an Audiology Room which has a very low internal noise level and is vibration isolated. The audiogram is obtained by detecting the levels at which specific tones can be heard and cease to be heard by the person being tested.

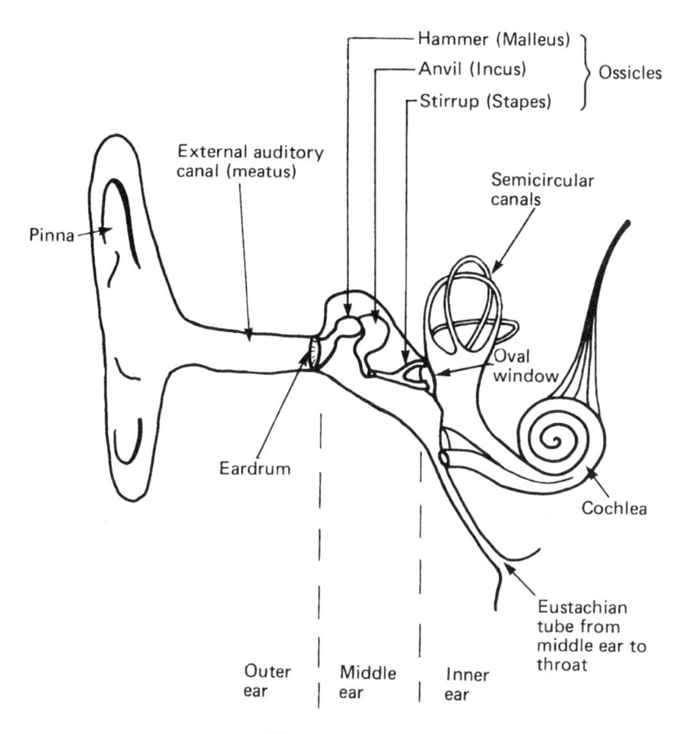

Figure 20.7 Diagram of human ear

20.5.1.2 Hearing loss

Noise-induced hearing loss (occupational hearing loss) is caused by over-stimulus of the receptor cells in the cochlea resulting in auditory fatigue. In its early stages it is often shown as an increase in the threshold of hearing – 'temporary threshold shift'. It may be accompanied by a ringing in the ears (tinnitus) which is indicative of temporary hearing damage. If exposure to high noise levels was continued, the result would be a 'permanent threshold shift' or noise-induced deafness. The effects of noise-induced deafness can be seen by the dip in the audiogram at 4 kHz, which deepens and widens with continuing exposure. Eventually the speech frequencies are affected (*Figure 20.8*), the sufferer loses the ability to hear consonants, and speech sounds like a series of vowels strung together.

Similar effects occur with age-induced hearing loss, which should always be considered when checking hearing impairment.

20.6 The equivalent noise level

In general the noise level in the community or inside a factory will vary with time. The equivalent noise level (L_{eq}) is defined as the notional steady noise level which, over a given period of time, would deliver the same amount of

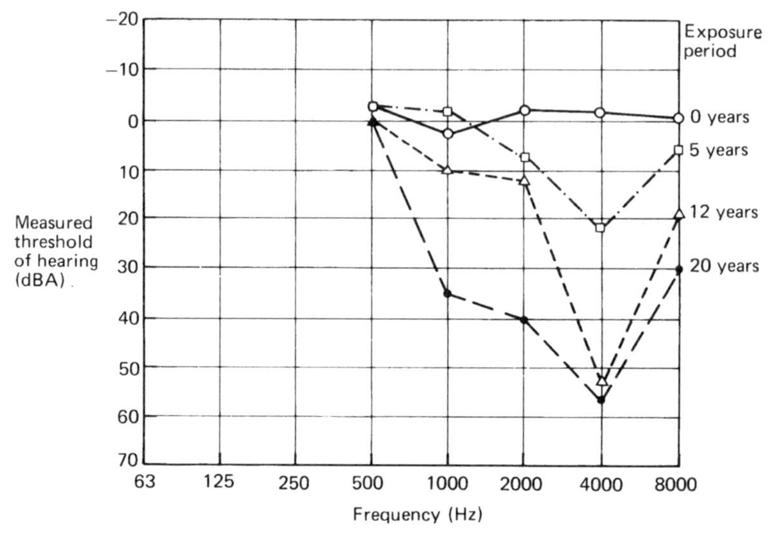

Figure 20.8 Typical audiograms showing progressive loss of hearing

sound energy as the fluctuating level. Thus to maintain the L_{eq} when SPL is doubled, i.e. increased by 3 dB, exposure time must be halved (*Table 20.2*). The equivalent noise level concept forms the basis of the exposure criteria used in the Noise at Work Regulations 1989[3] which calls it 'daily personal noise exposure' ($L_{EP.d}$). Where the fluctuation is not well defined the calculations can be done electronically using a dosimeter or statistical analyser[4]. Transient noises also require statistical analysis and those measurements often required are:

L_{10} – the noise level exceeded for 10% of the time (average peak).
L_{50} – the noise level exceeded for 50% of the time (mean level).
L_{90} – the noise level exceeded for 90% of the time (average background level).

Table 20.2. Exposure times vs. noise levels for an equivalent noise level of 90 dBA

Noise level (dBA)	*Exposure time (h)*
87	16
90	8
93	4
96	2
99	1
102	½
105	¼

20.7 Community noise levels

Noise generated in a factory may affect not only those employed in the factory but those living in the immediate neighbourhood, especially if machinery is run during night-time. Legislation has been brought into effect to give those affected the right of redress.

20.7.1 Control of Pollution Act 1974[5]

The Control of Pollution Act provides general legislation for limiting community noise as well as other pollutants. No specific limits are set, but the Act empowers the Local Authority to require a reduction in noise emission and impose conditions for noisy operations, e.g. specify a level of noise emission for a particular operation which must not be exceeded at certain given times. In certain instances they may not only set the limits but specify how they are to be met or how equipment is to be operated, referring to the relevant British Standard[6].

20.7.2 Method of rating industrial noise in the neighbourhood

Criteria that can be used to assess the reasonableness of complaints and for setting noise limits for design purposes are contained in a British Standard[7] which in itself does not recommend specific limits, but predicts the likelihood of complaints. These predictions are based on measured or predicted noise levels, corrections for the noise characteristic and measured or notional background noise levels.

20.7.3 Assessing neighbourhood noise

The following procedure is typical for evaluating the existing neighbourhood noise situation:

(1) Identify critical areas outside the factory by taking measurements over a reference period, as defined in the standard.
(2) Note which equipment in the factory is operating during the measurement period to help evaluate the major sources.
(3) Measure the L_{90} or background noise level. If the factory is always operating, the notional background noise level should be assessed using BS 4142[7]. Comparison of the factory noise level with the background level as explained in that British Standard, or with any limit set by the Local Authority will show whether further action is required.
(4) Where there is possibility of complaints, or complaints have been made, a noise control programme should be introduced to reduce the noise from the major sources identified in stage (2).

20.8 Work area noise levels

Legislation, in the form of the Noise at Work Regulations 1989[3] has been enacted to limit the risk of hearing damage to workpeople while at work. These regulations give effect to an EC directive[8] on the protection of workers from the risks related to noise at work. Under the regulations, employers are required to take certain actions where noise exposure reaches 'Action Levels'

First action level – $L_{EP.d}$ = 85 dB(A)
Second action level – $L_{EP.d}$ = 90 dB(A)
Peak action level – a peak sound pressure of 200 pascals.

Where employees are likely to be exposed to noise levels at or above the first action or peak action level, the employer must ensure that a noise survey is carried out by a competent person (reg. 4). Records of this assessment must be kept (reg. 5). Where a noise risk has been identified there is a requirement on the employer to reduce the emission to the lowest reasonably practicable level (regs 6 and 7).

If it is not reasonably practicable to reduce noise exposure levels to below 90 dB(A) $L_{EP.d}$ then the employer is required to provide personal hearing protection (reg. 8). An alternative to personal protection is the provision of hearing havens from which the worker can carry out his duties. Areas where there remains a hearing risk shall be designated 'ear protection zones' and be identified as such (reg. 9).

All hearing protection equipment must be kept in good order (reg. 10) and employees are required to use properly and look after any personal protective equipment issued to them. Considerable emphasis is placed on keeping employees informed (reg. 11) of:

- risks to their hearing,
- steps being taken to reduce the risk,
- procedures for obtaining personal protective equipment, and
- their obligations under the Regulations.

Regulation 12 modifies section 6 of HSW to impose a duty on manufacturers and suppliers to inform customers when the level of noise emitted by their product exceeds the first action level, i.e. 85 dB(A) $L_{EP.d}$.

20.9 Noise control techniques

Before considering methods of noise control, it is important to remember that the noise at any point may be due to more than one source and that additionally it may be aggravated by noise reflected from walls (reverberant noise) as well as the noise radiated directly from the source. With any noise problems there are the three distinct elements shown in *Figure 20.9* – source, path and receiver.

Having identified the nature and magnitude of any noise problem, the essential elements of a noise control programme are provided. Where a problem is evident, there are three orders of priority for a solution:

(1) Engineer the problem out by buying low noise equipment, altering the process or changing operating procedures.
(2) Apply conventional methods of noise control such as enclosures or silencers.
(3) Where neither of the above approaches can be used, the last resort of providing personal protection should be considered.

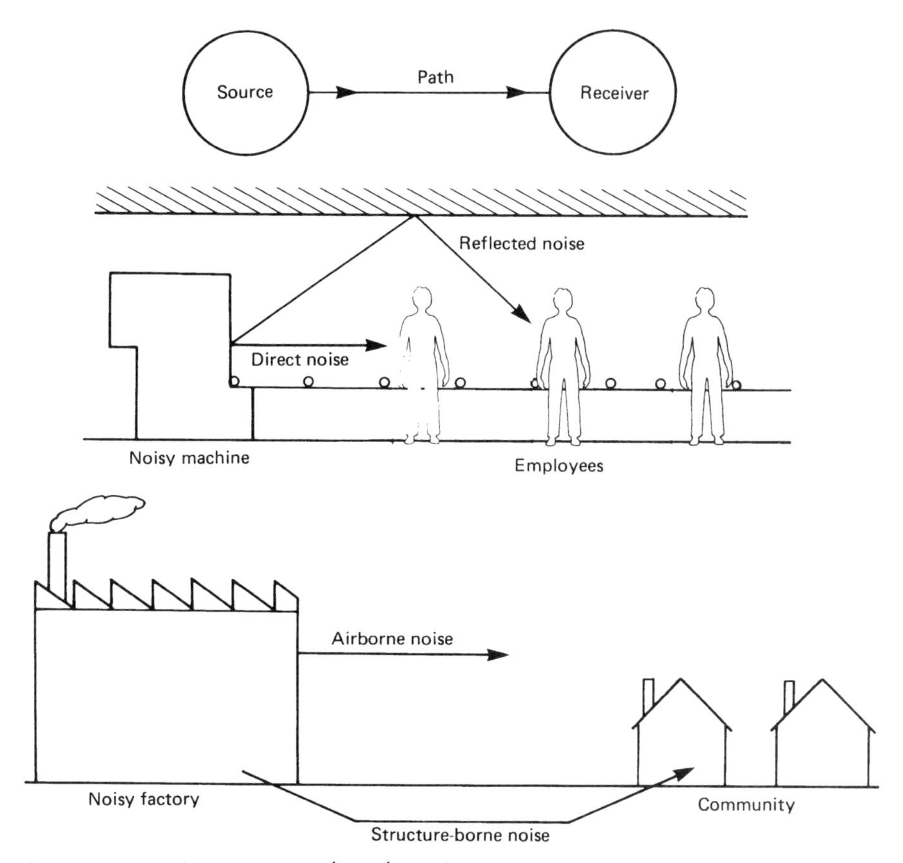

Figure 20.9 Noise source, path and receiver

20.9.1 Source

Although the control of noise at source is the most obvious solution, the feasibility of this method is often limited by machine design, process or operating methods. While immediate benefits can be obtained, this method should be regarded as a long-term solution.

20.9.2 Path

(a) *Orientation and location*
Control may be achieved by moving the source away from the noise sensitive area. In other cases where the machine does not radiate equally in all directions, turning it round can achieve significant reductions.

(b) *Enclosure*
Enclosures which give an attenuation of between 10 and 30 dBA[9] are the most satisfactory solution since they will control both the direct field and reverberant field noise components. In enclosing any source, the provision

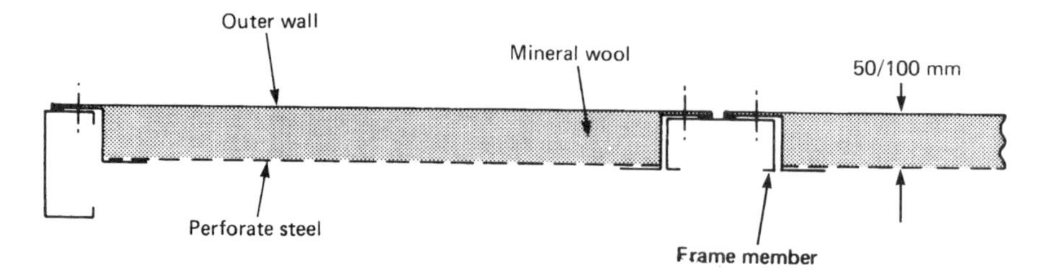

Figure 20.10 Noise enclosure. (Above) cross section through typical noise enclosure wall; (below) noise enclosure with access doors removed (courtesy Ecomax Acoustics Ltd)

of adequate ventilation, access and maintenance facilities must be considered. A typical enclosure construction is shown in *Figure 20.10*. The main features are an outer 'heavy' wall with an inner lining of an acoustically absorbent material to minimise reverberant build-up inside the enclosure. An inner mesh or perforate panel may be used to minimise mechanical damage.

The sound reduction, attenuation or insertion loss is defined as the difference in sound pressure level or sound power level before and after the enclosure (or any other form of noise control) is installed. The performance of the enclosure will be largely dependent on the sound reduction index (SRI) of the outer wall[10], assuming approximately 50% of the internal surface is covered with mineral wool or other absorption materials.[11] Typical values of sound reduction index for materials used for enclosures are shown in *Table 20.3*. Absorption coefficients are shown in *Table 20.4*.

When considering an enclosure for any item of equipment, it is essential to consider a number of other aspects as well as the noise reduction required. Ventilation may be required to prevent overheating of the

Table 20.3. Sound reduction indices for materials commonly used for acoustic enclosures

Material	*Octave band centre frequency: Hz*							
	63	*125*	*250*	*500*	*1 K*	*2 K*	*4 K*	*8 K*
22 g steel	8	12	17	22	25	26	25	29
16 g steel	12	14	21	27	32	37	43	44
Plasterboard	10	15	20	25	28	31	34	38
¼ in plate glass	12	16	19	21	22	36	31	34
Unplastered brickwork (110 mm thick)	22	31	36	41	45	(50)	(50)	(50)

Bracketed figures refer to practical installations rather than test conditions.

Table 20.4. Typical sound absorption coefficeints

Material	*Octave band centre frequency: Hz*							
	63	*125*	*250*	*500*	*1 K*	*2 K*	*4 K*	*8 K*
25 mm mineral wool	0.05	0.08	0.25	0.50	0.70	0.85	0.85	0.80
50 mm mineral wool	0.10	0.20	0.55	0.90	1.00	1.00	1.00	1.00
100 mm mineral wool	0.25	0.40	0.80	1.00	1.00	1.00	1.00	1.00
Typical perforated ceiling tile with 200 mm air space behind	0.30	0.50	0.80	0.95	1.00	1.00	1.00	1.00
Glass	0.12	0.10	0.07	0.04	0.03	0.02	0.02	0.03
Concrete	0.05	0.02	0.02	0.02	0.04	0.05	0.05	0.06

N.B. Concrete would normally be used for floor of enclosure.

equipment being enclosed. Where ventilation is required each vent should be silenced. Special consideration should be given to the access requirements of maintenance and production, and the designers should ensure that these requirements are considered at an early design stage. In selecting any form of noise control, care should be taken to ensure that the equipment will physically withstand an industrial environment especially if it is particularly hostile. It must be robust and be capable of being dismantled and reassembled.

(c) *Silencers*[12]

Silencers are used to suppress the noise generated when air, gas or steam flow in pipes or ducts or are exhausted to atmosphere. They fall into two forms:

(1) Absorptive, where sound is absorbed by an acoustical absorbent material.
(2) Reactive, where noise is reflected by changes in geometrical shape.

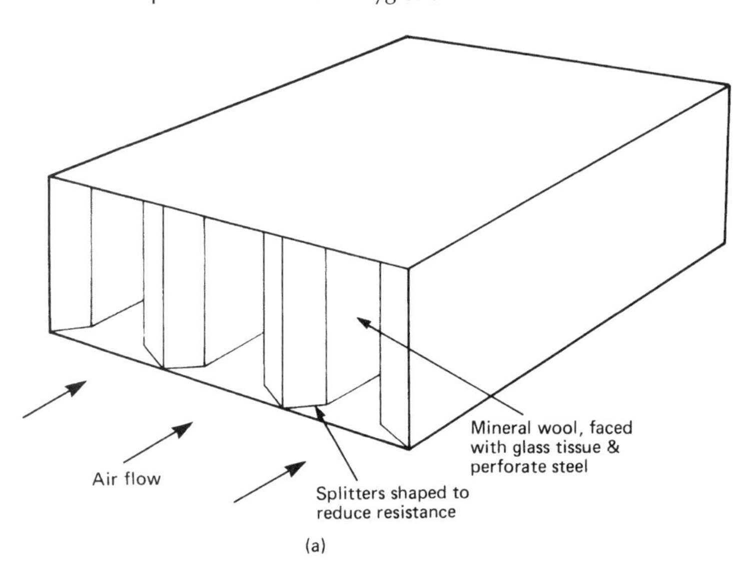

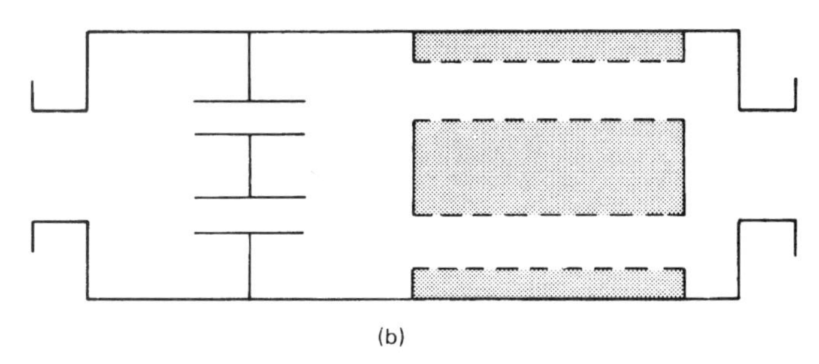

Figure 20.11 Typical silencers (a) absorptive splitter silencer; (b) combination reactive/absorptive silencer

Figure 20.11(a) shows a typical layout for an absorptive silencer, while *Figure 20.11(b)* shows a combination of the two types. The absorptive silencer normally has the better performance at higher frequencies, whereas the reactive type of silencer is more effective for controlling low frequencies.

The performance of splitter type of silencers is dependent on its physical dimensions. In general:

(1) Sound reduction or insertion loss increases with length.
(2) Low frequency performance increases with thicker splitters and reduced air gap.

Similarly for cylindrical silencers, the overall performance improves with length and the addition of a central pod. Performance would be limited by the sound reduction achievable by the silencer casing and other flanking paths. Typically no more than 40–50 dB at the middle frequencies could be expected without special precautions.

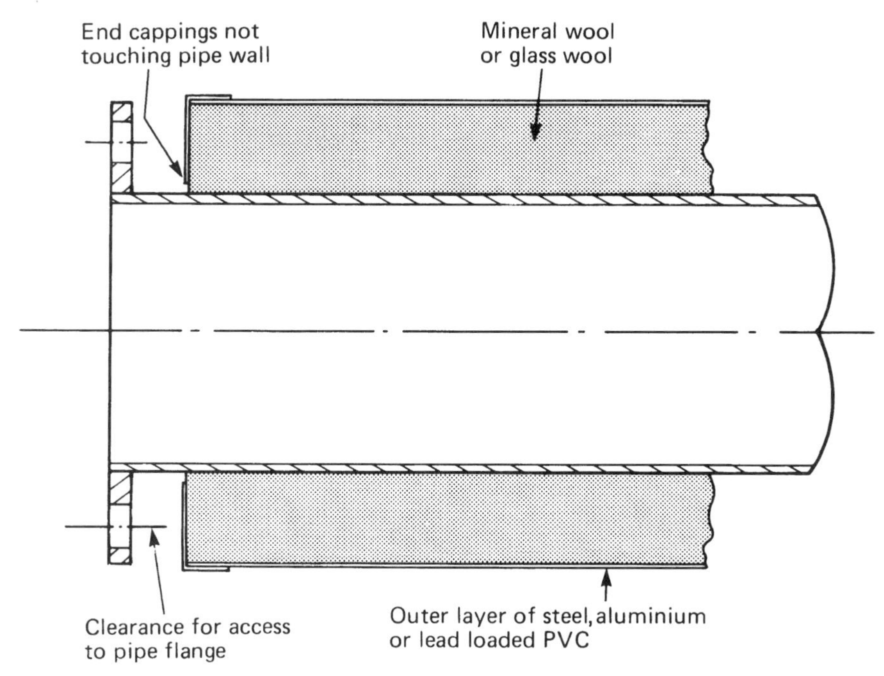

Figure 20.12 Pipe lagging

(d) *Lagging*[13]
On pipes carrying steam or hot fluids thermal lagging can be used as an alternative to enclosure and can achieve attenuations between 10 and 20 dBA, but it is only effective at frequencies above 500 Hz. The cross-section shown in *Figure 20.12* illustrates the main features of mineral wool wrapped around the pipes with an outer steel, aluminium or lead loaded vinyl layer. It is important that there is no contact between the outer layer and the pipe wall, otherwise the noise-reducing performance may be severely limited.

(e) *Damping*[14]
Where large panels are radiating noise a significant reduction can be achieved by fitting proprietary damping pads, fitting stiffening ribs or using a double skin construction.

(f) *Screens*
Acoustic screens (*Figure 20.13*) are effective in reducing the direct field component noise transmission by up to 15 dBA[15]. However they are of maximum benefit at high frequencies, but of little effect at low frequencies and their effectiveness reduces with distance from the screen.

(g) *Absorption treatment*
In situations where there is a high degree of reflection of sound waves, i.e. the building is 'acoustically hard', the reverberant component can dominate the noise field over a large part of the work area. The introduction of an

Figure 20.13 Acoustic screens (courtesy Ecomax Acoustics Ltd)

Figure 20.14 Acoustic absorption treatment showing suspended panels and wall treatment (courtesy Ecomax Acoustics Ltd)

acoustically absorbent material in the form of wall treatment and/or functional absorbers at ceiling height as shown in *Figure 20.14* will reduce the reverberant component by up to 10 dBA[11], but will not reduce the noise radiated directly by the source.

20.9.3 Personnel protection[16]

The two major methods of personnel protection are the provision of a quiet room or peace haven, and the wearing of ear-muffs or ear-plugs. The peace haven is similar in construction to an acoustic enclosure, and is used to keep the noise out. Ear-muffs or ear-plugs should be regarded only as the final resort to noise control. Their selection should be made with care having regard for the noise source, the environment and comfort of the wearer. Ear-

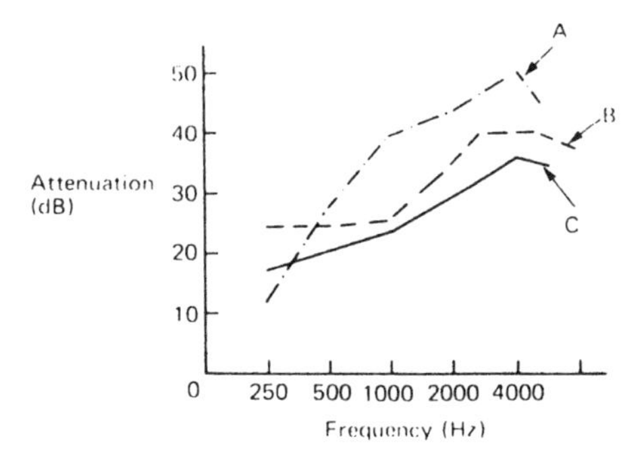

Figure 20.15 Attenuation characteristics of different hearing protectors (A: ear-muffs; B: plastic foam ear-plugs; C: acoustic wool ear-plugs)

plugs are only generally effective up to noise levels of 100–105 dBA while ear-muffs can provide protection at higher noise levels to meet a 90 dBA criterion, for noise received by the wearer. Comparative attenuation characteristics for various personal hearing protection devices are shown in *Figure 20.15*.

20.9.4 Effective noise reduction practices

A number of practical techniques can be used as part of normal day-to-day operational and maintenance procedures that will achieve significant reductions in the noise emitted, will cost nothing or very little to implement and can, additionally, give worthwhile savings in energy. Some of these techniques are listed below:

20.9.4.1 On plant

(1) Tighten loose guards and panels.
(2) Use anti-vibration mounts and flexible couplings.
(3) Planned maintenance with programme for regular lubrication for both oil and grease.

(4) Eliminate unnecessary compressed air and steam leaks, silence air exhausts.
(5) Keep machinery properly adjusted to manufacturer's instructions.
(6) Use damped or rubber lined containers for catching components.
(7) Switch off plant not in use, especially fans.
(8) Use rubber or plastic bushes in linkages, use plastic gears.
(9) Resite equipment and design-in noise control.
(10) Specify noise emission levels in orders, i.e. 85 dB(A) at 1 metre.
(11) Check condition and performance of any installed noise control equipment.

20.9.4.2 Community noise

(1) Keep doors and windows closed during anti-social hours.
(2) If loading is necessary, carry out during day.
(3) Vehicle manoeuvring – stop engine once in position.
(4) Check condition and performance of any noise control equipment and silencers.
(5) Turn exhaust outlets from fans away from nearby houses.
(6) Carry out spot checks of noise levels at perimeter fence, both during working day and at other times.

20.9.5 Noise cancelling

Recent experiments to cancel the effects of noise electronically by emitting a signal that effectively flattens the noise wave shape have achieved some success in low-frequency operation and situations where the receiver position and the source emission are well-defined. However, the industrial application of this technique is still in its infancy.

20.10 Vibration

Vibration can cause problems to the human body, machines and structures, as well as producing high noise levels.

20.10.1 Effect of vibration on the human body[17]

Low frequency vibration (3–6 Hz) can cause the diaphragm in the chest region to vibrate in sympathy giving rise to a feeling of nausea. This resonance phenomenon is often noticeable near to large slow speed diesel engines and occasionally ventilation systems. A similar resonance affecting the head, neck and shoulders is noticeable in the 20–30 Hz frequency region while the eyeball has a resonant frequency in the 60–90 Hz range.

The use of vibratory hand tools, such as chipping hammers and drills which operate at higher frequencies, can cause 'vibration induced white finger'. The vibration causes the blood vessels to contract and restrict the blood supply to the fingers creating an effect similar to the fingers being cold.

The effects of vibration on the human body will be dependent on the frequency, amplitude and exposure period and hence it is difficult to generalise on what they will be. However it is worthwhile remembering that in addition to the physiological effects vibration can also have psychological effects such as loss of concentration.

20.10.2 Machinery vibration

For machines that vibrate badly, apart from the increased power used and damage to the machine and its supporting structure, the vibrations can travel through the structure of the building and be radiated as noise at distant points (structure-borne noise).

Where the balance of the moving parts of a machine cannot be improved, vibration transmission can be reduced by a number of methods[18] of which the most commonly used are:

(1) Mount the machine on vibration isolators or dampers.
(2) Install the machine on an inertia block with a damping sandwich between it and the building foundations.

The method chosen will depend on the size and weight of the machine to be treated, the frequency of the vibration to be controlled and the degree of isolation required. Whichever form of vibration isolation (*Figure 20.16*) is

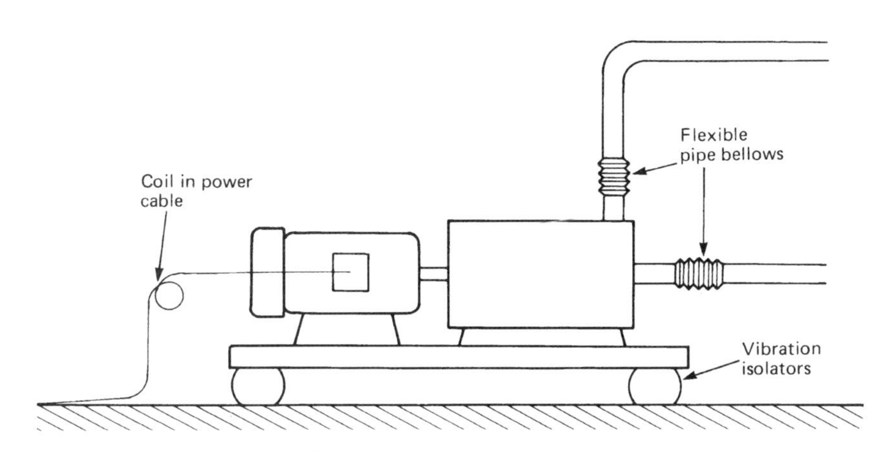

Figure 20.16 Vibration isolation

selected, care should be taken to ensure that the effect is not nullified by 'bridges'. For example, isolation of a reciprocating compressor set would be drastically reduced by rigid piping connecting it to its air receiver or distribution pipework, or by conduiting the cables to the motor. In severe cases rigid piping would fracture in a very short time.

20.11 Summary

The treatment of any noise source or combination of sources may use any of the control techniques individually or in combination. The selection of suitable measures will depend on:

(1) The type of noise field – whether dominated by the direct noise radiated from the machine or the reverberant field.
(2) The degree of attenuation required.
(3) Whether work area limits or community noise limits are to be met.
(4) Its cost effectiveness.

References

1. BS 3489:1962, Sound level meters (industrial grade); BS 4197:1967, A precision sound level meter, British Standards Institution, London
2. International Electrotechnical Commission, IEC Standard 651, Sound Level Meters, IEC, Geneva (or British Standards Institution, London) (1979)
3. H.M. Government, The Noise at Work Regulations 1989, SI 1989 No. 1790, HMSO, London (1989)
4. Hassall, J.R. and Zaveri, K., *Acoustic Noise Measurements*, Bruel and Kjaer, Naerum, Denmark (1979)
5. Control of Pollution Act 1974, part III, HMSO, London (1974)
6. BS 5228:1975, Code of practice for noise control on construction and demolition sites, British Standards Institution, London (1975)
7. BS 4142:1990, Method of rating industrial noise affecting mixed residential and industrial areas, British Standards Institution, London (1990)
8. European Economic Community, Directive No. 86/188/EEC on the protection of workers from the risks related to exposure to noise at work, Official Journal No. L137 of 24 May 1986, p. 28, HMSO, London (1986)
9. Warring, R.A. (Ed.), *Handbook of Noise and Vibration Control*, 509, Trade and Technical Press, London (1970)
10. Ref. 9, p. 474
11. Ref. 9, p. 460
12. Ref. 9, p. 543
13. Ref. 9, p. 249
14. Ref. 9, p. 595
15. Ref. 9, p. 504
16. Ref. 9, p. 571
17. Ref. 9, p. 112
18. Ref. 9, p. 586

Further reading

Blitz, J., *Elements of Acoustics*, Butterworth, London (1964)
Beranek, L. L., *Noise and Vibration Control*, McGraw-Hill, New York (1971)
Health and Safety Executive, Health and Safety at Work Series Booklet No. 25, *Noise and the Worker*, HMSO, London (1976)
Health and Safety Executive, Report by the Industrial Advisory Subcommittee on Noise, *Framing Noise Legislation*, HSE Books, Sudbury (undated)
Sharland, I., *Wood's Practical Guide to Noise Control*, Woods of Colchester Ltd, England (1972)
Taylor, R., *Noise*, Penguin Books, London (1970)
Webb, J. D. (Ed.), *Noise Control in Industry*, Sound Research Laboratories Ltd, Suffolk (1976)
Burns, W. and Robinson, D., *Hearing and Noise in Industry*, HMSO, London (1970)
Burns, W. and Robinson, D., *Noise Control in Mechanical Services*, Sound Attenuators Ltd, and Sound Research Laboratories Ltd, Colchester (1972)
Health and Safety Executive, *Noise 1990 and You*, HSE Books, Sudbury (1989)

Chapter 21

Workplace pollution, heat and ventilation

F. S. Gill

The solution of many workplace environmental problems, whether due to the presence of airborne pollutants such as dust, gases or vapours, or due to an uncomfortable or stressful thermal environment, lies in the field of ventilation. Ventilation can be employed in three ways:

(1) By using extraction as close to the source of pollution as possible to minimise the escape of the pollutant into the atmosphere. The extraction devices can be either hoods, slots, enclosures or fume cupboards coupled to a system of ducts, fans and air cleaners.
(2) By providing sufficient dilution ventilation to reduce the concentration of the pollutants to what is thought to be a safe level.
(3) By using air as a vehicle for conveying heat or cooling to a workplace to maintain reasonably comfortable conditions by employing air conditioning or a warm air ventilation system.

A flow of air which may be part of an industrial process can have a substantial effect upon the safety of the workplace by removing – or not – excessive heat, fume or dust. Such processes as ink drying, solvent collection, particulate conveying could fall into this category.

Before embarking upon the design for a ventilation system, it is necessary to assess the extent of the problem, that is, the amount of airborne pollution to be encountered in a workplace, and/or the degree of discomfort or stress expected from a thermal environment. Measurement and analysis techniques need to be devised and criteria and standards applied to the environment under consideration. Where measurement and analysis are concerned, the physics and the chemistry of the properties of the pollutant and its mode of emission need to be studied in such a way that a reliable and accurate assessment of the exposure of a worker can be made. As far as criteria and standards are concerned, medical evidence, biological research and epidemiological methods need to be applied to establish the relationship between the exposure and the long and short-term effect upon the human body of the worker taking into account the duration of exposure and the work rate. It can be seen, therefore, that many scientific skills require to be involved before a judgement can be made and a method of control devised.

21.1 Methods of assessment of workplace air pollution

Airborne pollutants can be divided roughly into three groups:

(1) dusts and fibres;
(2) gases and vapours;
(3) micro-organisms (bacteria, fungi etc.);

although some emissions from workplaces, for example oil mist from machine tools, could contain material from each group.

There is a wide range of techniques available to measure the degree of workplace pollution, some of which are described below. Firstly it is important to decide what information is required as the technique chosen will determine whether the concentration of airborne pollution is measured:

(a) in the general body of the workroom at an instant of time or averaged over the period of work, the latter being known as a time weighted average (TWA); or
(b) in the breathing zone of a worker averaged over the period of work; or
(c) in the case of dusts, as the total or respirable airborne dust (respirable dust is that which reaches the inner part of the lungs).

If a time weighted average is required, it might also be necessary to know whether any dangerous peaks of concentration occurred during the work period.

Owing to air currents in workrooms, pollutants can move about in clouds; thus concentrations vary with place and time and some statistical approach to measurement may be required. Also some workers move about from place to place in and out of polluted atmospheres and, whilst workplace concentrations might be high, operator exposure levels might be lower.

In addition, workers may be exposed to a variety of different pollutants during the course of a shift some being more toxic than others, so that the degree of exposure to each might be required to be known. When a mixture of pollutants occurs, the presence of one may interfere with the measurement of another. Therefore, the measurement of the degree of exposure of a worker must be undertaken with care. Occupational (or industrial) hygienists are trained in the necessary skilled techniques and should be called in to carry out the measurements.

21.1.1 Airborne dust measurement

The commonest method for measuring airborne dust is the filter method where a known volume of air is drawn through a pre-weighed filter paper (*Figure 21.1*) or membrane by means of a pump (*Figure 21.2*). The filter can be part of a static sampler located at a suitable place in the workroom or it can be contained in a special holder attached to a person as close to the face as possible, usually fixed to the lapel or shoulder strap of a harness and connected by tubing to the pump which is attached to the wearer's belt. At

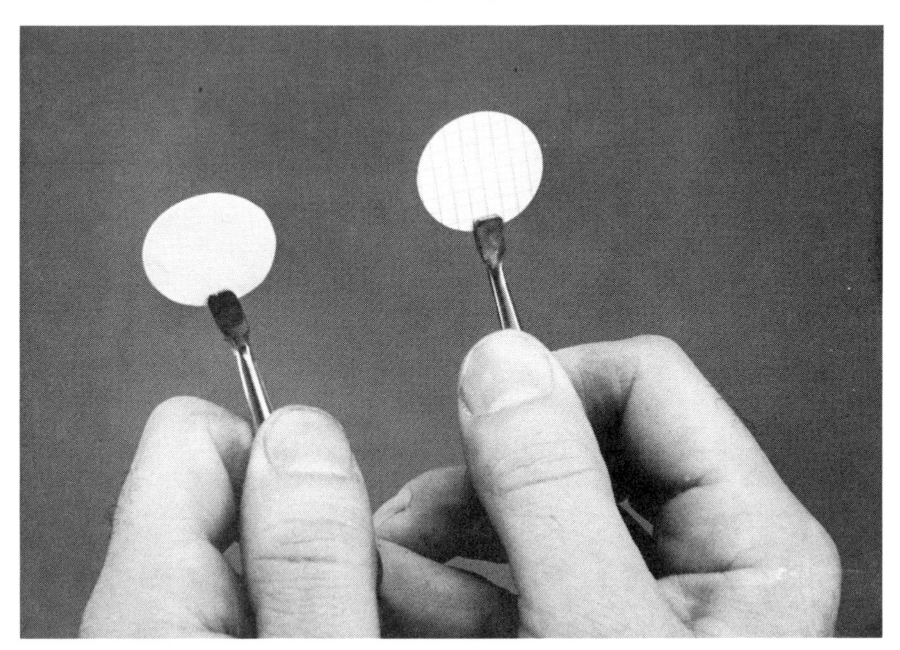

Figure 21.1 Dust sampling filters

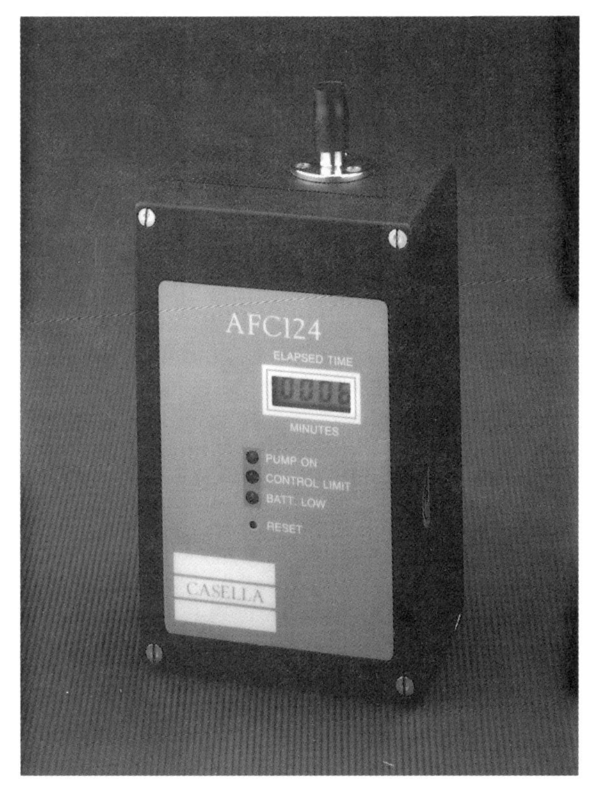

Figure 21.2 Personal dust sampling pump. (Courtesy Casella London Ltd)

the end of the sampling period the filter is weighed again and the difference in weight represents the weight of dust collected. This, divided by the total volume of air which has passed through the filter, gives the average concentration of dust over the period.

If a membrane-type filter made of cellulose acetate is used, it can be made transparent by the application of a clearing fluid, allowing the dust to be examined under a microscope and the particles or fibres counted if required. This is particularly important if fibrous matter such as asbestos is present.

Other filters can be chemically digested so that the residues can be further examined by a variety of chemical means. Types of filter are available that are suitable for examining the collected dust by X-ray diffraction, X-ray fluorescent techniques or by a scanning electron microscope. Thus, the choice of filter must be related to the type of analysis required. Weighing must be accurate to five decimal places of grams and, as air humidity can affect the weight of some filters, preconditioning may be required.

When on personal samplers the level of respirable dust is required, a device such as a cyclone is used to remove particles above 7 μm in diameter. Static samplers use a parallel plate elutriator for this purpose which allows the larger particles to settle so that only the respirable dust reaches the filter. With all separation devices the airflow rate must be controlled within close limits to ensure that the correct size fraction separation occurs.

Other techniques for static dust measurement include those which use the principle of measuring the amount of light scattered by the dust and one that uses a technique which measures the oscillation of a vibratory sensor which changes in frequency with the amount of dust deposited in it; another uses the principle of absorption of beta-radiation by the amount of dust deposited on a thin polyester film.

Further details of dust sampling and measurement techniques are given in refs. 1, 2 and 4.

21.1.2 Airborne gases and vapours measurement

The number of techniques available in this field is vast and the range is almost as wide as chemistry itself. Instruments can be used which are specific to one or two gases while others use the principle of infrared absorption and can be tuned to be sensitive to a range of selected gases. The principle of change of colour of paper or crystals is also used for specific gases and vapours; detector tube and impregnated paper samplers are of this type, but difficulties can be experienced if more than one gas is present as one may interfere with the detection of the other (*Figure 21.3*).

Techniques for unknown pollutants involve collecting a sample over a period of time and returning it to the laboratory for detailed chemical analysis. Collection can be in containers such as bladders, bags, cylinders, bottles, syringes or on chemically absorbent materials such as activated charcoal or silica gel. Those methods which use containers may lose some of the collected gases or vapours by adherence to the inside of the container. The chemical absorbers are usually contained in small glass or metal tubes

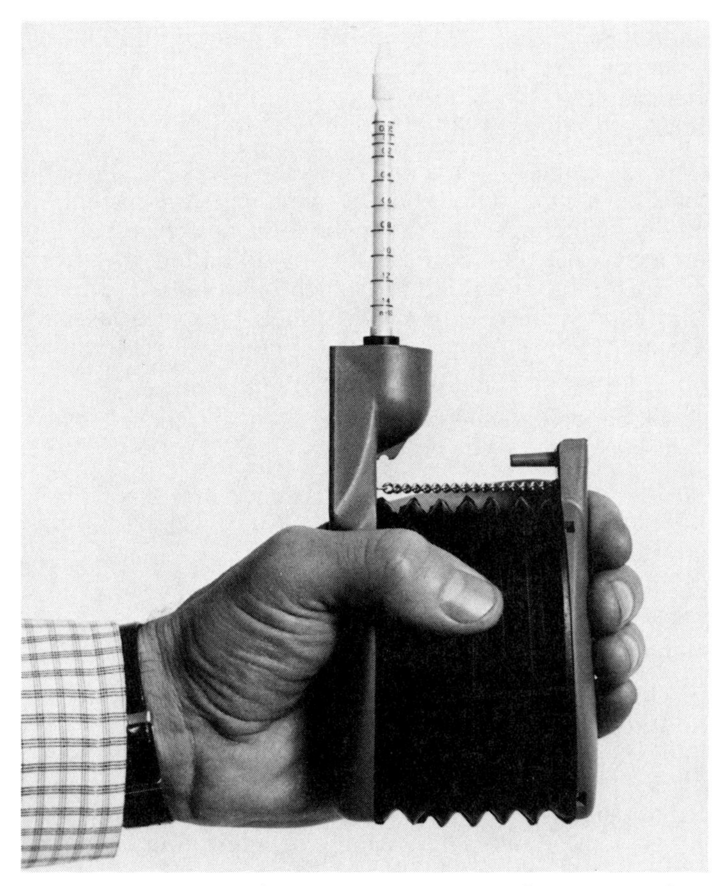

Figure 21.3 Draeger tube. (Courtesy Draeger Safety Group Ltd)

connected to a low-volume sampling pump and can be worn by a worker in a similar fashion to the personal dust samplers.

All these techniques require a good knowledge of chemistry if reliable results are to be obtained as many problems exist in the collection and analysis of gases and vapours[2,3,4].

21.2 Measurement of the thermal environment

Many working environments are uncomfortable owing to excessive heat or cold in one form or another; some can be so extreme that they lead to heat or cold stress occurring in the workers. When investigating these environments, it is important to take into account the rate of work and the type of clothing of the worker as these affect the amount of heat the body is producing and losing.

To obtain a correct assessment of a thermal environment, four parameters require to be measured together:

(1) The air dry bulb temperature.
(2) The air wet bulb temperature.
(3) The radiant temperature.
(4) The air velocity.

If any one of these is omitted, then an incomplete view is obtained. The sling psychrometer (sometimes known as the whirling hygrometer) will measure wet and dry bulb temperatures, a globe thermometer responds to radiant heat, and air velocity can be measured by an airflow meter or a katathermometer. There are several indices which bring together the four measurements and express them as a single value: some also take into account work rate and clothing. A number of these indices are listed below, and their values can be calculated or derived from charts.

(1) The Wet Bulb Globe Temperature index (WBGT) can be calculated from the formula:

    For indoor environments
    WBGT = 0.7 × natural wet bulb temperature + 0.3 × globe temperature

    For outdoor environments
    WBGT = 0.7 × natural wet bulb temperature + 0.2 × globe temperature + 0.1 × dry bulb temperature

(2) Corrected Effective Temperature index (CET) can be obtained from a chart and can take into account work rate and clothing.
(3) Heat Stress Index (HSI) can be calculated or obtained from charts and takes into account clothing and work rate, and from it can be obtained recommended durations of work and rest periods.
(4) Predicted four hour sweat rate (P4SR) can be obtained from charts and takes into account work rate and clothing.
(5) Wind chill index, as its name suggests, refers to the cold environment and uses only dry bulb temperature and air velocity but takes into account the cooling effect of the wind.

These five indices are considered in detail, including charts and formulae, in refs 5 and 6.

21.3 Standards for workplace environments

Authorities from several countries publish recommended standards for airborne gases, vapours, dusts, fibres and fume. The two main sources are the Health and Safety Executive in the UK who publish occupational exposure limits (OELs) in their Guidance Note Series No EH40[7] and the American Conference of Governmental Industrial Hygienists who publish threshold limit values (TLVs)[8] Both these sources update annually.

The occupational exposure limits are published in EH40. Essentially these standards are in two parts: maximum exposure limits (MELs) and occupational exposure standards (OESs), and they have a legal status under regulation 7 of COSHH[9].

21.3.1 Maximum Exposure Limits

Regulation 7(4) of COSHH requires that where an MEL is specified the control of exposure shall be such that the level of exposure is reduced as low as reasonably practicable and in any case below the MEL. Where short term exposure limits are quoted they shall not be exceeded. In the 1993 edition of EH40 there are 45 MELs quoted.

21.3.2 Occupational Exposure Standards

Regulation 7(5) of COSHH requires that where an OES is quoted the control of exposure shall be treated as adequate if the OES is not exceeded or, if exceeded, the employer identifies the reasons and takes appropriate action to remedy the situation as soon as is reasonably practicable. There are some 700 OESs published.

21.3.3 Units and recommendations

The standards in the above document are quoted in units of parts per million (ppm) and milligrams per cubic metre (mg m^{-3}). They are given for two periods: Long-term exposure which is an 8-hour time weighted average (TWA) value and a short-term exposure which is a 10 minute TWA. Certain substances are marked with a note 'Sk' which indicates that they can also be absorbed into the body through the skin, others have a 'Sen' notation indicating that they may possibly sensitise an exposed person.

It should be remembered that all exposure limits refer to healthy adults working at normal rates over normal shift duration. In practice it is advisable to work well below the recommended value, as low as one quarter, to provide 'a good margin of safety'.

21.4 Ventilation control of a workplace environment

As a result of the COSHH regulations there is a legal duty to control substances that are hazardous to health. The Approved Code of Practice (ACOP)[9] associated with these regulations sets out in order the methods that should be used to achieve adequate control. Extract and dilution ventilation are two of the methods mentioned. These regulations also require the measurement of the performance of any ventilation systems that control substances that are hazardous to health. The places where measurements are required to be taken are listed in para. 61 of the ACOP.

21.4.1 Extract ventilation

In the design of extract ventilation it is important to create, at the point of release of the pollutants, an air velocity sufficiently strong to capture and draw the pollutants into the ducting. This is known as the capture velocity and can be as low as 0.25 m/s for pollutants released gently into still air such as the vapour from a degreasing tank or as much as 10 m/s or more for heavy particles released at a high velocity from a device such as a grinding wheel. The capturing device can be a hood, a slot or an enclosure to suit the layout of the workplace and the nature of the work but the more enclosure that is provided and the closer to the point of emission it is placed, the more effective will be the capture.

Difficulty can be experienced with moving sources of pollution such as the particles from hand-held power saws and grinders. In these circumstances high velocity low volume extractors can be fitted to the tools using flexible tubing of 25–50 mm diameter to draw the particle-laden air to a cleaner which contains a high efficiency filter and a strong suction fan (*Figure 21.4*).

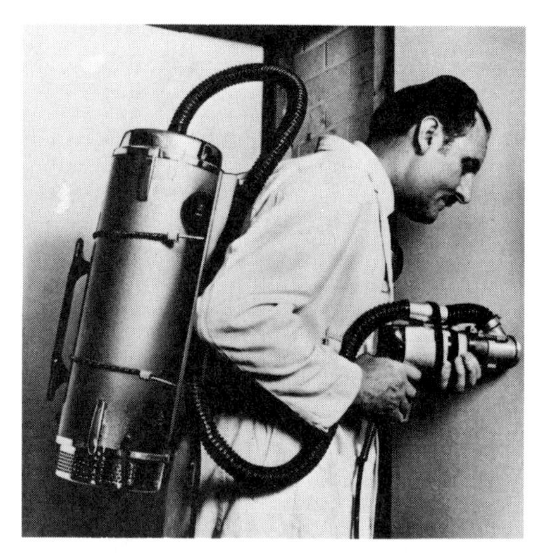

Figure 21.4 High velocity low volume extractor. (Courtesy B.V.C. Ltd)

Hoods attached to larger diameter flexible tubing can be used for extraction from the larger moving sources such as welding over wide areas, but owing to the higher weight of these devices some form of movable support system is required (*Figure 21.5*).

When siting a capture hood or slot, advantage should be taken of the natural movement of the pollutants as they are released. For example, hot substances and gases are lighter than air and tend to rise, thus overhead capture might be most suitable, whereas some solvent vapours when in concentrated form are heavier than air and tend to roll along horizontal surfaces, so capture points are best placed at the side. Care must be taken to ensure that all contaminants are drawn away from the breathing zone of the worker – this particularly applies to places where workers have to lean over or get close to their work. It is important to note that whenever extract ventilation is exhausted outside, a suitably heated supply of make-up air must be provided to replace that volume of air discarded.

There are established criteria for the design of extract systems[10].

Figure 21.5 Portable collecting hood. (Courtesy Myson Marketing Services Ltd)

21.4.2 Dilution ventilation

This method of ventilation is suitable for pollutants that are non-toxic and are released gently at low concentrations and should be resorted to only if it is impossible to fit an extractor to the work station. It should not be used if the pollutants are released in a pulsating or intermittent way or if they are toxic. The volume flow rate of air required to be provided must be calculated taking into account the volume of the pollutants released, the concentration permitted in the workplace and a factor of safety which allows for the layout of the room, the airflow patterns created by the ventilation system, the toxicity of the pollutant and the steadiness of its release[11,12].

Hourly air change rates are sometimes quoted to provide a degree of dilution ventilation. The volume flow rate of air in cubic metres per hour is calculated by multiplying the volume of the room in cubic metres by the number of air changes recommended. There are recommended air change rates for a range of situations[13].

21.5 Assessment of performance of ventilation systems

In addition to the testing of the airborne concentrations of pollutants, it is necessary, and indeed is a requirement of COSHH, to check airflows and pressures created in a ventilation system to ensure that it is working to its designed performance by measuring:

(1) Capture velocity.
(2) Air volume flow rates in various places in the system.
(3) The pressure losses across filters and other fittings and the pressures developed by fans.

The design value of these items should be specified by the maker of the equipment. Therefore, instruments and devices are required to:

(1) Trace and visualise airflow patterns.
(2) Measure air velocities in various places.
(3) Measure air pressure differences.

Air flow patterns can be shown by tracers from 'smoke tubes' which produce a plume of smoke when air is 'puffed' through them (*Figure 21.6*). For workplaces where airborne particles are released it is possible to visualise the movement of the particles by use of a dust lamp. This shines a strong parallel beam of light through the dust cloud highlighting the particles in the same way that the sun's rays do in a darkened room.

Figure 21.6 Smoke tube

Air velocities can be measured by a variety of instruments but vane anemometers and heated head (hot wire or thermistor) air meters are the most common. Vane anemometers (*Figure 21.7*) have a rotating 'windmill' type head coupled to a meter and are most suitable for use in open areas such as large hoods and tunnels. The heated head type of air meter (*Figure 21.8*) is more suitable for inserting into ducting and small slots and is more versatile than the vane anemometers except that it is unsuitable for use in areas where flammable gases and vapours are released. Most air flow measuring instruments require checking and calibration from time to time.

One instrument which requires no calibration but is only effective in measuring velocities above approximately 3 m/s is the pitot-static tube which, in conjunction with a suitable pressure gauge, measures the velocity component of the pressure of the moving air which can be converted to air velocity by means of the simple formula:

Figure 21.7 Vane anemometer. (Courtesy Air Flow Developments Ltd)

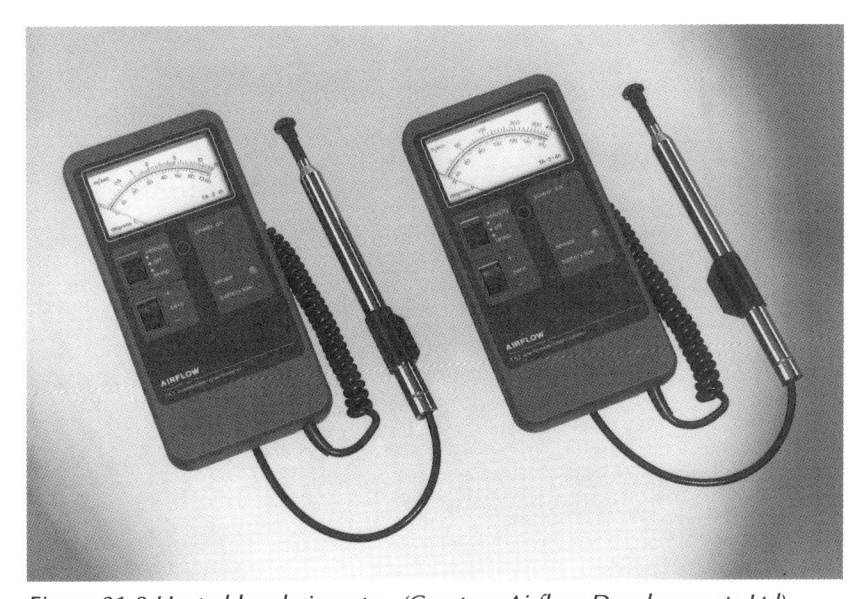

Figure 21.8 Heated head air meter. (Courtesy Airflow Developments Ltd)

$$p_v = \tfrac{1}{2} \rho v^2 \qquad \text{or} \qquad v = \sqrt{\frac{2p_v}{\rho}}$$

where p_v = velocity pressure (N/m^2 or Pa); ρ = air density (usually taken to be 1.2 kg/m^3 for most ventilation situations); and v = air velocity (m/s).

Pitot-static tubes are small in diameter and can easily be inserted into ducting.

All the above air velocity measuring instruments need to be placed carefully in an airstream so that their axes are parallel to the stream lines; any deviation from this will give errors.

Differences in air pressure can be measured by a manometer or U-tube gauges filled with water or paraffin, placed either vertically or, for greater accuracy, inclined. If the two limbs of the gauge are coupled by flexible plastic or rubber tubing to either side of the place to be measured, such as a fan or a filter, then the difference in height between the two columns of the tube indicates the pressure difference. Pressure tappings in ductings must be at right angles to the airflow.

Liquid-filled gauges are prone to spills and the inclusion of bubbles and before use must be carefully levelled and zeroed. Diaphragm pressure gauges avoid these problems but need to be checked for accuracy from time to time.

Airflow measuring techniques vary to suit the application[4].

References

1. Brief, R.S., *Basic Industrial Hygiene*, Section 9, Exxon Corporation (1975)
2. ACGIH, *Air Sampling Instruments*, American Conference of Governmental Industrial Hygienists, Cincinnati, Ohio (1972)
3. Thain, W., *Monitoring Toxic Gases in the Atmosphere for Hygiene and Pollution Control*, Pergamon, Oxford (1980)
4. Ashton, I. and Gill, F.S., *Monitoring for Health Hazards at Work*, Chapter 4, 'Ventilation', Blackwell Scientific Publications, Oxford (1992)
5. Gill, F.S., 'Heat, physics and measurement' chapter in *Occupational Hygiene* (Eds. Harrington, J.M., and Waldron, H.A.), Blackwell Scientific Publications, Oxford (1980)
6. Harrington, J.M. and Gill, F.S., *Occupational Health Pocket Consultant*, Blackwell Scientific Publications, Oxford (1992)
7. Health and Safety Executive, Guidance Note EH40, Occupational Exposure Limits, HSE Books, Sudbury (1993)
8. ACGIH, *Threshold Limit Values for Chemical Substances and Physical Agents in the Workroom Environment*, American Conference of Governmental Industrial Hygienists, Cincinnati, Ohio (1993)
9. Health and Safety Commission, *Control of Substances Hazardous to Health, Approved Code of Practice*, HSE Books, Sudbury (1988)
10. British Occupational Hygiene Society, Technical Guide No 7, *Controlling Airborne Contaminants in the Workplace*, Science Reviews Ltd., Leeds (1987)
11. Gill, F.S., 'Ventilation' chapter in *Occupational Hygiene* (Eds. Harrington, J.M. and Waldron, H.A.), Blackwell Scientific Publications, Oxford (1980)
12. ACGIH, *Industrial Ventilation*, 20th edn, American Conference of Governmental Industrial Hygienists, Cincinnati, Ohio (1988)
13. Daly, B.B., *Woods Practical Guide to Fan Engineering*, chapter 2, Woods of Colchester Ltd (1978)

Further reading

Ashton, I. and Gill, F. S., *Monitoring for Health Hazards at Work*, Blackwell Scientific Publications, Oxford (1992)

Chapter 22

Lighting

E. G. Hooper

22.1 Introduction

The part that lighting plays in ensuring a safe and healthy place of work is being increasingly recognised and specific requirements incorporated into legislation such as that concerned with the workplace[1], work equipment[2], docks[3], the use of electricity[4] and display screen equipment[5]. Most people prefer to work in daylight, making the best possible use of natural light. However, for work indoors, especially in large working areas, natural light does not usually give sufficient illuminance for the whole of the working day and it has therefore to be supplemented by an artificial source of lighting, usually electric lighting. The effects of glare, an excess of natural or artificial light, and of lighting deficiency are discussed below. Both extremes must be avoided.

22.2 Physics of the eye

The eyes are set in a deep, protective cavity in the skull and are further protected by upper and lower lids that also clean the front of the eyeball. The front of the eye comprises a white dense fibrous tissue shining through the translucent conjunctiva in the centre of which is the transparent cornea through which light passes into the eye. Behind the cornea is the iris which is coloured and has a hole in its centre called the pupil. At the back of the iris is the lens. The innermost part of the eye is the retina which contains the optic nerve ends linked with the brain. When light from an object falls upon the eye it is refracted at the cornea and passes through to the lens which causes further refraction and focus, producing an accurate, small, inverted image on the retina. In bright light, and if the object viewed moves closer, the pupil narrows by the muscular action of the iris causing the lens to become more convex.

If the object moves further away from the eye or the level of light is reduced the pupils widen or dilate and the lens becomes less convex.

The image formed in the retina, and also the effects caused on the retina by the different wavelengths of the observed colours, are converted into nerve impulses that are carried along the optic nerves to the brain where the impulses are reproduced as a recognisable picture of what is seen.

22.3 Eye conditions

The eye is a very delicate and sensitive structure and is subject to a number of disorders and injuries requiring skilled treatment: some of these disorders are mentioned briefly below.

Conjunctivitis is an inflamed condition of the conjunctiva of the eye caused by exposure to dust and fume and occasionally to micro-organisms.

Eye strain can be caused by subjecting the eye to excessively bright light, glare and is also a general term used for symptoms of refractive errors, uncorrected by spectacles, causing headaches, inflammation and general tiredness etc.

Accommodation is a term for the ability of the eye to alter its refractive powers and to adjust for near or distant vision. As the eye ages the lens loses its elasticity and hence its accommodation, thus affecting the ability to read and requiring corrective spectacles. In addition to this ageing process defects in accommodation can occur early in life, such as by the presence of conditions known as

(1) *Astigmatism* due to the cornea of the eye being unequally curved and affecting focus;
(2) *hypermetropia*, or long sight, in which the eyeball is too short; and
(3) *myopia*, or short sight, in which the eyeball is too long.

These defects can usually be corrected by spectacles.

Nystagmus is an involuntary lateral or up and down oscillating and flickering movement of the eyeball, and is a symptom of the nervous system observed in such occupations as mining.

Double vision is the inability of both eyes to focus in a co-ordinated way on an object usually caused by some defect in the eye muscles. It can be due to a specific eye injury, a symptom of some illness or to tiredness. It may be a momentary phenomenon or may last for longer periods.

Colour blindness is a common disorder where it is difficult to distinguish between certain colours. The most common defect is red/green blindness and may be of a minor character where red merely loses some of its brilliance, or of a more serious kind where bright greens and reds appear as one and the same colour – a dangerous condition in occupations requiring the ability to react to green and red signals or to respond to colour coding of pipework or electrical cables.

Temporary blindness may be due to some illness but it can occur in the following circumstances:

(1) Involuntary closure of the eyelids due to glare.
(2) Impairment of vision due to exposure to rapid changes in light intensity and to poor dark adaptation.

The act of seeing requires some human effort which is related to the environmental conditions. Even with good eyesight a person will find it difficult to see properly if the level of illumination is not adequate for the task involved, e.g. for the reading of small print or working to fine detail. But no standard of lighting, however well planned, can correct defective vision and anyone with suspected disability in this connection should be encouraged to undergo an eye test and, if advised, wear corrective spectacles.

22.4 Definitions[6]

The following terms are used in connection with illumination:

Candela (cd) is the unit of luminous intensity, i.e. the measure describing the power of a light source to emit light.

Candlepower, an obsolete term for the luminous intensity. Replaced by the candela.

Lumen (lm) is the unit of luminous flux used to describe the quantity of light emitted by a source or received by a surface.

Lux (lx) is the unit of illuminance and is equal to 1 lumen per square metre. Formerly, the term foot-candle (the number of lumens per sq ft (lm/ft^2) was used but the term is now being superseded by lux. N.B. 1 lm/ft^2 = 10.76 lx.

Illumination is the process of lighting an object.

Illuminance (E) is the luminous flux density at a surface measured in lux.

Service illuminance is the mean illuminance throughout the life of a lighting installation averaged over a working area.

Standard service illuminance is the service illuminance recommended for standard conditions.

Incandescent lamp is a lamp where the passage of current through a coiled filament raises its temperature to white heat giving out light. (This is the most common type of lamp used.) Oxidation within the lamp glass bulb is prevented by inert gas or vacuum sealing of the bulb. Higher efficiency incandescent lamps are produced by the introduction of iodine or bromine to the gas within the bulb, such as in the tungsten halogen lamps.

Electric arc lamp was formerly a lamp whose light was produced by the ionisation of a gap between two carbon electrodes, the resulting electric arc across the gap producing light. It also has other uses such as the production of ultraviolet rays used in setting certain printing inks.

Electric discharge lamps – electric current passing through certain gases produced an emission of light. Mercury vapour or mercury-halide lamps, tubular fluorescent, and sodium discharge lamps are examples.

Luminaire is a general term for all the apparatus necessary to provide a lighting effect. It includes all components for the mounting and protection of lamps and connecting them to a power supply, i.e. the whole lighting fitting.

22.5 Types of lighting

The selection of the source of light appropriate to the circumstances depends on several factors not least of which must be capital costs. But it is also important to consider efficiency, installation, running costs, and maintenance, life characteristics, size, robustness, and heat and colour output. The efficiency of any lamp can be calculated in terms of the light output per unit of electricity consumed and this is usually quoted in lumens per watt. Generally speaking, incandescent lamps are less efficient than discharge and fluorescent lamps.

Type of lamp	*Lumens per watt*
Incandescent lamps	About 15
Tungsten halogen	About 22
High pressure mercury	35–55
High pressure sodium	About 100

In any choice between incandescent and the other types of lamp the total lighting costs must take into account not only running costs but also installation and replacement costs. Incandescent lamps are much cheaper to buy and install, they give out light immediately they are switched on but they are more expensive than other types of lamp to run; they also require more frequent replacement because the filament is subjected to high temperatures. Discharge and fluorescent lamps cost more to install but their greater efficiencies and longer life make them more appropriate for general working interior lighting. Fluorescent lamps come on to full light output reasonably quickly but discharge lamps need some time to achieve maximum output depending upon the gas type and wattage.

In larger places of work the choice is often between discharge and fluorescent lamps. Where colour performance is important the mercury discharge lamp, with its normal bluish white appearance (although it can be colour-corrected to some extent), and the sodium lamp, with its rather warm golden effect, may not really be suitable and the choice is usually the tubular fluorescent lamp. A limitation of the fluorescent lamp is the restricted loading per point (i.e. more lamps are required per unit surface area) and in certain workshops where luminaire-mounting at heights is required (in overhead travelling crane workshops for example) the high pressure discharge lamp with its higher loading per point is often selected.

22.6 Levels of illuminance

The standard of illuminance required depends upon such things as the visual efficiency necessary for the tasks involved and general comfort and amenity requirements. The average illuminance out of doors in the UK is about 5000 lux on a cloudy day, but may be 10 times that on a sunny day. Inside a workplace, the illuminance from natural light at, say, a desk next to a window, will probably be only about 20% of the value obtaining outdoors. As working areas get further from windows the natural light produces illuminance values of perhaps only 1 to 10% of outdoor values so require supplementing by artificial lighting.

In normal practice, decisions should be based on the recommendations of the code for interior lighting produced by the Illuminating Engineering Society[7]. Typical values of standard service illuminance for certain locations and tasks are given below but for detailed information, for particular industries and tasks, reference should be made to the Illuminating Engineering Society. Guidance and advice can be obtained from an HSE publication[8], the Lighting Industry Federation[9] and the Electricity Association[10].

Although the term standard service illuminance represents levels that are good for general purposes, increases over the figures given may be necessary where high perception, low reflection or contrast are required or where the location is a windowless interior.

Location and task	*Standard service illuminance* (lx)
Storage areas, plant rooms, entrance halls, etc.	150–200
Rough machinery and assembling, conference rooms, typing rooms, canteens, control rooms, wood machinery, cold strip mills, weaving and spinning etc.	300–400
Routine office work, medium machinery and assembly etc.	500
Demanding work such as in drawing offices, inspection of medium machinery etc.	750
Fine work requiring colour discrimination, textile processing, and fine machinery and assembly, etc.	1000
Very fine work, e.g. hand engraving and inspection of fine machinery and assembly	1500

For a discussion on average illuminances, minimum measured illuminances and for maximum ratio of illuminances between working and adjacent areas see reference 6.

22.6.1 Maintenance of light fittings

Dust, dirt and use will progressively reduce the light output of lamps. Attention to good general cleaning and maintenance, and a realistic lamp replacement policy will help maintain the standard of illuminance.

22.7 Factors affecting the quality of lighting

The eye has the faculty of adjusting itself to various conditions and to discriminating between detail and objects. This visual capacity takes time to adjust to changing conditions as, for example, when leaving a brightly lit workroom for a darkened passage. So, sudden changes of illuminance and excessive contrast between bright and dark areas of a workroom should be avoided. A recent problem, resulting from the introduction of word-processors and other equipment using video screen displays, is the effect on eye discomfort and general wellbeing of viewing screens for extended periods of time.

22.7.1 Glare

Glare causes discomfort or impairment of vision and can be considered at three separate levels, viz. disability glare, discomfort glare, and reflected glare.

Disability glare, or dazzle, impairs the ability to see clearly without necessarily causing personal discomfort. The undipped headlamps of a car is an example of this.

Discomfort glare is a type of glare that causes visual discomfort without necessarily impairing the ability to see and may occur from unscreened windows in bright sunlight or when over-bright or unshaded lamps in the workroom are far too strong in terms of brightness for the workroom environment.

Reflected glare is the effect of light reflecting from a shiny or polished non-matt surface. The visual effect may be reduction of contrast, or distortion, and can be both irritating and, in certain workplaces, dangerous.

22.7.2 Glare indices

The Illuminating Engineering Society (IES)[7] has produced a glare index system for the numerical evaluation of direct glare. This has become the standard system in the UK and is also used in certain other countries.

The relationship between direct glare discomfort and the factors affecting it, viz. source brightness (B_s), background brightness (B_b), *apparent size of the source angle* (ω), and an index (p) representing the position of the source in relation to the direction of vision, is as follows:

$$\text{Glare index} = 10 \log_{10} \left[0.5 \times \text{constant} \sum \frac{B_s^{1.6} \ \omega^{0.8}}{B_b} \times \frac{1}{p^{1.6}} \right]$$

A set of comprehensive tables, based on this formula, has been produced by the IES giving glare indices for a wide range of situations, types of luminaires etc., and these should be referred to for specific advice. Examples of numerical values of maximum desirable glare indices range from about 19 for a general office to about 28 for industrial work. Figures above the recommended levels for a given location may lead to visual discomfort.

22.7.3 Protection from glare

The most common cause of glare results from looking directly at unscreened fluorescent tubes or incandescent lamps from normal viewing angles. Any form of diffuser fitted over the lamp, or a suitably placed reflector used as a screen will help to reduce the effect of glare from a lamp. The minimum screening angle below the horizontal should be about 20°. Reflected glare can only really be eliminated by changing the offending shiny surface for a matt one.

Glare from sunlight coming through windows can be reduced by using exterior or interior blinds but this reduces the amount of natural lighting.

22.7.4 Effect of shadow

Shadow will affect the amount of illumination, and its impact on people in working areas will depend on the task being performed, and on the

Figure 22.1 Factory lighting of correct illuminance, free from shadow and glare. (Courtesy Thorn Lighting Ltd)

disposition of desks, work benches etc. The remedy is to use large luminaires or to increase their number. *Figure 22.1* shows an example of factory lighting of the correct illuminance.

22.7.5 Stroboscopic effect

The earlier type of fluorescent tube was criticised because of stroboscopic effect which gives the illusion of motion or non-motion as the case may be. For example, it was thought that the pulsating effect of alternating current, operating at 50 Hz, produced a 'flicker' which when superimposed on rotating machinery would appear to indicate that the machinery was slowing down or even stationary. The danger of stroboscopic effect is self-evident but with modern fluorescent tubes the problem has been minimised by reduced flicker effect and should not cause problems. Wiring adjacent lighting tubes to different supply phases overcomes this effect.

22.7.6 Colour effect

The reflection of light falling on a coloured surface produces a coloured effect in which the amount of colour reflected depends upon the light source and the colour of the surface. For example, a red surface will only appear red if the incident light falling upon it contains red: under the pure yellow of sodium street lighting, for example, a red surface will appear brown. The choice of lamp is important if colour effect or 'warm' or 'cool' effect is required and can be as important a consideration as the degree of illuminance itself. Where accurate colour judgements have to be made the illuminance should be not less than 1000 lx.

22.8 Use of light measuring instruments

The human eye is unreliable as an indicator of how much light is present. For accurate results in the measurement of the illuminance at a surface it is necessary to use a reliable instrument. Light meters are available for this purpose.

A pocket light meter, normally adequate for most locations, is a photocell which responds to light falling upon it by generating a small electric current which deflects a pointer on a graduated scale measured in lux. Most light meters have a correction factor built into their design to allow for using a filter when measuring different types of light (daylight, mercury vapour lamps, fluorescent tubes etc.). The recommended procedure for taking measurements with a light meter is to:

(1) Cover the cell with opaque material until the pointer is at zero on the scale or zero adjustment in accordance with manufacturer's instructions.
(2) Allow a few minutes for the instrument to 'settle down' before taking a reading. A longer period will be required in the case of fluorescent tubes and discharge lamps which have just been switched on.
(3) Select the appropriate scale on the instrument, i.e. the one that gives the greatest pointer deflection within the limit of the scale. If the pointer moves to maximum a higher scale should be chosen.
(4) If readings are to be taken during daylight two readings are necessary:
    (a) with the lights on and with the window blinds drawn back so as to record the combined effect of natural and artificial light, and
    (b) with the same natural light conditions as in (a) but with the artificial lights switched off.

    The result required, i.e. the measure of the artificial light, is the difference between the two readings. If the two readings are large and approximately equal it will be necessary to re-check the artificial light reading after dark.

The level of illuminance measured should be checked against the appropriate standard service illuminance for the location and task. It is important to avoid great contrast and shadow, particularly where people are working or passing because of the visual adaptation problem of the eye. The correct use of a light meter is an important aid to establishing good levels of illuminance. However, to ensure accurate readings, the instrument should be kept in its case when not in use and away from damp and excessive heat.

References

1. H.M. Government, *Workplace (Health, Safety and Welfare) Regulations 1992*, regulation 8, HMSO, London (1992)
2. H.M. Government, *Provision and Use of Work Equipment Regulations 1992*, regulation 21, HMSO, London (1992)
3. H.M. Government, *The Docks Regulations 1988*, regulation 6, HMSO, London (1988)
4. H.M. Government, *The Electricity at Work Regulations 1989*, regulation 15, HMSO, London (1989)
5. H.M. Government, *Health and Safety (Display Screen Equipment) Regulations 1992*, the schedule, HMSO, London (1992)
6. BS 4727, Glossary of electrotechnical power, telecommunications, electronic, lighting and colour, Part 4, Terms particular to lighting and colour, British Standards Institution, London
7. Illuminating Engineering Society, *Code for Interior Lighting*, IES, London (1977)
8. Health and Safety Executive, *Lighting at Work*, Health and Safety: Guidance Booklet No. HS(G)38, HSE Books, Sudbury (1989)

Further reading

In addition to the above, numerous booklets and pamphlets on lighting for occupational premises and processes may be obtained from:

9. The Lighting Industry Federation, 25 Bedford Square, London WC1B 3HH
10. The Electricity Association, 30 Millbank, London, SW1.

Chapter 23

Ergonomics and human error

C. Mackay

23.1 Introduction

There have been a number of recent and well publicised accidents in which human error has played a prominent part. At Three Mile Island a combination of poor display design, bad maintenance and operator error led to the reactor core becoming uncovered and the release of radioactivity into the atmosphere. At the Ekofisk oil field incident a blowout preventer was installed upside down which resulted in an accident, following which widespread environmental pollution ensued. In both these cases the overall losses have been estimated in tens and hundreds of millions of pounds. These and similar incidents have brought sharply into focus the need to include an appraisal of human reliability and the factors which affect it when considering the safety performance of both large and small scale systems of work.

The study of the needs and requirements of human beings within working environments encompasses several disciplines. It can be argued, however, that the study of human reliability and the factors which affect it falls most easily within the remit of ergonomics. Ergonomics is the study of the relationship between man, the equipment with which he works and the physical environment in which this 'man-machine system' operates. The basic disciplines of psychology, physiology and anatomy are used to define and optimise the interrelationships between various aspects of the working environment. Ergonomics (usually referred to in the United States as human factors, human engineering or engineering psychology) obtained much of its impetus as an emerging discipline from work carried out during World War II which was concerned, amongst other things, with designing operational environments for flying personnel which were safe, efficient and reliable when looked at from the point of view of the operators' performance. However, the early roots of ergonomics can be traced back somewhat earlier to the work of F.W. Taylor[1] and the Gilbreths[2]. Their engineering approach to work design emphasised the need for production tasks to be broken down into basic individual elements requiring only simple motions or manipulations on the part of the worker, to produce the most economical and efficient performance. This approach has subsequently become incorporated into the procedure of task analysis which is widely used by ergonomics practitioners for the appraisal and description of man-machine relationships. A second example of work which led to the formation of ergonomic practices, and whose methods were incorporated into its early philosophy, is that of the Committee on the Health of Munitions Workers[3] (later the Industrial Fatigue Research Board) which investigated problems of fatigue in workers, a factor which was associated with low levels of morale and ill-health. Today military and aerospace applications continue to be a major user of ergonomic techniques but over the last quarter of a century such techniques have been increasingly used in the design of industrial and consumer products and have been applied to the design of the working and domestic environments.

23.2 Basic philosophy

The basic philosophy behind the ergonomic approach to safety is that of designing to take into account the limitations of the human operator. Man is endowed with a range of skills and abilities, some of which may reflect innate functional capacities such as those concerned with psychological processes like perception and memory, and physical skills such as manual dexterity (although showing inter-individual differences these can be thought of as 'fixed'), and skills which are learned, either in general terms, but also those acquired prior to or during the performance of specific work-related tasks (such as operating a fork-lift truck). Given these various capacities and skills the human being is able to operate satisfactorily within a given range of conditions. Departure from this range of optimum conditions may be associated with impairments in the operator's ability, often leading to degradation in task performance. Defining the optimum conditions for specific tasks is one of the main goals of fundamental research in ergonomics. Apart from degradations in human performance, the study of which is largely concerned with human error, severe departure from optimum conditions may be associated with impaired physical and mental health. It is in this area that ergonomics overlaps environmental hygiene and occupational medicine. The goals of the application of ergonomics are therefore two-pronged: first, to design equipment and environments which fit the individual's capacities and needs and by so doing allow him to perform effectively, and second, to ensure that design of work systems and environments does not violate requirements for physical and mental well-being, including the acceptability of the system to the operator and his level of general comfort whilst working.

23.2.1 The 'man-machine interface'

The basic approach of ergonomics, outlined above, is that of 'fitting the task to the man' as contrasted with the complementary approach of 'fitting the man to the task' which emphasises the requirements for selection and training. The term 'task' includes the totality of those aspects of the working environment which may influence the individual's effectiveness and comfort. A crucial aspect concerns the interaction of the man with the machinery and equipment with which he is working. Inherent in this interaction process is the transfer of either energy (power) or information between the man and the machine. This exchange takes place via an imaginary plane known as the 'man-machine interface'. This information passes from the machine to the man through the *display* elements of the interface and from the man to the machine through the so-called *control* elements of the interface. Here man and machine are combined in a closed loop feedback system. The nature of the system is shown schematically in *Figure 23.1*.

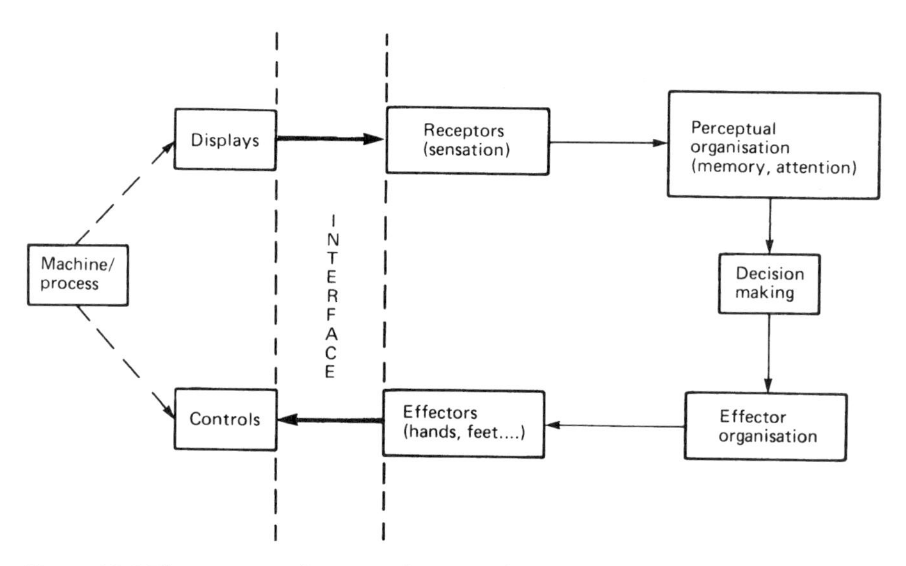

Figure 23.1 The man–machine interface, emphasising the performance of the human information processing system. Note that the broken lines underline that displays and controls are links between the operator and the machine

Figure 23.1 also incorporates the basic components of the human information processing system, the operation and limitations of which need to be taken into account when considering the design and effectiveness of particular man-machine systems. First of all, information about the status of the machine or plant must be compatible with the functional characteristics of the human sense organs (reception). Secondly, the information transmitted via the sensory receptors must be organised in such a fashion as to be easily interpreted by the perceptual elements in the human information processing system. (This aspect of structuring of displayed information is referred to as coding.) For example, one of the principal requirements for the interpretation of coded information is that it does not overload the operator's short-term memory capacity or make heavy attentional demands upon him.

The third stage at which information is handled in the system is the point at which the operator decides whether or not to act once the information displayed has been perceived and evaluated, and, if so, the nature of the action. The decision processes involved may be situation specific and are likely to vary in complexity. For our present purposes it is worth noting that the effectiveness of any decision will depend upon the quality and quantity of the information transmitted and whether the man is working under stress imposed by time constraints, emergency conditions or other pressures. Fourthly, implementing the decision requires the operator to organise controlling actions in a logically structured and sequenced fashion in order to harmonise with the control dynamics of the machine. The fifth and final aspect of the information processing loop therefore requires the manipulation of controls by effectors. It is when the design of the man-machine interface does not take the limitations of the human information processing system into account that occurrence of human error is increased. This aspect of ergonomics and safety will be returned to in more detail below.

23.2.2 Allocation of function

Another concept fundamental to man-machine system ergonomics is that of allocation of function. Simply, this term refers to those parts of a system that will be performed by the machine and the role to be played by man. The problem of allocation of function can be seen as one of defining the functional location of the interface referred to in the previous section. The solution is not simply one of separating functions appropriate to one or other of the two components, but must also be seen in the light of social, economic and political arguments. Thus the introduction of computerised and automated production systems may have implications for de-skilling and levels of employment which go far beyond consideration of an individual operator's role in the work process. Nevertheless in the design of most processes where the human operator is involved the problem of allocating functions must be resolved.

On what basis should this task allocation be decided? The most obvious way to separate functions, given that constraints such as cost-benefit have been considered, is on the relative capabilities and limitations of the two components. Such an approach was attempted by Fitts[4] in drawing up his now famous list itemising the relative advantages of men and machines. An updated version of the so-called 'Fitts List' as modified by Singleton[5] is shown in *Table 23.1*.

It becomes clear from an examination of this list that humans appear to surpass machines in detection, pattern recognition, flexibility, inductive reasoning and judgement. On the other hand, machines appear to surpass humans in speed, precision of response and application of sustained power, repetitive performance, short term memory, deductive reasoning and multi-channel performance. The human does however have one particular crucial ability: that of his inbuilt error-correction and error-monitoring facility, which coupled with his great flexibility and versatility means that he can often detect then act to minimise the consequence of error.

Unfortunately such a list suffers from at least three main disadvantages. First, with the rapid development of the micro-electronic technologies the relative advantages are constantly shifting in favour of the 'machine' (if task performance is the only criterion). Such lists tend therefore to become

Table 23.1. An updated version of the Fitts List (from Singleton[5])

Property	*Machine performance*	*Human performance*
Speed	Much superior Consistent at any level Large constant standard forces and power available	Lag one second 2 horse-power for about ten seconds 0.5 horse-power for a few minutes 0.2 horse-power for continuous work over a day
Consistency	Ideal for – routine, repetition and precision	Not reliable – should be monitored Subject to learning and fatigue
Complex activities	Multi-channel	Single channel Low information throughput
Memory	Best for literal reproduction and short term storage	Large store multiple access Better for principles and strategies
Reasoning	Good deductive Tedious to reprogramme	Good inductive Easy to reprogramme
Computation	Fast, accurate Poor at error correction	Slow Subject to error Good at error correction
Input	Can detect features outside range of human abilities	Wide range (10^{12}) and variety of stimuli dealt with by one unit, e.g. eye deals with relative location, movement and colour
	Insensitive to extraneous stimuli Poor pattern detection	Affected by heat, cold, noise and vibration Good pattern detection Can detect very low signals Can detect signal in high noise levels
Overload reliability	Sudden breakdown	Gradual degradation
Intelligence	None Incapable of goal switching or strategy switching without direction	Can deal with unpredicted and unpredictable Can anticipate Can adapt
Manipulative abilities	Specific	Great versatility and mobility

outdated quickly. Secondly, at best, such a list can only be used as a rough guide to aid the designer in the first stages of the allocation process. Unfortunately no adequate systematic methodology exists in which engineering data usually available (in a highly quantified form) can be reliably contrasted with comparable data on human performance. (A further problem is the lack of an adequate bank of human performance data except those obtained in very constrained and artificial situations.) Thirdly, and perhaps the most important (but least understood), is the need of individuals to utilise and develop their skills (at however low a level) rather than being allocated the remaining 'odd jobs' which cannot be successfully automated. A somewhat simplified example of the advantages and difficulties with the 'allocation of function' concept in a wider context is shown in *Figure 23.2*.

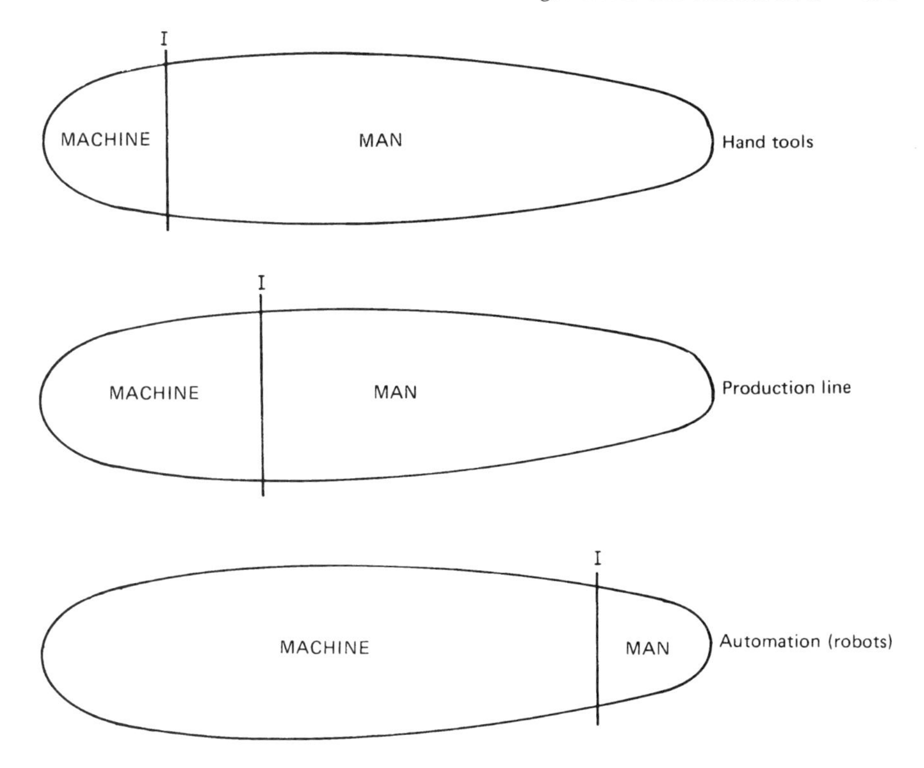

Figure 23.2 A schematic representation of the shift in the man–machine interface and change in allocation of function in different types of vehicle production (see text for details)

Consider the production of motor vehicles where three types of manufacturing process can be envisaged. In the first type the position of the 'interface' is to the left. Much of the manufacture of the car is based upon hand machining of parts and hand beating of panels allowing for utilisation of individual skills (often developed over many years) and a sense of creativity resulting from a finished product of high quality. In the second type of system a production line approach is adopted with pre-formed bodies and the extensive use of power tools. The man on the production line must work in a repetitive fashion assembling the same components as each partly finished vehicle passes along the line. Here, there is often under-utilisation of human skills and little opportunity for creativity. In the third, most recent and most 'advanced' system both the manufacture and assembly of the major parts of the car are undertaken by computer-controlled industrial robots. The function of the human operator is that of monitoring the flow of cars and maintaining faulty machines. His role is one of process controller. Thus differences in allocation of function produce three widely differing jobs, each emphasising different characteristics required of the human operator. Although the man-machine interface problems are quite different for each of the three jobs, ergonomic techniques can be used for their solution.

The usual engineering approach to allocation of function is one of *laissez-faire* – as much as possible being done by a machine at a reasonable cost and letting the human operator do the rest. This often leads to the human operator performing fragmented 'odd jobs' which bear little relationship to the concept of an integrated task. The description, due to Bowen[6] of 'monkey see, monkey do' is particularly appropriate here. A further analogy which is also illustrative is that of the human operator acting like a form of 'human-glue' which because of its almost infinite elasticity and flexibility holds together even the most badly designed system. This approach to allocation of function can however lead to problems of safety. An example is described below.

Example 1: Nail collating machine Nail collating machines are used for the production of cartridges for nail guns. The machine operates as follows. Loose nails are delivered into a hopper 5 m above floor level at the rear of the machine. Beneath the hopper a vibrator sorts the nails and a feed mechanism passes them onto a gravity fall conveyor. To ensure that the nails are aligned and all facing in the same direction, and for the removal of damaged nails, a circular wire brush and two photoelectric cells which operate the conveyor arm are used. Each nail is individually picked up by a chain which transports them to a forming head where they are brought together into a cartridge. A design fault on these machines means that the automatic alignment mechanism does not work satisfactorily and an operator has to be employed to ensure a continuous flow of nails and also to detect damaged nails. This is done by watching for damaged nails and feeling for them with finger tips as they pass along the conveyor. It was the operator's practice to lift out misaligned or damaged nails and flick them free in one motion. On one occasion, a reject nail flew forward unexpectedly and landed on the metal plate in front of the nip formed by the nail feed chain and the conveyor arm where the nails are discharged. As the operator attempted to move the nail his finger was drawn into the nip and damaged.

There is of course a requirement here for a simple but adequate guard to isolate the hazard and this simple expedient would of course prevent any recurrence of the accident. However this incident can also be seen as an example of poor machine design and poor allocation of function leading to human error. It also illustrates a pattern in a number of accidents: that of human error leading to risk-taking behaviour. Thus in the above example the operator was aware of the strict instruction not to approach the machinery whilst it was still in motion. However he felt that he could remove the damaged nail without isolating the machine and without creating a hazard. A brief discussion of human error and ways in which it can be minimised is given in the following section.

23.3 Human error

As a mathematical or engineering concept, error may be described as the occurrence of a discrepancy between an optimum and an achieved path, the degree of error being shown by either the *frequency* of the occurrence or the *extent* or *magnitude* of the discrepancy. Statistically, error can be defined as the extent to which a given set of variables is unable to account for the total amount of measured variation within a given system; that is, the behaviour of the system which cannot be accounted for.

In laboratory studies of human behaviour error is commonly used as an index of performance. A common finding in these studies is the marked trade-off between the speed of response and number of errors made. An example of this trade-off is shown in *Figure 23.3*. It is interesting to note that many of these approaches have been incorporated into concepts of human error as they apply to considerations of safety.

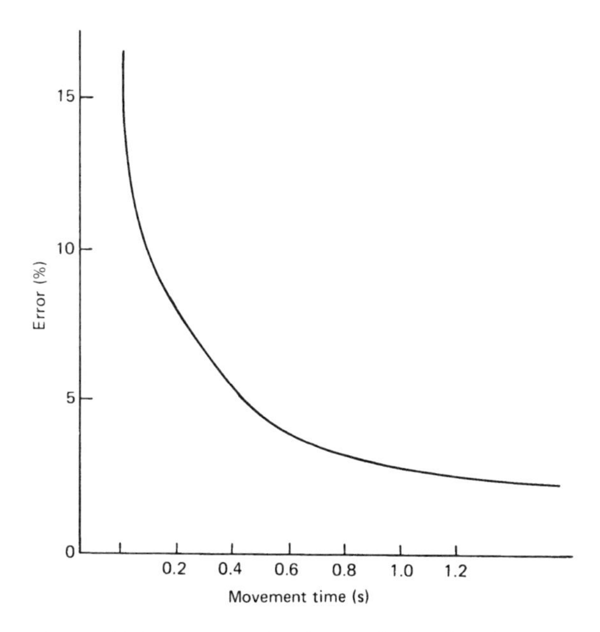

Figure 23.3 An example of the relationship between speed and error in a simple control task

Existing attitudes to human error are somewhat varied. The term is often used in a loose, colloquial sense to categorise accidents where behavioural factors have made a significant contribution to the cause. One school of thought assumes that once engineering or equipment 'failures' have been discounted, responsibility for an accident must, by definition, lie with the human component in the system, that is the accident was the result of a 'human failure'. Definitions of this type are used within a socio-legal context in the sense of allocating responsibility or blame. In the past, investigations of railway accidents and plane crashes have been conducted within this frame of reference. Another commonly held view of human error adopted by some managers and designers of equipment is that human error occurs because of an inadequacy or idiosyncrasy of motivation, skill, training or capacity. The remedy in this case lies in the use of 'zero defects programmes' which in their mildest form involve the use of safety propaganda such as posters but in their most punitive form involve retraining, demotion or job transfer as a means of minimising error. A further and related attitude to human error believes that human beings are essentially unreliable, the consequence of which is that errors are both inevitable and ineradicable. Solutions proposed by advocates of this approach emphasise the need for design of equipment which minimises both the occurrence and the consequences of human error and in its most draconian form advocates the 'designing-out' of the human operator altogether placing reliance on fully automatic systems. Where this is not possible, the reliance is placed on interlocking and self-checking systems so that a wrong action or sequence of actions cannot be performed.

Human error may be defined as a failure, on the part of the human operator, to perform an assigned task within specified limits of tolerance, with such limits generally being defined in terms of accuracy, sequence or time. This failure can be seen as a disruption or breakdown of information processing. Thus McCormick[7] states that human error occurs when the input-mediation-output process is impaired such as 'failure to perceive a stimulus, inability to discriminate among various stimuli, misinterpretation of the meaning of stimuli, not knowing what response to make to a particular stimulus, physical inability to make a required response and responding out of sequence'. Of particular concern will be, first, the extent to which the operator is clear as to what represents an error, and secondly, his confidence as to whether or not he has made an error.

It is possible to describe human error by a number of salient dimensions the most important of which are the type of error, its cause, the probability of its occurrence within a specified time and its consequences, the product of the latter two providing a crude estimate of 'risk'.

23.3.1 Human error and the ergonomics of displays and controls

The definition of human error proposed above emphasises the disruption of information processing linked to difficulties in communication between man and machine. Problems of misreading and misinterpretation of instrumentation and the incorrect activation of control elements may be considerable; even when the probability of error is low, the consequences of failure may be considerable. The flow diagram shown earlier (*Figure 23.1*) emphasises the need for optimum information flow in order to aid the operator in monitoring and subsequent decision making. Functionally, a good display is one which allows the best combination of speed, accuracy and sensitivity in transferring information from the machine to the man. General characteristics of the main types of display[8] are shown in *Table 23.2*.

Problems arise when the choice of display is inappropriate to the operator's task, where there are difficulties with the legibility of the display and where there are difficulties with display-control relationships. Apart from the design of the display itself, legibility is also affected by lighting conditions and workplace layout.

Just as the correct choice and design of display has a marked effect upon safety performance so has the use of appropriate control devices. Even though many industrial tasks are changing in content from those in which the man is involved in minute-to-minute manipulation of materials and tools, to those in which his role is one of passive monitor, there will always remain a need (particularly under emergency conditions) for the operator to achieve manual control of the system. It is important to ensure therefore that he is able to employ optimum sequencing and timing of control actions and for this reason the correct choice, design and arrangement of controls is essential.

Table 23.2. Functional characteristics of the main types of display. Based upon Grandjean[8]

Display type	*Functional characteristics*			
QUALITATIVE				
Auditory	For attracting immediate attention. Annunciators as warning devices.			
Visual	For representing three or more status conditions by the use of colour, shape, size or location coding.			
Representational	Provides operator with a model of the system. Good for showing spatial (temporal) relationships between variables. Static (maps). Dynamic (Process control – signalboxes).			
QUANTITATIVE	*Ease of reading*	*Precision*	*Detecting rate of change*	*Setting to a reading*
Analogue (dials and meters)				
Moving pointer	Acceptable	Acceptable	Very good	Very good
Moving scale	Acceptable	Acceptable	Acceptable	Acceptable
Digital	Very good	Very good	Poor	Acceptable

Choice of the most appropriate control for a specified task is determined by the type of control function desired, which in turn depends largely upon consideration of functional anatomy. Thus fingers and hands are best for precision and speed whilst arms and feet are best for control movements requiring force. An overview of different types of control[5,7–9] is given in *Table 23.3*, which also incorporates the main considerations which govern the choice of a particular display.

Apart from matching the type of control to the needs of the operator's task further factors are involved, particularly those relating to the *design* of the control itself. Take for example, the ease of identification and location of a particular control. Both are important in minimising error, and can be achieved by differences in shape, size or colour and by locating functionally different controls separately from each other. Controls can also be delineated by the use of labels and 'mimic' lines. Problems with the activation of controls often occur because they violate what are termed 'population stereotypes', where a whole population expect a specific response to a particular control activation or movement. For example, people *expect* a pointer on a dial to move to the right when a control knob is rotated in a clockwise direction, and, to most people, this instinctively means an *increase* in the variable under control. The average estimated probability of turning a control in a wrong direction is 0.0005. When the control violates a strong population stereotype the probability increases dramatically to 0.05. Under conditions of high stress the chances of turning the control the wrong way are even ($P = 0.5$). Under emergency conditions therefore, poor design increases the risk of error by a factor of a thousand. Similarly the error probabilities of mis-setting a multiposition indicator vary widely (0.0001 to 0.1) depending very largely upon the extent to which ergonomics are incorporated in the design of the equipment. Again these estimates are based upon data gathered in the context of nuclear power plant operation. However the problem of operator expectation and the behaviour of controls also occurs at a level which has consequences for the individual worker, as described in Example 2.

Table 23.3. Functional characteristics of various types of controls. [1]A = Activation; D = Setting a discrete variable; C = Setting a continuous variable; CC = Continuous control; D = Data entry. Adapted from Murrell[9] (1965), McCormick[7] (1970), Singleton[5] (1974) and Grandjean[8] (1980)

Type of control	*[1]Typical function*		*Comparative performance*			
			Speed	*Accuracy*	*Force*	*Range*
Hand push button	A		Good	Unsuitable	Very poor	Very poor
Toggle	A,D		Good	Unsuitable	Very poor	Very poor
Selector switch	D		Moderate	Moderate	Unsuitable	Moderate
Knob	D,C,CC		Very poor	Good	Unsuitable	Moderate
Thumbwheel	D,C,CC		Very poor	Good	Unsuitable	Moderate
Crank	C,CC	Small	Good	Poor	Very poor	Good
		Large	Good	Very poor	Good	Good
Handwheel	C,CC		Poor	Good	Moderate	Moderate
Lever	C,CC	Horizontal	Good	Poor	Poor	Poor
		Vertical	Good	Moderate	Good	Poor
Keyboard	D		Good	Moderate	Unsuitable	Poor
Foot push button	A,D		Moderate	Unsuitable	Very poor	Very poor
Foot pedal	C,CC		Good	Poor	Good	Very poor

Example 2: Shovel loading vehicle This example of control design and human error concerns the operation of a shovel loading vehicle. To move the vehicle in a forward or reverse direction the driver operated a floor-mounted, hand-operated gear stick type lever. Forward direction was achieved by pushing the lever forward; pulling the lever back made the vehicle reverse. The lever was automatically self-centering (see *Table 23.3*). Even with the throttle full open once the driver let go of the lever it returned immediately to the zero (upright) position and the vehicle stopped dead. This particular vehicle was withdrawn from service for essential maintenance, and was replaced by an ostensibly identical type. Unfortunately,

unknown to the suppliers of the loader and thus to the user, a small design modification had been made, such that driver control was no longer self-centering. No mention was made of this modification so with the throttle open, the vehicle on a slight incline and the lever in a forward position, the driver let go and jumped to the ground expecting the vehicle to stop. When it failed to do so but gathered speed instead, he ran after it down the slope, and, whilst trying to jump into the cab, fell under the back wheels of the vehicle and was killed instantly. Here we have an example of an accident with a multiple cause, where a major factor was operator expectancies being based upon the behaviour of the controls he had become accustomed to. As with the nail collating machine incident this is an example of the human operator, having made an error albeit caused by design changes, attempting to retrieve the situation and in so doing becoming involved in risk-taking behaviour.

23.4 Anthropometry and machine guarding

A measurement technique widely used in ergonomics is that of anthropometry. This may be defined as the science of the measurement of human bodily dimensions as they relate to the three-dimensional space in which the human operator works. Because the human is so flexible the frame can adapt, over time, to the most constrained and uncomfortable of working postures. However there are a number of costs that result from having to adapt to badly designed workspaces, costs such as injuries and strains from bad working posture, less spare capacity to deal with emergencies and a greater probability of errors and accidents, often as a result of postural fatigue. By careful task analysis, measurement, in terms of 'body size', of the population who will use the equipment, and by defining the reaching and seeing requirements for adequate performance, the ergonomist is able to describe the three-dimensional space which will be most comfortable for the operator and result in the most efficient performance.

Anthropometry is also a very powerful tool in defining the distances at which barriers should be erected from dangerous machinery, so as to give protection to the majority of a given population. It is in this way that Booth and Thompson[10] tested the applicability of the German Standard DIN 31 001 concerning reach over barriers. Using 200 male subjects and a specially constructed test rig they were able to demonstrate that the safety distances given in DIN 31 001, whilst protecting 95% of the German workforce, would not protect the same percentage of an equivalent UK population, at most barrier heights.

23.5 Manual handling

One way of classifying injuries is into those which are acute and those which are chronic. The acute type are those immediate injuries such as cuts, bruises and fractures or torn ligaments which usually arise from easily defined accidental causes such as falls and the 'struck by falling object' category. The second, chronic type of injury is often cumulative and is less easy to define than the acute variety. These latter often result in disorders such as tenosynovitis, epicondylitis and bouts of non-specific back pain. Back troubles are particularly painful and can lead to a severe impairment in mobility and life-style. They lead to long absences from work, may require extensive rehabilitation and are one of the main causes of early disability. Those in occupations such as labourers, farmers, porters and nurses are particularly at risk. Lundgren[11] questioned 1200 Swedish workers who had been absent because of back trouble. Those engaged in heavy work reported twice as much absenteeism from back complaints as those employed in light work. Kramer[12] has estimated that overall 20% of all absenteeism is caused by injury to discs and this accounts for half of all premature retirements. Further research (Davis and Stubbs[13,14]) has indicated that a close relationship exists between forces acting on the lower back and pressures generated in the abdominal cavity. They have found that those occupations in which workers frequently have abdominal pressures in excess of 100 mm Hg are those with significantly high incidences of back pain. Based upon observations on some 700 British male subjects, a guide[15] has been produced by the Materials Handling Research Unit, University of Surrey which gives force limits for lifting, pulling and thrusting. By using this guide, working practices can be designed so as not to exceed the maximum limiting intra-abdominal pressure (90 mm Hg), and thus, indirectly, reduce the likelihood of back injury.

23.6 New legislative requirements

Under Article 118 of the Treaty of Rome, Member States are obligated to incorporate into national laws the substance of health and safety directives. The rationale behind these obligations is the aim to achieve equivalent levels of protection for workpeople throughout the Community. In the UK six new Regulations, based on EC directives and made under HSW, came into force at the beginning of 1993. A basic feature of the directives, and now part of the Regulations, while not dissimilar from existing approaches, is the inclusion of a requirement for risk assessment as a step in risk reduction. The texts of the original directives also emphasised the importance of an ergonomic approach to risk reduction, particularly as regards the use of display screens and manual handling and, to a lesser degree, in the use of personal protective equipment.

The new Regulations came at a time when many countries were recognising the cost and impact of so-called ergonomic illnesses. These are ill-health conditions that stem from poor work organisation and bad machine and tool design that lead to chronic injury, mostly of insidious onset. Amongst those of principal concern are the prevalence and growing incidence of work-related musculoskeletal disorders. An exercise carried out on behalf of the HSE and based upon the Labour Force Survey found that in England and Wales the annual prevalence of work-related musculoskeletal disorders is approaching one million cases (predominantly back, shoulder and upper-limb disorders). Other self-reported disorders of high prevalence included headaches and eyestrain. Similar patterns of disease have been reported in North America and Scandinavia, and, no doubt, occur in other European countries.

Costs to the individual in terms of pain, disability and loss of livelihood; to the National Health Service for treatment; and to industry for insurance and sickness absence, make these top priority items in loss and accident prevention programmes. Some of these problems have been referred to in the earlier sections.

The emphasis on assessments in the new Regulations is an important starting point in the preventative process. The assessment obviously needs to be tailored to the likelihood of risk in a given situation and the extent of the assessment is covered in various guises in the new Regulations that require them to be *suitable* and *sufficient*. An ergonomic assessment of a nuclear power station control room is clearly of a different order from that of a simple repetitive task on a production line, but similar principles apply. A key feature must be, wherever possible, to involve individual workpeople in the assessment process. Indeed with respect to display screens, the requirement for an assessment of individual workstations can, in most cases, be met by the individual operator, provided training has been given in what to look for, and that any problems highlighted are tackled by the employer. Only with complex or non-standard workstations will recourse to specialist ergonomic advice be necessary.

More generally, approaches based upon ergonomic task analysis or records of work-related pain or discomfort, or a combination of both, will provide the basis for a risk assessment. The approach taken in the Manual Handling Operations Regulations is to emphasise the need to take into account factors other than just the weight of the load. There are many other characteristics of the load and, indeed, of the immediate environment in which the handling is to take place, that affect the likelihood of an injury occurring and these must be assessed systematically. Such assessments can be done with the aid of an ergonomic checklist, again not necessarily by an ergonomic specialist. There are a number of proprietary methods available or a scheme may be tailor-made based on existing guidance[16--18]. An advantage of a systematic approach is that individual consideration of specific risk factors often generate ideas for preventative solutions. A further advantage is that tasks which are inherently risky in terms of musculoskeletal or other ergonomic deficiencies are often organised inefficiently. Ergonomic intervention or redesign often, therefore, has multiple benefits.

23.7 Concluding remarks

This chapter on ergonomics has been necessarily brief and has concentrated on a few narrowly defined topics with the aim of drawing the reader's attention to some of the basic concepts in ergonomics and illustrating its use with examples where ergonomic techniques have had a significant impact in areas of occupational health and safety. Because of the selection of material many aspects of the subject have been excluded such as the design of protective clothing, the physical environment (noise, heating, lighting) and the design of jobs amongst others. The essentially multi-disciplinary nature of the subject means that information is often contained in specialist journals which range from medicine to engineering. There are, however, a number of useful source materials: these are listed below in the section concerned with further reading.

References

1. Taylor, F.W., What is scientific management? *Classics in Management*, rev. edn, American Management Association, New York (1970)
2. Gilbreth, F.B., Science in management for the one best way to do work, *Classics in Management*, rev. edn, American Management Association, New York (1970)
3. Committee on Health of Munition Workers, Output in relation to hours of work, *Interim Report on Industrial Efficiency and Fatigue*, Cmnd 8511, HMSO, London (1917)
4. Fitts, P.M., *Handbook of Experimental Psychology*, chapter on Engineering psychology and equipment design, John Wiley, London (1951)
5. Singleton, W.T., *Man-machine Systems*, Penguin Books, Harmondsworth (1974)
6. Bowen, M.H., Rational design, *Industrial Design*, **11** (1964)
7. McCormick, E.J., *Human Factors Engineering*, McGraw-Hill, New York (1976)
8. Grandjean, E., *Fitting the Task to the Man: An Ergonomic Approach*, Taylor & Francis, London (1980)
9. Murrell, K.F.H., *Ergonomics, Man and his Working Environment*, Chapman & Hall, London (1965)
10. Booth, R.T. and Thompson, D., Anthropometry and machinery guarding: An investigation into forced reach over barriers and through openings, Part 1: Reach over Barriers. *Journal of Occupational Accidents*, **3,** 45–56 (1980)
11. Lundgren, N.P.V., *Menschengerechte Gestaltung der Schwerabeit* (Industrial Organisation), **29,** 401–412 (1960)
12. Kramer, J., *Biomechanische Veranderungen im lumbalen Bewegungssegment*, Hippokrates, Stuttgart (1973)
13. Davis, P.R. and Stubbs, D.A., Safe levels of manual pressure for young males (1), *Applied Ergonomics*, **8** (3), 141–150 (1977)
14. Davis, P.R. and Stubbs, D.A., Safe levels of manual forces for young males (2), *Applied Ergonomics*, **8** (4), 219–226 (1978)
15. The Materials Handling Research Unit, University of Surrey, Guildford, *Force Limits in Manual Work*, IPC Science & Technology Press, Guildford (1980)
16. Health and Safety Executive, *Legislation Booklet No. L23, Manual Handling. Manual Handling Operations Regulations 1992. Guidance on Regulations*, HSE Books, Sudbury (1992)
17. Health and Safety Executive, *Health and Safety Guidance Booklet No. HS(G)60, Work related Upper Limb Disorders: A Guide to Prevention*, HSE Books, Sudbury (1990)
18. Pheasant, S., *Ergonomics, Work and Health*, Macmillan, Basingstoke (1991)

Further reading

Ergonomics, one of the official journals of The Ergonomics Society (published by Taylor & Francis) and *Human Factors* published by The Human Factors Society are the two principal, specialist journals in the area. They are devoted primarily to research topics. *Applied Ergonomics*, also an official journal of The Ergonomics Society, contains papers and articles more concerned with examples of the application of ergonomics, together with information of more general interest. The interested reader is referred to the books by Grandjean, McCormick, Murrell and Singleton already listed under References. The following titles may also be of interest.

Brown, B. C. and Martin, J. T. *Human Aspects of Man-machine Systems*, The Open University Press, Milton Keynes (1977)

McCormick, E. J. *Human Factors in Engineering and Design*, McGraw-Hill, New York (1976)

Morgan, C. T., Cook, J. S., Chapanis, A. and Lund, M. W. *Human Engineering Guide to Equipment Design*, McGraw-Hill, New York (1963)

Shackel B., *Applied Ergonomics Handbook*, IPC Science & Technology Press, Guildford (1974)

Surry, J. *Industrial Accident Research: A Human Engineering Appraisal*. Labor Safety Council, Ministry of Labor, Ontario (1974)

Van Cott, H. P. and Kinkade, R. G. (Eds.), *Human Engineering Guide to Equipment Design*. US Government Printing Office, Washington (1972)

Woodson, W. E. and Conover, D. W. *Human Engineering Guide for Equipment Design*. University of California Press, Berkeley (1966)

Part IV

General science

Much of the work undertaken by safety advisers requires an understanding of technical industrial processes. Even in a single factory unit the safety adviser may be called upon to advise on avoiding the hazards from a chemical reaction, guarding particular types of machinery, the standards of safe working to be expected of a building contractor, the precautions to be taken to prevent fire and the fire fighting equipment that should be provided, etc.

To carry out his duties effectively, the safety adviser should have an understanding of basic physics and chemistry and of the current safety techniques for reducing the risks associated with the more commonly met industrial processes. This Part considers some of these processes and the basic sciences from which they stem.

Chapter 24

Engineering science

J. R. Ridley

24.1 Introduction

In the construction of machines, plant and products, materials are selected because they have particular physical and chemical properties. Wood, metals, concrete, plastics and other substances all have their uses but there are limitations as to what they can do and how long they can do it. Properties may change with use, temperature, operating atmosphere, contamination by surrounding chemicals and for many other reasons. It is necessary to know the properties of the materials and how and why they have been used so that an assessment can be made of whether likely changes in the properties may give rise to hazards.

These properties stem from the chemical and physical characteristics of the different materials and substances used and their behaviour under certain conditions can determine the safety or otherwise of a process or operation. This chapter looks at some of the characteristics and properties of materials in common use, their application, circumstances of use and possible causes of hazards.

24.2 Structure of matter

Everything that we use in our work and daily life is made up of chemical substances, by themselves or in combination of one sort or another. Each substance consists of elements which are the smallest part of matter that can exist by itself. In its free state, an element comprises one or more atoms. When atoms combine together they form molecules of the element or, if different atoms combine, of compounds. The ratio in which atoms combine is determined by their combining power or valency.

Atoms are made up of three particles:

- protons which have a unit mass and carry a positive charge,
- neutrons which have a unit mass but carry no charge, and
- electrons which have no mass but carry a negative charge.

Table 24.1. List of elements giving symbol, atomic number and atomic mass

Name	*Symbol*	*Atomic no.*	*Atomic mass (relative)*
Actinium	Ac	89	(227)
Silver (argentum)	Ag	47	107.868
Aluminium	Al	13	26.9815
Americium	Am	95	(243)
Argon	Ar	18	39.948
Arsenic	As	33	74.9216
Astatine	At	85	(210)
Gold (aurum)	Au	79	196.9665
Boron	B	5	10.81
Barium	Ba	56	137.34
Beryllium (glucinum)	Be	4	9.0122
Bismuth	Bi	83	208.9806
Berkelium	Bk	97	(249)
Bromine	Br	35	79.904
Carbon	C	6	12.011
Calcium	Ca	20	40.08
Cadmium	Cd	48	112.40
Cerium	Ce	58	140.12
Californium	Cf	98	(251)
Chlorine	Cl	17	35.453
Curium	Cm	96	(247)
Cobalt	Co	27	58.9332
Chromium	Cr	24	51.996
Caesium	Cs	55	132.9055
Copper (cuprum)	Cu	29	63.546
Dysprosium	Dy	66	162.50
Erbium	Er	68	167.26
Einsteinium	Es	99	(254)
Europeium	Eu	63	151.96
Fluorine	F	9	18.9984
Iron (ferrum)	Fe	26	55.847
Fermium	Fm	100	(253)
Francium	Fr	87	(223)
Gallium	Ga	31	69.72
Gadolinium	Gd	64	157.25
Germanium	Ge	32	72.59
Hydrogen	H	1	1.0080
Hahnium	Ha	105	–
Helium	He	2	4.0026
Hafnium	Hf	72	178.49
Mercury (hydrargyrum)	Hg	80	200.59
Holmium	Ho	67	164.9303
Iodine	I	53	126.9045
Indium	In	49	114.82
Iridium	Ir	77	192.22
Potassium (kalium)	K	19	39.102
Krypton	Kr	36	83.80
Lanthanum	La	57	138.9055
Lithium	Li	3	6.941
Lawrencium	Lr	103	–
Lutetium	Lu	71	174.97
Mendelevium	Md	101	(256)
Magnesium	Mg	12	24.305

Table 24.1. *(Continued)*

Name	*Symbol*	*Atomic no.*	*Atomic mass (relative)*
Manganese	Mn	25	54.9380
Molybdenum	Mo	42	95.94
Nitrogen	N	7	14.0067
Sodium (natrium)	Na	11	22.9898
Niobium (columbium)	Nb	41	92.9064
Neodymium	Nd	60	144.24
Neon	Ne	10	20.179
Nickel	Ni	28	58.71
Nobelium	No	102	(254)
Neptunium	Np	93	(237)
Oxygen	O	8	15.9994
Osmium	Os	76	190.2
Phosphorus	P	15	30.9738
Protactinium	Pa	91	231.0359
Lead (plumbum)	Pb	82	207.2
Palladium	Pd	46	106.4
Promethium (Ilinium)	Pm	61	(145)
Polonium	Po	84	(210)
Praseodymium	Pr	59	140.9077
Platinum	Pt	78	195.09
Plutonium	Pu	94	(242)
Radium	Ra	88	226.0254
Rubidium	Rb	37	85.4678
Rhenium	Re	75	186.2
Rutherfordium (kurchatovium)	Rf	104	–
Rhodium	Rh	45	102.9055
Radon (niton)	Rn	86	(222)
Ruthenium	Ru	44	101.07
Sulphur	S	16	32.06
Antimony (stibium)	Sb	51	121.75
Scandium	Sc	21	44.9559
Selenium	Se	34	78.96
Silicon	Si	14	28.086
Samarium	Sm	62	150.4
Tin (stannum)	Sn	50	118.69
Strontium	Sr	38	87.62
Tantalum	Ta	73	180,9479
Terbium	Tb	65	158.9254
Technetium (masurium)	Tc	43	(99)
Tellurium	Te	52	127.60
Thorium	Th	90	232.0381
Titanium	Ti	22	47.90
Thallium	Tl	81	204.37
Thulium	Tm	69	168.9342
Uranium	U	92	238.029
Vanadium	V	23	50.9414
Tungsten (wolfram)	W	74	183.85
Xenon	Xe	54	131.30
Yttrium	Y	39	88.9059
Ytterbium	Yb	70	173.04
Zinc	Zn	30	65.37
Zirconium	Zr	40	91.22

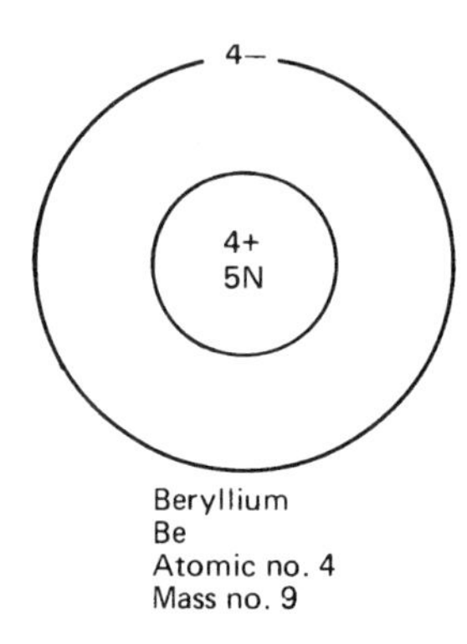

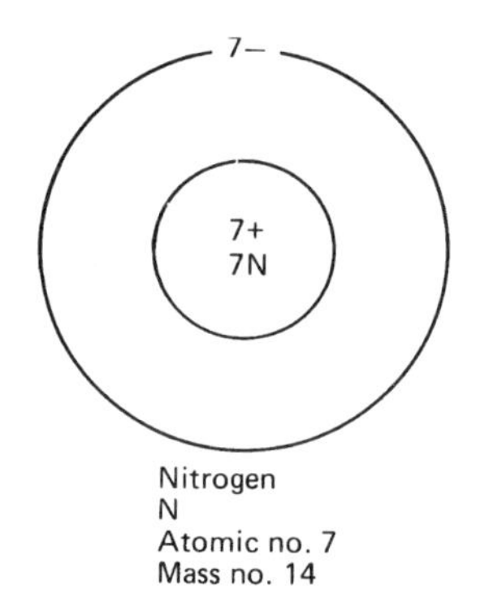

Figure 24.1 Atomic structures

The core or nucleus of the atom consists of protons and neutrons with electrons travelling in orbits around the nucleus (*Figure 24.1*). Elements normally have no overall charge since the number of protons is matched by an equal number of electrons. However, it is possible to upset this balance by removing either a proton or an electron resulting in the atom carrying a charge when it is said to be ionised.

In chemistry, atoms are given 'atomic numbers' which equal the number of protons or electrons in the atom. They are also given 'mass numbers' which equal the sum of the number of protons plus neutrons. The mass number is always equal to or greater than 2 $\times$ the atomic number. Some elements can occur in conditions where they have the same atomic number, and hence the same name, but different mass numbers; they are then known as isotopes and are generally referred to as nuclides. Very large heavy atoms, such as uranium, can be unstable and easily break down to produce smaller atoms with the production of particles or energy. These atoms are radioactive and provide the source of energy in nuclear reactors.

Approximately 100 different atoms have been identified and each has been given a name and a coded symbol which usually is the first one or two letters of its name: carbon – C, lithium – Li, titanium – Ti, etc. Exceptions in this coding system arise because when it was evolved in the early 1800s some chemicals were still known by their Latin names, such as copper (cuprum – Cu) and tin (stannum – Sn), see *Table 24.1*.

When atoms join together their molecular formulae are written as groups of atomic symbols to indicate the amount of those atoms present in the molecule.

Molecular formulae

Br_2	bromine
O_2	oxygen
H_2O	water
NaOH	sodium hydroxide (caustic soda)
$CBrF_3$	bromotrifluoromethane (BTM or Halon 1301)
HCHO	formaldehyde
$C_2H_5NO_3$	ethyl nitrate

Each compound has its own properties which may be vastly different from those of the constituent atoms. Atoms within a molecule cannot be separated unless the compound undergoes a chemical reaction.

A chemical reaction occurs when the atoms in molecules rearrange either by decomposing into smaller molecules or by joining with other atoms to form different molecules; in both cases the atoms reorganise themselves to form different structures. Endothermic reactions require the input of heat to make them happen whereas exothermic reactions occur with the evolution of heat.

$$2Na + 2H_2O = 2NaOH + H_2 + \text{heat}$$
$$2H_2 + O_2 = 2H_2O + \text{heat}$$

These chemical equations show in chemical shorthand, using the chemical codes, the rearrangement of atoms which occurs in these reactions, a balance of the number of atoms being maintained during the reaction. The molecular mass of molecules can be obtained by adding together the mass numbers of the constituent atoms.

Compounds which contain atoms of elements other than carbon, but including carbon dioxide (CO_2), carbon monoxide (CO) and the carbonates (e.g. calcium carbonate $CaCO_3$) are called inorganic chemicals. All other compounds which contain carbon atoms are known as organic chemicals.

Carbon is an unusual element; not only is it able to form simple compounds where one or two carbon atoms are joined to atoms of other elements, but carbon atoms can link together to form chains or rings of atoms. Almost all other atoms can be joined into these chains and rings to create millions of different organic compounds, from the comparatively simple ones consisting of one carbon with one other type of atom to the highly complex molecules with hundreds of linked carbon atoms joined with other different atoms. Organic chemicals include most of the solvents, plastics, drugs, explosives, pesticides and many other industrial chemical substances.

24.3 Properties of chemicals

Properties of chemicals are to a large extent determined by how the atoms are bonded.

24.3.1 Inorganic compounds

In some compounds one or more of the bonds joining the atoms are the result of an unequal sharing of electrons between the two atoms and these produce ionic compounds which are crystalline solids, usually with a high melting point. They are often soluble in water giving a solution which conducts electricity.

Many other compounds have bonds based on an equal sharing of electrons and so do not ionise. These compounds can be solids having low melting points, or liquids or gases. Usually they are not soluble in water unless they react with it. There are also many more compounds with types of bond intermediate between the two described and which exhibit

Table 24.2. Properties of a selection of inorganic compounds

Compound	*Formula*	*Properties*
Ammonia	NH_3	gas, b.p. –33°C, dissolves readily in water giving a basic solution
Carbon monoxide	CO	gas, b.p. –190°C, odourless, almost insoluble in water
Hydrogen chloride	HCl	gas, b.p. –85°C, dissolves readily in water giving hydrochloric acid
Hydrogen sulphide	H_2S	gas, b.p. –61°C, strong odour, burns in air
Hydrogen peroxide	H_2O_2	liquid, decomposes violently on heating, powerful oxidising agent
Stannic chloride	$SnCl_4$	liquid, b.p. 114°C, fumes, reacts rapidly with water
Sulphuric acid	H_2SO_4	liquid, decomposes at 290°C giving SO_3, strong acid, reacts violently with water
Aluminium silicate	$Al_2Si_2O_7$	solid, infusible, unreactive, clay silicate
Phosphoric acid	H_3PO_4	solid, m.p. 39°C or syrupy liquid, strong acid
Potassium chloride	KCl	solid, m.p. 770°C, very soluble in water, unreactive
Sodium hydroxide	$NaOH$	solid, m.p. 318°C, deliquescent, strong alkali

properties that relate to both types. *Table 24.2* lists some of the properties of a selection of inorganic compounds.

24.3.2 Metals

Metals are different in structure from both types of compounds, existing in the solid state as an ordered array of atoms held together by their electrons which circulate freely between them. Application of an electric potential across a metal allows the electrons to undergo a directional flow between the atoms making metals good electrical conductors. *Table 24.3* lists the properties of some typical metals and other elements.

With the exception of sulphur, stannic chloride and potassium chloride, all the elements and compounds listed in *Tables 24.2* and *24.3* present hazards to health.

24.3.3 Organic compounds

As most organic compounds contain a relatively large percentage of carbon and hydrogen atoms they are flammable and many are toxic. All living matter is constructed of complex interdependent organic chemicals and it is through interfering with the normal functioning of them that organic compounds constitute fundamental health and hygiene hazards.

Although there are very many organic compounds they can be grouped into a small number of classes according to their reactive properties. These broad groups are listed in *Table 24.4* which gives examples of compounds in each group.

Table 24.3. Properties of typical elements

Element	*Symbol*	*Properties*
Reactive metals		
Aluminium	Al	m.p. 660°C, good conductor, surface oxide formation resists attack by air or water
Barium	Ba	m.p. 850°C, soft, spontaneously flammable in air, reacts with water
Lithium	Li	m.p. 186°C, soft, burns vigorously in air, reacts with water
Less reactive metals		
Cobalt	Co	m.p. 1490°C, hard, not attacked by air or water
Iron	Fe	m.p. 1525°C, burns in oxygen, reacts slowly with water
Mercury	Hg	liquid, slowly attacked by oxygen, no reaction with water
Silver	Ag	m.p. 961°C, ductile, not attacked by oxygen or water
Non-metals		
Bromine	Br	dark red liquid, b.p. 59°C, very reactive, not flammable
Phosphorus	P	red form, m.p. 600°C or white form, m.p. 43°C, burns readily to P_2O_5, insoluble in water
Sulphur	S	yellow or white, m.p. 115°C, burns to SO_2, insoluble in water

Table 24.4. Examples of the main groups of organic compounds

Group	*Example*	*Formula*	*Use*
Aliphatic hydrocarbons	Methane	CH_4	natural gas
	Butane	C_4H_{10}	petroleum gas
Aromatic hydrocarbons	Benzene	C_6H_6	toxic solvent
	Toluene	$C_6H_5CH_3$	solvent
Halocarbons	Bromomethane	CH_3Br	fumigant
	Trichloroethane	CH_3CCl_3	solvent
Alcohols	Ethanol	C_2H_5OH	'alchol'
	Glycerol	$C_3H_5(OH)_3$	glycerine
Carbonyl compounds	Formaldehyde (methanal)	$HCHO$	fumigant
	Benzaldehyde	C_6H_5CHO	manufacturing
	Acetone	CH_3COCH_3	solvent
Ethers	Ethyl ether	$C_2H_5OC_2H_5$	anaesthetic
	Dioxan	$C_4H_8O_2$	solvent
Amines	Methylamine	CH_3NH_2	manufacturing
	Aniline	$C_6H_5NH_2$	manufacturing
Acids	Ethanoic acid	CH_3CO_2H	acetic acid
	Phthalic acid	$C_6H_4(CO_2H)_2$	manufacturing
Esters	Ethyl acetate	$CH_3CO_2C_2H_3$	Solvent
Amides	Acetamide	CH_3CONH_2	manufacturing
	Urea	$CO(NH_2)_2$	by-product

24.3.4 Acids and bases

Acids are compounds which dissolve in water to give hydrated hydrogen ions:

$$HCl + H_2O = H_3O^+ + Cl^-$$
$$H_2SO_4 + H_2O = H_3O^+ + HSO_4^-$$

Strong acids completely dissociate into ions in solution; weak acids only partially dissociate. A concentrated acid is one which is not diluted with water, and the terms *strong* and *concentrated* should not be confused.

Acids are corrosive in that they react with both metals and with body proteins. Acids are dangerous not just because of their acidity but they can be oxidising agents (HNO_3, $HClO_4$), violently reactive with water (H_2SO_4) and many are toxic. Phenol (C_6H_5OH) is one of the most dangerous acidic organic compounds.

Bases are of two types, solid alkalis such as metal hydroxides which dissolve in water to give hydroxide ions, and gases and liquids such as ammonia and the amines, which liberate hydroxide ions on reaction with water:

$$NaOH + H_2O = Na^+ + OH^- + H_2O$$
$$NH_3 + H_2O = NH_4^+ + OH^-$$

Some of the bases are toxic, many react exothermally with water and all are highly corrosive or caustic towards proteins. Alkalis spilled on the skin penetrate much more rapidly than acids and should be leached out with copious water and not sealed in by attempting neutralisation.

The reaction between an acid and a base is a vigorous, exothermic neutralisation forming a salt. The strength of acids and bases can be measured in terms of hydrogen ion concentration by the use of either meters or test papers, and it is expressed as a pH value on a scale from 0 (acid) to 14 (base). Pure water has a neutral pH of 7.

24.3.5 Air and water

Air and water deserve to be considered separately since they are ever present and are necessary for the operation of many processes and responsible for the degradation of many materials.

Air is a physical mixture of gases containing approximately 78% nitrogen, 21% oxygen and 1% argon. These proportions do not vary greatly anywhere on the earth but there can be additional gases as a result of the local environment: carbon dioxide and pollutants near industrial towns, sulphur fumes near volcanoes, water vapour and salts near the sea, etc.

Air can be liquefied and its constituent gases distilled off; liquid nitrogen (b.p. – 196°C) has many uses as an inert coolant, liquid oxygen (b.p. – 183°C) is used industrially in gas-burning equipment and in hospitals, and argon (b.p. – 185.7°C) is used as an inert gas in certain welding processes. Liquid oxygen is highly hazardous as all combustible materials will burn with extreme intensity or even explode in its presence. Combustion is a simple exothermic reaction in which the air provides the oxygen needed for oxidation. If the concentration of oxygen is increased the reaction will accelerate. This effect was experienced in the fire on HMS Glasgow[1].

Water is a compound of hydrogen and oxygen that will not oxidise further and is the most common fire extinguishant. However, caution must be exercised in its use on chemical fires since a number of oxides and metals react energetically with it, in some cases forming hazardous daughter products and in others producing heat and hydrogen which further exacerbate the fire.

24.4 Physical properties

All matter, whether solid, liquid or gas, exhibits properties that follow patterns that have been determined experimentally over the years and are now well established and proven. This section looks at some of the factors that influence the state of matter in its various forms.

24.4.1 Temperature

Temperature is a measure of the hotness of matter determined in relation to fixed hotness points of melting ice and boiling water. Two scales are universally accepted, the Celsius (or Centigrade) scale which is based on a scale of 100 divisions and the Fahrenheit scale of 180 divisions between these two hotness points. Because Fahrenheit had recorded temperatures lower than that of melting ice he gave that hotness point a value of 32 degrees. Converting from one scale to the other:

$$(^\circ F - 32) \times 5/9 = {}^\circ C$$
$$(^\circ C \times 9/5) + 32 = {}^\circ F$$

Man has long been intrigued by the theory of a temperature at which matter ceases to exist. This has never been reached but has been determined as being –273°C. The Kelvin or absolute temperature scale uses this as its zero, 0 K; thus on the absolute scale ice melts at +273 K.

Devices for measuring temperature include the common mercury in glass thermometer, thermocouples, electrical resistance and optical techniques.

24.4.2 Pressure

Pressure is the measure of force exerted by a fluid (i.e. air, water, oil, etc.) on an area and is recorded as newtons per square metre (N/m^2). With solids the term stress is used instead of pressure. Datum pressure is normally taken as that existing at the earth's surface and is shown as zero by pressure gauges which indicate 'gauge pressure' (i.e. the pressure above atmospheric). However, at the earth's surface the weight of the air of the atmosphere exerts a pressure of 1 N/m^2 or 1 bar. Beyond the earth's atmosphere there is no pressure and this is taken as the base for the measurement of pressure in absolute terms. Thus:

$$\text{gauge pressure} = \text{absolute pressure} - 1\ \text{N/m}^2$$
$$\text{or absolute pressure} = \text{gauge pressure} + 1\ \text{N/m}^2$$

An absolute vacuum (0 N/m^2 absolute) exists at the top of a mercury barometer, where the force of the atmospheric air outside the tube is equalled by the force exerted by the column of mercury inside the barometric tube.

Pressure can be measured by means of manometers which show the pressure in terms of the different levels of a liquid in a U-tube, by mechanical pressure gauges which record the differential effect of pressure forces on the inside and outside surfaces of a coiled tube, and electronic devices which measure the change of electrical characteristic of an element with pressure.

24.4.3 Volume

Volume is the space taken up by the substance. With solids which retain their shape, their volume can be measured with comparative ease. Liquid volume can be measured from the size of the containing vessel and the liquid level. Gases, on the other hand, will fill any space into which they are introduced, so to obtain a measure of their volume they must be restrained within a sealed container.

Each state of material reacts to changes of temperature and, to a lesser extent with solids and liquids, to changes of pressure, by increases or decreases in their volume and this fact can be made use of, or has to be allowed for, in many industrial processes and plant.

24.4.4 Changes of state of matter

At ordinary temperatures, matter exists as solid, liquid or gas and many substances change their state as temperatures change – for example, ice melts to form water at 0°C and then changes into steam at 100°C. The stages at which these changes of state occur are also influenced by the pressure under which they occur.

24.4.4.1 Gases

In gases the binding forces between the individual molecules are small compared with their kinetic energy so they tend to move freely in the space in which they exist. When heated, i.e. additional kinetic energy is given to them, they move much more rapidly and if restrained in a fixed volume impinge more energetically on the walls of the containing vessel, a condition that is measured as an increase in pressure. The relationship between temperature, pressure and volume of gases is defined by the general Gas Law:

$$\frac{PV}{T}\ \text{(initial)} = \frac{PV}{T}\ \text{(final)}$$

where P = absolute pressure, V = volume and T = absolute temperature.

Thus in a reaction vessel which has a fixed volume, if the temperature is increased, so the pressure will increase. If the reaction is exothermic and the temperature increase is not controlled there is a risk that the pressure in the vessel could rise above the safe operating level with consequent risk of vessel failure, a situation that may be met in chemical processes that use autoclaves and reactor vessels.

This general law applies with variation when gases are compressed in that the temperature of the gas rises. In air compressors where there is likely to be oil present the temperature of the compressed air must be kept below a certain level to prevent ignition of the contained oil. Conversely, when the pressure of a gas is decreased, the temperature drops, a condition that can be seen with bottles of LPG where a frost rime forms and where in cold weather there is a danger of the temperature of the gas dropping so low that the control valve freezes up.

Some gases can be compressed at normal temperature until they become liquids (e.g. carbon dioxide, chlorine, etc.), and can conveniently be stored in that state, while others, called permanent gases, cannot be liquefied in this way but are stored either as compressed gases (e.g. hydrogen, air, etc.) or under pressure in an absorbent (e.g. acetylene).

Air, carbon dioxide and a number of other gases which dissolve in water to some extent become more soluble as the pressure increases or the temperature decreases. Increase in temperature or decrease in pressure cause the dissolved gases to come out of solution, e.g. tonic water or fizzy lemonade. This is why hydraulic systems need venting. A similar condition arises with divers who surface too quickly and release dissolved gases from their blood causing the 'bends'.

The specific gravity (sp. gr.) of a gas is determined by comparing its weight with that of air and this has important implications at work. The charging of the batteries of fork lift and other trucks results in the generation of hydrogen (which has a sp. gr. of 0.07) which will either become trapped under any covers or lids left over the battery creating a serious explosion risk or will rise into the ceiling space where good ventilation is necessary to prevent a fire risk. At the other end of the scale carbon dioxide (CO_2, sp. gr. = 1.98) and hydrogen sulphide (H_2S, sp. gr. = 1.19) being heavier than air tend to sink to the floor and will fill the lower part of pits and storage vessels making entry hazardous. Similarly, the vapours of most flammable liquids are heavier than air and tend to collect in low places in the floor and will flow like water to the lowest point creating fire hazards possibly remote from the site of the leakage.

24.4.4.2 Liquids

In liquids, the kinetic energy of the molecules is sufficient to allow them to slide over each other but not sufficient for them to move at random in space since they are subject to the binding force of cohesion and remain a coherent mass with a definite volume. Thus the substance will flow to take up the shape of the container but cannot be compressed. When heated the kinetic energy of the molecules increases and the liquid expands until boiling point is reached when the cohesive forces can no longer hold adjacent molecules together, the liquid evaporates and behaves as a gas. If

Table 24.5. Properties of some flammable liquids and gases

Substance	*Sp. gr. of liquid*	*Sp. gr. of vapour*	*Boiling point, °C*	*Sp. heat of liquid at 20°C*	*Vapour pressure at 20°C in mmHg*	*OES long term ppm*
Acetaldehyde	0.78	1.52	21	–	760	100
Acetone	0.79	2.00	56.5	0.53	185	1000
Acetylene	–	0.91	–84	–	–	*
Ammonia	–	0.60	–33.4	–	–	25
Carbon disulphide	1.3	2.64	46	0.24	298	10 (MEL)
Carbon dioxide	–	1.98	–79	0.82	–	5000
Diethyl ether	0.71	2.56	35	0.54	442	400
Hydrogen	–	0.07	–253	–	–	*
Commercial propane	0.50	1.4–1.55	–45	1.53	9	*
Toluene	0.87	3.14	111	0.39	23	100

OES = occupational exposure standard.
MEL = maximum exposure limit.
* = asphyxiant, requires monitoring for oxygen content of atmosphere.

the temperature falls, i.e. the kinetic energy of the vapour is reduced, the vapour will revert to its liquid state by condensing. However, at temperatures well below the boiling point, some surface molecules gain sufficient energy to escape from the surface. This means that the liquid is constantly losing some of its surface molecules by vaporisation and the extent to which this is occurring is measured as vapour pressure (*Table 24.5*). Vapour pressure will increase as temperature rises.

In passing from the liquid to the vapour or gaseous state a considerable additional input of energy is required without raising the temperature. This energy is the 'latent heat of vaporisation' of the liquid and can be quite substantial (*Table 24.6*). This characteristic is made use of in fire fighting where a fine spray of water absorbs the heat of the flames to become steam and in so doing reduces the gas temperature to below that required to maintain combustion.

As liquids are heated they expand volumetrically and this effect has to be taken into account in the storage of liquids in closed vessels. It is normal to

Table 24.6. Heats of vaporisation and fusion

Substance	*M.P., °C*	*B.P., °C*	*Hv, kJ/kg*	*Hf, kJ/kg*
Aluminium	660	2450	10 500	397
Copper	1083	2595	4810	205
Gold	1063	2660	1580	64
Water	0	100	2260	335
Ethanol	–114	78	854	105
Oxygen	–219	–183	210	13.8

Hv = heat of vaporisation.
Hf = heat of fusion.

leave an air space above the liquid, the ullage, to allow for this expansion, and to fit pressure relief valves to release any excess pressure generated. An example of what can happen if these precautions are not taken was seen at Los Alfaques Camping Site at San Carlos de la Rapita in Spain in July 1978 when a road tanker carrying 23.5 tonnes of propylene exploded killing 215 people. The circumstances of the incident were that the unlagged tanker was filled completely with propylene during the cool of the early morning, leaving no ullage. The tank was not fitted with any form of pressure relief. As the tanker was driven in the heat of the sun, the tank and its contents heated up to a temperature where the vapour pressure of the contents exceeded atmospheric pressure and the propylene (an incompressible liquid), expanding at a greater rate than the tank, caused the tank to rupture as it was passing a holiday camp at 14.30 hours creating a BLEVE (boiling liquid expanding vapour explosion) that caused devastation over an area of five hectares.

The incompressibility of liquids is utilised in many ways in industry particularly in hydraulic power transmission. Two common applications are as power sources for machines and vehicles and in hydraulic cylinders. Normally the hydraulic medium is an oil and hazards can arise where leaks occur either as oil pools on a floor or from high-pressure lines as a fine mist which is highly flammable.

24.4.4.3 Solids

In a solid at room temperature the molecules are tightly packed together, have little kinetic energy and vibrate relative to each other. A solid has form and a rigid surface as a result of the strength of the cohesive forces between the molecules. As a solid is heated the vibrations of the molecules increase and the solid expands. Eventually a temperature is reached where the kinetic energy allows the molecules to slide over one another and the solid becomes a liquid.

Solids, and particularly metals, have a number of characteristics that are of great value in almost everything to do with modern standards of living. They have stability of form over a wide range of temperatures which enables them to be formed and machined into shapes, a measurable strength over a wide range of stresses and temperatures, and permanent physical characteristics such as rates of expansion and contraction with temperature.

The change of state from solid to liquid requires the input of a considerable quantity of heat – the latent heat of fusion – without raising its temperature (*Table 24.6*). This can easily be seen with melting ice where the temperature of the water remains at 0°C until all the ice has melted. At room temperatures some solids, such as solid carbon dioxide, do not pass through the liquid stage but change directly from a solid to a gas, a process known as 'sublimation', the liquid state only occurring at low temperature and/or high pressure. Organic substances such as wood also have no liquid stage but decompose on heating, rather than melting, reducing their very large molecules into smaller ones some of which are given off as vapours.

As solids are heated they expand at rates determined by their 'coefficients of expansion' that vary with the different substances. The heat needed to raise the temperature of different solids varies and can be determined from

Table 24.7. Physical properties of some solids

Material	*Density*	*Coefficeint of linear expansion*	*Specific heat*
	kg/m³	*μm/mK*	*kJ/kgK*
Aluminium	2700	24	900
Brass	8500	21	380
Cast iron	7300	12	460
Concrete (dry)	2000	10	0.92
Copper	8930	16	390
Glass	2500	3–8	0.84
Polythene	930	180	2.20
Steel	7800	12	480

its 'specific heat'. Problems can arise where dissimilar metals are welded together that have different specific heats and different coefficients of expansion. Allowance has to be made for differential expansions where different metals are moving in contact and have different coefficients of expansion, such as in bearings. However use is made of these differences in bimetal strips for temperature measurements and in thermostats.

Physical properties of some solids are shown in *Table 24.7.*

Solids, and particularly metals, have great strength and an ability to withstand high levels of stress and strain. By alloying metals they can be made to exhibit particular features to suit particular applications. Characteristics of metals can vary from very ductile lead and gold through high-strength high-tensile steels to brittle cast iron, each having its particular use. One exception is the metal mercury which is a liquid at room temperature.

24.5 Energy and work

There are two types of energy, kinetic and potential. Kinetic energy is involved in movement and may be available to do work; it can be present in a number of forms such as heat, light, sound, electricity and mechanical movement. Potential energy is stored energy and usually requires an agent to release it, as, for instance, the potential energy of a loose brick lying on a scaffold which needs to be kicked off to produce kinetic energy.

Heat is a form of energy where the degree of hotness of a material is related to the rate of movement of the atoms and molecules which make it up. Heat is transferred from one material to another by conduction, convection or radiation. Conduction occurs within and between materials that are in contact through the physical agitation of molecules by more energetically vibrating neighbours. Some materials are better conductors than others depending on the ease with which molecules can be made to vibrate. Convection is the conveyance of heat in gases and liquids by the heated fluid rising and heating any surface with which it comes into contact. Initial spread of fire within a building is most likely to be by convection and a method of control is by venting the hot gases to outside the building so preventing the lateral spread of fire. Radiation of heat occurs in infra-red emissions from a heat source which pass through the atmosphere by wave propagation.

If a force is exerted on an object, no energy is expended until the body moves when the force is converted into kinetic energy which is equal to the work done on the body by the force. Mechanical work is done whenever a body moves when a force is applied to it.

Work done (joules) = force (newtons) × distance moved (metres)

In all mechanical devices or machines the work done is never as great as the energy expended, as some of the energy is lost in overcoming friction. The ratio of work done to energy expended is the mechanical efficiency of the machine and is always less than one.

Power is the rate of doing work and is measured in joules per second:

1 joule per second = 1 watt of power
1 kWh = 3.6×10^6 joules
746 watts = 1 horsepower

The rate at which machines work is given in either horsepower or kilowatts.

Pressure is unreleased potential energy since when contained in a pressure vessel the fluid exerts equal and opposite forces on the vessel walls but no movement takes place. Pressure is independent of the shape of the vessel and is exerted at right angles to the containing surfaces. With gases the pressure is virtually the same throughout the containing vessel but with a liquid the pressure varies according to the depth, i.e. the weight of liquid above or below the point of measurement. Pressure can be converted into kinetic energy through the movement of a pneumatic or hydraulic cylinder.

24.6 Mechanics

Mechanics is that part of science that deals with the action of forces on bodies with much of the theory based on Newton's three Laws of Motion:

(1) Every body continues in its state of rest or of uniform motion in a straight line except in so far as it is compelled by external impressed force to change that state.
(2) Rate of change of momentum is proportional to the force applied and takes place in the direction in which the force acts.
(3) To every action there is always an equal and contrary reaction.

If a force is applied to a body resting on a plane, initially the body will not move because of the friction between itself and the plane. The force is resisted by an equal and opposite force due to friction. Once the applied force exceeds the friction force the body will move in the direction of the applied force. However, if the same force continues to act on the body its speed will accelerate because sliding friction is less than limiting (i.e. static) friction. Hence:

Before movement the force $F = W\mu_L$

where μ_L is the coefficient of static or limiting friction, and W is the reaction between the body and the plane (i.e. its weight = mass × gravity).

After movement the excess force $F_e = F - W\mu_s$

where μ_s is the coefficient of sliding friction.

This excess force overcomes the *inertia* of the body and causes it to accelerate at a rate f given in the formula

$$F_e = mf$$

This is why a sticking object will shoot away once it has been freed. Friction has an ambivalent role – it is necessary to enable us to walk, for cars to move and to enable us to control motion through the use of brakes; on the other hand in machines friction absorbs energy and makes the machine less efficient.

Bodies can possess potential energy as a result of the height at which they are located above a datum and that potential energy can be converted into kinetic energy by the object falling. Thus a mass m falling through a height h does work equal to

$$W = m \times g \times h$$
$$= \tfrac{1}{2} \times m\, v^2 \text{ its kinetic energy}$$

where v is its velocity.

For a body to move in the direction of the force applied to it, the line of that force must pass through the centre of gravity of the body. If the line of the force does not pass through the body's centre of gravity, the force will apply a turning moment to the body. Two equal and opposite forces acting on a body will apply a turning couple.

In machines forces act on component parts in overcoming friction, accelerating or decelerating or in transmitting power or loads. These forces can be tensile, compressive, shear, bending or torsional.

24.7 Strength of materials

For machines, plant and buildings to work and give an economic life, their component parts must be capable of resisting the various forces to which they are subject during normal working. Any load applied to a component induces stresses and strains in it. The type of stress depends on the manner in which the load is applied. Stress is force per unit area (N/m^2). Strain is the proportional distortion when subject to stress ($\delta l/l$) and is often quoted as a percentage. The ratio of stress to strain is constant and known as Hooke's Laws after its discoverer, thus:

$$\frac{\text{stress}}{\text{strain}} = \text{constant} = E$$

E is called Young's modulus or modulus of elasticity and its value depends on the material and the type of stress to which it is subjected.

Forces acting on a component can be tensile, compressive, shearing, bending or torsional. The effect of these is shown diagrammatically in *Figure 24.2*.

The calculation of tensile, compressive and shear stresses is relatively easy, but to obtain bending stresses it is necessary to know the moment of inertia of the section of the beam and in the case of torsion, although it is a shear stress, because it is spread along the length of the shaft its calculation is complex. With bending moments, a tensile stress is induced in the outer surface of the beam and an equal and opposite compressive stress occurs in the inner surface.

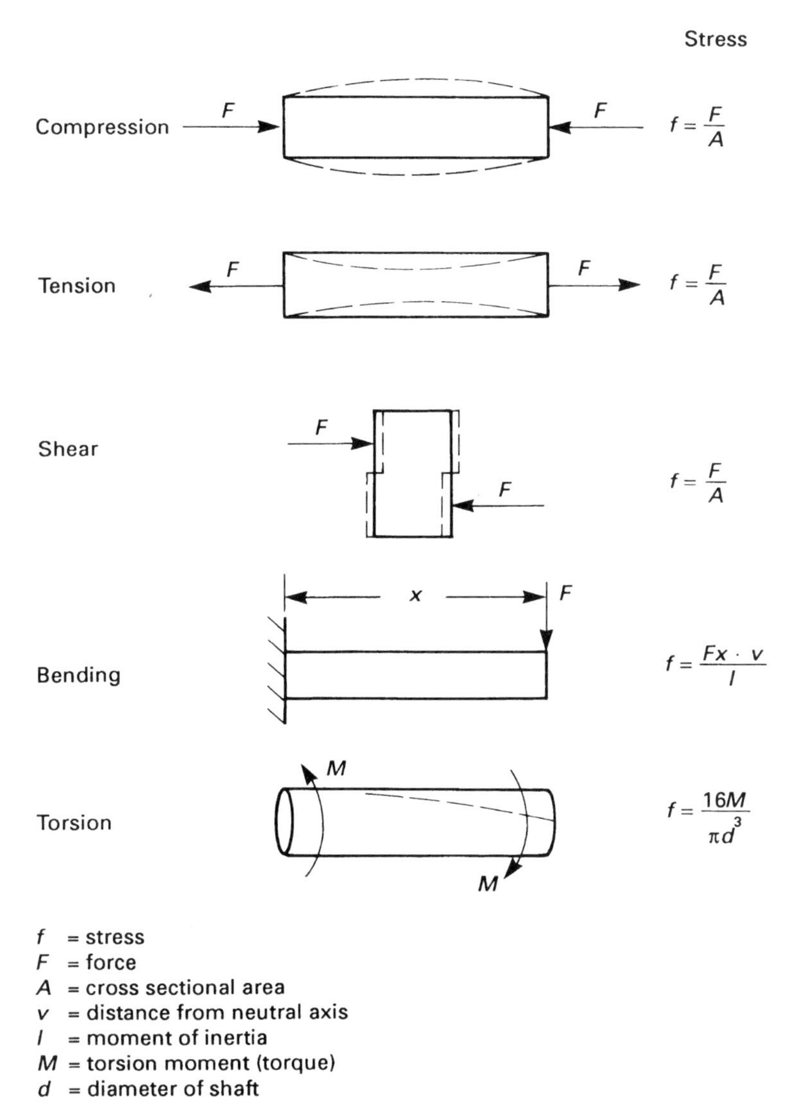

f = stress
F = force
A = cross sectional area
v = distance from neutral axis
I = moment of inertia
M = torsion moment (torque)
d = diameter of shaft

Figure 24.2 Forces producing deformation of a material

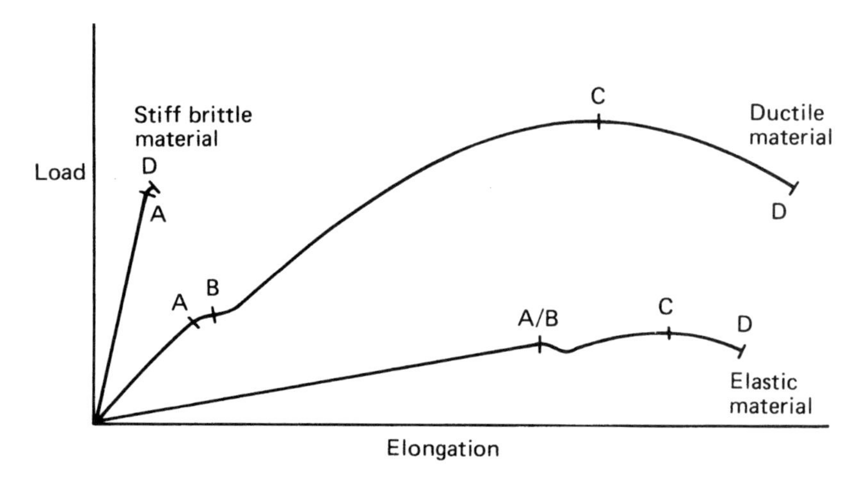

Figure 24.3 Behaviour of different materials under tensile load. A = elastic limit; B = yield point; C = ultimate tensile strength; D = breaking point

While each material exhibits particular physical characteristics, they all follow a broad pattern of behaviour as shown in *Figure 24.3*. Where an item has a load applied to it, provided that the strain induced does not go beyond the elastic limit, then when the load is removed the item will return to its original size, i.e. there will have been no permanent deformation. However, if the elastic limit is exceeded permanent deformation occurs and the characteristic of the material will have changed. Typical stiff brittle materials are cast iron, glass and ceramics which have no ductility and will fracture at the elastic limit. Ductile materials include mild steel and copper, and elastic materials include plastics.

24.8 Failures

Even in the best of designed machinery and plant failures can occur. When they do, the cause is normally investigated so that repetition can be prevented. Modes of failure relate to the type of stress under which the component has been working and the characteristic features of failures due to tension, compression, shear and torsion are well-known. Sometimes the failure is related more to the operating process than to the stress particularly when there is repeated stress cycling of the part and it can suffer fatigue failure.

In processes using certain chemicals stresses have been shown to make the parts prone to corrosive attack which can reduce their strength. Similarly failures have occurred where a chemical in contact with the part has affected its ability to carry stresses through causing embrittlement such as zinc embrittlement of stainless steel[2] and hydrogen embrittlement of grade T chain[3].

24.9 Testing

While there are quite extensive requirements for the testing of plant and equipment written into legal obligations, these tend to relate to the finished product. A great deal of testing takes place long before the product is manufactured and starts at the material producer who tests his product for quality and to ensure that it meets specification. Such materials testing is usually on specially shaped test pieces that are tested to destruction and by chemical analysis to ensure that the constituent materials are present in the correct amounts.

In the manufacture of plant and machinery and subsequent periodic inspections and tests, testing is normally non-destructive and a number of specialised techniques have been developed. Equipment is now available to enable visual inspections to be carried out of inaccessible places using fibre optics and remote-controlled television. A major aim of testing is to check the condition of the material of which the plant is constructed and to identify faults that cannot be seen by eye. Special detection techniques used to highlight the weaknesses or faults in the material include the use of magnetic particles, penetrant dyes, X-ray and gamma-ray sources, ultrasonic vibrations, microwave and infra-red rays.

24.10 Hydraulics

This side of engineering is concerned with fluids, both liquids such as water, oil, etc., and gases such as air and gas elements and compounds. Steam can be classified as a gas although it has characteristics that make it more hazardous than normal compressed gases. These two types of fluid exhibit disparate characteristics in that gases can be compressed while liquids are incompressible.

24.10.1 Compressible fluids

When gases are compressed, work is done on them and they acquire considerable potential energy as pressure. They also become hot and it is sometimes necessary to provide cooling to ensure that critical temperatures are not exceeded. Conversely when gas pressure is released, the gas does work or gives up energy and heat and becomes cold. This latter effect is the basis by which refrigerators work and it can be seen, even in the middle of summer, as rime on the outside of LPG cylinders.

Compression of gases and air to achieve high pressures with relatively low volumes is by positive displacement compressors of either the piston or rotary type, while for large volumes at lower pressures centrifugal compressors are used as for example in gas turbines. Gas flows can be measured by pressure drop across an orifice or by means of pitot tubes.

Compressed gases have a large amount of potential energy which can be utilised for actuating thrust cylinders. Once pressurised gas has been admitted to a cylinder, even if the supply is then cut off, the cylinder will continue to operate until all the energy of the gas has been absorbed in work done. A dramatic demonstration of the energy of compressed gases can be seen when a containing vessel ruptures with heavy sections being projected very considerable distances or the jet effect when a valve breaks off a gas cylinder. Because of the energy it contains, compressed air is not used for pressure testing of vessels except in certain very specific cases where the use of water is not practicable, e.g. in nuclear reactor vessels.

24.10.2 Incompressible fluids

Normally referred to as liquids, incompressible fluids have many advantages when used for power transmission in that, even at very high pressures, they do not in themselves contain a great deal of energy so that should a failure of a containing vessel or pipework occur the result is not catastrophic. These fluids also have the advantage that their flow and pressure can be controlled to a very fine degree which makes them ideal for use in applications where control is critical. They also have the advantage that they can be pumped at very high pressures and so permit the application of very large forces through hydraulic jacks and high torques through hydraulic motors. The power of a fluid jet has been harnessed in cutting lances where very high pressure liquid, normally water, is emitted from a very small orifice with sufficient force to cut through a variety of materials including steel. These lances have a number of applications but can be very dangerous so their use requires strict control.

The pumps used in handling these high-pressure liquids can suffer considerable damage from cavitation. Incompressible liquids will not compress, nor will they withstand tension; thus if the suction inlet to a pump is restricted the fluid will release any contained air to form cavities. This condition seriously affects the performance of the pump, can cause damage to its rotor and generates a great deal of noise. Gas or air entrained in a hydraulic fluid is detrimental to its effectiveness as a power transmission medium.

24.11 Summary

Although much of occupational safety is now recognised as being behavioural, engineering with its foundation in science still has a major role to play. Many of the principles learnt in the lecture room and science laboratory have application across a spectrum of work activities and have important connotations as far as occupational safety is concerned. This chapter has reviewed some of those applications – there are many more for the student to discover.

References

1. Health and Safety Executive, *Investigation report: Fire on HMS Glasgow, 23 September 1976*, HSE Books, Sudbury (1978)
2. Health and Safety Executive, *Guidance Note, Plant and Machinery series No. PM 13: Zinc embrittlement of austenitic stainless steel*, HSE Books, Sudbury (1977)

3. Health and Safety Executive, *Guidance Note, Plant and Machinery series No. PM 39: Hydrogen embrittlement of grade T chain*, HSE Books, Sudbury (1984)

Reading list

Basic Engineering Science

Titherington, D. and Rimmer, J. G., *Engineering Science*, Vol. 1 (1980), and Vol. 2 (1982), McGraw-Hill, Maidenhead

Deeson, E., *Technician Physics*, Longman, Harlow (1984)

Harrison, H. R. and Nettleton, T., *Principles of Engineering Mechanics*, Edward Arnold, London (1978)

Hannah, J. and Hillier, M. J., *Applied Mechanics*, Pitman, London (1971)

Houpt, G. L., *Science for Mechanical Engineering Technicians*, McGraw-Hill, Maidenhead (1970)

Properties of Materials

Peapell, P. N. and Belk, J. A., *BASIC Materials Studies*, Butterworth, London (1985)

Timings, R. L., *Engineering Materials*, Vol. 1, Longman, Harlow (1989)

Bolton, W., *Engineering Materials Technology*, Heinemann, Oxford (1989)

Charles, J. A. and Crane, F. A. A., *Selection and Use of Engineering Materials*, 2nd edn, Butterworth, London (1989)

Hanley, D. P., *Introduction to the Selection of Engineering Materials*, van Nostrand, Wokingham (1980)

Young, W. C., *Roark's Formulas for Stress and Strain*, 6th edn, McGraw-Hill, New York (1989)

Higgins, R. A., *Properties of Engineering Materials*, Hodder and Stoughton, Sevenoaks (1977)

Gordon, J. E., *The New Science of Strong Materials*, Pitman, London (1979)

Hydraulics

Turnbull, D. E., *Fluid Power Engineering*, Butterworth, London (1976)

Pinches, M. J. and Ashby, J. G., *Power Hydraulics*, Prentice Hall, Hemel Hempstead (1989)

Massey, B. S., *Mechanics of Fluids*, 6th edn, van Nostrand Reinhold, London (1989)

Basic Introductory Chemistry

Lewis, M. and Walter, G., *Thinking Chemistry*, Oxford University Press, Oxford (1980)

Jones, M., Johnson, D., Netterville, J. and Wood, J., *Chemistry, Man and Society*, Holt, Rinehart and Winston, Eastbourne (1980)

Hazardous Chemicals

Schieler, L. and Pauze, D., *Hazardous Materials*, van Nostrand Reinhold, Wokingham (1976)

Hazardous Chemicals, A manual for schools and colleges, Oliver and Boyd, Edinburgh (1979)

Chapter 25

Fire precautions

Dr P. Waterhouse

25.1 Introduction

The control of fire provided man with his first means of advancement. It broadened his food choice by enabling him to cook, it widened his living range by providing him with an external source of heat, it improved his tools by permitting him to extract and work with metals and it lengthened his day by giving him a source of light. Fire was so important to primitive man that he made it one of the four 'elements' – earth, fire, air and water – which made up his world. However man's discovery of fire also revealed to him its awesome destructive power. Primitive man used fire and worshipped it but he also went in fear and dread of its destructive nature.

Fire precautions are the measures taken in the provision of the fire protection in a building or in other situations to minimise the risk to the occupants, contents and structure from an outbreak of fire.

Fire prevention is the concept of preventing outbreaks of fire, of reducing the risk of fire spreading and of avoiding danger to persons and property from fire.

Fire protection deals with design features, systems or equipment in a building, structure or other fire risk situation, to reduce the danger to persons and property by detecting, extinguishing or containing fires.

25.2 Basic combustion chemistry

When a fire occurs, products of combustion, often called smoke, are evolved and heat and light are given off. Scientifically the process is known as combustion and the reaction that occurs is a gaseous exothermic oxidation reaction. Exothermic means that energy is given off when the reaction occurs, oxidation means that it is a reaction between the substance and oxygen and gaseous that it takes place in the gaseous state. In chemical terms the reaction is written as the following equation:

$$\text{substance} + \text{oxygen} \rightarrow \text{oxidation products} - \text{energy}$$

$$\text{e.g.} \quad CH_4 + 2O_2 = 2H_2O - 890\,\text{kJ}$$

Note: The evolution of energy is always expressed as a negative number. The amount is a characteristic of the reacting substance (the fuel).

Useful as it is to express combustion as a chemical reaction, the equation contains no information on the energy that is required to start it, how fast it will proceed once it has started and the conditions that are necessary to sustain it once it has started. For a reaction to take place it is necessary to put some energy into the mixture of reactants, the initiation energy, usually in the form of heat. Some mixtures require very little heat but for others a large amount is necessary. Often the heat input comes from heating the mixture but the heat can also arise from an electric discharge taking place in the mixture. In fire terms the initiation energy is known as the ignition energy and the source is called the ignition source.

The speed of the reaction is influenced by many factors such as the chemical nature of the combustible substance, the amount of mixing and the physical state of the substances but for an exothermal reaction increasing the temperature of the reactants will increase the speed of the reaction. The difference between a fire and an explosion is very largely the speed of the reaction that takes place once it has been initiated.

The conditions to sustain a combustion reaction once it has started are more difficult to explain in simple terms. At its simplest it is useful to imagine a chemical reaction occurring when two different molecules, A and B, collide, fuse together and then separate as two different molecules, C and D, with the release of some energy. If there are plenty of both A and B molecules around then collisions will occur frequently and the reaction will be seen as taking place. If however there are a large number of A molecules and very few of B then B will be quickly used up and the reaction will cease. Similarly if there are few A molecules but a large number of B, the A molecules will be quickly used up and the reaction will again cease. The measure of the number of molecules in a mixture is the concentration of the substance, the higher the number of molecules the greater the concentration. It follows therefore that it is not the fact that oxygen and a combustible material are present but rather the concentration in air (which contains the oxygen) of the combustible substance that is important. In particular there are two concentrations, an upper and a lower, beyond which no reaction will take place. In combustion terms these are known as the upper and lower flammable limits.

25.2.1 The fire triangle

The simple facts of combustion chemistry, reactants and ignition source, are commonly displayed as the fire triangle (*Figure 25.1*).

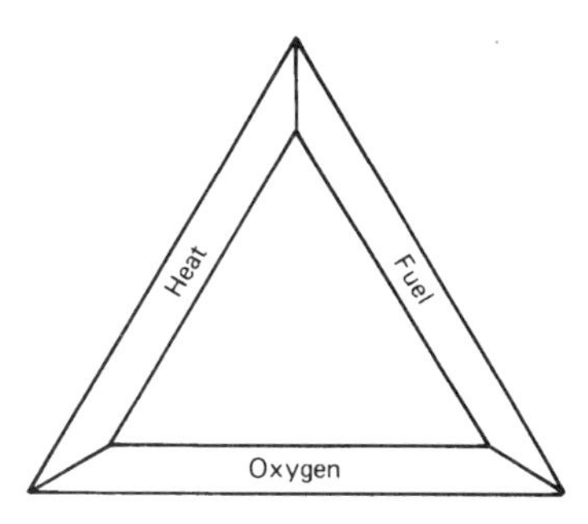

Figure 25.1 Fire triangle. The sides of the triangle represent the elements of combustion

It is important to note that the fire triangle contains no information on the speed of the reaction or of the amount of energy given off or of the upper and lower flammable limits.

25.2.2 Fuel

Almost all organic chemicals and mixtures whether in the solid, liquid or gaseous state are flammable so they are potential fuels. Examples of these include: wood, paper, plastics, fibres, petrol, oil, LPG, coal, and living tissue. Most but not all inorganic substances are not flammable and hence are not potential fuels. Exceptions to this are: hydrogen, sulphur, phosphorus, magnesium, titanium and aluminium.

25.2.3 Oxygen

The main source of oxygen in a fire is air which contains 21% oxygen. Other sources are oxidising agents and combustible substances which contain oxygen. Examples of the former include organic peroxides, hydrogen peroxide, sodium chlorate and nitric acid, and of the latter ammonium nitrate.

25.2.4 Means of ignition

The most common means of ignition is heat from an external source such as a burning match, a fire, a hot bearing, friction, incandescent sparks or the heat from a light bulb. Electrical discharges from the making and/or breaking of an electric circuit or from static electricity are also possible means of ignition.

25.2.5 Combustion characteristics

For liquids and gases it is possible to measure some combustion characteristics, i.e. the ignition energy and the flammable limits. When the measurements are carried out under standard conditions comparisons can be made and conclusions drawn.

There are three measures of ignition energy, the flash point, the fire point and the ignition temperature. These are defined as:

25.2.5.1 Flash point

The flash point is the minimum temperature at which sufficient vapour is given off to form a mixture with air which is capable of ignition under prescribed test conditions.

25.2.5.2 Fire point

The fire point is the lowest temperature at which the heat from the combustion of a burning vapour is capable of producing sufficient vapour to sustain combustion.

25.2.5.3 Ignition temperature

The ignition temperature is the lowest temperature at which the substance will ignite spontaneously.

The fuel/oxygen criteria are known as the lower and upper flammable (explosive) limits.

In all cases the flash point has a lower value than the fire point and the fire point has a lower value than the ignition temperature. The reason for this is that with the fire point the energy for ignition is provided from an external source and with the ignition temperature it is provided entirely by the heat contained in, i.e. the temperature of, the combustible substance.

The flammable limits are defined as:

25.2.5.4 Lower flammable (explosive) limit

The lower flammable (explosive) limit is the smallest concentration of flammable gas or vapour in air which is capable of ignition and subsequent flame propagation under prescribed test conditions.

25.2.5.5 Upper flammable (explosive) limit

The upper flammable (explosive) limit is the greatest concentration of flammable gas or vapour in air which is capable of ignition and subsequent flame propagation under prescribed test conditions.

It should be noted that, by virtue of the definition, flash point and fire point apply only to liquids while ignition temperature and flammable limits apply to both liquids and gases.

The standard tests for determining the flash point of a liquid are described in Schedules 1 and 2 of the Highly Flammable Liquids and Liquefied

Table 25.1. Properties of some flammable substances

Substance	*Flash point (°C)*	*Ignition temperature (°C)*	*Flammable range (% v/v in air)*
Acetic acid	40	485	4–17
Acetone	–18	535	2.1–13
Acetylene	–18	305	1.5–80
Ammonia	Gas	630	1.5–27
Benzene	–17	560	1.2–8
n-Butane	–60	365	1.5–8.5
Carbon disulphide	–30	100	1–60
Carbon monoxide	Gas	605	12.5–74.2
Cyclohexane	–20	259	1.2–8.3
Ether	–45	170	1.9–48
Ethanol	12	425	3.3–19
Ethylene	Gas	425	2.7–34
Hydrogen	Gas	560	4.1–74
Methane	Gas	538	5–15
Toluene	4	508	1.2–7
Vinyl chloride	–78	472	3.6–33

Petroleum Gases Regulations and are referred to in part IV of Schedule 1 of the Classification, Packaging and Labelling of Dangerous Substances Regulations. The properties of some flammable substances are given in *Table 25.1*.

25.3 The combustion process

Although easy to describe in chemical terms, the process of combustion is complex particularly for liquids and solids. Consider first a pool of liquid. Immediately above the liquid surface there is a layer of vapour, the concentration of which depends on the vapour pressure of the liquid at the temperature in question. If this is ignited and it continues to burn, the heat from the flame causes more of the liquid to evaporate and the burning is sustained. Close examination of the surface of the liquid shows that there is a thin gap between the liquid surface and the flame. In this gap the vapour concentration is above the upper flammable limit. With a solid there is no vapour above the solid to ignite so initially the heat from a flame raises the surface temperature of the solid which causes some decomposition of the surface layer with the evolution of a flammable vapour. This ignites, the heat from the flame causes more decomposition and the burning is sustained. Again, as with the liquid, close examination of the surface of the solid shows that here is a thin gap between the surface and the flame. In this gap the vapour concentration is above the upper flammable limit.

The chemical equation always shows that energy is released and that colourless gases are formed. The energy is actually the heat of the products of combustion and infrared radiation. In reality however, when combustion occurs there is a visible flame and there is smoke. The visibility of the flame arises from glowing particles, usually particles of carbon, which have not been fully oxidised to carbon dioxide. In addition to this visible radiation there are also infrared and ultraviolet radiations. Smoke is particulate matter where the particles are not sufficiently hot to glow and hence appear black. Smoke particles tend to be larger than the glowing particles in a flame and may include other decomposition substances. The visibility of the combustion flame differs with different substances, for example methanol burns with an almost invisible flame whereas toluene produces a dense black smoke. Generally the more air there is in a fire the less dense will be the smoke.

The hot products of combustion rise and will increase the temperature of everything with which they come into contact. Radiation is emitted uniformly in all directions – infrared radiation will heat any surface on which it falls. The intensity of the radiation diminishes as the square of the distance from the point of emission.

25.3.1 Flammable liquids and sprays

Flammable liquids are divided into three groups, depending mainly on the flash point of the liquid, which are given in Part 1 of CPL. A flammable liquid is one with a flash point between 21°C and 55°C. A highly flammable liquid is one with a flash point below 21°C. An extremely flammable liquid is one with a flash point below 0°C and a boiling point of less than 35°C. This division is important in the transport, use and storage of flammable liquids.

When a liquid is made into a spray, either deliberately through some form of atomiser or inadvertently when liquid under pressure leaks out through a small hole, the exposed surface area is increased which allows it to heat up more quickly. Also the unit mass of each droplet is reduced again allowing it to heat up more quickly thus increasing the evaporation rate. With the increase in evaporation, there is more fuel to burn in the vapour phase and so burning is enhanced. In fact burning liquid sprays display many of the characteristics of burning gases.

25.3.2 Flammable solids and dusts

Solids do not have the divisions of flammability that liquids do. Dusts, however, are finely divided solids and, as with sprays, they have an increased surface area and a decreased unit mass. Both of these features allow the particles to heat up more rapidly and so their rate of burning is increased. Dusts exhibit many of the burning characteristics of gases.

Table 25.2. Combustion characteristics of various dusts of particle size not more than 200 mesh

Substance	*Ignition temperature (°C)*	*Minimum spark ignition energy (mJ)*	*Lower explosive concentration (mg/m^3)*
Magnesium	520	40	200
Zinc	600	650	4800
Methyl methacrylate	440	15	200
Polystyrene	490	15	150
Coffee	160	85	500
Sugar	350	30	350
Coal	610	60	550
Wood floor	430	20	400

It is possible to measure the flammability limits of both dusts and sprays and to compare combustion characteristics under standard conditions (see *Table 25.2*).

25.3.3 Spontaneous combustion

A combustion characteristic which appears to be outside the usual range of characteristics is spontaneous combustion. This is the phenomenon of the apparently inexplicable bursting into flame or smoke of some substances such as oil-soaked rags or overalls, oil-soaked lagging, hay and straw. The explanation is simple: a slow oxidation of the oil gives off heat with the formation of substances with lower molecular weights which also have lower flash points. As the oxidation continues the temperature of the oxidising oil reaches the flash point of one of the degradation products and a fire occurs. The oxidation and generation of heat accelerate if the heat of oxidation cannot escape. The reaction rate is also increased when the oil is spread over a large surface. With hay and straw the initial source of heat is bacterial decomposition initiated by wet conditions and as hay and straw are good heat insulators the conditions for acceleration of the reaction are always present.

25.4 Classification of fires

Fires are commonly classified into four categories which take into account the type of substance that forms the fuel and the means of extinction.

25.4.1 Class A

Fires involving solid materials normally of an organic nature in which combustion occurs with the formation of glowing embers. Examples of Class A fires are wood, paper, coal and natural fibres. The most effective extinguisher agent is water either as a jet or spray. The mode of extinguishing is by cooling particularly of the glowing embers which propagate the fire.

25.4.2 Class B

Fires involving liquids or liquefiable solids which are subdivided into:

Class B(i)
Fires involving liquids which are soluble in water such as methanol and acetone. These fires can be extinguished with water spray, alcohol-resistant foam, light water, vaporising liquids ('Halons'), carbon dioxide and dry powder.

Class B(ii)
Fires involving liquids which do not dissolve in water such as petrol, oil, fats and waxes. These fires can be extinguished by foam, light water, vaporising liquids, carbon dioxide and dry powder.

25.4.3 Class C

Fires involving gases or liquefied gases in the form of liquid gas spillage or gas leaks. Examples are methane, propane, butane. Foam or dry chemical powder can be used on small liquefied gas spillages particularly in conjunction with water to cool the leaking container or spillage collector. When the fire involves leaking gas it can be extinguished by either isolating the supply or by inert gas injection. Direct flame extinguishment is difficult and probably counter-productive. If the flame is put out and the leak continues the possibility of re-ignition remains, occurring as a violent explosion. The extinguishers act by smothering.

25.4.4 Class D

Fires involving metals, such as magnesium and aluminium. Special dry powders must be used and no other extinguishers; these include powdered graphite, powdered talc, soda ash, limestone and dry sand. All these extinguishers work by smothering the fire.

25.4.5 Electrical fires

There is no classification called 'electrical' because the types of fires caused by live electrical equipment are covered by one or other of the recognised classifications. The electricity supply to live electrical equipment which is involved in a fire must always be isolated before the fire is tackled.

25.5 Ignition sources and their control

25.5.1 Sources of ignition

As heat is the most common source of ignition anything which will supply heat to a flammable substance is a potential ignition source. Direct heating sources such as steam and hot-water pipes, hot air, stoves and boilers, open fires, electrical heaters are just some of a long list of heat sources. Other common sources are friction heating and the hot surfaces of machines. Another source of ignition is the electrical discharge which can take place when a circuit is made or broken; these occur in switches and the commutators of motors. Electrical discharge can also take place when static electricity is discharged to earth. Frequent sources of fires are ovens, heating equipment etc. which have been left on after the end of work often in wooden huts; a close inspection of these premises must be carried out after the end of the working day and such equipment should be turned off unless there are specific instructions to the contrary. A similar problem arises when burning gear is used at the end of the working day and no time allowed for any hot area to cool down before workpeople leave. Again close inspection after the end of the working day is required. Where possible burning work should stop 15–30 minutes before the end of the working day.

For ignition to occur the heat from the source must be transmitted to the flammable substance by either conduction, convection or radiation. For conduction to occur the flammable substance must be in contact with the heat source or the conducting agency. Metals, as a group, are the best heat conductors followed by liquids and gases. Convection takes place through the atmosphere; the heat source raises the temperature of the air in contact with it, the air rises and passes its heat to the flammable substance. Radiation does not require an agency to carry heat to a flammable substance but transmits (radiates) heat in all directions. The heat radiated from a body decreases as the square of the distance from the body.

In order to control the radiation of heat it is important to maintain the maximum separation either by insulation or by distance between the source and the substance and in particular to ensure that there is no direct contact between the two. To minimise the convective risk it is important for the flammable substance not to be located directly over the heat source or for its vertical distance above a heat source to be the maximum possible.

25.5.1.1 Heating systems

Hot pipes and ducts should be lagged where they pass through walls, false ceilings and spaces to minimise the transfer of heat to unsupervised and potentially flammable parts of a building; as well as to conserve energy.

Stoves and portable heating appliances should be securely fixed on an incombustible base and surrounded by a guard rail or some such device to ensure that flammable materials cannot come in direct contact with the appliance.

25.5.1.2 Friction

Heating of a surface will occur when two materials, i.e. parts of a machine, in direct contact are moving relative to each other. The heat generated can be transmitted to a flammable substance by conduction and convection. The main method of dealing with this problem is to prevent the friction in the first place by providing proper lubrication and adjustment of the machine.

25.5.1.3 Hot surfaces

Hot surfaces occur on many machines used in manufacturing processes. Unintentional hot spots can be reduced to a minimum by good standards of maintenance. Where heat is an integral part of the process either insulation should be provided or measures taken to ensure that flammable materials are kept away from the parts at high temperature.

25.5.1.4 Electricity

Electricity, as such, is not a source of ignition, it is the arcing which occurs when a circuit is made or broken or the heat generated by the passage of electricity through a wire that forms the source of ignition. The energy of arcing discharge can be reduced by special measures and this approach is particularly important in the use of electrical equipment in potentially flammable atmospheres. The most common electrical heat source is an incandescent bulb from which heat can be transferred by convection to a flammable substance located above the bulb, by conduction to a flammable substance in contact with it or by radiation to adjacent materials (*Figure 25.2*). Incandescent bulbs should be surrounded by a light transparent guard to prevent direct contact with a flammable substance and flammable substances should not be kept or allowed to gather above them.

Discharge bulbs generate far less heat than incandescent bulbs and these should be used in preference.

Figure 25.2 Tungsten bulb too close to packing case creating fire risk. (Courtesy Fire Protection Association)

25.5.1.5 Static electricity

Static electricity is generated where there is relative motion and/or separation of two dissimilar materials. The charge can be transported or conducted some distance before it becomes great enough to cause a discharge which provides the source of ignition. The best means of control is to prevent the charge build up in the first instance and this can be achieved by providing a path to earth either by having one of the materials securely connected to earth or by creating a humid atmosphere. Other methods include devices using high voltages or ion exchange to effect the discharge. The static charge generated during the pouring of liquid hydrocarbons can be eliminated by providing earth bonding between both containers and not allowing the liquid to free fall.

25.5.1.6 Internal combustion engines

All internal combustion engines have hot surfaces and generate hot exhaust gases. In addition petrol engines require sparks to make them work and most engines have electrical equipment on them. The surface of the engine should be kept as cool as practicable as should the temperature of the exhaust gases. In places where there is a possibility of the presence of flammable gases or vapours then the use of petrol-driven engines should be banned. It should not be forgotten that flammable vapours may be sucked into internal combustion engines along with the air they require and this can cause the engine to run out of control.

25.5.1.7 Tools

Many tools generate hot surfaces in their use, others use heat for their operation and some give off incandescent particles in their operation. In circumstances where a flammable substance may be present, the use of such tools must be controlled. For burning, welding and grinding, the area in which the work has to be done should be clearly defined and surrounded by a non-flammable curtain, all flammable debris should be removed and standby firemen should be at the ready. When soldering irons are used insulated rests should be available and used.

25.5.1.8 Smoking

Fire statistics point to smoking as the most significant cause of fire. The real problems lie in the disposal of the source of ignition (the match) and the smouldering remains (stub end) when smoking ceases. The problem can be eliminated by banning smoking but this may drive it underground. The problem may be controlled by permitting smoking in designated areas where suitable ashtrays and waste bins are provided (*Figure 25.3*).

Figure 25.3 Smoking bay in high fire risk factory

25.5.2 Control of flammable substances

Since all organic and some inorganic substances are flammable it is virtually impossible to eliminate them from the workplace. A flammable substance inside a sealed, or specially designed, container can be considered as safe, but vapours will be released when it is used. The risk can be kept to a minimum if the amount of flammable substance in a workplace is kept as small as is reasonably practicable and certainly no more than is required for the day or shift.

25.5.2.1 Waste, debris and spillage

The fire statistics suggest that these are the most frequent sources of workplace fires. They can be controlled by maintaining high standards of housekeeping. Solid flammable waste should be put in a non-flammable container which should be emptied regularly. Spillages of flammable liquids should be contained using an adsorbent, then removed as soon as possible for safe disposal.

25.5.2.2 Gases and easily vaporisable liquids

All the time that these are contained within the equipment that is designed to hold them, whether it be pipeline, storage tank, reaction vessel, portable container, cylinder, etc., they do not present a fire risk. It is only when they are released to atmosphere either because their use requires it or inadvertently that they become a fire risk. Where the process requires them to be released, such as with burning equipment, the gases should be lit as soon as possible. The possibility of inadvertent release can be reduced by careful design of the equipment and training in strict systems of work. The prevention of accidental releases from leaking joints or pipe failure requires high standards of design, installation and maintenance of the equipment. Flanges, for example, should be enclosed so that any release can be contained and piped away. However since releases cannot be totally eliminated the potential sources of ignition in the area should be reduced to a minimum.

25.5.2.3 Flammable liquids

These present less of a fire risk than flammable gases because they have higher boiling points and hence higher ignition temperatures. Flammable liquids should be kept in properly designed containers in a flameproof store designated for the purpose. Electrical equipment in the store and the immediate area should be of the appropriate flameproof standard. Only the minimum amount of flammable liquid should be kept in the working area and even then it should be kept in a special fire-resistant cupboard. Special non-spill self-sealing containers should be provided for the transport and use of flammable liquids in the working area (*Figure 25.4*). Any spillage should

Figure 25.4 Safety can cut away to show stopper. (Courtesy Walter Page (Safeways) Ltd)

be contained by an absorbent and removed as quickly as possible. If the spillage is large it should be covered by a layer of foam and then removed. Flammable-liquid storage tanks should be surrounded by a bund to contain any spillage that may occur.

25.5.2.4 Flammable compressed gases and liquefied gases

The pressure cylinders which contain compressed flammable or liquefied gases should be stored in an upright position in the open air and protected from direct sunlight. Oxygen cylinders should never be stored next to flammable-gas cylinders. When in use compressed-gas cylinders should be secured in the upright position.

25.5.2.5 Liquid sprays

When a liquid is released through a small hole or slot the resultant spray has lower ignition requirements than the liquid since it has a much greater surface area in contact with the air. Hence it is a much higher fire and explosion risk than the bulk liquid.

25.5.2.6 Dusts

Dust is created, not only when solid material is worked on, but also whenever materials are moved or handled. A dust presents a higher fire risk than the solid because of its greater contact area with air with resultant lower ignition characteristics. A dust cloud has many of the fire characteristics of a gas. Accumulations of dust should be removed as quickly as possible by techniques which do not generate a dust cloud, i.e. damp sweeping or vacuum cleaning. The electrical equipment in areas where flammable solids and powders are handled or occur must be designed and maintained to the appropriate flameproof standard.

25.5.3 Products of combustion

When a substance burns it gives rise to four primary products: smoke, products of combustion, light and heat. Although smoke is one of the products of combustion because of its particular effects it is convenient to consider it separately. Light and heat are both electromagnetic radiations differing only in their wavelength, but their roles in a fire are such that they are best dealt with separately.

In the perfect state the process of burning changes the fuel into carbon dioxide, water vapour and the oxides of other elements in the fuel all or nearly all of which are colourless gases. In practice there is incomplete combustion in which not all the fuel is oxidised resulting in the production of carbon and other partially oxidised materials. Smoke is fine particulate carbon that is visible because it is black; its particles vary in size being far larger than dust particles. The smoke particles also have adsorbed on their surface some of the other products of combustion. In general the amount of smoke a fire produces is determined by the amount of air available for combustion: with less air more smoke is produced and with more air less smoke is produced. However some fuels, such as ethanol, burn without the formation of smoke and so the absence of smoke cannot be taken as an indicator that there is no fire.

The other products of combustion vary widely in their composition as they clearly depend on the fuel that is burning but they will always contain some carbon monoxide and carbon dioxide as well as water vapour. They will also contain some partial oxidation products of the fuel. The hot gas rising from a fire will always be a mixture of the products of combustion, oxygen that has not been consumed in the fire and nitrogen from the air. It is hazardous to health because it is hot, has a lower oxygen content than air and contains toxic substances.

Most fires give off light but often this is obscured by smoke. The light comes from glowing carbon particles and the less yellow the flame the higher its temperature. Some fires give off very little or no light at all and so the absence of light cannot be taken as an indicator that there is no fire.

Some of the heat that a fire produces is carried from the fire by the smoke and the products of combustion as they rise in the air. If these hot gases come in contact with other materials and substances some of that heat will be transferred causing a drying out of the material and a rise in its temperature with an increased risk of the material catching fire.

It is the combination of the hazardous nature of the products of combustion together with their high temperature which is the main cause of death in fires. The impenetrability of the smoke causes a state of panic which affects all but a few people. However because the products of combustion rise they generate a flow into the fire of cold air at lower level; this low layer of clean air can provide the means by which people can escape from a fire.

25.5.4 Fire spread

Four factors are involved in the spread of fire: conduction, convection, radiation and physical transport.

25.5.4.1 Conduction

If a solid body, a metal rod for instance, is heated at one end, the end remote from the source of heat will in time get hot because of the conduction of heat through the rod. In general metals are good conductors of heat; they have a high thermal conductivity. Therefore, if part of a steel beam is heated by a fire in time the whole of the beam will get hot and the heat may be sufficient to ignite flammable substances at a location remote from the original fire.

Depending on the characteristics of the solid and its insulation, there will be a gradual decrease in temperature throughout the solid with the highest temperature at the source of heat and the lowest at the most remote point.

25.5.4.2 Convection

Hot air rises because its density is less than that of the surrounding colder air. The upward movement of hot air is known as convection. As the hot air rises some of the heat is transferred to any object over which it passes as well as to colder air through which it passes. The same is true for liquids although the velocity of the upward current is less than that with air owing to the viscosity of the liquid.

In a fire the hot air is actually smoke and other products of combustion, the upward movement is known as the chimney effect and it causes cold air to feed into the base of the fire. If the smoke and products of combustion are prevented from escaping to the atmosphere, by a ceiling for instance, they spread underneath the ceiling heating it up. If the ceiling is of combustible material it may ignite. In addition, whether it is combustible or not, the ceiling will conduct some of the heat to the floor above which may cause further fires. Any natural chimney, such as a stair well, will act as a conduit for the smoke and products of combustion.

25.5.4.3 Radiation

A hot body radiates energy in the form of electromagnetic radiation, infrared radiation with wavelengths in the range 1 mm to 730 nm, which cannot be seen by the eye. As the temperature of the hot body rises then visible electromagnetic radiation is emitted, in the wavelength range 380 to 780 nm, as well as infrared. Ultimately a very hot body will emit ultraviolet radiation in the wavelength range 300 to 380 nm. When infrared radiation falls onto an object some of it is absorbed and the temperature of the object rises. Radiation is emitted uniformly in all directions and its intensity decreases with the square of the distance between the radiating source and the receiver, the inverse square law.

25.5.4.4 Physical transport

Fires can be spread by the collapse of hot or burning materials. They can be spread by the transport of burning debris in the updraught of the fire, by the flow of burning liquids either down a slope or by extinguisher jets literally pushing the burning liquid forward.

25.5.5 Combustion properties of building materials and construction

Buildings are rarely, if ever, made wholly of one material, so the fire risk on the structure can be complex. Common building materials include wood, steel, concrete, brick, stone, glass, plastics, plaster and finishing materials. Of these only wood, plastics and finishing materials actually burn in the fire. Steel expands due to the heat of the fire and becomes ductile at a temperature of 600°C. The effect of fire on concrete is to cause dehydration and contraction within the temperature range of 400–500°C; at a higher temperature, 800°C, the cement in the concrete begins to decompose leading to total disintegration of the concrete. Brick is essentially unaffected by fire other than surface flaking and so too is stone. Glass expands on heating and begins to melt at about 800°C. Plastic materials vary widely in their combustion characteristics but all will burn. Surface coatings, i.e. paint and wallpaper, are combustible and, because they form a thin coating, aid the rapid spread of fire.

Figure 25.5 illustrates the aftermath of fire in a textile mill.

All wood will burn, but in burning the ash formed often remains on the wood surface where it acts as an insulating coating thus slowing down the

Figure 25.5 Aftermath of fire in textile mill. (Courtesy Huddersfield Examiner)

combustion rate. Wood retains its strength when it is heated, so although it may burn, the structure it supports will continue to stand firm, much longer than a steel structure.

Asbestos-cement sheets break up very noisily in a fire because of the rapid decomposition of the cement. Bituminous-felt coverings burn vigorously in a fire and can cause rapid fire spread.

An important characteristic of fire in a building is that it will spread very rapidly in an upward direction, less rapidly horizontally and hardly at all downwards. Thus any openings in the floors of a building such as stair well, lift and hoist shafts will aid the spread of fire. Hence the need to minimise these floor openings and to enclose stair wells. A fire will remain in a closed room until the heat transfer through the walls or ceiling starts a secondary fire outside or the structural integrity of the room is destroyed. Hence the

Figure 25.6 Automatic fire ventilators on a roof discharging large quantities of heat and fume, so reducing the risk of fire spread. (Courtesy Colt International Ltd)

need to close up a room in which a fire is detected. Occasionally a fire in a closed room may go out because the available oxygen has been used up, but care must be taken before re-entering since the room and its contents will remain at a high temperature and the admission of air from a broken window or from a door being opened may cause re-ignition.

When a fire occurs in a large building, say a single-storey factory, and the products of combustion cannot escape, heat will build up in the roof space. The fire then may spread along the roof. If however the roof is opened by the use of smoke louvres, the products of combustion will escape and there will be little fire spread along the roof space (*Figure 25.6*).

False ceilings and shafts carrying services aid fire spread because they provide an easy passage for the spread of flame and the hot products of combustion both horizontally and vertically.

The basic fire prevention principle in all buildings is to divide up the buildings into compartments with effective fire and heat insulation so that a fire in any one compartment is prevented from spreading to others. This principle also applies to shafts, ducts and false ceilings all of which should have effective fire and smoke barriers built into them at suitable intervals.

25.6 Protection against fire

25.6.1 Structural precautions

The longer a building stands intact when engulfed by fire the higher are the chances that people can escape from it and the less is the consequential damage to its contents and to the business carried out within it.

The three key elements in structural fire resistance are insulation, integrity and stability. Insulation prevents the passage of heat by conduction through an item of structure such as a wall. Integrity refers to the prevention of the passage of flames and hot gases through the element of structure, while stability is concerned with the ability to resist collapse. As a general rule the minimum fire resistance that is required of a structure or element of structure is thirty minutes but in certain cases it could be several hours.

When the fire resistance of buildings is being considered attention must be given to a range of features that are effective in controlling fire spread.

26.6.1.1 Structural elements

Unless the structural members of the building have the necessary fire resistance they should be protected by means that provide the required fire resistance. This can be achieved by enclosing the structural member in concrete or in fire-resistant panels or insulation. The former method is to be preferred when there is the possibility of damage to the enclosing material.

25.6.1.2 Walls, doors and openings

The walls, glazing and door and other openings should be designed to contain a fire. Brick, concrete and other non-combustible wall materials will do this effectively. Normal glazing will collapse rapidly in fire though special glass can be used when the window frames need to have matching fire resistance. Doors serve two fire precaution purposes: they stop the spread of fire and they stop the spread of smoke. Wooden doors can be designed to provide fire resistance whereas normal metal doors will not. Careful thought must be given to the type and construction of doors where fire resistance is important. To prevent the spread of smoke the doors must be close fitting and ideally there should be no gap between the leaves or around the edges. Fire doors should either be self-closing or close automatically when fire occurs. Other openings in floors and walls such as for conveyors should be provided with a suitable fire-resistant door that closes automatically in the event of fire.

25.6.1.3 Roofs and floors

Roofs and floors can be affected by fire from below when heat and flames rise, and from a fire above by flame impingement. The outer surfaces of roofs and walls should wherever possible be non-flammable. A chimney effect can be created by the eaves of a pitched roof and hence draw the fire into the roof space.

25.6.1.4 Stairways and liftshafts

Stairways and liftshafts both extend upwards and communicate with all levels in a building and hence provide an effective path to allow the spread of fire throughout a building. All doors that communicate with the stairway, the column containing the liftshaft and the lift should have adequate fire resistance. The doors on lifts should close automatically in the event of fire and the lift should descend to the ground floor. All other doors should have thirty minutes fire resistance and should close automatically in the event of fire.

25.6.1.5 Cavities and voids

Cavities and voids are present in virtually all buildings and occur between the wall and the cladding, between the roof space of one building and the next and between a suspended ceiling and the floor slab above. These cavities often contain the services for the building. Also they are often interconnected so that once a fire has broken into the cavity or void it can spread throughout and may break back into a room or work area remote from the original fire. Cavities and voids should be separated by a fire-resistant partition to resist the spread of smoke and heat. In particular where a cavity occurs over a fire-resistant wall, the wall or an equivalent fire-resistant partition should be extended to the ceiling or roof of the cavity.

25.6.1.6 Cables

Buildings contain electricity cables for power, heating and lighting which commonly have PVC insulation. When exposed to fire this insulation will burn and the fire will spread along the surface of the cable. As the insulation burns smoke and acrid products of decomposition will be evolved, and the electrical insulation will decrease with the risk, if the cable is still live, of shorting. Cables which serve a key service such as emergency lighting or power cables to fire water pumps should be provided with special insulation or be of the mineral insulated type. It is possible for the cable itself to be the source of fire through overheating because it is overloaded. If cables are collected together and run in a service way or duct, heat and/or smoke detectors and automatic fire fighting devices can be fitted into the duct.

25.6.1.7 Compartmentisation

While open-plan buildings and stores have benefits from a business operating point of view, they encourage spread of flame in the event of fire. Wherever possible large spaces in buildings should be divided into smaller discrete compartments by fire-resisting walls which should be carried up to the ceiling or roof. Any breaks in these walls, for doors, conveyor ways, etc., should be provided with automatic fire doors.

25.6.1.8 Smoke control

Smoke is not only black and hazardous to health, it is also hot and where it accumulates its heat can be passed to the surrounding building, extending the fire. By placing ventilators in the roof of large open-span buildings which open when heat or smoke is detected, the build-up of smoke and heat can be prevented. The siting and size of the ventilators must be determined by the layout and size of the building. An additional benefit of these ventilators is that they create a chimney effect over the fire which draws the flames and smoke upwards reducing the lateral spread of the fire (*Figure 25.6*).

25.6.1.9 Plant

Buildings and premises may contain plant which needs special fire protection because of its nature. At one extreme these include storage tanks containing liquefied flammable gases which may require water drenching for cooling while at the other extreme cupboards containing vital records which should be metal cupboards with extensive heat insulation on the inner surface.

25.6.2 Fire detectors and alarms

The purpose of a fire detector is to detect the outbreak of fire before the means of escape is threatened. All workplaces should be provided with appropriate and adequate means for detecting fire which activate an alarm automatically.

There are a number of ways by which fire can be detected. The three principal fire detection methods involve sensing the heat, either as actual temperature or as rate-of-rise of temperature, the presence of flame or the presence of smoke. There is one other detection method which identifies a pre-fire condition and that is a flammable gas detector whose use is limited to places where flammable gases are likely to be present.

25.6.2.1 Heat detectors

There are three basic operating principles for heat detection: melting of a metal (fusion), expansion of a solid, liquid or gas, and electrical sensing.

25.6.2.1.1 Fusion detectors

The simplest application consists of an electrical circuit which contains a switch that is held in either the open or closed condition by a piece of low melting point alloy. When this melts in the heat of a fire, the switch is

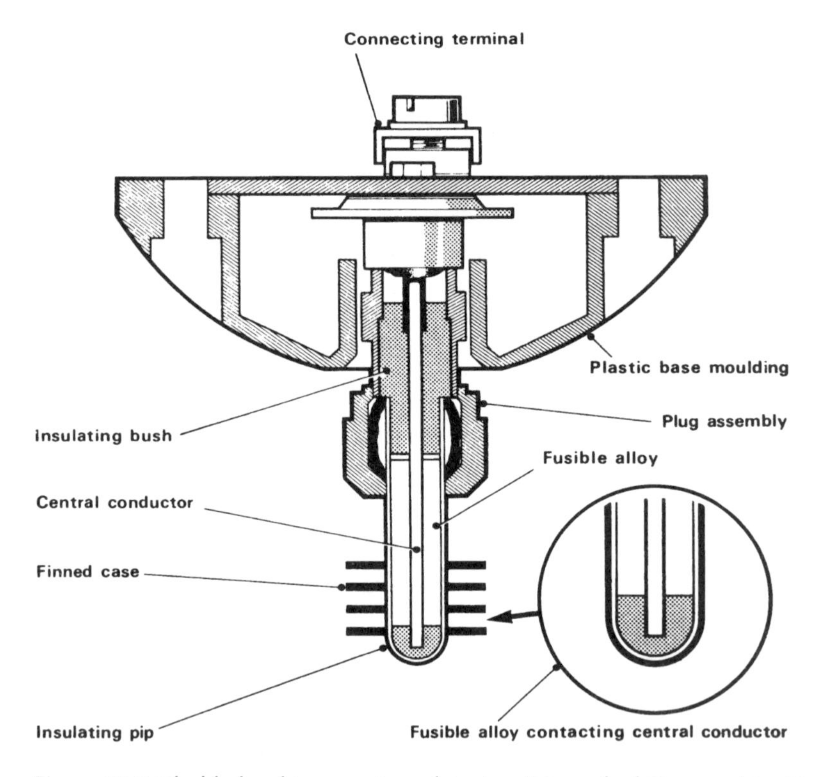

Figure 25.7 Chubb fixed temperature detector. (Manual of Firemanship. Courtesy Controller HMSO)

released and the circuit condition changes, i.e. it is either made or broken (*Figure 25.7*). A variation of this type of detector functions by the melting alloy running into a small cup to complete the alarm circuit.

25.6.2.1.2 Thermal expansion type

This type of heat detector utilises the fact that metals expand when heated to open an electrical contact and hence initiate the alarm. Bimetallic strip is used to enhance the amount of expansion obtained (*Figure 25.8*).

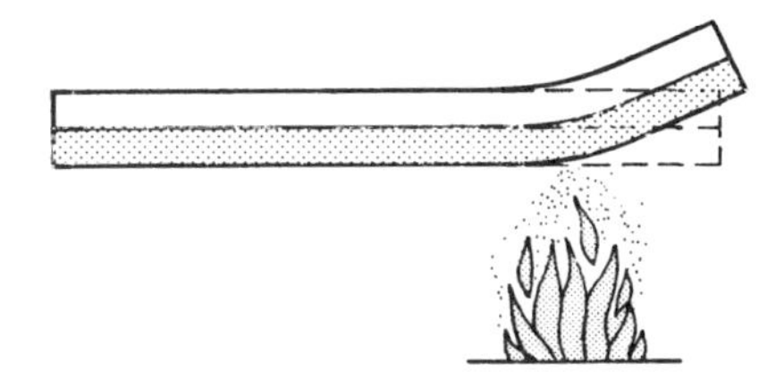

(a) Illustration of the effect of heat on a bimetal strip

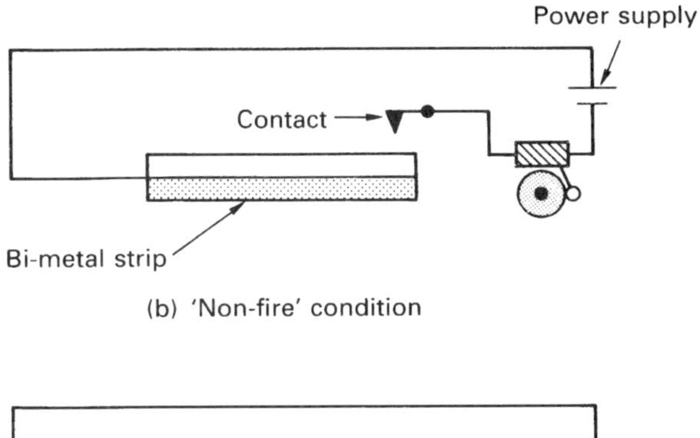

(b) 'Non-fire' condition

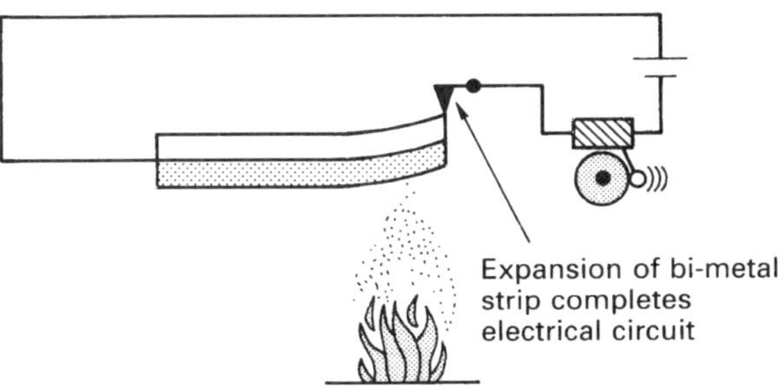

(c) 'Fire' condition

Figure 25.8 Illustration of a bimetal strip as a heat detector. (Manual of Firemanship. Courtesy Controller HMSO)

The rate-of-rise of temperature detector employs two bimetallic strips, one exposed and one partially insulated. For slow rates of rise, such as increases in ambient temperature, both strips warm up and expand at the same rate. However, if the rate of rise is high, as in the event of a fire, the exposed strip expands faster than the insulated one and trips the alarm circuit (*Figure 25.9*).

The coefficient of expansion of both gases and liquids is greater than that of solids, so potentially they are more sensitive as heat detectors. One such application is the quartzoid bulb in sprinkler systems which consists of a liquid-filled bulb blocking the outlet of the sprinkler head. The liquid expands more than the glass so when heated it will shatter on reaching a predetermined temperature. Once the bulb has broken, the sprinkler head will discharge.

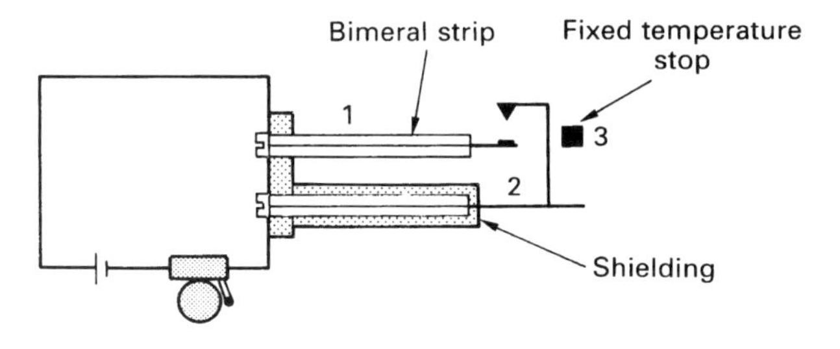

(a) 'non-fire' condition

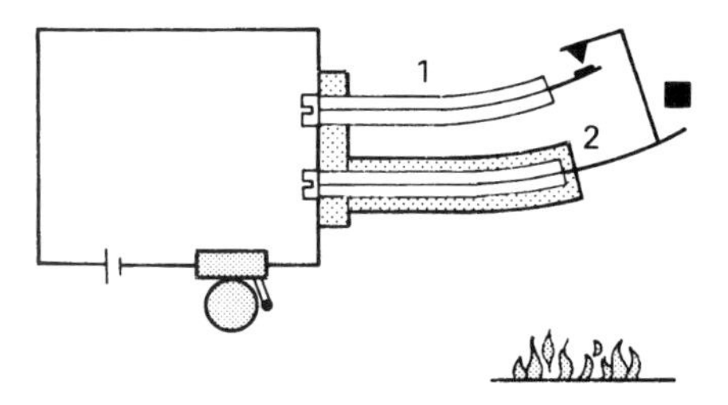

(b) slow rise in temperature

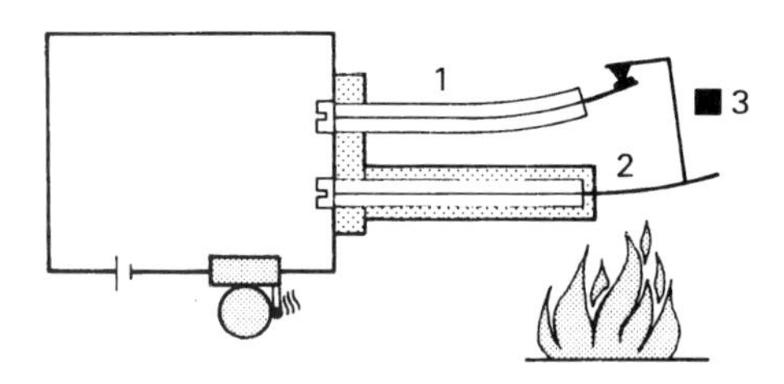

(c) rapid rise in temperature

Figure 25.9 Principle of the 'rate-of-temperature rise' detectors. (Manual of Firemanship. Courtesy Controller HMSO)

25.6.2.2 Combined heat and smoke detectors

In addition to light, fires emit infrared and ultraviolet radiations which are less affected by smoke than normal white light. By employing a photo-electric cell with infrared filters or an ultraviolet sensor, the presence of flames can be detected in spite of the smoke generated. The signal from the photoelectric cell or u.v. sensor can be arranged to trigger the alarm. These devices suffer from the disadvantage that they will react to infrared or ultraviolet radiations from other sources and hence are prone to spurious alarms. This disadvantage is overcome in a combined smoke and heat detecting system in which a modulated infrared beam is transmitted across

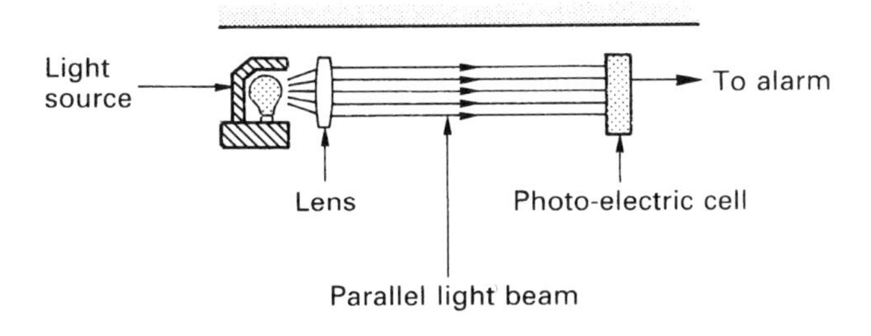

(a) 'Non-fire' condition

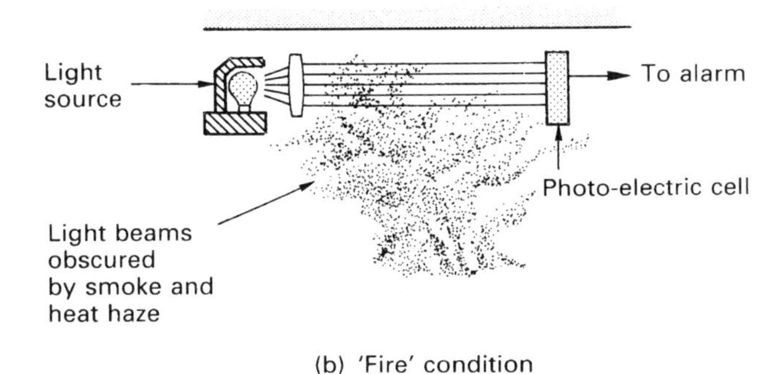

(b) 'Fire' condition

Figure 25.10 Combined heat and smoke detector. (Manual of Firemanship. Courtesy Controller HMSO)

the area to be protected. Smoke and heat haze interfere with the beam and this is detected in the receiving unit which initiates the alarm (*Figure 25.10*).

25.6.2.3 Smoke detectors

There are three types of smoke detectors utilising ionising radiations, light scatter and obscuration.

An ionisation detector works on the principle that ions are absorbed by smoke particles, thus when smoke enters a chamber which contains an ionised gas some of the ions are absorbed by the smoke particles and the current flow in the chamber is reduced. In practice an ionising source is used to maintain a predetermined level of ionisation in an open chamber. Ionised particles are attracted to the charged plates resulting in a flow of electricity. When smoke enters the chamber it attaches to the charged particles so fewer reach the plates and less electricity flows. This reduced flow of electricity is detected and the alarm triggered. The diagrams in *Figure 25.11 (a)* and *(b)* show this principle. A more sophisticated device employs two chambers, one sealed and one open, and compares the current flowing from them. When smoke enters the open chamber less current flows in that part of the circuit and when the difference in current flow reaches a predetermined level the alarm is activated.

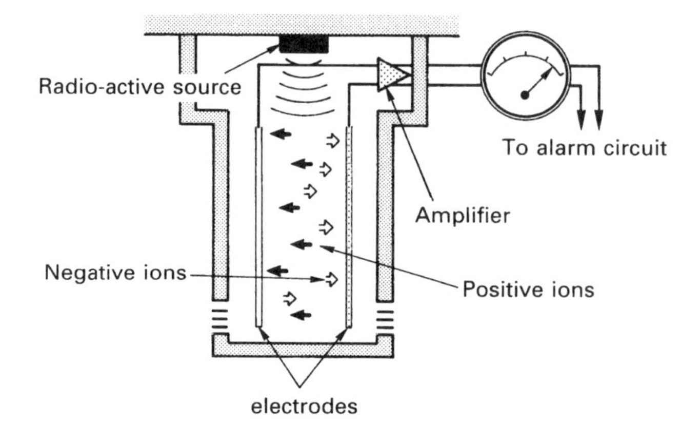

(i) 'non-fire' condition

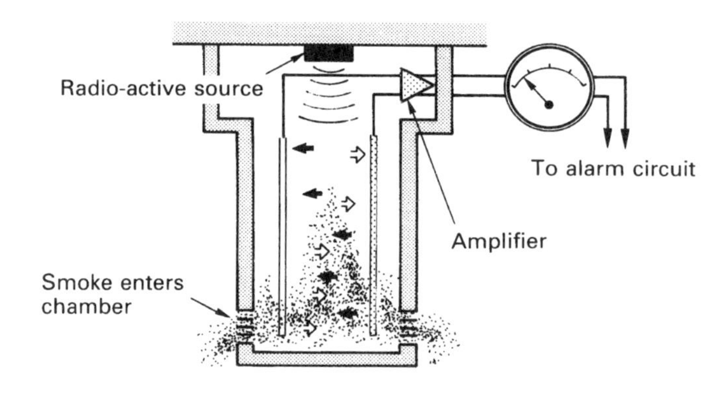

(ii) 'fire' condition

Figure 25.11(a) Diagram of smoke detector – ionisation type. (Manual of Firemanship. Courtesy Controller HMSO)

A light-scattering smoke detector works on the theory that smoke will scatter a beam of light. A photoelectric cell is fitted in a chamber at right angles to a light source and in the fire-free condition receives no light. However when smoke enters the chamber the light is scattered and some is detected by the cell which triggers the alarm. A typical photoelectric smoke detector is shown in *Figure 25.11(b)*. In an obscuration detector the light beam is directed at the photoelectric cell and when smoke enters less light is received by the cell and the alarm is raised. This principle is shown in *Figure 25.10*.

25.6.2.4 Flammable gas detector

A flammable gas detector is designed to measure the amount of flammable gas in the atmosphere and relate it to the upper and lower flammable limits.

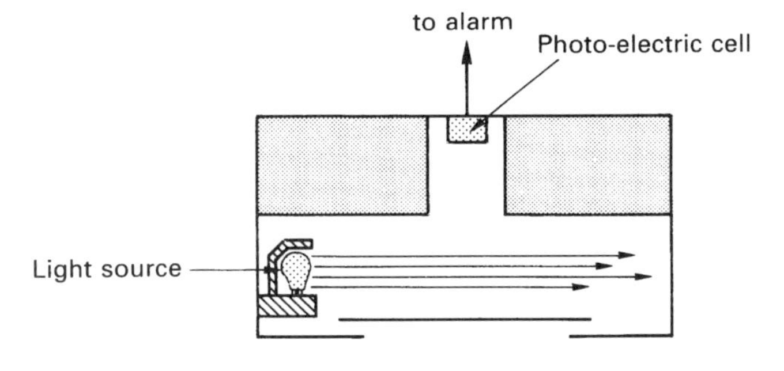

(i) 'non-fire' condition

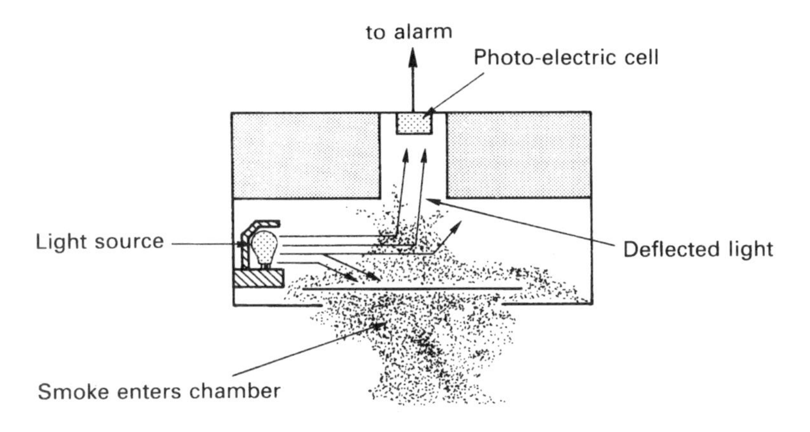

(ii) 'fire' condition

Figure 25.11(b) Diagram of smoke detector – optical light scatter type. (Manual of Firemanship. Courtesy Controller HMSO)

The gas mixture is drawn over a catalytic surface where oxidation, i.e. combustion, takes place. The combustion causes a rise in temperature of the surface which is measured by a decrease in its electrical resistance. The instruments have to be calibrated for the particular gas of interest but for petrol vapours pentane or heptane are used as the reference gas. The readings are usually displayed in terms of percentage of the lower explosive limit.

25.6.2.5 Detector installation

Three factors have to be taken into account when siting detectors, the distance between any point where a fire might occur and the detector, the area covered by the detector and the distance between detectors or detectors and walls.

Heat detectors should be sited so that the heat-sensitive element is between 25 and 150 mm below the ceiling or roof. Smoke detectors should normally be fixed in the highest point of, and between 25 and 600 mm below, any ceiling or roof. The distance between detector heads should not exceed 12 m in general and 18 m in corridors and they should where possible be fitted 6 m away from walls.

Detectors are intended to give an early warning of a fire so that the occupants of the premises can escape, therefore they should be fitted in all places where people work or congregate and along all escape routes. They should also be fitted in voids and cavities.

The wiring of a fire alarm system is as important as the detectors themselves. It should be protected from mechanical and fire damage and have an electrical supply which is either independent of the normal supply to the premises or is backed up by an independent supply. The control and indicator board apart from showing which detector has operated, should also show power supply faults, wiring faults and any other faults which may affect the security of the system.

25.6.2.6 Fire alarms

Any sort of warning device may be used as a fire alarm provided that it makes a distinctive sound which is audible in all parts of, and is understood by all those in, the premises.

Alarms may be initiated manually or automatically. Manually initiated alarms may be manually operated, such as bells and gongs, which are only suitable for use in small low-risk premises. Electrically operated alarms are to be preferred and should be considered as standard wherever possible. Alarm points should be fitted 1.4 m above floor level and at intervals such that no person need travel more than 30 m to reach one. Each alarm point should have the standard notice (see section 25.10.5) fixed nearby. Normally the alarm will be an audible bell, siren or hooter but it could be a distinctive sound or message over a public address system or, in some places frequented by the public, it could be an alarm which is recognised only by suitably trained staff who then pass a message to the public.

Automatic fire alarms of various types can be fitted which operate when a fire is detected. These have the advantage of raising the alarm in the event of fire in unmanned or unoccupied locations. Their setting can be critical to prevent spurious operation.

A display panel should be installed for both electrical and automatic fire alarms which shows the location of the alarm that has been initiated. Weekly tests should be made of the fire alarm so that people become familiar with the sound and so that any faults in the system can be detected and corrected. The fire alarm installation must be inspected and maintained on a regular and routine basis and a record made of all tests and inspections. This record should also include details of any faults, repairs and replacements. Inspections of the system should be carried out quarterly by a competent person. In high fire risk areas, serious consideration should be given to installing an automatic link from the display panel to the local fire brigade.

25.7 Extinction

Once a fire has started the control of its development and its ultimate extinction can be achieved by:

- reducing the amount of fuel available for combustion (fuel starvation),
- reducing the amount of oxygen available for combustion (smothering),
- removing the heat of the fire (cooling),
- direct interference with the oxidation reaction.

Often more than one of these features is employed in fighting a fire at any given time.

25.7.1 Fuel starvation

The simplest and most common example is turning off the tap on a gas cooker. If the fire involves burning gas or flammable liquid which is issuing from a pipe, isolating the supply will cause the fire to go out. Should the fire be in a flammable liquid storage tank or bund, removal of the liquid from below the fire will reduce the quantity of fuel available for, and hence the time of, burning. If the fire involves burning solids then removal of unburnt material will reduce the burning time or creating fire breaks in front of the advancing flame will prevent further development of the fire.

25.7.2 Smothering

Where a pool of flammable liquid is on fire, the oxygen supply can be reduced by covering the surface of the liquid with a layer of foam. On the other hand, a fire in a confined space can be extinguished by filling the space with an inert gas and so reducing the oxygen available for combustion. The same effect can be achieved by closing the doors and windows of a room which contains a fire and allowing the fire to consume the oxygen and hence self extinguish. Wrapping a fire blanket around a person whose clothes are on fire is another example of oxygen reduction. In the case where a leak of flammable gas is on fire, it is possible to inject an inert gas into the flammable gas upstream of the leak and so reduce the oxygen available in the flame zone. This effect can also be achieved by directing an inert gas into the flame. However, it is important to remember when extinguishing by oxygen removal: if the supply of oxygen is restored or the supply of inert gas fails then the fire may re-ignite if the heat source is still present. The technique of oxygen removal is of no use if the supply of oxygen comes from the substance that is burning, e.g. ammonium nitrate or a fire involving an oxidising agent.

25.7.3 Cooling

This is the commonest way of extinguishing a fire. A substance is put into the fire which readily absorbs heat thus preventing the fire itself from heating the

Figure 25.12 Use of large quantities of water to quench fire. (Courtesy Liverpool Daily Post and Echo PLC)

unburnt fuel up to ignition temperature. Water is the most common heat removal agent because it is cheap, readily available and has a high latent heat of vaporisation. Any cold substance can perform this task as can flame traps which dissipate the heat of the flame over a large area. With large fires, heavy jets are needed to provide the necessary quantity of water (*Figure 25.12*).

25.7.4 Interference with the reaction

Fire is propagated by means of a free radical chain reaction and so anything which will interrupt this chain will cause the fire to die. The 'halons' family of chemicals, which are chloro-, bromo-, fluoro-hydrocarbons, are very efficient at this. Spreading the heat over large surface areas, as in flame traps or with the powder of a dry powder extinguisher, also effectively removes the free radicals and hence results in a suppression of the fire.

25.8 Fire fighting

Fire fighting is the process of extinguishing a fire once it has started. All workplaces must be provided with adequate and appropriate means for fighting fires, the location of which must be suitably indicated and be so placed that it is readily available at all times. The equipment used for this purpose falls into two types: portable and fixed.

25.8.1 Portable fire extinguishers

A portable fire extinguisher is an appliance designed to be carried by hand, containing an extinguishing medium which can be expelled by the action of internal pressure and directed on to a fire. The maximum weight of a portable extinguisher in working order should be no more than 23 kg. They are intended for fighting small fires only.

Portable extinguishers should be manufactured to BS 5423: 1987 which also specifies the tests necessary to determine the rating of the extinguisher for both A and B class of fires. The size and number of extinguishers provided for any application are determined by an assessment of the size of fire that may need to be tackled. A requirement of this standard is the inclusion of controllable discharge devices. However many of the older single discharge type are still in service and these will remain acceptable until the end of their effective life.

There are a number of common features in the construction of all five types of portable extinguishers which are described in the British Standard, namely:

- Extinguishers should use gas or air pressure to eject the extinguishant and can be of either the stored pressure type or contain a high pressure gas cartridge.
- To operate them a sealing device must be pierced or broken.
- They must work without the need to turn them upside down.
- They must incorporate a device to show when they are discharged.
- They must incorporate a safety device to prevent inadvertent operation.
- They must have a means for controlling the discharge.

Portable fire extinguishers are divided into five groups according to the extinguishants they contain and, in the UK, these are colour coded.

Extinguishant	*Colour code*
Water	Red
Foam	Cream
Carbon dioxide	Black
Dry chemical powder	Blue
Vaporising liquid	Green

There are limitations to the classes of fires on which the different types of extinguisher can be used as indicated in *Table 25.3*.

25.8.1.1 Water-containing extinguishers

In these, water is the extinguishant and consequently they must have internal corrosion protection and the water will need anti-freeze in cold weather. The extinguishers must be operated in the upright position otherwise the gas or air will be released without expelling the water. The extinguisher will not work once the cylinder becomes depressurised.

Table 25.3. Classification of fires for portable extinguishers

Class of fire	*Description*	*Appropriate extinguisher*
A	Solid materials, usually organic, with glowing embers	Water, foam, dry powder, vaporising liquid and CO_2
B	Liquids and liquefiable solids:	
	(i) miscible with water, e.g. methanol, acetone	Water, foam (but must be stable on miscible solvents), CO_2, dry powder, vaporising liquid
	(ii) immiscible with water, e.g. petrol, benzene, fats and waxes	Foam, dry powder, vaporising liquid, CO_2
C	Gas or liquefied gas	–
D	Metals	–
–	Electrical equipment	Dry powder, vaporising liquids, CO_2 (not computers)

Note: Particular dry powders may be required for use on different classes of fires.

(a) *Soda-acid type*

These are the original form of water-containing extinguishers in which the cylinder was filled with an aqueous solution of sodium bicarbonate with, attached to the top of the cylinder, a small phial containing sulphuric acid. When the phial was broken, usually by means of a plunger, the acid mixed

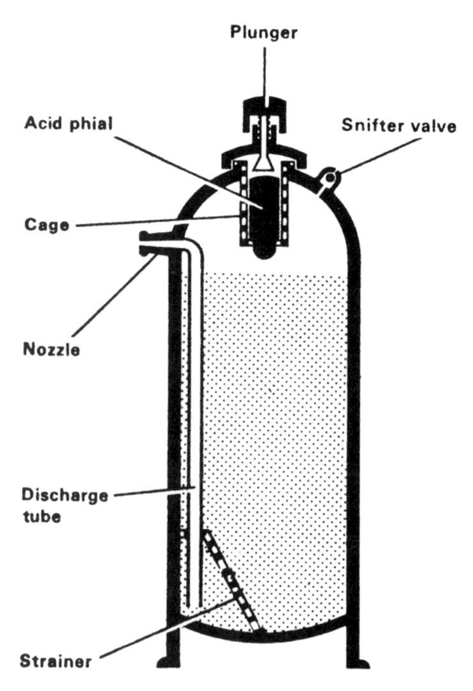

Figure 25.13 Section through a soda-acid extinguisher with plunger at the top. (Manual of Firemanship. Courtesy Controller HMSO)

with the bicarbonate solution generating carbon dioxide whose pressure expelled the solution (*Figure 25.13*). This method of generating gas pressure is obsolete and is being superseded by stored pressure or gas cartridge types.

(b) *Gas-cartridge type*
The cylinder is filled with water and a sealed container of liquid carbon dioxide is attached to the head cap assembly. When the seal of the carbon

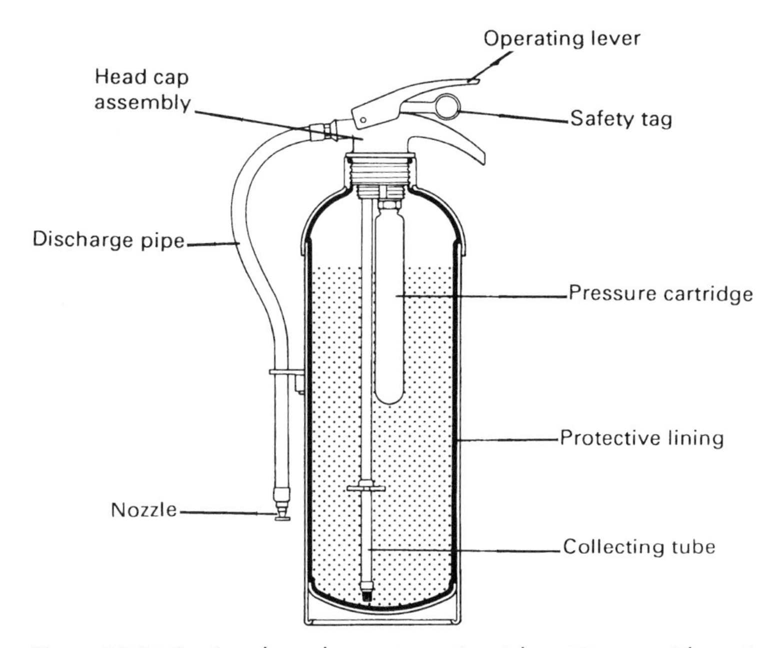

Figure 25.14 Section through a water extinguisher. (Courtesy Thorn Security Ltd)

dioxide container is pierced, the escaping gas pressurises the cylinder and the water is expelled. The discharge is controlled by the discharge control lever (*Figure 25.14*).

(c) *Stored-pressure type*
The cylinder, when filled with water, is sealed by the head cap assembly and then pressurised to between 10 and 15 bar with air. The extinguisher is actuated by means of the operating lever in the head cap assembly which punctures the seal and allows the water to be expelled. Again, the discharge can be controlled by means of a valve connected to the operating lever (*Figure 25.15*).

25.8.1.2 Foam extinguishers

Foam forming chemicals are corrosive so these extinguishers must have internal corrosion protection.

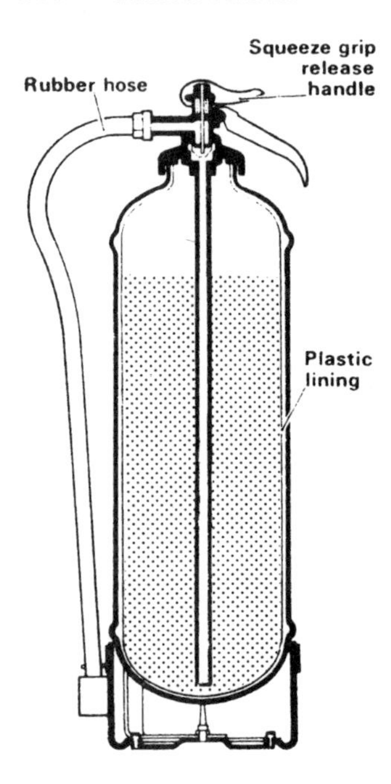

Figure 25.15 Section through a stored pressure type of water extinguisher. (Manual of Firemanship. Courtesy Controller HMSO)

(a) *Chemical foam*
In this type of extinguisher, which is now obsolete, the cylinder is filled with an aqueous solution of sodium bicarbonate which contains additional chemicals to enhance its foam forming characteristics. An inner cylinder containing an aqueous solution of aluminium sulphate is inserted into the cylinder. To operate, the extinguisher is inverted thus mixing the two chemicals which react to form the foam and also generates carbon dioxide gas to act as the expellant. Once operated the whole charge must be used.

(b) *Mechanical foam*
This type is so called because the foam is created by the mechanical effect as the foam solution is forced through a special nozzle on discharge. The foam solution is contained in the cylinder which is either sealed and pressurised to 10 bar or contains a cartridge of high-pressure carbon dioxide. When the extinguisher is operated, a seal is pierced and the liquid is forced out through a special self-aspirating nozzle at the end of the delivery hose. The flow of foam can be controlled by the operating lever in the head cap assembly (*Figure 25.16*). Various types of foam solution can be used depending on the fire risk: protein, fluor-protein, fluor-chemical and synthetic. Normally foam extinguishers must not be used on live electrical equipment, but one of the synthetic group, aqueous film forming foam (AFFF) has been certified for use at voltages up to 35 000 volts.

25.8.1.3 Dry powder extinguishers

The extinguishant in these is a very finely ground powder of sodium bicarbonate containing other chemicals to improve its fire fighting proper-

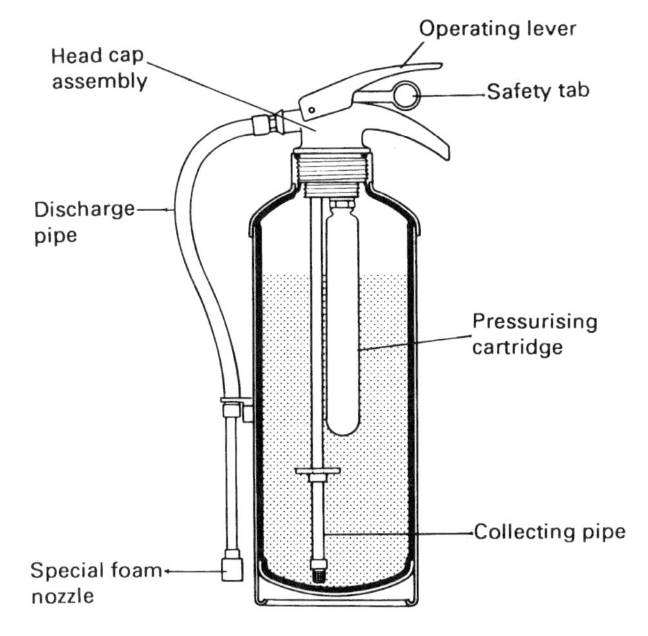

Figure 25.16 Section through a foam extinguisher. (Courtesy Thorn Security Ltd)

ties and to prevent it caking in the cylinder. These extinguishers suffer from the disadvantage that they are susceptible to dampness which can cause the powder to coagulate and either block or prevent the extinguisher from working properly. After discharge, the extinguisher needs to be thoroughly cleaned before being recharged.

(a) *Stored pressure type*
When the cylinder has been filled with the powder it is pressurised to 10 bar with dry nitrogen or air. Actuation is by means of the operating lever in the head cap assembly which also allows control of the discharge rate.

(b) *Gas cartridge type*
Attached to the head cap assembly is a sealed cartridge of liquid carbon dioxide which is pierced when the operating lever is depressed, so pressurising the cylinder. Control of discharge rate is by the operating lever (*Figure 25.17*).

25.8.1.4 Vaporising liquid extinguishers

Various types of vaporising liquids are available but all are halogenated hydrocarbons ('halons') and are toxic to some degree. In addition, the daughter products after contact with flames are more toxic and corrosive. If they are used in a confined space, it should be well ventilated before entry. Vaporising liquid extinguishers work by interfering with the combustion reaction but do not remove the heat, consequently if the supply of extinguishant runs out the fire can re-establish itself. In these extinguishers,

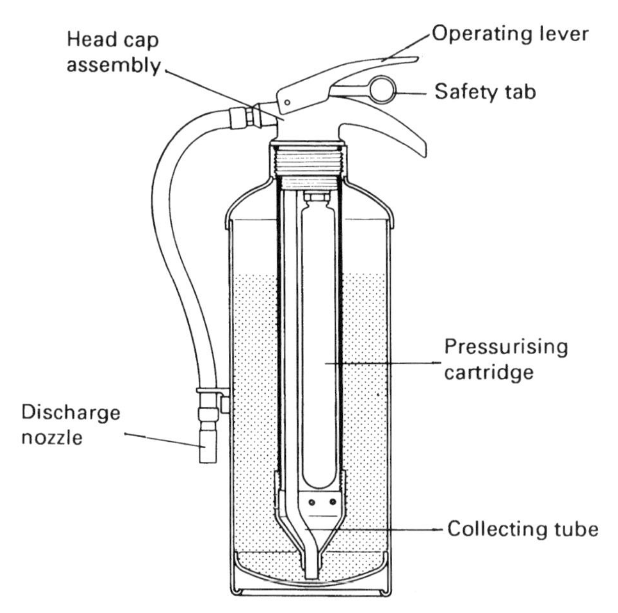

Figure 25.17 Section through a dry powder extinguisher. (Courtesy Thorn Security Ltd)

the halon liquid is sealed into the cylinder by the head cap assembly and the cylinder is then pressurised either to 10 bar with dry carbon dioxide or nitrogen or by piercing a gas cartridge when used. Actuation of the operating lever pierces the seal allowing the liquid to be expelled. The discharge rate can be controlled by the operating lever. Because of its adverse environmental effects production of the halon family of chemicals is being reduced.

25.8.1.5 Carbon dioxide extinguishers

As with the vaporising liquid extinguishers, this type works by smothering the fire but does not remove the heat so, if the flow of gas ceases, the fire can re-ignite. As carbon dioxide gas is heavier than air, it will collect in hollows and pits. It is also an asphyxiant so areas where it is used must be well ventilated before entry. The discharge of a carbon dioxide extinguisher is very noisy and the discharge horn can become cold enough to cause 'cold burns'.

The extinguisher itself consists of a pressure cylinder which contains liquid carbon dioxide at pressure. When the operating lever is depressed, the liquid is forced out in the form of carbon dioxide snow which turns into a gas in the heat of the fire (*Figure 25.18*).

25.8.1.6 Provision and maintenance of portable fire extinguishers

(a) *Provision*

Portable fire extinguishers are of little use unless the correct types are placed in sufficient quantities in the right places and are readily available at all times. They must be kept properly maintained and there should be sufficient

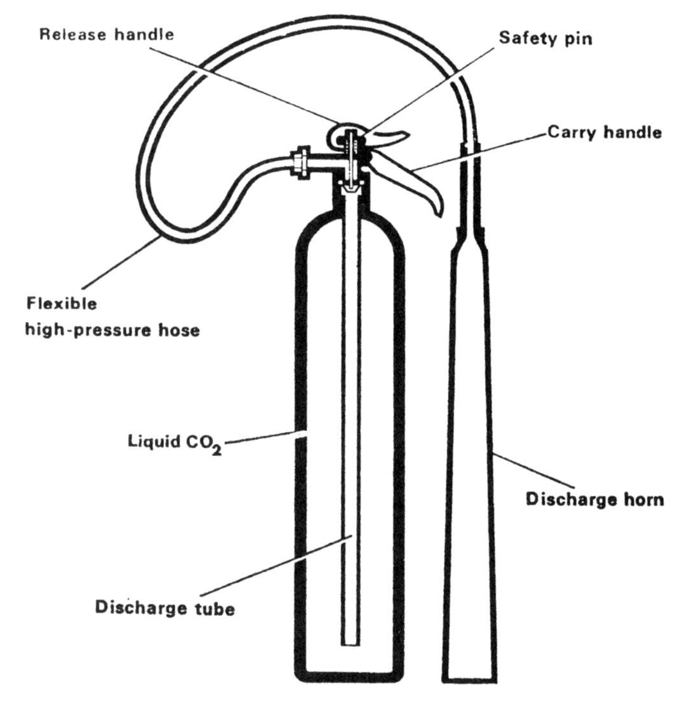

Figure 25.18 A carbon dioxide extinguisher showing the piercing mechanism, control valve and discharge horn. (Manual of Firemanship. Courtesy Controller HMSO)

numbers of persons trained to use them. They are suitable only for 'first aid' fire fighting and dealing with small fires.

(b) *Location*

Extinguishers should be located in conspicuous positions and always be accessible for immediate use. These positions should be clearly and distinctly marked with at least the standard sign. They should be on exit routes such as by doors, in corridors, on landings and in the same relative positions on different floors. If not floor standing, extinguishers should be either mounted on a wall with a bracket or standing on a shelf. The bracket should be fixed at a height no greater than 1 m and the shelf should be 0.75 m above the floor. Portable extinguishers can be grouped into fire points which should be situated no more than 30 m apart (*Figure 25.19*).

The number and type of fire extinguishers required is dependent on the level of fire risk and the size of the premises. Fire fighting capacity is indicated by either an A rating for carbonaceous material fires or a B rating for liquid fuel fires. The actual rating is determined by tests specified in BS 5423:1987.

(c) *Maintenance*

The cylinders of all fire extinguishers are pressure vessels and so require regular inspection and testing. This work should be carried out by someone

Figure 25.19 Clear unobstructed fire point. (Courtesy Reed Medway Sacks Ltd)

who has been trained and is suitably competent. It is usual for the supplier of the extinguishers to carry out this work. A monthly inspection should be carried out on all extinguishers, spare cartridges and replacement charges by a responsible person. The spare cartridges and replacement charges should be kept readily available so that the extinguishers can be recharged and returned to service as quickly as possible after use. On an annual basis a full maintenance should be carried out by a competent and trained person in accordance with the maker's instructions and the date of the check recorded on the extinguisher.

25.8.2 Fixed fire-fighting equipment

Fixed fire-fighting equipment can be split into two main categories:

(a) manually operated for general protection, including hose reels for use by the occupants of the premises, and hydrants and foam installations for use of the fire brigade;.
(b) automatically operated systems that provide either general protection such as sprinklers, or special risk protection for particular types of plant and equipment, i.e. main frame computers, high fire risk production processes, etc.

25.8.2.1 Manually operated fixed installations

(a) *Hose reel*

A hose reel is a coil of 25 mm ID flexible hose held on a suitable metal former and not more than 35 m long. It is connected directly to a fire main either by way of a manual shut-off valve or an automatic valve which is opened when the hose is pulled out. The end of the hose is fitted with a nozzle to control the water flow and spray pattern. Hose reels should be positioned near room exits or near staircase landings and if they are located in cupboards these should be marked with the standard sign (*Figure 25.20*).

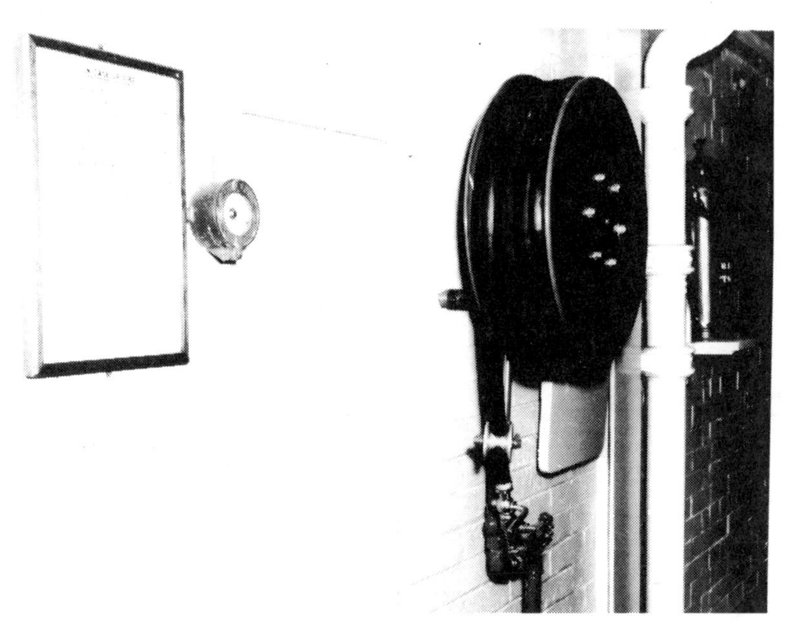

Figure 25.20 Manual fire alarm point, with hose reel. (Courtesy London Fire Brigade)

Sufficient hose reels should be provided so that no part of the building is more than 6 m from a nozzle when the hose is fully extended. The water supply to a hose reel should be sufficient to ensure a flow rate of 24 l per min at the nozzle and a minimum jet range of 6 m. Where there are two or more reels from the same supply there should be enough water to ensure 24 l per min when two reels are tested simultaneously.

(b) *Hydrant systems*

There are two types, dry rising mains and wet rising mains. The former consist of 100 mm ID vertical steel pipe rising through the building fitted with landing valves at each floor and having an inlet valve at street level onto which the fire brigade can connect their pumps. Wet rising mains are similar except they are charged with water at mains pressure either by direct connection into the mains or by pump from storage tanks. At each landing the valves should have standard hose fittings to enable the fire brigade to attach their hoses. Both dry and wet risers must only be used by the fire brigade or a fully trained works brigade. Dry risers are suitable for buildings up to 60 m high, wet risers should be used in taller buildings, and in both cases they should normally be situated inside a ventilated lobby or a staircase enclosure.

25.8.2.2 Automatically operated installations

25.8.2.2.1 Automatic sprinkler systems

Sprinkler systems are designed to provide an automatic means for detecting and extinguishing or controlling a fire by water in its earlier stages. They are suitable for the protection of most industrial and commercial buildings and although they are primarily designed to protect property they have a role in the protection of life.

The system comprises a series of water mains, usually overhead, to which sprinkler heads are fitted at intervals. As heat from the fire reaches them individual sprinkler heads operate at predetermined temperatures releasing a deluge of water over the fire below. The advantages of sprinkler systems are, that as only the heads in the vicinity of the fire operate, water damage is reduced to a minimum, they are automatic in operation, they are not impaired by smoke or fumes and are not hindered by the blockage of access routes.

For the purposes of sprinkler systems the Fire Offices Committee (FOC) have classified the fire hazards of building according to *Table 25.4*.

Sprinkler systems have been graded by the FOC according to the number and type of water supply:

Grade I — Duplicate water supply or one 'superior' water supply provided the total number of heads served does not exceed 2000 with not more than 200 on each separate risk.
Grade II — One 'superior' water supply and no limit on heads supplied.
Grade III — Single water supply from town mains or automatic pump supply.

Superior water supplies comprise town mains, automatic pump supply, elevated private storage, gravity tank, pressure tank.

Four types of sprinkler systems are available: wet-pipe, dry-pipe, alternate and pre-action. A wet-pipe system can only be used in premises not subject to frost or ambient temperatures above 70°C. A dry-pipe system is only suitable in premises subject to continuous temperatures below 0°C or above 70°C. An alternate system which enables either wet or dry operation to be selected is suitable in premises subject to freezing during cold weather. Finally a pre-action system, which is a system with a separate heat or smoke detector that allows water into the sprinkler pipework before the first head operates, should only be used where none of the others is suitable.

There are six parts to a sprinkler system: sprinkler heads, control valves, gauges and meters, alarms, pipework and water supplies.

The sprinkler head is an automatic water valve fitted with a deflector plate to give a specific spray pattern. The valve is sealed by a heat-sensitive element which is designed to burst at a predetermined temperature.

Table 25.4. Classification of fire hazards

Class	*Brief description*	*Sub-division*		*Some examples*
Extra light	Non-industrial occupancies where the combustibility of the contents is low	Nil		Hospitals Hotels Office buildings Schools
Ordinary	Commercial and industrial occupancies involving processing and storage of mainly ordinary combustible materials	Group I (Light ordinary)		Restaurants Cement works
		Group II (Medium ordinary)		Bakeries Engineering works Motor garages Most shops
		Group III (high ordinary)		General warehouses Clothing factories Large shops
		Group III (Special)		Film studios Distilleries
Extra high	Commercial and industrial occupancies having normal fire load but subject to special risk factors	(a) Process risks		Aircraft hangars Foam plastics manufacturers Paint manufacturers
		(b) High-piled storage risks	Category I	Clothing Groceries
			II	Non-foamed plastics Wood furniture
			III	Foamed plastics Wood wool
			IV	foamed plastics off-cuts

Sprinkler heads are available in three nominal sizes and are used according to *Table 25.5*.

The heat-sensitive element is either a specially designed solder seal or a heat-sensitive bulb (*Figure 25.21*). The temperature at which the water is released is indicated by a colour code on the yoke arm for the solder type or by the colour of the liquid in the bulb for the bulb type. This temperature, which should be not less than 30°C above the highest ambient temperature

Table 25.5. Sprinkler head sizes

Nominal size of sprinler head (mm)	*Hazard class*
10	Extra light
15	Ordinary or extra light
20	Extra high

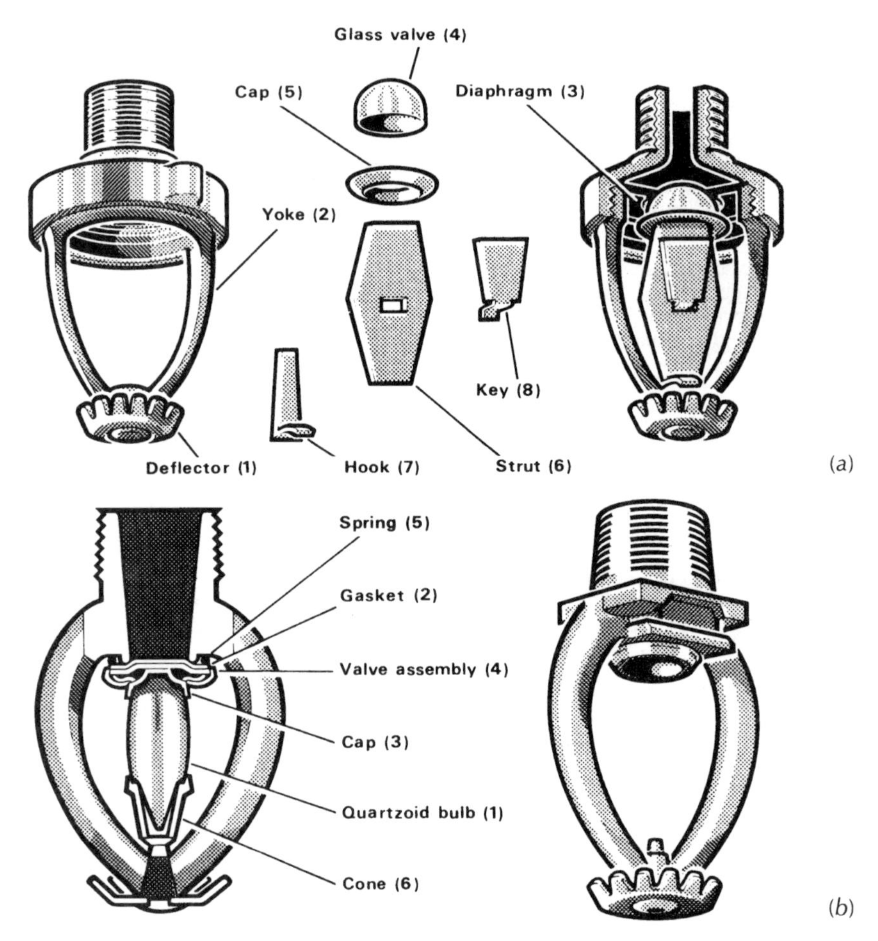

Figure 25.21 Typical sprinkler heads (a) a fusible solder type sprinkler head; (b) bulb type sprinkler head. (Manual of Firemanship. Courtesy Controller HMSO)

Table 25.6. Sprinkler head colour codes

Soldered type	*°C*	*Colour of yoke arm*
	68–74	Uncoloured
	93–100	White
	141	Blue
	182	Yellow
	227	Red
Bulb type	*°C*	*Colour of bulb liquid*
	57	Orange
	68	Red
	79	Yellow
	93	Green
	141	Blue
	182	Mauve
	204–260	Black

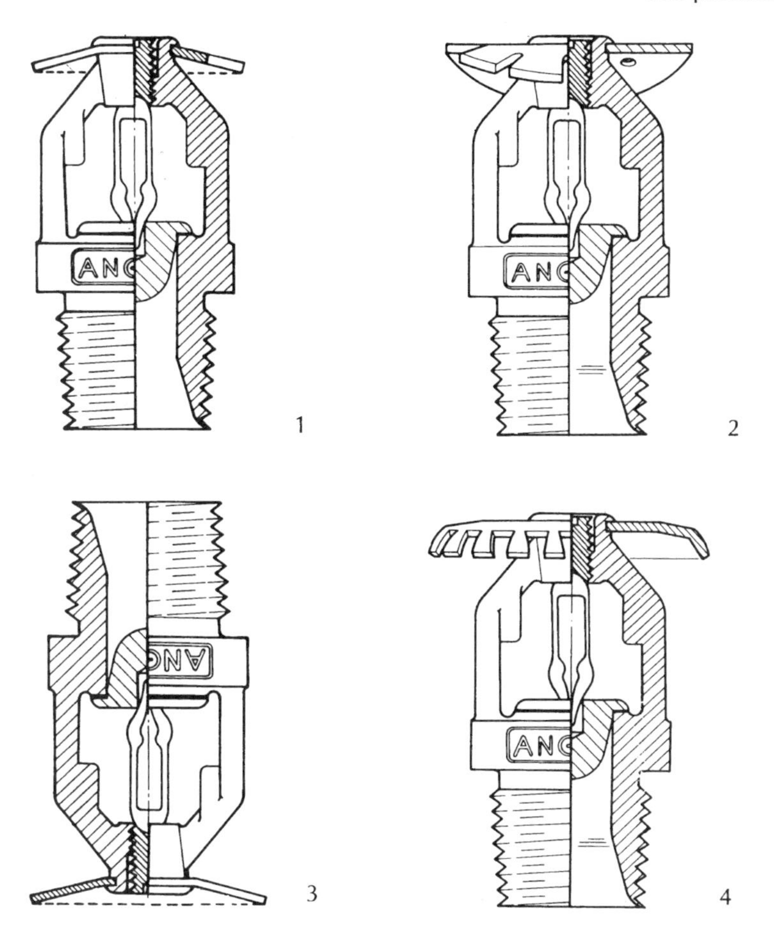

Figure 25.22 Types of sprinkler head deflector plates: (1) conventional; (2) sidewall; (3) pendant spray; (4) upright spray. (Courtesy Angus Fire Armour Ltd)

conditions in the area concerned, is specified according to the colour of the head given in *Table 25.6*.

The sprinkler head deflector plate is designed to provide three types of spray patterns, conventional, spray and sidewall, and for the first two types the head may be upright or pendant (*Figure 25.22*). The conventional spray provides a spherical discharge pattern with some of the water being thrown up towards the ceiling. With the spray type there is a hemispherical discharge below the sprinkler with little or no water reaching the ceiling. For the sidewall pattern the sprinklers are sited close to the wall and most of the water is deflected away from it.

25.8.2.2.2 Drencher systems

Ensuring adequate fire fighting provisions where large quantities of liquid or gaseous flammable materials are handled, e.g. chemical plants, tank farms, loading and off-loading installations, requires very large quantities of water.

When a fire occurs in these situations it is often far more important to provide adequate cooling of the installation than it is to extinguish the fire. It is usual to design for a minimum water flow rate of 15 to 20 m^3 per minute. Up to 2500 m^3 per hour may be used for a 5 to 6 hour period in very large fires. Such vast amounts of water call for excellent supply arrangements and when used can cause extensive flooding on the site if attempts are not made to contain or direct the disposal of the fire water. It should be remembered that fire water is often contaminated by chemicals and its release into natural water courses may cause environmental damage.

Water is provided at a pressure of 6 to 7 bar through a ring main usually buried to ensure it does not freeze and to protect it against accidental damage. Hydrant points are fitted in the ring main to surround the installation. Spray heads are fitted above vessels to provide cooling and protect the shell of the vessel from the effects of heat. Other spray heads may be fitted to provide a water curtain or to spray cooling water onto plant structure, other vessels or pipelines. Measures need to be taken to prevent the flow of water into nearby natural watercourses.

25.8.2.2.3 Fixed foam installations

Foam installations are provided where there is a risk of flammable liquid fires, e.g. in storage tanks and bunds, for confined spaces where access is difficult and for the total flooding of high-risk spaces.

Fire fighting foam is classified in two ways – its expansion characteristics and its composition. *Table 25.7* shows the expansion characteristics.

The six different types of foam concentrate have different fire fighting characteristics.

Table 25.7. Expansion characteristics of foam

Type of foam	*Expansion ratio*
(i) Low expansion (LX)	Ratio up to 50:1, but usually in the range of 5:1 to 15:1
(ii) Medium expansion (MX)	Ratio between 50:1 and 500:1, but usually in the range 75:1 to 150:1
(iii) High expansion (HX)	Ratio between 500:1 and 1500:1, but usually in the range 750:1 to 1000:1

(i) *Protein.* A solution of hydrolysed proteins generally used at 4% concentration for LX foam. It makes a stiff foam which is effective on most hydrocarbon fuels but not on miscible liquids. It is slower to extinguish fires than most fluid foams, keeps well when properly stored, does not corrode equipment if thoroughly washed out, and is the cheapest concentrate.

(ii) *Fluoroprotein.* A protein foam concentrate containing fluoro-chemical additives generally used at 4% concentration for LX foam. It has properties similar to protein, forms a more fluid foam, is quicker at extinction but is more expensive.

(iii) *Fluoro-chemical.* A solution of fluoro-chemicals generally used at 3–6% concentration for LX foam. It forms a very fluid foam, is a rapid extinguishant and is effective on hydrocarbon liquids and on partially miscible liquids. The undiluted foam may affect the skin and so any contamination must be washed off. It is the most expensive foam.

(iv) *Synthetic.* A solution of detergents generally used at 1.5–3% concentration for LX, MX and HX foam. Effective on low-boiling hydrocarbon liquids, it is very fluid and therefore rapid in action but the foam is readily broken down by heat. Gentle application is essential. It has good storage properties but is more expensive than protein.

(v) *Alcohol resistant.* Usually protein foams with additives to stabilise the bubbles in the presence of miscible solvents and generally used at 4–6% concentration for LX foam. It has similar properties to protein and is the only foam that can be used on miscible liquids. A high application rate is required.

(vi) *Chemical.* This type of foam is normally used only in portable fire extinguishers.

Foam making

There are two parts to the making of foam – the mixing of the foam concentrate with water and the aeration of the foam solution. The most common method of mixing foam concentrate with water is an in-line inductor in which as water flows a venturi device sucks in the concentrate. Foam is made in either a foam branch pipe or foam generator. In the branch pipe the flow of foam solution is made turbulent by baffles which cause air to be sucked in and the foam is made in a second set of baffles. In the foam generator, both concentrate and air are introduced at the same time. With HX foam large quantities of air are required so the air is blown into the unit by fans before the concentrate is introduced.

25.8.2.2.4 Carbon dioxide systems

Carbon dioxide can be used on both Class A and Class B fires and it can also be used on Class C where the risk of re-ignition after extinguishing is low. It is ideal for use on fires involving live electrical apparatus. However, its use does create risks to people since its OES is 5000 ppm which is considerably less than the 30% concentration in air that is needed for effective fire fighting, its density is greater than air so it collects in low places, rapid expansion of the gas on discharge causes an enormous drop in temperature and its discharge may generate static electricity.

Three types of system are available: total flooding, local application, and hand-held hoseline, with two modes of operation – automatic or manual. The choice of the system depends on the operating requirements, nature of the risk, the location and degree of enclosure of the risk and the risk to operators.

(i) *Total flooding systems*

These are automatic systems that are suitable for surface fires involving flammable liquids, solids or gases, deep-seated fires involving solids subject to smouldering and fires in electrical installations. The system should be automatic in operation and the area to be protected should not be manned during normal operation. When people have to work in the area, the system must be arranged for manual operation.

For the purposes of deciding the minimum carbon dioxide requirements it is usual to calculate on the basis of 34% concentration in the space to be protected. *Table 25.8* gives the requirements for a range of space volumes.

Table 25.8. CO_2 requirements for different spaces (assuming a 34% concentration)

Volume of space (m^3)	*Volume factors (kg/m^3)*	*Calculated miminum CO_2 required (kg)*
<4	1.15	–
4–14	1.07	4.5
14–45	1.01	16.0
45–126	0.90	45.0
126–1400	0.80	110.0
>1400	0.74	1110.0

The gas can be supplied from either a bulk storage tank or from banks of cylinders connected through a common manifold. There should be reserve capacity in case the fire re-asserts itself after the first extinguishing. *Table 25.9* gives the carbon dioxide requirements for typical high-risk plant.

(ii) *Local application system*

This system may be automatic or manually operated and is used for surface fires of Class B materials and for some Class A materials and is chosen when total flooding is undesirable, total flooding requirements cannot be met or as an adjunct to sprinkler installations. Potential areas of use include dip tanks, printing presses, textile machinery, food processing and process machinery. In designing a local application system the potential location of each fire must be identified and the nominal space determined for each. A useful guide is to assume a volume which is 600 mm longer than each of the actual dimensions and assume a discharge rate of 16 $kg/m^3/min$. The quantity of carbon dioxide available should be sufficient to maintain the full discharge rate for all areas of the system for a minimum period of 30 seconds.

(iii) *Hand-held hoseline*

This system should be manually operated and should be supplemented by portable fire extinguishers or other fire-fighting equipment. It must not be

Table 25.9. CO_2 requirements for particular plant

Risk	*Design concentration (%)*	*Flooding factors (kg/m^3)*
Electrical equipment	50	1.35
Storage of records and archives	65	2.00
Dust collectors	75	2.70

used as a substitute where a fixed carbon dioxide system is justified. With hand-held hoselines, which should have controllable discharge, there should be sufficient gas to permit discharge for at least one minute with all hoselines operating simultaneously. The use of hand-held hoselines will bring those using them near to the hazards of carbon dioxide and this requires special considerations in training and the system of operation adopted.

25.9 Fire risk

There are three categories of fire risk applicable to places of work; high, normal and low. The factors which lead to a workplace or parts of it being assessed as being 'high risk' include:

(a) whether the workplace is used for sleeping accommodation;
(b) whether the materials present are either easily ignited or, when alight, are likely to cause the rapid spread of fire, smoke or fumes;
(c) the presence of unsatisfactory structural features;
(d) whether there are certain parts of the workplace which could present a greater risk of fires occurring and developing than elsewhere; and
(e) any special consideration which may have to be given to occupants, either those who work in the workplace or others present.

Most workplaces will fall within the 'normal risk' category where the factors to be considered are:

(f) any outbreak of fire is likely to remain localised or is likely to spread only slowly;
(g) there is little risk of any part of the structure or contents of the building taking fire so readily or producing large quantities of smoke or fumes as to be a threat to life safety; and
(h) the nature of the occupancy does not require a higher risk classification.

A workplace may be assessed as of 'low risk' where the risk of fire occurring or of fire, smoke or fumes spreading is negligible. Such a workplace may be a heavy engineering shop or one where the process is entirely a wet one and non-combustible materials predominate.

25.10 Means of escape in case of fire

The principles on which means of escape provisions are based are that persons, regardless of the location of the fire, should be able to proceed safely along a recognised escape route, by their own unaided efforts, to a place of safety. Separate considerations apply to disabled persons. The characteristics of an escape route include:

(a) there should be an alternative means of escape from most situations;
(b) the distance that a person should travel to a place of safety is dependent upon the risk assessment of the workplace;

(c) where direct escape to a final exit is not possible, a place of relative safety, such as a protected escape route, should be reached within a reasonable travel distance;
(d) unless a place of safety can be reached within a reasonable travel distance the escape routes will need to be protected from the effects of fire;
(e) escape routes should be wide enough to cater for the number of occupants, e.g. not less than 1.05 m in width, and should not reduce in width; and
(f) from every room, storey or building there should be a sufficient number of available exits of adequate width.

The following are not acceptable as means of escape in case of fire: lifts (except where specially designated), escalators, portable or similar ladders, self rescue apparatus and lowering devices.

25.10.1 Travel distance

Suitable travel distances, in metres, are given in *Table 25.10.*

Table 25.10 Escape travel distances (in meters)

Type of premises		*Factory*			*Shop*	*Office*
Category of risk	high	normal	low	high	low	normal
Within a room or enclosure						
One direction only	6	12	25	6	12	12
More than one direction	12	25	50	12	24	25
From any point in an inner room to the exit from the access room	6	12	25	6	12	12
Total travel distance						
One direction only	12	25	45	12	18	18
More than one direction	25	45	60	25	30	45

25.10.2 Exits

More than one exit is required when the number of occupants of the room exceeds 60, when the travel distance to the only exit exceeds the value given in *Table 25.10* or when the room contains a high fire risk.

The minimum size for an exit should be 750 mm and where more than one exit is provided they should be more than 45° apart when viewed from the farthest point in the room.

25.10.3 From the site of fire to a place of safety

(a) Gangways and passages
Gangways and passageways in rooms from which persons need to escape in case of fire should be so arranged and kept that there is free passage at all times to the exit.

(b) Corridors
Corridors which form part of the escape route should be classed as a protected route, i.e. they should be so designed and constructed that they provide 30 minutes of protection from fire. Where escape is in one direction only, the route must be a protected route. Corridors should be at least 1.05 m wide and, where in factories and shops they exceed 30 m in length and in offices 45 m, they should be subdivided by fire doors.

(c) Stairways
It is good practice to provide at least two stairways in any building. Stairways should be at least 750 mm wide. All stairways should be separated from the remainder of the building by fire-resisting barriers and fire doors such that the stairwell is enclosed.

(d) Doors
Doors on the escape route should open in the direction of travel, each leaf should open not less than 90° and when open they should not obstruct any escape route. Fire doors protecting an escape route should be fitted with smoke seals and provide 30 minutes fire resistance. Devices to retain the doors open are permitted provided they allow the door to close automatically in the event of a fire. Doors used as a means of escape must be kept unlocked at all times when people are in the building. Panic bolts and latches are permitted to ensure security but in no case should a door be fastened in such a way that it cannot be opened easily and immediately from the inside without the use of a key.

25.10.4 Signs, notices and lighting

The appropriate signs and notices to BS 5378 and BS 5499 should be used and include: emergency exit sign, direction signs and suitable and appropriate signs and notices on doors, i.e. 'Push bar to open'. Where there is a danger that a door which is a fire exit may become obstructed then a conspicuous 'Fire escape – Keep clear' notice should be displayed on the appropriate face of the door.

All common escape routes, including stairways, should be provided with normal lighting which incorporates a suitable system of control to enable people to move along the route from areas that do not enjoy natural daylight and during the hours of darkness. Escape lighting should be provided in workplaces that operate during the hours of darkness, where there is underground or windowless accommodation, internal stairways without natural light and extensive internal corridors. Generally the need for escape lighting arises more frequently in shops than in factories or offices.

25.10.5 Emergency plans

The aim of an emergency plan is to ensure that, in the event of a fire, all employees are sufficiently familiar with escape procedure and the fire safety arrangements to take the correct action and evacuate the premises safely. The plan should specify the fire responsibilities of the managers and other nominated responsible persons. It should also detail the actions for:

(a) operating the fire warning system;
(b) calling the fire brigade;
(c) evacuating the occupants to predetermined assembly points in a safe place;
(d) attacking the fire if it is safe to do so;
(e) stopping machinery and plant and isolating power supplies where appropriate;
(f) closing doors; and
(g) liaising with the fire brigade on arrival.

When the fire brigade arrives on site they should be told: the location and extent of the fire, whether any persons are trapped or missing, any known special risks and the location of fire hydrants or other water supplies. The fire brigade should be encouraged to make regular visits to the site to familiarise themselves with the layout of premises and the risks posed.

The evacuation procedure should follow a pre-arranged plan which is regularly rehearsed and updated. It should identify the assembly points and the arrangements for checking that all employees and others on the site are accounted for. Annual fire drills should be carried out to test the emergency procedure and ensure that employees know what to do. Following a fire evacuation of a building, it should not be re-entered without the permission of the fire brigade.

A simple fire instruction, should be displayed at suitable locations in the workplace and typically contain the following information:

Fire Instructions

If you discover a fire –

1. Raise the alarm.
2. Attack the fire using the fire extinguishers provided but not putting yourself at risk.

On hearing the alarm –

3. Leave the building by the shortest route.
4. Go to your assembly area which is . . . .

Do NOT stop to collect belongings.

Do NOT re-enter the building until told to do so.

This notice must be rectangular with white lettering on a blue background as it is a mandatory notice (see BS 5378).

25.10.6 Instruction and training

No matter how good the provisions for fire prevention and protection it will not be effective if the people concerned are not given the necessary instruction and training. Every employee should be given instruction, training and refresher training in:

(a) the means of escape in case of fire;
(b) the action to be taken in case of fire;
(c) the location and method of operating equipment provided for fighting fires; and
(d) the location and use of fire alarms.

Fire marshals and other persons identified in the emergency plan must receive additional instruction and training appropriate to their responsibilities. This should include the checking of workplaces to ensure that they have been evacuated, the carrying out of roll calls at the assembly points and liaison with the fire brigade.

25.10.7 Maintenance and record keeping

All fire fighting equipment, fire detectors and alarms must be maintained regularly and kept in efficient working order. Records must be kept of such maintenance which should include details of any work carried out on the equipment. Similarly, records must be kept of all instruction, training and refresher training given to employees, and of all fire drills. The latter should include recommendations to remedy any defects found.

All records must be retained for a period of at least three years from the date on which they were made.

25.11 Legal requirements

Like all areas of occupational activities, fire has its legal requirements, both of a detailed and a general nature. The Fire Services Act 1947 places on the local authority the duty to create and maintain an efficient fire fighting force and outlines the fire fighting procedures, the powers of firemen and fire authorities and other matters necessary to ensure an efficient and speedy fire service. It also created Fire Prevention Officers.

25.11.1 The Fire Precautions Act 1971

The specific requirements contained in the Fire Precautions Act 1971 (FPA) have been considerably modified and extended by the Fire Safety and Safety of Places of Sport Act 1987 (the '87 Act) and it is necessary to refer to the two Acts when considering an employer's fire obligations.

No specific requirements in respect of fire, fire prevention, or fire precautions are contained in the HSW Act although under section 78 the Fire Precautions Act 1971 was extended to include places of work. This Act took over the fire clauses of the Factories Act, and Offices, Shops and Railway Premises Act, and introduced the requirement for a fire certificate rather than a means-of-escape certificate.

Under s. 1 of the FPA six classes of premises are designated:

(a) Premises providing sleeping accommodation, e.g. hotels, boarding houses.
(b) Premises providing treatment or care, e.g. hospitals, old people's homes.
(c) Premises providing entertainment.
(d) Premises providing teaching, research.
(e) Premises permitting access to the public.
(f) Premises used as a place of work.

Only premises in parts (a) and (f) must have a fire certificate before they can be brought into use and, except for one group in class (f), the local fire authority is responsible for the issue of a fire certificate. The issue of a fire certificate for places of work with high fire risk is the responsibility of the Health and Safety Executive. The issue of a fire certificate is conditional on the premises having a specified standard of fire safety.

25.11.2 Application for a fire certificate

This has to be made on Form FP1 (Rev) to the local fire authority. It is an offence to use premises not covered by a fire certificate except where an application has been made though the certificate may not have been issued nor refused. Before issuing a certificate the fire authority may inspect the building, ask for more information or plans and ask for remedial work to be carried out.

25.11.3 Issue of a fire certificate

The fire certificate will specify the following:

(a) The use or uses of the premises it covers.
(b) The means of escape in case of fire, usually indicated on a plan of the building.
(c) The means for ensuring the safety and effectiveness of the means of escape such as fire and smoke stop doors, emergency lighting and direction signs.
(d) The means of fighting fire for use of persons on the premises.
(e) The means of raising the alarm.
(f) Particulars of explosive or highly flammable liquids stored and used on the premises.

The fire certificate may also impose requirements relating to:

(g) The maintenance of the means of escape and keeping it free from obstruction.

(h) The maintenance of other fire precautions.
(i) The training of people and the keeping of records.
(j) Limitation on numbers of persons in the premises.
(k) Any other relevant fire precautions.

Once issued the certificate must be kept on the premises. Any proposed alterations to the premises which will affect any of the conditions specified in the fire certificate must be notified to the fire authorities before they are put in hand, otherwise an offence occurs. Fire certificates which were issued under the FA and for OSRP remain in force under the FPA.

A number of Orders were made under the FPA which rationalised the requirements of section 6, contents of a fire certificate. Three of these were the Fire Precautions (Factories, Offices, Shops and Railway Premises) Order 1976 which applied to medium and large factories, offices and shops, the Fire Precautions (Hotels and Boarding Houses) Order 1972 and the Fire Precautions (Non-certificated Factory, Office, Shop and Railway Premises) Order 1976, which removed the requirement for a fire certificate from smaller industrial premises/offices. In a Regulation made under the HSW Act, the Fire Certificates (Special Premises) Regulations 1976, the responsibility for the issue of a fire certificate for certain 'high risk' premises was transferred from the fire authority to the HSE.

25.11.4 The Fire Safety and Safety of Places of Sport Act 1987

This Act has had such a significant effect on the FPA that the two must be read together to obtain a view of the current legislation. In addition to amending the FPA, this '87 Act modified the powers of the Fire Authorities to enable them to specify in a designation order premises which were assessed as constituting a low risk and exempting them from the requirement to have a fire certificate. In these circumstances, occupiers become subject to a statutory duty to have adequate means of escape and fire fighting equipment, and to maintain them. A code of guidance is to be issued by the Secretary of State.

The '87 Act further extended the power of the Fire Authorities to enable them to enforce statutory obligations by the issue of Improvement and Prohibition Notices and by prosecution. The procedures regarding notices are similar to those under the HSW Act.

25.11.5 The Fire Certificates (Special Premises) Regulations 1976

These regulations apply to types of premises (14 are listed) in which identified 'high-risk' chemicals are used or stored, buildings subject to the Explosives Act 1875, surface buildings associated with mines and quarries, buildings associated with nuclear activities and 'construction huts'. Essentially the regulations, which do not add any requirements not already specified in the FPA, transfer the responsibility for enforcement from the local fire authority to the HSE.

25.11.6 The Fire Precautions (Places of Work) Regulations 199- (proposed)

These Regulations are being made under s. 12 of the Fire Precautions Act 1971 and are intended to give effect to the fire requirements of the EC framework directive no. 89/391/EEC. They will apply to all places of work including those to which the 1971 Act applies.

Separate regulations will contain requirements on:

- risk assessments and emergency plans,
- precautions relating to means of escape in case of fire,
- mean of fire fighting,
- means of giving warning in case of fire and for detecting fire,
- maintenance,
- storage of combustible refuse,
- supervision of maintenance and construction work,
- instruction and training in fire precautions, and
- keeping of records of maintenance work and instruction, training, etc.

Extensive guidance will be provided with these Regulations which will be based on the Guide to Fire Precautions in Existing Places of Work that require a Fire Certificate.

25.12 Liaison with the fire brigade

Under the Fire Services Act 1947, the local authority have a duty to provide an effective fire service, but the fire brigade cannot be effective if it cannot get to the fire or find its way round a site. Employers can do a number of things to maximise the effectiveness of the brigade.

(a) Ensure that there is adequate access for the brigade vehicles, which are large and heavy. The standard width is 3.66 m, height varies from 3.66 m to 3.96 m, and the weight lies between 10.2 and 18.3 tonnes. When these vehicles enter a factory site they may need to get under pipebridges and other over-road obstructions. Other factors relating to the buildings themselves determine how close the brigade need to get with their appliances and access arrangements need to be sorted out with the brigade before a fire occurs.

(b) Liaise with the brigade and arrange visits to the site so that they can familiarise themselves with the layout of the premises and the risks posed. It gives them pre-knowledge so that on call out they can work out their plan of attack before they reach the site. Regular visits should be arranged to ensure the brigade's knowledge of the site is kept up-to-date.

(c) On arriving at a site, the brigade need to be directed to the location of the fire so it is important that someone, preferably known to the brigade, is detailed to meet them at the gate and direct or guide them through the site.

Bibliography

Statutes and Statutory Instruments available from HMSO, London:
The Fire Services Act 1947 (as amended by the Acts of 1953 and 1959)
The Fire Precautions Act 1971
The Fire Safety and Safety of Places of Sport Act 1987
The Petroleum (Consolidation) Act 1928 (as modified by the Petroleum Spirit (Motor Vehicles etc.) Regulations 1929)
The Highly Flammable Liquids and Liquefied Petroleum Gases Regulations 1972
Fire Certificates (Special Premises) Regulations 1976
Fire Precautions (Factories, Offices and Railway Premises) Order 1976
Fire Precautions (Non-Certified Factories, Offices and Railway Premises) Order 1976
Fire Precautions (Hotels and Boarding Houses) Order 1976
Fire Precautions (Modifications) Regulations 1976
Fire Precautions (Places of Work) Regulations 199-
Safety Signs Regulations 1980
The Classification, Packaging and Labelling of Dangerous Substances Regulations 1984
Guidance on Statutes:
Guide to fire precautions in existing places of work that require a fire certificate: factories, offices, shops and railway premises
Code of practice for fire precautions in factories, offices, shops and railway premises not required to have a fire certificate.
Guidance on fire precautions in places of work.
Fire safety at work.

Further reading

Books
King R W and Magid J, *Industrial Hazard and Safety Handbook*, Butterworth, London (1979)
Handley W, *Industrial Safety Handbook*, 2nd edn, McGraw-Hill Book Company (UK) Ltd, Maidenhead (1977)
Lees F P, *Loss Prevention in the Process Industries*, Vols 1 & 2, Butterworth, London (1980)
Underdown G W, *Practical Fire Precautions*, 2nd edn, Gower Press, Teakfield Ltd, Farnborough (1979)
Wharry D M and Hurst R, *Fire Technology, Chemistry and Combustion*, Institution of Fire Engineers, Leicester (1974)
Palmer K N, *Dust Explosions and Fires*, Chapman & Hall, London (1973)
Field, P, *Explosability Assessment of Industrial Powders and Dusts*, HMSO, London (1983)
Dewis, M. and Stranks, J., *Fire Prevention and Regulations Handbook*, Royal Society for the Prevention of Accidents, Birmingham (1989)
Portsmouth University, *Safety Technology Unit Module ST4: Fire and Explosion*, Portsmouth (1988)
Everton, A., Holyoak, J. and Allen, D., *Fire Safety and the Law*, 2nd edn, Paramount Publishing Ltd, Borehamwood (1990)
Todd, C. *Croner's Guide to Fire Safety*, Croner Publications Ltd, Kingston upon Thames (1992)

Other publications (listed according to publisher)

Home Office, Manual of Firemanship, HMSO, London:
Book 1 Elements Elements of combustion and extinction
2 Fire brigade equipment
3 Fire extinguishing equipment
5 Ladders and mobile appliances
6 Breathing apparatus and resuscitation
7 Hydraulics and water supplies
8 Building construction and structural fire protection

9 Fire protection of buildings
10 Fire brigade communications and mobilising
11 Practical firemanship 1
12 Practical firemanship 2

Other parts of the Manual are currently under review but the old parts Nos 6B (1948), 6C (1962) and 7 (1944) remain effective.

Fire Offices Committee, London
Classification of Fire Hazards in Buildings
Rules for Automatic Sprinkler Systems

Fire Protection Association, London
Compendium of Fire Safety Data:
Vol 1 Management of fire risks (including organisation of fire safety)
2 Industrial and process fire safety (including occupancy fire safety)
3 Housekeeping and general fire precautions (including nature and behaviour of fire)
4 Information sheets on hazardous materials
5 Fire protection equipment and systems
6 Buildings and fire (design guide and building products)

British Standards Institution, London
BS 476 Fire tests on building materials and structures
Parts 1 & 2 withdrawn
Part 3:1975 External fire exposure roof test
Part 4:1970 Non-combustibility test for materials
Part 5:1968 Ignitability test for materials
Part 6:1968 Fire propagation tests for materials
Part 7:1971 Surface spread of flame
Part 8:1972 Test methods and criteria for the fire resistance of elements of building construction
BS 3116 Automatic fire alarms in buildings
Part 1:1970 Heat sensitive (point) detectors
Part 4:1974 Control and indicating equipment
BS 4422 Glossary of terms associated with fire
Part 1:1969 The phenomenon of fire
Part 2:1971 Building materials and structures
Part 3:1972 Means of escape
Part 4:1975 Fire protection equipment
Part 5:1976 Miscellaneous terms
BS 4547 1972 Classification of fires
BS 5306 Code of Practice for fire extinguishing installations and equipment on premises
Part 1:1976 Hydrants systems, hose reels and foam inlets
Part 2:1979 Sprinkler systems
Part 3:1980 Portable fire extinguishers
Part 4:1979 Carbon dioxide systems
Part 5, Section 5.1:1982 Halon 1301 total flooding systems
BS 5378 Safety signs and colours
BS 5423: 1987 Specification for portable fire extinguishers
BS 5499 Fire safety signs, notices and graphic symbols
Part 1:1984 Specification for fire safety signs
BS 5839 Fire detection and alarm systems in buildings
Part 1:1980 Code of Practice for installation and servicing
BS 5908: 1980 Code of Practice for fire precautions in chemical plant
BS 6020 Instruments for the detection of combustible gases
BS 6266: 1982 Code of Practice for fire protection for electronic data processing installations (formerly CP 95)

Health and Safety Executive publications available from HSE Books, Sudbury
Guidance Notes:
CS 1 Industrial use of flammable gas detectors
CS 3 Storage and use of sodium chlorate

CS 4 The keeping of LPG in cylinders and similar containers
CS 6 The storage and use of LPG on construction sites
CS 8 Small scale storage and display of LPG at retail premises
CS 11 Storage and use of LPG at metered estates
CS 21 The storage and handling of organic peroxides
EH 7 Petroleum based adhesives in building operations
EH 9 Spraying of highly flammable liquids
GS 3 Fire risk in the storage and industrial use of cellular plastics
GS 16 Gaseous fire extinguishing systems: precautions for toxic and asphyxiating hazards
GS 19 General fire precautions aboard ships being fitted out or under repair
GS 20 Fire precautions in pressurised workings
GS 21 Assessment of the radio frequency ignition hazard to process plants where flammable atmospheres may occur
GS 40 Loading and unloading of bulk flammable liquids and gases at harbours and inland waterways
PM 25 Vehicle finishing units: fire and explosion hazards
PM 58 Diesel engined lift trucks in hazardous areas

Guidance Booklets:
HS(G) 3 Highly flammable materials on construction sites
HS(G) 4 Highly flammable liquids in the paint industry
HS(G) 5 Hot work: welding and cutting on plant containing flammable materials
HS(G) 11 Flame arresters and explosion reliefs
HS(G) 22 Electrical apparatus for use in potentially explosive atmospheres
HS(G) 34 Storage of LPG at fixed installations
HS(G) 41 Petrol filling stations: construction and operation
HS(G) 50 The storage of flammable liquids in fixed tanks (up to 10 000 m^3 total capacity)
HS(G) 51 The storage of flammable liquids in containers
HS(G) 64 Assessment of fire hazards from solid materials and the precautions required for their safe storage and use: a guide for manufacturers, suppliers, storekeepers and users

Occasional Papers Series:
OP5 Electrostatic ignition: hazards of insulating materials – 1982

Chapter 26

Safe use of machinery

J. R. Ridley

26.1 Introduction

This chapter is concerned with the identification of mechanical hazards and the design of safeguards to eliminate or reduce the risks from those hazards. It has long been recognised that the majority of accidents involving machinery could have been prevented if reasonable practical precautions had been taken. Many of the preventable accidents are caused by the failure of employers to provide safeguards, or to provide training in the correct method of work; others are associated with the failure of the workpeople to use the guards or the method of work they have been taught.

The design of guards is critical to ensuring their use. Inconvenient and defeatable guards or those with unnecessarily complex methods of operation are more likely to cause, rather than prevent, accidents. It is, therefore, incumbent upon the designer to ensure that his design is not only effective in guarding against the identified hazards but also is acceptable to those who have to use them, thus ensuring that they will, in fact, be used. It is a serious shortcoming that the formal training of many designers does not cover effective techniques for guarding machinery. A result of this is that guards are treated as 'add-ons' after the machine has been designed, rather than incorporated as an inherent part of the design. Add-on guards are, inevitably, expensive and always look what they are – afterthoughts.

It is vital to appreciate that efforts to prevent machinery accidents require more than high guarding standards; they also require an appropriate package of behavioural and managerial measures underpinned by relevant law.

26.1.1 Changes in legislative arrangements

The past two decades have seen major changes in approaches to the problem of providing safer workplaces with the onus of responsibility moving towards the provider; in the case of the workplace – the employer, and in the case of equipment – the designer and supplier. The greatest change occurred on 1 January 1993 with the coming into effect of the European free market with its aim of harmonising safety standards across all

Member States of the European Community (EC). It has tackled this problem on two fronts: firstly new machinery and equipment must conform with the requirements contained in the Machinery Directive[1] before it may be put on the EC market and secondly all existing work equipment must conform with the Work Equipment Directive[2] by 1 January 1997. The nub of both these directives is compliance with harmonised health and safety standards to provide evidence of conformity with the directive's requirements.

While these two directives call for slightly different levels of safety which may give rise to dual safety standards it is an interim arrangement. In the longer term more uniform and higher standards will be achieved throughout the EC. In the UK, the requirements of the Machinery Directive have been incorporated into the Supply of Machinery (Safety) Regulations 1992[3] (SMSR) and those of the Work Equipment Directive in the Provision and Use of Work Equipment Regulations 1992[4] (PUWER). Like HSW, these regulations lay down the results to be achieved and rely on standards to specify the means by which conformity occurs. The standards concerned will be European harmonised standards (EN standards) drawn up by the European standard making bodies Comité Européen de Normalisation (CEN) and Comité Européen de Normalisation Électrotechnique (CENELEC).

Although these directives have resulted in changes to laws and standards concerned with the safe use of machinery, many of the new requirements are based on or are similar to existing UK practices. The techniques and practices developed in the UK, and largely contained in BS 5304: 1988[5] will, in most cases, give conformity with these new requirements. It is with those techniques and practices that this chapter is concerned.

26.2 Sources of information

Currently the most important publication on machinery guarding in the UK is the British Standard 5304:1988 'Code of Practice for Safety of Machinery'[5] which contains a comprehensive list of reference sources. However this standard will eventually be overtaken by a series of separate European harmonised (EN) standards that are in preparation to support the Machinery Directive. In a number of cases, sections of BS 5304 have been used as the basis for these EN standards. EN standards have been categorised into three types:

Type A standards – generic safety standards covering basic concepts and principles of safety design.

Type B standards – covering standards that apply to a group of machines:

type B1 standards deal with particular safety aspects such as safe distances, noise, etc.

type B2 standards specify safety related devices such as two hand controls, pressure sensitive devices, etc.

Type C standards are machine specific standards concerned with particular types or groups of machines.

Other information is published by HSE in the form of Advisory Committee Reports, Approved Codes of Practice, Guidance Notes and Health and Safety booklets, details of which are given in an HSE publication[6]. The HSE's

National Industry Groups can be consulted for information and advice on particular problems. A number of industry associations have prepared recommendations, in consultation with the HSE, on the safeguarding of machinery relevant to their operations. These are listed in the HSE publication[6]. Further information is also available in an open learning format[7].

Over the years a body of case law has established interpretations for words and phrases met in health and safety laws and activities. Cases such as *John Summers & Sons Ltd* v. *Frost*[8] that established the absolute nature of the duty to guard dangerous parts of machinery and *Uddin* v. *Associated Portland Cement Manufacturers Ltd*[9] concerning the concept of safe by position. It is unlikely that these, and the many other, interpretations will be changed or abandoned under the new EC based health and safety laws.

In the USA, OSHA has published 'Concepts and techniques of machine safeguarding'[10] which contains relevant and interesting suggestions but reflects American practices, some of which may not be acceptable in the UK.

26.3 Strategy for selecting guards

When considering the provision of safeguards for a machine or machinery, a strategy should be developed that encompasses all the foreseeable aspects that need to be looked at. Before the work of designing and manufacturing the safeguarding measures commences a plan of action should be prepared which, usefully, could be in the form of a pro forma or check-list of the items to be considered. To be confident that no important matter had been overlooked or forgotten the plan should be followed step by step.

Background information on the operating parameters and physical characteristics of the machine is necessary to provide a base on which decisions can be made on which technique of guarding or type of guard to employ. These details should include:

- its physical dimensions,
- method of drive and power requirements,
- limitations of speed, pressure, temperature, etc.,
- materials being processed or handled,
- access requirements especially for setting, adjustments and maintenance,
- environmental factors such as dust, fumes, noise, and
- operating requirements.

Thus an overall picture of any constraints or limitations on the design of the guard can be obtained which will provide a pointer to the most effective measures that will give the highest level of protection.

The action plan should incorporate the following sequential steps:

(1) identify the hazards,
(2) assess the risks from those hazards,
(3) remove the hazards wherever possible or reduce the risks to a minimum,

(4) select and provide safeguards,
(5) decide if any and what other precautions are necessary,
(6) inform operators of the hazards and instruct them in the safe methods of operating the machine,
(7) review the proposals and revise as necessary,
(8) design and install, and
(9) monitor the effectiveness of finished guard or safeguarding system and modify if necessary.

These steps are considered in more detail below.

26.4 Risk assessment

Risk assessment is defined in an EN standard[11] as a logical process made up of four elements and aimed at determining the likelihood of injury arising from the use of machinery with a view to eliminating or reducing that risk to an acceptable minimum. The four elements are:

(1) intended use of the machine,
(2) identifying the hazards,
(3) estimating the risks, and
(4) evaluating the effectiveness of the measures taken.

By examining these facets in a systematic way, the most effective measures necessary to achieve adequate levels of safety can be identified, and once installed, their effectiveness can be monitored.

26.4.1 Use of machines

In carrying out a risk assessment at the design stage it will be necessary to determine all possible uses to which the machine can be put – the proper and the improper – and allow for them. Whereas for an existing machine currently in service and performing a particular function, only the risks associated with that function need be considered. In making the assessment, the behaviour of the operator should be included in the considerations as should any known or possible behaviour which deviates from the laid down method of working, i.e. the foreseeability factor.

26.4.2 Identifying the hazards

A number of approaches to identifying and describing the dangerous parts of machinery have been used which stem from a study of the record of injury causing accidents. A more analytical and objective approach was taken in BS 5304[5] which has been followed in the European harmonised standard EN 292 part 1[12] and which considers three factors:

- the phases of the machine's life,
- the circumstances giving rise to injury, and
- the hazards that can cause injury.

26.4.2.1 Phases of machine life

This factor is important where the provision of guards or safeguard measures is considered at the initial design stage but it cannot be ignored when assessing existing machines. Safeguard arrangements should apply for the life of the machine so when conceiving a guard or safeguarding system it is important that all phases of the machine life are considered including:

(i) construction,
(ii) transport,
(iii) installation,
(iv) commissioning,
(v) start up, operation and shut down,
(vi) adjustment,
(vii) setting,
(viii) cleaning,
(ix) maintenance, and
(x) dismantling and disposal.

Inevitably conflicts will arise between the guarding requirements of different phases of the machine life so an order of priority will need to be worked out taking account of those facets that present the highest risk – not necessarily the greatest hazard but allowing for the frequency and duration of exposure. Examples of particular areas where priorities need to be given include cleaning for automatic food processing plants, operating (ergonomic aspects) for manually operated machines, maintenance of plant processing hazardous substances, etc. The aim of this review being to minimise or reduce to an acceptable level, the risks from any operation at any point in the machine life.

26.4.2.2 Machine situations giving rise to risk

In all phases of a machine's life work needs to be carried out on it. This work entails operators, fitters and engineers approaching the machine, both when it is stopped and when it is running, where they can move into an area of increased risk. Knowledge of the situations and circumstances that can occasion injury can assist in identifying the risks and ensure realistic risk assessments. The circumstances giving rise to injuries include:

(i) contact or entanglement with the machinery,
(ii) trapping between machine parts, the machine and material or fixed structures,
(iii) contact or entanglement with the material being processed,
(iv) being struck by ejected parts of the machine or the material being processed,
(v) unexpected start up, and
(vi) failure of a machine part due to corrosion or fatigue.

26.4.2.3 Hazards from machinery

In the operation of machinery, hazards can arise as a result of the machine's movement, the state of its parts or its failure to contain components or process material. Over the vast range of types and sizes of machines in use, the hazards met can be summarised in the following list:

- crushing,
- shearing,
- cutting or severing,
- entanglement,
- drawing-in or trapping,
- impact,
- stabbing or puncture,
- friction or abrasion, and
- high pressure fluid ejection.

To assist in identifying hazards the following five-question checklist can be used:

(1) *Traps:* Are there components whose movements could lead to trapping points?
- where the limbs are drawn into a trap (an 'in-running nip' point)
- where the limbs (etc.) are simply trapped by a closing or passing movement

(2) *Impact:* Are there parts which by their speed of movement could cause injury if a person gets in the way?

(3) *Contact:* Are there parts which because they are sharp, abrasive, hot, cold or electrically-live could injure as a result of touching them? (Contact injuries may involve both moving and stationary machinery)

(4) *Entanglement:* Is there a chance that hair, rings, gloves, clothing etc. could become entangled in moving machinery?

(5) *Ejection:* Is there a risk that machine components or material being worked upon will be thrown out of the machine?

A pictorial classification of dangers is presented in *Figures 26.1–26.6*.

Two further points should be observed when making the checks. The first is that some dangers may result from a combination of two or more of the

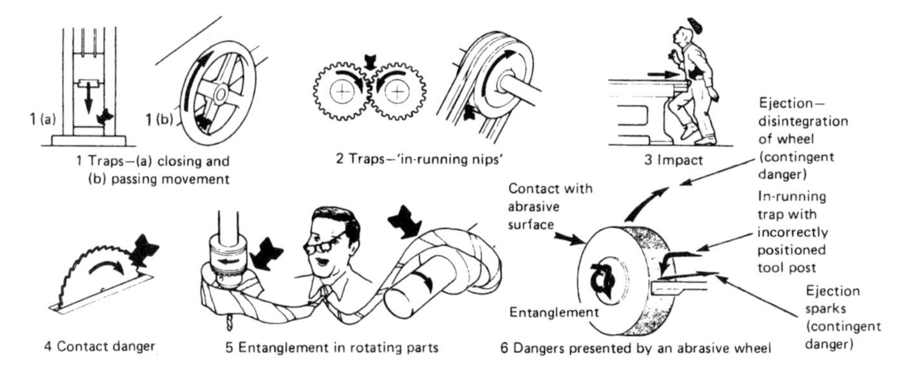

Figures 26.1–26.6 Classification of dangers. (Courtesy Engineering)

above list; for example, entanglement leading to a trap, or to contact. Secondly, consideration should be given not just to normal operation (continuing dangers), but also to the direct consequences of machine malfunction (contingent dangers, which follow directly from component failure).

26.4.2.4 Build-up of danger

In the assessment of risk it is useful to separate what might be called the 'build-up of danger'[13]. Thus the close approach to machinery in motion (perhaps involving the improper or necessary removal of a guard) represents a progressive build-up of danger. Once a person is immediately adjacent to the point of danger, the situation can be described as one of 'imminent danger'. The build-up stage may be associated with shortcomings in the system of work, risk-taking or an essential need for access. The condition of imminent danger may lead to an accident and injury depending upon the rigour with which any permit-to-work system is enforced or it may be the result of a human error on the part of the operator concerned. An important point is that the behavioural factors which determine the likelihood of

Table 26.1 Build-up of danger

(1) *The agent of danger*
 How could the harm occur (traps, etc.; continuing and contingent dangers)?
 What is the likely range of injury severities?
 How far are the hazards obvious?

(2) *Build-up of danger*
 The *need* for approach (access) leading to imminent danger
  What is the frequency, proximity, and duration of access?
  What are the circumstances of access (machine running; machine off/isolated)?
  Is the access required for operational or informal reasons?
 The *ease* of approach (access) leading to imminent danger.
  How far is access restricted by guarding etc.?
  To what extent can access be facilitated by removing or defeating safeguards?

(3) *Imminent danger*
 Probability of human errors.
 Have the personnel at risk a clear perception of the danger and the skills to avoid injury?
 What is the probability of human error?

intentional risk taking are quite different from the factors which lead to unintended errors. A number of the factors which determine the level of risk are shown in *Table 26.1*.

26.4.3 Estimating the risk

In general terms, risk is regarded as the likelihood that an event will occur in a set of foreseen circumstances. However, in an occupational risk assessment regimen a definitive definition is needed to provide a measure, either quantitative or qualitative, on which a decision regarding the priority for action can be based. Such a definition is given in an EN[11] standard as:

Risk is a function of:

severity of possible harm such as the extent of likely injury,

and *exposure to the hazard* in terms of duration and frequency,

and *the probability of an injury* being received or harm done, in which note is taken of factors such as reliability of equipment, accident history and comparison with similar operations,

and *the inability to provide suitable protection* on technical grounds against the hazard or the ease of defeating the precautions provided.

There are few, if any, reliable data available for most types of machinery to enable a quantitative estimate to be made of the risks, so in those cases the assessment will be essentially subjective and qualitative. However, a number of techniques have been developed, such as hazard and operability studies (HAZOPS), fault tree analysis (FTA), failure mode and effect analysis (FMEA) and the DELPHI technique (see chapter 10), which can be applied, particularly to complex plant, to give a reasonably accurate indication of the risks in a quantifiable form. There is also a growing body of reliability data for a whole range of component items of machinery and plant which will enable a quantitative value to be put on the risk.

26.4.4 Evaluating the measures taken

Essentially this is a technique whereby the effectiveness of the safeguarding measures taken is questioned. It involves identifying salient features of the measures and then establishing the extent to which they provide the protection required. Both the mechanistic and the behavioural aspects are reviewed. Typical questions posed could include:

Is the guard effective in preventing access to the dangerous parts?
Do the interlocks stop the machine before the operator can reach dangerous parts?
Are the interlock switches robust enough for the service and can they be defeated?

Raafat[14] has developed a widely applicable monogram for machinery risk assessment.

26.5 Risk reduction

Essentially reduction of risk can be achieved by technical, procedural or behavioural measures, or most notably by a mixture of all three, as shown in *Table 26.2*. Each machine or process should be studied and arrangements for its safe operation developed in the context of the criteria set out in the Table.

Table 26.2. Risk reduction checklist

(1) *Technical measures*
(to make the machinery inherently safer)
Dangerous parts: eliminate, or reduce capcity to cause harm:
- reduce speed of movement;
- reduce forces, torques, pressures, and inertia;
- reduce or eliminate rough or sharp surfaces and projections;
- reduce, or when appropriate, increase the size of clearances/gaps;

Dangerous parts: seek ways of making the dangers more obvious
Machine and guard design:
- reduce *need* for access (e.g. remote lubrication/adjustment)
- improve component and system design (reduce probability of failure and/or failure-to-danger)

Guard design:
- reduce *ease* of access to imminent danger (e.g. safeguards which allow safe access, but prevent access at other times)

Machine controls: improve ergonomic layout to reduce unintended errors

(2) *Procedural measures*
(to make the work tasks safer)
Planned maintenance and inspection of machines and guards
Systems of work (which minimise needs for dangerous access)
Permit-to-work procedures (to formalise precautions for imminent danger)

(3) *Behavioural measures*
(to ensure that personnel wish, or are obliged, to work safely)
Training: basic skills; systems and procedures; knowledge of hazards
Instruction and supervision
(Note that the absence of 'risk taking' is only attainable when the inconvenience of defeating the safeguard or system exceeds the benefits which accrue for so doing.)

26.6 Description and selection of safeguards

It should be remembered that guards are only required where the designer has failed to achieve intrinsic safety through not avoiding trapping and similar dangers. For example the spokes shown in *Figure 26.1(b)* could be filled in to eliminate a passing motion trap. However, *Figure 26.7* demonstrates how an inherently safe jump-out roller may be prevented from working in certain foreseeable circumstances. A fixed guard as shown in *Figure 26.7* may be the better option.

The considerations for choosing guard material are: strength, stiffness, durability (to cope with both continuing and contingent dangers), and the possible effect of the guard material on machine reliability. A solid guard may create problems of cooling and visibility; there are both operational and safety reasons for a clear view of the danger area. An important factor in non-solid or mesh guard material is the size of the opening in relation to its distance from the dangerous part.

26.6.1 Fixed Guards

A fixed guard has no moving parts and is defined[15] as a guard that is kept in place permanently by welding or by fasteners that can only be released with

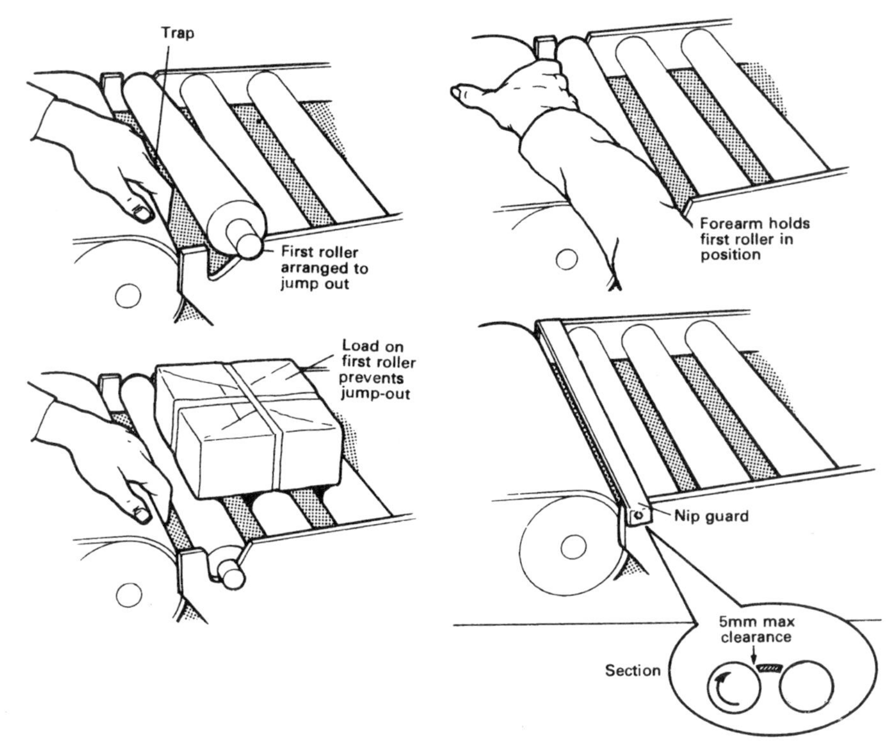

Figure 26.7 A jump-out roller may not be inherently safe – a fixed nip guard may provide a better standard of safety. (Courtesy Engineering)

the use of a tool. A more flexible definition could be 'a guard attached to a machine by a simple fixing method which is not linked to the controls, motion or hazardous condition of the machine' since this permits the use of moving parts to provide access for material or utilising a telescopic device.

A fixed guard should, when fitted, not be capable of being displaced casually, so the method of fixing is important.

Where fixed guards have to be removed periodically, removal time should be kept to a minimum by using fixing devices that are not awkward to undo. Ideally the fixing devices should be of the captive type or there is the risk that they will get lost when removed making the guard less effective when it is replaced. The fixing used is likely to be determined by the application – the sort of machine to which it is fitted and the particular industry – and by whether the guard is never moved or may have to be moved for setting, adjustment or clearing blockages. *Table 26.3* lists a number of fixing devices in order of preference. Traditional padlocks have little value as fixing devices if the keys are not kept under close supervision and control. Common sense, training and good supervision are factors that make the use of the simpler fixing devices more acceptable.

Careful design of fixed guards is required to prevent traps being created between the guard and moving (but otherwise safe) parts of the machine.

Table 26.3. Means of fixing guards

Fixing method	*Acceptability*
Quick-prelease catches, wing nuts, deadlock bolts	Not acceptable
'Secured by fastening requiring the use of a tool': Cheese-head screws Phillips/Pozidrive screws Hexagon nuts/bolts Socket headed screws Countersunk socket screws Deadlock latches (Budget lock) Key locks Padlock (key only available to authorised personnel in approved circumstances)	↓ Increasing acceptability ↓
Riveted or welded	Only suitable where there are no foreseeable reasons for access

26.6.2 Distance guard

A distance guard is simply a barrier sited at an appropriate distance from the danger. The degree of risk being faced will determine whether a fixed rail or fence is necessary; in the latter case, the distance from the dangerous part will determine the opening or mesh sizes, or vice versa. Guidance on safe distances is given in Appendix A of a BS[5] and two EN standards[16,17].

26.6.3 Adjustable guards

Adjustable guards comprise a fixed guard with adjustable elements that the setter or operator has to position to suit the job being worked on. They are widely used for woodworking and toolroom machines. Where adjustable guards are used, the operators should be fully trained in how to adjust them so that full protection can be obtained.

26.6.4 Self-adjusting guards

Self-adjusting guards prevent access to the danger except when the guard is forced open by the passage of the work. They usually incorporate a spring-loaded pivoted element.

26.6.5 Fixed enclosing guard

The provision of a fixed guard enclosing the whole of a dangerous machine achieves a high standard of guarding. Work can be fed to the machine through an opening in the guard by manual or automatic device arranged so that the operator cannot reach the dangerous parts – a feed tunnel is an example given in BS 5304[5]. By lining the enclosing guard with acoustic material additional protection from noise can be achieved.

In summary, fixed guards have many attractive features. Their disadvantage, compared for example with interlocking guards, is that access is barred equally when the machine is safe as when it is in a dangerous condition.

26.6.6 Interlocked guards

Where guards need to be moved or opened frequently and it is inconvenient to fix them, they can be interlocked mechanically, electrically, pneumatically, etc., to the machine controls. Two basic criteria must be observed: until the guard is closed the machine should not be capable of being started; and the machine should be brought to rest as soon as the guard is opened. Where there is a run down time, the guard may need to be fitted with a delay release mechanism. The interlock must connect to the start controls which should be arranged to prevent uncovenanted motion.

Interlocked guards can be hinged, sliding or removable but the integrity of the design of the interlocking mechanisms is crucial. The mechanisms must be reliable, capable of resisting interference and the system should fail safe.

Interlocking guards are provided to give ready access while ensuring the safety of the operator. However, there are circumstances that may require the machine to be moved when the guard is open, i.e.: for setting, cleaning, removing jams, etc. Such movement is only permitted under the following circumstances:

(1) As part of a 'permit-to-work' system.
(2) On true inch control.
(3) On limited inch control with each movement of the producing part not exceeding 75 mm (3 in) at a predetermined minimum speed.
(4) If for technical reasons the machine or process cannot accept intermittent movement, a continuous movement is permitted provided it is only at a predetermined minimum speed and is controlled by a hold-on switch, release of which causes the machine to stop immediately.

The choice of interlocking method will depend on power supply and drive arrangement to the machine, the degree of the risk being protected against and the consequences of failure of the safety device. The system chosen should be as direct and as simple as possible. Complex systems can be potentially unreliable, have unforeseen fail-to-danger elements and are often difficult to understand, inspect and maintain.

Interlocking mechanisms may be divided into those which lock the guard with the source of power, and those which lock with respect to the motion itself. An EN standard[18] reviews different power and control interlocks. The methods described may be used separately or in combination to achieve an effective interlocking system. Typical elements of interlocking systems are considered below, but it must be appreciated that the effectiveness is as dependent on the integrity of the control circuit – whether electrical, pneumatic, hydraulic, etc. – as on the switch arrangement itself. For normal safe operation, the electrical circuit is complete, or pneumatic/hydraulic circuits are pressurised; actuation of the interlock either breaks the electrical circuit or exhausts the pneumatic/hydraulic circuits.

26.6.6.1 Types of interlocks

(a) Direct manual switch or valve interlocks (*Figure 26.8*) where the switch or valve controlling the power source cannot be operated until the guard is closed, and the guard cannot be opened at any time the switch is in the run position.

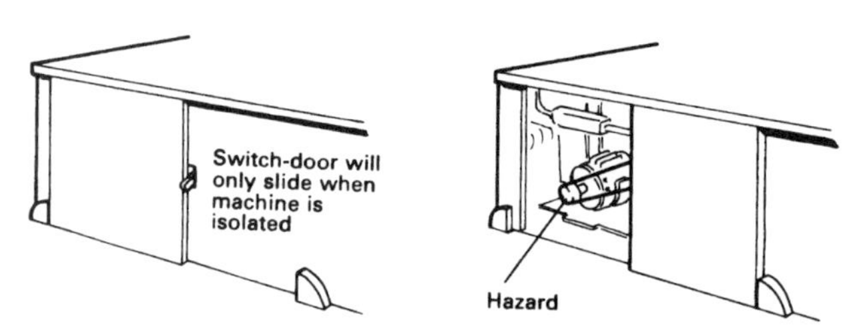

Figure 26.8 Direct manual switch interlock. (Detail design is crucial to ensure an effective arrangement.) (Courtesy Engineering)

(b) Mechanical interlocks provide a direct linkage from the guard to the power or transmission control. The most common application is on power presses (*Figure 26.9*) which are described in detail in an HSE report[19].

(c) Cam-operated limit switch interlocks are versatile, effective and difficult to defeat. They can be rotary (*Figure 26.10*) or linear (*Figure 26.11*) and in each case the critical feature is that in the safe operating position the switch is relaxed, i.e. the switch plunger is not depressed. Any movement of the guard from the safe position causes the switch plunger to be depressed, tripping the controls and stopping the machine. Interlock switches should work in the 'positive mode' (*Figure 26.11(b)*); negative mode operation is not acceptable (*Figure 26.11(a)*). However for certain high risk areas, a combination of positive and negative mode is recommended (*Figure 26.11(c)*) and this arrangement can incorporate a switch failure monitoring circuit as shown in *Figure 26.12*. The type of electrical switch used in interlocking is important. They must be of the positive make-and-break type (known as 'limit' switches) that fail to safety and have contacts capable of carrying the maximum current in the circuit. The principle of operation of such switches is shown in *Figure 26.13*. Micro-switches relying on leaf spring deflection for contact breaking are not acceptable.

(d) Trapped key interlocks (key exchange system) work on the principle that the master key, which controls the power supply to the machine through a switch at the master key box, has to be turned OFF before the keys for individual guards can be released. The master switch cannot be turned to ON until all the individual keys are replaced in the master box. Each individual key will enable its particular guard to be opened, releasing the

Figure 26.9 Interlocking guard, mechanically linked to a crankshaft arrestor fitted to a power press. (Courtesy J. P. Udal Company)

guard key which the operator should take with him when he enters the machine. The individual key is trapped in the guard lock until the guard is replaced and locked by the guard key. A diagram of a typical installation is shown in *Figure 26.14*.

(e) Captive key interlocking involves a combination of an electrical switch and a mechanical lock in a single assembly. Usually the key is attached to the movable guard. When the guard is closed, the key locates on the

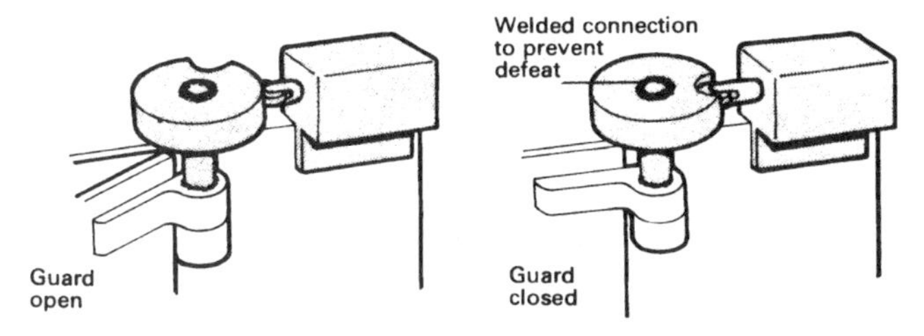

Figure 26.10 Cam activated limit switch for a hinged guard (positive) operation. (Courtesy Engineering)

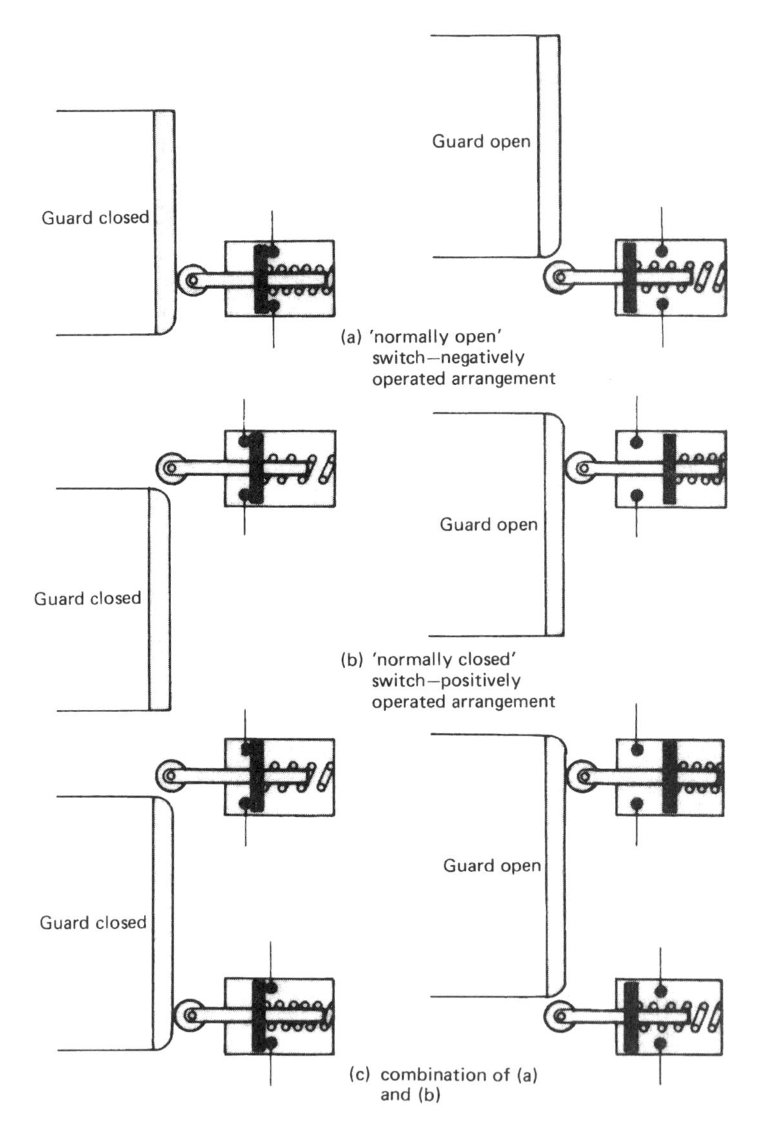

Figure 26.11 Mounting and layout of limit switches for a sliding guard. (From BS 5304:1988)

switch spindle. First movement of the key mechanically locks the guard shut and further movement actuates the electrical switch to complete the safety circuit (*Figure 26.15*).

(f) Magnetic switches with the actuating magnet attached to the guard have the disadvantage that they can be defeated easily. However, types using a shaped magnet have been accepted and in use for many years (*Figure 26.16*). Normal reed switches are not acceptable unless they incorporate special fail safe and current limiting features. Reed switches are often

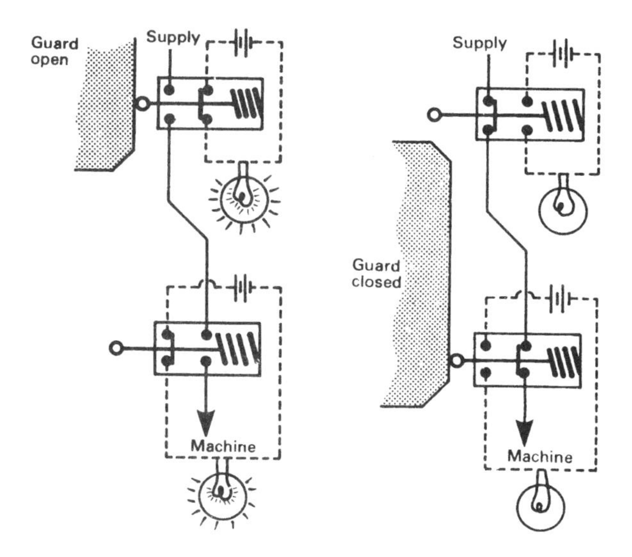

Figure 26.12 Limit switch failure monitoring circuit. (Courtesy Engineering)

Figure 26.13 Limit switch with cover removed showing the principle of fail-safe operation. (Courtesy Dewhurst & Partner p.l.c.)

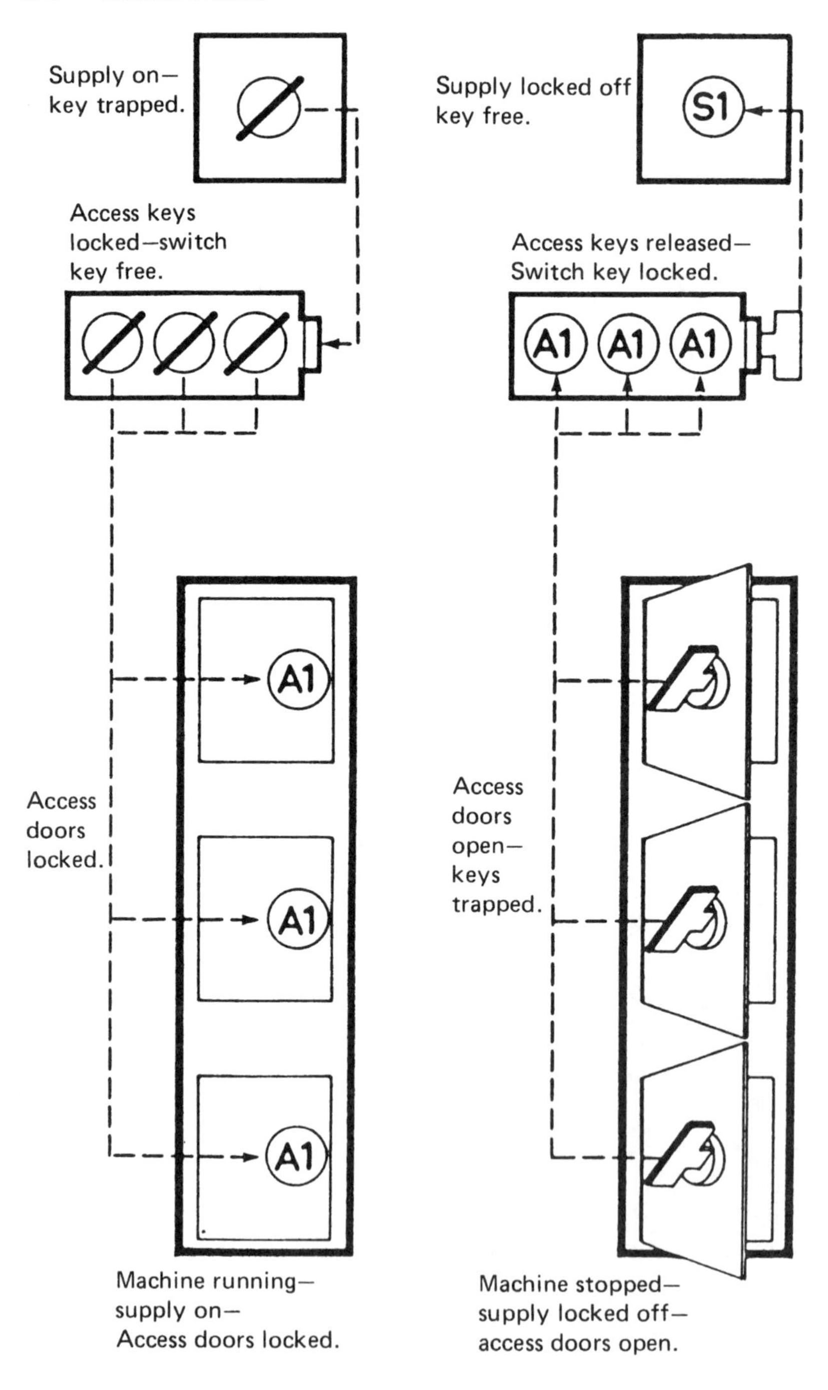

Figure 26.14 Diagrammatic arrangement of a key exchange system applied to a machine with a number of access doors required to be open at the same time. (Courtesy Castell Safety International Ltd)

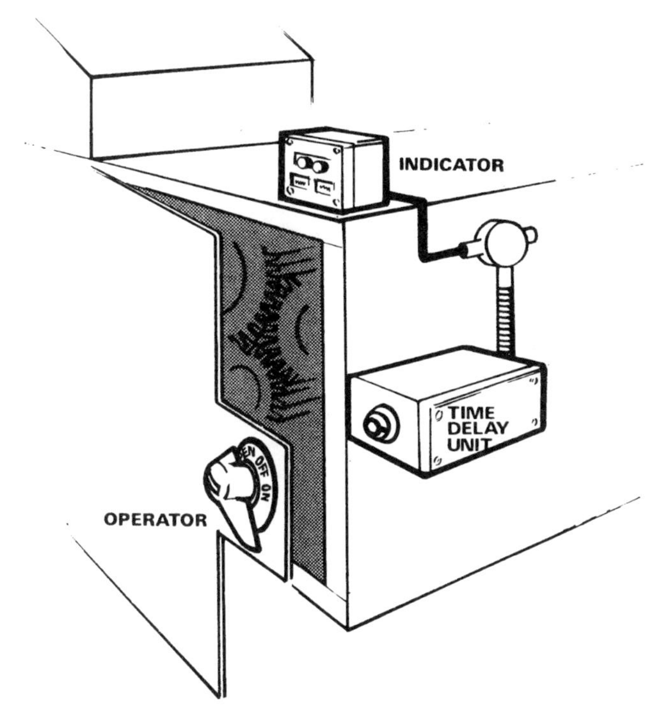

Figure 26.15 Captive key safety switch with electrical time delay release device. (Courtesy Unimax Switch Ltd)

encapsulated and have application in flammable atmospheres. A similar type of switch relies on inductive circuits.

(g) Time delay arrangements are necessary when the machine being guarded has a large inertia and, consequently, a long run down time on stopping. An electro-mechanical device is shown in *Figure 26.17* where the first movement of the bolt trips the machine, but the bolt has to be unscrewed a considerable distance before the guard is released. A solenoid-operated bolt can also be used in conjunction with a time delay circuit that is actuated from both the machine controls and the trip circuit.

(h) Mechanical scotches are required on certain types of presses to protect the operator when reaching between the platens. These scotches can be linked to the guard operation so that they are automatically positioned each time the guard is opened.

26.6.7 Automatic guard

This type of guard closes automatically when the machine cycle is initiated and is arranged so that the machine will not move until the guard is in the safe position. The machine then commences its movement automatically. Where trapping points occur as the parts of the guard come together,

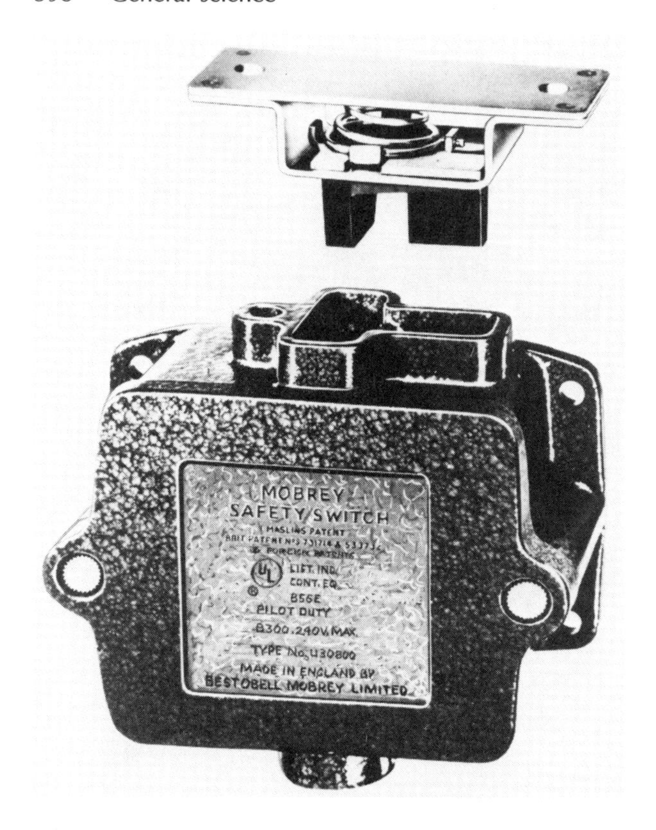

Figure 26.16 Magnetic switch with shaped magnet. (Courtesy Bestobell Mobrey Ltd)

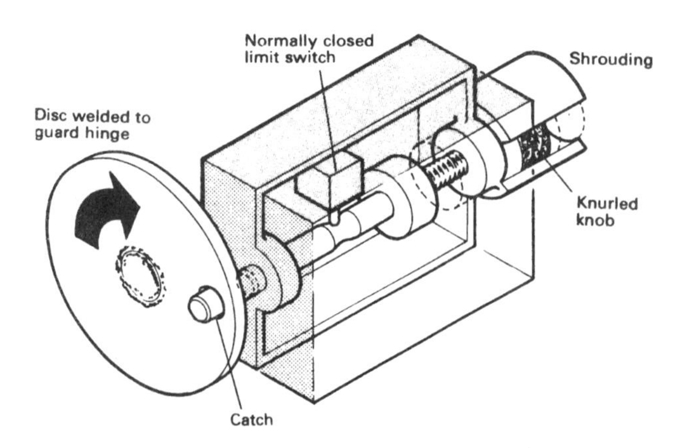

Figure 26.17 Time delay for limit switch; a shroud prevents the knurled knob from being spun too quickly. (Courtesy Engineering)

interlocked trip devices should be fitted. A version of this type of guard moves across the work area as the machine operation is initiated, removing any part of the operator that is in that area. This type is sometimes known as a 'sweep-away' guard.

26.6.8 Trip devices

A trip device ensures that an approach to a dangerous part beyond a safe limit stops or reverses the machine (*Figure 26.18*). These devices should be arranged so that they are actuated involuntarily by the movement of the limb into the danger area. Trip devices include: trip bars and wires, sensitive probes (for radial drills), photoelectric devices and pressure-sensitive strips and mats. In use it is important that trip devices are properly adjusted and

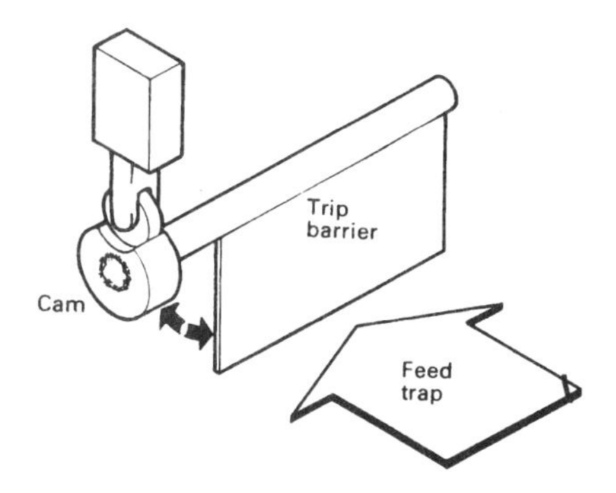

Figure 26.18 Hinged barrier with cam-operated limit switch. (Courtesy Engineering)

that the machine's brake is in good working order. Emergency stop wires, such as those mounted adjacent to conveyors, and emergency stop buttons require an operator to actuate them; while they have their value, they afford much less protection than a well maintained trip device. The safe distance of trip barriers and electrosensitive screens from the dangerous part is largely a function of the speed at which a person can move into the danger area and the rapidity with which the dangerous parts can be brought to rest. An EN standard[20] gives data on the approach speeds of parts of the body.

26.6.9 Two-hand control devices

Two-hand controls are used on machines on cyclic operations where the work is placed in the machine and the machine struck on. They are applicable only to machines with a single operator and the control buttons must be positioned more than a hand-span apart. The control circuit should be arranged so that both controls must be activated simultaneously to start the cycle and that both controls must be released after each cycle before the next cycle can be initiated. Release of either button during the dangerous part of the cycle must stop or reverse the machine movement. Details are specified in an EN standard[21].

26.6.10 Control guards

On some small machines, such as forming presses and gold blocking machines, in which the guard has to be moved out of the way to place and remove the work, the guard can be arranged so that as it is moved to the safe position it actuates the machine cycle, and all the time the machine is moving the guard is locked in position. This type of guard also has application on some large panel presses.

26.7 Anthropometric aspects of machinery guarding

26.7.1 Reach-through openings

While the prime object of all types of machinery guard is to prevent those operators at risk from making contact with the dangerous parts, there is often a need to see through the guard to check on the workpiece, and in some cases there is a need to feed through the guard the material being worked on. In both cases, the gaps provided must not interfere with the integrity of the guard. A number of standards exist that give guidance on the size of these openings and the one currently used in the UK is that given in BS 5304[5]. Further data are given in an EN standard[22]. In his 1975 Report[23], the Chief Inspector of Factories recommended the following formula:

$$y = \frac{x}{12} + 6, \quad \text{where } y = \text{gap in mm and } x = \text{distance from danger in mm.}$$

The above equation is primarily intended to give the depth of gap for wide slots, but it can be applied to square and round openings also. The sizes it gives fall within those recommended in BS 5304 for elongated, square and round openings.

26.7.2 Reach-over barriers

The most frequently quoted figures are those provided in the German Standard DIN 31001:1976[24], but this recommendation appears to allow a substantial amount of unsafe access to dangerous parts of machinery. Booth and Thompson[25] found that male subjects could reach substantially beyond the distances referred to in the German standard and their findings are presented in *Table 26.4*. In the absence of coherent standards, distance fences should be located further from the danger point than the maximum reach figures quoted in the table. A trip rail or wire along the top of a distance fence can provide additional protection for those who lean over it.

26.8 Summary checklist for machinery safety

The provision of the high standard of safeguarding when linked with appropriate systems of work, permit-to-work systems, and safeguard maintenance programmes is the most effective method of preventing machinery accidents and the following seven questions present a useful checklist for considering whether the safeguards are appropriate.

Table 26.4. Forced reach distances over barrier rails (from Booth and Thompson)

Summary of results of reach profile of 95th percentile male: Horizontal reach (mm)

Distance of hazard point from floor (mm)	*Height of edge of safety feature (barrier) (mm)*						
	2200	*2000*	*1800*	*1600*	*1400*	*1200*	*1000*
2400	*100**	*100*	*100*	*100*	*100*	*100*	*100*
	308	425	534	486	591	701	761
	+208	+325	+434	+386	+491	+601	+661
2200	*250*	*350*	*400*	*500*	*500*	*600*	*600*
	321	528	699	718	794	920	994
	+71	+178	+299	+218	+294	+320	+394
2000		*350*	*500*	*600*	*700*	*900*	*1100*
		537	785	824	920	1071	1171
		+187	+285	+224	+220	+171	+70
1800		–	*600*	*900*	*900*	*1000*	*1100*
		477	826	890	993	1189	1295
		+477	+226	–10	+93	+189	+195
1600		–	*500*	*900*	*900*	*1000*	*1300*
		271	835	898	1026	1267	1397
		+271	+335	–2	+125	+267	+97
1400			*100*	*800*	*900*	*1000*	*1300*
			764	869	1026	1318	1456
			+664	+69	+125	+318	+156
1200				*500*	*900*	*1000*	*1400*
				788	976	1322	1484
				+288	+76	+322	+84
1000				*300*	*900*	*1000*	*1400*
				637	889	1310	1482
				+337	–11	+310	+82
800					*600*	*900*	*1300*
					740	1273	1449
					+140	+373	+149
600					–	*500*	*1200*
					422	1160	1373
					+422	+660	+173
400						*300*	*1200*
						1116	1313
						+816	+113

* *100* = Safety distance (reach distance + safety allowance) recommended by DIN 31 001.
308 = Reach distance (+ 3 S.D.) found be experiment.
+ 208 = Distance by which experimental reach exceeded (+) or failed to reach (–) DIN standard.

(1) Does the safeguard when in its correct position and when working properly totally prevent approach to dangerous parts?
(2) Is the safeguard reasonably convenient to use (i.e. does it interfere with either the speed or quality of the work); are there foreseeable reasons why it is necessary to override the safeguard?
(3) Can the guard be defeated or is it susceptible to misuse?
(4) Will the safeguard cope with foreseeable machine failure?
(5) Are the components of the safeguard reliable; and do they fail-safe?

(6) Is the safeguard straightforward to inspect and maintain?
(7) Are the instructions for safe use of the machine and safeguard adequate to cope with all foreseeable dangers in use?

References

1. EC, *Directive on the approximation of the laws of Member States relating to machinery*, No. 89/392/EEC, EC, Luxembourg (1989)
2. EC, *Directive concerning the minimum safety and health requirements for the use of work equipment by workers at work*, No. 89/655/EEC, EC, Luxembourg (1989)
3. Department of Trade and Industry, *The Supply of Machinery (Safety) Regulations 1992*, HMSO, London (1992)
4. Health and Safety Executive, *The Provision and Use of Work Equipment Regulations 1992*, HMSO, London (1992)
5. British Standards Institution, *Code of Practice for Safety of Machinery*, BS 5304:1988, BSI, London (1988)
6. Health and Safety Executive, *Publications in Series; List of HSE/C publications*, Health and Safety Executive, Sheffield and local offices, updated twice yearly
7. Booth, R.T., Waring A. and Raafat, H.M.N. *Machinery Safeguarding*, Safety Technology Unit ST1, Portsmouth University, Portsmouth (1989)
8. Summers (John) & Sons Ltd *v.* Frost, (1955) AC 740; (1955) 1 All ER 870
9. Uddin *v.* Associated Portland Cement Manufacturers Ltd (1965) QB 582; (1965) 2 All ER 213
10. United States Department of Labour, *Concepts and Techniques of Machine Safeguarding*, OSHA 3067, US Government Printing Office, Washington USA (1980)
11. British Standards Institution, *Safety of Machinery – Risk Assessment*, BS EN 1050, BSI, London (to be published)
12. British Standards Institution, *Safety of Machinery – Basic Concepts, General Principles for Design, Part 1: Basic Terminology, Methodology*, BS EN 292, BSI, London (to be published)
13. Surry, J., *Industrial Accident Research*, University of Toronto, Department of Industrial Engineering, 22 (1974)
14. Raafat, H.M.N., 'Risk assessment and machinery safety', *Journal of Occupational Accidents*, **11**, pp. 37–50 (1989)
15. British Standards Institution, *Safety of Machinery – General Requirements for the Design and Construction of Guards (Fixed, Movable)*, BS EN 953, BSI, London (to be published)
16. British Standards Institution, *Safety of machinery – Safety Distances to Prevent Danger Zones being Reached by Upper Limbs*, BS EN 294, BSI, London (to be published)
17. British Standards Institution, *Safety of Machinery – Safety Distances to Prevent Danger Zones being Reached by Lower Limbs*, BS EN 811, BSI, London (to be published)
18. British Standards Institution, *Safety of Machinery – Interlocking Devices with and without Guard Locking; General Principles for Design*, BS EN 1088, BSI, London (to be published)
19. Health and Safety Executive, *Power Press Safety*, Standards prepared by the Joint Standing Committee on safety in the use of power presses, HSE Books, Sudbury (1979)
20. British Standards Institution, *Safety of Machinery – Hand/Arm Speed; Approach Speed of Parts of the Body*, BS EN 999, BSI, London (to be published)
21. British Standards Institution, *Safety of Machinery – Two-hand Control Devices*, BS EN 574, BSI, London (to be published)
22. British Standards Institution, *Safety of Machinery – Human Body Dimensions (7 parts)*, BS EN 547, BSI, London (to be published)
23. Health and Safety Executive, *Health and Safety, Manufacturing and Service Industries*, HSE Books, Sudbury (1975)
24. Deutsches Institut für Normung, *Safety Design of Technical Equipment, Safety Devices, Definitions, Safety Distances for Adults and Children*, DIN 31001 (1976)
25. Booth, R.T. and Thompson, D., Anthropometry and machinery guarding: An investigation into forced reach over barriers and through openings, part 1 – Reach over barriers, *Journal of Occupational Accidents*, **3**, 45 (1980)

Chapter 27

Mechanical handling

J. R. Ridley

27.1 Internal powered trucks

Internal powered trucks refer to those trucks that are employed for internal operations within a factory or works and which have proved to be one of the most efficient means in industry of moving materials. They are highly mobile, very flexible in use and can be fitted with a variety of attachments. Because of their mobility and adaptability care must be exercised in their use. Throughout industry there are numerous types and sizes of powered trucks and many of the features that contribute to their safe operation are common to all. Broadly these features divide into three sections:

(1) Safe operating conditions.
(2) Hazards to be avoided.
(3) Operating techniques and training.

27.1.1 Conditions for the safe operation of trucks

To maximise the performance and utilisation of a powered truck, the conditions under which it operates must be suitable. Bad housekeeping, poor plant layout and wrong vehicle for the job are some of the features that can prevent its full potential being achieved. The more important of these features include:

27.1.1.1 The floor

The floor must be of suitable construction and capable of withstanding the point loading of powered truck wheels as well as the stacks or racks placed on it. It should be flat although a maximum gradient of 1/60 for drainage purposes is acceptable. Any gullies should have bridging plates and the floor should be maintained in good condition and free from rubbish.

27.1.1.2 Storage and stacking areas

These areas should be laid out so that there is adequate room to manoeuvre the trucks, especially in the rack aisles. Truck manufacturers will give the minimum width of stacking aisle required for their trucks and a simple

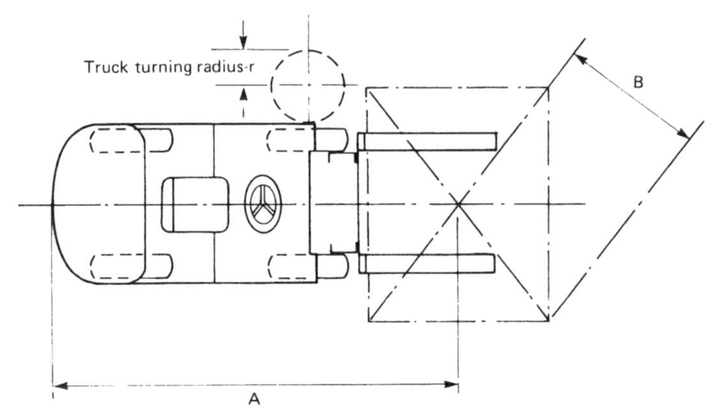

Aisle width $W = A + B + 10\%$

Example: if $A = 1800$ mm
$B = 800$ mm
aisle width $W = (1800 + 800) + 10\%$
$= 2600 + 10\% = 2860$ mm

Where trucks have a large turning circle (e.g. i/c engined trucks) the inner turning radius 'r' must be added, thus aisle width $W_1 = (A + B + r) + 10\%$

Figures 27.1 Width of stacking aisle for fork trucks

method for calculating it is given in *Figure 27.1*. Blind corners should be avoided and, if necessary, passing places provided. Stacking areas should be clearly defined.

With pallet racking, due allowance must be made for pallet overhang when calculating aisle widths (*Figure 27.2*). There should always be a space

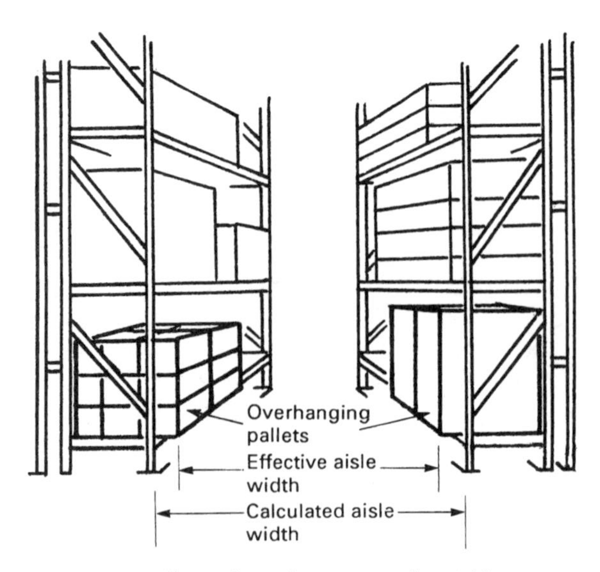

Figure 27.2 Effect of overhang on aisle width

of at least 100 mm (4 in) between the top of a load and the racking transom above to allow safe lifting of the load during removal. This also provides additional visibility at corners and intersections.

27.1.1.3 Lighting

The position of lighting, natural or artificial, in stacking areas is more important than the quality, with lighting over gangways being better than over stacks. Guidance on the quality and type of lighting is given in the Illuminating Engineering Society's code[1] and an HSE guidance booklet[2]. Legislative requirements for adequate lighting are contained in regulation 21 of PUWER[3].

27.1.1.4 Loading and unloading areas

Where loading docks are used, they should be provided with dock levellers and if the trucks need to drive onto the vehicle, the condition of the platform should be checked and the vehicle wheels chocked. When loading or unloading from the ground, adequate space should be allowed between and around vehicles (*Figure 27.3*).

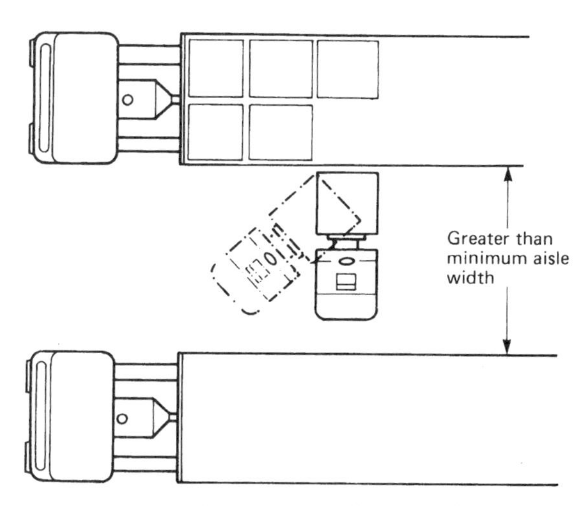

Figure 27.3 Gap between vehicles when being loaded

Trestles or mobile support frames placed under the long overhang of free-standing trailers will prevent them from tipping (*Figure 27.4*). Wheel scotches must be used where the ground is sloping and the trailer parking brakes must be applied. Reliance should not be put on vacuum or air brakes.

27.1.1.5 Ramps and slopes

The maximum gradient of a ramp or slope used by trucks should not exceed 1:10 except with the manufacturer's written agreement and trucks should not be driven across a slope.

Figure 27.4 Trestle to support long overhang of parked trailers during loading

27.1.1.6 Battery charging areas

Areas for charging traction batteries should be separate from the main work area and have a roof region that is well ventilated. Charging bays should be designated as NO SMOKING/NO NAKED LIGHTS areas. Good lighting is required to read the charger instruments. For interchangeable batteries, adequate lifting and moving facilities should be provided. Eye washing facilities must be available.

Figure 27.5 Separation of truck and pedestrian gangways

27.1.1.7 Pedestrians

Ideally, pedestrians should be excluded from trucking areas. However, where this is not possible, separate clearly marked pedestrian gangways should be provided and trucks should give right-of-way to pedestrians (*Figure 27.5*).

27.1.1.8 Trucks

Care should be taken to match the size, type and capacity of the truck to the job it is to do. In service, the manufacturer's maintenance schedules should be followed with any necessary repairs being carried out promptly. High-lift trucks should have an overhead guard and those used to load trailers and lorries should have a guard behind the driver (*Figure 27.6*).

Figure 27.6 High lift fork truck with load rest, overhead and rear guards. (Courtesy Hyster Europe Ltd)

27.1.1.9 Truck attachments

Many attachments are available for powered trucks and they can greatly extend the trucks' flexibility, utilisation and efficient operation. Those available include:

(1) Load rests which should be provided on high-lift trucks and be arranged to support the highest load (*Figure 27.6*).
(2) Side shift is a facility for limited side movement of the load to assist in its accurate placement.

Figure 27.7 Fork truck fitted with load clamp

(3) Clamps which are used for bales, unit loads and reels should have a non-return valve fitted in the hydraulic circuit such that full clamping pressure is maintained on failure of the hydraulic system. Release of the clamp should be by a separate control requiring positive action.
(4) Load clamps are used for restraining loosely stacked loads whilst being carried on the truck (*Figure 27.7*).
(5) Working platforms to give access to heights. The platform design and procedures for its use should follow the advice in the HSE's Guidance Note[4].

27.1.2 Hazards in truck operations

27.1.2.1 Overturning

This often occurs when manoeuvring with the load elevated or at too high a speed. Other causes are:

(1) Sudden braking.
(2) Wheels striking obstruction.
(3) Use of forward tilt with elevated load.
(4) Driving down ramp with load preceding truck.
(5) Striking overhead obstruction when reversing.
(6) Turning on or crossing ramps.
(7) Load shifting.
(8) Wheels sinking into unsuitable floor or ground.
(9) Overloading.

27.1.2.2 Overloading

The maximum safe load is shown as the RATED CAPACITY defined by a weight and distance, e.g. 1000 kg at 500 mm load centre, or 2240 lb at 20 in load centre. The load centre is measured from the heel of the forks to the

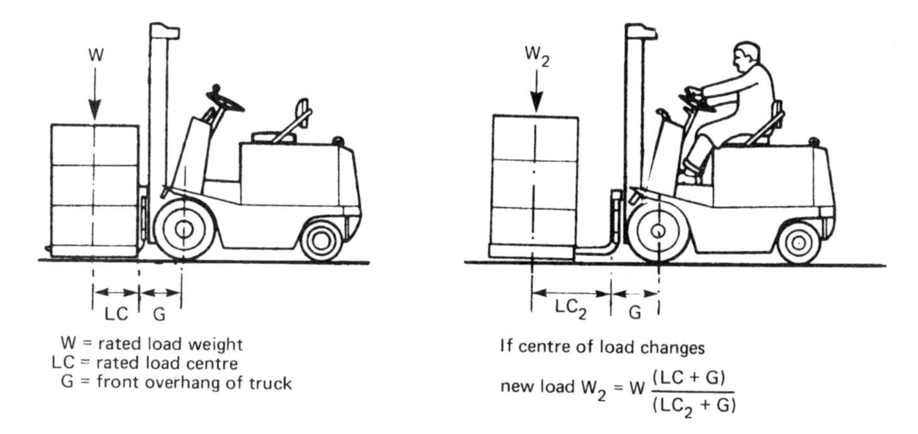

Figure 27.8 Rated load of fork truck

centre of the load (*Figure 27.8*). If the load centre increases, the carrying capacity decreases. On high-lift trucks the manufacturers specify the safe loads at various heights.

27.1.2.3 Collisions

Where collisions with parts of the building or racking occur, extensive and costly damage can be caused. Similarly, injuries caused by being hit by a truck can be very serious.

27.1.2.4 Floor level hazards

With the exception of rough terrain trucks, manufacturer's specifications are usually based on operations on FIRM, LEVEL FLOORS. The existence of gullies, potholes and debris on the floor should not be tolerated.

27.1.2.5 Overhead obstructions

Pipes, cables and other fixtures installed within the height that a truck mast can reach have a high damage potential, and avoiding them reduces the operator's efficiency.

27.1.2.6 Loss of loads

Load displacement occurs because recognised methods of load restraint are not used. Every load should be supported to its full height by a load rest (*Figure 27.6*). Careful use of backward tilt will prevent the load falling forward. Bonded stacking and load ties of paper, cardboard, thin timber battens, etc. at vertical intervals between individual levels of a stack will help to stabilise a load. Shrink or stretch wrapping or cardboard inserts can stabilise loose loads (*Figures 27.9* and *27.10*).

Figure 27.9 Bonded and wrapped pallet load

Figure 27.10 'Tied' pallet load. (Courtesy Spicers Ltd)

27.1.2.7 Truck failures

Failure of the truck can be prevented by regular checks being made and this will also ensure that the truck remains safe to operate. Particular attention should be paid to the following:

(1) Brakes.
(2) Steering.
(3) Chains.
(4) Hydraulic system.
(5) Condition of forks.
(6) Other attachments.
(7) Horn.
(8) Wheel nuts.
(9) Tyres.
(10) Electric wiring insulation.
(11) Fuel supply systems on i/c engined trucks.
(12) Cleanliness of power unit, i.e. free from dust, oil, etc.

27.1.2.8 Explosions and fire

On battery electric trucks, hydrogen is generated during charging and can be ignited by a battery spark to cause a damaging explosion. Explosions on electric trucks can be caused by the contact of live cables with the metal of the truck frame and the shorting of battery terminals by metal tools or lifting tackle. Battery terminals should have suitable insulating covers. Any cover put over the battery during charging should allow the free passage upwards of any hydrogen generated in the cells.

A recurring cause of fires on i/c engine powered trucks is lack of basic cleanliness. Dust accumulates on the engine which becomes hot, igniting the dust. Leaking fuel pipes are another frequent cause of fires.

27.1.2.9 Passengers

Passengers should not be carried on a powered truck unless there is a properly fitted and suitable passenger seat. Provision should be made for passenger security during truck movement.

27.1.3 Safe operating techniques and training

The provision of good operating conditions for trucks and the elimination of hazards is no guarantee that the trucks will be operated safely. This will always depend on the knowledge and skill of the operator in following recognised safe methods of operation. All truck operators must be trained in these methods. The training of operators of rider operated trucks must comply with that laid down in the HSE's Code of Practice[5].

27.2 Cranes

Perhaps the most commonly used piece of handling equipment is the crane, which over the years has been developed to meet highly specialised applications, with the result that there is now a great range of types and sizes in use in industry, the docks and on construction sites. As a result of accidents in the past, a body of legislation has grown up which covers the construction and use of cranes. The major statutes relating to the construction and use of existing cranes are the Factories Act 1961, the Construction (Lifting Operations) Regulations 1961 and the Docks Regulations 1988[6]. However all new equipment must carry the EC mark as evidence that it conforms to the requirements contained in the Machinery Directive[7] and embodied in the Supply of Machinery Safety Regulations 1992 (SMSR)[8]. Conformity requires the compilation of a 'technical file' on the equipment which includes records of static and dynamic tests at specified 'test coefficients', i.e. working factors of safety. The periodic examinations required by s. 27 of FA 1961 still apply.

In each case there is a general requirement that the equipment shall be of good construction, sound material, adequate strength and free from patent defect. In addition there are requirements that the equipment is inspected regularly and that it is properly used.

There are a number of common techniques that apply to the safe operation of every type of crane and these are dealt with below.

27.2.1 Identification and capacity

The manufacturer must issue a test certificate for every crane he produces, identifying it and specifying the safe working load (SWL) which must be clearly marked on the crane structure. In large organisations which have their own system of plant identification it should relate to the manufacturer's certificate.

27.2.2 Maintenance

Apart from statutory inspections, cranes should be inspected regularly by the user's own staff and preventative maintenance schedules followed. Records should be kept of these checks and of any work done, especially if it is in response to a statutory inspection requirement.

27.2.3 Safety measures

To protect both the operator and the crane itself, a number of safety devices are incorporated in the design of the crane and these include:

27.2.3.1 Overtravel switches

To prevent the hook or sheave block from being raised right up to the cable drum, a robust limit switch should be fitted to the crab or upper sheave block. Checks of this limit switch should be included in routine inspections.

27.2.3.2 Protection of bare conductors

Where bare pick-up conductors are used to carry the power supply they must be shielded from accidental contact particularly if near cabin access. Suitably worded notices, e.g. WARNING – BARE LIVE WIRES, should be posted on the walls or building structure. The power supply isolating switch should be provided with means for locking-off during maintenance work.

27.2.3.3 Controls

The controls of cranes, whether cabin, pendant or radio, should be clearly identified to prevent inadvertent operation. On overhead electric travelling (OET) cranes with electric pendant controls the directions of travel should be unambiguously marked. Controls should be of the 'dead-man' type.

27.2.3.4 Load indicators

Load indicators are required to be fitted to jib cranes and can be used with benefit on all cranes. There are two types:

(1) A load/radius indicator that shows the radius at which the crane is working and the safe load at that radius. It must be clearly visible to the driver.
(2) An automatic safe load indicator that warns, either visually or audibly, that the load is approaching the maximum safe level or that it has exceeded that value.

Where a load indicator is used for work under the Construction (Lifting Operations) Regulations 1961 it should be of an approved type.

27.2.3.5 Safety catches

Wherever practicable, crane hooks should be fitted with safety catches[9] to prevent slings, chains, ropes, etc. from 'jumping' off the hook (*Figure 27.11*).

27.2.3.6 Emergency escape

Where, on travelling cranes, access to the cab is not an integral part of the crane, suitable escape equipment should be provided to enable the driver to reach the ground quickly and safely in an emergency.

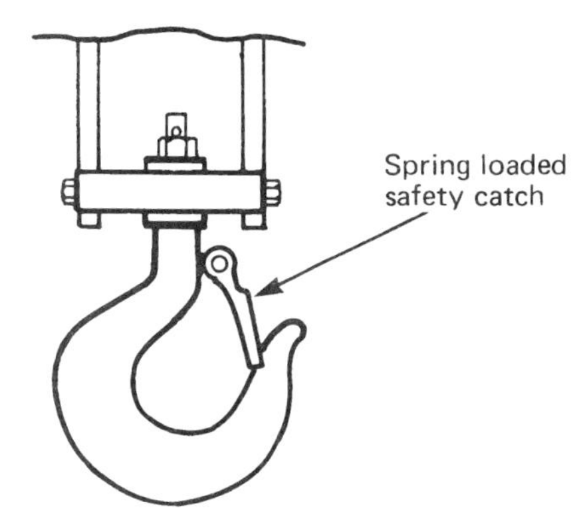

Figure 27.11 Crane hook with safety catch. (Source: Dickie and Short[9])

27.2.3.7 Access

Safe means of access should be provided to enable:

(1) the driver to reach his operating position;
(2) the necessary inspections and maintenance work to be carried out safely.

27.2.3.8 Operating position

The arrangement of the driver's cab should ensure:

(1) a clear view of the operating area and loads;
(2) all controls are easily reached by the driver without the need for excessive movement of arms or legs;
(3) all controls are clearly marked as to their function and method of operation.

27.2.3.9 Passengers

No one, other than the driver, should be allowed on the crane when it is operating unless there is a special reason for being there and it has been authorised. 'Riding the hook' is prohibited but should it be necessary to carry persons, the properly designed and approved chair or cradle should be used.

27.2.3.10 Safe working load

All cranes should be marked with their safe working load which must never be exceeded except for test purposes. If there is any doubt of the weight to be lifted, advice should be sought.

27.2.4 Crane operations

With many cranes including overhead electric travelling, mobile jib and construction tower cranes, the safe moving of loads relies on team effort involving the driver, slinger and sometimes a separate signaller (or banksman). Only one person, the signaller or if there is no signaller the slinger, should give signals to the driver and these should be clearly understood by both. The basic signals[10] are shown in *Figure 27.12*.

Slingers, signallers and drivers should be properly trained – in the latter two cases this is a requirement under the Construction (Lifting Operations) Regulations 1961 – medically fit and of a steady disposition.

27.2.5 Safe operating techniques for cranes

There are certain basic techniques that must be practised to ensure the safest possible operation of the crane. These techniques include:

(1) The driver carrying out a brief check of the crane at the beginning of his shift. The items checked may vary in different factories, but they should be agreed locally.
(2) Loads should not be left hanging on the hook at the end of a shift.
(3) Those working above floor level who are likely to be struck by the crane or its load must be warned.
(4) The crane must not approach nearer than 20 ft to anyone working on the track.
(5) Loads should not be carried over people.
(6) Loads must be lifted vertically – the crane must not be used to drag a load.
(7) Power to the crane should be switched off when the crane is left unattended.

Mobile jib cranes present extra hazards in use and the British Standard Code of Practice[11] should be followed. Particular points requiring attention include:

(8) The travelling height of the crane with jib lowered and raised should be marked on the crane.
(9) Overhead obstructions or hazards should be clearly identified and marked with the maximum clearance height.
(10) Care should be taken to ensure that no one is trapped by counter-balance weights, etc., as the crane slews.
(11) The condition of the ground should be checked before lifting with a mobile crane and it should not approach close to excavations.
(12) Tandem lifts should be avoided if possible. If they must be undertaken, it must be by a fully trained gang under the strict control of a competent person.

Detailed advice is given by Dickie and Short[12] and note should be taken of an HSE report on three crane accidents[13].

27.2.6 Lifting accessories

'Lifting accessories' are defined in SMSR as any component or equipment used to join the load to the lifting hook and 'separate lifting accessories' are

Figure 27.12 Crane signals. (BS 5744:1979)

Table 27.1. Maximum safe working loads for alloy steel chain slings. (Courtesy Parsons Chain Company)

KUPLEX GRADE T WORKING LOAD LIMITS
Based on a mean stress of 200 N/mm^2 (MPa)

Size	Single Leg Slings	Endless Slings	Two Leg Slings				
				30°	60°	90°	120°
mm	tonnes	tonnes	tonnes	tonnes	tonnes	tonnes	tonnes
7	1.5	2.25	3.0	2.9	2.6	2.1	1.5
10	3.2	4.8	6.4	6.1	5.5	4.5	3.2
13	5.4	8.1	10.8	10.4	9.3	7.6	5.4
16	8.0	12.0	16.0	15.4	13.8	11.3	8.0
19	11.5	17.25	23.0	22.2	19.9	16.2	11.5
23	16.9	25.3	33.8	32.6	29.25	23.9	16.9
26	21.6	32.4	43.2	41.7	37.4	30.5	21.6
32	32.0	48.0	64.0	61.8	55.4	45.2	32.0

those items that 'help to make up a slinging device, such as eyehooks, shackles, rings, eyebolts, etc.' All lifting accessories must be tested to an overload specified in the Regulations before being put to use. In addition s. 26 of FA 1961 remains in effect and requires that all chains, ropes and lifting tackle used for lifting persons, goods or materials shall be of good construction, sound material, adequate strength, free from patent defect and certified as to its safe working load. In any establishment covered by the Act a register must be kept of all chains, ropes and lifting tackle. There are certain exceptions with regard to fibre rope slings. Strict control should be exercised in the use and storage of lifting tackle.

27.2.6.1 Chain slings

Chains prepared for slinging purposes are in the form of 2, 3, or 4 legged chain slings each with its own hook and connected at the other end to one common lifting ring. No chain should be passed round sharp corners nor should it be tied in knots to shorten it. The safe working load of a chain sling varies with different angles of the legs as shown on the chart (*Table 27.1*). However a British Standard[14] recommends the rating and marking of lifting gear.

27.2.6.2 Wire rope slings

Made of a length of wire rope, the slings have a loop or eye formed at each end. The recommended methods for forming the eyes are to loop the wire round a rope thimble and lay it back on itself, securing the end in position

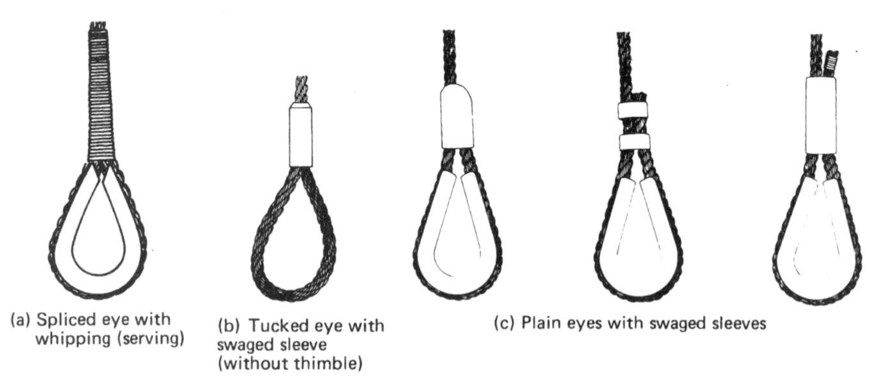

Figure 27.13 Types of wire rope eyes. (Source: Dickie and Short[9])

by either an Admiralty or 'cross-tuck' splice or a swaged sleeve or ferrule. These are shown[9] in *Figure 27.13*. Every wire sling should carry an identifying tag which also gives its safe working load. Wire rope slings are very vulnerable to misuse. They should be checked frequently and have a thorough examination every six months. Typical faults include kinking, bird caging and needling. Charts should be displayed showing the safe working loads of wire rope slings for different leg angles (*Table 27.2*).

Table 27.2. Safe working loads for slings constructed from ropes with fibre cores (BS 1290: 1983)

Rope diameter	*SWL*									
	Single-leg slings		*Multi-leg slings*							
	Single-part terminated by ferrules or splices	*Double-part endless or double-part grommet*	*Leg angle $0° \leq \alpha \leq 90°$ $0° \leq \beta \leq 45°$*				*Leg angle $90° < \alpha \leq 120°$ $45° < \beta \leq 60°$*			
			Two-leg		*Three- or four-leg*		*Two-leg*		*Four-leg*	
			Single part leg	*Double-part leg*	*Single part leg*	*Double-part leg*	*Single part leg*	*Double-part leg*	*Single part leg*	*Double-part leg*
A	*B*	*C*	*D*	*E*	*F*	*G*	*H*	*I*	*J*	*K*
mm	t	t	t	t	t	t	t	t	t	t
5	0.276*	0.414*	0.386*	0.579*	0.579*	0.869*	0.276*	0.414*	0.414*	0.621*
6	0.398	0.597	0.557	0.836	0.836	1.2	0.398	0.597	0.597	0.895
7	0.542	0.813	0.759	1.1	1.1	1.7	0.542	0.813	0.813	1.2
8	0.762	1.1	1.0	1.5	1.6	2.3	0.762	1.1	1.1	1.6
9	0.962	1.4	1.3	1.9	2.0	2.9	0.962	1.4	1.4	2.1
10	1.2	1.8	1.7	2.5	2.5	3.8	1.2	1.8	1.8	2.7
11	1.4	2.1	1.9	2.9	2.9	4.4	1.4	2.1	2.1	3.1
12	1.7	2.5	2.4	3.5	3.5	5.2	1.7	2.5	2.5	3.7
13	2.0	3.0	2.8	4.2	4.2	6.3	2.0	3.0	3.0	4.5
14	2.3	3.5	3.2	4.9	4.8	7.3	2.3	3.5	3.4	5.2
16	3.0	4.5	4.2	6.3	6.3	9.4	3.0	4.5	4.5	6.7
18	3.8	5.7	5.3	8.0	8.0	11.0	3.8	5.7	5.7	8.5
19	4.3	6.4	6.0	8.9	9.0	13.4	4.3	6.4	6.4	9.6
20	4.7	7.1	6.6	9.9	9.8	14.9	4.7	7.1	7.0	10.6
22	5.7	8.6	8.0	12.0	11.9	18.0	5.7	8.6	8.5	12.9
24	6.8	10.2	9.5	14.3	14.3	21.4	6.8	10.2	10.2	15.3
26	8.0	12.0	11.0	16.8	16.8	25.2	8.0	12.0	12.0	18.0
28	9.3	14.0	13.0	19.6	19.5	29.4	9.3	14.0	13.9	21.0
32	12.1	18.2	16.9	25.5	25.4	38.2	12.1	18.2	18.1	27.3
35	14.5	21.8	20.3	30.5	30.4	45.8	14.5	21.8	21.7	32.7
36	15.4	23.1	21.5	32.3	32.3	48.5	15.4	23.1	23.1	34.6
38	17.1	25.7	23.9	36.0	35.9	53.9	17.1	25.7	25.6	38.5
40	19.0	28.5	26.6	39.9	39.9	59.8	19.0	28.5	28.5	42.7

* SWLs of less than 1.0 t are normally cited in kilograms.

27.2.6.3 Fibre rope slings

Probably the most versatile type of sling and also the most abused, fibre rope slings can be made of natural or man-made fibres. Where intended for use in a corrosive or hot atmosphere, the maker's advice should be sought on the type to be used. A table giving safe working loads for fibre rope slings (*Table 27.3*) should be displayed where they are used. Because of their

Table 27.3. Maximum safe working loads up to the angle illustrated for man-made fibre rope slings. (Courtesy Bridon Fibres & Plastics)

Rope Dia mm	Sisal Tonnes	Grade 1 Manila Tonnes	Sisal Tonnes	Grade 1 Manila Tonnes	Sisal Tonnes	Grade 1 Manila Tonnes	Sisal Tonnes	Grade 1 Manila Tonnes
12	0·09	0·10	0·18	0·19	0·15	0·17	0·22	0·24
14	0·13	0·14	0·26	0·27	0·22	0·24	0·32	0·34
16	0·19	0·22	0·38	0·43	0·34	0·38	0·48	0·54
18	0·24	0·28	0·48	0·56	0·42	0·49	0·60	0·70
20	0·34	0·39	0·69	0·78	0·60	0·69	0·86	0·98
24	0·54	0·61	1·1	1·2	0·95	1·1	1·4	1·5
28	0·77	0·88	1·5	1·8	1·3	1·5	1·9	2·2
32	1·0	1·2	2·1	2·4	1·8	2·1	2·6	3·0
36	1·4	1·5	2·8	3·1	2·4	2·7	3·5	3·9
40	1·7	2·0	3·5	3·9	3·0	3·4	4·3	4·9
44	2·2	2·4	4·4	4·9	3·8	4·3	5·5	6·1
48	2·6	3·0	5·2	6·0	4·6	5·2	6·6	7·5

The safe working load quoted above is the maximum allowable for the size, and quality of rope stated, used in the defined lifting mode. In adverse operating conditions it is necessary for the user to reduce the safe working load quoted to a value appropriate to those conditions.

susceptibility to misuse fibre ropes and slings should be examined frequently for signs of damage – it is prudent to given them a cursory check each time they are used.

27.2.6.4 Eyebolts

Although included as separate lifting accessories under SMSR, eyebolts are not classified as lifting tackle under the FA 1961, but they are covered under the Construction (Lifting Operations) Regulations 1961. In use, eyebolts should be screwed fully home and any lift should be vertically only. They should not be subjected to lateral or sideways pulls. The differing screw thread sizes used on eyebolts have created hazards on which advice is given by the HSE[15]. Eyebolts must be of forged construction and not welded.

27.2.6.5 Shackles

Shackles should be of forged construction and can have plain or threaded pins. Only the original size and type of pin should be used.

27.2.6.6 Special lifting equipment

Where repeated lifts of a particular load are made, special lifting beams and equipment such as spreader beams (*Figure 27.14*) may be used.

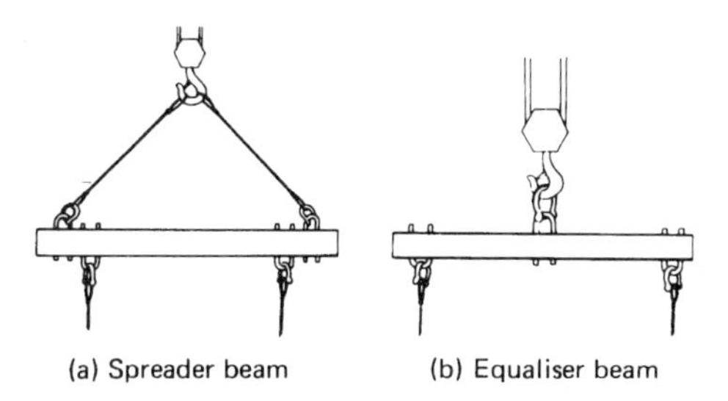

Figure 27.14 Spreader and equaliser beams. (Source: Dickie and Short[9])

27.2.6.7 Heat treatment

Chains, rings, hooks, shackles and swivels should be heat-treated at specified intervals unless they are made of materials which will be adversely affected by the high temperatures. The exemptions to this requirement are listed in Redgrave's Health and Safety[16].

27.3 Conveyors

Probably the most widely used mechanical handling equipment, there are three basic types:

(1) Belt conveyors,
(2) Roller conveyors,
(3) Screw conveyors,

each of which have many variations but there are certain common safety arrangements that, with adaptation, apply to all types.

27.3.1 Belt conveyors

The main risks from plain belt conveyors, whether flat or troughed, are at the head and tail pulleys which should be guarded or enclosed for a distance of at least 1070 mm (42 in) from the trapping point (*Figures 27.15* and *27.16*).

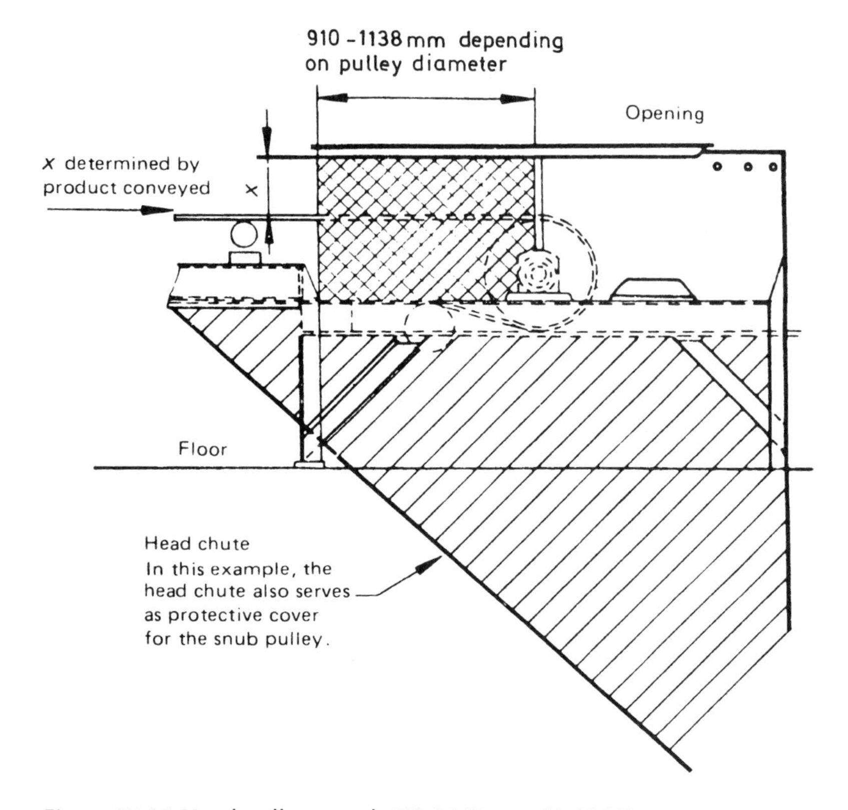

Figure 27.15 Head pulley guard. (BS 5667:part 19:1980)

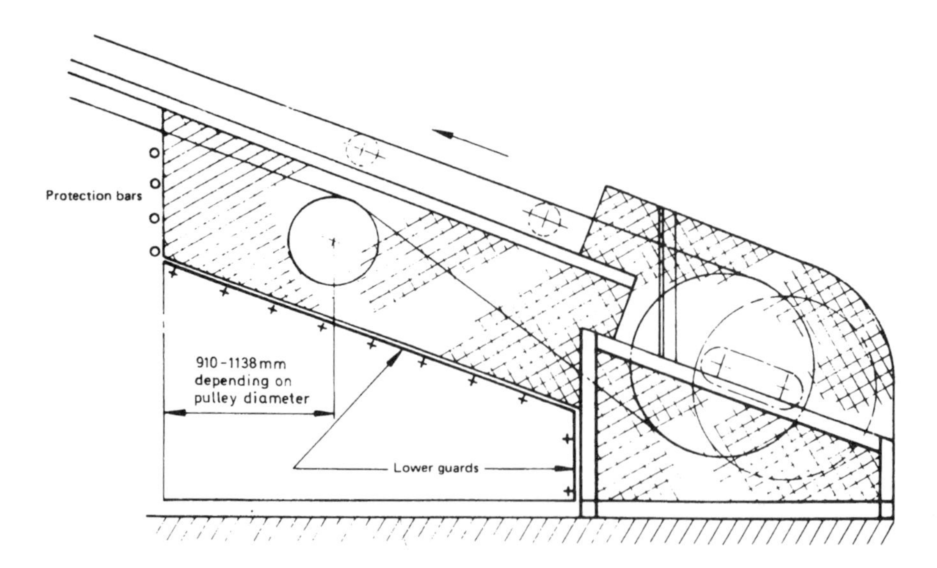

Figure 27.16 Tail pulley guard. (BS 5667:part19:1980)

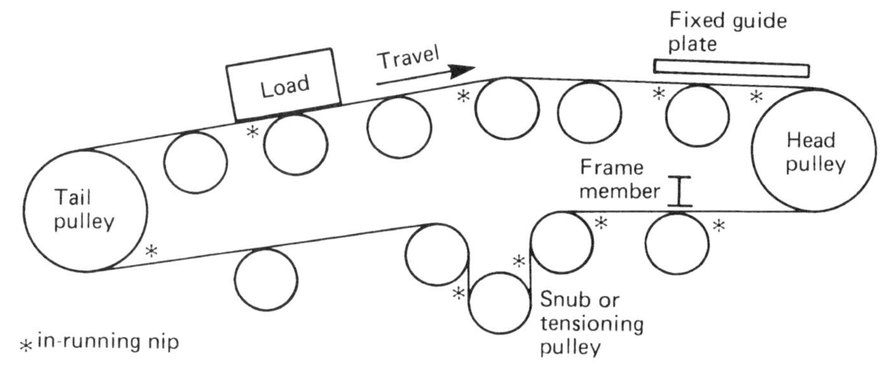

Figure 27.17 Diagrammatic layout of belt conveyor showing in-running nips

Additional trapping points occur wherever the belt changes direction as it passes over a pulley and also on straight runs as it passes over idler pulleys if the weight of the material being carried is too great to lift (*Figure 27.17*). Feed hoppers, guideplates and the conveyor frame may also contribute to creating nips by restricting the clearance above a roller or pulley. Belt conveyors may be guarded by enclosing the whole of the sides or by fitting individual guards to each nip.

On long conveyor runs it may be necessary to construct bridges to provide access to the opposite side of the conveyor. Long runs of conveyors present particular guarding problems and an acceptable solution is by the provision of a trip wire along the whole length of the machine (*Figure 27.18*) which

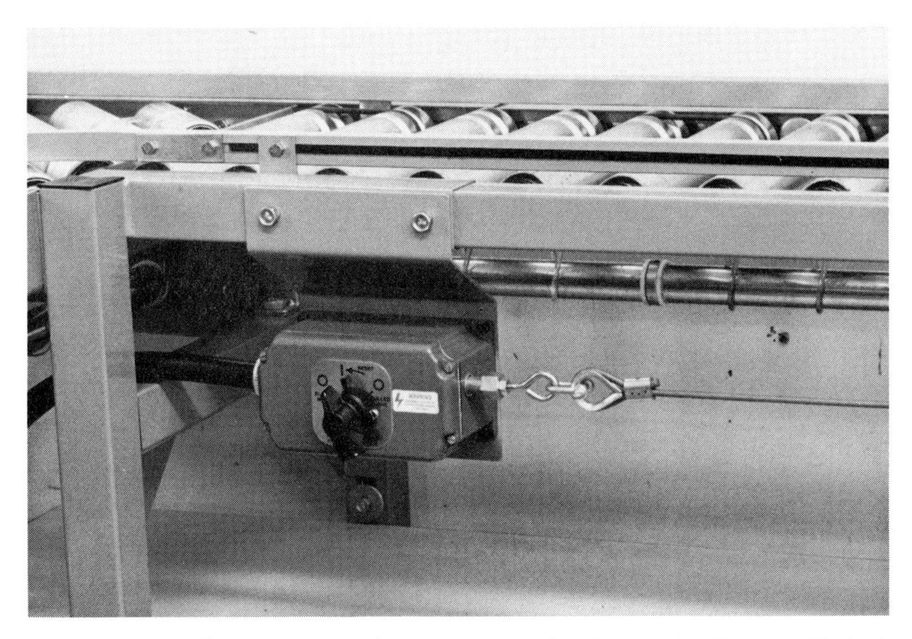

Figure 27.18 Roller conveyor with emergency grab wire switch. (Courtesy Craig & Derricott Ltd)

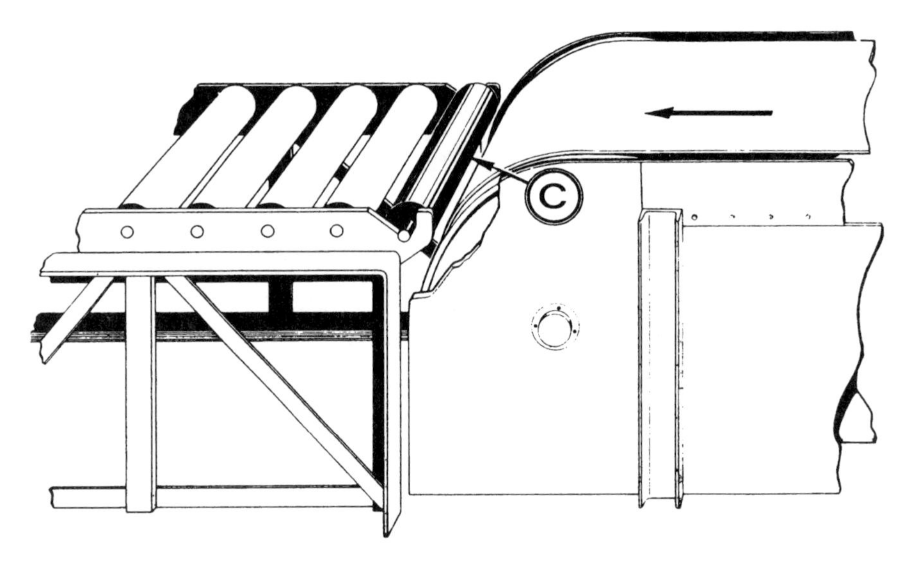

Figure 27.19 *Preventing free running roller trap. (Courtesy HSE)*

must be placed close to the belt so that it is immediately available in an emergency. If work is transferred from a belt conveyor to a free running roller conveyor, a trapping point can be formed between the belt and the first roller. This can be overcome by mounting the spindle of the roller in an open vee slot (item C in *Figure 27.19*).

27.3.2 Roller conveyors

There are three basic types of roller conveyors in use although there are many variations of each type. The basic types have:

(1) All rollers power driven.
(2) Some rollers power driven.
(3) All rollers free running.

Where every roller is power driven (by individual belt or chain) there are no in-running nips between adjacent rollers but the driving mechanism will need to be guarded.

On some conveyors, alternate rollers are power driven and this arrangement creates an in-running nip between each pair of rollers which should be guarded (*Figure 27.20*). Keeping these guards in position and in good repair may prove time-consuming and expensive, a problem which is eliminated if all the rollers are driven. There is a similar problem when a belt is fitted over the rollers but only driven by the end rollers. Frequently the belts do not cover the full width of the rollers and in-running nips can occur between the drive and adjacent roller and between the exposed part of the rollers and the belt (*Figure 27.21*).

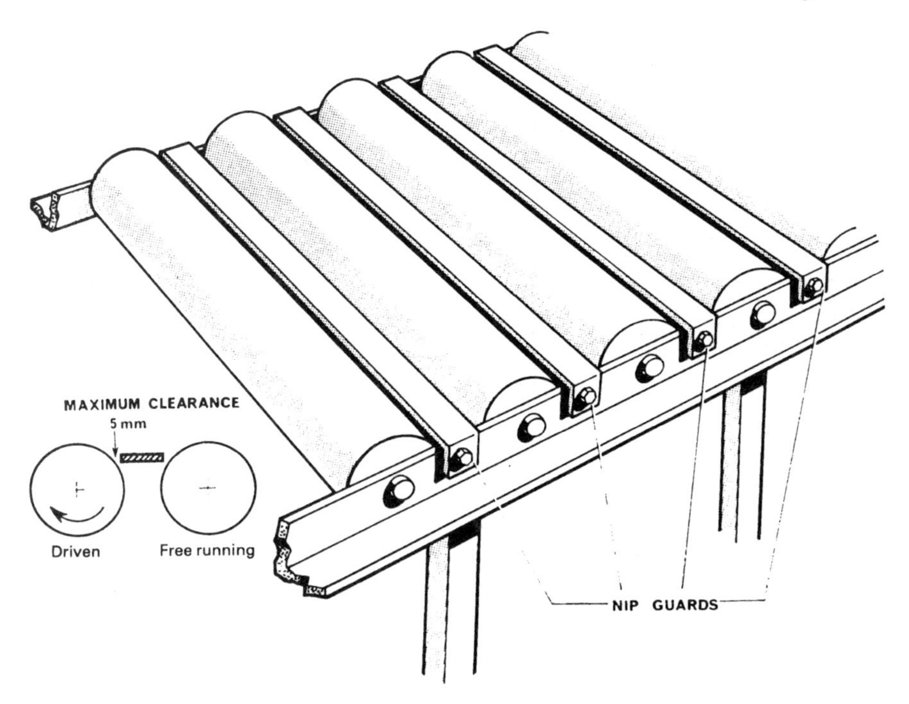

Figure 27.20 Guards between alternate drive-rollers

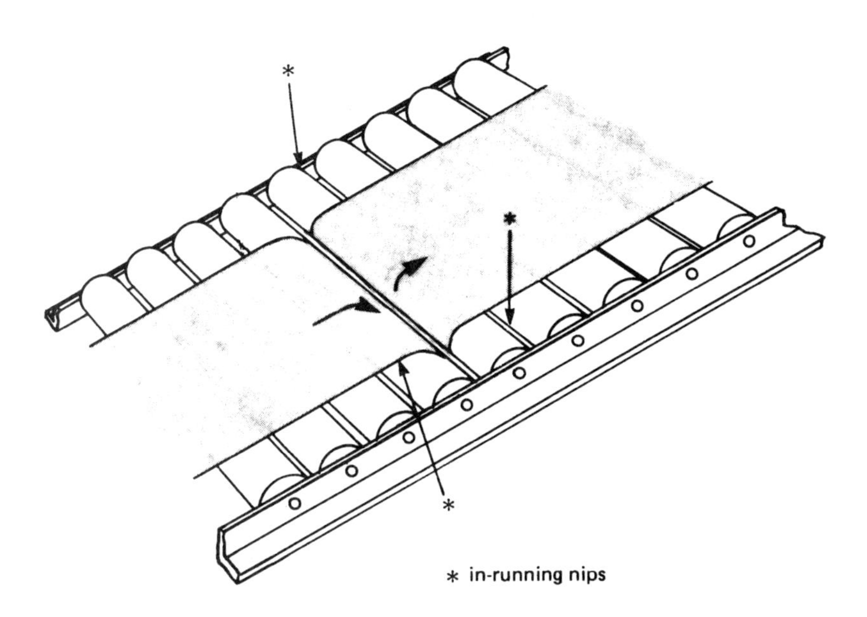

Figure 27.21 Nip points on roller conveyor with belts

Figure 27.22 Walkways between and across roller conveyors

While there is no intake risk between the rollers of a free running roller conveyor system, hazards do arise when people attempt to walk on them. Walkways should be provided between adjacent conveyors with crossing gangways at regular intervals along the conveyor length (*Figure 27.22*).

27.3.3 Screw conveyors

Because of the injury potential of these conveyors, they should be guarded so that when in use there is no possible means of access to them (*Figure 27.23*). Repairs and maintenance should only be undertaken when the drive motor is locked-off.

27.3.4 Other conveyor equipment

There is now a variety of equipment available for use with conveyor systems, such as pivoting corner units, right angle transfer units, etc. Where these are power driven, particularly on automatic or remote control, they should be fenced to prevent unauthorised access. For maintenance,

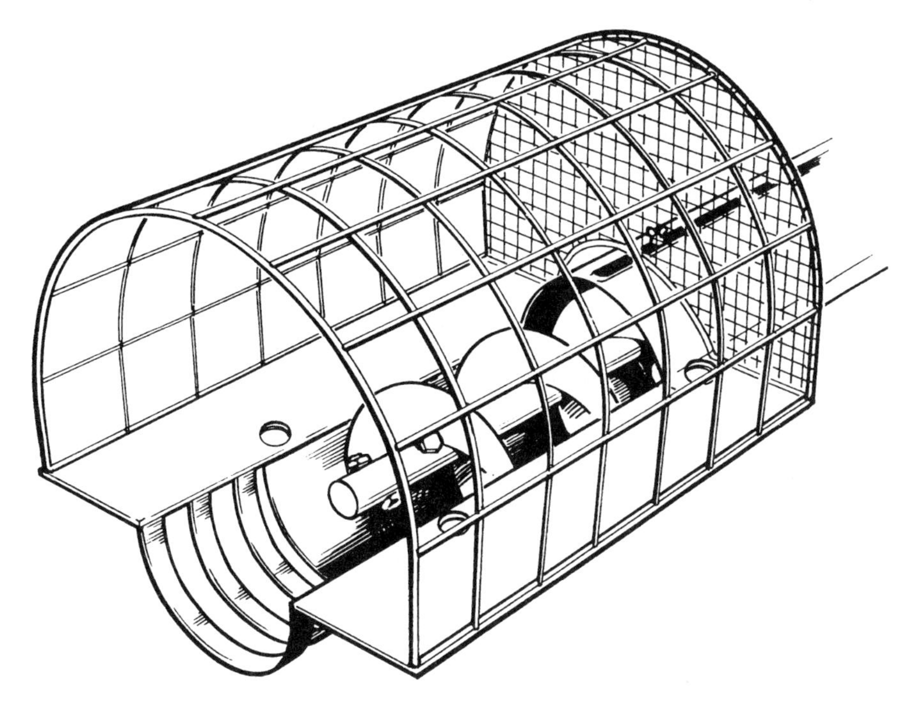

Figure 27.23 Screw conveyor guarding. (Courtesy HSE)

adjustment or repairs, a safe system of work should be laid down and followed involving the locking-off of the main drive before work commences.

27.4 Conclusion

The mechanical handling of materials plays an essential role in industry with the benefits from it being enormous; however any laxity in its use can prove very costly in terms of damage to plant and product, and, worse, of injury to employees. Whether the equipment is mobile, as with cranes and powered trucks, or is fixed in position as conveyors are, agreed safe systems of work and procedures for using it must be followed to reduce any risk to a minimum and to ensure the best utilisation possible.

References

1. Illuminating Engineering Society, *Code for Interior Lighting,* IES, London (1977)
2. Health and Safety Executive, *Health and Safety Guidance Booklet No. 38, Lighting at Work,* HSE Books, Sudbury (1987)
3. Health and Safety Executive, *The Provision and Use of Work Equipment Regulations 1992,* HMSO, London (1992)
4. Health and Safety Executive, *Working platforms on fork lift trucks,* Guidance Note No. PM28, HSE Books, Sudbury (1981)
5. Health and Safety Commission, *Approved Code of Practice no: COP 26, Rider Operated Lift Trucks – Operator Training,* HSE Books, Sudbury (1989)

6. HM Government, *The Docks Regulations, 1988*, S.I. 1988 No. 1655, HMSO, London (1988)
7. EEC, *Council Directive of 14 June 1989 on the approximation of the laws of Member States relating to machinery*, EC, Luxembourg (1989)
8. Department of Trade and Industry, *The Supply of Machinery (Safety) Regulations 1992*, HMSO, London (1992)
9. Dickie, D.E., *Lifting Tackle Manual*, (Ed. Douglas Short), Butterworths, London (1981)
10. BS 5744 Code of practice for the safe use of cranes, British Standards Institution, London
11. BS CP 3010 Safe use of cranes (mobile cranes, tower cranes and derrick cranes). British Standards Institution, London
12. Dickie, D.E., *Crane Handbook*, (Ed. Douglas Short), Butterworths, London (1981)
13. Health and Safety Executive, *Management's Responsibilities in the Safe Operation of Mobile Cranes – Report on 3 Crane Accidents*, HSE Books, Sudbury (1980)
14. BS 6166 Rating of lifting gear for general purposes, British Standards Institution, London
15. Health and Safety Executive, Guidance Note No. PM16, *Eyebolts*, HSE Books, Sudbury (1980)
16. Fife, I. and Machin, E.A., *Redgrave, Fife and Machin: Health and Safety*, 2nd edn, note attached to s. 26 of Factories Act 1961, Butterworths, London (1993)

Further reading

Fife, I. and Machin, E. A., *Redgrave's Health and Safety*, 2nd edn, 79–98, Butterworths, London (1993)

King, R. W. and Magid, J., *Industrial Hazard and Safety Handbook*, pp. 567–603, Butterworth-Heinemann, Oxford (1979)

British Standards Institution, London:

BS 3726:1978 Methods for stability testing of counterbalanced fork lift trucks.
BS 4338:1968 Rated capacities of fork lift trucks.
BS 4430:1969, parts 1 and 2, Recommendations for the safety of industrial trucks.
BS 4436:1978 Methods for stability testing of reach trucks and straddle trucks.
BS 5667:1980, parts 1–19. Continuous mechanical handling equipment – Safety requirements.
BS 5744:1979 Code of practice for safe use of cranes.

Health and Safety Executive, the following publications which are available from HSE Books, Sudbury:

Health and Safety Series Booklet No. HS(G)6 *Safety in Working with Lift Trucks* (1981), *Transport Kills* (1982)
Legislation Series Booklet No. L20, *A Guide to the Lifting Plant and Equipment (Records of Test and Examination etc.) Regulations 1992*
Guidance Notes: General Series
GS9 *Road transport in factories and similar workplaces* (1992)
GS39 *Training of crane drivers and slingers* (1986)
Guidance Notes: Plant and Machinery Series
PM7 *Lifts: thorough examination and testing* (1982)
PM8 *Passenger carrying paternosters* (1977)
PM24 *Safety in rack and pinion hoists* (1981)
PM26 *Safety at lift landings* (1981)
PM27 *Construction hoists* (1981)
PM34 *Safety in the use of escalators* (1983)
PM42 *Excavators used as cranes* (1984)
PM43 *Scotch derrick cranes* (1984)
PM45 *Escalators: periodic thorough examination* (1984)
PM46 *Wedge and socket anchorages for wire ropes* (1985)
PM54 *Lifting gear standards* (1985)
PM55 *Safe working with overhead travelling cranes* (1985)
PM58 *Diesel engined lift trucks in hazardous areas* (1986)
PM63 *Inclined hoists used in building and construction work* (1987)

Chapter 28

Electricity

E. G. Hooper

28.1 Alternating and direct currents

28.1.1 Alternating current

An alternating current (ac) is induced in a conductor rotating in a magnetic field. The value of the current and its direction of flow in the conductor depends upon the relative position of the conductor to the magnetic flux. During one revolution of the conductor the induced current will increase from zero to maximum value (positive), back to zero, then to maximum value in the opposite direction (negative) and, finally, back to zero again having completed one cycle. A graph plotted of the value and direction of the induced current follows a standard sine wave. This alternating current is produced at power stations in alternators at a frequency of 50 hertz (i.e. 50 rotations of the conductor per second).

28.1.2 Direct current

Direct current (dc) has a constant positive value above zero and flows in one direction only, unlike ac. A simple example of direct current is that produced by a standard dry battery.

DC is really ac which has its positive or negative surges rectified to provide the uni-directional flow. In a dc generator the natural ac produced is rectified to dc by the commutator.

DC is little used in standard electricity distribution systems but is used in industry for special applications. Although there are slight differences in the effects under fault and shock conditions with dc when compared with ac, it is a safe rule to apply the same general safety precautions for the treatment of electric shock.

28.2 Electricity supply

Alternating current electricity is generated in thermal power stations (coal-, gas- and oil-fired) and also at nuclear power stations under the control of the privatised generating and distributing companies. It is then transmitted by way of overhead lines at 400 kV (i.e. at 400 000 volts) (*Figure 28.1*), 232 kV

Figures 28.1 400 kV suspension towers on the National Grid's Sizewell–Sundon 400 kV transmission line. (Courtesy National Grid)

or 132 kV to distribution substations where it is transformed down to 33 kV or 11 kV for distribution to large factories, or transformed down to 240 V for use in domestic and commercial premises and the smaller factories.

The Area Electricity Boards, of which there were 12 in England and Wales and which have become separate distribution companies under privatisation, are responsible for the distribution systems. However, in Scotland there are two all-purpose companies dealing with generation and transmission in addition to distribution. Although a growing number of places of work – such as the chemical and paper industries–generate their own electricity, the UK electricity supply industry supplies over 20 million industrial, commercial and domestic consumers who consume over 230 000 million units of electricity every year.

Electricity is received by most industrial and commercial consumers at 415/240 V, ac, the standard distribution and utilisation range. However, consumers with large energy requirements often find it more economical and convenient to receive electricity at 11 kV and to transform it down to the lower values as required.

In all cases, however, it is essential that the consumer's electrical installation and all apparatus, both fixed and portable, meets good design, construction, protection, maintenance and use standards. The British Standards Institution[1] issues standards and codes giving guidance on electrical safety matters and the Wiring Regulations published by the Institution of Electrical Engineers[2] give excellent guidance on standards for electrical wiring and installations. However, it is important to appreciate the legal obligations relating to the safe use of electricity and electrical machinery at places of work.

The Health and Safety at Work etc. Act 1974 is an enabling Act providing a legal framework for the promotion of health and safety at all places of work. Although the Act says nothing specific about electricity it does require, amongst other things, the provision of safety systems and methods of work; safe means of access, egress and safe places of employment; and adequate instruction and supervision. These requirements have wide application but they are, in general terms, also relevant to the safe use of electricity. But for specific advice on the electrical legal requirements we must turn to a set of Regulations[3] made under the HSW Act.

28.3 Statutory requirements

28.3.1 The Electricity at Work Regulations 1989

Until they were revoked on 1st April 1990, the main statutory requirements for the generation, transformation, conversion, switching, controlling, regulating, distribution and use of electrical energy in factory premises, building operations, etc. were contained in the Electricity (Factories Act) Special Regulations 1908 and 1944. These regulations (normally referred to as the Electricity Regulations), although made under the Factory Act 1901, were continued in force under the Factories Act 1961 and under the Health and Safety at Work etc. Act 1974. An explanatory memorandum not only outlined the Regulations but gave advice as to their interpretation. However, on 1st April 1990, a new set of Regulations, the Electricity at Work Regulations 1989[3], came into effect to replace the 1908 Regulations and extended cover to all places of work.

The new Regulations specify fundamental principles of electrical safety while at the same time allowing flexibility to accommodate technical developments. Whereas certain of the requirements of the new Regulations apply to all industries, a number of them apply only to mines. To assist in the application and comprehension of them the Health and Safety Executive have issued three supporting documents:

(1) a Memorandum of Guidance[4], and
(2) two Approved Codes of Practice[5,6] dealing respectively with the use of electricity in mines and in quarries.

The Memorandum of Guidance referred to above gives technical and legal guidance on the Regulations and provides a source of practical help. Similarly, the two Codes of Practice provide essential advice for mines and quarries.

Contained in the Electricity at Work Regulations are 33 separate regulatory requirements which place firm responsibilities on employers, the self-employed, managers of mines and of quarries and employees to comply as far as they relate to matters within their control. Additionally, employees have a duty to co-operate with their employers so far as is necessary for the employers to comply with the Regulations.

Topics covered by the Regulations include:

(a) Systems, work activities and protective equipment (Reg. 4).
(b) Strength and capability of electrical equipment (Reg. 5).
(c) Adverse and hazardous environments (Reg. 6)
(d) Insulation, protection and placing of conductors (Reg. 7)
(e) Earthing and other suitable precautions (Reg. 8)
(f) Integrity of 'referenced' conductors (Reg. 9)
(g) Connections (Reg. 10)
(h) Means for protecting from excess current, for cutting off supply and isolation (Regs. 11 and 12)
(i) Precautions for work on dead equipment and on or near live conductors (Regs. 13 and 14).
(j) Working space, lighting and access (Reg. 15).
(k) Persons to be competent to prevent danger and injury (Reg. 16).
(l) Regulations applicable to mines only (Regs. 17–28).

28.3.2 Status of Regulations

Certain of the Regulations are subject to a 'so far as is reasonably practicable' clause which means that a decision has to be based on a balance between the risk and the cost of eliminating it.

The rest of the Regulations are 'absolute' which means that the requirements must be met regardless of cost. However these absolute Regulations are subject to the defence provisions of Regulation 29 which applies in criminal proceedings and provides a defence for the accused if he can show that he took all reasonable steps and exercised due diligence to avoid committing the offence.

28.4 Voltage levels

Unlike the 1908 Regulations which were based on voltage banding (low, medium, high and extra high voltage) the new Regulations apply equally to all voltages and to all electrical systems and equipment.

28.5 Electrical accidents

Electricity is a safe and efficient form of energy and its benefits to mankind as a convenient source of lighting, heating and power are obvious. But, if electricity is misused, it can be dangerous – a statement of the obvious, but one which must be made so as to keep the matter in proper perspective.

In the UK, every year, up to 50 people may be killed, at work, as a result of an electrical accident. In addition, up to a thousand or so are injured. These figures, considering the widespread use of electricity in industry and when compared with the numbers killed and injured as a result of other types of accident, are relatively small. Nevertheless, a knowledge of electricity is essential because the injuries are usually more severe than those resulting from other causes, and also because misuse of electricity or poor maintenance is responsible for a large proportion of fires.

28.6. The basic electrical circuit

For an electrical current to do its job of providing lighting, heating and power, it must move safely from its source, through the conducting path and back from whence it came. In short, electric current requires a suitable circuit to assist its flow without danger. The circuit must be of suitable conducting material, e.g. copper, covered with a suitable insulating material (to stop the current 'leaking' out) such as PVC or rubber.

For an electrical current (measured in amperes) to flow in a circuit it requires pressure (measured in voltage). As it flows it encounters resistance from the circuit and apparatus and this characteristic is measured in ohms.

This relationship between volts, amps and ohms is brought together in the famous Ohm's Law known to most schoolboys. Thus, to put it simply, the current in a circuit is proportional to the voltage driving it and inversely proportional to the resistance it has to overcome:

$$\text{amps} = \frac{\text{volts}}{\text{ohms}} \tag{1}$$

or

$$\text{ohms} = \frac{\text{volts}}{\text{amps}} \tag{2}$$

or

$$\text{volts} = \text{amps} \times \text{ohms} \quad (3)$$

From Ohm's Law may be derived other equations such as the relationship between power (measured in watts) and the voltage and current. Thus:

$$\text{watts} = \text{volts} \times \text{amps} \quad (4)$$

$$= \text{amps}^2 \times \text{ohms} \quad (5)$$

$$= \frac{\text{volts}^2}{\text{ohms}} \quad (6)$$

28.6.1 Impedance

As an alternating current passes around a circuit under the action of an applied voltage it is impeded in its flow. This may be due to the presence in the circuit of resistance, inductance or capacitance, the combined effect of which is called the impedance and is measured in ohms.

In a pure resistance circuit the applied voltage has to overcome the ohmic value of the resistance as is the case for direct current (see equations (1) to (6) above).

If, however, the circuit contains inductance, such as the presence of a coil, the alternating magnetic field set up by the alternating current will induce a voltage in the coil which will oppose the applied voltage and cause the current to lag vectorially behind the voltage (up to 90° where the circuit contains pure inductance only). This property is called reactance and is measured in ohms. Sometimes a circuit may contain a condenser: the applied voltage 'charges' the condenser and the effect is such as to cause the current vectorially to lead the voltage. This property is also called reactance. Now most circuits contain resistance, inductance and capacitance in various quantities, and the effect of impedance is found as follows:

$$\text{impedance}^2 = \text{resistance}^2 + \text{reactance}^2$$

where the reactance is calculated from the characteristics of the impedance and capacitance and a term known as the angular velocity. It is beyond the scope of this section to go further into these relationships and readers are referred to a standard textbook on electricity[7] and to BS4727[8].

28.7 Dangers from electricity

It has been said that properly used electricity is not dangerous but out of control it can cause harm, if it passes through a human body, by producing electric shock and/or burns. Electricity's heating effect can also cause fire but we will now deal with the electric shock phenomenon.

28.7.1 Electric shock

If a person is in contact with earthed metalwork or is inadequately insulated from earth then, because the human body and the earth itself are good conductors of electricity, they can form part of a circuit (albeit an abnormal one) through which electricity, under fault conditions, can flow. How can such fault conditions occur?

If, for any reason, there was a breakdown of insulation in a part of an electric circuit or in any apparatus such as, say, a hand-held metal-cased electric drill, it is conceivable that current would flow external to the closed circuit of the conductor of the drill, if a path were available. For example, the metalwork of the drill may be in contact with a live internal conductor at the point of insulation breakdown (an example of indirect contact). Or, take the example of someone working at a switch or socket outlet from which the cover had been removed before the electricity supply had been isolated. In such cases the person concerned could touch live metal or a live terminal and, if the conditions were right, would thereby cause an electric current to flow through his body to earth (an example of direct contact). If the total resistance of the earth fault path were of a sufficiently low value, the current could kill or maim.

Electric shock is a term that relates to the consequences of current flow through the body's nerves, muscles and organs and thereby causing disturbance to normal function. Owing to a current's heating effect the body tissue could also be damaged by burns. A particular danger with electric shock from alternating current is that it so often causes the person concerned to maintain an involuntary grip on the live metal or conductor (particularly hand-held electric tools) and this prolongs current flow. Death could occur when the rhythm of the heart is disturbed such as to affect blood flow and hence the supply of oxygen to the brain, a condition that is known as ventricular fibrillation. Unless skilled attention is given immediately, by, say, the proper application of oral resuscitation, ventricular fibrillation can be irreversible. However, it is still fortunately the case that most electric shock victims recover without permanent disability or lasting effect.

Although the effect of a direct current shock is generally not as dangerous as with ac (there is no dangerous involuntary grip phenomenon for example), it is recommended that similar precautions against shock be taken.

The severity of an electric shock is the product of the current value and the time for which it flows through the body. Thus a 50 mA shock current (i.e. 0.05 A) could probably flow through a body, without much danger, for up to 4 seconds; whereas a 500 mA current (0.5 A) flowing for only 50 ms (0.05 s) could be fatal.

Personal sensitivity to electric shock varies somewhat with age, sex, heart condition, etc., but for an average person the relationship between shock current, and time for which the body can accommodate it, is given by a formula of the following kind:

$$\text{current} = \frac{116}{\sqrt{\text{time}}}$$

where the current is measured in milliamps (mA) and the time is measured in milliseconds (ms). Above the duration of one heart beat a lower current threshold is recommended.

The maximum safe 'let go' current is less than 10 mA, whereas 20 mA to 40 mA directly across the chest could arrest respiration or restrict breathing; currents above 500 mA flowing for as little as 500 mS can be fatal. However, even 'safe' currents at the level of about 5 mA to 10 mA could still cause a minor shock sensation and cause someone to fall if working at a height.

From all this it will be concluded that at normal mains voltage of 240 V, and given the average value of resistance of a human body at 1000 Ω, the current flowing through the body would, from equation (1), be a maximum of

$$\frac{240}{1000} = 0.24 \text{ A or } 240 \text{ mA}$$

– a dangerously high value. Under normal circumstances there will be additional resistances (or impedances as they are more correctly called) such as, for example, the resistance of the circuit, the earth electrode, and any footwear worn. It must also be remembered that body resistance varies from person to person depending upon biological, environmental and climatic conditions. But even so, given the very small value of current that could cause harm, all possible sources of contact with live electric parts must be avoided. This brings us to the precautions to be taken.

28.8 Protective means

28.8.1 Earthing and other suitable precautions

For the prevention of danger where it is 'reasonably foreseeable' that a conductor (other than a circuit conductor) may become charged with electricity, earthing or other suitable protective means shall be taken. With earthing, metalwork forming part of an electrical installation or of apparatus, such as cable armouring, conduit, metal enclosures of switchgear, and frames and metalwork of motors and transformers, should be adequately bonded to a suitable electrode connection with earth. The resistance of the bonding and connection with earth must be of such low value as to allow sufficient current to flow to earth from the fault so as to 'blow' the fuse or operate the circuit breaker protecting the circuit in question. It is important to ensure that not only is there good initial design and construction of the earthing arrangements, but also that regular and frequent inspections and tests are carried out by a competent person.

Any flexible metallic covering of conductors should not, of itself, be the only earth, i.e. there must be a separate earth wire in the conductors in addition to the flexible metallic covering.

A British Standard Code of Practice[9] gives useful advice on the subject of earthing as do the IEE Wiring Regulations[2].

28.8.2 Precautions to be taken when working

When work is to be carried out on a part of a circuit or piece of electrical equipment, certain precautions need to be taken to protect the worker concerned from electrical danger. The electricity supply should first of all be switched out and locked off to ensure that the circuit or apparatus being worked upon is effectively electrically isolated and cannot become live.

Using a suitable voltage detecting instrument, that part of the circuit to be worked on should be checked to ensure that it is dead before work is allowed to commence. If the risk is high a Permit-to-Work system should be initiated requiring additional precautions, such as earthing, the posting of caution and danger notices etc., as described below. Although Regulation 14 permits live working, this is subject to very strict conditions. Normally dead working should always be the preferred choice.

28.8.3 Insulation

Mention has already been made of the need to ensure that electrical conductors etc. are adequately insulated. Insulating material has extremely high resistance values such as virtually to prevent electric current flow through it. The principle of insulation is used when work has necessarily to be carried out at or near uninsulated live parts. Obviously such parts should always be made dead if at all possible. If this cannot be done then properly trained people, competent to do the work, can make use of insulating stands, mats or screens, or rubber insulating gloves to protect them from electric shock. Periodic examination of all such insulating devices is necessary.

28.8.4 Fuses

A fuse is essentially a strip of metal, placed in a circuit, of such size as would melt at a predetermined value of current flow and therefore cut off the current to that circuit. Obviously a *properly rated fuse* is a most useful precaution because, in the event of abnormal conditions such as a fault, when excess current flows, the fuse would 'blow' and protect the circuit or apparatus from further damage. To operate effectively and safely the fuse should be placed in the live conductor and never in the neutral conductor, otherwise even with the fuse blown or removed, parts of the circuit, such as switches or terminals, could still be live. Fuses come in various sizes with different construction characteristics and degrees of protection. Good practice advises that every fuse must be so constructed, guarded and placed as to prevent danger from such things as overheating and the scattering of hot metal when it blows. Modern cartridge fuses, the simplest variant of which is contained in the standard 13 amp fused plug, are well constructed to meet these requirements but it is difficult to tell at a glance if they have 'blown'. Simple battery continuity tests are available for easy checking and for the larger industrial sizes of cartridge fuse an automatic indication is provided.

Overfusing, that is to use a fuse rating higher than that of the circuit it is meant to protect, is dangerous because in the event of a fault a current may flow to earth without blowing the fuse. This could endanger workpeople and the circuit or apparatus concerned. In addition it could result in the cable carrying an excessive current leading to considerable overheating with the risk of fire.

28.8.5 Circuit breakers

A circuit breaker, although more expensive than a fuse, has several advantages for excess current circuit protection. The principle of operation is that excess current flow is detected electromagnetically and the mechanism of the breaker automatically trips and cuts off electricity supply to the circuit it protects. Circuit breakers are also available to detect earth leakage current and, indeed, units are available that detect both over-current and earth leakage currents and thereby give very good circuit protection.

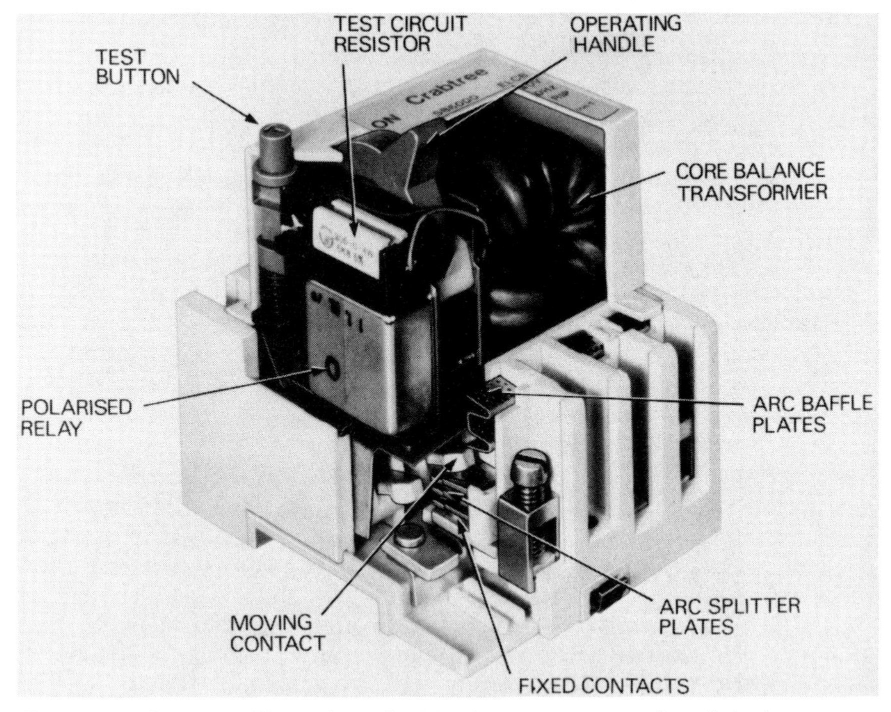

Figure 28.2 Cut-away illustration of a 30 mA current operated earth leakage circuit breaker. (Courtesy Crabtree Electrical Industries Ltd)

The majority of electric shock injuries occur when the body acts as conductor between line and earth. A general level of protection against such shocks is provided by the inclusion of a current sensitive earth leakage circuit breaker in the supply line. A typical example is shown in *Figure 28.2*.

28.8.6 Work near underground cables and overhead lines

Work near underground electricity cables and overhead electric lines has caused many serious and fatal accidents over the years. The precautions to be taken are dealt with in two HSE guidance notes[10,11].

28.9 Competency

Unlike the 1908 Regulations with their use of such terms as 'authorised persons' who had specific duties, the new Regulations recognise only one class of person who may be engaged in any work activity where technical knowledge or experience is necessary to prevent danger or injury. That person is one who must have such knowledge or experience (or is under appropriate supervision). In other words, every one has to be competent in relation to the work he is doing and having regard to the degree of supervision he is under at a particular time.

28.10 Permits-to-Work

Permits-to-Work are an essential prerequisite to the commencement of certain classes of work involving special danger to people. As regards work on electrical apparatus that has a high degree of risk, such as those operating at the higher voltages, a Permit-to-Work, of an agreed format, should be used as a form of declaration by an authorised person to the person in charge of the work, that the apparatus to be worked on is dead, properly isolated from all live conductors, has been discharged of electricity, is efficiently connected to earth and that work can safely commence. The use of proper voltage testing devices to ensure that the item to be worked upon is dead is an essential safety check. It is the duty of the person issuing the Permit-to-Work to ensure that the necessary safety precautions detailed in the Permit have been carried out, and that the person receiving the Permit is fully conversant with the nature and extent of the work to be done. Proper arrangements for issue, receipt, clearance and cancellation of Permits-to-Work are essential.

A typical Permit-to-Work for use on electrical apparatus is given on pages 640 and 641.

28.11 Static electricity

When two dissimilar bodies or substances meet, electrons pass from one to the other at the surface contact area. When the bodies separate, particularly if they are of an insulating material, a difference of potential occurs across the separating medium which manifests itself as static electrical charge. Such an effect can, for example, occur when there is separation of liquid from a pipe, powder from a tray or paper from a reel. In the movement of flammable liquid or a powder or dust with fine particles, static electricity can be generated that can give rise to sparking of sufficient energy to ignite the vapour or dust. The precautions to be taken to prevent fire and explosion as a result of static electricity depend upon the nature of the materials and process[12]. Examples of the approach to the more common problems are as follows:

(1) All pipework and containers used for conveying flammable liquids should be effectively bonded together and earthed so that any static

MODEL FORM OF PERMIT-TO-WORK – FRONT

NAME OF FIRM

PERMIT-TO-WORK

1. ISSUE NO. . . . .

To ....

I hereby declare that it is safe to work on the following Apparatus, which is dead, isolated from all live conductors and is connected to earth:–

....

....

ALL OTHER APPARATUS IS DANGEROUS

Points at which system is isolated ....

....

....

Caution Notices posted at ....

....

....

The apparatus is efficiently connected to earth at the following points

....

....

Other precautions ....

....

....

The following work is to be carried out ....

....

....

Signed ....
being an Authorised Person.

Time .... Date ....

MODEL FORM OF PERMIT-TO-WORK – BACK

2. RECEIPT

I hereby declare that I accept responsibility for carrying out the work on the apparatus detailed on this Permit-to-Work and that no attempt will be made by me, or by the men under my control, to carry out work on any other apparatus.

Signed ....

Time .... Date ....

Note: After signature for the work to proceed this Receipt must be signed by and the Permit-to-Work be retained by the person in charge of the work until the work is suspended or completed and the Clearance section has been signed.

---

3. CLEARANCE

I hereby declare that the work for which this Permit-to-Work was issued is now *suspended/completed, and that all men under my charge have been withdrawn and warned that it is no longer safe to work on the apparatus specified on this Permit-to-Work, and that gear, tools and additional earthing connections are all clear.

Signed ....

Time .... Date ....

*Delete word not applicable.

---

4. CANCELLATION

This Permit-to-Work is hereby cancelled.

Signed ....

being an Authorised Person possessing authority to cancel a Permit-to-Work.

electricity produced is immediately discharged to earth before it builds up to a dangerous voltage.

(2) Workshop atmospheres where flammable solvents are used for spreading as in, say, material proofing processing, can be artificially humidified. Where practicable specialised radioactive static eliminators or earthed metal combs near the charged material etc. can also be used. It is also a wise precaution to ensure adequate ventilation such as to keep the gas/air mixture well below the lower explosive limit of the flammable solvent concerned.

(3) High speed rotating flat belts and pulleys are known to produce dangerous static charges and should not be used at or near flammable solvents. Where there are difficulties, special conducting belts, belt dressings and earthing of drive and pulley shafting can help to eliminate the build-up of static electricity.

(4) Certain processes, such as electrostatic paint spraying, make use of the characteristics of static electricity and special precautions against solvent ignition are required[13].

28.12 Electrical equipment in flammable atmospheres

28.12.1 Explosive and flammable atmospheres

The techniques to be adopted to prevent danger when using electrical equipment in the vicinity of potentially explosive or flammable atmospheres have changed over the years. The new Regulations recognise these changes and in Regulation 6, which deals with various kinds of adverse or hazardous environments including the effects of the weather, temperature, pressure, wet, dusty and corrosive conditions (as well as exposure to flammable or explosive dusts, vapours or gases), the requirement is that electrical equipment which may 'reasonably foreseeably' be exposed to such conditions, shall be of such construction, or as necessary protected, so as to prevent (so far as is reasonably practicable) danger arising from such exposure.

28.12.2 Construction of equipment for use in flammable or explosive atmospheres

The construction of electrical equipment to be used where a flammable or explosive atmosphere is likely to occur must be such as to prevent ignition of that atmosphere. The selection and installation of such equipment are detailed and specialised matters requiring expert knowledge. Guidance may be obtained from HSE publication HS(G)22[14], BS 5490[15], BS 5345[16] and BS 5501[17].

28.12.3 Classification of hazardous areas

In industry, with the exception of mining, areas that are hazardous, so far as flammable gases and vapours are concerned, are classified according to the probability of occurrence of explosive concentrations of gas or vapour. These classifications, called zones, are as follows:

Zone 0 is a zone in which a flammable atmosphere is continuously present or for long periods.
Zone 1 is a zone in which flammable atmosphere is likely to occur in normal working.
Zone 2 is a zone in which flammable atmosphere is unlikely to occur except under abnormal conditions and then only for a short time.

The particular zone determines the types of protection required for electrical equipment in use in that zone. Ideally, the prime method of protection should be to exclude electrical apparatus from any hazardous area. However, where this is not practical or economic, the next consideration should be whether the electrical apparatus can be segregated by fire-resistant impermeable barriers. Where the installation of electrical apparatus in hazardous areas is unavoidable, the following types of protection may be used according to the circumstances:

28.12.4 Type N equipment

Type N equipment is so constructed that it will not ignite explosive gas in normal conditions. Such equipment is designed for use in Zone 2 areas.

28.12.5 Type 'e' equipment

Type 'e' equipment, also known as 'increased safety' equipment, employs a protection method to electrical apparatus that does not, in normal operation, produce sparks. Examples are transformers and squirrel cage motors. Type 'e' equipment may be used in Zone 1 areas.

28.12.6 Pressurising and purging

Pressurising is a method used whereby clean air, drawn from outside, is blown at a pressure slightly above atmospheric into the room or enclosure to maintain the atmosphere at a pressure sufficiently high to prevent ingress of the surrounding potentially flammable atmosphere.

Purging is a method whereby a flow of air or inert gas of sufficient quantity is maintained in a room or enclosure to reduce or prevent the flammable atmosphere occurring.

28.12.7 Electrical fittings

Electrical installations used in flammable atmospheres require special consideration as regards design, construction and installation of the metallic cable sheathing, cable armouring or conduit, junction boxes, cable gland sealing, wiring etc. Particular attention should be given to the earthing arrangements. This is a specialised subject for which expert opinion should be sought.

28.12.8 Marking of equipment

The standards for electrical equipment for use in flammable atmospheres are affected by the harmonisation processes of the EC, CENELEC (European Committee for Electrotechnical Standardisation) and the IEC (International Electrotechnical Committee).

BS 5345[16] details basic requirements for the selection, installation and maintenance of apparatus for use in potentially explosive atmospheres and various parts of this standard deal with different types of equipment and should be referred to for information additional to that given below.

Electrical equipment for use in flammable atmospheres should be marked both with a group number, which relates to the relative ease with which flammable gases may be ignited by electrical sparking from the equipment, and by a temperature classification which relates to the maximum surface temperature of the equipment.

Group I equipment is used in the mining industry and Group II equipment in other industries. Letters A, B or C markings following the Group number indicate the suitability of the equipment for various gases subject also to the marking for the appropriate temperature classification.

Thus representative gases for the various groups are as follows:

Equipment Group	*Gas*
I	Methane
IIA	Propane
IIB	Ethylene
IIC	Hydrogen

Temperature classification	
T1	450°C
T2	300°C
T3	200°C
T4	135°C
T5	100°C
T6	85°C

28.12.9 Intrinsically safe systems

In a circuit where the amount of electrical energy available to cause a spark is below that necessary for igniting flammable vapour or gas, the equipment is considered to be intrinsically safe. It is particularly suitable for use in instrumentation, remote control etc. Intrinsically safe equipment to category IA may be used in any Zone except that in which there are contacts that could spark. It should not be used in Zone 0. Category IB equipment may be used in all zones except Zone 0, See BS 5958[12].

28.12.10 Flameproof equipment

Flameproof equipment is regarded as safe for use when exposed to the risk of explosive atmosphere for which certification has been given. Electrical apparatus, defined as flameproof, has an enclosure that will withstand an internal 'explosion' of the flammable vapour or gas in question which may enter the enclosure. The joints of the enclosure which are designed with clearance gaps to prevent a build-up of internal pressure also prevent any internal 'explosion' igniting vapour or gas surrounding the equipment. The surface temperature of the enclosure must be below the ignition temperature of the vapour or gas in question, hence the importance of the temperature classification referred to above. Flameproof enclosures are primarily intended for use in Zone 1 or Zone 2 classification but *not* in Zone 0.

28.13 Portable tools

Portable electric tools are a convenient aid to many occupational activities. However, the necessity to use flexible cables to supply electricity to the tool introduces hazards. For example, such cables are often misused and abused resulting in damaged insulation and broken or exposed conductors. The tool itself could also become unsafe if, say, its metalwork became charged with electricity due to a fault. Constant care and adequate maintenance and storage are essential to safe use.

Of the hand-held power operated electrical tools, the most common are the electric drill and portable grinding wheel. The major safety requirement for portable tools is contained in Regulation 13 which requires that:

(1) The flexible cables (or cords as they are increasingly called) must be connected to the source of supply by proper connections and connectors (plugs and so on). This means not only that connectors are designed to appropriate British Standards, but that the cable is properly connected up and the sheath firmly clamped.
(2) The metalwork of portable apparatus should be efficiently earthed. This is best effected by using an earth core of the flexible cable. Flexible conduit is not considered suitable for providing an earthing connection. An alternative to a straight earth connection is the use of a 110 V supply with centre tapped to earth. In this circuit a suitable transformer is used with the mid-point of its secondary 110 V winding connected to earth; thus the effective shock voltage is reduced to 55 V – a relatively safe level. In fact, it has been estimated that if this recommendation for reduced voltage for portable apparatus was universally adopted in the UK, the fatal accidents due to electricity would probably be halved.
(3) Portable apparatus should be controlled by efficient means, or easily accessible, within reach and near the apparatus for cutting off the electricity supply. For 240 V tools the plug itself, if it can be pulled out of the socket easily enough and without danger, will meet this requirement. However, for voltages above 240 V an isolating switch will be required.

It is essential that portable apparatus is properly maintained, with the earth core of the flexible cable and associated earth connections being tested regularly by a competent person using a substantial current (special test sets are available) to ensure continuity and strength of the earthing.

Equipment, to the relevant British Standard, of the double-insulated or all-insulated types need not be earthed[18].

28.14 Residual current devices

Additional back-up protection can be provided by residual current devices that ensure that in the event of an earth fault the current is cut off before a fatal shock is received. This form of protection works on the principle of monitoring any differential between (i) the current entering a circuit to supply power to the portable apparatus and (ii) the current returning to the supply point. For normal safe operation this current differential is zero but if there is a fault, such as leakage to earth, a differential current occurs which the device rapidly senses, tripping its contacts to cut off the supply to the apparatus.

28.15 Maintenance

It is a requirement of the Electricity Regulations that all electrical apparatus and conductors shall, amongst other things, be adequately maintained. Maintenance does not simply mean general care and cleanliness but implies a system for regular inspection and testing to ensure serviceability. The frequency of maintenance and the competency of those doing it depends upon the type of equipment, its location and the work it is called upon to do. Records should be kept of all maintenance and the results of any testing.

28.16 Conclusion

However well an electrical installation or apparatus is designed and constructed, however well it is operated or worked upon, however adequate are the systems and methods of work and the protective clothing and safety equipment provided and used, safety precautions are incomplete unless suitable arrangements are made for proper maintenance and testing by competent people.

The current 16th edition of the IEE Wiring Regulations[2] is based on internationally agreed installation rules and is an excellent guide to electrical installation, design and practice, with considerable impact on the application of fundamental electrical safety requirements to installations in and about buildings. Although the IEE Wiring Regulations have no legal status as such, compliance with them will, in general, satisfy the requirements of the new Regulations[3] where the latter apply to electrical installations in buildings. For this reason, and also because they introduce new internationally agreed terminology which, in time, will presumably be in common use among the electrical fraternity, the IEE Wiring Regulations[2] merit special study.

Reference should also be made to the British Standards Institution's Standards and Codes of Practice listed in the BSI Yearbook[1], some of which are referred to above, and to the manufacturer's instructions. Given attention to such matters, and to the proper training, instruction and supervision of people who work at or near electrical equipment, electricity should continue to be a safe, convenient and efficient source of power.

References

1. British Standards Institution, *British Standards Yearbook*, BSI, London (latest edn)
2. Institution of Electrical Engineers, *Regulations for Electrical Installations*, 16th (or latest) edn, IEE, London (1991)
3. H.M. Government, *The Electricity at Work Regulations 1989*, S.I. 1989 No. 635, HMSO, London (1989)
4. Health and Safety Executive, *Memorandum of Guidance on the Electricity at Work Regulations 1989*, Health and Safety Series Booklet HS(R)25, HSE Books, Sudbury (1989)
5. Health and Safety Executive, *Approved Code of Practice No. COP 35, The Use of Electricity in Quarries*, HSE Books, Sudbury (1989)
6. Health and Safety Executive, *Approved Code of Practice No. COP 34, The Use of Electricity in Mines*, HSE Books, Sudbury (1989)
7. Hughes, E., *Electrical Technology*, Longmans, London (1978)
8. British Standards Institution (London), *Glossary of electrotechnical, power, telecommunications, electronics, lighting and colour terms*, BS 4727 –
   Part 1 – Terms common to power, telecommunications and electronics
   Part 2 – Terms particular to power engineering
   Part 3 – Terms particular to telecommunications and electronics
   Part 4 – Terms particular to lighting and colour
9. BS CP 1013:1965, *Earthing*, British Standards Institution, London (1965)
10. Health and Safety Executive, *Avoidance of Danger from Overhead Electricity Lines*, Guidance Note GS 6, HSE Books, Sudbury (1991)
11. Health and Safety Executive, *Avoiding Danger from Underground Services*, Health and Safety Guidance Booklet No. HS(G)47, HSE Books, Sudbury (1989)
12. British Standards Institution, BS 5958, *Code of Practice for control of undesirable static electricity*, BSI, London
13. British Standards Institution, BS 6742, *Electrostatic painting and finishing using flammable materials*, BSI, London
14. Health and Safety Executive, *Electrical Apparatus for Use in Potentially Explosive Atmospheres*, Health and Safety Series Booklet No. HS(G)22, HSE Books, Sudbury (1984)
15. British Standards Institution, BS 5490: 1977 (1985), *Specification for classification of degrees of protection provided by enclosures*, BSI, London (1985)
16. British Standards Institution (London), *Code of practice for the selection, installation and maintenance of electrical apparatus for use in potentially explosive atmospheres (other than mining applications or explosive processing and manufacture)*, BS 5345: Part 1, *Basic requirements for all parts of the code*
17. British Standards Institution, BS 5501, *Electrical apparatus for potentially explosive atmospheres*, BSI, London
18. British Standards Institution (London), *Hand held electric motor-operated tools*, BS 2769:1984

Chapter 29

Statutory engineering inspections

E. S. Long

29.1 Introduction

The industrial revolution began in the late 18th century with the mechanisation of the textile industry and subsequent major developments in mining, transport and industrial production, based upon Britain's rich mineral resources such as coal and iron ore, and the use of steam power.

The great industrial towns such as Manchester began to expand with steam power providing the impetus, and this great human exploit was subsequently to make Manchester the world's first industrial city. The industrial conurbation within a 10 mile radius of central Manchester contained in excess of 50 000 boilers – the largest concentration of steam boilers in the world – supplying power to textiles, engineering and to other industries of the period.

The demand by industry for ever-higher boiler operating pressures was outstripping the ability of engineers to meet it safely. As a consequence there was great public concern over boiler explosions which were occurring around Greater Manchester at an alarming rate. The boilers were literally blowing to pieces, causing multiple deaths and injuries.

From 1851 the famous Manchester engineer, Sir William Fairbairn, began arguing for periodic inspection of boiler plant. An advocate of high pressure, in the interests of economy, he decided that the explosions arose from avoidable mechanical causes which could be located in time by carrying out periodic inspections. So in 1854 he enlisted the help of two other eminent Manchester men – Henry Houldsworth, master cotton spinner, and the celebrated engineer, Sir Joseph Whitworth – to form an Association for boiler inspections. This became known as the Manchester Steam Users' Association.

The Boiler Explosions Act 1882 made the reporting of all boiler explosions compulsory and required enquiries to be conducted by Board of Trade surveyors as to the causes and circumstances of the explosions.

At this stage there was still no statutory requirement to have boilers inspected, but, as the 20th century approached, the evidence of over 1000 enquiries held under the Boiler Explosions Act clearly demonstrated the value of regular thorough inspections by competent engineers. The result, included in the Factories and Workshops Act of 1901, was a requirement that steam boilers be subjected to periodic thorough internal examinations by competent persons, and so became the first piece of legislation requiring an item of engineering plant to be inspected, and laid the foundations for the extensive and varied provisions of present day UK law. Electrical equipment inspections and insurance became a prominent part of the insurance portfolio towards the end of the 19th century, with lifts and cranes from early in the 20th century. Dust extraction plant (now referred to as local exhaust ventilation) and power presses followed much later.

29.2 Legislation

While HSW is the principal Act governing the health and safety of employees and others in the workplace, it does not contain any detailed reference to the inspection of plant but relies on those parts of the FA'61 that are still in effect and on Regulations made under either the HSW or FA.

Plant that is subject to periodic statutory inspections includes lifting plant for both goods and people and associated ancillary components, boilers, pressure vessels, power presses, extract ventilation plant and some electrical installations in special situations. A list of the principal Regulations containing plant inspection requirements is given in Table 29.1.

29.3 Pressure vessels

The position regarding the statutory inspection of pressure vessels is changing from the requirements developed under the FA to new requirements contained in new Regulations made under the HSW. Since these two sets of requirements will run concurrently for the transition period up to 1 July 1994 they are dealt with separately below.

29.3.1 Requirements of the Factories Act 1961

These requirements, which have developed with time, tend to relate to particular pieces of plant or operating situations. Specific requirements are covered by FA ss 32 to 38 supported by the Examination of Steam Boilers Regulations 1964 and the Examination of Steam Boilers Reports Order 1964. In addition there is reference to steam boilers and air receivers in the Quarries (General) Regulations 1956 and the Shipbuilding and Ship Repairing Regulations 1960. These will continue in effect until 1 July 1994 unless a user opts to comply with the new Regulations before that date.

29.3.1.1 Boilers

The definition of a boiler is given in FA s 38 as 'Any closed vessel in which for any purpose steam is generated under pressure greater than atmospheric, and includes any economiser used to heat water being fed to any such vessel, and any superheater used for heating the steam at constant pressure'.

Table 29.1. Schedule of principal statutory inspection provisions

Statute		*Class of plant*	*Report form or register number*
The Factories Act 1961	Sect. 27	Cranes & Lifting machines	See SI 1963 No. 1382
ditto	Sect. 22	Hoists & lifts	F54
ditto	Sect. 26	Chains, ropes and tackle	See S.R. & O. 1938 No. 599
Construction (Lifting Operations) Regulations 1961,	Reg. 28	Lifting appliances (machines)	Form 91, Part 1, Sect. E
ditto	Reg. 40	Chains, ropes & tackle	Form 91, Part 2, Sect. J
ditto	Reg. 46	Hoists	Form 75 & Form 91.H
Docks Regulations 1988,		Lifting machinery	–
ditto		Lifting tackle	–
Shipbuilding & Ship Repairing			See SI 1961 No. 530
Regulations 1960,	Reg. 34	Lifting appliances (machines)	
ditto	Reg. 37	Chains, ropes & tackle	See SI 1961 115 & 116
Quarries (General) Regulations 1956,	Reg. 13	Cranes & Lifting	M & Q Form No. 237
[But note the Lifting Plant and Equipment (Records of Test and Examination) Regulations 1992, regarding legislation on report forms or registers which is being phased in]			
For existing pressure plant until July 1994:			
The Factories Act 1961 Examination of Steam Boilers Regulations 1964 Examination of Steam Boilers Reports (No. 1) Order 1964	Sect. 33	Steam boilers, Economisers, Superheaters, ovens, and hot plates, etc.	F55, F55A F56, F57 F62, F62A F62B
Factories Act 1961,	Sect. 36	Air receivers	F59 & F60.
Factories Act 1961,	Sect. 35	Steam receivers	F58, F58A, F60, F58B
Quarries (General) Regulations 1956	Reg. 18	Boilers	283–286
ditto	Reg. 22	Air receivers	238 & 239
For new pressure plant and all pressure plant after July 1994:			
Pressure Systems and Transportable Gas Containers Regulations 1989	Reg. 9	All pressure plant	Engineer/surveyor report
Control of Substances Hazardous to Health Regulations 1988	Reg. 9	Local exhaust ventilation plant and Dust and fume extraction plant	Engineer/surveyor report
Electricity at Work Regulations 1989		Electrical installations	Engineer/surveyor report

NOTE: The above schedule does not claim to be exhaustive. There are many statutes requiring more frequency but less thorough inspections; inspections and tests before commissioning; and some requiring users to draw up their own schemes of examination and implement them (e.g. Quarries (Ropeways and Vehicles) Regulations 1958, Reg. 14). There are also many statutory report forms additional to those specified above, mainly in connection with particular certificates of exemption.

Any reference to a steam boiler also includes all fitments and attachments. These requirements apply to steam raising plant only and not to hot water boilers used in central heating installations.

All existing boilers should have a maximum permissible working pressure which for a new boiler is the pressure specified by the manufacturer, but for existing boilers is the pressure specified on the last report of thorough examination. This is the lift-off pressure of the safety valve although it is normal to run the boilers at 90–95% of this to minimise continual leakage from the safety valves.

All boilers are required to have a safety valve so adjusted as to prevent the boiler being worked at a pressure greater than the permissible working pressure, a suitable stop valve, steam pressure gauge, water level gauge, means for attaching a test pressure gauge and unless externally fired be provided with a fusible plug or efficient low water alarm device. The boiler must also carry a distinctive number or mark to identify it especially where more than one boiler operates in parallel. Certain of these provisions do not apply to economisers or superheaters.

Boilers should be thoroughly inspected by a competent person before being taken into use for the first time and subsequently once every 14 months. In the case of certain water-tube boilers, waste-heat boilers and heat exchangers that are of fusion welded or solid forged construction the period of inspection is 26 months.

The examination of a steam boiler is in two parts:

(1) A thorough examination, both internally and externally by a competent person, when it is cold, after the interior and exterior have been properly prepared in accordance with one or more of the following provisions:
   (a) Opening up and cleaning out and scaling of the fire and water sides including the removal of all access doors from manholes, mudholes and handholes.
   (b) Opening out for cleaning and inspection of fittings including pressure parts of automatic controls, safety valves and water gauges, and blow down valve.
   (c) In the case of water-tube boilers, the removal of drum internal fittings as required by the competent person.
   (d) In the case of shell-type boilers, the dismantling of firebridges if made of brick, and furnace protective brickwork at specified periods.
   (e) All brickwork, baffles and coverings must be removed for the purpose of the thorough examination, to the extent required by the competent person, but in any case to expose the headers, seams of shells and drums:
      (i) not less frequently than once every six years in the case of steam boilers in the open, or exposed to the weather or damp, and
      (ii) not less frequently than once every ten years in the case of every other steam boiler.
(2) The boiler (including economiser and superheater if fitted) must be thoroughly examined by a competent person when it is under normal steam pressure. This examination should be made on the first occasion when the steam is raised after the examination when the boiler is cold,

and must consist of ensuring that the safety valve is adjusted so as to prevent the boiler being worked at a pressure greater than the maximum permissible working pressure. It should then be locked off to prevent unauthorised tampering with the setting. The operation of the pressure gauges, water gauges, float controls, alarms and cut-outs must also be checked.

It is possible for the two parts of the thorough examination to be carried out by two different competent persons.

Whilst the examination of the boiler, like any other item of statutory plant, is based upon a visual inspection, it may be supplemented by: hammer testing to establish material integrity/extent of surface corrosion, material thickness drilling, proving a clear waterway through the tubes, withdrawal of sample tubes for evidence of deterioration of wall thickness, internal material detection of cracking/flaws by ultrasonic, radiographic, magnetic particle or dye penetrant testing, or by hydraulic testing. These tests may be undertaken by specialists but the results must be made available to the competent person.

Entry into any boiler, which is one of a range of two or more, is forbidden unless all inlet pipes through which steam or hot water could enter the boiler have been disconnected, or the valves controlling entry of steam or hot water have been closed and locked. Particular attention should be paid to blow-down valves which discharge into a common line or sump, where special safety procedures must be followed[1]. Suitable precautions should also be taken to ensure that the boiler is free from dangerous fumes and has adequate ventilation.

The purpose of the examination is to ascertain the material condition of the boiler with particular reference to any defects which could affect the continued safe working of the boiler at its current maximum permissible working pressure. During the examination when cold the competent person will be checking for the following types of typical defects:

(1) *External* – Wastage from corrosion due to leakage at joints between fittings and shell, tubes and tubeplates, or at seams in rivetted boilers; wastage of manhole, mudhole and handhole joint seatings; general physical damage.
(2) *Fire side* – Wastage in combustion chamber plates due to leaking tubes, stay tubes or stays; furnace flame impingement damage; erosion of material by fast moving gases and/or entrained particles – particularly where gases are hottest and changing direction (i.e. combustion chamber ends of tubes in horizontal shell-type boilers); overheating damage due to scale or sludge build-up, or water shortage; furnace bulges and combustion chamber crown bulges.
(3) *Water side* – Chemical damage causing wastage, pitting, thinning and necking of material due to oxygen or other chemical corrosion; mechanical damage caused by expansion and contraction, and grooving at plate flange bends, evidence of 'peaking' at the longitudinal weld caused during the manufacturing process, cracking at the tube/tube plate joints, and other slowly developing fatigue cracks at a variety of locations.

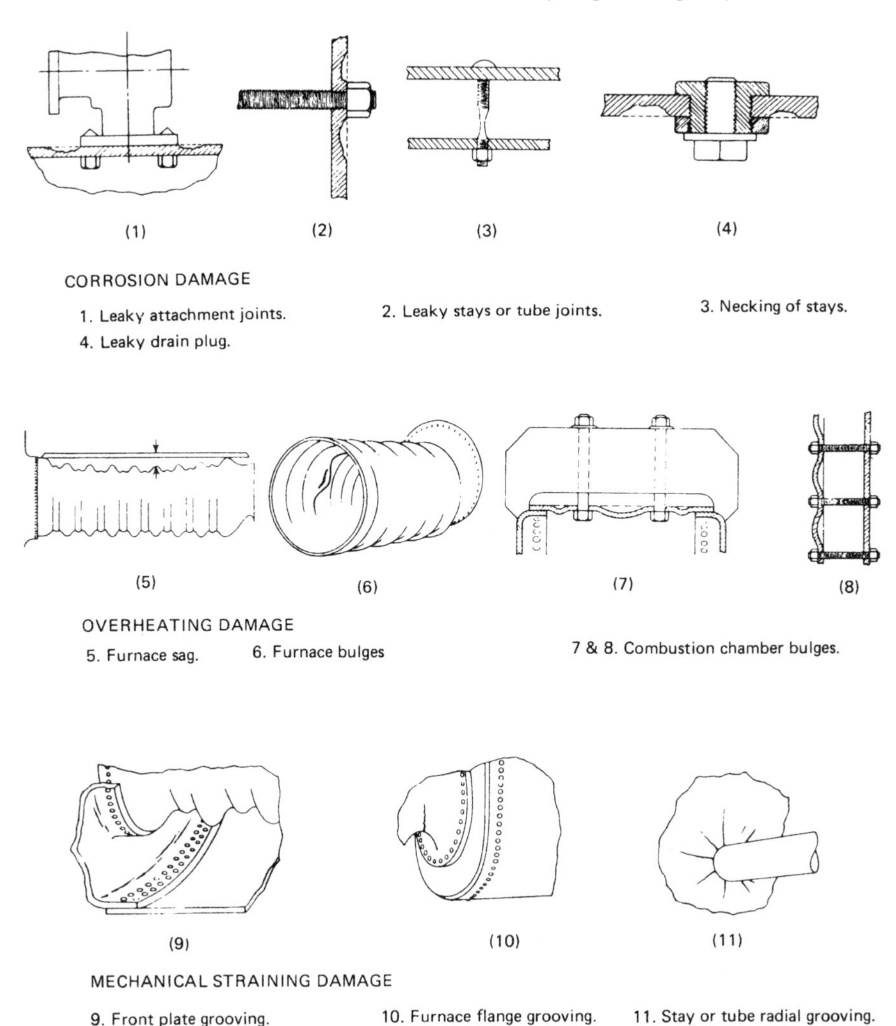

Figures 29.1 Some typical examples of different types of boiler defects. (Courtesy Milton and Leach[2])

Examples of typical defects are shown in *Figure 29.1*.

Reports for both parts of the examination must be made on prescribed forms and include the prescribed particulars contained in the Examination of Steam Boilers Reports (No. 1) Order 1964, Schedules 1 (cold examination) and 2 (normal working pressure). The report must be signed by the competent person making the examination and in addition be countersigned by the chief engineer of the company employing the competent person.

When the examination reveals defects which affect the safe working of the boiler at the current maximum permissible working pressure, the report must make recommendations on conditions to be met before the boiler can be returned to operation. The conditions may include certain defined repairs, or the reduction of the maximum permissible working pressure. In both cases a copy of the report of examination must be forwarded to the local Health and Safety Executive office within 28 days.

Where a defect has been found in the boiler and subsequent repairs carried out, the competent person should satisfy himself that the repair work has in fact been carried out satisfactorily before the boiler is released for service.

29.3.1.2 Steam and air receivers

These are defined in ss 35 and 36 of FA as:

(1) *Steam receiver* – Any vessel or apparatus (other than a steam boiler, steam container, steam pipe or coil, or part of a prime mover) used for containing steam under pressure greater than atmospheric pressure.
(2) *Steam container* – Any vessel (other than a steam pipe or coil) constructed with a permanent outlet into the atmosphere, and through which steam can be passed at approximately atmospheric pressure, for the purpose of heating, boiling, drying, evaporating or other similar purpose.
(3) *Air receiver* – Any vessel for containing compressed air and connected to an air compressing plant, a fixed vessel for containing compressed air or compressed exhaust gases used for the purpose of starting an internal combustion engine, a fixed or portable vessel used for the purpose of spraying by means of compressed air, any paint, varnish, lacquer or similar substance, or any vessel in which oil is stored and from which is forced by compressed air.

There are other types of vessels and tanks, not covered by the definition of air receivers, from which fluids are blown by compressed air, and whilst non-statutory at present, it is advisable to subject such vessels to similar inspection procedures.

The safe working pressure is the legal term specifically applied to steam receivers, although also used in the case of air receivers. In the case of a new receiver, it means that pressure specified by the manufacturer, and in the case of a receiver which has been examined in accordance with the statutory provisions, that specified in the report of the last examination.

Both the steam receiver and the air receiver must be provided with a suitable reducing valve or other automatic appliance to prevent the safe working pressure being exceeded if the vessel is not suitable for the pressure of the source of supply, a pressure gauge, a safety valve with a vent to atmosphere and, where more than one vessel is in use in the factory, each must bear a separate distinguishing mark. In the case of the steam receiver only, it must have in addition a suitable stop valve. Air receivers must be fitted with a drain plug, provided with a suitable manhole, handhole or other means which will allow the interior to be thoroughly cleaned and have the safe working pressure marked on it.

The steam receiver and its fittings should be thoroughly examined by a competent person, as far as the construction permits, at least once in a period of 26 months. In the case of the air receiver that is of solid drawn construction the competent person making the examination may specify in writing a period exceeding 26 months, but not exceeding 4 years, within which the next examination must be made. Also, if the air receiver is so constructed that the internal surface cannot be thoroughly examined, a suitable hydraulic test of the receiver may be carried out instead of an internal examination.

Unlike boilers, the manner of the examination of these vessels is not detailed in the legislation. It is simply required that they be thoroughly examined and, in the case of air receivers, that they are thoroughly cleaned as well.

Reporting is on a prescribed form and the competent person will need to have examined the vessels internally and externally, checked that the required fittings are provided, properly adjusted and in good working order, and that the vessels are clearly marked with an identification number and safe working pressure where necessary.

Where a defect is found, there is no requirement to forward copies of the reports on steam or air receivers to the Health and Safety Executive.

The most common defect found in these types of vessels is thinning or pitting of the shell plates due to corrosion on the inner surface and this is normally countered by reducing the safe working pressure, which is calculated from plate thicknesses and other data ascertained at the examination. The safe working pressure must be specified in the report.

Section 37 of the Factories Act allows particular classes or types of boiler, steam or air receiver to be exempt or excepted from the requirements of Sections 32 to 36.

29.3.2 Requirements of the Pressure Systems and Transportable Gas Containers Regulations 1989

These Regulations have been made under HSW and deal with broad objectives to be achieved in the operation of complete pressure systems at all places of work, as opposed to dealing with specific requirements for specific pieces of plant under FA 1961.

Pressure systems are defined as:

(a) a system comprising one or more pressure vessels of rigid construction, any associated pipework and protective devices;
(b) the pipework with its protective devices to which a transportable gas container is, or is intended to be, connected; or
(c) a pipeline and its protective devices which contains, or is liable to contain, a relevant fluid, but does not include a transportable gas container.

Relevant fluid is defined as:

(a) steam;
(b) any fluid or mixture of fluids which is at a pressure greater than 0.5 bar above atmospheric, and which fluid or mixture of fluids is:
    (i) gas, or
    (ii) a liquid which would have a vapour pressure greater than 0.5 bar above atmospheric pressure when in equilibrium with its vapour at either the actual temperature of the liquid or 17.5 degrees Celsius; or

(c) a gas dissolved under pressure in a solvent contained in a porous substance at ambient temperature and which could be released from the solvent without the application of heat.

Since July 1990, the Regulations have placed obligations on anyone who manufactures or constructs a new pressure system, and anyone who repairs or modifies a new or existing pressure system or part of it, to ensure that no danger will arise when it is operated within the safe operating limits specified for that plant. The Regulations revoke ss. 32 to 38 of FA 1961, and no pressure plant may be put into operation until a written scheme for its periodic examination has been prepared by a competent person.

The written scheme of examination is perhaps one of the most important features of the new Regulations and is required to be compiled before a pressure system can be operated. Incorporated in the scheme must be details of the pressure vessels, protection devices and pipework where a defect in them could give rise to danger. In addition, it must specify the nature and frequency of examinations and the measures necessary to prepare the system for safe examination over and above the precautions that the user or owner of the system would reasonably be expected to take.

The onus for ensuring that these requirements are met lies with the user of an installed pressure system or the owner or filler of a mobile system – as opposed to the employer or occupier under the FA.

A report of the periodic examination by the competent person must be given to the user or owner of the system within 28 days, but only if there is imminent danger from the continued operation of the system is a copy to be provided to the enforcing authority. Copies of the periodic reports must be kept, presumably for the life of the system, but they can be kept in a computer.

For existing pressure plant, the requirements of these Regulations regarding a programme of examinations and the carrying out of the examinations do not come into effect until 1 July 1994, although the user of an existing system can opt to comply at an earlier date.

These Regulations are supported by Approved Codes of Practice[3] which are accompanied by Guidance Notes. The Approved Codes of Practice outline, in relative terms, the qualifications to be held by a competent person or authority to enable them to compile a written scheme for a major, intermediate or minor system, the size of the system being specified in bar-litres.

29.4 Lifting and handling plant

The range of what has generically become known as lifting and handling plant has expanded over the years in both scope and sophistication, with many innovative designs of plant now being used in industry and for various handling processes. Because of the very nature of this type of plant, together with its versatility, it is used in many industrial locations, and as such a number of Acts and Regulations make provision for it. These comprise: the Factories Act 1961; the Offices, Shops and Railway Premises (Hoists and Lifts) Regulations 1968; the Miscellaneous Mines Order 1956; the Miscellaneous Quarries Order 1956; the Shipbuilding and Ship Repairing Regulations 1960; the Docks Regulations 1988; the Loading and Unloading of Fishing Vessels Regulations 1988, and lastly, but by no means least, the Construction (Lifting Operations) Regulations 1961.

A report has to be made following each examination and the possibility of rationalising the many reporting forms used for lifting plant is being reviewed with the object of reducing the number of report forms and test certificates, and so simplifying the administrative procedures.

29.4.1 Lifts and hoists

There are many variations on the fundamental design concepts for hoists and lifts, but basically they are of four main classes: (1) electric, incorporating winding gear and a traction sheave drive or winding drum, (2) hydraulic, either direct or indirect acting, (3) manual service type, and (4) continuous paternoster type lifts; each of which may be classified as passenger or goods lifts only, or a combination of both. Continuous lifts (known as paternosters)[4] have no doors and the passengers step into and out of the car as it passes the landing. These are normally to be found in hospitals and colleges where the rapid mobility of staff and students is required.

The principal legislation covering hoists and lifts is ss. 22 to 25 of the FA 1961, and the Offices, Shops and Railway Premises (Hoists and Lifts) Regulations 1968 (OSRP Regulations) which define a hoist or lift as an appliance which incorporates a platform or cage where the direction of movement is restricted by a guide or guides. In the case of service lifts, the generally accepted definition, in addition to the above, is that the maximum floor area of the cage is 1.0 m^2 and its maximum height is 1.2 m. A greater height may be allowed if the floor area is sub-divided into compartments.

Every hoist or lift should be of good mechanical construction, sound material and adequate strength, and should be properly maintained. In the case of lifts constructed after 1937 in factory premises, and after 1968 in offices, the gates of the lift must be fitted with interlocks, usually of the electro-mechanical pre-locking type and of fail safe design, so that neither the cage nor landing gate can be opened unless the cage or platform is at the landing, and that the cage or platform cannot be moved unless the gates are closed.

This provision does not apply to manual service lifts or continuous lifts. Where the hoist or lift is used for carrying persons, as opposed to 'goods only' lifts, it must be fitted with: (a) an automatic limiting device to prevent the cage or platform over-running, (b) at least two suspension ropes separately anchored, each capable of taking the maximum working load and weight of the cage etc., and (c) safety gear that will support the cage with its maximum working load in the event of the suspension ropes failing. Section 48 of the Factories Act states that all lifts constructed after 1938 must have a fire proof enclosure.

The frequency of thorough examinations for a hoist or lift by a competent person under the FA and OSRP Regulations is six months. In the case of continuous lifts and manual service lifts the frequency is every 12 months.

It should be noted that escalators are non-statutory items of plant but, under the Health and Safety at Work Act, owners would be well advised to have them inspected at regular frequencies.

The results of the inspection must be reported on a prescribed form and must give details of design and construction, maintenance, parts inaccessible, defects, observations and the maximum working load which must also be displayed in the lift car. Where defects are found a copy of the report must be sent to the enforcing authority (HSE or LA), depending upon whether the lift is located in an office, in factory premises or in a commercial enterprise.

Under the Hoists Exemption Order 1962, a number of types of hoist or lift are exempted from certain of the Regulations, depending upon the design. The main examples are: lifts descending from pavement level to the basement of a building, mobile and fixed lifts used for loading and unloading goods directly from vehicles where the maximum height of travel is restricted, electrical and mechanical service lifts where either the landing and cage entrances are fitted with lattice type gates or the lift cage has no door and moves in a flush lined shaft, and hoists used solely for lifting materials directly onto a machine.

The purpose of the examination is to ascertain whether the lift can continue to be used with safety up to its designated operating limit for the period of time until the next inspection and, if not, to specify any repairs, renewals or adjustments necessary to ensure this. The person making the examination must have regard to all factors affecting safety and take into consideration both the mechanical and electrical components of the installation.

The safety of passengers in the lift relies on the proper operation of many components in the installation but, in the case of an electrical lift, particular attention should be paid to the following:

(1) Winding gear, especially the condition of the traction sheave and brake friction pads. The gearing should be opened up every ten years to ascertain the condition of the worm and worm wheel.
(2) Suspension ropes and fixings, checking for surface wear, broken wires, splintering, corrosion, fluctuations in diameter and unequal tensions in the ropes. Because pre-formed ropes are used, each wire has little or no strain energy and when breakage occurs the wire still maintains its helical form and will not spring out proud of the outer surface of the rope. As such wire breakages are very difficult to detect. Particular attention should be paid to ropes that are over ten years old.
(3) Safety gear. All passenger lifts must be fitted with safety gear which will arrest and support a fully loaded car in the event of suspension rope failure. The safety gear linkages, attachments, ropes etc. should be carefully checked for security and freedom of movement at each examination. The safety gear must also be tested every four years.
(4) Control equipment and the overload protection devices.
(5) Landing gate locks must be checked internally every two years.
(6) Upper and lower limit switches should be checked, and the slack rope switch on drum-type winding gear.
(7) Integrity of guide fixings.

There is no legal requirement to test a lift prior to commissioning, although clients may require the competent person to witness the pre-commissioning tests carried out by the manufacturers or contractors. The types of tests that would normally be carried out include:

(a) Motor current and speed checks under no load, balanced load, contract load and with a proof load of 110 per cent of the contract load when travelling both up and down.
    In the case of a hydraulic lift, pressure readings should be taken in the above modes of operation, and a hydraulic relief valve proving test carried out.
(b) Overload protection relay checks for the main drive motor and the door operation motor.
(c) Testing of the over-speed governor and measurement of the stopping distance of the cage when the safety gear is applied.
(d) Insulation resistance testing on the lift motor, power wiring and safety circuits, and earth continuity testing of electrical components – this value should not exceed 0.5 ohms. It must be noted that where the controls comprise electronic circuitry, insulation resistance tests must not be carried out.
(e) Testing of all safety devices such as car-door interlocks, ultimate travel limits, phase failure/phase reverse relays, slack rope switch (if fitted), broken selector tape switch, car top controls, safety gear switch and emergency lowering device. In the case of hydraulic lifts, the anti-creep device and low-pressure valves should be checked.
(f) Checking of the run-by clearances of the car and counterweight and the accuracy of car levelling at the landings.
(g) Testing of safety controls such as the overload cut out on the car loading device, car emergency alarm/telephone/lighting, kinetic energy of lift doors when closing and the effectiveness of photo-electric protection device.
(h) Performance tests for 30 minutes with contract load, noting the temperature rise of main motor, winding gear lubricant and bearings, etc.

29.4.2 Cranes and lifting machines

The types of cranes and lifting machines in current usage are vast, but the following are those in common usage: tower cranes, overhead travelling cranes, mobile cranes, portal cranes, derrick cranes and goliath cranes; and in the case of lifting machines: electric hoist-blocks, manual pulley or chain blocks, fork trucks, swinging jibs and electric teagle hoists.

Lifting machines are defined differently, depending upon the industry and which particular piece of legislation applies. For example, s. 27 of the FA defines a crane as a 'lifting machine', and includes in the definition a crab, winch, teagle, lifting block (manual or electric), gin wheel, transporter or runway (sometimes referred to as a trolley). The Construction (Lifting Operations) Regulations 1961 define a 'lifting appliance' as including shear legs, excavator, dragline, piling frame, aerial ropeway or overhead runway in addition to the above, and under the Shipbuilding and Ship Repairing

Regulations 1960 a 'lifting appliance' also includes a derrick. The Miscellaneous Mines Order and the Miscellaneous Quarries Order define a 'lifting machine' as simply a crane, crab or winch. Finally, the Docks Regulations 1988 define the 'lifting appliance' more generally as a stationary or mobile appliance, including anchorages, fixings and supports when used on dock premises for suspending, raising or lowering or moving loads, and so includes fork trucks, chain blocks, lifting cradles, cranes, straddle carriers and access lifting platforms. A winch used solely for horizontal movements of loads is not a lifting appliance.

It should be noted that the scope of the new Docks Regulations is wider than the revoked Docks Regulations 1934, and now refers to all places where 'dock operations' are carried out including neighbouring land and buildings upon the land, used for landing, storage, sorting, inspection, checking, weighing or handling of the goods.

Whatever the strict legal definition of the lifting machine, all parts of the fixed or mobile working gears, including anchorages, must be of good construction, sound material, adequate strength, free from patent defect and properly maintained. The Loading and Unloading of Fishing Vessels Regulations 1988 and the Docks Regulations 1988 also imply that it must be of good design and fit for the purpose for which it is to be used.

A lifting machine must be tested and thoroughly examined by a competent person and have its safe working load (above 1 tonne in the case of the Construction Regulations) marked on it before being taken into use for the first time. In the case of the Docks Regulations 1988 and the Loading and Unloading of Fishing Vessels Regulations 1988, this requirement applies also to plant that has been modified, altered and also following a major repair that is likely to affect its strength or stability. Under the Construction Regulations, this testing requirement is rather more complex. Generally, any crane, crab or winch must be tested and thoroughly examined every four years, and all pulley blocks, gin wheels and shear legs should be tested before use. In addition to this, every time a crane is erected on site, or after partial dismantling and re-erection on another part of the site, removal or adjustment involving the anchorage and/or ballast, also any major repairs, or modifications to members in the direct line of stress of a crane, crab or winch and affecting its strength or stability, or any substantial alteration to a pulley block, gin wheel, shear legs greater than 1 tonne, it must be tested and thoroughly examined. This provision to test a crane after erection on site would apply to such as a tower crane or very large mobile crane of the maxi-lift type and not to the normal size of mobile cranes. As it is unlikely that such cranes will be on site for four years, these cranes would be tested at a greater frequency. In the case of a lifting appliance on board ship, under the Docks Regulations, this must have been tested within the last five years.

All lifting appliances must be thoroughly examined by a competent person every fourteen months under the Factories Act, the Construction Regulations, the Miscellaneous Mines and Miscellaneous Quarries Orders, and every twelve months under the Shipbuilding and Ship Repairing Regulations, the Loading and Unloading of Fishing Vessels Regulations and the Docks Regulations. In addition, the Docks Regulations stipulate that more frequent inspections can be specified by the competent person when deemed necessary. Also, under the Miscellaneous Mines and Miscellaneous

Quarries Orders a crane, grab or winch, if once dismantled or out of regular use for more than two months, should be inspected by a competent person before going back into use. Moreover, the Loading and Unloading of Fishing Vessels Regulations require the lifting appliance to be visually inspected by a responsible person before loading/unloading commences, and the Construction Regulations require all lifting appliances to be inspected by the operator at least once every week and the results reported on a prescribed form.

The Report of thorough examination must be on a prescribed form. Where the machine has a defect and cannot continue to be used with safety, then a copy of the report must be sent to the Health and Safety Executive within 28 days. Under the Lifting Machines (Particulars of Examinations) Order 1963, reports must contain information such as the identification mark, safe working load, particulars of defects, observations, name of the competent person and the date of examination. The Docks Regulations also stipulate that the reports are to be retained for at least two years; this practice is, however, recommended in all cases, regardless of which legislation is applicable.

The provision for establishing the maximum safe working load that can be lifted at a particular jib radius of the crane has become a complex issue under various pieces of legislation. Under the Factories Act, a mobile crane with a varying jib radius must be fitted with a safe load indicator, or a table of safe working loads used in conjunction with a pendulum type indicator mounted on the jib such that it is clearly visible to the operator and shows the radius of the crane hook. Again the Shipbuilding and Ship Repairing Regulations simply require a radius indicator for cranes with derricking jibs with corresponding indication of safe working loads. The Construction Regulations are slightly more demanding in the sense that all mobile, tower and derricking cranes above 1 tonne must be fitted with an approved type of automatic safe load indicator. Where this safe load indicator has been wholly or partially dismantled it must be tested by a competent person before the crane is taken back into service and the results reported on a prescribed form. The requirements under the Docks Regulations are somewhat more onerous in the sense that they specify three separate situations for various types of plant. Firstly, a load/radius indicator is required for such as tower/cantilever cranes of the hammer-head type; secondly, a load/angle indicator for mobile cranes operating with fly jib or locked manual extension; and thirdly, an approved automatic safe load indicator for mobile cranes where the safe working load is greater than 1 tonne.

Regarding the safe working load, the Construction Regulations and the Docks Regulations state that it is that load specified in the latest certificate of test obtained from the manufacturers or suppliers, or the latest report of thorough examination. This latter point is relevant when the competent person has derated the lifting machine owing to certain mitigating circumstances. It should be noted at this stage that, whilst not quoted in the legislation, the safe working load (SWL) will in most cases be the same as the working load limit (WLL). However, if a lifting machine has been derated because of the heavily corrosive environment for example, in which the equipment is being used, then the SWL stated on the equipment will be less than the WLL depending upon the amount of deration, thus building into the equipment a further factor of safety.

The philosophy underlying any inspection of lifting machines should entail a check on all items in the direct line of stress. Commencing at the hook this would comprise the following major components, where appropriate, but is by no means exhaustive:

(1) Hook return block, jib head pulleys, lifting rope and rope anchorage points and fixings. Condition of luffing ropes, bridle ropes, compensating pulleys, wedge type socket fixings and bulldog grip fixings.
(2) Jib and mast structures. General condition of main structural members (chords), bracings, struts, ties, jib foot pin, in terms of deformation and impact damage, and a visual inspection of the welds.
(3) Crane chassis structure, including the 'A' frame in the case of a mobile crane and loose or failed bolts in the crane slewing ring.
(4) Machine winding gear.
(5) Control, including any derricking limits etc; particular attention being paid to any interlocking arrangements between separate control functions.
(6) Safe working load indicators. This can vary from a simple notice of SWL's and pendulum type load/radius indicators on the jib, through to a dynamometer incorporating a load cell on the jib with a display readout in the operator's cab, to sophisticated microprocessors incorporating automatic safe load indication and overload alarms. The competent person must satisfy himself that the type of safe load indicator fitted conforms to the relevant statutory requirements, and that the indicator is accurate for the configuration of the crane rigging.

Generally, the testing procedure would comprise an initial inspection of the lifting machine, a functional test during full operation of the crane, application of the safe working load and the proof load, a measurement of deflections of the structure (where appropriate), and a re-inspection to ensure permanent deformation has not taken place.

In the case of mobile cranes, the test load depends upon the date of manufacture of the crane. For those manufactured before March 1981, the amount the test load exceeds the SWL depends upon the size of crane:

(1) Up to 20 tonnes, the test load is the SWL plus 25 per cent.
(2) From 20 up to 50 tonnes, the test load is the SWL plus 5 tonnes.
(3) Over 50 tonnes, the test load is the SWL plus 10 per cent.

For mobile cranes manufactured after March 1981, regardless of size, the crane must be subjected to a dynamic test with a load of the SWL × 1.1, where the crane is moved through all its motions, plus a static test of the SWL × 1.25 where the load is simply raised slightly off the ground.

Additional special tests are required in the case of cranes with manual extensions and for fly jib extensions.

Under the Construction Regulations the safe load indicator also requires to be tested and reported upon and, in addition, in the case of a tower crane an anchorage test needs to be carried to and reported upon.

Higher proof loads should be applied for stability tests on cranes, but in practice would very rarely be carried out.

29.4.3 Miscellaneous lifting equipment

Section 26 of the Factories Act defines lifting tackle as chains, ropes, rings, hooks, shackles and swivels, and the Construction Regulations and Shipbuilding and Ship Repairing Regulations extend this definition to include links, plate clamps and eyebolts. Regardless of the strict definitions quoted in the Statutory Instruments, the term lifting tackle/gear refers to any items of ancillary equipment used in a 'lifting system'. The Docks Regulations sum this up with its definition that lifting gear is a means by which a load can be attached to a lifting appliance, but which does not form an integral part of that appliance or load.

Again the equipment must be of good construction, sound material, adequate strength and free from patent defect. The Loading and Unloading of Fishing Vessels Regulations and the Docks Regulations extend this to include good design, properly installed and assembled, and properly maintained. All items must be separately identified with the safe working load clearly marked, and in the case of multi-legged slings under the Factories Act a table of safe working loads displayed in the distribution store.

The safe working load means that specified on the last test certificate, or report of examination, or as specified by the manufacturer/supplier. The equipment must be supplied with a test certificate before being used under any of the relevant Statutory Instruments, and re-tested after any substantial repair or modification.

In the case of the Factories Act and Construction Regulations, lifting tackle must be thoroughly examined every six months by a competent person. Under the Shipbuilding and Ship Repairing Regulations, lifting gear and chains must also be examined every six months, but in the case of wire ropes the frequency is every three months; where the examination of the rope reveals that a wire of the rope or rope sling has broken, then the examination frequency must be monthly until such time as the rope is discarded. The Loading and Unloading of Fishing Vessels Regulations and the Docks Regulations state twelve monthly examination frequency, or a shorter period as specified by the competent person; the more frequent examination could be carried out by a responsible person on site.

Chains, links and shackles of wrought iron construction (where still in use) must be annealed every fourteen months to comply with the Factories Act, Construction Regulations and the Shipbuilding and Ship Repairing Regulations; where they are manufactured from 0.5 inch bar or smaller the frequency is increased to six monthly.

Under the Construction Regulations, hooks used on cranage must be fitted with a safety catch, or the hook must be of such a shape as to prevent the possibility of the sling becoming displaced.

Some of the Statutory Instruments such as the Construction Regulations and the Shipbuilding and Ship Repairing Regulations outline certain discard-criteria for wire ropes, such that if in any length of ten diameters the total number of visible broken wires exceeds 5 per cent of the total number of wires in the rope, then the rope should be rejected.

The proof load parameters for testing lifting tackle are complex, particularly with regard to the Shipbuilding and Ship Repairing Regulations, and the Docks Regulations, and the reader is referred to the Regulations for further details.

The Shipbuilding (Reports on Chains and Lifting Gear) Order 1961 and the Shipbuilding (Reports on Ropes and Rope Slings) Order 1961 give details of what is required regarding the contents of the reports. Under the Docks Regulations the reports should be kept for at least two years, which is a reasonable period of time regardless of any Statutory Instrument.

It must be noted that where a defect is found in the equipment, a copy of the report does not need to be sent to the Health and Safety Executive.

Lifting tackle is not only used as ancillary equipment to a lifting operation, but will also be an integral part of the lifting machine, such as lifting ropes, hooks etc. on a statutory item such as a crane, or a non-statutory item such as a fork lift truck or a motor vehicle lifting table. In the latter cases, even though the items are non-statutory, the ropes/chains are statutory items and require to be inspected every six months.

The examination is to assess the suitability of the equipment for continued use, and the following are some typical defects of lifting tackle: stretching, distortion, corrosion, fatigue cracking, wear between adjacent chain links, broken wires in ropes and degradation of the internal fibre core of the rope etc. The degree of 'acceptable damage' is a matter of judgement on the part of the competent person in many respects, as the dividing line between a defect and an observation can be extremely arbitrary. Also, lifting tackle may appear to be relatively simple items of engineering equipment and, as such, abused in many instances. Notwithstanding this, the stresses imposed on an item of lifting equipment during normal use are highly complex, so it is constructed with substantial factors of safety to withstand shock loads and maltreatment. In addition, wire ropes are a subject in themselves and generally have a factor of safety of about 5 to 1. However, they can deteriorate rapidly if abused, or if left outside in sub-zero temperatures, or subjected to elevated temperatures from a manufacturing process, or used over salt baths, etc.

29.4.4 The Lifting Plant and Equipment (Records of Test and Examination) Regulations 1992[5]

These Regulations are not concerned with the actual tests and examinations of pressure systems but with revoking all references to 'register', 'certificate' and 'report' in earlier legislation. In addition, they revoke the requirement to report on a 'prescribed form' and also remove any requirement to provide a written report except where required to do so by an HSE inspector. Instead, they provide for the keeping of a 'record' in a form agreed by both the competent person carrying out the test/examination and the plant owner. This should enable future records to be kept by electronic means.

29.5 Power presses

A power press or press brake is defined under the Power Press Regulations 1965 as a machine which is power driven and which embodies a flywheel and a clutch mechanism and is specifically used for working cold metal. Presses working hot metal, hydraulic presses, and metal-cutting guillotines are not included. There is also a number of exemption certificates in force which cover machines used for compacting metal powders, machines with a stroke not exceeding 6 mm, punching and shearing machines, straightening machines, upsetting machines, riveting machines, stapling machines; this list is not exhaustive but a representative sample.

The Regulations deal with the appointment of competent persons for thorough inspections, examination and testing of the press and its safety devices, and the procedures to be adopted where defects are found.

The Regulations specify two types of inspection. The first, under Regulation 7, is an inspection and test by a responsible person, such as a suitably trained press toolsetter, which must be carried out during the first four hours of every eight-hour shift in the factory, and after setting, re-setting or adjustment of the tools. It is a visual inspection and a functional test of the safety devices, and the responsible person making the inspection and test must sign a certificate stating that the safety devices are in efficient working order, or inform management that they are not. In the latter case, the press must be taken out of use.

The second type of inspection, under Regulation 5, is a thorough examination and test of the power press and its safety devices carried out by a competent person every twelve months on fixed guard machines, and every six months on all other types that incorporate some form of interlocked or automatic guard. The inspection will always involve close examination of guard linkages, interlock mechanisms, guard control mechanisms, clutch and brake condition and performance, and anything else that could affect safety at the tools. It may also involve opening up clutch mechanisms for close inspection.

The report of examination by the competent person must be made on a prescribed form, F2197, and include a note of any defect that requires to be repaired, either immediately or within a specified period of time, to ensure safe operation of the machine. The owner must be informed immediately, both verbally and by official handwritten note. In addition, a copy of the report of examination must be forwarded by the competent person to the local offices of the Health and Safety Executive within 14 days. If the repairs are required immediately, the press must be taken out of service until the repair work has been completed.

Where remedial action has been taken to repair a defect found during the examination, an accurate record must be kept of the date and the actual repair work carried out.

Since these Regulations came into effect in 1965 they have provided stringent requirements regarding safety in the use of power presses. As a result, this type of machinery will not be greatly affected by the Provision and Use of Work Equipment Regulations 1992[6]. However, the general duties of these Regulations require:

- selection and suitability of equipment to ensure that the work can be done safely,
- work equipment to be properly maintained,
- authorisation of persons to use any equipment with special risks, and

– information, instruction and training to be given to operators and their supervisors and managers,

which place additional obligations on the users of power presses.

29.6 Local exhaust ventilation

Local exhaust ventilation (LEV) equipment is intended to control mechanically the emission of contaminants such as dust and fumes that are given off during a manufacturing process or in a chemical laboratory. Normally this is done as close to the point of emission as possible using a stream of air to remove the airborne particulate matter and transport it to where it can be safely collected for ultimate disposal. The physical layout and setting of LEV equipment is critical for it to work effectively, and comparatively minor alterations can affect its performance. It is, therefore, important that LEV equipment[7] should be properly designed, manufactured, installed, operated and maintained.

The main elements of the equipment would normally comprise:

(1) captor hood,
(2) exhaust ducting,
(3) extraction fan, and
(4) filter/collecting bags.

In addition, the following items would be regarded as LEV equipment:

(5) parts of machinery such as integral machine casing or guarding which has a dual purpose of controlling and venting emissions from the process to atmosphere;
(6) vacuum cleaners when permanently connected to an exhaust system and fitted to portable tools;
(7) flues from a furnace/oven when the plant is producing hazardous or toxic fumes – but not when the flue only serves to create a draught for a combustion process;
(8) fume cupboards in chemical laboratories; and
(9) low volume/high velocity (lv/hv) extraction for cutting processes.

The major legislative requirements concerning the examination and testing of LEV plant are contained in the Control of Substances Hazardous to Health Regulations 1988 (COSHH) and its supporting Approved Codes of Practice[8,9].

Where an assessment has identified an exposure level above the occupational exposure standard (OES) or maximum exposure limit (MEL) for the substance, then under a hierarchy of preferred controls, LEV equipment must be installed wherever practicable, as opposed to providing personal protective equipment, as the means for reducing the employee's exposure. Such equipment must be properly used by the operator and visually inspected for obvious defects. Further, the employer must ensure that the equipment is maintained and the statutory inspections are carried out. As a part of this inspection it may be necessary to monitor the working environment by air sampling to ensure that the plant is continuing to operate effectively.

COSHH has revoked those parts of the Regulations that contained requirements regarding the examination and testing of LEVs and replaced them by the following general requirements listed in Schedule 3 of the Regulations:

(1) All LEVs other than those specific processes mentioned below – every 14 months.
(2) Processes in which blasting is carried out in, or incidental to, the cleaning of metal castings, in connection with their manufacture – monthly.
    (It must be noted that these monthly inspections refer to castings only. In the case of blasting other items the inspection frequency is 14 months.)
(3) Processes, other than wet processes, in which metal articles (other than gold, platinum, or iridium) are ground, abraded or polished using mechanical power, in any room for more than 12 hours in any week – every 6 months.
(4) Processes giving off dust or fumes in which non-ferrous metal castings are produced – every 6 months.
(5) Jute cloth manufacture – every month.

Where plant is old, the frequency of inspections may need to be increased.

COSHH does not extend to include asbestos or lead which are covered by extant regulations.

The Control of Asbestos at Work Regulations 1987 requires, *inter alia*, that every employer shall take control measures necessary to protect the health of employees from inhaling asbestos. Details of the pressure and air velocity tests and inspection procedures to be used are given in an Approved Code of Practice[10] which provides for three levels of inspection, i.e. weekly inspections by a responsible person, an examination of new and substantially modified plant (referred to as Part 1 examination), and six monthly inspections by a competent person (referred to as Part II examination).

The competent person may enlist the assistance of specialists to carry out certain tests, but he must make a report within 14 days of the examination. The report of Part I tests should include technical details of the plant.

The Control of Lead at Work Regulations 1980 requires employers to provide such control measure, other than by the use of respiratory equipment or protective clothing, as will prevent the exposure of his employees to lead. An Approved Code of Practice[11] outlines the weekly inspections to be carried out by a responsible person and requires that the annual examination and test covers the condition of the LEV plant, static pressures and air velocities at various points and a check that the lead dust or fume is being effectively controlled. Although the Code does not require it, it would be prudent for the annual examination to be carried out by a competent person.

Apart from the visual inspection of the LEV equipment, a statutory thorough examination under COSHH would include some of the following tests:

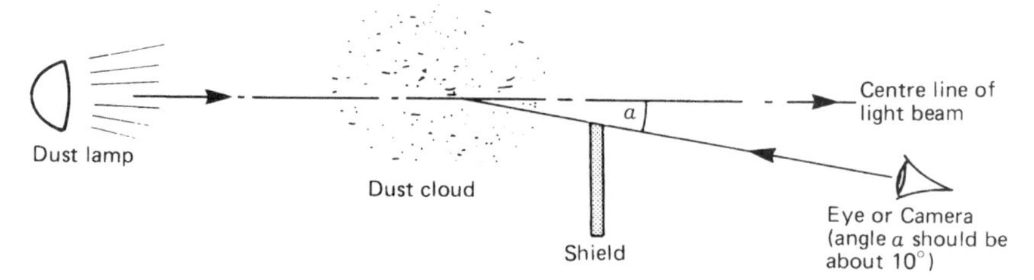

Figure 29.2 Use of dust lamp to see or photograph dust

(1) Tyndall (dust) lamp (*Figure 29.2*) and/or fuming sulphuric acid test;
(2) static pressure behind the captor hood (N.B. this is perhaps the most important test as this will determine whether the performance of the LEV equipment has altered);
(3) air velocities across the face of the captor hood;
(4) centreline velocity of the small duct between the captor hood and the main exhaust ducting;
(5) velocity and static pressure at the main exhaust ducting;
(6) static and total pressures at the inlet and outlet of the exhaust fan; and
(7) static pressure at the inlet and outlet of the dust collector and filters to obtain the differential pressure – particularly required for systems that recirculate the filtered air back into the workplace.

29.7 Electrical installations

The periodic inspection and testing by a competent person of electrical plant and installations as a statutory requirement arises only under the Cinematograph (Safety) Regulations 1955 which require inspection and testing by a competent person every twelve months.

However, Local Authorities may impose licensing conditions requiring petrol filling stations (under the Petroleum (Consolidation) Act 1928), launderettes, sports grounds and buildings used for public entertainment, to be inspected, tested and certified periodically (see also below). The Home Office Model Conditions, on which most Local Authority licensing conditions are based, specify annual inspections, testing and certification that the installations comply with the current (16th) edition of the Institution of Electrical Engineers (IEE) Regulations[12].

In Scotland compliance with the IEE Regulations is a requirement of the Building (Scotland) Act 1959.

More importantly, the Electricity at Work Regulations 1989 refer to design parameters such as the circuit's strength and capability, insulation, earthing, protective devices and the precautions to be taken when working on electrical equipment. All systems must be maintained, as far as is reasonably practicable, so as to prevent danger. A technique to ensure proper maintenance is through programmed preventative maintenance of which inspection and testing are essential parts. The frequency of maintenance is not specified, but practical experience should be a useful indicator. Factors to consider when determining the frequency of maintenance include mechanical wear and tear, impact damage, corrosion, excessive electrical loading, ageing and environmental conditions.

The Inspection and Testing Guidance Notes to the IEE Regulations suggest the following periods between inspections:

domestic premises	10 years
commercial premises	5 years
educational premises	5 years
hospitals	5 years
industrial works	3 years
public buildings	1 year
(i.e. cinemas, leisure complexes, restaurants, hotels etc.)	
special installations	1 year
(i.e. fire alarms, launderettes, petrol filling stations, etc.)	

In the case of testing class I, IIA, IIB and III portable electrical equipment reference should be made to an HSE Guidance Note[13]. Portable electrical tools should be inspected each time they are returned to stores and, with heaters, fans and business equipment, tested annually, although under certain hostile operating conditions testing should be more frequent.

The testing of electrical installations and equipment should include:

- polarity,
- earth fault loop impedance/earth continuity,
- insulation resistance,
- operation of devices for isolation and switching,
- operation of residual current devices,
- visual inspection of fuse rating,
- relay and starter contact erosion, and
- wiring integrity.

Portable equipment should be tagged to identify it and indicate its inspection status. This can provide the data for a register which can be used to maintain a record of inspection dates. Similar records should be kept of inspections and examinations of fixed installations.

29.8 General considerations

A number of common issues arise from the various legislative requirements for examination and inspection of plant and equipment.

29.8.1 Competent person

In general where the law requires particular items of plant to be thoroughly examined and tested it requires that the work be carried out by a 'competent person'.

There is no precise guidance in the statute or in subsequent case law as to what constitutes 'competence'. However, the competency of a person to carry out particular examinations or tests is a matter of fact on which the occupier or owner of the statutory equipment must be satisfied. In the event of legal proceedings it will require to be demonstrated in court that the person chosen was indeed competent to carry out the statutory surveys.

An often quoted definition is that 'the competent person should have such practical and theoretical knowledge and actual experience of the type of machinery or plant which he has to examine, as will enable him to detect defects or weaknesses which it is the purpose of the examination to discover, and to assess their importance in relation to the strength of the machinery or plant in relation to its function'. It is not sufficient for the person making the examination to be able to detect faults, he must also, from his knowledge and experience, be able to assess their seriousness.

Table 29.2 shows a chart of plant typically examined by a competent person.

The competent person need not necessarily be from an independent authority, but can be an internal appointment of a suitably qualified person by the company concerned, but in exercising his responsibilities for carrying out examinations he must be separated from any other functions that might cause a clash of interests. The competent person, whether employee or from a specialist organisation, must be allowed to act objectively and in a professional manner.

For complex plant and equipment it is doubtful whether any one individual would have sufficient knowledge and expertise to carry out the full examination on his own. In such circumstances, the support of a team of suitably qualified specialists may be needed.

The Code of Practice on Pressure Systems[3] considers that competence should be related to the size, complexity and hazard associated with the plant concerned and suggests different degrees of qualification for minor, intermediate and major pressure systems. The largest employers of competent persons, the engineer-surveyors, are the specialist engineering inspecting authorities associated with insurance companies, but there is an increasing number of small inspection companies providing 'competent person' inspection services who have no interest in insurance. While there is no requirement in law to have any item of plant, whether subject to statutory inspection or not, insured, it is prudent to do so. Some large companies – particularly in the steel and petrochemical industries – have their own in-house inspection departments.

The onus of responsibility for compliance with the statutory inspection obligations has been placed firmly with the company or organisation that owns the plant, and the enforcing authorities have always taken a consistent and firm line on this point. Thus, if a specialist inspection organisation is engaged to carry out all statutory examinations and it fails to do so, either properly or within the statutory period of time, the inspecting organisation may be in breach of contract, but it is the company that owns the plant that is held responsible for the breach of the particular legislation.

However, new Regulations[14] recognise the fact that a person, other than the owner and not employed by him, may, by his actions or failure to act, cause a breach resulting in an offence and offer the owner a defence if he identifies the person who caused the breach and can prove that he – the owner – exercised all due diligence to avoid the breach. Nevertheless, under s. 36 of HSW the owner may still be liable.

Under certain regulations there is a requirement for a 'responsible person' to carry out routine daily or weekly inspections of plant. That responsible person could be the employee who regularly operates the plant or equipment provided he has been trained in the checks to be carried out.

29.8.2 Patent and latent defects

When referring to requirements to be met by various lifting and boiler plant, the phrase is used in the FA that 'the machine shall be of good construction, sound material, adequate strength', and sometimes the words 'and free from patent defect' are added. While no definition is given of this latter phrase, its meaning seems to relate to the design of the equipment. Similarly, latent – which means hidden – would seem to refer to the material used to construct the item. The only guidance on this is from *McNeil* v. *Dickson and Mann (1957)* S.C. 345 where it was held that it was irrelevant whether a defect was patent or latent since the words did not qualify the major requirement 'to be of good construction, sound material and adequate strength'.

In considering whether a machine is of 'good mechanical construction', regard must be given to the stresses imposed upon the item when used for its intended purpose and to any additional stresses that could conceivably be imposed upon the item during normal use, due to impact loading, excessive vibration, inertia forces or temperature (both elevated or sub-zero).

'Sound material' means in fact material which is inherently intact and without internal defect, and not material which merely appears to be sound.

Whether plant is of 'adequate strength' must be related to its ability to withstand the stresses and strains met in its intended or foreseeable use.

29.8.3 Thorough examination

A thorough examination is a detailed visual examination, both stationary and under working conditions, carried out as carefully as the conditions permit in order to arrive at a reliable conclusion as to the safety of the plant or machinery and hence an assurance that the operation of the plant will be safe until the next statutory inspection.

Where the competent person needs to satisfy himself as to the condition of internal component parts he can require the plant to be dismantled. In addition, whenever considered necessary the visual examination can be supplemented by non-destructive testing of components to determine their internal or surface condition without causing any detrimental change in the material.

Table 29.2 indicates the plant that should be inspected regularly for both statutory and economic reasons.

29.8.4 Reporting

The completion of a report of an examination is a statutory requirement in respect of certain types of plant and it can be made by separate reports or by annotating the register of plant, or a combination of both.

Table 29.2. Chart of plant covered by competent person examinations and inspections. (Courtesy: British Engine Insurance Ltd.)

PLANT INSPECTION CHART

REPRODUCED BY KIND PERMISSION OF

Head Office Longridge House Manchester M60 4DT Tel: 061-833 9282

This chart is intended as a guide only and deals with categories of plant in general use. Whilst correct to the best of our knowledge, no legal liability or responsibility can be accepted for the information contained. It is not legally authoritative and in cases of doubt reference should be made to the relevant legislation (copies available from H.M. Stationery Office) or the appropriate enforcing authority.

TYPE OF PLANT / TYPE OF LOCATION	Bakeries	Breweries	Building Contractors	Churches & Halls	Clothing Factories	Department Stores & Shops	Docks	Dry Cleaners	Educational Establishments	Engineering Works	Factories/Workshops not otherwise specified	Farms	Flats	Food Manufacturers & Canneries
Aerial Cable & Ropeways			C14											
Air Receivers	A26	A26	A26		A26	A26	A26	A26		A26	A26			A26
Cranes & Lifting Machinery	A14	A14	C14		A14		B12	A14		A14	A14			A14
Dust Extraction & Ventilation Equipment	R	R					R			R	R	R		R
Economisers & Superheaters	A14	A14	A14		A14		A14	A14		A14	A14			A14
Electrical Installations and Earthing Systems														
Excavators			C14											
Fork Lift Trucks	A6	A6	C6		A6		B12			A6	A6			A6
Hot Water Boilers														
Lifts and Hoists	A6	A6	C6		A6	L6	A6	A6	L6	A6	A6			A6
Lifting Gear	A6	A6	C6		A6		B12	A6		A6	A6			A6
Power Presses & Press Brakes										K6	K6			K6
Refrigerating and Air Conditioning Plant														
Steam Boilers and Bakers Steam Tube Ovens	A14	A14	A14		A14	A14	A14	A14		A14	A14			A14
Steam Pressure Vessels	A26	A26	A26		A26		A26	A26		A26	A26			A26
Suspended Access Equipment (for Window Cleaning etc)														
Vehicle Lifting Tables														

HOW TO USE THIS CHART

A6 A square containing a letter or number denotes that the item of plant indicated when situated at the type of location is subject to statutory requirement for inspection. The letter indicates the legislation applicable as set out in the Key to Legislation. The number indicates the maximum monthly interval at which the plant should be inspected by a competent person.

A shaded square denotes that inspection is recommended on the grounds of safety and cost effective use of plant.

KEY TO LEGISLATION

- **A** Factories Act, 1961
- **B** Docks Regulations, 1988
- **C** Construction (Lifting Operations) Regulations, 1961
- **D** Ship Building & Ship Repairing Regulations, 1960
- **E** Coal and Other Mines (Steam Boilers) Regulations, 1956
- **F** Miscellaneous Mines (General) Regulations, 1956
- **G** Code of Practice — The Use of Electricity at Mines
- **H** Quarries (General) Regulations, 1956
- **J** Cinematograph (Safety) Regulations, 1955
- **K** Power Press Regulations, 1965
- **L** Offices, Shops and Railway Premises (Hoists and Lifts) Regulations, 1968
- **M** Quarries (Ropeways and Vehicles) Regulations, 1958
- **N** 1969 Model Code Regulations (adopted by most L.A.) Main Act — Petroleum (Consolidation) Act, 1928
- **P** Code of Practice — The Use of Electricity at Quarries
- **R** Control of Substances Hazardous to Health Regulations

Garages – Vehicle Workshop and/or Filling Stations	Hotels, Restaurants and Public Houses	Laundries	Millers	Mines – Coal, Ironstone, Shale & Fireclay	Mines – Other	Office Buildings	Printing Works	Quarries	Shipbuilding yards and Boat Builders	Supermarkets	Theatres & Cinemas	Timber Merchants	Warehouses	TYPE OF LOCATION / TYPE OF PLANT
								M14						Aerial Cable & Ropeways
A26	A26	A26	A26	E26	F26		A26	H26	A26	A26		A26	A26	Air Receivers
A14		A14	A14		F14		A14	H14	D12			A14	A14	Cranes & Lifting Machinery
R		R	R	R	R			R	R			R		Dust Extraction & Ventilation Equipment
	A14	A14	A14	E14	F14			H14	A14				A14	Economisers & Superheaters
N12				G6	G6			P6			J12			Electrical Installations and Earthing Systems
														Excavators
A6		A6	A6		F6		A6		D6			A6	A6	Fork Lift Trucks
														Hot Water Boilers
A6	L6	A6	A6			L6	A6		A6	L6		A6	A6	Lifts and Hoists
A6		A6	A6				A6		D6			A6	A6	Lifting Gear
									K6					Power Presses & Press Brakes
														Refrigerating and Air Conditioning Plant
A14	A14	A14	A14	E14	F14	A14	A14	H14	A14			A14	A14	Steam Boilers and Bakers Steam Tube Ovens
A26	A26	A26	A26	E26	F26	A26	A26	H26	A26			A26	A26	Steam Pressure Vessels
														Suspended Access Equipment (for Window Cleaning etc)
A6														Vehicle Lifting Tables

Statutory reports have to be in a special format, sometimes on prescribed forms which must contain certain particulars specified by legislation. The reports of an examination usually indicate the condition of the plant, e.g. 'in good order', list the defects requiring correction or place restrictions on the use of the plant because of its condition. Where defects are found a copy of the report has to be sent to the HSE within a prescribed time limit, usually 28 days. Defects can affect the safe working of the plant or machinery and when identified must receive the attention specified in the report within the time limit given therein.

It is common practice for the engineer-surveyor to make observations in his report to draw attention to other matters of a less serious nature. These observations are advisory and do not affect the continuing safe operation of the plant and machinery.

The reports of the statutory inspections must be kept readily available for inspection by the enforcing authorities. Where needed as working documents, photocopies of the report should be used so that the original can be kept as a master copy in a central file.

29.9 Conclusion

The chapter has endeavoured to cover the principal statutory inspection requirements in the UK that are likely to be of concern to occupational safety advisers. Certain areas have not been covered, such as the inspections of gasholders under the Factories Act, excavations and builders' hoists under the Construction Regulations nor the slightly differing requirements in Eire, Northern Ireland, the Channel Islands and the Isle of Man.

It should not be assumed that because there is no statutory requirement to periodically inspect a particular type of plant or machine, that it need not be so inspected. Section 2 of HSW places a duty on employers to provide and maintain plant in a safe condition and to provide a safe place to work generally. To meet this general obligation it is prudent to carry out regular inspections and tests on a range of plant and machines and to record the results. This is underlined by the increasing number of industry-produced Codes of Practice, such as for injection moulding machines, die-casting machines and concrete pumping booms, which recommend inspections and tests as being the best practical means of ensuring continuing compliance with HSW.

References

1. Health and Safety Executive, *Guidance Note No. PM 60, Steam boiler blowdown systems,* HMSO, London (1987)
2. Milton, J.H. and Leach, R.M., *Marine Steam Boilers,* 4th edn, Butterworth, London (1980)
3. HSE, (a) *Approved Code of Practice, COP 37, Safety of pressure systems,* (1989); (b) *Approved Code of Practice, COP 38, Safety of transportable gas containers,* (1990). HSE Books, Sudbury
4. HSE, *Guidance Note No. PM 8, Passenger carrying paternosters,* HSE Books, Sudbury (1977)
5. Health and Safety Executive, *The Lifting Plant and Equipment (Records of Test and Examination etc.) Regulations 1992,* HMSO, London (1992)

6. Health and Safety Executive, *The Provision and Use of Work Equipment Regulations 1992*, HMSO, London (1992)
7. HSE, Health and Safety: Guidance Booklets Nos.
   HS(G)37, *Introduction to local exhaust ventilation* (1987);
   HS(G)54, *The maintenance, examination and testing of local exhaust ventilation* (1990)
   HSE Books, Sudbury.
8. HSE, *Legislation booklet L5, Control of substances hazardous to health: Control of carcinogenic substances*, HSE Books, Sudbury (1993)
9. HSE, *Approved Code of Practice, COP 30, Control of substances hazardous to health in fumigation operations*, HSE Books, Sudbury (1988)
10. HSE, *Legislation booklet L27, Control of asbestos at work*, HSE Books, Sudbury (1993)
11. HSE, *Approved Code of Practice, COP 2, Control of lead at work*, HSE Books, Sudbury (1985)
12. The Institution of Electrical Engineers, *Regulations for Electrical Installations*, 16th edn, IEE, London (1991)
13. Health and Safety Executive, Guidance Notes: Plant and Machinery Series No. PM32, *Safe use of portable electrical apparatus (electrical safety)*, HSE Books, Sudbury (1990)
14. *The Pressure Systems and Transportable Gas Containers Regulations 1989, Reg. 23*, HMSO, London (1989)

Further reading and references

General

Sinclair, T. Craig, *A Cost-Effective Approach to Industrial Safety*, HMSO, London (1972)

Legal

Fife, I. and Machin, E. A., *Redgrave Fife and Machin; Health and Safety*, Butterworth-Heinemann, Oxford (1993)

Munkman, J., *Employer's Liability at Common Law*, 11th edn, Butterworth, London (1990)

Pressure vessels

Jackson, J., *Steam Boiler Operation: Principles and Practices*, 2nd edn, Prentice Hall, London (1987)

Robertson, W. S., *Boiler Efficiency and Safety*, MacMillan Press, London

Brown, Nickels and Warwick, *Periodic Inspection of Pressure Vessels*, (A.O.T.C.) I. Mech. E. Conference, London (1972)

British Standards Institution, London:

- BS 470:1984 Specification for inspection, access and entry openings for pressure vessels
- BS 709:1983 Methods of destructive testing fusion welded joints and weld metal in steel
- BS 759 (Pt 1):1984 Specification for valves, gauges and other safety fittings for application to boilers, and to piping installations for and in connection with boilers
- BS 1113:1989 Specification for design and manufacture of water tube steam generating plant (including superheaters, reheaters and steel tube economisers)
- BS 1123:1987 Specification for safety valves, gauges and other safety fittings for air receivers and compressed air installations
- BS 2790:1989 Specification for the design and manufacture of shell boilers of welded construction
- BS 2910:1986 Methods for radiographic examination of fusion welded circumferential butt joints in steel pipes
- BS 5169:1975 Specification for fusion welded steel air receivers
- BS 5500:1988 Specification for unfired fusion welded pressure vessels
- BS 6244:1982 Code of Practice for stationary air compressors

ANSI/ASME Boiler and pressure vessel code

- Sec. I – Power boilers
- Sec. VII – Recommended guidelines for the care of power boilers

HSE Guidance Notes, HSE Books, Sudbury

- GS 4 Safety in pressure testing (1976)

GS 5 Entry into confined spaces (1977)
GS 20 Fire precautions in pressurised workings (1983)
PM 5 Automatically controlled steam and hot water boilers (1977)
HS(G)29 Locomotive boilers (1986)
HS(R)30 *A guide to the Pressure Systems and Transportable Gas Containers Regulations 1989*
PM 60 Steam boiler blowdown systems (1987)

Lifting and handling plant
Phillips, R. S., *Electric Lifts*, Pitman, London (1973)
Dickie, D. E., *Lifting Tackle Manual*, (Ed. Douglas Short), Butterworth, London (1981)
Dickie, D. E., *Crane Handbook*, (Ed. Douglas Short), Butterworth, London (1981)
Dickie, D. E., *Rigging Manual*, Construction Safety Association of Ontario (1975)
Associated Offices Technical Committee (A.O.T.C.) *Guide to the testing of cranes and other lifting machines*, 2nd edn, AOTC, Manchester (1983)
British Standards Institution, London:
BS 466:1984 Specification for power driven overhead travelling cranes, semi-goliath and goliath cranes for general use
BS 1757:1986 Specification for power driven mobile cranes
BS 2452:1954 Specification for electrically driven jib cranes mounted on a high pedestal or portal carriage (high pedestal or portal jib cranes)
BS 2573 (Pt 1): 1983 Specification for the classification, stress calculations and design criteria for structures
BS 2573 (Pt 2): 1980 Specification for the classification, stress calculations and design of mechanisms
BS 2853: 1957 Specification for the design and testing of steel overhead runway beams
BS 4465: 1986 Specification for design and construction of electric hoists for both passengers and materials
BS 5655 (10 parts) covering safety of electric and hydraulic lifts, dimensions, selection and installation, control devices and indicators, suspension eyebolts, guides, and the testing and inspection
BS 5744: 1979 Code of Practice for the safe use of cranes
ISO 4309 Wire ropes for lifting appliances – Code of Practice for examination and discard
ISO 4310 Cranes – test code and procedures
HSE Guidance Notes, HSE Books, Sudbury
PM 3 Erection and dismantling of tower cranes (1976)
PM 7 Lifts: thorough examination and testing (1982)
PM 8 Passenger carrying paternosters (1987)
PM 9 Access to tower cranes (1979)
PM 24 Safety at rack and pinion hoists (1981)
PM 26 Safety at lift landings (1981)
PM 27 Construction hoists (1981)
PM 34 Safety in the use of escalators (1983).
PM 43 Scotch derrick cranes (1984)
PM 45 Escalators: periodic thorough examination (1984)
PM 54 Lifting gear standards (1985)
PM 55 Safe working with overhead travelling cranes (1985)
PM 63 Inclined hoists used in building and construction work (1987)
HS(G)19 Safety in working with power operated mobile work platforms (1982)
HS(G) 23 Safety at power operated mast work platforms (1985)
L20 *A guide to the Lifting Plant and Equipment (Records of Test and Examination etc.) Regulations 1992*

Power presses
Joint Standing Committee on Safety in the Use of Power Presses:
Safety in the use of power presses, HSE Books, Sudbury (1959)
Power press safety; Safety in material feeding and component ejection systems, HSE Books, Sudbury (1984)
British Standards Institution, London:
BS 4656 (Pt 34):1985 Specification for power presses, mechanical, open front
BS 6491: Pt 1:1984 Electro sensitive safety systems for industrial machines
Pt 2:1986 Particular requirements for photo-electric sensing units

HSE Guidance Notes, HSE Books, Sudbury
PM 41 Application of photo-electric safety systems to machinery (1984)

Local exhaust ventilation plant
Industrial Ventilation, American Conference of Government Industrial Hygenists, Cincinatti, Ohio
Principles of Local Exhaust Ventilation, Report of the Dust and Fume Sub Committee of the Joint Standing Committee on Health, Safety and Welfare in Foundries, HSE Books, Sudbury (1975)
Relevant Standard:
BS 6540 (Pt 1):1985 Methods of test for atmospheric dust spot and efficiency, and synthetic dust weight arrestance
HSE Guidance Notes, HSE Books, Sudbury
MS 13 Asbestos (1988)
EH10 Asbestos – Exposure limits and measurements of airborne dust concentrations (1988)
EH 25 Cotton dust sampling (1980)
EH 28 Control of lead: air sampling techniques and strategies (1986)
EH 30 Control of lead: pottery and related industries
HS(G) 37 Introduction to local exhaust ventilation (1987)

Electrical installations
Institution of Electrical Engineers, *Regulations for Electrical Installations*, 16th edn, London (1991)
British Standards Institution, London:
BS 2754:1976 Construction of electrical equipment for protection against electric shock
BS 2771 (Pt 1):1986 Specification for general requirements (of electrical and electronic equipment of machines)
BS 4444:1980 Guide to electrical earth monitoring
BS 5958 (Pt 1):1987 Code of Practice for control of undesirable static electricity – general considerations
BS 5958 (Pt 2):1983 Code of Practice for control of undesirable static electricity – recommendations for particular industrial situations
BS 6233:1982 Methods of test for volume resistivity and surface resistivity of solid electrical insulating materials
HSE Guidance Notes, HSE Books, Sudbury
GS 6 Avoidance of danger from overhead electrical lines (1980)
PM 29 Electrical hazards from steam/water pressure cleaners etc. (1988)
PM 32 The safe use of portable electrical apparatus (1983)
PM 38 Selection and use of electric hand lamps (1984)

Chapter 30

Safety on construction sites

R. Hudson

The construction industry has always been plagued with an abundance of reportable accidents coming to an all time high of over 45 000 accidents in 1966 with, over the previous decade, an average of 250 persons killed each year. These appalling figures occurred in spite of a considerable volume of safety legislation aimed at improving safe working in the construction industry.

Many of the causes of these accidents are reflected in the detailed requirements of the relevant Regulations[1] which lay down the preventative measures to be taken. This chapter looks at the safety legislation for the construction industry and some of the techniques for meeting the required safety standards.

30.1 Construction accidents

An indication of the size and seriousness of the problem can be obtained by considering past accident records.

30.1.1 Accident statistics

Each year the HSE publishes a report[2] containing data on fatal and major accidents and respective incidence rates. In the period 1987/88–1991/92 figures for fatal accidents in the construction industry compared with manufacturing industries were:

	1987/8	1988/9	1989/90	1990/1	1991/2
Manufacturing industries	99 (5)	94 (11)	108 (10)	88 (14)	67 (7)
Construction industry	103 (55)	101 (50)	100 (65)	96 (37)	82 (22)

(numbers in brackets are for persons other than employees, i.e. sub-contractors, members of the public, etc.)

However, numbers employed in the two industries differ widely and have varied considerably over the years. To put the figures in perspective the fatal accident incidence rate for employees gives a better measure of the problem.

	1987/8	1988/9	1989/90	1990/1	1991/2
Manufacturing industries	1.9	1.8	2.1	1.8	1.4
Construction industry	10.3	9.9	9.4	9.3	9.2

While overall there has been a reduction in fatal accidents this is accounted for mainly by the smaller workforce of a less buoyant industry rather than improved standards. As the incidence rate shows, the construction industry is, in comparison with manufacturing, consistently some six times more dangerous. As some 70% of the accidents investigated could have been prevented by management action[3] this continues to be an unacceptable situation.

30.1.2 Types of accidents

Analysis of both fatal and major accidents reported[4] for the year 1989/90 gives a good indication of the problem areas, *Table 30.1*. While the numbers vary from year to year the pattern remains fairly constant with 'falls from height' accounting for some 40% of major injuries and 50% of fatalities.

Table 30.1. Causes of construction site fatal and major injuries 1989–90 (Fatalities recorded in brackets)

Classification		*Fatal and major injuries*
Contact with moving machinery or material being machined		83 (2)
Struck by moving, flying or falling objects		521 (9)
Struck by moving vehicles		143 (14)
Struck against something fixed or stationary		93
Injured while handling, lifting or carrying		195
Slip, trip or fall on same level		516
Falls from heights		
– up to and including 2 m	466 (3)	
– over 2 m	742 (49)	
– not known	117 (1)	
– Total all heights		1325 (53)
Trapped by something falling or overturning		66 (13)
Drowning or asphyxiation		7 (3)
Exposure to or contact with harmful substance		51
Exposure to fire		12 (1)
Exposure to an explosion		8
Contact with electricity or an electrical discharge		107 (5)
Injuries by an animal		0
Other kinds of accident		51
Injuries not classified by kind		2
Total:		3180 (100)

30.2 Safe working in the industry

In considering safe working and accident prevention in the construction industry, this chapter will follow broadly the progression of a construction operation. All stages should be adequately planned making allowance for the incorporation of safe systems of work.

Planning has been the province of the main contractor but with the coming into effect of the Construction (Design and Management) Regulations[5] (CONDAM) this responsibility will be extended. Under these Regulations the client will have an obligation to appoint a competent planning supervisor for the project. In many instances this role will be filled by professional advisers such as architects or engineers who act on behalf of the client.

The planning supervisor will be required to:

- ensure the design includes adequate information about the design and materials to be used where they might affect the health and safety of those carrying out the construction work;
- prepare a health and safety plan to be included with the tender documentation which sets out the arrangements on the project for ensuring the health and safety of all persons at work, how the works will be managed as far as health and safety is concerned, and how compliance will be monitored;
- advise on the appropriate financial and time allowance to enable contractors to comply with the duty imposed by the relevant statutory provisions while carrying out the construction works; and
- prepare and deliver to the client a health and safety file on the as-built structure which the client retains for reference during subsequent construction works on the structure.

The client will also be required to appoint a competent principal contractor for the project.

The principal contractor must for his part:

- adopt and develop as appropriate both the health and safety plan and health and safety file;
- ensure the health and safety plan is followed by all persons on the site; and
- co-ordinate the activities of others on the site and ensure that all co-operate in complying with the relevant statutory provisions that affect the works.

For these purposes the principal contractor can give directions or establish rules for the management of the construction works as part of the health and safety plan. Such rules must be in writing and be brought to the attention of all affected persons.

Finally, one of the main provisions of these Regulations defines the responsibility which designers, such as architects, have for health and safety during the construction stages. Designers will have to ensure, so far as is reasonably practicable and provided the structure conforms to their design, that persons building, maintaining, repairing, re-painting, redecorating or cleaning the structure are not exposed to risks to their health and safety. In addition, the designer must ensure that included in the design documentation is adequate information about the design and materials used, particularly where they may affect the health and safety of persons working on the structure.

Basically, these requirements place designers of buildings and structures, such as architects, under similar obligations to those who design articles and substances, whose obligations are contained in s. 6 of HSW.

As work gets under way, the principal contractor, who has responsibility for the construction phase of the project, has to ensure that all those employed are properly trained for their jobs. Under HSW, now amplified by the Management of Health and Safety at Work Regulations 1992[1] (MHSWR), the employer is required to provide training in specific circumstances, i.e. on joining an employer, when work situations change and at regular intervals.

In addition, specific job training is prescribed in numerous statutory provisions such as the Abrasive Wheels Regulations 1970 (for mounting abrasive wheels) and the Provision and Use of Work Equipment Regulations 1992[1] (PUWER) (adequate training in the use of work equipment).

In an industry increasingly reliant upon the use of sub-contractors, the main contractor retains the onus for health and safety on site. This onus can extend to training employees of sub-contractors where their activities may affect the health and safety of the employees of the main contractor and of the sub-contractor himself (ss. 2 and 3 of HSW). This will be more clearly defined by CONDAM which will require the principal contractor to ensure that other employers on the construction work provide their employees with appropriate health and safety training when they are exposed to new or additional risks due to:

- changes of responsibilities, i.e. promotion,
- use of new or changed work equipment,
- new technology, or
- new or changed systems of work.

The provision of information is also an essential contribution to reducing risks to health and safety. As with training, ss. 2 and 3 of HSW apply to both main and sub-contract employers in relation to informing each other of risks, within their knowledge, arising out of their work. The decision in *Regina* v. *Swan Hunter Shipbuilders Ltd*[6] clarified this in respect of 'special risks'. In this case, a number of fatalities resulted from a fire in a poorly ventilated space in a ship which had become enriched with oxygen due to an oxygen supply valve being left open by a sub-contractor's employee. An intense fire developed when another contractor struck an electric arc to do some welding. Swan Hunter were well aware of the fire risk associated with oxygen enrichment and provided detailed information for their own, but not sub-contractors' employees. Swan Hunter were prosecuted under s. 2 of HSW and convicted for failing to ensure the health and safety of their own employees by not informing the employees of sub-contractors of special risks which were within its, Swan Hunter's, knowledge, i.e. from fires in oxygen enriched atmospheres. This decision has been overtaken by MHSWR (reg. 9) which requires employers who share a workplace to take all reasonable steps to inform other employers of any risks arising from their work. Further, CONDAM will place a duty on the principal contractor to inform other contractors of the risks arising out of or in connection with the works and ensure that those sub-contractors inform their employees of:

- risks identified by the contractor's own general risk assessment,
- the preventative and precautionary measures that have to be implemented,
- any serious or imminently dangerous procedures and the identity of any persons nominated to implement those procedures, and
- details of the risks notified to him by the principal or another contractor.

30.2.1 Notification of construction work

Where, on any site, a 'building operation or work of engineering construction'[7] is to be undertaken which is likely to take longer than six weeks, the contractor must, before any work commences, notify the HSE, using form F10, of the nature of the work being undertaken. However, whether there is a need to notify or not, full compliance with all the requirements of the relevant safety legislation is necessary.

Under the CONDAM Regulations a number of changes will occur to the notification procedure. Firstly, the definition of 'building operation or work of engineering construction' will be replaced by the broader term 'construction work'. Ultimately this term will replace the phrase 'building operation or work of engineering construction' in existing legislation so extending the scope of such legislation as the Construction Regulations to a wider range of activities.

Secondly, the notification responsibility will become that of the client, unless there is no project client when it will rest with the principal contractor. The parameters for the notification will also change to a minimum construction phase of 30 days involving more than 500 man-days of construction work.

Apart from the overall obligation placed on both the main and sub-contractor employer by HSW, more extensive requirements specific to the industry are contained in legislation known collectively as the 'Construction Regulations'. This chapter is mainly concerned with the practical application of those Regulations to construction sites.

30.2.2 The Construction (General Provisions) Regulations 1961

30.2.2.1 Responsibilities

The responsibility for complying with the requirements of these Regulations is placed on the contractor and every employer engaged in construction work with details being outlined in Reg. 3. To be a contractor or employer of workmen undertaking a building operation is not a narrow term applying only to main contractors. All involved in the work are encompassed. Thus, in *Williams* v. *West Wales Plant Hire Co.*[8], a plant hire company providing a mobile crane with their own employee as operator was held to be a vicarious 'contractor' or 'employer' and as such subject to the Construction Regulations.

Unless it is stated otherwise, the duties under these Regulations are an 'absolute one' so that if the work cannot be done as required by these regulations, it should not be done at all. Duties are also placed on employees by Reg. 3(2) to co-operate in the carrying out of the regulations and to report to their employers, supervisor or the safety supervisor any defects discovered in the plant or equipment in use.

Every contractor who employs more than 20 persons, whether or not they are all on one site, must appoint as safety supervisors one or more suitably qualified persons with experience of the type of work being undertaken (Reg. 5). They need not be employed full-time, but their safety duties must take priority over any other work they may undertake (Reg. 6).

The appointment of a safety supervisor has been reinforced by Regulation 6 of MHSWR which places a duty on all employers to have access to competent assistance to develop and implement procedures to enable the employer to comply with the 'relevant statutory provisions'. The requirement for a safety supervisor, while still a legal requirement, will be revoked when CONDAM comes into effect.

30.2.2.2 Safety in excavations

In any excavation, shaft or tunnel more than 1.2 m (4 ft) deep where there is a risk of material collapsing or falling, proper timbering, trench sheeting, etc., must be used to safeguard operatives. Protection is also required for those engaged in timbering or sheeting operations and there must be an adequate supply of suitable timber. This Regulation (Reg. 8) does not apply to excavations etc. with sloping sides where there is no risk of earth collapse or of men being struck by material falling from a height of more than 1.2 m (4 ft). These safe alternative trenching techniques are:

(1) *Battering the sides,* i.e. cutting the sides of the excavation back from the vertical to such a degree that fall of earth is prevented.
(2) *Benching the sides.* The sides of the excavation are stepped to restrict the fall of earth to small amounts. Maximum step depth 1.2 m (4 ft).

Figure 30.1 shows typical examples of these trenching techniques.

In tunnelling operations, it is recognised that physical conditions may be met over which the employer has no control and against which it is not reasonably practicable to provide protection (Regs. 8(2); 12(2); 16(2)).

Inspection (Reg. 9) must be made:

(a) every day that men are working in any type of excavations more than 1.2 m (4 ft) deep where there is any risk of material collapsing or falling;
(b) at the beginning of every shift, in the case of:
   (i) tunnel working faces,
   (ii) working ends of trenches more than 1.98 m (6 ft 6 in) deep, and
   (iii) at base or crown of a shaft.

No record need be kept of these inspections. However a thorough examination of every type of excavation must be carried out:

(a) after explosive charges have been fired;
(b) after any damage to timbering, trench sheeting etc., or after any fall of earth or collapse of material;
(c) in any case, every seven days.

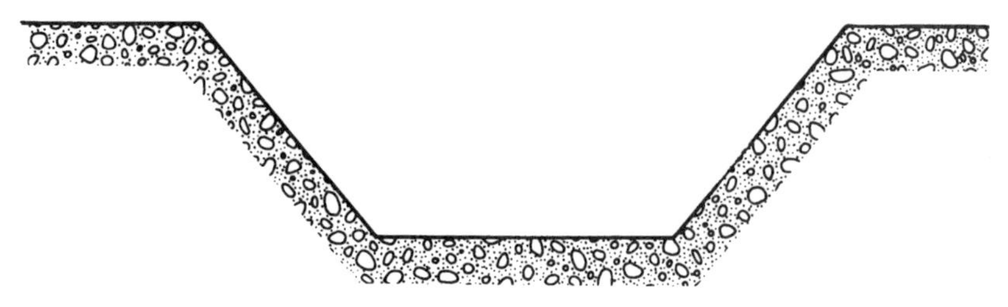

(a) Battering the sides

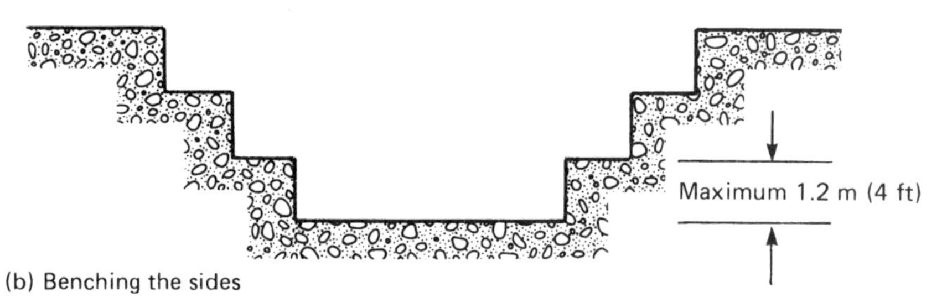

(b) Benching the sides

Figure 30.1 Safe trenching methods without the use of timber

A record of these examinations must be made in the official register; Form 91 (Part 1 Section B).

All material used for timbering or sheeting must be inspected before use and material found defective must not be used (Reg. 10). Timbering and sheeting must only be erected, altered or dismantled under competent supervision and whenever practicable by experienced operatives. All timbering and sheeting must be properly constructed and maintained in good order. Struts and braces must be fixed so that they cannot be accidentally dislodged. In addition, in the case of a coffer-dam or caisson, all materials must be examined and only if found suitable should they be used (Reg. 17).

If there is risk of flooding in any excavation, shaft, tunnel (Reg. 11), coffer-dam or caisson (Reg. 16), ladders or other means of escape must be provided.

When excavating in close proximity to existing buildings or structures, be they permanent or temporary, there is a requirement to give full consideration to their continued stability. This is intended to protect persons employed on site. However under the Health and Safety at Work Act this responsibility is extended to the safety of the public, i.e. those not employed on the site, and may relate to private dwelling houses, public buildings or public rights of way. It is particularly important when excavating near scaffolding.

Where any existing building or structure is likely to be affected by excavation work in the vicinity, shoring or other support must be provided to prevent collapse of the building or structure (Reg. 12). Examples of trench shoring are given in *Figures 30.2, 30.3* and *30.4*.

Excavations, shafts or pits more than 1.98 m (6 ft 6 in) deep near which men work or pass, must be protected at the edge by guardrails or barriers or must be securely covered (Reg. 13). Guardrails, barriers or covers may be temporarily moved for access or for movement of plant or materials but must be replaced as quickly as possible.

Where the excavation is not in an enclosed site and is accessible to the public the standard for protection is more onerous. Even the most shallow depressions should be fenced so that members of the public are not exposed to risks to their health and safety.

Materials, plant, machinery etc. must be kept away from the edge of all excavations, shafts and pits to avoid collapse of the sides and the risk of men falling in, or material falling on men (Reg. 14).

While work is being carried out in a coffer-dam or caisson, an inspection must be carried out at the beginning of each day or shift, and a thorough examination conducted at least once every seven days, after explosives have been used, or where substantial damage has occurred due to falling materials (Reg. 18). The result of the examinations must be recorded in Form 91 (Part 1 Section B), a copy of which should be held on site. Thorough daily and shift inspections should be carried out and while they do not have to be recorded, it is prudent to do so.

Severe weather conditions such as heavy rain, or where timber has become wet then followed by a hot dry spell, could so affect shoring,

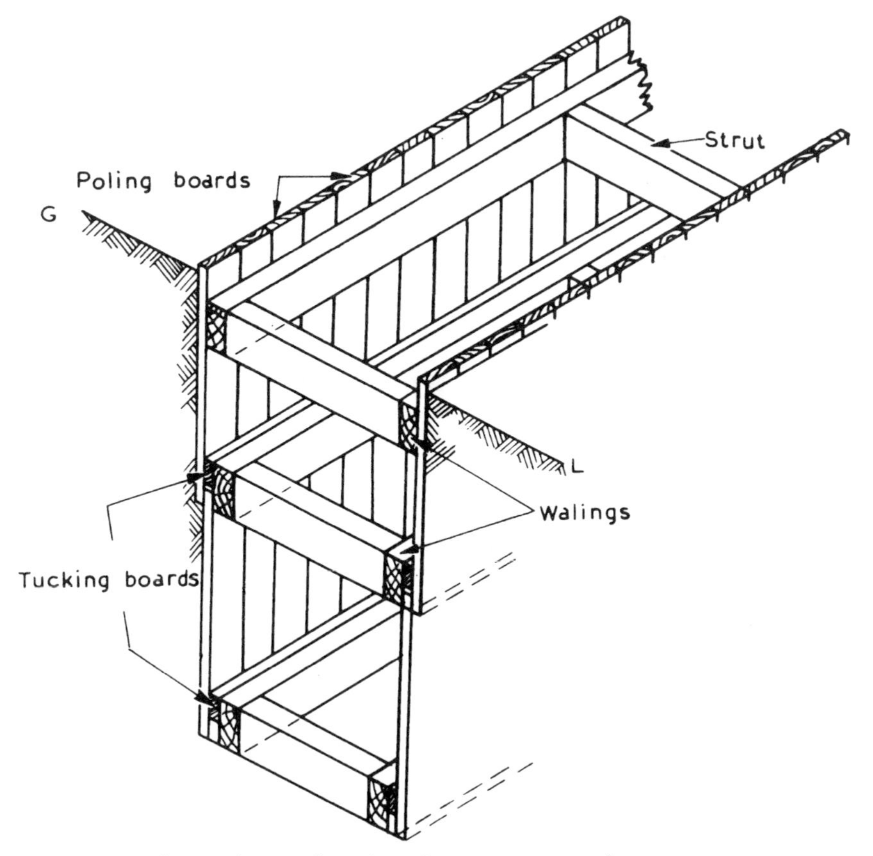

Figure 30.2 Close poling with tucking frames. (BS 6031)[9]

timbering etc. as to cause it to become dangerous. In these circumstances it is prudent to carry out a thorough examination. Guidance on the construction of trenches, pits and shafts is given in the British Standard CP 6031 Code of Practice for Earthworks[9].

On sites where mobile machinery such as tippers, diggers, rough terrain fork lift trucks, etc. are used, special care should be taken to ensure that operators are fully aware of the stability of their machines and of the maximum slope on which they can be safely used. Particular attention should be paid to the condition of the ground and whether it is capable of bearing the vehicle weight. Information on safe ground conditions and angles of tilt can be obtained from the machine manufacturers.

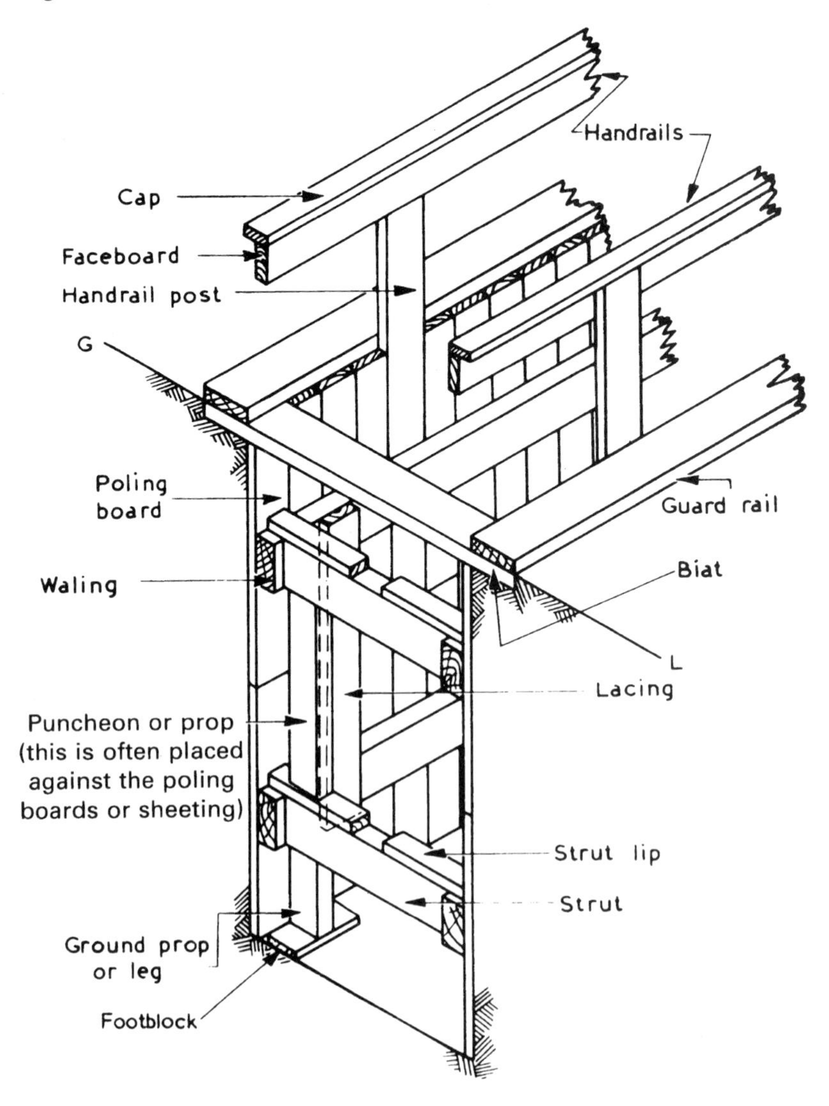

Figure 30.3 Typical single or centre waling poling frame. (BS 6031[9])

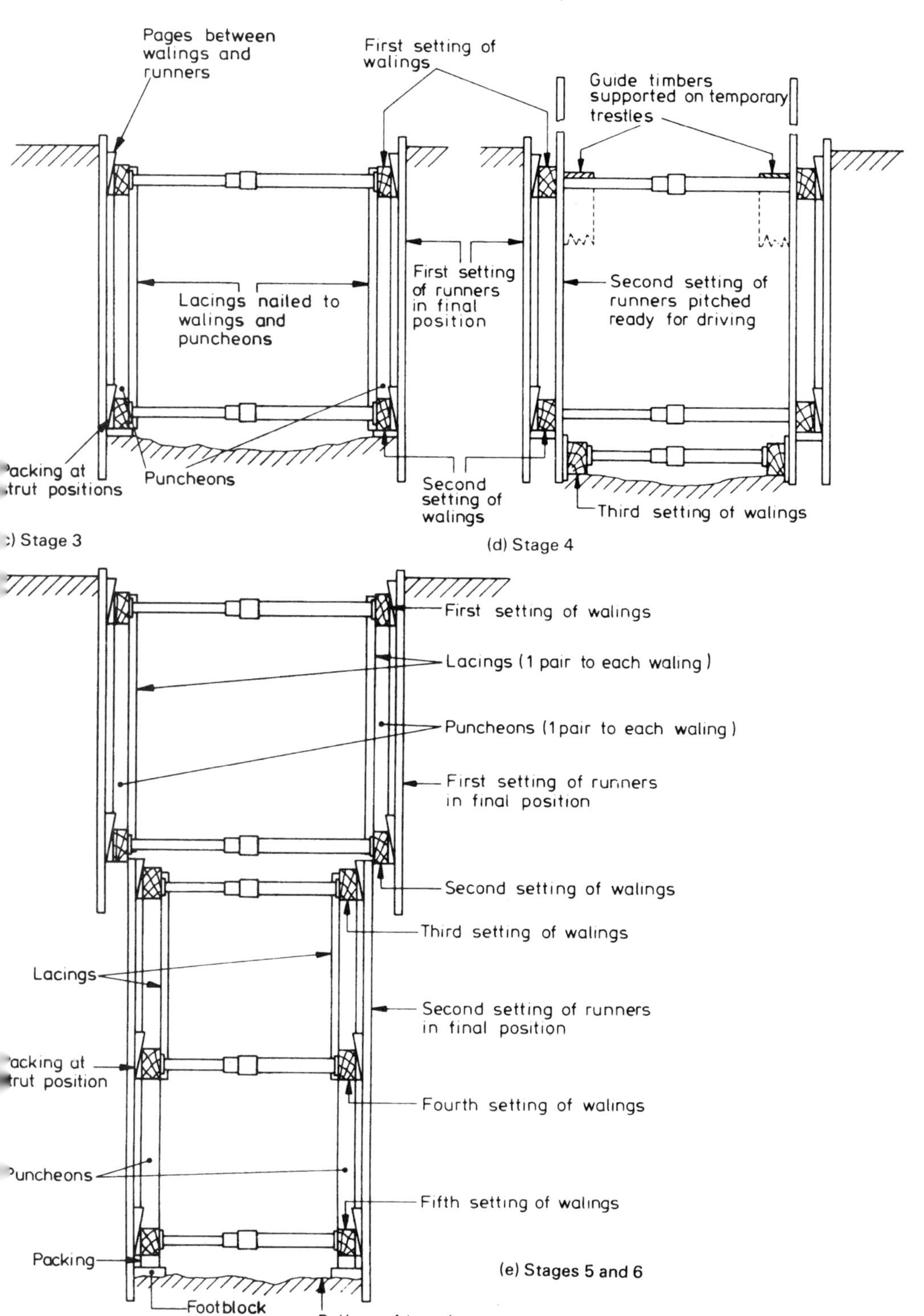

Figure 30.4 Trench excavation using steel trench sheets as runners. (BS 6031)[9]. Note Stages 1 and 2 are shown in the British Standard

Where[9] overhead cables cross the line of excavations, particular care must be taken in the selection of the type of plant to be used and precautions taken[10] to ensure that the equipment does not or cannot touch live high voltage conductors (Reg. 44). Underground cables, be they high or low voltage, telephone or television links, together with gas piping, present a more difficult problem which, in the main, rests with the excavating contractor. Advice is given in an HSE publication[10] but the contractor should approach each of the service authorities asking for accurate information on the actual location, run and depth of their services. This should be in writing, and the information supplied by the authority should preferably be marked on a drawing; ideally the authority should authorise the drawing as correct. The location should be confirmed using devices for locating cables and other services and the route of the service marked on the surface. Services should be carefully exposed by hand-dig methods to verify their precise location and depth before mechanical means are employed. Mechanical equipment such as excavators should not be used within 0.5 m of the suspected cable until its route has been specifically located.

Knowledge of the whereabouts of underground services is also necessary where heavy plant or vehicles are used since many such services are at shallow depth and can be damaged by the sheer weight of equipment. Although injury is not likely to result, considerable cost to the contractor could be involved.

Careful planning, including the selection of the correct plant and equipment is essential, for both safety and economic reasons, when carrying out excavation work. For example, the correct size of excavator can act as a crane eliminating the need to bring extra specialised plant onto site. However, when using an excavator in this manner the machine must comply with either all the requirements of the Construction (Lifting Operations) Regulations 1961 that apply to cranes or to the less onerous but restrictive Certificate of Exemption[11] which exempts an excavator from certain requirements of the Regulations applicable to cranes but imposes alternative conditions. However, under the Exemption the excavator is limited to operations immediately associated with excavation work, i.e. lifting pipes into an excavation. It may not be used for lifting items from the back of a lorry into a storage compound or subsequently moving them to an excavation.

Before work is started on a construction site there are a number of matters that should be checked, from both a prudent and statutory point of view:

(1) Provide site security – particularly to stop children getting in.[12]
(2) Investigate the nature of the ground before excavations begin and decide the form that the support work will need to take and ensure that adequate supplies of sufficiently strong materials are available. Special precautions may be needed where trenches pass near adjacent roads or buildings.
(3) Locate all public services, water, gas, electricity, telephone, sewers etc., and avoid if possible; if not, take necessary precautions.
(4) Provide material for barriers and authorise traffic notices.
(5) Provide adequate lighting.
(6) Position spoil heap at a distance not less than the depth from the edge

of excavation. Tip if possible on blind side of excavator to ensure operator has visibility when swinging back to trench excavation.

(7) Provide personal protective equipment.
(8) Provide sufficient ladders of suitable length, strength and type.
(9) Query necessity for bridges and gangways.
(10) Take note of all overhead services, the arrangements made for their protection and the safety of all working in their vicinity.

30.2.2.3 Mechanical plant and portable tools

Prior to 1 January 1993 machinery, whether power driven or not, used on a construction site had to be effectively guarded under the Construction (General Provisions) Regulations 1961 where the requirements for both existing machinery (Reg. 42) and new machinery (Reg. 43) were specific, referring mainly to fencing shafts.

Since 1 January 1993 however, all machinery for use at work is subject to the Provision and Use of Work Equipment Regulations 1992[1] (PUWER) with its requirements for guarding dangerous parts, concerning controls and associated matters. Existing machinery, however, has until 1 January 1997 to conform during which period the Construction (General Provisions) Regulations will continue to apply.

The general principles of guarding are contained in BS 5304[13] but this standard is being replaced by a number of harmonised European (EN) standards. Other publications[14] are available that give advice on the safety and care of site plant and equipment.

Where guards are removed to enable maintenance work to be carried out, they must be replaced before the machinery is returned to normal work.

Portable tools are used extensively on construction sites and commonly suffer damage. Arrangement should be made for regular checks on the condition of portable tools, paying particular attention to the integrity of electrical insulation on the tool itself and on the lead and to damage to rotating parts remembering that the Electricity at Work Regulations 1989 require such equipment to be properly maintained, involving regular inspection and testing. In the case of compressed air equipment, the Pressure Systems and Transportable Gas Container Regulations 1989 also have effect.

Under PUWER, not only the operators of any plant and machinery, being work equipment, but their supervisors and managers must be trained:

- in the correct method of use,
- on the risks such use may involve, and
- on the precautions to be taken.

In the case of cranes, capstans, winches, woodworking machines and mechanically propelled vehicles, there are legal obligations for them to be so. Operators of these machines must be over 18 years old unless being trained and under the direct supervision of a competent person.

30.3 Site hazards

The extensive use of temporary or semi-permanent wiring on construction sites, the rough usage that equipment gets, the hostile conditions under which it is used and, often, the lack of knowledge of those using the equipment contribute to the high risk potential of the use of electricity. Compliance with the Electricity at Work Regulations[15] will reduce the hazards, which broadly can be divided into three categories:

(1) Electrocution.
(2) Fire.
(3) Glare.

30.3.1 Electrocution

There are three operations that carry the highest risk of electrocution: use of portable tools, striking a buried cable, and cranes and diggers making contact with overhead power lines.

Portable tools are used extensively on sites and maintaining them and their connecting cables in good repair is a critical factor of their safe use. Electrocution occurs when the body acts as the conductor from a power line and earth, often because the earth connection on the tool has broken. All portable tools must be securely earthed or be of double or all insulated construction and the plug on the lead must be correctly fused. Unfortunately it is frequently difficult to keep track of every item, so reliance has to be placed on the person using them.

Protection for 240 V supplies can be obtained by the use of residual current devices (RCDs) either in the supply circuit or on the connection to the particular equipment. In addition, on construction sites 240 V supplies should be carried by armoured, metal sheathed or other suitably protected cables.

The greatest protection from electrical shock on construction sites is through the use of a reduced voltage system. Where portable hand held tools are used this is essential and the system recommended is 110 V ac with the centre point of the secondary winding tapped to earth, so that the maximum voltage of the supply above earth will be 55 V which normally is reckoned to be non-fatal. A further alternative is to use a low voltage supply at 24 V, but in this case, because of the low voltage, the equipment tends to be heavier than with higher voltages. The risk of electrocution from power tools is increased if they are used in wet conditions.

Electrocution from striking an underground cable can be spectacular when it occurs and the most effective precaution is to obtain clearance from the local Electricity Board or the factory electrical engineer that the ground is clear of cables. If doubt exists, instruments are available that enable underground cables to be traced.

All too often overhead power cables cross construction sites and are a potential hazard for cranes and mechanical equipment. If the supply cannot be cut off, suspended warning barriers should be positioned on each side and below the level of the cable and drivers warned that they may only pass under with lowered jib.

30.3.2 Fire

Usually caused through overloading a circuit, frequently because of wrong fusing. Repeated rupturing of the fuse should be investigated to find the cause rather than replacing the blown fuse by a larger one in the hope that it will not blow. A second cause can be through water getting into contact with live apparatus and causing a short circuit which results in overheating of one part of the system. Where electrical heaters are used on sites, they should be of the non-radiant type, i.e. tubular, fan or convector heaters. Multi-bank tubular heaters used for drying clothes should be protected to prevent clothing, paper etc. being placed directly on the tubes. This protection can be achieved by enclosing the heaters in a timber frame covered with wire mesh.

30.3.3 Glare

Not usually recognised as a hazard, glare can prevent a crane driver from seeing clearly what is happening to his load and it can cause patches of darkness in accessways that prevent operators from seeing the floor or obstacles. Electric arc welding flash can cause a painful condition known as 'arc-eyes', so welding operations should be shielded by suitable flame-resistant screens. Floodlights are designed to operate at a height of 6 m or more and must never be taken down to use as local lighting as the glare from such misuse could create areas of black shadow and may even cause eye injury. Floodlights should not be directed upwards since they can dazzle tower crane drivers.

30.3.4 Dangerous and unhealthy atmospheres

Conditions under which work is carried out on construction sites is largely dictated by the weather, ranging from soaking wet to hot, dry and dusty and suitable protection for the health of the operators has to be provided. However, there is also a considerable range of substances[16] and working techniques now in use that have created their own hazards and a number of these are considered below.

30.3.4.1 Cold and wet

Cold is most damaging to health when it is associated with wet, as it is then very difficult to maintain normal body temperature. Being cold and wet frequently and for substantial periods may increase the likelihood of bronchitis and arthritis and other degenerating ailments. The effects of cold and wet on the employee's health and welfare can be mitigated by three factors: food, clothing and shelter. Where practicable, shelter from the worst of the wind and wet should be provided by sheeting or screens. The accommodation which has to be provided 'during interruption of work owing to bad weather' could also be used for warming-up and drying-out breaks whenever men have become cold, wet and uncomfortable.

30.3.4.2 Heat

Excessive heat has tended to be discounted as a problem on construction sites in the UK, and cases of heat exhaustion which do occur during heat-waves are often attributed to some other quite irrelevant cause. Common forms of heat stress produce such symptoms as lassitude, headache, giddiness, fainting and muscular cramp. Sweating results in loss of fluids and salt from the body, and danger arises when this is not compensated for by increased intake of salt and fluids. If the body becomes seriously depleted it can lead to severe muscular cramps.

30.3.4.3 Dust and fumes

Despite the general outdoor nature of the work, construction workers are not immune from the hazards of airborne contaminants. Although natural wind movement will dilute dust and fumes throughout the site, operatives engaged on particular processes may have a dangerous concentration in their immediate breathing zones unless suitable extraction is provided. This is particularly relevant for work in shafts, tunnels and other confined spaces where forced draught ventilation may have to be provided.

Certain processes commonly met on construction sites create hazardous dusts and fumes. Typical are:

Cadmium poisoning from dust and fumes arising from welding, brazing, soldering or heating cadmium plated steel.

Lead poisoning resulting from inhalation of lead fumes when cutting or burning structures or timber that has been protected by lead paint.

Silicosis due to inhaling siliceous dust generated in the cleaning of stone structures, polishing and grinding granite or terrazo.

Carbon monoxide poisoning caused by incomplete combustion in a confined space or from the exhausts of diesel and petrol engines.

Metal fume fever from breathing zinc fumes when welding or burning galvanised steel.

Each of these hazards would be eliminated by the provision of suitable and adequate exhaust ventilation or, in the case of silicosis, by the provision of suitable breathing masks. In each case food should not be consumed in the area, and the medical conditions can be exacerbated through habitual smoking. A good standard of personal hygiene is also an important factor in maintaining good health on site.

30.3.4.4 Industrial dermatitis

The use of an increasing range of chemical based products on sites poses a potential health risk to those who handle them unless suitable precautions are taken. The complaint is neither infectious nor contagious, but once it develops the sufferer can become sensitised (allergic) to the particular chemical causing the complaint and will react to even the smallest exposure. All chemical substances supplied to sites should carry instructions for use on the label and if the precautions recommended by the maker are followed little ill-effect should be experienced.

Barrier creams may be helpful but suffer the disadvantage of wearing off with rough usage or being washed off by water. Effective protection is provided by the use of industrial gloves and, where necessary, aprons, face masks etc. Again good personal hygiene is important and the use of skin conditioning creams after washing is beneficial.

30.3.4.5 Sewers

Before any work is carried out in sewers, the atmosphere must be tested to ensure that it is not deficient in oxygen. Because of the gases generated by decaying matter, smoking in sewers should be prohibited. Where the atmosphere is foul, respirators or breathing apparatus, as appropriate, should be worn. Due consideration must be given to preventing the onset of 'Weil's Disease', a 'flu-like disease which if untreated can have serious or fatal outcome. It is transmitted in rat's urine and enters the body through breaks in the skin or, more rarely, by ingestion of contaminated food.

30.3.5 Vibration-induced white finger

The vibrations from portable pneumatic drills and hammers can produce a condition known as white finger or Raynaud's phenomenon in which the tips of the fingers go white and feel numb as if the hand was cold. Anyone showing these symptoms should be taken off work involving the use of these drills or hammers and found alternative employment.

30.3.6 Ionising radiations

There are two main uses for radioactive substances that give off ionising radiations on construction sites. Firstly, tracing water flows and sewers where a low powered radioactive substance is added to the flow and its route followed using special instruments. Only authorised specialists should be allowed to handle the radioactive substance before it is added to the water. Once it is added, it mixes rapidly with the water and becomes so diluted as not to present a hazard.

The second application is in the non-destructive testing of welds where a very powerful gamma (γ) source is used. Because of its penetrating powers and the effects its rays have on human organs, very strict controls must be exercised in its use. The relevant precautions are detailed in Regulations[17] whose requirements must be complied with.

30.3.7 Lasers

Beams of intense light, they are radiations but do not ionise surrounding matter. Hazards stem from the intensity of the light which can burn the skin, and, if looked into, can cause permanent damage to eyesight. Ideally, lasers of classes 1 or 2 should be used as these present little hazard potential. Class 3A lasers give rise to eye hazards and should only be used in special cases under the supervision of a laser safety adviser. Class 3B and above generally should not be used on construction work, but if the necessity arises only adequately trained persons should operate them. When eye protection is assessed as being necessary, the type supplied must be certified as providing the required attenuation for the laser being used[18].

30.3.8 Compressed air work

The health hazards of work in compressed air and diving are decompression sickness ('the bends') and aseptic bone necrosis ('bone rot'). Both these illnesses can have long-term effects varying from slight impairment of mobility to severe disablement. The protective measures, including decompression procedures, are laid down in the Work in Compressed Air Special Regulations[19] and these should be observed in conjunction with the medical code of practice[20] for work in compressed air. Where diving work is involved, the Diving Operations at Work Regulations 1981[1] apply.

30.4 The Construction (Working Places) Regulations 1966

30.4.1 Access

Although there is a trend in the construction industry towards specialised plant to meet a particular need, the most common material at present employed to provide access scaffolding is scaffold tube and couplers. Large-scale or difficult projects are best carried out by experts but there is a very large amount of scaffold erection of the smaller type in short-term use which can be quickly and safely erected by craftsmen who are to work on them, provided they have been trained in the basic techniques and requirements of these Regulations and the British Standard Code of Practice[21].

As with all structures, a sound foundation is essential. Scaffolds must not be erected on an unprepared foundation. If soil is the base it should be well rammed and levelled and timber soleplates at least 225 mm (9 in) wide and 40 mm (1½in) thick laid on it so that there is no air space between timber and ground.

The standards should be pitched on baseplates 150 mm × 150 mm (6 in × 6 in) and any joints in the standards should occur just above the ledger. These joints should be staggered in adjacent standards so that they do not occur in the same lift. Ledgers should be horizontal (Reg. 13), placed inside the standards and clamped to them with right-angle couplers. Joints should be staggered on adjacent ledgers so that they do not occur in the same bay.

Decking (Reg. 25) will generally be 225 mm × 40 mm (9 in × 1½in) boards and each board should have at least three supports (although the Regulations state a maximum span between supports on a 40 mm (1½in) board should not exceed 1.52 m (5 ft) the British Standard[22] recommends that they do not exceed 1.2 m (4 ft)). Boards are normally butt jointed but may be lapped if bevel pieces are fitted or other measures taken to prevent tripping. A 40 mm (1½in) board should extend beyond its end support by between 50 mm and 150 mm (2 in–6 in).

Guardrails must be fitted at the edges of all working platforms (Reg. 28) at a height between 920 mm and 1.15 m (3 ft–3 ft 9 in) above the platform level. Toeboards and/or barriers at least 150 mm (6 in) high must also be fitted to prevent materials from falling from the platform. The space between toeboards and guardrails must not exceed 760 mm (30 in).

Ladders must stand on a firm level base and must be secured at the top and bottom so that they cannot move. All ladders must extend at least 1.07 m (3 ft 6 in) beyond the landing level (Reg. 32). To preserve them they may be treated with a clear preservative or be varnished but must not be painted. All rungs must be sound and properly secured to the stile.

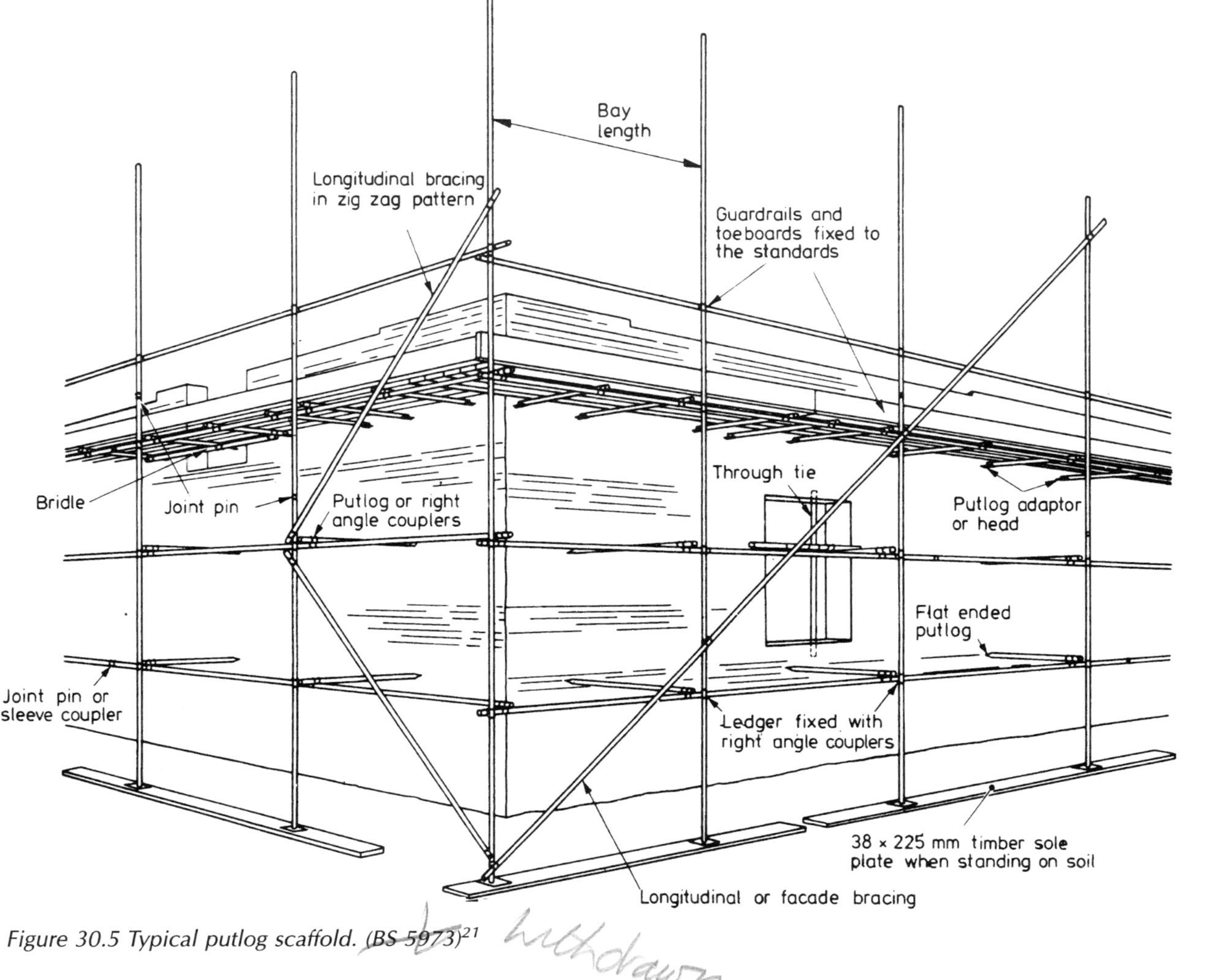

Figure 30.5 Typical putlog scaffold. (BS 5973)[21]

Unless properly designed to stand on their own, all scaffolds must be sufficiently and effectively anchored to the building or structure by ties (Reg. 15) which are essential to ensure stability of the scaffold. Before using a scaffold, the employer has a duty to arrange for it to be inspected by a competent person, then ensure that it is inspected every seven days and an entry made in the official register form F91 (Part 1 Section A). All scaffold material must be kept in good condition and free from patent defect. Damaged equipment should be stored separately (Reg. 10) and identified as 'damaged' or destroyed. Metal scaffold tubes and fittings and timber scaffold boards should comply with the appropriate British Standard[22,23]. *Figure 30.5* shows a simple scaffolding structure where only an outside row of standards are used to support the platforms, with putlogs fixed into brickwork joints.

30.4.2 Mobile towers

A mobile access tower[24] (*Figure 30.6*) is a tower formed with scaffold tube and mounted on wheels. It has a single working platform and is provided with handrails and toeboards. It can be constructed of prefabricated tubular frames and is designed to support a distributed load of 30 lb/ft^2. The height of the working platform must not exceed three times the smaller base dimension and no tower shall have a base dimension less than 4 ft. Rigidity of the tower is obtained by the use of diagonal bracing on all four elevations and on plan. Castors used with the tower should be fixed at the extreme corners of the tower in such a manner that they cannot fall out when the tower is moved and shall be fitted with an effective wheel brake. When moving mobile towers great care is essential. All persons, equipment and materials must be removed from the platform and the tower moved by pushing or pulling at the base level. Under no circumstances may mobile towers be moved by persons on the platform propelling the tower along.

30.5 The Construction (Lifting Operations) Regulations 1961

These Regulations apply to all equipment used for lifting on construction sites and include fixed, mobile and travelling cranes, hoists used for both goods and passengers and also the ropes, chains, slings, etc. that support the load being lifted.

There are certain requirements that are common to all this equipment in that they must be of sound construction and free from patent defect (Regs. 10, 21, 34, 47). When erected the equipment must be properly supported and secured (Regs. 11, 18) and that ground conditions are such as to ensure stability (Regs. 19, 20). Erection must be under competent control (Reg. 25) and in the case of derricking cranes follow certain procedures (Reg. 33).

All lifting equipment must be examined regularly (Regs. 10, 28, 34, 46) with safe means of access provided for those carrying out the inspection (Reg. 17). The safe working load must be clearly indicated (Regs. 29, 45) and jib cranes must have an automatic safe load indicator (Reg. 30) which must be tested. The specified safe working load must not be exceeded (Reg. 31).

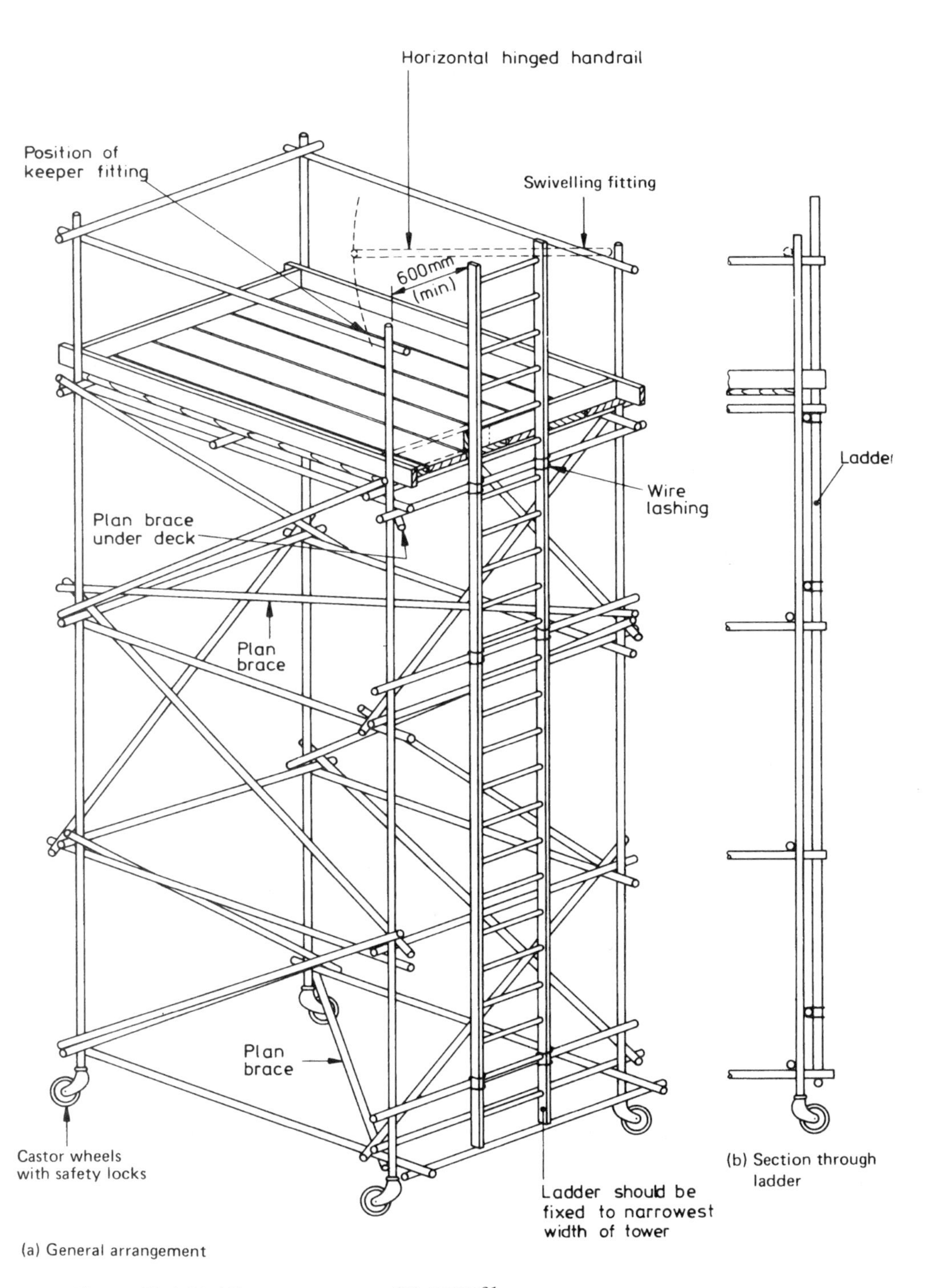

Figure 30.6 Mobile access tower. (BS 5973)[21]

The positioning of travelling or slewing cranes should be such that a clear passageway, 0.6 m (2 ft), is ensured at all times (Reg. 12). Where drivers or banksmen require platforms that are more than 2 m (6 ft 6 in) above an adjacent level, suitable guardrails must be provided (Reg. 13) and a cab with safe access should be provided for drivers exposed to the weather (Reg. 14). Any communication between banksman and driver must be clear (Reg. 27). *Figure 27.12* (p. 616) shows the visual signals in common use[25].

Lifting tackle such as chains, rings, hooks, shackles, etc. must not be modified by welding unless by a competent person and followed by a test (Reg. 35). Hooks should have a safety clip (Reg. 36) and slings must not be used in such a way that is likely to damage them (Regs. 37, 38, 39). Chains, ropes and lifting gear must be thoroughly examined by a competent person every six months (Reg. 40) and chains and hooks of the appropriate material are to be heat-treated every 14 months (Reg. 41).

A hoist is defined (Reg. 4) as '. . . a lifting machine, whether worked by mechanical power or not, with a carriage, platform or cage the movement of which is restricted by a guide or guides, but does not include a lifting appliance used for the movement of trucks or wagons on a line or rails' and special requirements attach to the protection of hoistways, platforms and cages (Reg. 42) and to their operation (Reg. 43). Special precautions apply where people are carried by lifting equipment (Regs. 47, 48) and to the securing of loads (Reg. 48).

Records must be kept of the examinations of lifting equipment (Reg. 50).

To facilitate compliance with these requirements, check lists can be used for the different items of lifting gear and tackle and the following are typical lists that apply to the various statutory requirements under this section of the Construction Regulations.

30.5.1 Checklists

30.5.1.1 Mobile cranes

Prior to work commencing ensure a competent 'lifting co-ordinator' has been appointed:

When was the crane selected, and what information was available/used at the time?

Has the selected crane been supplied?

Check that the ground is capable of taking the loads (outriggers/crane/load/wind). If in doubt get ADVICE from specialist departments/firms.

Ensure that the approach and working area are as level as possible.

Ensure that the area is kept free of obstructions – minimum 600 mm (2 ft) clearance.

Ensure that the weights of the loads are known, and that the correct lifting gear is ordered/available.

Ensure that there is a competent, trained banksman available.

Check that there are no restrictions on access, i.e. check size(s) of vehicles, etc.

Ensure that the work areas are adequately lighted.

Check that the Plant Department/Hirer has provided information re the cranes, etc.

Whilst work is in progress:
Check that there is an up-to-date test certificate.
Check that the daily/weekly inspections are being carried out and entered in Form 91 (Part 1).
Ensure that the crane is operating from planned/approved positions only.
Ensure that the banksman is working in the correct manner.
Ensure that the correct lifting gear is being used.
Ensure that outriggers are being used, and are adequately supported.
Check that the safe load/radius indicator is in working order.
Check that the tyres/tracks are at the correct pressure and in good, clean condition.
Check that the crane is kept at a safe, predetermined distance from open excavations, etc.
Check that, when travelling, the load is carried as near to the ground as possible and that hand lines are being used.
Check that when travelling on sloping ground the driver changes the radius to accommodate the moving of the load.
Check that loads are not being slewed over persons and that persons are not standing or walking under the load.

30.5.1.2 Automatic safe load indicator

All cranes with a lifting capacity in excess of 1 ton must be fitted with an approved type of automatic safe load indicator (Construction (Lifting Operations) Regulations 1961).

It is the responsibility of the operator to:

(a) determine the type of indicator fitted;
(b) determine how the adjustments are made;
(c) ensure that it is correctly adjusted for the various lifting duties;
(d) ensure that the electrical circuit is tested for serviceability;
(e) take immediate action when an overload is indicated.

The signals given by the indicator take the form of coloured lights, a dial indicator or both and a bell.

Green/white – Indicator adjusted for 'free' duties
Blue – Indicator adjusted for 'blocked' duties
Amber – Maximum safe load being approached
Red – Overload condition reached.

The red light will be supported by a bell to give an audible warning of overload.

30.5.1.3 Hoists, static and mobile – safe working checklist

(a) Erect the hoist in a suitable position.
(b) Make the hoistway as compact as possible.
(c) Hoistway to be efficiently protected by a substantial enclosure at least 2 m (6 ft 6 in) high.
(d) Hoist gates – guards to be at least 2 m (6 ft 6 in) high.
(e) Engine or motor must also be enclosed to a height of 2 m (6 ft 6 in) where practicable.

(f) Make sure that no one can come into contact with any moving part of the hoist.
(g) Enclosures at the top may be less than 2 m (6 ft 6 in) but in no case less than 0.9 m (3 ft) providing that no one can fall down the hoistway and that there is no possibility of anyone coming into contact with any moving part.
(h) All intermediate gates will be 2 m (6 ft 6 in) unless this is impractical, i.e. confined space, etc.
(i) The construction of the hoist shall be that it can only be operated from the upmost working position at any one time.
(j) It shall not be operated from inside the cage (unless designed for the purpose).
(k) The person operating the hoist must have a clear and unrestricted view of the platform throughout.
(l) The safe working load shall be plainly marked on every hoist platform and this load must not be exceeded.
(m) No person shall ride on the hoist (unless so designed), and a notice to this effect must be exhibited on the hoist so that it can be seen at all levels.
(n) Every hoist must be fitted with an efficient automatic device to ensure that the platform does not overrun the highest point for which it is intended to travel.
(o) Every hoist must be fitted with an efficient device which will support the platform and load in the event of the failure of the ropes or lifting gear.
(p) All movable equipment or plant must be scotched, to prevent its displacement while in motion.
(q) All materials will be so placed as to prevent displacement.
(r) Gates should be kept closed on all landing stages. Every person using hoists must close landing place gates immediately after use. (This is a statutory duty imposed upon the person actually using the hoist but the employer also has the duty of seeing that the regulation is obeyed, and the employer's representative on site is the General Foreman or Site Agent.)
(s) Landing stages should be kept free from materials and plant.
(t) No person under the age of 18 must be allowed to operate or give signals to operator.
(u) Only a competent person should operate the hoist.
(v) Signals to be of distinct character; easily seen or heard by person to whom they are given.
(w) Every hoist must be inspected once a week and a written entry should be made weekly in Form 91 (Part I Section C).
(x) Every hoist should be examined every six months by a competent person and certificated.

30.5.1.4 Chains, rope slings and lifting gear – safe working checklist

Prior to commencing work:

(a) Examine the slings provided and check that the 'thorough examination' has been carried out and recorded.

(b) Determine and clearly mark the Safe Working Loads for all slings.
(c) Ensure that the correct and up-to-date copies of the Sling Chart and Safe Working Load Tables are available, when using multi-leg slings.
(d) Ensure that a copy of the correct crane signals is available.
(e) Ensure that a suitable rack is available for storing slings, etc. not in use. N.B. Wire ropes should be stored in a dry atmosphere.
(f) Ensure that the weights of loads to be lifted are known in advance, and that load weights are clearly marked.
(g) Find out the type of eye bolt fitted to the load, in advance, to ensure that the correct equipment, shackles/hooks/lifting beams, is available on site.

Whilst work is in progress:
(h) Ensure that the 'right' techniques are being used.
(i) Ensure that the copies of the Sling Chart and the Safe Working Loads Tables are being used, where necessary.
(j) Ensure that the correct crane signals are being used, and that signals are given only by 'approved' banksmen.
(k) Ensure that regular inspections of the equipment are being carried out.
(l) Ensure that unfit slings are destroyed, or at least removed from site.
(m) Stop persons 'hooking back' onto the legs of slings.
(n) Ensure that slingers understand that 'doubling up' the sling does NOT 'double up' the Safe Working Load: avoid this practice if possible.
(o) Limit the use of endless wire rope slings.
(p) Ensure that wire rope slings are protected from sharp corners of the loads, by suitable packings.
(q) Prevent/stop slings/ropes from being dragged along the ground.
(r) Ensure that the hooks used for lifting are NOT also carrying unused slings.
(s) Ensure that the crane hook is positioned above the load's centre of gravity.
(t) Ensure that the load is free before lifting and that all legs have a direct load.
(u) Ensure that the load is landed onto battens to prevent damage to slings.
(v) Ensure that a sling is NOT passed through more than one eye bolt.
(w) Ensure that 'snatch' loading does NOT take place.
(x) Ensure that NO ONE rides on a load that is being slung.

Additional guidance on the standards to be achieved is given in British Standards[25–27] and in HSE Guidance Notes[28].

30.6 The Construction (Health and Welfare) Regulations 1966

Under the Regulations, every contractor or employer has a duty to provide certain health and welfare facilities for his own employees who must have proper access to them. Because a number of contractors may be working on the site some of the facilities may be shared (Reg. 4) or alternatively arrangements may be made by the contractor to use the facilities offered by adjacent factories. Such arrangements should be agreed in writing. When such agreements are terminated, to prevent confusion it is advisable to give notice in writing.

If facilities are shared with another employer or contractor on the site, then the one who provides the facilities or equipment must:

(1) in deciding what facilities to provide, assume that he employs the total number of men who are to use the facilities: e.g. say own employees = 50, other employees = 40 – therefore, for the purposes of providing facilities, assume that he employs 90;
(2) keep a register (Form F2202) showing the facilities to be shared and the names of the firms sharing them. It should be available for examination by those allowed to use the facilities and by inspectors;
(3) give to each firm sharing the facilities a certificate (Form F2202 Part A) detailing what facilities are to be shared.

Contractors or employers on the same site can jointly appoint the same man to take charge of first-aid and ambulance arrangements.

30.6.1 Facilities to be provided on site

Clearly marked 'FIRST AID' boxes must be provided and put in the charge of a responsible person whose name must be displayed near the box. After assessing the level of risk, the availability of emergency services and other matters detailed in the Regulations[29] and the Approved Code of Practice[30], it may be necessary for the responsible person to be a trained and certificated first aider.

Where there is a large workforce on a site a suitably staffed and equipped first-aid room should be provided. However, where a large workforce is divided into several dispersed working groups or the location of the site makes access to places of treatment outside it difficult, the needs of such a site may be better met by the provision of first-aid equipment and trained first-aiders at different parts of the site.

Regardless of the number of employees there must be at least one first-aid box on site, and provision should be made for every employee to have reasonably rapid access to first-aid.

Construction workers are frequently exposed to the weather and facilities must be provided to store the clothes they do not wear while working, to warm themselves, and to dry their clothing when not in use. In addition, a supply of drinking water must be available and suitable arrangements for warming and eating food.

Where anyone is employed on a site for more than four hours, suitable washing facilities must be provided (Reg. 12) and toilets, accessible from all workplaces on the site, must be under cover, partitioned from each other, have a door with fastening, be ventilated and provided with lighting. They must not open directly into workrooms or messrooms and must be kept clean (Reg. 13). Separate conveniences must be provided for men and women. For new sites these facilities must conform to the Workplace (Health, Safety and Welfare) Regulations 1992[1] (WHSWR) while existing sites have until 1 January 1996 to conform.

30.7 Other relevant legislation

30.7.1 Offices, Shops and Railway Premises Act 1963

This Act applies to all regular offices and site offices (which are in use for more than six months if movable or more than six weeks if fixed) in which people work for more than a total of 21 hours per week (self-employed man and family-only staff excepted). It applies equally to stairs, passages, landings, entrances, exits, store rooms, yards and any associated canteen. In very general terms the Act as it applies to construction sites requires:

(a) All premises to be kept properly clean.
(b) Each office worker to have $3.7\,m^2$ ($40\,ft^2$) of floor space.
(c) Offices to be maintained at more than 16°C (60.8°F) and be adequately lit.
(d) Reasonably accessible sanitary conveniences to be provided, kept clean, lit and ventilated, with separate accommodation for males and females.
(e) Washing facilities with at least running warm water, soap and towels.
(f) A sufficient supply of good drinking water and drinking vessels together with means of washing up.
(g) Suitable office seating arrangements.
(h) Eating facilities.

The WHSWR supersede the Act for all new regular offices while existing offices have until 1 January 1996 to comply. Site offices will be subject to similar requirements (to be contained in Regulations due to come into effect in 1994) until which time the Act remains in effect.

30.7.2 Personal protective equipment

The requirements to be met in the application and use of this equipment are laid down in the Personal Protective Equipment at Work Regulations 1992[1] (PPER).

The general assessment required by MHSWR should identify the hazards, the extent of the risks faced and enable the necessary preventative and precautionary measures to be decided. If personal protective equipment is considered appropriate, PPER sets out the steps to be taken in the process of selecting suitable and effective equipment which the employer has to provide to his employees.

These Regulations revoke early regulations made under FA 1961, such as the Protection of Eyes Regulations 1974, and have marginally modified subsequent regulations that include provisions for personal protective equipment, such as the Noise at Work Regulations 1989, The Construction (Head Protection) Regulations 1989, etc., which continue to apply.

30.7.3 The Construction (Head Protection) Regulations 1989[1]

Under these Regulations employers are required to provide, maintain and replace, as necessary, suitable head protection for their employees and others working in the areas over which they have control. They must ensure that the head protection is worn unless there is no risk of a head injury occurring other than by falling over. To be 'suitable' head protection must be:

- designed, so far as is reasonably practicable, to provide protection against foreseeable risks of injury to the head to which the wearer may be exposed,
- adjustable so that it can be made to fit the wearer, and
- suitable for the circumstances in which it is to be used.

Persons who have control over a site, such as management contractors, may make written rules concerning the wearing of head protection by anyone, employees and others (with the exception of Sikhs wearing turbans), working on that site, and should make arrangements for enforcing those rules. Employees are required to take care of equipment and to report cases of damage, defect or loss.

30.7.4 Fire Certificates (Special Premises) Regulations 1976

(a) These Regulations were introduced as a result of the operation of the Health and Safety at Work Act 1974 and take effect from 1st January 1977. Most temporary site buildings used as offices or workshops at building operations and works of engineering construction were previously subject to certain fire precautions provisions, under the Offices, Shops and Railway Premises Act, or occasionally under the Factories Act. Health and Safety Executive Inspectors already inspect building sites for general inspection purposes, and by including these premises into the new Regulations the Health and Safety Executive now becomes responsible for fire precautions at them. It is not intended to issue a Fire Certificate for every temporary site building, however, and the Regulations are designed to exclude small site buildings conditionally from the certification procedure. Some large site buildings, or those with special risks, will require a Certificate.

(b) A Fire Certificate will be required for any building or part of a building which is 'constructed for temporary occupation for the purposes of building operations or works of engineering construction, or, which is in existence at the commencement there of any further such operations'. But a Fire Certificate will not be required for these buildings if all of the following conditions are complied with:
   (i) Not more than 20 persons are employed at any one time in the building or part of the building.
   (ii) Not more than 10 persons are employed at any one time elsewhere than on the ground floor.
   (iii) No explosive or highly flammable material is stored or used in or under the building.
   (iv) The building is provided with reasonable means of escape in case of fire for the persons employed there.
   (v) Appropriate means of fighting fire are provided and maintained and so placed as to be readily available for use in the building.

(vi) While anyone is inside, no exit doors may be locked or fastened so that they cannot be easily opened from inside.
(vii) If more than 10 people are employed in the building, any doors opening on to any staircase or corridor from any room in the building must be constructed to open outwards unless they are sliding doors.
(viii) Every exit opening must be marked by a suitable Notice.
(ix) The contents of every occupied room in the building must be arranged so that there is a free passageway for everyone employed in the room to a means of escape in case of fire.

(c) To obtain a Fire Certificate application must be made to the HSE (Construction) office for the area in which the site is located, using form F2003 to give the required particulars, which are listed below. The HSE will then carry out an inspection of the site buildings. In practice, all final exit doors and any door opening on to a staircase or corridor from any room in the building should open outwards. Fire extinguishers to be hung on wall brackets adjacent to the final exit, with the top of the extinguisher 1.07 m (3 ft 6 in) from the floor. A fire blanket should be hung on a wall bracket adjacent to the cooker in the canteen with the top of the blanket 1.52 m (5 ft 0 in) from the floor. LPG cylinders should be sited outside huts and have an isolating valve at the cylinder and an ON/OFF control valve as near as practicable to the heater.

Particulars required on form F2003 when applying for a Fire Certificate include:

(a) Address of premises.
(b) Description of premises selected from those listed in Schedule 1 to the Regulations.
(c) Nature of the processes carried on, or to be carried on, on the premises.
(d) Nature and approximate quantities of any explosive or highly flammable substance, kept, or to be kept, on the premises.
(e) Maximum number of persons likely to be on the premises at any one time.
(f) Maximum number of persons likely to be in any building of which the premises form part at any one time.
(g) Name and address of any other person who has control of the premises.
(h) Name and address of the occupier of the premises.
(j) If the premises consist of part of the building, the name and postal address of the person or persons having control of the building or other part of it.

30.7.5 Food Hygiene (General) Regulations 1970

The Food Safety Act 1990 is now one of the main pieces of legislation concerning food hygiene. It is an enabling Act allowing specific subordinate legislation (Regulations) to be made as necessary. An example is the Food Premises (Registration) Regulations 1991 which require all places where food is served, including site canteens, to be registered with the local authority. More specifically, the 1970 Regulations still apply to site canteens that are regarded as fixed premises and require that articles or equipment with which food is likely to come into contact must be non-absorbent, capable of being cleaned and kept in good condition. People handling food must be clean, must not spit or smoke in rooms in which food is handled, and must keep any cut or abrasion covered with a waterproof dressing. They must wear clean, washable overclothing and neck and head covering. There must be accommodation for their outdoor clothing and footwear.

No containers or wrapping may be used that could contaminate the food (e.g. newspaper). If a person handling food suffers any infection of the digestive system, the Medical Officer of Health must be notified immediately. An adequate supply of hot and cold water and hand wash basins must be supplied for the use of food handlers, and separate sinks with hot and cold water for preparing vegetables and for washing equipment. No toilet may connect direct with a room used for preparing or eating food. Food rooms must be adequately lit and ventilated and walls, floors, doors, windows, ceilings, woodwork etc., must be kept clean and in good repair. Waste must not be allowed to accumulate.

Where the site buildings are not regarded as fixed premises, kitchens and canteens are subject to the Food Hygiene (Markets, Stalls and Delivery Vehicles) Regulations 1966 and the Food Hygiene (Markets, Stalls and Delivery Vehicles) Amendments Regulations 1966 under which the requirements are virtually the same as those discussed above.

30.7.6 Petroleum Consolidation Act 1928
Petroleum Spirit (Motor Vehicles etc.) Regulations 1929
Petroleum Mixtures Order 1929

Broadly, the Act and Regulations deal with petroleum spirit and mixtures whether liquid, viscous or solid. Petroleum means substances giving off a flammable vapour at a temperature of less than 23°C (73°F). It will be appreciated that this is less than body heat so there is a need for strict control on their uses. By prior arrangement with the local authority, up to 60 gallons may be kept on site without licence but this is subject to a review annually. Where quantities in excess of 60 gallons are to be kept, a petroleum licence must be obtained from the Petroleum Officer.

30.7.7 Mines and Quarries Act 1954

The requirements of this Act apply to sites, or parts of sites, that fall within the description of a quarry, as defined in s. 180 of the Act, the most common example being a borrow pit used in motorway construction, although it can also apply where old quarries are being in-filled with spoil. Enforcement of this Act is split between the Factory Inspectorate who cover building on the surface, and the Mines and Quarries Inspectorate who deal with work within the quarry itself.

Notification of the commencement and termination of work in a quarry must be made on Mines and Quarries form 213. The owner of the quarry is required to appoint a manager and a deputy manager.

30.7.8 Asbestos

Asbestos is a strong, durable, non-combustible fibre, and these physical properties make it ideal as a reinforcing agent in cement, vinyl and other building materials, e.g. vinyl floor tiles, bath panels, cold water cisterns, roof felts, corrugated or flat roof sheets, cladding sheets, soffit strips, gutters and distribution pipes. It is also a good insulant and has been used for protecting structures from the effects of fire.

Asbestos has been used extensively in the past and may be found in many forms in existing buildings including: industrial wall and roof linings; internal partitions; duct and pipe covers; suspended ceilings; fire doors and soffits to porch and canopy linings. It also occurs in plant rooms and boiler houses and as asbestos coatings and insulating lagging on structures and pipework.

Work with asbestos products is governed in the UK by specific legislation:

The Asbestos (Licensing) Regulations 1983
The Asbestos (Prohibitions) Regulations 1985
The Asbestos Products (Safety) Regulations 1985
The Control of Asbestos at Work Regulations 1987
The Control of Asbestos in the Air Regulations 1990

The main purpose of the 1983 regulations is to control the manner in which companies and the self-employed carry out work which involves the disturbance of asbestos insulations or coatings. All persons carrying out such work must be in possession of an HSE licence and receive regular medical examinations. Wherever asbestos products are encountered the regulations should be complied with, using advice given in HSE guidance notes[31]. Useful information is also available from the Asbestos Information Centre Ltd.

Control limits for asbestos dust in the working environment are set out in an Approved Code of Practice[32]. These control limits are not intended to represent safe levels of airborne dust but the upper limits of permitted exposure. There is still a statutory duty to reduce exposure to the lowest level that is reasonably practicable.

Where asbestos containing materials have to be stripped, its disposal is governed by special regulations[33].

Acknowledgement

Grateful thanks are due to the Construction Health and Safety Group, Chertsey, for their assistance in preparing this chapter.

References

1. Statutory Instruments relevant to construction work:
   The Construction (General Provisions) Regulations 1961 (SI 1961 No. 1580)
   The Construction (Lifting Operations) Regulations 1961 (SI 1961 No. 1581)
   The Construction (Working Places) Regulations 1966 (SI 1966 No. 94)

The Construction (Health and Welfare) Regulations 1966 (SI 1966 No. 95)
The Woodworking Machines Regulations 1974 (SI 1974 No. 903)
The Electricity at Work Regulations 1989
The Abrasive Wheels Regulations 1970 (SI 1970 No. 535)
The Reporting of Injuries, Diseases and Dangerous Occurrences Regulations 1985 (SI 1985 No. 2023)
The Highly Flammable Liquids and Liquefied Petroleum Gases Regulations 1972 (SI 1972 No. 917)
The Diving Operations at Work Regulations 1981 (SI 1981 No. 399 as amended by SI 1990 No. 966 and SI 1992 No. 608)
The Work in Compressed Air Special Regulations 1958 (SI 1958 No. 61 as amended by SI 1960 No. 1307 and SI 1973 No. 36)
Noise at Work Regulations 1989 (SI 1989 No. 1790)
Construction (Head Protection) Regulations 1989 (SI 1989 No. 2209)
Lifting Plant and Equipment (Records of Test and Examination, etc.) Regulations 1992 (SI 1992 No. 195)
The Management of Health and Safety at Work Regulations 1992 (SI 1992 No. 2051)
The Provision and Use of Work Equipment Regulations 1992 (SI 1992 No. 2932)
The Manual Handling Operations Regulations 1992 (SI 1992 No. 2793)
The Personal Protective Equipment at Work Regulations 1992 (SI 1992 No. 2966)
The Workplace (Health, Safety and Welfare) Regulations 1992
All published by HMSO, London.

2. Health and Safety Executive, Manufacturing Services and Industries – Annual Report (published each year), HSE Books, Sudbury
3. Health and Safety Executive, *Blackspot Construction, A study of five years fatal accidents in the building and civil engineering industries*, HSE Books, Sudbury (1988)
4. H.M. Government, *The Reporting of Injuries, Diseases and Dangerous Occurrences Regulations 1985* (SI 1985 No. 2023) HMSO, London (1985)
5. Health and Safety Commission, *The Construction (Design and Management) Regulations 1993*, HMSO, London (1993)
6. Regina *v.* Swan Hunter Shipbuilders Ltd and Telemeter Installations Ltd [1981] IRLR 403
7. Fife, I. and Machin, E.A., *Redgrave, Fife and Machin: Health and Safety*, Butterworths, London (1993)
8. Williams *v.* West Wales Plant Hire Company and others, [1984] All ER 353
9. British Standards Institution, BS 6031, *Code of Practice for earthworks*, BSI, London (1981)
10. Health and Safety Executives,
    Health and Safety: Guidance Booklet No. HS(G)47, *Avoiding danger from underground services*, (1989)
    Health and Safety: Guidance Note No. GS6, *Avoidance of danger from overhead electrical lines*, (1991)
    HSE Books, Sudbury
11. Health and Safety Executive, *Construction (Lifting Operations) Regulations 1961 – Certificate of Exemption: CON(LO)/1981/2* (General), *Excavator, loaders and combined excavator/loaders used as cranes*, HMSO, London (1981)
12. Health and Safety Executive, *Health and Safety Guidance Note No. GS7, Accidents to children on construction sites*, HSE Books, Sudbury (1989)
13. BS 5304 Code of Practice for safety of machinery, British Standards Institution, London (1988)
14. Health and Safety Executive, Guidance Notes:
    No. MS 15 Welding
    No. PM 17 Pneumatic Nailing and Stapling Tools
    No. PM 14 Safety in the Use of Cartridge Operated Tools
    No. PM 5 Automatically Controlled Steam and Hot Water Boilers
    No. PM 1 Guarding of Portable Pipe Threading Machines
15. H.M. Government, *The Electricity at Work Regulations 1989*, HMSO, London (1989)
16. H.M. Government, *The Control of Substances Hazardous to Health Regulations 1988 (as amended by SI 1991 No. 2341 and SI 1992 No. 2382)*, HMSO, London (1988)
17. H.M. Government, *The Ionising Radiations Regulations 1985*, HMSO, London (1985)
18. British Standards Institution, BS 7192, *Radiation, safety of laser products*, BSI, London

19. H.M. Government, *The Work in Compressed Air Special Regulations 1958*, HMSO, London (1973)
20. *Medical Code of Practice for Work in Compressed Air*, Report No. 44, 3rd edn, CIRIA, London (1988)
21. BS 5973, Code of Practice for access and working scaffolds and special scaffold structures in steel, British Standards Institution, London
22. BS 2482, Specification for timber scaffold boards, British Standards Institution, London
23. BS 1139 – Metal scaffolding
    Part 1: Specification for tubes for use in scaffolding
    Part 2: Specification for couplers and fittings for use in tubular scaffolding
    Part 4: Specification for prefabricated steel splitheads and trestles
    British Standards Institution, London
24. Health and Safety Executive, *Health and Safety: Guidance Notes No. GS 42, Tower Scaffolds*, HSE Books, Sudbury (1987)
25. BS 5744, Code of Practice for safe use of cranes (overhead/underhung travelling and goliath cranes, high pedestal and portal jib dockside cranes, manually operated and light cranes, container handling cranes and rail mounted low carriage cranes), British Standards Institution, London
26. CP 3010, Safe uses of cranes (mobile cranes, tower cranes and derrick cranes), British Standards Institution, London, (1972) (See also ref. 27)
27. BS 7121, parts I–IX, *Safe use of cranes*, British Standards Institution, London
28. Health and Safety Executive, Guidance Notes:
    PM 3 Erection and Dismantling of Tower Cranes
    PM 9 Access to Tower Cranes
    PM 8 Passenger Carrying Paternosters
    PM 16 Eyebolts
    PM 27 Construction Hoists
    GS39 Training of Crane Drivers and Slingers
    HSE Books, Sudbury
29. H.M. Government, *The Health and Safety (First Aid) Regulations 1981*, HMSO, London (1981)
30. Health and Safety Executive, *Approved Code of Practice No. COP 42, First aid at work. Health and Safety (First Aid) Regulations 1981. Approved Code of Practice and Guidance*, HSE Books, Sudbury (1990)
31. Health and Safety Executive, Guidance Notes:-
    EH35 Probable asbestos dust concentrations at construction processes.
    EH36 Work with asbestos cement
    EH37 Work with asbestos insulating board
    EH40 Occupational exposure limits
    HSE Books, Sudbury
32. Health and Safety Executive, *Legislation booklet L27, The control of asbestos at work. The Control of Asbestos at Work Regulations 1987, approved Code of Practice*, HSE Books, Sudbury (1988)
33. H.M. Government,
    The Environmental Protection Act 1990,
    The Control of Pollution (Special Waste) Regulations 1980,
    HMSO, London

Further reading

Dickie, D. E., *Crane Handbook*, (Ed.: Douglas Short), Butterworth, London (1981)
Dickie, D. E., *Lifting Tackle Manual*, (Ed.: Douglas Short), Butterworth, London (1981)
Construction Safety Manual, Construction Safety, Crawley, Sussex
Dickie, D. E., Ed. Hudson, R., *Mobile Crane Manual*, Butterworth, London (1985)
King, R. and Hudson, R., *Construction Hazard and Safety Handbook*, Butterworth, London (1985)

Chapter 31

Safe use of chemicals

S. Bradley

31.1 Introduction

All chemicals can be handled safely! This applies to even the most toxic and flammable substances known. Conversely, to mishandle them is to court danger.

A chemical incident has the potential to affect many people, not only the operator, nearby workers and others on the site, but also members of the public. The results of some incidents, such as at Seveso and Bhopal, have been of epic proportions and were truly catastrophic. Such events carry a massive price tag in terms of human suffering and loss of life as well as the cost of rebuilding the plant and loss of business. All incidents, regardless of scale, damage the reputation of the organisation and, indeed, the whole of the industry.

31.2 Hazards and risks

The words 'hazard' and 'risk' arise frequently and are often misunderstood. However, they do have precise meanings and should not be confused:

HAZARD – is the inherent property of a substance to cause harm.
RISK – is the likelihood of a hazardous substance actually causing harm.

For example, concentrated sulphuric acid is undoubtedly hazardous. However, if it is in a bottle, properly stoppered, on a shelf in a locked cupboard, then the risk is minimal.

The safe storage and usage of chemicals requires the highest level of competence and professionalism from not only the qualified chemist but also from the engineer, manager, safety adviser and the workpeople. To reduce the likelihood of an incident and to minimise its effects should one occur, it is essential that the hazards and risks associated with the storage and use of the chemicals on the site are known and understood.

31.3 Discovering the hazards

Authoritative guidance in relation to health hazards associated with various substances is given in the Control of Substances Hazardous to Health Regulations 1988[1] (COSHH) and its amendments where 'substances hazardous to health' are defined as:

(a) those substances specified as very toxic, toxic, corrosive or irritant in the Classification, Packaging and Labelling of Dangerous Substances Regulations 1984 (CPL)[2];
(b) a substance which has been given a maximum exposure limit (MEL) or an occupational exposure standard (OES)[3];
(c) a micro-organism which creates a hazard to health;
(d) substantial concentrations of dust; and
(e) any substance not mentioned above but which creates a comparable hazard.

Some fourteen hundred or so of the more common hazardous substances have been classified under the Chemical (Hazard Information and Packaging) Regulations 1993[2] (CHIP). The details are given in an associated Approved Supply List[4]. In essence, substances with any of the following properties are hazardous to some degree:

(a) explosive
(b) oxidising
(c) flammable
(d) highly flammable
(e) extremely flammable
(f) toxic
(g) very toxic
(h) harmful
(i) corrosive
(j) irritant
(k) carcinogenic (causes cancer)
(l) mutagenic (causes inherited changes)
(m) teratogenic (causes harm to the unborn)
(n) micro-organisms that create a hazard to health
(o) substantial concentrations of dust
(p) radioactive materials
(q) any substance not mentioned above which creates a comparable hazard.

31.3.1 Information sources

All packaged dangerous substances are required by CPL to be labelled with a label that includes a hazard warning sign, an indication of the various risks and the necessary first aid treatment if contact occurs.

A second source of information is the supplier's Safety Data Sheet which the supplier is obliged to give to his customers. General guidance on the sort of information a supplier should give are outlined in Guidance Booklet HS(G)27[5]. Also, for the more commonly used substances there are many reference books that give comprehensive technical data[6--8].

It is likely that there is expertise in the user organisation since, if the substance has been ordered for use within the process, the person initiating the order must have a good knowledge of its characteristics.

For substances that are not bought in but which are produced and used exclusively within an organisation, such as reaction intermediates, then in-house expertise should be available. If there are no hazard data for a substance, e.g. a new research molecule, and it is not possible to estimate its characteristics from similar known molecules then that substance must be treated as toxic until proved otherwise.

31.4 Risk assessment

An essential feature of the measures to reduce the risk from hazardous substances is the requirement in COSHH to carry out risk assessments. These assessments should review all the chemicals or substances present to obtain an indication of the likely risk they represent to the health of the employees.

There are two broad approaches when assessing risks, quantitative and qualitative.

31.4.1 Quantitative risk assessment

This technique, which is employed extensively in the nuclear industry and the major chemical industries, uses the manipulation of numerical data, normally derived from equipment reliability data, to develop a quantitative measure of a risk. Because it is highly sophisticated and mathematically complex it is not widely adopted elsewhere.

Another approach, which has been pioneered by the HSE, is based on vapour dispersal characteristics. Computer programmes model the dispersion of various sized releases taking into account the physical characteristics of the vapour, the release rates, wind speeds, etc. These models produce concentration figures which, when compared with toxicological or flammability criteria, enable risk contours to be developed. The HSE has considerable expertise in this field particularly with chlorine and LPG (propane and butane). There are several readily available computer programmes suitable for desk-top computers which may be useful when assessing the risks in a major hazard site. Information on this technique is given in a number of publications[9–11].

From a quantitative point of view, if the risk of an event, such as a fatality, is less than 10^{-6} per year, i.e. less than a one in a million chance of an occurrence in any specific year, then it is considered acceptable. However, if the risk is greater than 10^{-4} per year, i.e. more than a one in ten thousand chance, it is generally regarded as unacceptable.

31.4.2 Qualitative risk assessment

An opposite approach to quantitative risk assessment is a subjective and qualitative risk assessment. It is based on inspection, process and chemical knowledge and experience. This sort of assessment needs to be carried out by experienced and competent people who are conversant with the hazards of the substances and exactly how those substances are and should be handled. Being subjective, this approach cannot be mathematically validated but it is by far the most practical method. However, it does require a good experienced assessor who needs to think laterally, to look beyond the obvious and have good interrogative skills to tease out crucial information which is often assumed to be common knowledge or which is wrongly

Risk Assessment - British Sugar plc

Site:		**Area:**
Operations covered by this Assessment:		
Maximum No. of people exposed: **Frequency & durations of exposure:**		
Hazards		
Actions already taken to reduce the risk		
Assessment of residual risk:		
Further actions required:		
Signed: **Position:**	**Date:**	**Review Date**

Figure 31.1(a) Risk assessment form. (Courtesy British Sugar plc)

considered to be unimportant. While there are a number of commercially available inspection/audit systems, there is no system which has universal application, each having to be tailored to the particular circumstances.

Examples of the documentation used in two risk assessment systems are shown in *Figures 31(a)* and *31(b)*. They are used to provide a simple, workable and manageable approach to general risk assessment under MHSWR and COSHH. The assessments are carried out by a small expert

BRITISH SUGAR PLC

COSHH ASSESSMENT AREA: ______ SITE: ______

SUBSTANCES/MAIN COMPONENTS		STOCK CODE:	
TASK/EXPOSURE POINTS	HAZARD:	EH 40 EXP LIMIT:	
FREQUENCY/DURATION OF EXPOSURE		NO. OF PEOPLE EXPOSED:	
CONTROL MEASURES	W. I. REFERENCES:		
MONITORING REQUIRED: YES/NO	RISK: LOW MEDIUM HIGH		
ACTION REQUIRED		RESPONSIBILITY	BY

SIGNATURE:	DATE FOR NEXT ASSESSMENT:

Figure 31.1(b) COSHH assessment form. (Courtesy British Sugar plc)

team involving the safety adviser, production staff and operators. Each site is split into a number of discrete manageable units based either on location or process. All the chemicals and activities in each unit are considered as are foreseeable maintenance activities.

Assessment of exposure of employees should be carried out for each substance used except where they can be grouped into categories having similar generic characteristics. In the first instance, the assessment can be subjective but this should be confirmed by monitoring the exposure under actual working conditions. Where the more hazardous substances are used, this monitoring should be repeated more frequently.

The qualitative risk assessment, whatever the method employed, will have produced a list of actions to be taken. There needs to be a structured and logical system to rank these different actions in a priority order. One method relies on a semi-quantitative approach of assessing the potential risks by looking at the health aspects. It is based on five parameters:

(a) potential of the hazard to harm health,
(b) estimated exposure concentration,
(c) frequency of exposure,
(d) duration of exposure, and
(e) numbers of workpeople exposed.

Each of these parameters is assigned a numerical rating in which low exposure gets a low rating and high exposure gets a high rating. Each action point identified by the risk assessment is analysed for each of the above five parameters and the results summed to give a final total. A similar scheme is described in section 10.1.3. This total gives a measure of the risk factor and provides a relative ranking or priority. It is not an absolute measure but assists in indicating those activities that need to be addressed first.

31.4.3 Other aspects

When carrying out a risk assessment on operations or processes, potential risks to other employees, visitors, contractors, maintenance personnel and also members of the public must not be overlooked.

Similar consideration must be given to environmental aspects of risk such as air pollution, ground pollution, noxious emissions etc., and this should be extended to cover the waste, either solid, liquid or gaseous, that is an integral by-product of every manufacturing process that uses chemicals. The risks posed in this area must be treated with the same level of importance as the risks resulting from the raw material or finished product.

31.5 Risk management

Having determined the hazards of the substances and assessed the risks it is necessary to decide on a course of action. Firstly the objective to be achieved must be established. While nothing can be absolutely and unequivocally safe and totally free from risk, the aim must be to achieve a standard that reduces to a minimum the risks from exposure and at the same time is economically feasible. Thus a balance must be achieved between the risk and the costs of protecting against that risk, see *Edwards* v. *National Coal Board (1949)*[12].

In evolving a strategy the health and safety of the workforce is not the only facet that needs to be considered. A number of faults or failures can occur to the plant which may not expose the workers to risk to their health but could put the plant out of action and so jeopardise future employment – they also must be considered.

Any strategy must accommodate these non-risk faults and failures which can include:

- the flammability of the process material;
- monitoring the state of reactions – particularly in enclosed vessels – to ensure that they remain within specified limits and do not become dangerous;
- the effect of loss of containment, leakage or spillage not only on those working in the plant but the wider air pollution issues that might affect the local community;
- the amount of waste produced by the process – how to minimise it – how to dispose of it; and
- measure to ensure that emissions remain within acceptable environmental limits.

The following sections concentrate on those facets of risk management aimed at preventing or minimising the adverse effects of chemicals on the health and safety of the workforce, namely:

(i) preventing exposure,
(ii) minimising exposure,
(iii) training and information,
(iv) monitoring, and
(v) keeping records.

31.5.1 Preventing exposure

Many extremely hazardous substances can be employed quite safely in manufacturing processes provided that at all times they are contained within the plant and not allowed to escape into the working environment. This assumes automated and bulk handling into storage, thence from storage to process by transfer either in sealed containers or through enclosed and dedicated pipelines. An essential feature of containment is a high level of operating integrity in the plant, instrumentation and control systems. These features in turn demand high standards of plant maintenance.

31.5.2 Minimising exposure

For each substance assessed measures must be developed to ensure that under all foreseeable operating conditions no one, whether employee, visitor, contractor or member of the public, is exposed to any hazardous substance in a concentration that may injure their health. A number of actions are possible and in order of priority they are:

(1) Substitution by a less hazardous substance or by the same substance in a less hazardous form.
(2) Minimise emissions.
(3) Partial enclosure with local exhaust ventilation.
(4) Local exhaust ventilation.
(5) Personal protective equipment.

In each of these cases a number of aspects need to be considered.

(a) Substitution
This is the best and most effective safeguard but it must be realistic in terms of process need. Therefore:

(i) It is necessary to check with the chemist and process controller that the substitute proposed will produce an end-product of the required specification.
(ii) The cost of the substitute must not be such as to make the process uneconomical or commercially unviable.
(iii) The substitute itself must be a safer substance than the original it replaces.
(iv) In the manufacturing process there must be no side effects or daughter products that nullify the benefit of the change.

(b) Minimise emissions
The plant, processes and systems of work should be designed to minimise the generation of hazardous dusts, fumes, micro-organisms, etc. The process must also limit the area of contamination in the event of spills and leaks.

(c) Partial enclosure with local exhaust ventilation
If the process requires access into parts of the plant so as to prevent total containment, then the enclosure of the plant must be to the greatest degree possible and local exhaust ventilation provided with exhaust vented to a safe place to reduce exposure to an absolute minimum.

(d) Local exhaust ventilation
Used where enclosure is not practicable, the design of the capture systems and discharge points must ensure that the greatest amount of the hazardous substance is prevented from escaping into the general workplace and is discharged at a safe height and position outside the normal workplace (see also section 21.4.1.) In some cases the use of chemical scrubbing should be considered.

(e) Personal protective equipment
This is used only as a last resort or in cases of very low risk. It can be very effective provided:

- the equipment is suitable for the hazard to be faced;
- it is acceptable to and effective on the wearer;
- it does not interfere with the functions to be carried out by the person wearing it;
- it is of an approved type; and
- the wearing of it does not generate additional but different hazards.

In addition, the equipment must be properly looked after and maintained in good working conditions. Ideally PPE should be issued to an individual, but if this is not possible then adequate sanitary cleaning must be carried out between each use. (See also section 31.8.3.)

31.5.3 Training and information

Essential to any strategy is the need to provide those handling or using the substances with information about the substances and training in the techniques necessary to minimise the risks faced. Thus training and instruction should be given in:

(a) Characteristics of the substances to be handled, the hazards they present and the action to be taken if exposure or contamination occurs.
(b) The system of work which must be followed to ensure:
   (i) that the substance is correctly used in the process,
   (ii) the safe transfer from store to process plant,
   (iii) the safe handling of the finished product, and
   (iv) safety in the event of a leakage or spillage.
(c) Any special plant or equipment which is to be used, with particular emphasis on recognising deviations from normal operation or performance.
(d) The proper use of PPE, its correct application, cleaning and care, and the reporting of any damage or faults in it.
(e) Understanding the results obtained from monitoring checks – in the quality of the product, on concentrations in the atmosphere and any measured exposures.

31.5.4 Monitoring

Where there is likely to be a health risk from exposure to a substance, the concentrations in the work area should be measured regularly and the likely exposure to all those working there evaluated. In carrying out the evaluations it may be necessary to make adjustments for the effectiveness (or reducing effectiveness) of any PPE that is worn. This is particularly important where the substance involved has been given an MEL. Where monitoring shows an exposure above the OES, corrective action should be initiated to bring actual exposure down to as low a level as possible.

With certain substances that are listed in COSHH there is a requirement that those exposed should have the facility of health surveillance on request.

The condition of the plant that contains hazardous substances should be monitored to ensure that it is being properly maintained, is not deteriorating and is in fact working at optimum efficiency.

31.5.5 Records

Because of the long-term effects that exposure to hazardous substances can have on health, it is essential that adequate records are kept of all those working in areas or on processes where they are likely to be exposed, any exposure or contamination that occurs and the results from monitoring of both the plant and the work area. Finally records of health surveillance must be kept for 5 to 40 years depending on the substance involved and the degree of exposure.

31.6 Legislative requirements

After the Flixborough incident in 1974[13] the HSE appointed a committee of experts, the Advisory Committee on Major Hazards, to consider the health and safety problems posed by major chemical sites and to make recommendations. This they did in three reports which identified a need for three basic elements of control:

(i) Identification of the site.
(ii) The location of the site.
(iii) An assessment of the potential hazards arising.

It had been the intention to implement these three recommendations through a single set of regulations. However, following two major incidents at Seveso and Manfredonia in Italy in 1976, the European Community adopted a directive on major accident hazards. This resulted in the preparation of two slightly conflicting sets of regulations, the Notification of Installations Handling Hazardous Substances Regulations 1982 (NIHHS)[14] and the Control of Industrial Major Accident Hazards Regulations 1984 (CIMAH)[15] with its amendments. In addition, continuing concern about the effects on workers of substances which they have to handle in the course of their work resulted in, amongst other things, COSHH and The Chemical Hazard (Information and Packaging) Regulations 1992 (CHIP).

31.6.1 The Notification of Installations Handling Hazardous Substances Regulations 1982

The main thrust of the regulations is to identify those sites which handle or store more than specified quantities of hazardous substances and it has resulted in the compiling of a central register of all major chemical and other potentially dangerous sites. The regulations required only that the sites be notified and placed no further obligations on the employers. Guidance on the application of these regulations is given in an HSE publication[16].

31.6.2 The Control of Industrial Major Accident Hazards Regulations 1984

These are the principal regulations which govern major hazard sites in the UK and incorporate the requirements contained in the EEC Directive No. 82/501/EEC[17] (often referred to as the Seveso Directive). The regulations place duties on the owners of hazardous sites to demonstrate safe operation and to notify any major accidents which occur. For those activities covered by the various schedules, which list the more dangerous ones, the sites have to prepare a report on their activities and their likely impact on the surrounding area (referred to as the Safety Case). Further they have to prepare an on-site emergency plan and provide information to the local authority who are themselves required to prepare an off-site emergency plan. Finally, information has to be given to the local population who may be affected by the site operations. Guidance on the application of these regulations and on the preparation of emergency plans is given in HSE publications[18,19].

These regulations had a significant impact on the chemical industry in general causing it to be much more overt in its operations. The amendment in 1990 widened the application of the regulations to include aggregated amounts of categories of substances rather than relating only to single substances. Thus if a site contains a total of more than 200 tonnes of various substances, which in themselves were not listed but were classified as toxic, then the site becomes subject to CIMAH.

31.6.3 The Chemical (Information and Packaging) Regulations 1993[2] (CHIP)

This major piece of legislation replaces CPL with the object of increasing the protection of people and the environment from the ill effects of chemicals. It does this by requiring suppliers:

(a) to identify the hazards of the chemicals which they supply,
(b) to give information about the hazards to the people who are supplied, and
(c) to package the chemicals safely.

These are known as the *supply requirements.* A supplier is someone who supplies chemicals as part of a transaction, and includes manufacturers, importers and distributors. Similar duties are imposed on those who consign chemicals for transport by road. These are known as the *carriage requirements.* It should be noted that the term 'chemical' covers pure chemicals, such as ethanol, and also preparations or mixtures of chemicals such as paints and pharmaceutical compounds.

The full package of legislation and supporting guidance for CHIP consists of:

(i) the Regulations[2];
(ii) the Approved Supply List[4];
(iii) the Approved Carriage List;
(iv) a guide to the packaging of explosives for carriage[20];
(v) the Approved Guide to the Classification and Labelling of Dangerous Substances and Preparations for Supply[21], and
(vi) the Approved Code of Practice on the classification and labelling of dangerous substances for carriage by road in tankers, tank containers and packages[22];
(vii) the Approved Code of Practice on the packaging of dangerous substances for carriage by road[23]; and
(viii) the Approved Code of Practice on safety data sheets.

One of the most important requirements of CHIP is for safety data sheets to be provided wherever dangerous chemicals are supplied for use at work. The sheet must contain sufficient information about the chemical to enable the recipient to take the necessary precautions to protect his health. The Approved Code of Practice gives detailed advice on the information to be given in the safety data sheets. While the safety data sheet can be invaluable in helping employers to carry out a COSHH assessment, it is not a substitute for an assessment. The data sheets must describe the hazards of the chemicals but only the user can assess the probability of danger (the risk) arising in the workplace.

If a dangerous chemical is supplied in a package, the package must carry a label informing anyone who handles it or uses the chemical about its hazards and the precautions to be taken. For workpeople the label should supplement the information provided by the employer; for others it is an important way of obtaining precautionary information.

The label on the package must contain details of:

(a) the supplier,
(b) the chemical,
(c) the category of danger, and
(d) appropriate risk phrases and safety phrases.

Risk phrases summarise the main danger of the chemical, e.g. 'may cause cancer' or 'toxic by inhalation'. Safety phrases tell the user what to do, or what not to do, e.g. 'keep away from children' or 'do not empty into drains'. A warning symbol is also required on most labels, e.g. a skull and crossed

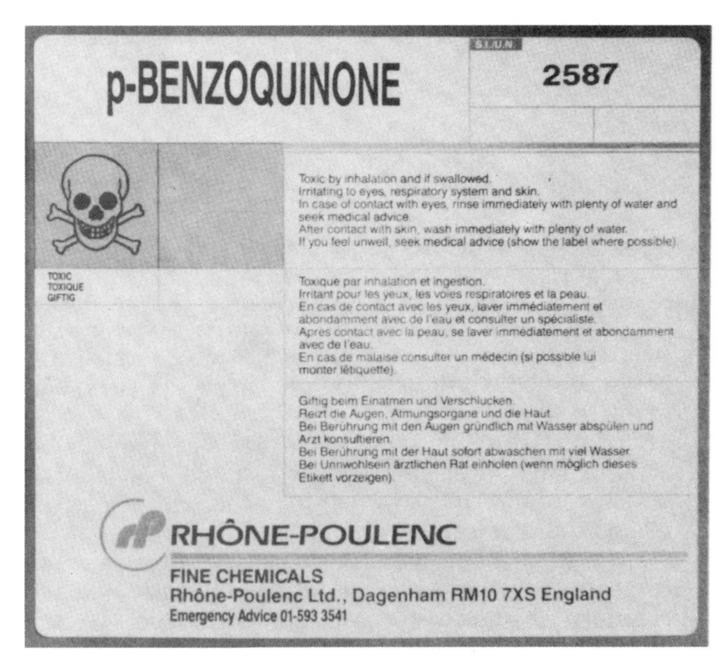

Figure 31.2 An example of a label for hazardous substance package showing warning sign. (Courtesy Ciba-Geigy Plastics)

bones or a picture of an explosion. Suppliers are responsible for ensuring that the label carries the correct information. CHIP also requires labels to be provided to warn those who handle packages of chemicals during transit on a public road. These transit labels are similar to but not as comprehensive as the package labels, concentrating more on the immediate information needed should a vehicle accident occur. If necessary the transit label can be combined with the package label to avoid duplication. An example is shown in *Figure 31.2*.

Finally, CHIP requires chemicals to be packaged safely and in a manner to withstand the foreseeable conditions of supply and carriage.

Many of the requirements of CHIP are similar to those of CPL, but additional items cover requirements from various EC directives.

31.6.4 The Control of Substances Hazardous to Health Regulations 1988

These regulations as amended provide for a complete package of duties on health issues relating to the use, handling and storage of hazardous substances. In this, they place onerous duties on the employer as well as obligations on the employee.

Hazardous substances are defined in Reg. 2 supplemented by reference to those substances that have been given a Maximum Exposure Limit (MEL) and an Occupational Exposure Standard (OES) in Guidance Note EH 40[3].

Under these regulations employers are required to:

(1) Establish which of the substances they use are hazardous, what those hazards are and, if a new substance is proposed, what hazards it presents.
(2) Before using a substance, to assess the potential risk to their employees.
(3) Design control measures to prevent employees being exposed to dangerous concentrations. Regulation 7 allows three priorities of control:

    (i) prevention of exposure;
    (ii) maintaining exposure levels to below the OES; and
    (iii) control of exposure using PPE.

    Employees must not be exposed to levels above the MEL but if this occurs the employer must investigate and take corrective action to prevent a repetition.
(4) Provide suitable control measures and ensure that they are properly used and looked after. Where these measures include plant or equipment, it must be properly maintained, regularly examined and tested and records kept of such tests and examinations.
(5) Where there is a risk that employees are liable to be exposed to dangerous concentrations of a substance, that exposure must be monitored, recorded and the records kept for between 5 and 30 years depending on the circumstances of exposure.
(6) Where certain hazardous substances, listed in Reg. 11, are used, health surveillance should be carried out.

(7) Ensure that employees are:
   (a) given information about the substances being used;
   (b) trained in the hazards those substances present;
   (c) instructed in the procedures and techniques necessary for the safe use of those substances and any precautions that should be taken;
   (d) provided with information on the results of monitoring and health surveillance, and
   (e) given access to the risk assessment.

A major feature of COSHH is the duty to carry out assessments of the health risks associated with hazardous substances. There is no universally accepted approach to assessing risks and each company has to do what is appropriate in the circumstances of their particular operations. An example of an assessment form is shown in *Figure 31.1(b)*. Many companies have found it helpful to have a team carrying out these COSHH assessments; the team consisting of safety adviser with representatives from the chemists, production staff, engineers, operators and safety representatives. The assessment should consider the risks from not only those chemicals that are brought in to be stored, used or worked on in the workplace but also substances that are given off during the process or work activity as well as the finished product, residues, waste, scrap, etc.

In preparing for an assessment, it is vital to discover where and in what circumstances the substances are stored, used, handled, generated, released, etc. It is important to observe people handling the substances as opposed to relying on the assumption that all work is carried out according to instructions. It is also necessary to observe and monitor the effectiveness of the measures taken to control and minimise exposures since these are part of the overall system of work. A knowledge of the effect that the various substances have on the body and of the routes of entry, such as inhalation, ingestion, skin absorption, etc., is important. Measured exposure levels should be compared with published exposure limits. The number of people exposed and the duration of exposure should also be noted.

The assessment should be recorded to document the conditions found and the actions that were judged to be necessary to reduce the risk. If conditions change within the plant or process, the COSHH assessment should be reviewed and may need to be repeated. Advice and guidance on carrying out risk assessments are contained in HSE publications[24–26].

In endeavouring to meet the obligations posed by these regulations, it is helpful to break the work into manageable parts, remembering the less visible people such as cleaners, maintenance staff, storekeepers, etc.

31.6.5 The Environmental Protection Act 1990[27]

This piece of legislation is the most profound development in British environmental law this century and contains many fundamental changes to pollution control regimes that are likely to have far-reaching effects on industry. The Act is split into nine parts with Parts 1, 2 and 3 having the largest implications for industry. It is an enabling Act that specifies broad environmental objectives to be achieved but relies on subsidiary regulations to provide the details of the requirements.

31.6.5.1 Part 1: Integrated pollution control and air pollution control

This part deals largely with the administration of the Act outlining the authorisation of the bodies for enforcing it and the powers vested in them. It considers enforcement in two parts:

(i) Integrated pollution control (IPC) by H.M. Inspectorate of Pollution (HMIP).
(ii) Air pollution control (APC) by local authorities (LAs).

The Environmental Protection (Prescribed Processes and Substances) Regulations 1991 allocate the enforcement of standards between HMIP and LAs. The large technically complex processes and industries are enforced by HMIP while the remainder of premises are enforced by LAs. Advice is provided by some 100 guidance notes produced by HMIP.

IPC deals with the environmental aspects of a process at all stages aiming to achieve the least environmentally damaging combination of options, hence minimising pollution. By contrast, APC deals only with emissions into the atmosphere, although the LAs who enforce it are required to seek and abide by the decisions of HSE on matters affecting human health, and of the National Rivers Authority where discharges to water courses are involved. This is a more limited approach than IPC and relates to processes having a supposedly lower pollution potential. *Figure 31.3* shows a modern scrubbing plant for cleaning vapour discharges.

Owing to the enormity of the task of granting authorisations, the application period has been extended to January 1996. In the interim period, processes must meet current environmental standards. After January 1996 each authorisation will state the standard to be achieved. Failure to meet that standard will attract prosecution with, on conviction, penalties similar to those under HSW.

The conditions in an IPC authorisation cover emissions to water, land and air and extend to the prevention of persistent offensive odours at or beyond the boundary of the premises. The standards demanded by the authorisations will require the application of the *best available techniques not entailing excessive costs* (BATNEEC). This is a new enforcement concept which has yet to be tested in the courts, but it is seen as equivalent to *so far as is reasonably practicable* of HSW.

The Act empowers Inspectors to issue Enforcement and Prohibition Notices which must specify the remedial steps to be taken. There is an appeal procedure against such notices to the Secretary of State who will appoint an arbiter.

Included in this part of the Act is a power to recover the cost of making the authorisations and of enforcement – on the principle of 'the polluter pays'. Fees will be charged for initial applications and for the annual renewal of the authorisations.

31.6.5.2 Part 2: Waste disposal[28–30]

Waste is any substance which constitutes a scrap material, an effluent, other unwanted surplus substance or any substance or article that needs to be disposed of as being broken, worn out, contaminated or otherwise spoiled.

Figure 31.3 Modern vapour-scrubbing plants to ensure compliance with atmospheric emission standards. (Courtesy Rhône-Poulenc-Rorer)

Certain substances can be classified by regulations as *Special Wastes*, e.g. solvents, toxic chemicals, etc.

A novel feature of this Act is the introduction of the concept of a 'duty of care' which is placed on all involved in the production and handling of waste. Five aspects are listed:

(1) to prevent the keeping, treatment or disposal of waste without a licence or in breach of a licence;
(2) to prevent the escape of waste;
(3) to transfer waste only to an authorised person;
(4) to ensure that there is clear information and labelling of the waste; and
(5) to retain documentary evidence.

The major responsibility for ensuring that these duties are fulfilled rests with the creator of the waste.

Thus the producers of waste have an obligation to prevent the escape of waste along the whole length of the disposal chain through the use of reputable disposal companies and registered hauliers. Transporters of waste have to be registered with the County Council for the area in which they are based. Producers must ensure that waste is carried by authorised carriers and disposed of at facilities which have planning permission, are licenced and that the licence allows that type of waste to be deposited. Producers must also satisfy themselves that the waste can be stored securely without loss of containment and that the method of disposal is appropriate. There must be a tight control on the documentation for the movement of the waste to demonstrate fulfillment of the duty of care. Producers must periodically audit each stage of the waste disposal operation. The penalties for non-compliance are similar to those under HSW.

The standards for the keeping and disposal of waste are being significantly tightened with land-fill licence conditions becoming more stringent, usually by requiring better leachate control, methane monitoring, aquifer monitoring, etc. More frequent inspections will be carried out by inspectors of the Environmental Protection Agency which will take over the waste regulation function currently carried out by the County Councils. As with IPC, recovery of the cost of enforcement and administration of waste control is introduced. Regulations may be made to extend the definition of 'special wastes' to encompass many of the known hazardous substances in current use. Other regulations may provide for a list of exemptions from licensing for substances even though they are still classified as special waste.

31.6.5.3 Part 3: Statutory nuisances

A statutory nuisance is the emission of smoke, fumes, gases, dust, steam, smells, other effluvia and noise at a level which is judged to be prejudicial to health or a nuisance to the community or anyone living in it. The legal interpretation is somewhat flexible.

Where a local authority is satisfied that either a statutory nuisance exists or is likely to occur or recur, it can serve an Abatement Notice requiring the abatement, prohibition or restriction of the occurrence or recurrence. The notice will specify the date by which the notice is to be complied with and may also specify the remedial action to be taken. Appeals against the notice can be made within 21 days. Failure to comply with the notice can attract a fine on conviction of up to £20 000. It is a defence to a prosecution to demonstrate that the best practicable means had been used to prevent or counteract the effects of the nuisance. A private individual can start a proceeding in a Magistrates' Court and, if the court is satisfied that a nuisance exists, it can require the accused to abate or terminate it. The court may also specify the remedy to be followed.

31.6.5.4 Summary

Governmental and public pressure is demanding a significant tightening of environmental standards. Much of the onus for achieving this will fall on industry which may be involved in considerable expense in carrying out the work necessary to meet the new standards demanded.

31.7 Plant design

When designing a plant to handle hazardous substances, the main aim must be for containment and process control. There are well established techniques for checking the integrity of containment – by simple gas leakage tests or by monitoring for escape of the substance. In many cases the assessments and monitoring carried out under COSHH will achieve this. However what this checking does not do is to indicate whether the degree of process control is adequate and this raises the question of process safety.

31.7.1 Process safety

This section relates to the safe design of process plant which starts with the chemist developing a chemical reaction or series of reactions that produce the required end-product. Initially the proposed process will be tried out in the laboratory in gram quantities. When the feasibility of the process is proved a pilot-scale plant will be built which will carry out the synthesis in kilogram quantities to establish the manufacturing viability of the process. At this scale of operation the chemist will not only be looking for product purity and quality but also at some of the physical chemistry aspects such as heats of reaction, temperature rises, rate of pressure rises, temperature of first exotherm, etc.

The information gathered from the pilot-plant study is compiled into a process dossier which provides the data for the chemical engineer to use in the design of the full-scale production plant. It is at this stage that the safety adviser should become involved.

The initial design will be in the form of engineering line diagrams which show the vessels, pumps, interconnecting pipework, etc., but not the detailed instrumentation or control systems. Sufficient information should be available to enable a preliminary safety analysis to be carried out drawing on the expertise of the process chemist, process engineer, control systems engineer, production engineer and the safety adviser. The object of this analysis is to identify possible hazards arising from the proposed plant and to define the measures necessary to eliminate or reduce them to an acceptable level. Some of the more obvious hazards to be looked for will be low or high pressures, low and high temperatures and highly exothermic reactions which could cause a runaway reaction with potentially catastrophic results.

In the analysis, each process step is looked at for deviations in temperature, pressure, level, etc., and the possible effects of physical problems such as pipe blockage, valve failure, corrosion, etc. All the details of this analysis should be recorded. On the findings of this analysis, the chemical engineer can refine his design and produce detailed plans including pipework layouts and instrumentation (P and I) diagrams.

Once the detailed plans have been prepared, a more detailed safety analysis can be carried out which could be in the form of a Hazard and Operability Study (HAZOP)[31,32].

A HAZOP study requires a multi-disciplinary approach by a team made up of technical specialists, i.e. chemical engineer, chemist, production manager, instrumentation engineer, safety adviser, etc. It is co-ordinated by a leader who guides the systematic investigation into the effects of the various faults that could occur and their effects. The success of this study depends heavily on the quality of the leader and the positive and constructive attitude of the team members. It is essential that the team have all the basic data plus line diagrams, flow charts etc.

In the study each logical unit of the plant, e.g. a reactor with all its inlets and outlets, is examined in turn and the team leader uses questioning guide words such as 'what happens if there is . . . none, more of, less of, part of, more than, . . . etc.' For example, looking at 'more of' in relation to pressure in a vessel, the possible cause could be a blocked pressure relief vent with the consequence of vessel failure due to excessive pressure building up – the proposed corrective action could be to include high-pressure alarms or duplicate venting. A well conducted HAZOP study should eliminate 80–85% of the major hazards thereby reducing the level of risk in the plant. By carrying out a further study at a later stage when the detailed design is finalised further hazards can often be eliminated.

31.7.2 Plant control systems

Many small simple relatively low hazard plants are fully manually operated, but with increasingly complex plant and especially where increasing hazards, risk, criticality and quality needs are met there is a tendency to introduce automated controls using electronic control systems[33–35]. However, because a plant is computer-controlled does not necessarily make it safer. To achieve optimum levels of safe operation, any computer programmes should be devised jointly by the software specialist and the production staff. All operational requirements must be discussed, otherwise the software designer may make assumptions which could result in faulty or even dangerous operation of the plant. Not only must the programme work satisfactorily during routine operations, it must be designed to accommodate plant failures and any testing or checking during or following maintenance.

Once the computer programme has been written, and before introducing chemicals into the plant, it must be challenged in all possible situations to ensure that it matches operational requirements. Any software changes must be fully described and recorded.

When computer control systems are implemented, operators tend to rely on them completely to the extent that there is a risk that the operators forget how to control the plant. This can be critical in an emergency and it may be prudent to switch the computer off occasionally thereby forcing the operators to control the plant manually (if possible).

A plant should not be provided with too many instruments since this can confuse the operator and prove counter-productive. However sufficient instrumentation is needed to enable the operators to know what is going on inside closed vessels, pipes, pumps, etc.

There is no universal formula for control systems and each production unit must devise a control strategy for each plant based on the hazards, risks and systems of work. A small batch plant consisting of two chemical reactors having a mixture of manual and automatic controls is shown in *Figure 31.4*.

31.7.3 Further safety studies

Having carried out a HAZOP study on the plant and incorporated its findings into the design, it is prudent to carry out a further review during the commissioning period to check that the design modifications have produced the desired results. This is necessary since final details of the physical installation are often left to the installing engineers to decide and these could produce unforeseen hazards. Finally, once the plant is commissioned and operational there should be routine safety checks carried out on a regular basis.

31.7.4 Plant modifications

Plant modifications, even apparently simple ones, can have major consequential effects[13]. It is crucial that the plant is not modified without proper authorisation and the completion of an analysis of possible effects. This can be achieved by the use of a 'process change form' which gives the reasons for the change. Use of such a form also ensures a degree of control on the modifications made, especially if it has to be sanctioned by senior technical specialists such as the process engineer, safety adviser, production manager and maintenance manager. There needs to be clear guidance as to when the process change form has to be used so that there can be no misunderstanding.

31.8 Safe systems of work

Since human beings are necessary in the operation of chemical plants there is the likelihood of errors being made that could result in hazards. This can be reduced to a minimum by introducing formal written systems of work.

Figure 31.4 Two chemical reactors having manual and automatic controls. (Courtesy Rhône-Poulenc-Rorer)

31.8.1 Instructions

There should be detailed operating instructions for every chemical plant. These can be on three levels: the first is for the plant operator giving specific instructions on which valves to turn, how much chemical X to add at which stage in the process, what temperature rises are acceptable, etc. Included should be information on the hazard data of the substances involved and any first aid and emergency procedures. The plant operator must be thoroughly instructed and trained in the operation of the plant – these plant instructions can provide a useful foundation for that training.

The second level of instruction is more detailed and should contain a synopsis of the chemistry and potential problems, such as exotherms, that may be met. It should be aimed at the chemist or engineer in charge of the plant. Finally a process dossier should be provided that gives information about every item of the plant, the substances to be used and how the plant is to be operated and this would be the major reference document for use by the process engineers.

31.8.2 Permits-to-work

A system of permits-to-work is required whenever the work to be carried out is sufficiently hazardous to require strict control of those doing it. This can occur when maintenance and other non-routine work is being carried out in a chemical plant or for any normal operation where the risks faced make clear and unequivocal instruction necessary for the safety of the operators.

The essential elements of a permit-to-work scheme are:

(a) Full explanation to the workpeople involved of the hazards of the plant and the chemicals.
(b) The work to be carried out is properly detailed and understood by both sides.
(c) The area in which the work is to be carried out is clearly identified and made safe, or the hazards are highlighted.
(d) The 'owner' of the area, who must be competent and responsible, must sign the document stating that he is satisfied that any necessary isolation, blanking off, etc., has been completed and it is safe for workmen to enter the area.
(e) The workmen must sign the permit to say that they fully understand the work that is to be carried out, the limitations of access and the hazards and potential risks to be faced.
(f) Any monitoring required before, during and after the work must be specified and the results noted on the document.
(g) When the work is completed, the workmen must sign off the permit to say they have completed the specified work and left the plant in a suitable state to return to operations.
(h) The 'owner' of the area must sign to accept that the work has been completed and that he is now back in charge.

The format of permit-to-work documents will be determined by the particular type of work being undertaken, but all such documents should cover the eight points listed above. Different types of permit will be needed to cover general work, hot work, vessel entry, excavations and high-voltage electrical work. For the scheme to work it is essential that the procedure is properly and fully followed. The responsibility for ensuring this rests with the employer, or the person he appoints to oversee the work. The circumstances under which a permit-to-work scheme is to be used must be fully understood as must the responsibility of each of the participants. This applies to contractors as well as employees and they must be fully instructed in its operation before being allowed to start work.

31.8.3 Protective clothing

In many manufacturing processes, the wearing of protective clothing is necessary to protect the health and safety of the operator. Under the Personal Protective Equipment at Work Regulations 1992[36] (PPER) duties are placed on every employer to provide suitable personal protective equipment (PPE) where the risk to their employees' health and safety has not been adequately controlled by other means. To be effective, PPE must:

(a) be appropriate for the risks involved and the conditions where the exposure to the risk may occur;
(b) take account of ergonomic requirements and the state of health of the person who has to wear it;
(c) be capable of fitting the wearer correctly after adjustment; and
(d) be effective in preventing or adequately controlling the risks without increasing the overall risk.

The type of PPE chosen should be determined by the findings of an assessment of the risks to health and safety which it has not been possible to avoid by other means. The PPE must be maintained in good working order and clean. Suitable accommodation should be provided for storing it when not in use and adequate information, instruction and training must be given to those who are required to wear it.

These Regulations apply to all work situations except those covered by:

(i) The Control of Lead at Work Regulations 1980
(ii) The Ionising Radiations Regulations 1985
(iii) The Control of Asbestos at Work Regulations 1987
(iv) The Control of Substances Hazardous to Health Regulations 1988
(v) The Noise at Work Regulations 1989
(vi) The Construction (Head Protection) Regulations 1989

where types of PPE specific to the risks faced are specified. An example of good protective clothing to provide protection against contact and dust is shown in *Figure 31.5*.

31.9 Transport

Chemicals are transported either in discrete packages or in bulk, often by tanker. Careful selection must be made of the material and construction of

Figure 31.5 Protective clothing to provide protection against contact and dust. (Courtesy Rhône-Poulenc-Rorer)

the package or container: for example, a mild steel road tanker will not last long if used to transport 20% sulphuric acid whereas it is perfectly satisfactory for carrying fuel oil. Where the substance is packaged in small quantities, the packaging and labelling must comply with CHIP. In addition, when the substances are to be carried on the public highway the packages must comply with the Road Traffic (Carriage of Dangerous Substances in Packages, etc.) Regulations 1986 (PGR)[37] and the vehicle must carry a suitable warning sign[38]. These Regulations, backed by an Approved Code of Practice[39] and guidance[40] cover the suitability of the vehicle, the information to be available about the substance being carried and the instruction and training of the drivers. The signs required by the Regulations are the orange rectangles at the front and rear of vehicles transporting more than 500 kilogrammes of dangerous substances.

When chemicals are carried in bulk the requirements for vehicle signing are specified in the Dangerous Substances (Conveyance by Road in Road Tankers and Containers) Regulations 1981[41] which are supported by Approved Codes of Practice[42,43] and guidance notes[44,45].

Both these Regulations for transport in packages and in bulk have similar requirements for the proper stowage of the load, adequate instruction and training for drivers, information on the load being carried and proper labelling of the vehicles. A review of the safe handling of chemicals and dangerous substances in road transport operations is contained in a chemical industry publication[46].

31.10 Storage

The HSE has issued a number of guidance notes on the storage of specific substances such as chlorine[47], LPG[48], highly flammable liquids[49], etc. Further guidance can be obtained from the distributive trade association[50] which offers practical advice on standards of storage.

Figure 31.6 shows a storage tank for non-flammable liquids.

An overriding principle in the storage of chemicals is that they should not be adversely affected by other adjacent substances or operations. An HSE note gives useful guidance on this[51].

31.10.1 External storage of flammable liquids

A common problem met in factories using chemicals is where to store flammable solvents. Whether in drums or in bulk, outside storage has the advantages that it:

(a) allows the dispersion of any flammable vapours from vents, leaks, spillages, etc.,
(b) minimises the sources of ignition, and
(c) provides secure storage to minimise damage to containers.

External storage of chemicals and hazardous substances requires particular facilities which for drums include:

(a) an impervious hard standing, i.e. a concrete plinth,
(b) spillage retention facilities of a bund with drainage to a special settlement sump,
(c) a separation of at least 4 metres from boundaries, buildings and sources of ignition, and
(d) a ramp to allow trolleys and fork lift trucks easy access for handling the drums.

In the case of bulk storage, the tank should be mounted above ground level and be provided with:

(a) A contents gauge arranged for local or remote reading.
(b) A pressure/vacuum relief valve, or vent to allow the tank to breathe during filling and emptying.

Figure 31.6 Storage tank for non-flammable liquids. (Courtesy Rhône-Poulenc-Rorer)

(c) A bund capable of retaining 110% of the tank capacity.
(d) Secure attachment to the ground to prevent movement due to high winds.
(e) Adequate separation distances from other tanks and from sources of ignition.
(f) Fixed installation for fire fighting which can be sprinkler, drench or foam.

31.10.2 Warehousing

The first step when chemicals and substances are to be stored is to know what the substances are and their characteristics. This will enable storage to be arranged so that for incompatible substances there is either a suitable separation distance, an intermediate fire wall or a mutually compatible substance between them to provide separation. The effects of storing incompatible chemicals in the same area have been described in HSE reports[52,53].

Fire detection and sprinkler systems are beneficial but if used with high bay racking consideration should be given to incorporating inter-rack sprinklers which counteract the chimney effect of a fire. Where sprinklers are fitted care should be taken that no water-reactive chemicals are stored. The advice of the local fire prevention officers, sprinkler suppliers, insurance surveyors and safety advisers should be sought on storage and rack layouts and on the arrangement of the sprinklers.

Following an incident at Basle, Switzerland, in which contaminated fire fighting water from a warehouse fire entered the Rhine and caused severe ecological damage, the chemical industry and storage companies have been studying better methods of run-off water control. Their recommendations are published in a booklet[50].

31.11 Laboratories

The use of chemicals in laboratories poses totally different problems from those met in production facilities. The scale is much smaller, the equipment is generally more fragile and while the standard of containment for bench work is often less, the skill and knowledge of the workforce is very high.

Work in quality control laboratories is normally repetitive using closely defined analytical methods. Research laboratories are far wider in the scope of the reactions they investigate and the equipment they use, sometimes dealing with unknown hazards. The principal hazards met in laboratories are fire, explosion, corrosive and toxic attacks. A limit should be specified for the total amount of flammables allowed in a laboratory at any one time. A useful guide could be the HFL Regulations which permit only enough for the day's work.

Hazardous and potentially hazardous reactions should be carried out in a fume cupboard. The effectiveness of extraction in the fume cupboard should be checked regularly. A well ordered and tidy fume cupboard is shown in *Figure 31.7*. These cupboards should not be used for extra storage space.

An effective way to improve safety in laboratories is to insist on a high standard of housekeeping and a minimising of the inventory of chemicals in use. Scientists are notorious for hoarding samples for analysis at 'some time in the future'. Laboratory safety is a very wide subject and there are a number of publications giving excellent guidance[54–56]. Many of the larger companies produce their own practical guidance and are pleased to supply copies.

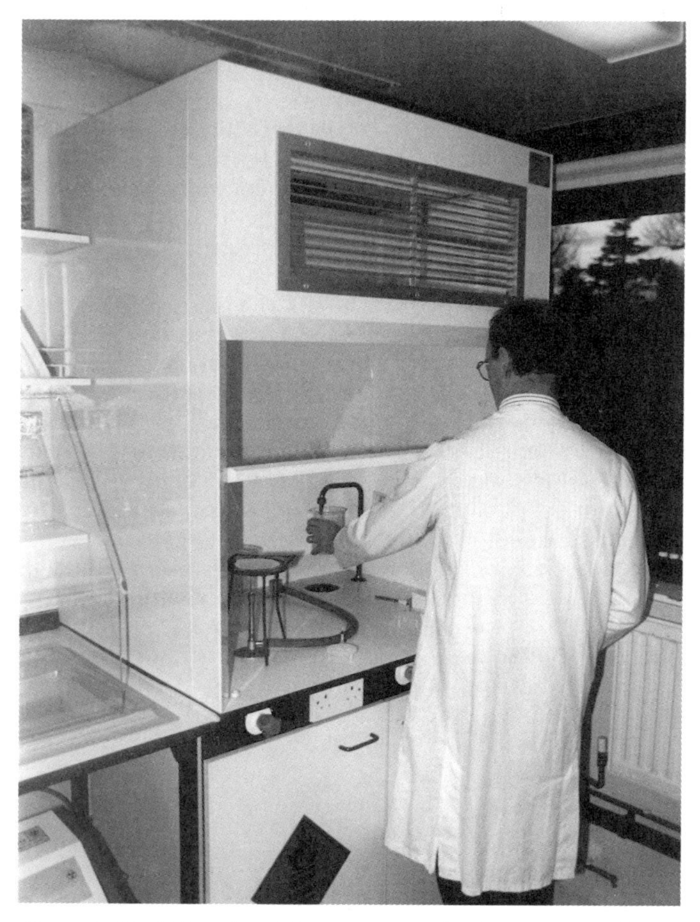

Figure 31.7 Well ordered fume cupboard. (Courtesy British Sugar plc)

31.12 Emergency procedures

The Management of Health and Safety at Work Regulations 1992 (MHSWR) imposes on employers an explicit duty to have in place effective procedures to be followed in the event of serious or imminent danger to people at work. The CIMAH Regulations also require affected manufacturers to prepare on-site emergency plans. In addition, CIMAH requires employers to co-operate with the local authority in developing off-site emergency plans. Irrespective of these statutory requirements it is prudent for every user and storer of hazardous substances to prepare an emergency plan to cover all reasonably foreseeable events such as fire, major spillage or toxic release. The plans can be at two levels, one for the immediate production or storage area and the second for the site as a whole taking account of the likely effects on the local community.

It is very important that employees and the local emergency services know exactly and unambiguously what to do should an incident occur. The

Dangerous Substances (Notification and Marking of Sites) Regulations 1990 require that the entrances to sites are labelled such that the emergency services have pre-warning that there are hazardous chemicals on site. Additionally, the Planning (Hazardous Substances) Regulations 1992 require notification to the local authority of the amounts of hazardous substances held on site. A clear drawing or sketch showing the layout of the site should be available for the emergency services. It should also contain details of the buildings and highlight fire extinguishers, emergency exits, spillage control equipment, etc. All employees should be properly instructed, fully trained and rehearsed in those emergency plans. The local emergency services should be encouraged to familiarise themselves with the site.

Where there is a potential for a major emergency, which would involve the local emergency services and local authority, there must be an agreed plan of action to co-ordinate all the services including managers and employees on the site with their specialised knowledge of the site and its processes. The emergency plans should include a list of emergency contacts including such bodies as the local authority, the water authority, the factory inspectorate, the pollution inspectorate, the police, etc.

Advice and guidance on preparing emergency plans are contained in publications by The Society of Industrial Emergency Services Officers (SIESO)[57] and the Chemical Industries Association (CIA)[58]. It must be emphasised that all emergency plans must be regularly practised and reviewed – there is no substitute for actually doing it!

31.13 Conclusions

This chapter has summarised very briefly some of the health and safety problems faced when using chemicals. There are many laws and a great deal of guidance available to assist in ensuring high standards of protection of the health and safety of all those who may be affected by operations using chemicals. Those with responsibilities in this area are recommended to study the laws and guidance when implementing precautionary measures.

References

1. *The Control of Substances Hazardous to Health Regulations 1988* (as amended by SI 1991 No. 2341 and SI 1992 No. 2382), HMSO, London
2. *The Chemical (Hazard Information and Packaging) Regulations 1993*, HMSO, London (1993)
3. Health and Safety Executive, *Guidance Note EH 40, Occupational Exposure Limits*, HSE Books, Sudbury (1993 but updated annually)
4. Health and Safety Executive, *Approved Supply, Information approved for the classification and labelling of dangerous substances and preparations for supply*, HSE Books, Sudbury (1993)
5. Health and Safety Executive, *Guidance Booklet No. HS(G)27, Substances for use at work: the provision of information*, HSE Books, Sudbury (1989)
6. Sax, N.I., *Dangerous Properties of Industrial Materials*, 7th edn, Van Nostrand Reinhold (1989)
7. *The Merck Index*, 10th edn, Merck & Co. Inc. (1983)
8. Bretherick, L., *Handbook of Reactive Chemical Hazards*, Butterworth, London (1979)

9. Pape, R.P. and Nussey, C., 'A basic approach for the analysis of risks from major toxic hazards', *Assessment and Control of Major Hazards,* I. Chem. E. Symposium, Manchester (April 1985)
10. Jagger, S., 'Development of CRUNCH: A dispersion model for continuous releases of a denser-than-air vapour into the atmosphere', SRD., UKAEA Report R229 (1983)
11. Davies, P. and Hymes, I., 'Chlorine toxicity criteria for hazard assessment', *The Chemical Engineer,* 30 (June 1985)
12. Edwards *v.* National Coal Board [1949] 1 KB 704
13. Health and Safety Executive, *Investigation report: Flixborough disaster, [1974],* HSE Books, Sudbury (1975)
14. Health and Safety Executive, *The Notification of Installations Handling Hazardous Substances Regulations 1982,* HMSO, London
15. Health and Safety Executive, *The Control of Industrial Major Accident Hazards Regulations 1984,* HMSO, London
16. Health and Safety Executive, *Regulation Booklet No. HS(R)16, Guide to the Notification of Installations Handling Hazardous Substances Regulations 1982,* HMSO, London
17. European Commission, *Directive No. 82/501/EEC, Council Directive on Major Accident Hazards of Certain Industrial Activities,* European Commission, Luxembourg (1982)
18. Health and Safety Executive, *Regulation Booklet No. HS(R)21, Guide to the Control of Industrial Major Accident Hazards Regulations 1984,* HMSO, London (1985)
19. Health and Safety Executive, *Guidance Booklet No. HS(G)25, Control of Industrial Major Accident Hazards Regulations 1984: further guidance on emergency plans,* HMSO, London (1985)
20. Health and Safety Executive, *Legislation Booklet No. L13, A Guide to the Packaging of Explosives for Carriage Regulations 1991,* HSE Books, Sudbury (1991)
21. Health and Safety Executive, *Approved guide to the classification and labelling of substances and preparations dangerous for supply,* HSE Books, Sudbury (1988)
22. Health and Safety Executive, *Approved Code of Practice No. COP 19, Classification and labelling of dangerous substances for conveyance by road in tankers, tank containers and packages,* HSE Books, Sudbury (1988)
23. Health and Safety Executive, *Approved Code of Practice No. COP 40, Packaging and labelling of dangerous substances for conveyance by road,* (revision 2), HSE Books, Sudbury (1990)
24. Health and Safety Executive, *COSHH Assessment: A step by step guide to assessment and the skills needed for it,* HSE Books, Sudbury (1988)
25. Health and Safety Executive, *Legislation Booklet No. L5, Control of substances hazardous to health and control of carcinogenic substances. Approved Codes of Practice,* 3rd edn, HSE Books, Sudbury (1992)
26. Health and Safety Executive, *Approved Code of Practice No. COP 30, Control of substances hazardous to health in fumigation operations,* HSE Books, Sudbury (1988)
27. *The Environmental Protection Act 1990,* HMSO, London
28. *The Control of Pollution Act 1974* (as amended by *The Environmental Protection Act 1990*), HMSO, London (1990)
29. Health and Safety Executive, *The Control of Pollution (Special Waste) Regulations 1980,* HMSO, London (1980)
30. Health and Safety Executive, *The Control of Pollution (Supply and Use of Injurious Substances) Regulations 1980,* HMSO, London (1980)
31. Kletz, T.A., *Hazop & Hazan: Notes on the identification and assessment of hazards,* The Institution of Chemical Engineers, Rugby (1986)
32. *A guide to hazard and operability studies,* Chemical Industries Association, London (1987)
33. Health and Safety Executive, *Occasional Paper No. OP2, Microprocessors in industry: Safety implications of the uses of programmable electronic systems in factories,* HSE Books, Sudbury (1981)
34. Health and Safety Executive, *Programmable electronic systems in safety related applications: No. 1, An introductory guide,* HSE Books, Sudbury (1987)
35. Health and Safety Executive, *Programmable electronic systems in safety related applications: No. 2, General technical guidelines,* HSE Books, Sudbury (1987)
36. Health and Safety Executive, *The Personal Protective Equipment at Work Regulations 1992,* HMSO, London (1992)
37. *The Road Traffic (Carriage of Dangerous Substances in Packages, etc.) Regulations 1986,*

amended by the Road Traffic (Carriage of Dangerous Substances in Packages, etc.) (Amendment) Regulations 1989, HMSO, London (1989)
38. *Approved substance identification numbers, emergency action codes and classifications for dangerous substances conveyed in road tankers and tank containers*, 3rd edn, HMSO, London (1989)
39. Health and Safety Executive, *Approved Code of Practice: Operational Provisions of the Road Traffic (Carriage of Dangerous Substances in Packages, etc.) Regulations 1986*, HSE Books, Sudbury
40. Health and Safety Executive, *Regulations Booklet No. HS(R)24, Guide to the Road Traffic (Carriage of Dangerous Substances in Packages, etc.) Regulations, 1986*, HSE Books, Sudbury (1987)
41. Health and Safety Executive, *The Dangerous Substances (Conveyance by Road in Road Tankers and Tank Containers) Regulations 1981*, HMSO, London (1981)
42. Health and Safety Executive, *Approved Code of Practice No. COP 11, Operational provisions of the Dangerous Substances (Conveyance by road in road tankers and tank containers) Regulations 1981*, HSE Books, Sudbury (1981)
43. Health and Safety Executive, *Approved Code of Practice No. COP 14, Road tanker testing: examination, testing and certification of the carrying tanks of road tankers and of tank containers used for the conveyance of dangerous substances by road*, HSE Books, Sudbury (1981)
44. Health and Safety Executive, *Regulations Booklet No. HS(R)13, Guide to the Dangerous Substances (Conveyance by Road in Road Tankers and Tank Containers) Regulations 1981*, HSE Books, Sudbury (1981)
45. Health and Safety Executive, *Guidance Booklet No. HS(G)26, Transport of dangerous substances in tank containers*, HSE Books, Sudbury (1986)
46. 'Hauliers Safety Audit' booklet, Chemical Industries Association, London (1986)
47. Health and Safety Executive, *Guidance Booklet No. HS(G)40, Chlorine from drums and cylinders*, HSE Books, Sudbury (1987)
48. Health and Safety Executive, *Guidance Note, Chemical series No. CS 4, Keeping of LPG in cylinders and similar containers*, HSE Books, Sudbury (1986)
49. Health and Safety Executive, *Guidance Note, Chemical Series No. CS 2, Storage of highly flammable liquids*, HSE Books, Sudbury (1977)
50. British Distributors and Traders Association, *Warehousing of Chemicals Guide*, London (1988)
51. Health and Safety Executive, *Guidance Note, Chemical series No. CS 17, Storage of packaged dangerous goods*, HSE Books, Sudbury (1986)
52. Health and Safety Executive, *Investigation report* [unnumbered], *Fire and explosion at B and R Hauliers, Salford, 25th September 1982*, (ISBN 0 11 883702 8), HSE Books, Sudbury (1983)
53. Health and Safety Executive, *Investigation report* [unnumbered], *Fire and explosions at Cory's Warehouse, Toller Road, Ipswich, 14th October 1982*, (ISBN 0 11 883785 0), HSE Books, Sudbury (1985)
54. Bretherick, L., *Hazards in the Chemical Laboratory*, 4th edn, The Royal Society of Chemistry, London (1986)
55. Weston, R., *Laboratory Safety Audits and Inspections*, Institute of Science Technology, London (1982)
56. *Safe Practices in Chemical Laboratories*, The Royal Society of Chemistry, London (1989)
57. The Society of Industrial Emergency Services Officers, *Guide to Emergency Planning*, Paramount Publishing Ltd., Borehamwood (1986)
58. The Chemical Industries Association, *Recommended Procedures for Handling Major Emergencies*, London (1976)

Endnotes

The HSE has published the following guidance on CHIP:

Approved carriage list: Information approved for the classification, packaging and labelling of dangerous substances for carriage by road.

Approved supply list: Information approved for the classification and labelling of substances and preparations dangerous for supply.

Chip guides:
No. 1. The complete idiot's guide to CHIP.
No. 2. How to classify substances.
No. 3. How to classify preparations, part I.
No. 4. How to classify preparations, part II.
No. 5. Supply labelling.

Appendices

Appendix 1

The Institution of Occupational Safety and Health

The Institution of Occupational Safety and Health (IOSH) is the leading professional body in the United Kingdom concerned solely with matters of safety and health. Its growth in recent years reflects the increasing importance attached by employers to safety and health for all at work. The Institution provides a focal point for safety practitioners in the setting of professional standards, their career development and for the exchange of technical experiences, opinions and views.

Increasingly employers are demanding a high level of professional competence in their safety advisers, calling for them to hold recognised qualifications and have a wide range of practical expertise. These are evidenced by Corporate Membership of the Institution for which proof of a satisfactory level of academic knowledge of the subject reinforced by a number of years of practical experience in the field is required.

The recognised academic qualification is that provided by successful completion of the Diploma Examination of the National Examination Board in Occupational Safety and Health (NEBOSH).

For those whose day-to-day responsibilities include a safety function but who do not wish to acquire a professional level of qualification in safety and health, obtaining a NEBOSH National General Certificate will provide a broad but basic knowledge of the subject and will allow them to apply for the non-corporate Associate grade of membership.

Further details of requirements for membership can be obtained from the Institution.

Appendix 2

The National Examination Board in Occupational Safety and Health (NEBOSH)

The Board is an independent body which is recognised in the UK as setting, through its Diploma level syllabus and examinations, suitable academic standards for those who wish to become qualified practitioners in the field of health and safety.

Additionally, it caters for those managers, supervisors and part-time safety advisers who have a health and safety function as one of their day-to-day responsibilities by providing a Certificate examination which ensures an adequate basic knowledge of the subject.

The Board publishes comprehensive guides to each of its examinations. These include the full syllabus, intended learning outcomes, and further essential information for candidates and their tutors. Copies of these guides are available from NEBOSH. The syllabi are copyright to NEBOSH and summaries of them are reproduced below by kind permission of the Board.

Because of rapid changes that take place in safety and health laws and practices, the syllabi are kept under constant review and candidates for the examinations are advised to refer to the current NEBOSH guide before undertaking a course of study.

Outline syllabus for the National Diploma in Occupational Safety and Health

The Diploma is a vocational qualification for all those who aspire to responsible positions as safety practitioners in the general field of occupational safety and health. The syllabus, with its case study, has been designed to develop the ability of the candidate to draw on a broad spectrum of knowledge when tackling safety and health problems met in working life. It is intended to equip them to:

- be able to identify the full range of hazards present in the workplace;
- understand the principles and practices for the control of risks and to apply those principles to specific problems;
- appreciate the implications of the social, economic and legal context in which safety and health problems have to be tackled.

The examination is divided into modules:

Module A: Risk management
Module B: Health and safety law
Module C: Occupational health and hygiene
Module D: Safety technology
Module E: Case study

Successful completion of the examination in each of the five modules is necessary before the Diploma will be awarded. The Diploma meets the academic requirements for Corporate Membership of the Institution of Occupational Safety and Health.

Module A: Risk Management

This module is intended to give the candidate the ability to control hazards, particularly to individuals, based on:

identification of the nature of the hazard;
evaluation of the magnitude of the risk;
selection of appropriate methods for eliminating or reducing the hazard and minimising the consequences if such events occur;
evaluation of the success or failure of safety efforts; and
an awareness of their own limitations, and a knowledge of the sources of information and assistance which are available.

The module is concerned with the techniques which can be applied for the control of all the main types of risk: and in particular with methods of general application for improving systems and human reliability. The module is also concerned with the ways in which changes in organisational structure and behaviour may be used to improve the control of work-based risks. Within the module there is emphasis on the integration of safety and health with all management functions of the organisation. Candidates should be aware that safety has to be managed and is, therefore, amenable to management techniques such as the setting of objectives, the allocation of resources, implementation of techniques of control and accountability.
The candidate will be expected to succeed through a knowledge and understanding of:

1 *Identification of Hazards to People and Evaluation of Risk*

a Definitions of hazard, risk, danger.

b Main types of hazard (immediate physical injury, long-term physical injury, long-term damage to health). Models of accident causation: sequence of events leading to injuries or dangerous occurrences.

c Data on injuries and ill-health
Use of data on actual injuries and ill-health to identify hazards and evaluate size of risk.
Problems in obtaining data on ill-health of people outside the workplace.
Information available from national, industry and insurance company sources.
Information available within an organisation: systems for recording, storing, retrieving, and analysing data: significance of data on sick absence, damage accidents and product complaints.
Analysis of data by standard statistical techniques: patterns of random events: confidence limits: frequency and incidence rates.

d Analysis of potential hazards
Job safety analysis: workplace inspection procedures: safety sampling.
Investigation and analysis of hazardous incidents.
Probabilistic risk assessment, data sources, systematic risk identification methodologies (e.g. HAZOPS): assessment methodologies (fault trees and event trees): estimation of consequences.
Toxicology and epidemiology in identification of long-term health and environmental hazards.
Analysis of hazards at the planning and design stage.

e Risks from human failure
Contribution of behavioural science to accident control.
Limitations of human performance (intelligence, perception, memory, expectancy): main types of human error.
Degradation of human performance (length and distribution of periods at work: stress: psychological effects of the physical environment).
Variations in individual performance (age, experience, personality, stresses outside work, ethnic and cultural background).

f Cost-benefit analysis
Methods of evaluating the costs of injuries and ill-health.
Estimation of the costs of injury or ill-health to the sufferer: to the employer: to society.

Methodological problems: 'cost of life': costing of hazardous incidents, damage accidents and product failures.
Costing of control measures: hardware: system changes.

2 *Perception of Risk and Responses to Risk*

a Reactions of individuals to risk to themselves.
Process of recognition of danger and its consequences: attention mechanisms.
Effectiveness of verbal, written and pictorial warnings.
Reaction to perceived risk: stimulation/acceptance/aversion: factors affecting this.
Ability of individuals to estimate the degree of risk.
Economic factors affecting acceptance of risk.
Revealed and expressed preferences.
Effects of group responses and of expected behaviour.

b Reaction of society to risk
Ability of people to estimate or rank different types of risk.
Relation between perceived risk and acceptability.
Attitude to risks involving catastrophic events: and to long-term health risks compared with sudden death.
Reactions of the media, governments and local communities to risk. Variations between communities.

3 *Methods of Controlling controlling Risk*

a Elimination of hazards and control of risks.
Designing-out hazards and minimising specific risks at the planning or design stage or later by process change: by equipment change: or by changes in the system of work. Problems in estimating new risks introduced by such changes.

b Improvements in systems reliability
Systemic approaches to complex problems. Systems analysis and design: reasons for failure. Holistic and reductionist approaches to analyses of complex failure.
Basic reliability theory: maintainability: switching and control elements: methods of estimating failure probability and implications for safety.
Quality assurance techniques. British and international standards. Product and service reliability.
Maintenance systems: planned maintenance: asset register.

c Human-systems interface
Fundamentals of interface design: principles of ergonomics.
Stress conditions related to poor machine, equipment or system design.
Design of control equipment and systems to suit individual ability as an information processor and controller.

d Improvements in human reliability
Techniques for assessing and predicting human reliability.
Selection and initial and continuing training of workers to reduce the risk of injuries from human failure: identification of training needs: individual and group training techniques.
Provision of information on hazards and reasons for appropriate systems of work.
Definition and design of messages: choice and use of media.
Control of human behaviour by rules. Statutory requirements and their enforcement by the employer.
Organisational rules and their enforcement.
Permit to work systems: conditions for effectiveness.
Problems arising from changes in process, equipment and systems of work and legal requirements. Methods of overcoming resistance to change.

e Safety organisation: reduction in risk
Safety policies: conditions for effectiveness in an organisation: main elements of such policies: differences related to size of organisation, to management style and technology in use.

Role and function of management: difference between role and function: organisational control systems: vertical/lateral managements: effect of payment and reward systems: interaction of health and safety goals with other management goals.
Essential features of good safety organisation: relationship of safety professionals to line management and to the workforce. Potential role conflicts and their resolution.
Role of the organised workforce. Safety representatives and their effectiveness. Safety Committees. Joint consultation. Bargaining, negotiation, consultation and participation in health and safety matters.
Problems in relations with contracting firms and individuals.
External pressures on an organisation and its response:
Official enforcement agencies. Relations between Inspectors and management, workers and safety professionals. Healthy and unhealthy types of relationship.
Industrial Tribunals.
From contractual relationships with purchasers of goods and services, with suppliers, with main contractors in construction activities.
Trade Unions (as opposed to the workforce and their representatives).
Insurance companies: financial and technical pressures.

f Safety organisation: minimising consequences
Arrangements for reducing the risk of injury and minimising the consequences of dangerous occurrences.
Arrangements for maintaining an orderly system of activity in the organisation: first aid and medical facilities: emergency use of protective equipment: internal emergency procedures, fire and emergency drills and evacuation procedures.
Emergency planning for events which could affect the general public. Liaison with the fire service, the police, local authorities and government agencies. Works emergency plans.
Communication and control systems. Testing the system.
Arrangements for reducing the cost to the organisation of injuries and damage. Insurance as a tool of risk management.
Investigation and analysis of hazardous incidents whether involving injury or not.
Principles of such investigations.

4 *Assessment of Success/Failure*
Analysis of statistics of injuries, ill-health, and hazardous incidents: identification of significant trends.
Monitoring the environment (internal or external) and assessment of improvement and/or deterioration.
Setting of objectives in specific terms and evaluation of performance.
Monitoring and testing systems including safety audits: principles of design and requirements for effectiveness.
Data on total costs of injuries and damage accidents: assessment of trends. Systems of loss control.
Review of the effectiveness of systems of evaluation and control.

Module B: Health and Safety Law

This module is concerned with the general structure of the law as it applies to health and safety at the workplace. It is intended to give candidates a detailed knowledge of the Health and Safety at Work, etc. Act 1974 and of the arrangements for enforcing its provisions through the courts and industrial tribunals. It is also intended to provide an understanding of the structure of subordinate legislation made under the Act, and of the status of earlier legislation such as the Factories Act 1961 and the Offices, Shops and Railway Premises Act 1963. An understanding of Regulations and Approved Codes of Practice is gained by a detailed study of the enactments and codes themselves.
This module is also intended to provide candidates with an understanding of the main features of other safety and health legislation, such as that concerned with the supply of products to consumers and pollution controls, as well as a background knowledge of other legislation that has to be taken into account when operating safety and health policies in any organisation.
The candidate will be expected to understand:

1 *Judiciary System*

Main types of court likely to deal with safety and health issues in England and Wales, Scotland and Northern Ireland.
Relationship between such courts including appeal procedures.
Main elements of procedure in such courts: identity of court personnel.
Industrial Tribunals: composition, function and powers: appeal procedures: relationship to Courts: procedures in Tribunals.
European Community Courts: function and powers: relationship to national courts.
Nature and use of precedent in legal actions.
Legal terminology and its use in courts and Acts of Parliament.

2 *Civil Liability*

Nature and development of common law and its main features in relation to safety and health matters.
'Common duty of care': main features as developed by courts and as affected by statute law: application to safety and health in important relationships –
employer/employee
manufacturers – suppliers/users – consumers
occupiers of premises/workers – visitors – others
employers – employees in relation to nuisance and trespass.
Civil actions for damages in cases of injury: negligence: assessment of damages: defences.
'No fault' liability systems: advantages and disadvantages.

3 *Statute Law and European Community Legislation*

Procedures for passing new health and safety legislation: White Papers: Green Papers: Consultative Documents: Stages of Parliamentary procedure.
Detailed legislation: procedure for approval and coming into force: amendment of existing Statutory Provisions by regulations under the Health and Safety at Work etc. Act.
European Community Legislation: main elements of procedure with examples from safety and health Directives. Enforcement in UK.
Enforcement of health and safety legislation. Prosecutions of bodies corporate and individuals: enforcement authorities and enforcement offices – duties and powers: rights of accused persons: effects of HSWA Sections 17 and 40 on prosecution procedures. Enforcement notices: under HSWA: under other health and safety legislation.

4 *Health and Safety at Work etc Act 1974 and Associated Legislation*

Persons and activities covered by the Act: those not so covered: activities formally covered but which the Health and Safety Commission has been directed to leave to other legislation.
General duties set out in Sections 2–9. Meaning of 'so far as is reasonably practicable'. Relationship between the general duties and specific requirements.
Relevant statutory provisions. Status of earlier health and safety legislation listed in Schedule 1. Extant provisions of the Factories Act 1961; Offices, Shops and Railways Premises Act 1963 and other Acts applicable to the employer's premises/business.
Regulations made under Section 15. Procedures by the Commission in preparing proposals to Ministers. What may be included in Regulations.
Approved Codes of Practice issued under Section 16. Procedures for drafting and approval. Approval of other documents (e.g. British Standard Specifications). Distinction between Approved Codes of Practice and other documents called 'Codes of Practice' and Guidance Notes issued by the Commission or Executive.
Understanding the structure and application of Regulations and associated Approved Codes of Practice by study of selected examples, viz –
Safety Representatives and Safety Committees Regulations:
Reporting of Injuries, Diseases and Dangerous Occurrences Regulations:
Classification, Packaging and Labelling of Dangerous Substances Regulations:
Protection of Eyes Regulations:
Abrasive Wheels Regulations.
Enforcement authorities authorised under Section 18. Division between local authorities and HSE. Other enforcement authorities for specific provisions. Administrative division of enforcement between different Inspectorates within HSE.

Inspectors appointed under the Act: their powers and status including rights as agents to enforce legislation other than HSW and its relevant statutory provisions.
Improvement and Prohibition Notices (Sections 21–24). Circumstances in which they can be issued. Appeals procedure. Results of appeals.
Prosecutions under the Act. Penalties on conviction. Prosecution of Directors and Managers: of people failing to take reasonable care for their own safety.
Sources of information about new legal requirements and guidance and decisions in important legal cases.

5 *Other Health and Safety Legislation*
Fire Safety: Main provisions of the Fire Precautions Act 1971. Responsibilities of Fire Authorities. Fire Certificates. Responsibilities concerning fire hazards which remain with HSE (Fire Certificates (Special Premises) Regulations 1976). Fire prevention legislation. (See also Module D, Safety Technology: Fire)
Consumer Safety: Main provisions of the Consumer Protection Act 1987. General duty to supply safe goods. Relationship to HSW. Consumer products subject to special Regulations. Safety of Gas Appliances. Enforcement Procedures. Banning of imports. Prohibition of sale of specific products. Product Liability.
Control by central, regional and local authorities in aspects of:
- offensive trades
- statutory nuisances
- atmospheric pollution
- noise
- water pollution
- industrial hazardous wastes
- hygiene in catering establishments and food shops
- pest control.

Dangerous Substances: manufacture, use, storage, labelling, transportation. Statutory controls.
Safety of Buildings: Building Control Regulations in respect of new buildings. Control of dangerous structures under Public Health Act.
Freight Transport: Licensing of fleet operators, managers and drivers of road freight vehicles: use of works vehicles on public roads: weight limits and distribution of loads.

6 *Other Legislation bearing on the Work of Safety Practitioners*
Legislation which may subsequently influence or constrain the application of health and safety policies in an organisation.
Industrial Relations Legislation: status of trade union officials (shop stewards): Contracts of Employment: unfair/wrongful dismissal: disciplinary arrangements.
Anti-discrimination legislation (sex and race): relevance to selection and training; warning notices; protective equipment. Special protection for women in terms of reproductive cycle.
Law of Contract: as it relates to relations between employers and employees: suppliers and customers: owners/occupiers of premises and visitors: unfair contract terms in relation to civil and statutory liability for possible injury.
Insurance requirements: compulsory insurance for Employers Liability: third party insurance against road accidents: NI sickness, injury and disease benefits.

7 *Legislation about Provision of Information*
Provisions either requiring information to be provided or requiring information to be kept confidential.
Provisions of the HSW. Information on manufacturers, importers and suppliers of goods (Section 6): Section 20(2)(j): Sections 27–28: rights of safety representatives: provision of information –
(1) by operators of hazardous installations;
(2) before placing new products on the market.
Disclosure for purposes of civil liability proceedings.
Confidentiality of individual medical reports: trade secrets: Law of defamation: privilege.
Disclosure of personal details held on computers.
Access to papers of local authorities.

8 *Case Law*
Identity and background of certain important cases applicable to the preceding topics.

Module C: Occupational Health and Hygiene

This module is intended to equip the candidate to identify and control the main types of specific hazard which are likely to result in impairment of the health and well being of workers (and in some cases the general public) arising from work activities. General principles are illuminated by a study of a limited range of typical important chemical, physical and biological hazards plus the special legal requirements relating to the specific hazards. The candidate will be expected to be able to relate general legal requirements such as the 'general duties' under the Health and Safety at Work etc Act and general systems of risk management to the hazards dealt with in this module.

The candidate will be expected to understand:

1 *General Aspects of Occupational Health*
Main types of occupational health risk: chemical, physical, biological. Classification of dangerous substances. Main routes of attack on the body.
General strategy for control. Identification of a hazard. Assessment of risk. Choice of control measures (including: process change, substitution, segregation, engineering controls, reduced time exposure, personal protective equipment).
Nature of exposure limits.
Role and interfaces of medicine, nursing, occupational hygiene, environmental engineering and toxicology in the control of health hazards.
Statutory requirements for recording and reporting cases of illnesses and diseases.

2 *Chemical Health Hazards*
Main forms of chemical attack on the body: toxic (including carcinogens), corrosive and dermatitic. Main routes of entry to absorbing areas: absorption: target organs and systems: natural body defences.
Control strategies for typical important toxic substances: lead, benzene, trichlorethylene, isocyanates. Signs and symptoms of ill-health conditions: measurement of absorption. Main features of Regulations and Codes of Practice for these substances.
Control strategies for known or 'suspected' carcinogens. 'No threshold' concept. Definitions of 'suspected carcinogens'. Strategies for typical important carcinogens: asbestos, tar, vinyl chloride monomer, MBOCA. Main features of Regulations and Codes of Practice for these substances.
Control strategies for typical corrosive and irritant substances: chromium compounds, ammonia. Main features of legal requirements for these substances.
Control strategies for substances causing acute or chronic respiratory disorders. Silica: asbestos: carbon monoxide. Main features of legal requirements for these substances.

3 *Physical Health Hazards*
Main forms of physical stressors: noise, temperature, light: vibration and repetitive strain: radiations. Main effects on the body.
Control strategies for these physical stressors. Main legal requirements.

4 *Biological Health Hazards*
Main forms of biological agent, ie zoonoses, which can cause harm to the body: bacilli, viruses, leptospira, etc.
Control strategies for typical biological ill-health conditions: Weil's Disease; air conditioning/humidifier fever; anthrax; Legionnaire's Disease; glanders; brucellosis; aspergillosis; viral hepatitis; AIDS.

5 *Toxicology, Epidemiology and Pathology*
Main methods of recognising and evaluating health hazards: their uses and limitations, viz:
- Experimental toxicological data on substances (and its interpretation particularly to managers and workers).

- Epidemiological studies – methodology and interpretation. Problems in establishing past exposures of workers. Application of results to a particular workplace. Practicability of such studies in a single organisation.
- Information from doctors, nurses and general practitioners about occupationally related illnesses.

6 *Environmental and Biological Monitoring*

Main methods of detection, sampling and measurement of hazardous substances or physical stressors in the work and general environment: continuous monitoring: spot sampling: personal samplers.

Main types of instrumentation for collection and analysis of:
airborne particulates (behaviour, collection and analysis);
airborne gases;
noise (levels: octave band analysis: noise 'dose');
radiation (counters: dose meters: cumulative doses).

Limitations of information collected and difficulties in interpretation.

Methods of biological monitoring. Ethics of such monitoring. Confidentiality of results. Use as a check on environmental monitoring. Blood and urine samples: X rays: lung, liver and renal function: nerve conditions: performance tests.

Main statutory requirements for environmental and biological monitoring.

7 *Environmental Engineering*

Main methods of controlling hazards at work by engineering:

- Air contamination control by dilution and local exhaust ventilation. Problems of design and efficiency: fan, trunking and filtration systems: location of exhaust outlets: maintenance: monitoring of performance.
- Acoustic control: identification of sources of noise: enclosure of noise: muffling: re-design of noise producing parts.
- Re-design of systems of work to avoid contact with harmful chemicals.
- Methods of control of thermal and lighting environment to avoid damage to health and to achieve good standards of comfort.

8 *Personal Protective Equipment*

Main types of equipment available for protection against health hazards.

Performance of equipment: performance test theory: calculations of theoretical performance: performance in practice (fit: removal for short periods: maintenance).

Circumstances in which personal protective equipment is appropriate: acceptability to workers: systems of operation and maintenance: assessment of effectiveness.

The relevant statutory provisions.

Module D: Safety Technology

This module is intended to equip the student to identify the hazards and control the risks from a range of situations which could result in immediate physical injury to workers (and in some cases the general public).

The module is applicable to the construction, installation, use, maintenance, modification and removal of buildings, plant and equipment and the storage, handling and transport of goods and materials.

In addition it covers the safe operation of processes which could result in the release of toxic, corrosive, flammable and explosive substances.

It will enable the candidate to:

appreciate the relevance of the properties of structures, materials and chemicals to the range of hazards examined;

appreciate the engineering, ergonomic and operational principles relevant to safety in the design, layout, use and maintenance of plant, equipment and buildings;

use this knowledge to establish safe places of work and safe systems of work.

The module is not intended to equip students to carry out detailed design of safe machinery or structures. Rather it is to equip them to know what questions need to be asked and what calculations need to be done, communicate effectively with engineers, designers and architects, command their confidence and assess their approach to safety issues.

1 *Maintaining Integrity of Machinery, Plant and Structures*
Identification of causes and consequences of failure of components or structures;
main modes of causes and consequences of failure of materials: metals and concrete;
main modes of failure: fractures (brittle, ductile and fatigue): buckling: corrosion: wear;
main causes of failure: sources of stress: stress concentrations: environmental conditions: material composition;
main consequences of failure: failure of equipment: collapse of structures: explosion in pressure systems: release of hazardous substances.
Preventive measures:
Specification of materials and components to match service conditions.
Quality assurance systems (v. British Standards).
Planned maintenance.
Non-destructive testing: assessment of significance of defects found.
Avoiding repetition of failures:
Investigatory methods to establish causation: forensic examination of failed components.
Application to some typical situations:
cranes and lifting tackle
falsework
pressurised systems.
Main current legal requirements:
general
for the typical situations above.

2 *Safe Use of Machinery*
Identification of main types of hazard:
contact (with moving parts, cutting edges, abrasive surfaces);
trapping and entanglement;
ejection of parts of machine or materials.
Preventive strategies:
elimination of hazards by redesign of machinery;
reduction of risk by:
safeguarding measures
systems of work
changes in operator behaviour.
Use of simple safeguarding devices:
fixed and adjustable interlocked guards
trip devices
perimeter fencing (control of access by barriers).
Use of advanced guarding systems:
interlocks and trip devices for pneumatic and hydraulic machines;
application of risk assessment methodologies and reliability engineering principles to complex guarding systems.
Special problems in the safeguarding of robotic and computer-controlled systems: software failures.
Ensuring safety during maintenance and adjustment of machinery.
Anthropometrics of machinery safeguards.
Action when supplying plant and machinery to other people: guarding requirements: provision of information.
Application to some typical situations: abrasive wheels: conveying equipment: robots.
Main current legal requirements:
general
for typical situations as above.

3 *Safety in Hazardous Processes*
Identification of main hazards arising in certain chemical processes from:
high/low pressure: high/low temperature;
use of toxic, corrosive, flammable or explosive chemicals;

exothermic reactions;
release of toxic gases: behaviour of heavy and light gases;
release of flammable or explosive liquid or gas;
release of toxic chemicals to ground, water or air.
Principles of safe design and operation. Precautions during plant modification. Special considerations with computer-controlled processes. Training of operators.
Establishment of safe systems for the handling and transport of hazardous chemicals: for the disposal of chemical wastes.
Application to typical situations:
thermal cracking
gas/liquid separation
liquid/solid separation.
Main current legal requirements.

4 *Safety in Movement of People and Goods*
Main causes of injury in workplaces from movement of people:
slips and falls on level ground;
falls from heights (stairs: ladders: scaffolding: working on roofs: working over water).
Measures to reduce risk of injury:
redesign of workplaces: circulation routes, floor finishes, stairs;
systems of work above ground: inspection of equipment.
Main causes of injury in workplaces from movement of goods:
manual handling: back injuries and falls;
falls of goods on to people (stacking: falls from vehicles: overhead movement of goods);
impact of vehicles on people (forklift trucks: loading and unloading of goods vehicles).
Measures to reduce risk of injury:
restriction of manual handling to acceptable loads: introduction of mechanical handling systems;
segregation of movement of people and vehicles;
systems of work for loading and offloading.
Main current legal requirements.

5 *Safe Use of Electricity*
Identification of electrical hazards:
effect of electricity on the human body
common causes of electric shock.
Principles of electrical safety:
isolation: insulation: reduced voltage: earthing, fuses and circuit breakers: systems of work: maintenance and operation of electrical systems.
Safe working with high voltage systems.
Static electricity: minimising build-up and dissipation.
Use of electrical equipment in flammable atmospheres: and in hazardous areas.
Overhead and buried cables.
Hand-held electrical tools.
IEE Regulations: nature and status.
Main current legal requirements.

6 *Fire: Prevention, Fighting and Escape*
Identification of fire hazards in buildings:
sources of ignition
factors affecting the spread of fire
smoke and toxic fumes.
Reduction of fire risk by design of structures and choice of materials: effect of fire on structures, buildings, finishes and furnishings.
Measures to reduce the risk of injury in a fire:
fire detection and alarm systems;
active fire fighting equipment and use: training;
passive fire fighting systems;
means of escape in case of fire. Human behaviour in fires: fire drills.
Application to some typical situations:
department store

paint factory
store for flammable liquids and gases.
Main current legal requirements.

Module E: The Case Study

Health and safety in the workplace does not separate into discrete and isolated areas but is an inter-weaving of a range of disciplines including those covered by this syllabus. In his day-to-day work a safety and health practitioner has to deal with situations that, for their resolution, require lateral thinking, drawing on techniques and practices from a number of different disciplines. This module, for which there is no separate syllabus, embraces the full spectrum of material covered by modules A to D and is intended to assess the candidate's ability to deal effectively with real life problems.

Syllabus for the National General Certificate in Occupational Safety and Health

The Certificate is intended as a qualification for newly appointed Safety Advisers who have no previous experience, for people such as managers and supervisors whose executive responsibilities include safety and health, and for safety representatives and safety committee members.
The intended learning outcome is that candidates will be able to:

1. identify the common hazards and unsafe practices likely to be found in workplaces and know the appropriate preventative and protective measures to be taken;
2. advise on appropriate action to minimise the risk of fire in the workplace and assist in the development of procedures to deal with fires;
3. monitor management's policies, procedures and performance in ensuring health and safety in the workplace;
4. assist in the preparation and regular review of statements of health and safety policy;
5. describe how human and organisational factors can affect health and safety performance;
6. use proactive and reactive strategies to prevent or control the risk of accidents and ill-health;
7. specify the health and training needs of the young, inexperienced and newly-appointed persons;
8. communicate effectively on health and safety matters;
9. prepare internal memoranda, letters and formal reports on health and safety;
10. identify sources of guidance and advice on health and safety matters;
11. maintain an information retrieval system and disseminate information when needed;
12. liaise with safety committees and facilitate relations with worker representatives;
13. investigate events that led, or might have led, to death, injury, ill-health or a dangerous occurrence: prepare formal reports and make logical and cost effective proposals to prevent a repetition;
14. maintain records of events at 13;
15. identify the main requirements of health, safety and welfare legislation and approved codes of practice which apply to the workplace;
16. use reference material to identify the requirements of health and safety legislation and to ensure sound health and safety practices;
17. select appropriate safety monitoring techniques (audits, surveys, inspections, tours and sampling) for a given work situation; and
18. undertake workplace inspections, note any hazards and the way they are safeguarded, safe and unsafe working practices, and recommend appropriate and cost-effective remedial action.

The holder of a Certificate is eligible to join IOSH in the non-corporate grade of Associate (which does not allow the use of designatory letters after the name).
The Certificate syllabus, which is practical and aimed at work-related experience rather than being theoretical, is in three parts and the following is a summary of the more detailed syllabus set out in the NEBOSH Certificate guide.

Part A1 – The identification and control of hazards

1. Hazards and safe systems of work associated with:
   - the use of work equipment including machinery, appliances and hand tools;
   - manual handling operations;
   - pedestrian operated trolleys and vehicles used in the workplace;
   - maintenance work;
   - construction and demolition work;
   - the use of access equipment;
   - the use of electricity and electrically operated tools and equipment; and
   - the activities of contractors and sub-contractors.
2. Fire prevention measures and actions in the event of fire:
   - main fire hazards in workplaces;
   - fire precautions and fire prevention measures; and
   - actions to be taken in the event of a fire.
3. Controlling health hazards:
   - main types of occupational health risks: physical, chemical and biological;
   - classification of dangerous substances: main routes of entry to the body;
   - general strategies for the identification, measurement, evaluation and control of health hazards;
   - monitoring exposure to airborne contaminants;
   - nature of exposure limits; and
   - prevention and control strategies.
   - relevant statutory provisions.

Part A2 – Safety and health management

4. Methods of controlling risk
   - statement of health and safety policy: main elements and conditions for effectiveness in an organisation;
   - proactive and reactive strategies in accident prevention;
   - legal requirements for ensuring a safe and healthy work environment;
   - human factors which influence compliance with health and safety practices;
   - health and safety training; and
   - relevant statutory provisions.
5. Effective communication
   - effective written and oral communication;
   - sources of information and advice on health and safety matters;
   - selection and distribution of information on health and safety matters; and
   - joint consultation on health and safety issues.
6. Investigation of accidents and other incidents and the maintenance of records
   - investigation of accidents, dangerous occurrences, near misses and work-related ill-health;
   - preparation of internal records and the use and interpretation of national accident data; and
   - relevant statutory documents.
7. Compliance with general legislation
   - the main features of common law and its relation to health and safety at work;
   - the powers of courts and tribunals in dealing with breaches of health and safety legislation;
   - the Health and Safety at Work, etc. Act 1974
     - general duties of employers, employees and others
     - statements of health and safety policy
     - approved codes of practice and guidance notes
     - the role of enforcement authorities
     - powers of inspectors
     - the structure and application of Regulations and associated Approved Codes of Practice and guidance notes.

Part B – The practical assessment

8. Inspections of workplaces
   - the various types of inspections; and
   - identification of hazards and appropriate remedial action.

Appendix 3

List of official safety guidance booklets published by HMSO, London

A complete list of HSE publications is contained in an HSE Library and Information Services booklet, Publications in Series: List of HSC/HSE publications (see latest issue).
Note: some of these publications that relate to information on and packaging of chemicals and substances are being reviewed in the light of the Chemicals (Hazard Information and Packaging) Regulations 1993.

APPROVED CODES OF PRACTICE ISSUED UNDER SECTION 16 OF THE HEALTH AND SAFETY AT WORK ETC. ACT 1974

COP 1 Safety representatives and safety committees. (1988) (The Brown Book)
COP 2 Control of lead at work: approved code of practice: (revised June 1985) (in support of SI 1980 No. 1248)
COP 3 Work with asbestos insulation, asbestos coating and asbestos insulating board. Revised approved code of practice. (1988)
COP 6 Plastic containers with nominal capacities up to 5 litres for petroleum spirit: Requirements for testing and marking or labelling (in support of SI 1982 No. 830). (1982)
COP 7 Principles of good laboratory practice: Notification of New Substances Regulations (1982) (in support of SI 1982 No. 1496)
COP 11 Operational provisions of the Dangerous Substances (Conveyance by road in road tankers and tank containers) Regulations 1981
COP 14 Road tanker testing: examination, testing and certification of the carrying tanks of road tankers and of tank containers used for the conveyance of dangerous substances by road (in support of SI 1981 No. 1059)
COP 15 Zoos: safety, health and welfare standards for employers and persons at work: approved code of practice and guidance note. (1985)
COP 16 Protection of persons against ionising radiation arising from any work activity: The Ionising Radiations Regulations 1985: approved code of practice. (1985)
COP 17 Notice of approval of the code of practice for the operational provisions of the Road Traffic (Carriage of Dangerous Substances in Packages etc.) Regulations 1986, by the Health and Safety Commission. 1987: Approved Code of Practice. (1987)
COP 18 Dangerous substances in harbour areas: The Dangerous Substances in Harbour Area Regulations 1987: approved code of practice. (1987)
COP 19 Classification and labelling of dangerous substances for conveyance by road in tankers, tank containers and packages. (Dangerous substances (Conveyance by Road in Road Tankers and Tank Containers) Regulations 1981). (Classification, Packaging and Labelling of Dangerous Substances Regulations 1984). (1988)
COP 20 Standards of training in safe gas installation: approved code of practice. (1987)

COP 21 Control of asbestos at work: The Control of Asbestos at Work Regulations 1987. (1988)
COP 22 Classification and labelling of substances dangerous for supply. Notification of New Substances Regulations 1982. Classification, Packaging and Labelling of Dangerous Substances Regulations 1984. (1988)
COP 23 Exposure to radon: the Ionising Radiations Regulations 1985. Approved Code of Practice. (1988)
COP 24 Preventing accidents to children in agriculture: approved code of practice and guidance notes. (1988)
COP 25 Safety in docks: Docks Regulations 1988: Approved code of practice with regulations and guidance. (1988)
COP 26 Rider operated lift trucks – operator training: approved code of practice and supplementary guidance. (1988)
COP 27 Explosives at quarries. Quarries (Explosives) Regulations 1988. Approved code of practice. (1989)
COP 28 Safety of exit from mines underground workings. Mines (Safety of Exit) Regulations 1988. Approved code of practice. (1989)
COP 29 Control of substances hazardous to health: Control of carcinogenic substances. Control of Substances Hazardous to Health Regulations 1988. Approved Code of Practice. (1988)
COP 30 Control of substances hazardous to health in fumigation operations: Control of Substances Hazardous to Health Regulations 1988. Approved Code of Practice. (1988)
COP 31 Control of vinyl chloride at work. Approved Code of Practice. (1988)
COP 32 First aid on offshore installations and pipeline works. Offshore Installations and Pipeline Works (First-Aid) Regulations 1989. Approved Code of Practice with Regulations and guidance. (1990)
COP 33 Transport of compressed gases in tube trailers and tube containers. Dangerous Substances (Conveyance by Road in Road Tankers and Tank Containers) Regulations 1981. Approved Code of Practice. (1989)
COP 34 The use of electricity in mines. Electricity at Work Regulations 1989. Approved Code of Practice. (1989)
COP 35 The use of electricity at quarries. Electricity at Work Regulations 1989. Approved Code of Practice. (1989)
COP 36 Carriage of explosives by road. Road Traffic (Carriage of Explosives) Regulations 1989. Approved Code of Practice with regulations and guidance. 1989
COP 37 Safety of pressure systems. Pressure Systems and Transportable Gas Containers Regulations 1989. Approved Code of Practice (1990)
COP 38 Safety of transportable gas containers. Pressure Systems and Portable Gas Containers Regulations 1989. Approved Code of Practice (1990)
COP 40 Packaging and labelling of dangerous substances for conveyance by road (revision 2) (1990)
COP 41 Control of substances hazardous to health in the production of pottery. Approved Code of Practice (1990)
COP 42 First aid at work. Approved Code of Practice and guidance (1990)

OTHER CODES OF PRACTICE

Code of practice for reducing the exposure of employed persons to noise. 1972
Code of safe practice at fairs. 1984
Code of safe practice at fairs: amusement device log book
Code of safe practice at fairs: record of daily inspection of passenger carrying device
Code of safe practice at fairs: technical annex. 1988
Pesticides: code of practice for the safe use of pesticides on farms and holdings. (1990)

AUTHORISED AND APPROVED LISTS

Approved supply list. Information approved for the classification and labelling of substances and preparations dangerous for supply. (1993)
Information approved for the classification, packaging and labelling of dangerous substances for supply and conveyance by road . . . , 3rd edn (1990)

Road tanker approved list. Approved substance identification numbers, emergency action codes and classifications for dangerous substances carried by road tankers and tank containers. (1992)

HEALTH AND SAFETY AT WORK BOOKLETS (still in print)
11 Guarding of Hand-fed Platen Machines
12 Safety at Drop Forging Hammers
45 Seats for Workers in Factories, Offices and Shops

This series of booklets is gradually being superseded by the Health and Safety series (HS(R) and HS(G)) booklets and guidance notes.

LEGISLATION SERIES BOOKLETS explaining and interpreting legislation.
(Note: these L series booklets will gradually supersede Approved Codes of Practice and the HS(R) series of booklets).
L1 A guide to the Health and Safety at Work etc. Act 1974, 4th edn (1990)
L2 Power presses. The Power Press Regulations 1965 and 1972 (1991)
L4 Woodworking Machines Regulations 1974. Guidance on the regulations (1991)
L5 Control of substances hazardous to health and control of carcinogenic substances. Control of Substances Hazardous to Health Regulations 1988. Approved Code of Practice, 4th edn (1993)
L6 Diving operations at work. The Diving Operations at Work Regulations 1981 as amended by the Diving Operations at Work (Amendment) Regulations 1990 (1991)
L7 [see also COP 23] Dose limitation – restriction of exposure: Additional guidance on regulation 6 of the Ionising Radiations Regulations 1985. Approved Code of Practice – part 4 (1991)
L8 The prevention or control of legionellosis (including legionnaires disease). Approved Code of Practice (1991)
L9 The safe use of pesticides for non agricultural purposes. Approved Code of Practice (1991)
L10 A guide to the Control of Explosives Regulations 1991 (1991)
L11 A guide to the Asbestos (Licensing) Regulations 1983 (1991)
L13 A guide to the Packaging of Explosives for Carriage Regulations 1991 (1991)
L14 Classification and labelling of dangerous substances for carriage by road in tankers, tank containers and packages (1993)
L15 Packaging of dangerous substances for carriage by road (to be published)
L16 Design and construction of vented non-pressure road tankers used for the carriage of flammable liquids. Road Traffic (Carriage of Dangerous Substances in Road Tankers and Tank Containers) Regulations 1992. Approved code of practice (1993)
L17 Design and construction of road tankers used for the carriage of carbon disulphide. Road Traffic (Carriage of Dangerous Substances in Road Tankers and Tank Containers) Regulations 1992. Approved code of practice (1993)
L18 Design and construction of vaccuum insulated road tankers used for the carriage of non-toxic deeply refrigerated gases. Road Traffic (Carriage of Dangerous Substances in Road Tankers and Tank Containers) Regulations 1992. Approved code of practice (1993)
L19 Design and construction of vacuum operated road tankers used for the carriage of hazardous wastes. Road Traffic (Carriage of Dangerous Substances in Road Tankers and Tank Containers) Regulations 1992. Approved code of practice (1993)
L20 A guide to the Lifting Plant and Equipment (Records of Test and Examination etc.) Regulations 1992 (1992)
L21 Management of Health and Safety at Work Regulations 1992: Approved Code of Practice (1992)
L22 Work Equipment Provision and Use of Work Equipment Regulations 1992. Guidance on regulations (1992)
L23 Manual handling. Manual Handling Operations Regulations 1992. Guidance on regulations (1992)
L24 Workplace health, safety and welfare. Workplace (Health, Safety and Welfare) Regulations 1992. Approved Code of Practice and guidance (1992)
L25 Personal Protective Equipment at Work Regulations 1992: guidance on regulations (1992)

L26 Display screen equipment work. Health and Safety (Display Screen Equipment) Regulations 1992. Guidance on regulations (1992)
L27 The Control of Asbestos at Work. Control of Asbestos at Work Regulations 1987. Approved code of practice, 2nd edn (1993)
L28 Work with asbestos insulation, asbestos coating and asbestos insulating board. Control of Asbestos at Work Regulations 1987. Approved code of practice, 2nd edn (1993)
L29 A guide to the Genetically Modified Organisms (Contained Use) Regulations 1992
L30 A guide to the Offshore Installations (Safety Case) Regulations 1992: guidance on regulations (1992)
L31 A guide to the Public Information for Radiation Emergencies Regulations 1992 (1993)
L37 Safety data sheets for substances and preparations dangerous for supply (1993)
L38 Approved guide to the classification and labelling of substances and preparations dangerous for supply (1993)
L42 Shafts and winding in mines. Approved Code of Practice on the Mines (Shafts and Winding) Regulations 1993 (1993)

Booklets in the Health and Safety series explaining Regulations (HS(R)) and giving guidance (HS(G)) are listed below.

HS(R) SERIES BOOKLETS explaining Regulations
2 A guide to agricultural legislation
4 A guide to the 1963 OSRP Act
12 A guide to the Health and Safety (Dangerous Pathogens) Regulations 1981
13 A guide to the Dangerous Substances (Conveyance by Road in Road Tankers and Tank Containers) Regulations 1981
14 A guide to the Notification of New Substances Regulations 1982
15 Administrative guidance on the implementation of the European Community 'Explosive Atmospheres' Directives (76/117/EEC and 79/196/EEC) and related directives
16 A guide to the Notification of Installations Handling Hazardous Substances Regulations 1982
17 A guide to the Classification and Labelling of Explosives Regulations 1983
18 Administrative guidance on the application of the European Community 'Low Voltage' Directive (72/23/EEC) to electrical equipment for use at work in the United Kingdom
21 A guide to the Control of Industrial Major Hazards Regulations 1984
23 A guide to the Reporting of Injuries, Diseases and Dangerous Occurrences Regulations 1985
24 Guide to the Road Traffic (Carriage of Dangerous Substances in Packages etc.) Regulations 1986
25 Memorandum of guidance on the Electricity at Work Regulations 1989
26 Guidance on the legal and administrative measures taken to implement the European Communities Directives on lifting and mechanical handling appliances and electrically operated lifts
27 A guide to Dangerous Substances in Harbour Areas Regulations 1987
28 A guide to the Loading and Unloading of Fishing Vessels Regulations 1988
29 Notification and marking of sites. The Dangerous Substances (Notification and Marking of Sites) Regulations (1990)
30 A guide to the Pressure Systems and Transportable Gas Containers Regulations 1989 (1990)

HS(G) SERIES BOOKLETS giving guidance
1 Safe use and storage of flexible polyurethane foam
2 Poisonous chemicals on the farm
3 Highly flammable materials on construction sites
4 Highly flammable liquids in the paint industry
5 Hot work: welding and cutting on plant containing flammable materials
6 Safety in working with lift trucks
7 Container terminals: safe working practice
8 Fabric production: safety in the cotton and allied fibres industry
9 Spinning, winding and sizing: safety in the cotton and allied fibres industry
10 Cloakroom accommodation and washing facilities
11 Flame arresters and explosion reliefs

12 Off-shore construction health, safety and welfare
13 Electrical testing: safety in electrical testing
14 Opening processes: cotton and allied fibres
15 Storage of liquefied petroleum gas at factories
16 Evaporating and other ovens
17 Safety in the use of abrasive wheels: adjusting ring to compensate for wheel wear
18 Portable grinding machines: control of dust
19 Safety in working with power-operated mobile work platforms
20 Guidelines for occupational health services
21 Safety in the cotton and allied fibres industry: cardroom processes
22 Electrical apparatus for use in potentially explosive atmospheres
23 Safety at power operated mast work platforms
24 Guarding of cutters of horizontal milling machines
25 Control of Industrial Major Accident Hazards Regulations 1984 (CIMAH); further guidance on emergency plans
26 Transport of dangerous substances in tank containers
27 Substances for use at work: the provision of information
28 Safety advice for bulk chlorine installations
29 Locomotive boilers
30 Storage of anhydrous ammonia under pressure in the UK: spherical and cylindrical vessels
31 Pie and tart machines
32 Safety in falsework for in situ beams and slabs
33 Safety in roofwork (supersedes Guidance Note GS 10: Roofwork, prevention of falls)
34 Storage of LPG at fixed installations
35 Catering safety: food preparation machinery
36 Disposal of explosive waste and the decontamination of explosive plant
37 Introduction to local exhaust ventilation
38 Lighting at work
39 Compressed air safety
40 Chlorine from drums and cylinders
41 Petrol filling stations: construction and operation
42 Safety in the use of metal cutting guillotines and shears
43 Industrial robot safety
44 Drilling machines: guarding of spindles and attachments
45 Safety in meat preparation: guidance for butchers
46 A guide for small contractors: Site safety and concrete construction
47 Avoiding danger from underground services
48 Human factors in industrial safety
49 The examination and testing of portable radiation instruments for external radiations (1990)
50 The storage of flammable liquids in fixed tank (up to 10 000 m^3 total capacity) (1990)
51 The storage of flammable liquids in containers (1990)
52 The storage of flammable liquids in fixed tanks (exceeding 10 000 m^3 total capacity) (1991)
53 Respiratory protective equipment: a practical guide for users (1990)
54 The maintenance examination and testing of local exhaust ventilation (1990)
55 Health and safety in kitchens and food preparation areas (1990)
56 Noise at work. Noise assessment, information and control. Noise guides 3 to 8 (1990)
57 Seating at work (1991)
58 Evaluation and inspection of buildings and structures (1990)
60 Work related upper limb disorders: a guide to prevention (1990)
61 Surveillance of people exposed to health risks at work (1990)
62 Health and safety in tyre exhaust premises (1991)
63 Radiation protection off site for emergency services in the event of a nuclear accident (1991)
64 Assessment of fire hazards from solid materials and the precautions required for their safe storage and use: a guide for manufacturers, suppliers, storekeepers and users. Prepared by the Health and Safety Executive, the Home Office and the Scottish Office (1991)
65 Successful health and safety management (1991)
66 Protection of workers and the general public during the development of contaminated land (1991)

67 Health and safety in motor vehicle repair (1991)
70 The control of legionellosis including legionnaire's disease (1991)
71 Storage of packaged dangerous substances (1992)
72 Control of respirable silica dust in heavy clay and refractory processes (1992)
73 Control of respirable crystalline silica in quarries (1992)
74 Control of silica dust in foundries (1992)
76 Health and safety in retail and wholesale warehouses (1992)
77 COSHH and peripatetic workers (1992)
78 Container packing: guidance for those packing and transporting dangerous goods in CTUs for carriage by sea (1992)
81 Fairgrounds and amusement parks: a code of safe practice (1992)
83 Training woodworking machinists (1992)
84 Shopping trolleys: safe system of work guidance (1992)
85 Electricity at work: safe working practices (1993)
86 Veterinary medicines. Safe use by farmers and other animal handlers (1992)
89 Safeguarding agricultural machinery: moving parts (1992)
91 A framework for the restriction of occupational exposure to ionising radiation (1992)
93 The assessment of pressure vessels operating at low temperature (1993)
96 The costs of accidents at work (1993)
97 A step by step guide to COSHH assessment (1993)
108 CHIP for everyone (1993)

GUIDANCE NOTES issued by the Health and Safety Executive are grouped in the following five series:

General series

GS 2 Metrication of construction safety regulations
GS 3 Fire risk in the storage and industrial use of cellular plastics
GS 4 Safety in pressure testing
GS 5 Entry into confined spaces
GS 6 Avoidance of danger from overhead electrical lines
GS 7 Accidents to children on construction sites
GS 8 Articles and substances for use at work – guidance for designers, manufacturers, importers, suppliers, erectors and installers
GS 9 Road transport in factories
GS 10 Roofwork: prevention of falls
GS 11 Whisky cask racking
GS 12 Effluent storage on farms
GS 13 Reporting of accidents to pupils and students
GS 15 General access scaffolds
GS 16 Gaseous fire extinguishing systems: precautions for toxic and asphyxiating hazards
GS 17 Safe custody and handling of stock bulls on farms and at artificial insemination centres
GS 18 Commercial ultra-violet tanning equipment
GS 19 General fire precautions aboard ships being fitted out or under repair
GS 20 Fire precautions in pressurised workings
GS 21 Assessment of radio frequency ignition hazard to process plant where flammable atmospheres may occur (see also BS 6656:1986)
GS 23 Electrical safety in schools
GS 24 Electricity on construction sites
GS 25 Prevention of falls to window cleaners
GS 26 Access to road tankers
GS 27 Protection against electrical shock
GS 28/1 Safe erection of structures. Part 1: Initial planning and design
GS 28/2 Safe erection of structures. Part 2: Site management and procedures
GS 28/3 Safe erection of structures. Part 3: Working places and access
GS 28/4 Safe erection of structures. Part 4: Legislation and training
GS 29/1 Health and safety in demolition work. Part 1: Preparation and planning
GS 29/2 Health and safety in demolition work. Part 2: Legislation
GS 29/3 Health and safety in demolition work. Part 3: Techniques

GS 29/4 Health and safety in demolition work. Part 4: Health hazards
GS 30 Health and safety hazards associated with pig husbandry
GS 31 Safe use of ladders, step ladders and trestles
GS 32 Health and safety in shoe repair premises
GS 34 Electrical safety in departments of electrical engineering
GS 35 Safe custody and handling of bulls on farms and similar premises
GS 36 Safe custody and handling of bulls at agricultural shows, markets and similar premises off the farm
GS 37 Flexible leads, plugs, sockets and other accessories
GS 38 Electrical test equipment for use by electricians
GS 39 Training of crane drivers and slingers
GS 40 The loading and unloading of bulk flammable liquids and gases at harbours and inland waterways
GS 41 Radiation safety in underwater radiography
GS 42 Tower scaffolds
GS 43 Lithium batteries
GS 46 In situ timber treatment using timber preservatives: health, safety and environmental precautions
GS 47 Safety of electrical distribution systems on factory premises (1991)
GS 48 Training and standards of competence for users of chain saws in agriculture, and forestry (1990)
GS 49 Pre-stressed concrete (1991)
GS 50 Electrical safety in places of entertainment (1991)
GS 51 Facade retention (1992)

Chemical safety series
CS 1 Industrial use of flammable gas detectors
CS 2 The storage of highly flammable liquids
CS 3 Storage and use of sodium chlorate and other similar strong oxidants
CS 4 The keeping of LPG in cylinders and similar containers
CS 5 The storage of LPG at fixed installations
CS 6 The storage and use of LPG on construction sites
CS 7 Odorisation of bulk oxygen supplies in shipyards
CS 8 Small scale storage and display of LPG at retail premises
CS 9 Bulk storage and use of liquid carbon dioxide: hazards and precautions
CS 10 Fumigation using phosphine
CS 11 Storage and use of LPG at metered estates
CS 12 Fumigation using methyl bromide (bromomethane)
CS 15 Cleaning and gas freeing of tanks containing flammable residues
CS 16 Chlorine vaporisers
CS 17 Storage of packaged dangerous substances
CS 18 Storage and handling of ammonium nitrate
CS 19 Storage of approved pesticides: guidance for farmers and other professional users
CS 21 The storage and handling of organic peroxides (1992)

Plant and machinery series
PM 1 Guarding of portable pipe-threading machines
PM 2 Guards for planing machines
PM 3 Erection and dismantling of tower cranes
PM 4 Safety at high temperature dyeing machines
PM 5 Automatically controlled steam and hot water boilers
PM 6 Dough dividers
PM 7 Lifts: thorough examination and testing
PM 8 Passenger-carrying paternosters
PM 9 Access to tower cranes
PM 10 Tripping devices for radial and heavy vertical drilling machines
PM 13 Zinc embrittlement of austenitic stainless steel
PM 14 Safety in the use of cartridge operated tools

PM 15 Safety in the use of timber pallets
PM 16 Eyebolts
PM 17 Pneumatic nailing and stapling tools
PM 18 Locomotive boilers
PM 19 Use of lasers for display purposes
PM 20 Cable-laid slings and grommets
PM 21 Safety in the use of woodworking machines
PM 22 Training advice on the mounting of abrasive wheels
PM 23 Photo-electric safety systems
PM 24 Safety at rack and pinion hoists
PM 25 Vehicle finishing units: fire and explosion hazards
PM 26 Safety at lift landings
PM 27 Construction hoists
PM 28 Working platforms on fork lift trucks
PM 29 Electrical hazards from steam/water pressure cleaners etc.
PM 30 Suspended access equipment
PM 31 Chain saws
PM 32 Safe use of portable electrical apparatus (electrical safety)
PM 33 Safety of bandsaws in the food industry
PM 34 Safety in the use of escalators
PM 35 Safety in the use of reversing dough brakes
PM 36 Weld defect acceptance levels for in-service non destructive examination of set-in endplate to furnace and shell connections of shell boilers
PM 37 Electrical installations in motor vehicle repair premises
PM 38 Selection and use of electric handlamps
PM 39 Hydrogen embrittlement of grade T chain
PM 40 Protection of workers at welded steel tube mills
PM 41 The application of photo-electric safety systems to machinery
PM 42 Excavators used as cranes
PM 43 Scotch derrick cranes
PM 45 Escalators: periodic thorough examination (1984)
PM 46 Wedge and socket anchorages for wire ropes
PM 47 Safe operation of passenger carrying amusement devices – the Waltzer
PM 48 Safe operation of passenger carrying amusement devices – the Octopus
PM 49 Safe operation of passenger carrying amusement devices – the Cyclone Twist
PM 51 Safety in the use of radio-frequency dielectric heating equipment
PM 52 Safety in the use of refuse compaction vehicles
PM 53 Emergency private generation: electrical safety
PM 54 Lifting gear standards
PM 55 Safe working of overhead travelling cranes
PM 56 Noise from pneumatic systems
PM 57 Safe operation of passenger carrying amusement devices: the Big Wheel
PM 58 Diesel engined lift trucks in hazardous areas
PM 59 Safe operation of passenger carrying amusement devices – the Paratrooper
PM 60 Steam boiler blowdown systems
PM 61 Safe operation of passenger carrying amusement devices – the Chair-O-Plane
PM 62 Expanded polystyrene moulding machines
PM 63 Inclined hoists used in building and construction work
PM 64 Electrical safety in arc welding
PM 65 Worker protection at crocodile (alligator) shears
PM 66 Scrap baling machines
PM 68 Safe operation of passenger carrying amusement devices – Roller Coasters
PM 69 Safety in the use of freight containers
PM 70 Safe operation of passenger carrying amusement devices – Ark/Speedways
PM 71 Safe operation of passenger carrying amusement devices – Water Chutes
PM 72 Safe operation of passenger carrying amusement devices – Trabant (1990)
PM 73 Safety at autoclaves (1990)
PM 75 Glass reinforced plastic vessels and tanks: advice to users (1991)
PM 76 Safe operation of passenger carrying amusement devices – Inflatable Bouncing Devices (1991)
PM 77 Fitness of equipment used for medical exposure to ionising radiation (1992)

Medical series

MS 3 Skin tests in dermatitis and occupational chest disease
MS 4 Organic dust surveys
MS 5 Lung function
MS 6 Chest X-rays in dust diseases
MS 7 Colour vision
MS 8 Isocyanates – medical surveillance
MS 9 Byssinosis
MS 10 Beat conditions and tenosynovitis
MS 12 Mercury – medical surveillance
MS 13 Asbestos
MS 15 Welding
MS 16 Training of offshore sick-bay attendants ('rig-medics')
MS 17 Biological monitoring of workers exposed to organo-phosphorus pesticides
MS 18 Health surveillance by routine procedures
MS 19 The control of lead at work: medical surveillance
MS 20 Pre-employment health screening
MS 21 Precautions for the safe handling of cytotoxic drugs
MS 22 The medical monitoring of workers exposed to platinum salts
MS 23 Health aspects of job placement and rehabilitation – advice to employers
MS 24 Health surveillance of occupational skin disease (1991)
MS 25 Medical aspects of occupational asthma (1991)

Environmental hygiene series

EH 1 Cadmium – health and safety precautions
EH 2 Chromium – health and safety precautions
EH 4 Aniline – health and safety precautions
EH 5 Trichloroethylene: health and safety precautions
EH 6 Chronic acid concentrations in air
EH 7 Petroleum based adhesives in building operations
EH 8 Arsenic – toxic hazards and precautions
EH 9 Spraying of highly flammable liquids
EH 10 Asbestos – exposure limits and measurement of airborne dust concentrations
EH 11 Arsine – health and safety precautions
EH 12 Stibine – health and safety precautions
EH 13 Beryllium – health and safety precautions
EH 14 Level of training for technicians making noise surveys
EH 15 Threshold limit values (superseded by EH 40)
EH 16 Isocyanates: toxic hazards and precautions
EH 17 Mercury – health and safety precautions
EH 18 Toxic substances: a precautionary policy
EH 19 Antimony – health and safety precautions
EH 20 Phosphine – health and safety precautions
EH 21 Carbon dust – health and safety precautions
EH 22 Ventilation of the workplace
EH 23 Anthrax: health hazards
EH 24 Dust and accidents in malthouses
EH 25 Cotton dust sampling
EH 26 Occupational skin diseases: health and safety precautions
EH 27 Acrylonitrile: personal protective equipment
EH 28 Control of lead: air sampling techniques and strategies
EH 29 Control of lead: outside workers
EH 30 Control of lead: pottery and related industries
EH 31 Control of exposure to polyvinyl chloride dust
EH 32 Control of exposure to talc dust
EH 33 Atmospheric pollution in car parks
EH 34 Benzidine based dyes. Health and safety precautions
EH 35 Probable asbestos dust concentrations at construction premises
EH 36 Work with asbestos cement

EH 37 Work with asbestos insulating board
EH 38 Ozone: health hazards and precautionary measures
EH 40 Occupational exposure limits (updated yearly)
EH 41 Respiratory protective equipment for use against asbestos
EH 42 Monitoring strategies for toxic substances
EH 43 Carbon monoxide
EH 44 Dust in the workplace: general principles of protection
EH 45 Carbon disulphide: control of exposure in the viscose industry
EH 46 Man-made mineral fibres
EH 47 The provision, use and maintenance of hygiene facilities for work with asbestos insulations and coatings
EH 48 Legionnaire's disease
EH 49 Nitrosamines in synthetic metal cutting and grinding fluids
EH 50 Training operatives and supervisors for work with asbestos insulation and coatings
EH 51 Enclosures provided for work with asbestos insulation, coatings and insulating board
EH 52 Removal techniques and associated waste handling for asbestos insulation, coatings and insulating board (1989)
EH 53 Respiratory protective equipment for use against airborne radioactivity (1990)
EH 54 Assessment of exposure to fume from welding and allied processes (1990)
EH 55 The control of exposure to fume from welding, brazing and similar processes (1990)
EH 56 Biological monitoring for chemical exposures in the workplace (1992)
EH 57 The problems of asbestos removal at high temperatures (1993)
EH 58 The carcinogenicity of mineral oils (1990)
EH 59 Crystalline silica (1992)
EH 60 Nickel and its inorganic compounds: health and safety precautions (1991)
EH 62 Metalworking fluids – health precautions (1992)
EH 63 Vinyl chloride: toxic hazards and precautions (1992)
NB: to be read in conjunction with the COSHH Approved Codes of Practice: L5 and COP 31
EH 64 Occupational exposure limits: criteria document summaries. Synopses of the data used in setting occupational exposure limits (1992)
NB: summaries should be read in conjunction with the COSHH Regulations 1988, associated Approved Codes of Practice, and guidance publications EH 40 and EH 42
EH 65/1 Trimethylbenzenes: criteria document for an occupational exposure limit (1992)
EH 65/2 Pulverised fuel ash: criteria document for an occupational exposure limit (1992)
EH 65/3 N-N-dimethylacetamide: criteria document for an occupational exposure limit (1992)

OCCASIONAL PAPERS

OP 2 Microprocessors in industry: safety implications of the uses of programmable electronic systems in factories (1981)
OP 3 Managing safety: a review of the role of management in occupational health and safety (1981)
OP 4 Dust in London Underground (1982)
OP 5 Electrostatic ignition: hazards of insulating materials (1982)
OP 6 Underground cable damage survey November 1980–July 1981 (1983)
OP 8 Training for hazardous occupations: a case study of the Fire Service (1984)
OP 11 Measuring the effectiveness of HSE's field activities (1985)

In addition, a number of reports have been prepared by Industry Advisory Committees giving advice specific to operations in those particular industries. These reports are listed in the List of Publications obtainable from HSE Books, Sudbury.

Appendix 4

List of abbreviations

AAL	Authorised and Approved List
ac	alternating current
ACAS	Advisory, Conciliation and Arbitration Service
ACGIH	American Conference of Governmental and Industrial Hygienists
ACOP	Approved Code of Practice
ADS	Approved dosimetry service
AFFF	Aqueous film forming foam
ALA	Amino laevulinic acid
All ER	All England (Law) Reports
APAU	Accident Prevention Advisory Unit
BLEVE	Boiling liquid expanding vapour explosion
CBI	Confederation of British Industry
cd	candela
CD–ROM	Compact Disk – Read Only Memory
CEC	Commission of the European Community
CEN	European Committee for Mechanical Standardization
CONDAM	The Construction (Design and Management) Regulations 1993
CENELEC	European Committee for Electrotechnical Standardisation
CET	Corrected Effective Temperature
CHASE	Complete Health and Safety Evaluation
CHIP	The Chemicals (Hazard Information and Packaging) Regulations 1993
CIA	Chemical Industries Association
CIMAH	Control of Industrial Major Accident Hazards Regulations 1984
COREPER	Committee of Permanent Representatives
COSHH	Control of Substances Hazardous to Health Regulations 1988
CPA	Consumer Protection Act 1987
CPL	Classification, Packaging and Labelling of Dangerous Substances Regulations 1984
CRE	Commission for Racial Equality
dc	direct current
dB	decibels
DPP	Director of Public Prosecutions
DSER	The Health and Safety (Display Screen Equipment) Regulations 1992
DSS	Department of Social Security
E	Illuminance
EAT	Employment Appeal Tribunal
EcoSoC	Economic and Social Committee
E(E)C	European (Economic) Community
EMAS	Employment Medical Advisory Service
EOC	Equal Opportunities Commission
EP	European Parliament
EPA	The Environmental Protection Act 1990

eV	electron volt
FA	Factories Act 1961
FAFR	Fatal accident frequency rate
FEP	free erythrocytic porphyrin
FMEA	Fault modes and effect analysis
FOC	Fire Offices Committee
FPA	Fire Precautions Act 1971
FTAB	Fault tree analysis
HAZAN	Hazard Analysis
HAZOP	Hazard and Operability Study
hfl	Highly flammable liquid
HSC	Health and Safety Commission
HSE	Health and Safety Executive
HSI	Heat Stress Index
HSW	Health and Safety at Work etc. Act 1974
ICRP	International Commission on Radiological Protection
ID	Internal diameter
IEC	International Electrotechnical Committee
IEE	Institution of Electrical Engineers
IES	Illuminating Engineering Society
IFR	Injury frequency rate
IOSH	Institution of Occupational Safety and Health
IQ	Intelligence quotient
IRLR	Industrial relations law report
ISRS	International Safety Rating System
J.P.	Justice of the Peace
JSA	Job Safety Analysis
LA	Local Authority
$L_{EP,d}$	Daily personal noise exposure
LEV	Local exhaust ventilation
L.J.	Lord Justice
lm	lumen
LPG	Liquefied petroleum gas
LVHV	Low volume/high velocity
lx	lux
MEL	Maximum Exposure Limit
MHOR	The Manual Handling Operations Regulations 1992
MHSWR	The Management of Health and Safety at Work Regulations 1992
MPL	Maximum Potential Loss
M.R.	Master of the Rolls
NADOR	Notification of Accidents and Dangerous Occurrences Regulations 1980
NAIR	National Arrangements for Incidents involving Radioactivity
NDT	Non-destructive testing
NEBOSH	National Examination Board in Occupational Safety and Health
NI	Northern Ireland (Law Reports)
NIG	National Industry Groups
NIHHS	Notification of Installations Handling Hazardous Substances Regulations 1982
NIJB	Northern Ireland Judgements Bulletin
NIOSH	National Institute for Occupational Safety and Health, USA
NLJ	Northern Ireland Legal Journal
npf	Nominal protection factor
NR	Noise Rating
NRPB	National Radiological Protection Board
OEL	Occupational Exposure Limit
OES	Occupational Exposure Standard
OET	Overhead electric travelling (crane)
OSHA	Occupational Safety and Health Act (or Administration), USA
OSRP	Offices, Shops and Railway Premises Act 1963
Pa	Pascal
PCB	Polychlorinated biphenyl
PCT	Polychlorinated terphenyl

PGR	Road Traffic (Carriage of Dangerous Substances in Packages, etc.) Regulations 1986
ppm	Parts per million
P4SR	Predicted four hour sweat rate
PPER	The Personal Protective Equipment at Work Regulations 1992
ptfe	Polytetrafluoroethylene
PTS	Permanent Threshold Shift
PUWER	The Provision and Use of Work Equipment Regulations 1992
PVC	Poly-vinyl chloride
QA	Quality assurance
RGN	Registered general nurse
RIDDOR	Reporting of Injuries, Diseases and Dangerous Occurrences Regulations 1985
RM	Resident magistrate
RoSPA	Royal Society for the Prevention of Accidents
RPA	Radiation Protection Adviser
RPS	Radiation Protection Supervisor
RRP	Recommended retail price
SC	Sessions Cases (Scotland)
SEN	State enrolled nurse
SIESO	Society of Industrial Emergency Services Officers
SLT	Scots Law Times
SMSR	The Supply of Machinery (Safety) Regulations 1992
SPL	Sound pressure level
SRN	State registered nurse
SSP	Statutory sick pay
SWL	Safe working load
TLV	Threshold Limit Value
TTS	Temporary Threshold Shift
TUC	Trades Union Congress
TWA	Time Weighted Average
UKAEA	United Kingdom Atomic Energy Authority
v.	versus
VAT	Value-added tax
VCM	Vinyl chloride monomer
WBGT	Wet Bulb Globe Temperature
WHSWR	The Workplace (Health, Safety and Welfare) Regulations 1992
WLL	Working load limit
WLR	Weekly Law Reports
YTS	Youth training scheme

Appendix 5

Organisations providing safety information

Institution of Occupational Safety and Health, The Grange, Highfield Drive, Wigston, Leicester LE18 1NN 0533 571399

National Examination Board in Occupational Safety and Health, NEBOSH House, Newton Lane, Wigston, Leicester LE18 3PP 0533 888858

Royal Society for the Prevention of Accidents, Cannon House, The Priory Queensway, Birmingham B4 6BS 021 200 2461

British Standards Institution, 2 Park Street, London W1A 2BS 071 629 9000
For purchase of Standards: British Standards Institution, Newton House, 101 Pentonville Road, London N1 9ND

Health and Safety Executive, Enquiry Point, St. Hugh's House, Trinity Road, Bootle, Liverpool L20 3QY 051 951 4000
or any local offices of the HSE

HSE Books, PO Box 1999, Sudbury, Suffolk CO10 6FS 0787 881165

Employment Medical Advisory Service, St Hugh's House, Trinity Road, Bootle, Liverpool L 20 3QY 051 951 4381

Fire Protection Association, Aldermary House, Queen Street, London EC4N 1TJ 071 248 5222

Institution of Fire Engineers, 148 New Walk, Leicester LE1 7QB 0533 59171

Medical Commission on Accident Prevention, 35–43 Lincolns Inn Fields, London WC2A 3PN 071 242 3176

The Asbestos Information Centre Ltd, PO Box 69, Widnes, Cheshire WA8 9GW 051 420 5866

Chemical Industry Association, 93 Albert Embankment, London SE1 071 735 3001

Institute of Materials Handling, Cranfield Institute of Technology, Cranfield, Bedford MK43 0AL 0234 750662

National Institute for Occupational Safety and Health, 5600 Fishers Lane, Rockville, Maryland, 20852, USA

Noise Abatement Society, PO Box 8, Bromley, Kent BR2 0UH 081 460 3146

Department of Energy, Thames House South, Millbank, London SW1P 4QI

Electrical Engineering Inspectorate	071 211 4447
Atomic Energy Safeguards Office	071 211 3054
Gas Safety	071 211 7163
Offshore Petroleum Safety	071 211 5344

Department of the Environment, Waste Disposal Directorate, Romney House, 43 Marsham Street, London SW1P 3PX

Domestic waste	071 212 6216/6224
Industrial waste	071 212 8666
Commercial waste	071 212 6287
Hazardous wastes	071 212 6086

Home Office, 50 Queen Anne's Gate, London SW1A 9AT

Fire Services Inspectorate	071 213 4311
Library and Information Services	071 213 4277

Department of Trade and Industry
- Consumer Safety Unit, 10–18 Victoria Street, London SW1H 0NN 071 215 3322
- Manufacturing Technology Division, 151 Buckingham Palace Road, London SW1W 9SS 071 215 5000

Department of Transport
- Railways Inspectorate, 2 Marsham Street, London SW1P 3EB 071 212 3434
- Road and Vehicle Safety Directorate, 2 Marsham Street, London SW1P 3EB 071 212 4143/4274/3014

Advisory, Conciliation and Arbitration Service (ACAS), 11–12 St James' Square, London SW12Y 4LA 071 214 6000

Council of Independent Inspecting Authorities (CIIA), 54 Hagley Road, Birmingham B16 8QP 021 455 9876

Health Education Council, 78 New Oxford Street, London WC1A 1AH 071 631 0930

National Radiological Protection Board (NRPB), Harwell, Didcot, Oxfordshire OX11 0RQ 0235 831600

Northern Ireland Office
- Health and Safety Inspectorate, 83 Ladas Drive, Belfast BT6 9FJ 0232 701444
- Agricultural Inspectorate, Dundonald House, Upper Newtownwards Road, Belfast BT4 3SU 0232 65011 ext: 604
- Employment Medical Advisory Service, Royston House, 34 Upper Queen Street, Belfast BT1 6FX 0232 233045 ext: 58

Commission of the European Communities, Information Office, 8 Storey's Gate, London SW1P 3AT 071 222 8122

British Safety Council, National Safety Centre, Chancellor's Road, London W6 9RS 071 741 1231/2371

Confederation of British Industry, Centre Point, 103 New Oxford Street, London WC1A 1DU 071 379 7400

For most countries, the names and addresses of organisations concerned with safety and health matters can be found in: The Health and Safety Directory, Kluwer Publishing Ltd, Brentford (1988).

Appendix 6

Table of Statutes, Regulations and Orders

Appendix 7

Table of Cases

Index

Safety sign	Meaning	Standard wording	Safety sign	Meaning	Standard wording
	Smoking is prohibited	No smoking		Caution risk of explosion	Explosive material
	Smoking and naked flames prohibited	No naked flame		Directional arrow	
	Caution risk of fire	Highly flammable material			

Safety signs that have been internationally agreed. (Courtesy: British Standards Institution)